The Handbook of the Psychology of Communication Technology

Handbooks in Communication and Media

This series aims to provide theoretically ambitious but accessible volumes devoted to the major fields and subfields within communication and media studies. Each volume sets out to ground and orientate the student through a broad range of specially commissioned chapters, while also providing the more experienced scholar and teacher with a convenient and comprehensive overview of the latest trends and critical directions.

The Handbook of Children, Media, and Development, *edited by Sandra L. Calvert and Barbara J. Wilson*
The Handbook of Crisis Communication, *edited by W. Timothy Coombs and Sherry J. Holladay*
The Handbook of Internet Studies, *edited by Mia Consalvo and Charles Ess*
The Handbook of Rhetoric and Public Address, *edited by Shawn J. Parry-Giles and J. Michael Hogan*
The Handbook of Critical Intercultural Communication, *edited by Thomas K. Nakayama and Rona Tamiko Halualani*
The Handbook of Global Communication and Media Ethics, *edited by Robert S. Fortner and P. Mark Fackler*
The Handbook of Communication and Corporate Social Responsibility, *edited by Øyvind Ihlen, Jennifer Bartlett and Steve May*
The Handbook of Gender, Sex, and Media, *edited by Karen Ross*
The Handbook of Global Health Communication, *edited by Rafael Obregon and Silvio Waisbord*
The Handbook of Global Media Research, *edited by Ingrid Volkmer*
The Handbook of Global Online Journalism, *edited by Eugenia Siapera and Andreas Veglis*
The Handbook of Communication and Corporate Reputation, *edited by Craig E. Carroll*
The Handbook of Media and Mass Communication Theory, *edited by Robert S. Fortner and P. Mark Fackler*
The Handbook of International Advertising Research, *edited by Hong Cheng*
The Handbook of the Psychology of Communication Technology, *edited by S. Shyam Sundar*

The Handbook of the Psychology of Communication Technology

Edited by

S. Shyam Sundar

WILEY Blackwell

This edition first published 2015

Registered Office
John Wiley & Sons, Ltd, The Atrium, Southern Gate, Chichester, West Sussex, PO19 8SQ, UK

Editorial Offices
350 Main Street, Malden, MA 02148-5020, USA
9600 Garsington Road, Oxford, OX4 2DQ, UK
The Atrium, Southern Gate, Chichester, West Sussex, PO19 8SQ, UK

For details of our global editorial offices, for customer services, and for information about how to apply for permission to reuse the copyright material in this book please see our website at www.wiley.com/wiley-blackwell.

Library of Congress Cataloging-in-Publication Data

The handbook of the psychology of communication technology / edited by S. Shyam Sundar.
pages cm
Includes bibliographical references and index.
ISBN 978-1-118-41336-4 (cloth)
1. Communication and technology–Psychological aspects. 2. Information technology–Psychological aspects. 3. Communication–Psychological aspects. I. Sundar, S. Shyam.
P96.T42H366 2015
303.48′33–dc23

2014026814

A catalogue record for this book is available from the British Library.

Cover image: © Peshkova / Shutterstock

Set in 10/12.5pt Galliard by SPi Publisher Services, Pondicherry, India

1 2015

Dedicated to Cliff Nass (1955–2013), teacher, advisor, mentor, friend, and a pioneering force in the study of psychology of communication technology

Contents

Biographical Notes

Cristina Botella is Full Professor of Clinical Psychology at Universitat Jaume I, Spain. Her main research line is the design and testing of clinical applications based on information and communication technologies for the treatment of emotional disorders. She has published over 100 papers and participated in various research projects granted by national as well as European funds.

Jeremy N. Bailenson is Founding Director of VHIL and an Associate Professor in the Department of Communication at Stanford. His main area of interest is the phenomenon of digital human representation. His findings have been published in over 70 academic papers in the fields of communication, computer science, education, and psychology.

Juana Bretón-López is an Assistant Professor of Psychological Treatments at the Universitat Jaume I, Spain. Her main research interests are anxiety disorders or behavioral addictions and the application of new technologies to treat these. She has published over 27 papers and participated in several research projects.

Pam Briggs holds a Research Chair at Northumbria University, where she explores identity, trust, privacy, and security and inclusion issues in social media. Her current projects address trust issues in peer-to-peer health care, the influence of social media in message dissemination during pandemics, location-based services and technologies, and cybersecurity identity management and digital personhood.

Judee K. Burgoon is a Professor and Director of Human Communication Research, Center for the Management of Information at the University of Arizona and a Distinguished Visiting Professor with the Department of Communication at the University of Oklahoma. She is the most prolific female scholar in communication in the twentieth century.

Jenna L. Clark is a Graduate Student and Royster Fellow at the University of North Carolina at Chapel Hill. Her research explores how individuals' perceptions of online relationships (perceived reality) can affect the outcomes of those relationships.

Antonios Dakanalis is currently Post-Doc Researcher at the University of Pavia, Pavia, Italy. His research interests are related to the concept of self-objectification and its role in the etiology of eating disorders. More, he is interested to explore the potential of positive technologies in the field of well-being and health care.

Kathleen M. Cumiskey is Associate Professor at the College of Staten Island, City University of New York. She is an expert in the social consequences of mobile technology. Her latest publications focus on the role that mobile phones play in changing people's perception of their Social and emotive environments. She is the coeditor (with Larissa Hjorth) on a volume published by Routledge entitled: *Mobile Media Practices, Presence, and Politics: The Challenge of Being Seamlessly Mobile* (2013).

Nicole B. Ellison (Ph.D., University of Southern California) is an Associate Professor in the School of Information at the University of Michigan. Her research focuses on how individuals and groups use online tools to initiate, maintain, and benefit from their social relationships.

Andrew J. Flanagin is Professor in the Department of Communication at the University of California at Santa Barbara. His research focuses on the ways in which information and communication technologies structure and extend human interaction, with particular emphasis on the processes of organizing and information evaluation and sharing.

Azucena García-Palacios is Professor of Abnormal Psychology at Universitat Jaume I, Spain. One of her main line of research is the design and testing of clinical applications based on information and communication technologies for the treatment of emotional disorders. She has published over 50 papers and she has participated in several research projects.

Melanie C. Green is an Assistant Professor of Communication at the University of Buffalo (SUNY). Her research examines the reasons why individuals choose online or ersatz interactions, and the consequences of these actions.

Mark D. Griffiths is Professor of Gambling Studies at the Nottingham Trent University. Dr. Griffiths has published over 450 refereed journal papers mainly in the area of behavioral addiction and has won 14 national and/or international awards for his work.

Jamie Guillory (Ph.D., Cornell University) is a Postdoctoral Fellow at the University of California, San Francisco, in the Center for Tobacco Control Research & Education (http://tobacco.ucsf.edu/users/jeg258). She studies the influence of technology on emotions, social behavior, and health using big data, experiments, surveys, and mobile Apps (https://sites.google.com/site/jamieguillory/home).

Anouk den Hamer (VU University Amsterdam) is a Ph.D. student in Media Psychology and Lecturer in Statistics and Methods at VU University Amsterdam. She holds a B.A. in Media Studies and an M.A. in Communication Science. She studies victimization and the role of media use in cyberbullying behavior among adolescents.

Jeffrey T. Hancock is a Professor in the Department of Communication and Cochair of the Department of Information Science at Cornell University. His work is concerned with how social media affect psychological and interpersonal processes, with a particular emphasis on understanding how language can reveal psychological and social dynamics.

Brandon Van Der Heide is an Assistant Professor of communication at Michigan State University. His primary interests include investigating how impression formation and social influence processes are changed when people interact using communication technologies and how these communication processes affect areas of practical interest such as electronic commerce.

Laura Hoffmann is doing her Ph.D. on social robotics. She is working as a Research Associate at the Department of Social Psychology: Media and Communication at the University of Duisburg-Essen, Germany. Her research interests lie in the area of social psychology and media psychology, with a particular focus on human–robot interaction (https://www.uni-due.de/sozialpsychologie/Hoffmann.shtml).

Johan F. Hoorn (D.Litt., D.Sc.) is Senior Associate Professor of Communication Science at VU University, Amsterdam. He is the principal investigator and project leader of Services of Electro-Mechanical Care Agencies (SELEMCA) directed at the creation of android robots in the care domain, so called "Caredroids."

Yan Huang is a doctoral student at the Pennsylvania State University. Her current research focuses on the psychological effects of media technologies. She is especially interested in how technological affordances influence the processes and outcomes of online persuasion in a variety of contexts such as health communication, advertising, and e-commerce.

Haiyan Jia (PhD, Penn State University) is a post-doctoral scholar at the Penn State University in the College of Information Sciences and Technology. Her research interest primarily focuses on the social and psychological effects of communication technology ranging from Web to mobile apps to smart objects. She also researches technology use in cross-cultural and international communication contexts.

Benjamin K. Johnson (M.A., Michigan State University) is a doctoral candidate at the School of Communication, The Ohio State University. His research is focused on selective exposure in new media settings, especially as it relates to social comparison and impression management (http://www.comm.ohio-state.edu/component/comprofiler/userprofile/106.html).

Sriram "Sri" Kalyanaraman is a Professor of journalism at the University of Florida's College of Journalism and Communications. His primary research focuses on persuasion and attitude change in online environments, particularly as informed by customized and interactive messages. He currently serves as co-editor of Media Psychology.

Cody O. Karutz is the Manager of Stanford University's Virtual Human Interaction Lab (VHIL). His main areas of interest are how virtual reality technologies and devices can be utilized to create new educational platforms.

Silvia Knobloch-Westerwick (Ph.D., University of Music, Theater, and Media, Hanover, Germany) is Professor at the School of Communication at The Ohio State University and Director of the Graduate Studies Program. She serves as Editor of Media Psychology. Her research examines media uses and effects, with a focus on selective exposure in the contexts of political communication, health communication, news, entertainment, and communication technologies (http://www.comm.ohio-state.edu/sknoblochwesterwick). Her latest monograph is titled "Choice and Preference in Media Use: Advances in Selective Exposure Theory and Research" (http://www.routledge.com/9780805855159).

Nicole C. Krämer (Ph.D., University of Cologne) has been leading the team of Social Psychology: Media and Communication as a Professor at the University of Duisburg-Essen, Germany. Her research interests include social aspects of human technology interaction and computer-mediated communication (https://www.uni-due.de/sozialpsychologie/ kraemer.shtml).

Elly A. Konijn (Ph.D., Utrecht University, the Netherlands) is Professor of Media Psychology at VU University Amsterdam (http://www.fsw.vu.nl/en/departments/communication-science/staff/konijn/), Former Editor of Media Psychology and Chair of ICA's Information Systems. Her research focuses on media use among adolescents, emotions and reality-fiction perceptions, virtuality, and empathic robots. Application areas are human well-being and health, media education, and creative industry (e.g., http://crispplatform.nl/projects/selemca).

Daria J. Kuss is a Lecturer in Applied Psychology at the Birmingham City University, UK. Dr. Kuss has published widely in many international journals and book chapters in the area of cyberpsychology and online addiction, with particular expertise in online gaming addiction.

Robert LaRose, Ph.D. (University of Southern California, 1979), is a Full Professor in the Department of Media and Information at Michigan State University. His area of interest is the use and effects of new media.

Eun-Ju Lee (Ph.D., Stanford University) is Professor in the Department of Communication at Seoul National University. Her research focuses on social cognition and social influence in computer-mediated communication. She has served as Associate Editor of *Human Communication Research* and *Journal of Communication*.

Debra A. Lieberman, Ph.D., is a media researcher and lecturer at the University of California, Santa Barbara, where she teaches in the Department of Communication, directs funded research projects, and directs the Center for Digital Games Research (www.cdgr.ucsb.edu). Her research focuses on processes of learning and health behavior change with interactive media and games.

Rich Ling, is the Shaw Foundation Professor of New Media at Nanyang Technological University. He is also employed by Telenor in Oslo, Norway, has been the Pohs Visiting Professor of communication studies at University of Michigan. His publications include *The Mobile Connection* (2004), *New Tech, New Ties* (2008), *Taken for Grantedness* (2012), and over 50 articles on social consequences of mobile communication.

Fabrizia Mantovani, Ph.D., is Professor of Communication at the University of Milano Bicocca, Milan, Italy. She is also Senior Researcher of the Centre for Studies in Communication Sciences (CESCOM) and member of the international research network on "Methodology for the Analysis of Social Interaction" (MASI).

Miriam J. Metzger is Professor in the Department of Communication at the University of California at Santa Barbara. Her interests lie at the intersection of media, information technology, and trust, centering on how information technology alters our understandings of credibility, privacy, and the processes of media effects.

Kristine L. Nowak (Ph.D., Michigan State University, 2000) is an Associate Professor and Department Head in the Department of Communication at the University of Connecticut. Her research examines human–computer interaction and how the use computer media influences interactions as well as perception, attribution, and information processing. She is particularly interested in how people use different types of avatars in computer-mediated interactions and how those avatars influence self-concept as well as person and message perception. Her work has been published in several journals including the *Journal of Computer Mediated Communication, Computers and Human Behavior, Journal of Communication, Media Psychology*, and *Presence: Teleoperators and Virtual Environments.*

Soo Youn Oh (M.A., Seoul National University) is a doctoral student in the Department of Communication at Stanford University. Her research interests include computer-mediated communication, human–computer interaction, media multitasking, and the emotional dynamics of virtual communities.

Jorge Peña, Assistant Professor at UC Davis, examines how avatar appearance and design features unconsciously affect communication in virtual settings. He is assessing how military training technologies impact trainees' cognitive and behavioral readiness. In partnership with Portuguese researchers, Dr. Peña is designing and evaluating a virtual environment for international collaborations.

Xanthe S. Plaisier (Ph.D. candidate, VU University Amsterdam) is currently pursuing her Ph.D. in Media Psychology at the Department of Communication Science at VU University Amsterdam. She holds an M.A. in Communication Science and an M.A. in Developmental Psychology. Her research focuses on adolescents' use of immoral, risky, and antisocial media content as a function of their developmental stage.

Tom Postmes is Professor of Social Psychology at the University of Groningen. He studies human behavior in virtual groups and communities, in crowds, and also in organizations and "normal" teams. Central to his work is the influence of personal and social identity: the relation of the individual to the group.

Astrid M. Rosenthal-von der Pütten (Ph.D., University Duisburg-Essen) works as a Research Associate at the University of Duisburg-Essen, Faculty of Engineering, Department for Social Psychology: Media and Communication. Her research interests include social effects of artificial entities, especially the uncanny valley, linguistic alignment with robots and virtual agents, and crisis communication in social media (https://www.uni-due.de/sozialpsychologie/vonderpuetten.shtml).

Soledad Quero is Professor of Clinical Psychology at Universitat Jaume I, Spain. Her main research is the adaptation and validation of CBT programs for different emotional disorders and the application of communication and information technologies to improve psychological treatments. She has published over 50 papers and she has participated in several research projects.

Artemio Ramirez, Jr., is an Associate Professor in the Department of Communication at the University of South Florida. His research focuses on the intersection between interpersonal and computer-mediated communication. Recent projects include examinations of shifts between communication modalities, social media use by older populations, and textual harassment among college students.

Giuseppe Riva, Ph.D., is Director of the Interactive Communication and Ergonomics of NEw Technologies—ICE-NET—Lab. at the Catholic University of Milan, Italy. He is also Professor of General Psychology and Communication Psychology and member of the steering committee of the Ph.D. program in Psychology at the same university.

Rosa Maria Baños is Full Professor of Psychopathology at University of Valencia, Spain. Her main research activity is focused on the application of technologies to Clinical Psychology for the treatment of mental disorders and promoting well-being. Dr. Baños has published over 100 papers and has participated in various research projects granted by national as well as European funds.

Elizabeth Sillence is a Senior Lecturer in Psychology at Northumbria University, UK, and a founding member of the PaCT Lab (http://www.pactlab.org.uk/). Her research investigates trust, social exchange, and decision making within e-health environments http://www.northumbria.ac.uk/researchandconsultancy/refprofiles/lizsillence/).

Russell Spears is a Faculty Professor at the University of Groningen. His research focuses on intergroup relations, social identity, and the role of group emotions in a range of group processes. He has researched how such processes operate in contexts of computer-mediated communication, as affected by characteristic the features of these contexts (e.g., anonymity, isolation).

Marloes Spekman is a Ph.D. student in Media Psychology within SELEMCA (http://crispplatform.nl/projects/selemca) at the Department of Communication Science of the VU University Amsterdam. Her project focuses on the role of emotion in perceiving and interacting with healthcare robots. Her research interests include affect and media use, serious gaming, pathological gaming, and health communication.

S. Shyam Sundar (Ph.D., Stanford University) is Distinguished Professor and Founding Director of the Media Effects Research Laboratory (http://www.psu.edu/dept/medialab) in the College of Communications at Penn State University. His research investigates social-psychological effects of technological features such as modality, interactivity, and navigability in digital media interfaces (http://comm.psu.edu/people/individual/s.-shyam-sundar).

Leila Takayama (Ph.D., Stanford University) is a Senior Researcher at Google[x], a Google lab that aims for moonshots in science and technology. Prior, she was a Research Scientist and Area Manager for human–robot interaction at Willow Garage. Her research investigates the intersection of ubiquitous computing and embodied cognition (http://www.leilatakayama.org).

Jolanda Veldhuis (Ph.D. 2014, VU University Amsterdam) is an Assistant Professor at the Department of Communication Science at VU University Amsterdam, the Netherlands. Her research interests include media effects on body image perceptions among adolescents, peer influence, communication strategies to negotiate the effects of idealized body exposure, and applications in the fields of health communication and media education.

Arun Vishwanath is Associate Professor of Communication at the University at Buffalo. His research focuses on the cognitive, affective, and behavioral determinants of technology adoption. His research looks at how individuals conceptualize new innovations and the role of interpersonal and mass-mediated messages in the adoption and utilization of innovations.

Jessica Vitak (Ph.D., Michigan State) is an Assistant Professor in the College of Information Studies at the University of Maryland. Her research examines how the technical structure of SNSs shapes users' experiences, as well as how these technologies help individuals maintain relationships with a variety of connections.

T. Franklin Waddell is a Ph.D. candidate at Pennsylvania State University. His research examines the psychology of person perception in the context of computer-mediated communication. In addition, he also investigates users' social responses to robots, avatars, and related interface agents.

Joseph B. Walther is the Wee Kim Wee Professor in Communication Studies at Nanyang Technological University's Wee Kim Wee School of Communication and Information, in Singapore. He is a Fellow in the International Communication Association and was a 2013 Fulbright Scholar at the Amsterdam School of Communication Research. He has developed several original theories and numerous empirical studies on CMC and interpersonal relations; see http://cmcresearch.org/walther/.

Axel Westerwick (Ph.D., University of Technology, Aachen, Germany) is Assistant Professor Clinical at the School of Communication at The Ohio State University. His research investigates information seeking on the Internet, online information credibility, and the use of communication technology.

Kevin B. Wright (Ph.D., University of Oklahoma) is a Professor in the Department of Communication at George Mason University. His main research interests include new technologies and health information-seeking, social support and health outcomes, and stress and burnout among healthcare workers.

Bartosz W. (Bart) Wojdynski is an Assistant Professor in the Grady College of Journalism and Mass Communication at the University of Georgia. His research focuses on technological and design variables in digital content presentation, and how they interact with individual differences to influence visual attention and information processing.

Preface

From websites and social media to mobile devices and robotics, the last 20 years have witnessed an explosion in the design, deployment, and diffusion of information and communication technologies. Together, these technologies have fundamentally changed the nature and scope of our media use as well as our communications with each other.

Scholars in communication, psychology, and related fields have been at the forefront of this revolution, contributing new theories and conducting exciting empirical studies to discover unique aspects of human psychology when interacting with communication technology. Insights gained from this scholarship have advanced our knowledge about not only human-technology interactions but also informed design of new technologies.

A nuanced understanding of user psychology is critical for anyone interested in the role played by communication technologies in a variety of domains, from individual health to interpersonal interactions to organizational decision-making to online mass communications. The 25 chapters in this book provide a range of insights that go beyond simply cataloging how human users have responded to new communication technologies and focus on how the aspects of the technology interact with aspects of human psychology, thereby explaining the continued fascination with—and success of—tools such as Smartphones, Games, and Social Networking Apps. Across the chapters, it is clear that a wide array of technological variables interact with a number of human variables in order to make such success stories possible. The interplay of these factors is quite complex, and researchers provide a series of parsimonious accounts to help us understand such phenomena.

The three theoretical overviews in the opening part of the handbook provide analytical frameworks for academic research on the role played by communication technologies in interpersonal communication, group communication, and mass communication, respectively. The next part addresses a core question underlying our interactions with media technologies—how do we orient to technologies as sources of communication? The human tendency to assign agency to technological entities has resulted in a number of significant psychological outcomes, including greater socialness in human–computer and human–robot interactions. The third part of the book identifies the technological affordances that play a role in social interactions, particularly online and mobile communications, among users in a variety of contexts, from dating sites to online courses. Part IV examines the psychology behind the actual use of communication technologies, from adoption and diffusion to overuse and addiction. Part V addresses user experience of digital media by analyzing how users exercise control over their online interactions, including how they selectively read and assess the credibility of online content. The final part of the book focuses on health technologies, a dominant domain of application of the research on the psychology of communication technology, with chapters focusing on online health information and the use of new technologies for preventive health as well as therapy.

The authors of the chapters in this handbook are all leaders in their respective research domains and were especially invited to synthesize their corpus of work for this volume and use it as a vehicle for articulating their agenda for research. As a result, each chapter serves as a go-to place for the philosophy, theory, and methodology of a particular researcher or group of researchers. All authors were asked to (a) cite their influences, (b) outline their assumptions, (c) state their broad thesis in theoretical terms, (d) explicate the central concepts involved, (e) derive representative hypotheses for empirical testing, (f) describe methods used in general terms, (g) highlight major findings across studies, including new conclusions being drawn, (h) discuss theoretical implications of these findings, (i) brainstorm design and other practical implications of those findings, and (j) provide directions for future work in the area. As a result, the chapters represent summative accounts and the latest thinking in the major streams of research at the intersection of psychology and communication technology. They serve as useful reference work for graduate students and other scholars interested in this area of research. They also provide roadmaps for stimulating future research on psychological aspects of communication technologies.

S. Shyam Sundar

Acknowledgments

I would like to thank all the authors for their contributions to this volume. Given the unique focus of this handbook on synthesizing their own work, the authors had to make a special effort to articulate their respective programs of research within a limited time. I appreciate their willingness and enthusiasm to embark on this project with me.

I would also like to acknowledge the substantial institutional support provided to me by the College of Communications at Penn State University, including the assignment of graduate assistants who have helped me with this compilation. Since the inception of the project, four assistants—Hyang-Sook Kim, Xue "Yuki" Dou, Haiyan Jia, and T. Franklin Waddell—have managed the logistics pertaining to different phases of the book.

I am particularly thankful to members of my own lab group. Saraswathi Bellur, Eun Go, Jiangxue "Ashley" Han, Yan Huang, Anne Oeldorf-Hirsch, Eunhwa Jung, Hyunjin Kang, Guan-Soon Khoo, Jeeyun Oh, Akshaya Sreenivasan, T. Franklin Waddell, Qian Xu, Mu Wu, and Bo Zhang have been instrumental in inspiring me to undertake this venture and provided support at various stages, from preparing the prospectus to pitching potential authors to providing feedback on chapters.

The staff at Wiley has also been a major source of support and encouragement. I would like to thank Elizabeth P. Swayze, Senior Editor, Communications and Media Studies, for pursuing me to edit this handbook, Julia Kirk for being an especially responsive Project Editor, and editorial assistant Allison Kostka for handling the contracts and related paperwork so that the authors could focus on the content of the book.

Finally, and most importantly, my thanks go to my family—my wife Savitha Kolar and our son Shreyas Kolar Sundar—for their understanding and support as I took this project on top of my already overfull plate of research, teaching, and service activities. Their sacrifice, kindness, and good cheer have been instrumental in helping me in all my academic endeavors, including the editing of this book.

Part I

Theoretical Overviews

1

Interpersonal and Hyperpersonal Dimensions of Computer-Mediated Communication

Joseph B. Walther[1], Brandon Van Der Heide[2], Artemio Ramirez, Jr.[3], Judee K. Burgoon[4], and Jorge Peña[5]

[1] *Nanyang Technological University, Singapore*
[2] *Michigan State University, MI, USA*
[3] *University of South Florida, Tampa, FL, USA*
[4] *University of Arizona, Tucson, AZ, USA*
[5] *University of California, Davis, CA, USA*

The social information processing theory of computer-mediated communication (CMC) was the first of several theoretical models of interpersonal interaction online to explain how individuals and groups formed impressions and developed relational communication via text-based electronic communication. Prior to its introduction, the predominant theoretical approaches to CMC predicted that the relative lack of nonverbal cues in CMC compared with face-to-face (FtF) communication would reduce the socioemotional quality of communication online. The social information processing (SIP) theory, in particular, articulated assumptions about the CMC medium, the relationships between nonverbal and verbal cue systems, and users' adaptation to media that represented a significant departure from other prevalent models at the time it was introduced.

Premises of the social information processing theory became the basis for several models to follow, including the hyperpersonal model of CMC. Both approaches to CMC focus on message qualities and how the characteristics of the CMC channel (such as the lack of most nonverbal cues, and, at times, the asynchronous nature of messaging systems) interact with interpersonal goals and strategies, resulting in systematic patterns of interaction via mediated channels. They each involve a high degree of human agency and depict how users appropriate the channel and its technological characteristics to suit their communicative purposes. They depict CMC users as more or less creative and opportunistic rather than as passive adopters of a relatively restricted medium. This chapter reviews the development, status, and

The Handbook of the Psychology of Communication Technology, First Edition. Edited by S. Shyam Sundar.

future of the social information processing theory and hyperpersonal model of CMC, and their potential contributions to our knowledge about new media in interpersonal relations.

The Social Information Processing Theory of CMC

To understand these models and their potential contributions fully, it is useful to consider the emergence of SIP in its historical theoretical context.

The theoretical landscape prior to SIP

Although it would not be until the mid-1990s that the Internet was available to the general public, CMC started its adoption in a few professional, educational, and recreational venues in the late 1970s and 1980s. As it appeared in high-tech businesses and proprietary services like CompuServe and Prodigy, in universities and dial-up bulletin-board systems (see Rapaport, 1991), researchers, journalists, and early adopters began to ask how CMC might change communication and how its changes might affect the social processes to which it may be put (e.g., Hiltz & Turoff, 1978; Johansen, Vallee, & Spangler, 1979; Rheingold, 1993).

The new medium was text-based, rather than multimodal. That is, compared with FtF communication, and even to telephone conversations, there were no nonverbal cues accompanying the written messages in CMC. Early on, research considered this characteristic a likely culprit that would make CMC devoid of social and interpersonal richness. The SIP and hyperpersonal models would turn this concern on its head, so to speak, first by addressing how users overcome the lack of nonverbal cues in making their messages sufficiently personal over time, and later, by explaining the actual advantages that accompany the flexibility of communicating via language and only language.

Although CMC sent written messages across vast distances almost instantaneously, it also featured the ability to "store-and-forward" messages asynchronously. Most commonly seen in email, and now text messaging and social network systems, asynchronous communication means that one individual can post a message and it is retained in the CMC system until its intended reader(s) examine it at another time, at their convenience. This characteristic, too, departed from FtF and telephone messaging. Some observers suggested that asynchronous messaging would make it impossible for communicators to make coherent sense of a series of messages and responses (see, e.g., McGrath, 1990). The hyperpersonal model of CMC, in particular, would argue how asynchronous communication and the perceived control over message construction can actually be advantageous and facilitate more desirable messages and enhanced communicative control (Schouten, Valkenburg, & Peter, 2007).

When CMC was new and first being studied, then, much was expected, but much doubt about its interpersonal potential also accompanied the early theories and research about the new medium. Would CMC simply facilitate communication without any particular change, and obviate the need to schedule (or travel to) FtF meetings (Rockart & DeLong, 1988; cf. Vallee et al., 1975)? Would communicating with others remotely, without nonverbal cues, dehumanize its users (see Orcutt & Anderson, 1977)?

To answer these and more middle-range questions about the effects of interactive media, researchers appropriated established theories from teleconferencing research and developed original theories focusing on CMC per se for the purpose of predicting and explaining the likely effects of online interaction, primarily in large and small groups. These positions, as it turned out, were consistent with many positions in nonverbal communication research. The earliest theoretical positions argued that since the vast majority of our emotional expressions relied on the exhibition and detection of nonverbally encoded messages, text-based messaging without nonverbal cues must therefore lack socioemotional expression. Some theorists went farther to suggest that CMC, without the warmth of natural human communication, would lead users to antagonism and hostility with one another. Empirical research, primarily experimental, seemed to support these predictions.

The assumptions and propositions of the "cues-filtered-out" approaches to CMC (see Culnan & Markus, 1987) have been summarized in numerous publications. Their essential position is that nonverbal cues facilitate a number of functions related to identifying who others are, what their individual personalities are, how they express emotion, and what their utterances mean. As though nonverbal cues have a unique capacity to accomplish these functions, the general argument of these approaches is that CMC, without nonverbal cues, is impoverished or is incapable of supporting these communicative functions (for review, see Walther, 2010). Small group interaction research long held that in order to be successful, groups need to support both task and socioemotional communication. Task communication is the facts, opinions, ideas, and arguments that group members need to surface in order to inform their decision making. Socioemotional messages convey charisma, humor, agreement, and interpersonal regard, and are also considered critical in order for groups to have effective and satisfying conversations. Research concerned itself with the question of whether and how well CMC could support both these aspects of small group interaction.

Empirical support for these positions often involved experiments comparing small groups communicating by CMC or FtF methods for a limited time, and content analysis of transcripts with which to compare verbal communication in both settings. Many such experiments appeared to support the cues-filtered-out perspective.

At about the same time as these theoretical and research-based interpretations of the effects of CMC appeared in the management, information systems, engineering, and psychology literature, stories of a quite different nature appeared, sometimes in academic outlets, and at other times in the popular press. Anecdotes described shy youngsters who found friends online, and remained online, who had never experienced the kinds of best friends as they did in cyberspace. Spontaneous romances arose via text, surprising their participants with emotional intensity, and appalling their friends. Case studies of high-tech firms showed that internal networks were being used as much for play as for work (Ord, 1989; Steinfield, 1986), for exchanging movie reviews as a hobby as much as distributing parking rules to employees. The cases did not fit the theories, and the experiments were at odds with the anecdotes.

Although anecdotal examples should not be definitive scientifically, they seemed to reflect the experiences of a growing number of CMC users, while the theories and research seemed to map on to the suspicions about the medium among those who had not engrossed in it as much.

One other departure from the cues-filtered-out approach appeared in a theory of managerial media selection. It was originally referred to, also, as the social information processing theory of CMC. The term, in this case, came from work by Salancik and Pfeffer (1978), who had argued that managers' perceptions of organizational artifacts are influenced by the social information one's coworkers generate in regard to them. That is, managers perceive things due in part to the communication reflecting others' perceptions of those same things. Fulk, Steinfield, and colleagues (1987) applied this approach to organizational members' perception of the richness of email. They argued, and later demonstrated empirically, that workers viewed email's expressiveness based not only on their own apprehension of email's capacity, and not only due to its actual features, but to a significant extent their perceptions were affected by the opinions and email-oriented behaviors of other individuals who shared strong sociometric ties to one another. Fulk and colleagues soon renamed their model a *social influence theory* of CMC (e.g., Fulk, Schmitz, & Steinfield, 1990).

Social information processing

The other social information processing theory of CMC, introduced in the 1992 *Communication Research* article by Walther, was the first formal theory among several that would soon emerge to suggest predictions and explanations about the relational potential of online communication alternative to the cues-filtered-out approach. The introduction of the theory attempted several objectives: to reflect a fundamental but relatively obscure set of paradigmatic assumptions about the relationship of verbal and nonverbal message cues and the relevance of this relationship for the translation of affective expression from physical to verbal behavior when physical behavior is obviated; to be able to account both for the impersonal communication findings of numerous laboratory studies and to account for the anecdotal accounts of relationship development online, through identification of specific factors with the potential to moderate the effect of the medium on its users' communication; and to articulate new theoretical propositions capable of generating testable hypotheses and a new view of CMC. SIP specified a new set of assumptions about what people do when they communicate using different channels, and how they respond to a severe reduction of nonverbal cues by CMC.

The theory seeks to explain how, with time, CMC users are able to accrue impressions of and relations with others online that achieve the level of development that is expected through offline communication. It was developed in light of certain philosophies of communication that are useful in understanding the positions laid out by SIP.

A functional approach to communication

Social information processing theory arose among a community of researchers who espoused a functional approach to communication research. This approach offered some fundamental assumptions about communication and how to approach its analysis. It focused on how communicators undertook the processes of social influence, impression formation and management, information processing, and relational communication, across settings and contexts. In rigorous research on nonverbal communication, rather than focus on what certain behaviors or cues meant per se or what specific cues were the sole conveyors of specific effects, the functional approach subscribed to an alternative idea: Any of a number of nonverbal behaviors could contribute to the communication of different communicative functions depending on the activation of other nonverbal and verbal cues. Moreover, any communicative message could potentially be conveyed through alternative combinations of cues, and that, typically, the accomplishment of communication functions involved the combination of verbal as well as nonverbal cues. This latter notion does not suggest that verbal cues denote content while nonverbal cues connote affect (cf. Watzlawick, Beavin, & Jackson, 1967). Rather, message elements from different cue systems may be duplicative, complementary, contradictory, or otherwise intertwined with one another (Burgoon, Buller, & Woodall, 1989). As Walther and Ramirez (2009, p. 289) would later reflect, "No single nonverbal behavior, or specific set of behaviors, had a monopoly on the conveyance of social meanings. To study communication was to study the fluid encoding and decoding of complex interactions of appearance, proximity, touch, and other cues, while communicating verbally, and as relational contexts varied."

Several prior studies were particularly informative in translating these precepts into the domain of CMC. One was a study by Donohue, Diez, Stahle, and Burgoon (1983) that reported a clever experiment examining whether communicators whose personal space is encroached, who cannot physically retreat, respond to the incursion through psychologically distancing verbal behavior. Indeed, subjects who were crowded by a confederate reduced the ***verbal immediacy*** of their speech. This dynamic demonstrated a functional interchangeability among nonverbal and verbal cue systems for the management of interpersonal immediacy. When nonverbal responses were constrained, communicators accommodated through language variations to accomplish what would otherwise have been done through physical behaviors.

With this background, the cues-filtered-out notions of CMC—which suggested that nonverbal cues held a monopoly on the expression and detection of identity, affect, and meaning—were antithetical to the functional approach to communication generally and the notion of verbal/nonverbal interchangeability. Others had speculated that CMC users might get used to the medium and learn to improve its emotional expressiveness with experience (Hiltz & Turoff, 1981; see also Carlson & Zmud, 1994). A functional approach suggested that communicators already had the skills to write expressively when a FtF or phone conversation were unavailable, and that this skill should readily port to electronic writing.

From this perspective, the name of the SIP theory suggests that the study of communication generally, and CMC in particular, should focus on how individuals process social information (as well as instrumental, or task-related information) using whatever cue systems they have with which to do so. The use of the term social information processing was consistent with research in cognitive social psychology that focused on cues that lead to variations in person perception (for review, see Wegner & Vallacher, 1977).

Assumptions and propositions

The SIP theory specifies several premises that explain how CMC can affect impressions and relational communication. Although it explicitly recognizes that CMC is devoid of the physical nonverbal cues that accompany FtF communication, it does not recognize that this provides an incapacity or a motivational deterrent to develop impressions. Its first assumption was that communicators seek to develop relationships with others no matter what medium they use. It proposes that communicators use whatever cues they have available to them in order to generate and apprehend interpersonal (as well as instrumental) messages. Therefore, when they cannot employ nonverbal cues to do so, individuals may adapt the encoding and decoding of social information (i.e., personal, socioemotional, or relational messages) into text. Although many readers have interpreted this argument to refer to emoticons (typed-out smiles, frowns, and other faces; Derks, Bos, & von Grumbkow, 2007), the theory focuses primarily on language content and style as more important conduits of interpersonal information. A later study established that, when both emoticons and verbal messages appeared in email, emoticons accounted for less variance in inferences than did language variations (Walther & D'Addario, 2001). But the first argument of SIP theory is that individuals translate their social message encoding and decoding into verbal behavior when, as in CMC, nonverbal cues are unavailable, and that in doing so they can achieve as much impression development and relational depth as do people communicating FtF.

The second major argument is that, despite CMC's capacity to convey personal and socioemotional content, the exchange of information among communicators operates at a different rate than FtF communication. That is, because the language channel carries fewer messages per instant than FtF utterances (since FtF utterances are accompanied by a variety of vocalic, kinesic, proxemic, and appearance cues), CMC users' ability to achieve levels of impression and relational definition equivalent to FtF interaction are expected to require more exchanges. Generally this will require more time, especially when CMC is asynchronous, but the difference should pertain to real-time CMC as well. CMC users need time to compensate for the slower rate in order to accumulate sufficient information with which to construct cognitive models of partners and to emit and receive messages with which to negotiate relational status and definition.

Empirical support

Empirical tests of SIP initially examined the second proposition, looking at the interaction effect of time and medium on impression development (Walther, 1993) and relational communication (Walther & Burgoon, 1992). Small groups communicated via asynchronous CMC addressing three decision-making tasks over 6 weeks, or via FtF meetings three times, 2 weeks apart. The impression development results were most clear: After the first task, FtF participants had strongly developed impressions of their partners, whereas CMC participants' impressions were significantly less developed. The results after one time interval only perfectly replicated the cues-filtered-out prediction. But after the second interval, CMC participants' impressions continued to develop, and after the third, there was no difference in impression development between the CMC and FtF conditions. Scores on relational communication measures generally reflected greater positivity over time, although the pattern of differences and similarities in relative levels of each by medium was less clearly pronounced.

These patterns have been replicated in a number of studies by other researchers. It is noteworthy that some of these investigations used synchronous CMC chat systems repeatedly over several days (rather than asynchronous CMC over a period of weeks) and achieved results that strongly supported SIP theory's predictions of development and improvement of impressions and relations over time (Hian, Chuan, Trevor, & Detenber, 2004; Wilson, Straus, & McEvily, 2006). A meta-analysis of published studies existing at the time indicated that CMC experiments that had restricted users' time online averaged less positive socioemotional communication than did studies that did not restrict CMC use (Walther, Anderson, & Park, 1994).

Research has also addressed the first major theoretical contention of SIP, the translation of affective messages from nonverbal to verbal behavior. Ironically, by the time the first direct empirical examination of this proposition was first submitted for publication review in 2004, the premises of SIP theory had diffused so strongly that one journal reviewer argued that the mechanism examined in this study had already been well established. (As a later discussion will review, this theoretical premise is now strongly accepted in some quarters, although it remains strongly challenged elsewhere.) This study, which was the first of its kind, compared the verbiage that dyad members exchanged in an experimental CMC chat setting to the verbiage, kinesic cues, and vocalic cues generated in FtF dyads, in which one dyad member had been prompted by researchers to behave in a strongly friendly or unfriendly manner (Walther, Loh, & Granka, 2005). Researchers did not instruct these ad hoc confederates how to signal positive or negative effect, but to do so as the participants wished. Researchers then analyzed recordings of the FtF confederates in three ways: For verbal behavior, they transcribed the conversations and subjected them to coding content-level strategies; for kinesic cues, they had coders rate the participant's movement by watching videorecordings without any sound; and for vocalic analyses, coders listed to recordings using a low-pass content filtering systems that obfuscated the verbiage while playing the vocal tones, pitch variety, pause/phonation sequences, and other vocal behaviors aside from the verbal content. Likewise, coders analyzed the verbal transcripts from the CMC chats.

Results indicated that CMC transcripts presented as strong a degree of affective difference as did FtF recordings. In the CMC conditions, various verbal cues emerged, ratings on which accounted for over 90% of the variance in perceived liking and immediacy. Some of these behaviors included the manner in which individuals expressed disagreement with partners, either by praising partners' ideas and agreeing with them, or using more bluntly dismissive and disagreeable phrases. In FtF settings, unsurprisingly, vocalic cues conveyed the greatest proportion of variance in affective expression, followed by kinesic behaviors; although some verbal behaviors differed between the liking and disliking conditions in the FtF settings, the verbal behaviors as a whole did not account for a significant degree of emotional expression when compared statistically with the groups of vocalic and kinesic cues. However, the verbal behaviors in CMC accounted for as much variance in affective expression as did the multimodal cues in the FtF setting. Despite its centrality to the empirical support for SIP theory, there have been few replications of this research (cf. Hancock, Gee, Ciaccio, & Lin, 2008; Hancock, Landrigan, & Silver, 2007).

Another important link in the theoretical chain of SIP pertains to self-disclosure, a form of messaging with a long history of association with the development of impressions and the development of relationships. Independent researchers more or less simultaneously found similar results with respect to the impact of online disclosure relative to FtF. In CMC, disclosures and personal questions comprise greater proportions of the total utterances in dyadic conversations among strangers than they do in comparable FtF discussion (Joinson, 2001; Tidwell & Walther, 2002). Moreover, CMC users' disclosures and questions are more intimate than those of FtF counterparts.

Continuing impact of SIP

The tenets of SIP appear to have had a meaningful impact on the study of CMC and the field of communication more generally. Although the SIP theory originally emerged before the World Wide Web, when CMC systems were entirely text-based, and although its currency has been questioned in an era of multimodal social media, a variety of factors indicate that its potency as a model of CMC remains strong. Text-based messaging continues to increase on a variety of platforms. Although teens' use of Instant Messaging may or may not be as popular as it once was, teens in particular are among the most frequent users of texting via smartphones (Lenhart, 2012). The Radicati Group (2011), a corporate technology company, estimates that a typical corporate employee sends or receives 105 email messages per day. To the extent that SIP is connected to the use of text-based CMC, its utility persists.

The SIP approach has generated numerous studies extending its focus on language, its use, and its particular potency in online discourse. Hancock (2004) established that communicators are more explicit in their expression of irony online than in FtF encounters. LaRose and Whitten (2000) used the framework to examine instructors' immediacy in online courses, and O'Sullivan, Hunt, and Lippert (2004) examined verbal immediacy as a code that encouraged affiliation via educational websites.

The foundational studies about online disclosure have influenced a variety of new lines of research, including its frequency and accuracy in the use of online date-finding systems (Gibbs, Ellison, & Heino, 2006) and the role of CMC in the development of one's psychological sense of self among adolescents. In this latter regard, a series of studies by Valkenburg, Peter, and colleagues (for review, see Valkenburg & Peter, 2009) has established that teens' online disclosure can lead to improvements in their psychosocial development, in that, presuming that teens disclose more online than offline, their friends offer more reinforcement for the concepts of self that adolescents explore in online chat discussions. Their level of benefit is related to the control they believe they have in using these systems, and the feedback they get from their real-life friends via online channels.

New extensions to SIP continue to emerge. Recent work suggests that different message contents in CMC vary in the distinctiveness and relevance they offer readers, depending on the context and goal of those readers. As a result, readers place different weight on contextually distinct pieces of social information. In one study, participants formed strong extraversion or intelligence judgments of a target only when the context propelled them to try to identify message senders in terms of these specific personality characteristics. When the same cues were distinctive but not contextually relevant (or vice versa), they impacted impression judgments significantly less (Van Der Heide, 2009). Future research on information-seeking goals and the differential value of social information cues has the potential to replace SIP's original assumptions that CMC users garner impressions holistically and inductively. Research on contextual influences on the interpretation of social and relational information in CMC has already established that readers interpret text-based self-disclosures exchanged via Facebook as being more intimate when they are transmitted privately rather than shared publicly (Bazarova, 2012). Turning to the social information value of photographic information (see Westerman, Van Der Heide, Klein, & Walther, 2008), D'Angelo and Van Der Heide (in press) demonstrated that observers evaluate physicians who post a professional photograph of themselves on Facebook as significantly more credible than those who post the same photograph on a WebMD.com profile. The researchers concluded that when social information is normative (and not particularly distinctive) it has less impact on social judgments than social information that stands out.

Challenges to SIP

The impact that SIP has had has not gone without challenges. One line of contest arose from disagreements about CMC's potential to convey personal information rather than group-based impressions. Game studies presents an opportunity for SIP research that has been relatively underexplored to date. Another challenge has come, and continues to surface, from the perspective that text-based CMC is communicationally impoverished compared with multimodal alternatives.

A strong critique of SIP appeared in a number of papers advancing an alternative approach to CMC, the social identity model of deindividuation effects, or SIDE model (for review, see Spears, Lea, & Postmes, 2000). Authors connected with the SIDE model have suggested that data-analytic results from SIP-related research does not support SIP (when indeed a pure SIP effect was overridden in an disordinal interaction), to more extreme criticism dismissing the entire enterprise of online interpersonal relations as asystematic, charging that the study of interpersonal communication online was misguided and had set back the study of CMC (Postmes & Baym, 2005). The SIDE model formerly argued, as an alternative, that when CMC users communicate with others who they do not see, they experience deindividuation, or a suspension of individual identity. That is, CMC users who communicate without seeing each other and without visually witnessing each other's individual appearance characteristics experience *visual anonymity.* As a result, according to SIDE theory, an overall sense of interpersonal anonymity pervades CMC use. In this individual identity vacuum, when a group to which such a communicator belongs becomes salient to the CMC user due to contextual clues (such as the name of the communication site, the purpose to which communication takes place, due to unconscious cueing, or due to instructions from an experimenter, CMC users gravitate toward the group identity and relate to other CMC conversants on that basis. Rather than relate interpersonally, that is, as individuals, CMC users relate on the basis of in-group/out-group dynamics. According to SIDE, members' attraction to the group supersedes what otherwise might have been, as SIP would argue, attraction to other individual group members. A good deal of evidence supports the SIDE model (for review, see Walther & Carr, 2010).

Careful readers will detect that a basic disagreement between the SIDE and SIP approach deals with anonymity. Anonymity is a central construct in SIDE theory, but it is more or less dismissed in SIP. The SIDE model relies on a causal leap from the existence of visual anonymity of CMC to a sense of depersonalization. On the other hand, SIP theory admits there is visual anonymity online, but holds that CMC users get to know each other inter-individually despite the lack of visual cues. Moreover, SIP acknowledges that there are a number of other individual identifiers in much CMC, such as unique usernames, or self-descriptions, not to mention style and language differences, that readily mark individuals as different from one another online, and become especially salient over time.

In their later works, SIDE theorists have come to agree that CMC users do get to know each other online, over time, as individual people. Specifically, some SIDE advocates—major critics of SIP among them—have embraced the potential necessity of the development of interpersonal impressions, identifications, and relationships online over time and the exchange of text-based messages as part of a process of developing a group identity in CMC groups (Postmes, Spears, Lee, & Novak, 2005).

But rather than suggesting that the two theories are converging, it is most likely the case that each theory has its place depending on the boundary conditions imposed by the nature of the setting, and especially the time frame within which online communication may take place. A recent article suggested that the study of CMC was badly in need of more consideration of theoretical boundaries, so that, rather than dismiss one theory as superior to others, we may identify circumstances in which one theory simply does not apply and another may (Walther, 2009). For instance, in settings where CMC users have time-constrained (or no actual) interaction with one another, anticipate no future interaction, and have been prompted to seek common group characteristics, there is every reason SIDE dynamics may result, whereas the full impact of SIP cannot be expected to obtain. Likewise, when CMC users encounter comments left by other users who are similar to them in some obvious respect, yet are visually anonymous, and do not interact, SIDE adheres (e.g., Walther, DeAndrea, Kim, & Anthony, 2010). Where CMC users experience long-term associations, however, interpersonal attraction tends to increase over time even when CMC groups are prompted to look for group identification (Rogers & Lea, 2004).

The SIP approach has had minimal impact on the study of online games, but more could be done in this area. One significant contribution has been in the work of Peña and Hancock (2006), who established that unacquainted gamers playing battle games exchange a considerable proportion of socioemotional communication. Klimmt and Hartmann (2008) provide an extensive review of the interpersonal dynamics at various levels of relationships within online games, many of which involve a great deal of social interaction. For instance, not only do unacquainted gamers become acquainted, they also join clans or other social groupings in which players develop a sense of one another's personal and sociotechnical capabilities, and come to trust and regard one another. As game studies tend to be cross-sectional rather than longitudinal, the opportunities to examine SIP's precepts in these environments have been minimal, but can occur in future research.

It is also the case that SIP continues to be challenged by researchers who maintain that more bandwidth (i.e., more communication channels) is superior to text-based CMC. These positions do not generally offer rationales that differ substantively from the cues-filtered-out positions that were popular in the 1980s. Empirical studies accompanying such challenges generally find that telephone communication, FtF interaction, or avatars generate more positive socioemotional responses than does text-based CMC (e.g., Epley & Kruger, 2005; Okdie, Guadagno, Bernieri, Geers, & Mclarney-Vesotski, 2011). However, rather than accept that SIP is simply incorrect, it is important to note (as most such research efforts do not) that in no case have these studies included the boundary conditions that SIP requires. That is, no such study has included long-term interaction or even the anticipation of future interaction. In a sense, they corroborate SIP's contention that, when CMC users operate in a time-limited context with no anticipation of future interaction, their behavior is expected (by SIP) to reflect the impersonal qualities that are also predicted by the cues-filtered-out positions. Only when such parameters are relaxed is SIP expected to adhere, and when those conditions are met, SIP appears to be relatively accurate.

One recent study sought to estimate the role that alternative channels in social network sites such as Facebook or the Dutch Hyves system offered users in terms of uncertainty reduction about another person: Antheunis, Valkenburg, and Peter (2010) argued that social network sites provide an abundance of asynchronous and unintrusive biographical, pictorial, and sociometric information about other people, and that such forms of social information are more accessible than interactive, text-based CMC should offer. Their results indicated, however, the interactive communication remained the information-seeking strategy that reduced uncertainty the most strongly, despite the appeal and use of other modalities. Despite the massive popularity of Facebook, its most frequently used features remain text-based messaging in the form of wall postings, status updates, commenting, and private messaging using synchronous or asynchronous modes. Despite the perseverance of plain-text CMC, the prevalence of multimodal CMC compel us to ask what happens to disclosure and other relational messages when they occur within interfaces with additional modalities. If pictures and videos and such do not replace interactive exchanges, do they provide interactive exchange an interpersonal running start, so to speak? That is, do conversations go deeper, faster, when they are grounded against a profile and/or photos? These are important questions for future research.

Hyperpersonal CMC

The aspect of SIP that has become a foundation point for other conceptual models of CMC is its assumption that CMC users encode and decode language cues in the service of impressions and relational goals, and that different levels of effort accompany different media's capacity to transact communication that supports these goals. The hyperpersonal model of CMC (Walther, 1996) extends the first of these assumptions into processes in which impressions and relational states exceed rather than simply meet what is expected to occur in parallel, FtF settings. The hyperpersonal model proposes a set of four concurrent routines that, together, seek to explain CMC's support of relationships with relatively greater desirability and intimacy than occur in offline counterparts.

Receivers

When receiving messages from others in CMC, an individual may tend to exaggerate perceptions of the message sender. Absent the physical and other cues that FtF encounters provide, rather than fail to form an impression, receivers fill in the blanks with regard to missing information. This often takes the form of idealization if the initial clues about another person are favorable. The original hyperpersonal model drew on SIDE theory (Lea & Spears, 1991) in formulating how receivers make over-attributions of similarity when communicating under conditions of visual anonymity, if contextual cues suggest that a conversational partner shares some salient social identity with the receiver. A revision of the hyperpersonal model has broadened its view to consider that receivers' exaggerated impressions may derive from stereotyped perceptions of others personality characteristics or person prototypes (Walther, 2006).

A recent study shows how simple these kinds of overattributions based on language can be (Spottswood, Walther, Holmstrom, & Ellison, 2013). An experiment presented sample postings from an online social support system, where one individual ostensibly complained about being stood up for a date, and a variety of responses followed. These responses' language varied in the degree of person-centeredness they exhibited, that is the degree they reflected sympathy and acknowledged the original poster's feelings rather than giving advice or telling him what to do or think. In offline social support, user-centered messages are more normatively associated with females, and they are not accepted by receivers when male support-givers offer them. In this study, in some cases, person-centered messages were accompanied by usernames for which the gender was indeterminate (e.g., b0k_choi and zy523). Indeed, when the message form was more person-centered, observers attributed the sender's sex to be female, whereas when the messages appeared in low person-centered form, observers interpreted them as coming from male online helpers.

Senders

Text-based CMC facilitates selective self-presentation. Online users can transmit only those cues they desire, and need not reveal physical, environmental, or undesirable speech behaviors. By constructing messages that portray themselves in desirable ways, they contribute to the idealized perceptions receivers develop of them.

One of the most interesting applications of this component of the hyperpersonal model appears in studies of deceptive self-presentation, particularly in online dating systems. Toma, Hancock, and Ellison (2008) described how facets of CMC facilitate deception by explicit reference to the hyperpersonal model. They argue that the development of online dating profiles takes place using asynchronous communication tools, allowing users to plan and write their online self-presentations more deliberately than they would FtF. Moreover, as Toma and Hancock (2011) suggested, online profiles are editable, allowing users to rewrite and revise their profiles to make them more appealing, a notion that leads to the next aspect of the hyperpersonal model.

Channel

The third dimension of the hyperpersonal model is characteristics of the channel and how CMC as a medium contributes to the deliberate construction of favorable online messages. One part of the channel factor focuses on the mechanics of the CMC interface, suggesting that users exploit the ability to take time to contemplate and construct messages mindfully, as alluded to in the dating/deception research cited earlier. In many CMC applications (especially asynchronous systems), users may take some time to create optimally desirable messages without interfering with conversational flow, unlike the effects of response latencies in FtF interaction. The hyperpersonal model further suggests that CMC users capitalize on the ability to edit, delete, and rewrite messages, in order for messages to reflect intended effects, before sending them. The introduction of the model further suggested that CMC users may redirect cognitive resources into enhancing one's messages; without the need to pay attention to the physical behaviors of one's conversational partner or oneself, or to the ambient elements where one is physically located when communicating (in contrast to these demands on attention in FtF conversations), CMC users can focus their attention on message construction to a greater extent than they could FtF. Research found that CMC users in an asynchronous discussion system who believed they were writing to an attractive partner exhibited more editing (backspaces, deletions, replacements) composing their messages than did those who believed they were writing to less attractive partners (Walther, 2007). The degree of editing also corresponded to the degree of relational affection that outside coders found in the recorded messages.

A recent study extended the idea of CMC users planning their messages for optimal affective impact by examining how CMC users avail themselves of web-based resources in order to plan and devise the composition of their exchanges (Walther, Van Der Heide, Tong, Carr, & Atkin, 2010). In this study, pairs of strangers were told they would be discussing what the best hamburger restaurant was in the area. One of the partners was either instructed to try and get the other partner to like him or her, or to get the partner to dislike him or her. More often in the dislike than like condition, the CMC user looked up the partner's favorite hamburger before the chat started in order to find (bad) things to say about it during the discussion. These users expressed more disagreements and negative opinions during the chat as well. Users who had been asked to be more friendly were more spontaneously agreeable.

One side effect of these discussion strategies is that the friendly and unfriendly chat users came to experience a shift in their own attitudes as a result of the discussion. This finding is consistent with theories of counter-attitudinal advocacy, cognitive dissonance, and self-perception, and so the exact psychological mechanism underlying the effect warrants greater study. Nevertheless, it is apparent that, in order to get a partner to like oneself in CMC, people express their interpersonal affinity by means of sharing or conflicting attitudes about some target of the discussion. In FtF conversations, we expect liking and disliking to be expressed using vocalic and kinesic cues primarily, signaling one's affective orientation to one's partner quite directly. In CMC, without nonverbal cues to reflect affect, users focus on something else more visible—their stated opinions and their consistency with their partners'. But doing so not only affected their attitudes about the topic. It also affected their perceptions of their partners, in that disagreeable participants came to think of their partners as less physically desirable and socially attractive than did those who tried to get their partners to like them. These kinds of distortions seem unlikely to take place in FtF interaction, but that notion awaits empirical verification.

Feedback

The hyperpersonal model of CMC suggested that the enhancements provided by idealization, selective self-presentation, and channel effects reciprocally influenced matters, forming a feedback system by which the CMC intensified and magnified the dynamics each component of the model contributes. That is, when a receiver comes upon a selectively self-presented message and idealizes its source, that individual may respond in such a way that reciprocates and reinforces the partially modified personae, reproducing, enhancing, and potentially exaggerating them. The manner by which the dynamics of these reciprocated expectations modify the participant's character may reflect the process of behavioral confirmation.

Behavioral confirmation (Snyder, Tanke, & Berscheid, 1977) describes how one individual's impression about a target partner leads the first individual to behave toward the target, and how that individual's behavior actually alters the responses of the target person. The original behavioral confirmation study involved male subjects who were shown photos priming them to believe that their upcoming female telephone interaction partners were physically attractive or unattractive (even though the actual partners were not really those depicted in the photos but were randomly selected female subjects). Not only did this expectation affect the male's involvement, it affected the female's personality-related responses as well, as revealed in outside raters' evaluations of the females' personalities based on audio-recordings of their conversations. The hyperpersonal model argued idealized impressions of online partners lead CMC users to respond based on that biased impression, and that those responses influence the partner's behavior toward the expected impression. In this way feedback intensifies hyperpersonal effect, bringing together receiver idealization, sender selective self-presentation, and manipulation of channel.

Recent research has added to our understanding of how these cycles affect the development of hyperpersonal relations. An experiment involving dyads who were about to chat via CMC, where, in one condition, experimenters led one dyad member to expect that his partner was in an unpleasant mood; although, the experimenter, bad moods are malleable. Subjects in another condition were told that their partner had an unpleasant personality, which—the experimenter reminded the subject—is pretty permanent. In actuality the partners were naive subjects whose moods and personalities were normal. When the chat was over, subjects who expected bad moods found that their partners had changed (Walther Kashian, Jang, & Shin, in press). Similar studies show definitively that these changes are due to the conversational efforts of the expecting subjects (Tong & Walther, in press). Yet, in this study, the subjects attributed the partner's mood elevations not to their own efforts, but to the conclusion that their partners liked them personally; and this effect did not occur among subjects who believed their partners to have an unpleasant personality. Obviously the subject's attributions were false. The targets behaved nicely because the subjects coaxed them into doing so. But the (misplaced) attribution of personal liking becomes the (false) basis for mutual attraction in such settings. It does not take much, and we are relatively unaware of the things we ourselves do, to affect others' moods online, and rather than recognize our subtle interpersonal influence on others, we think others really like us.

Future research will continue to push these models forward as they deal with the benefits and constraints of new media forms such as social media and similar platforms. When does an online interaction lead CMC users to "Google" their partner, and does the text in historical archives tell a user enough about a partner, or is continued interaction required? Does access to a group partner's picture on Facebook satisfy the desire for impression formation, or does it give CMC users a head start for conversations that can then probe deeper and reveal more about the participants, perhaps in less time than before the advent of online photos and "about me" sections of Facebook? How much does one learn about friends of friends online through their postings to each other's status updates, and how much must they interact directly with one another? New media platforms seem to encourage more SIP and hyperpersonal questions for future study, rather than fewer of them.

Conclusion

At the National Communication Association's 2012 conference, a panel discussed the contributions of SIP theory to communication research in general, in light of its 20th year after publication. This chapter concludes with a review of some of these qualities.

It is an unusual theory in that it specifies its assumptions explicitly, which offers researchers a clear view with which to argue about its premises and its logic.

It is a theory with an explicitly process view of communication, with its emphasis on time and its longitudinal perspective on change.

It reinforced the notion of cue substitutability, in which nonverbal and verbal cues may comprise equivalent forms of communication. In doing so, it was one of few theories to emphasize a functional rather than structural view of nonverbal communication.

Its considerations of cues, rate, and time reflect the principle of equifinality, akin to a systems theory perspective on communication, in which various different combinations and interactions among system attributes routes lead to similar end-states.

It connected the independent principle of interactivity into a theoretical framework.

CMC today is a whirlwind of applications and tools that operate on our desktops, laptops, and hand-held devices. Researchers are beginning to recognize that contemporary social interactions are not conducted through one medium or another but often through a variety of channels. Research has yet to conceptualize what this means for people. Even those who urge researchers to take a "multimodal perspective" have not offered methods by which to do so. At the same time, the social goals that individuals and groups undertake may exert considerable influence over what people do online and how they do it. The SIP and hyperpersonal models assume that social goals are undeterred by interactive media, and to study CMC is to study the way people pursue these goals using new channels that present challenges as well as opportunities to the enhancement of social information-based activities. Advancing technology makes it easier to connect, have fun, and stay in touch. By making some communications easier, they reinforce that other communications require more effort, but the SIP and hyperpersonal perspective suggest that there is a payoff for the extra effort that some of these communication technologies require. Communication was never completely easy and it has not always been done FtF. Letters and phone calls were the mainstays of communication in close relationships in the past, and as these technologies diffused so did concerns over their depersonalizing ill-effects. Just as one form led to love letters and another led to all-night banter, new technologies transform but do not ultimately impede relationships. To study CMC from the SIP and hyperpersonal perspective is to observe the latest in these transformations.

References

Antheunis, M. L., Valkenburg, P. M., & Peter, J. (2010). Getting acquainted through social network sites: Testing a model of online uncertainty reduction and social attraction. *Computers in Human Behavior*, *26*, 100–109.

Bazarova, N. N. (2012). Public intimacy: Disclosure interpretation and social judgments on Facebook. *Journal of Communication*, *62*, 815–832.

Burgoon, J. K., Buller, D. B., & Woodall, W. G. (1989). *Nonverbal communication: The unspoken dialogue*. New York: Harper & Row.

Carlson, J. R., & Zmud, R. W. (1994). Channel expansion theory: A dynamic view of media and information richness perceptions. In D. P. Moore (Ed.), *Academy of management: Best papers proceedings 1994* (pp. 280–284). Madison, WI: Omnipress.

Culnan, M. J., & Markus, M. L. (1987). Information technologies. In F. M. Jablin, L. L. Putnam, K. H. Roberts, & L. W. Porter (Eds.), *Handbook of organizational communication: An interdisciplinary perspective* (pp. 420–443). Newbury Park, CA: Sage.

D'Angelo, J. D., & Van Der Heide, B. (in press). The formation of physician impressions in online communities: Negativity, positivity, and non-normativity effects. *Communication Research*.

Derks, D., Bos, A. E. R., & von Grumbkow, J. (2007). Emoticons and social interaction on the Internet: The importance of social context. *Computers in Human Behavior*, *23*, 842–849.

Donohue, W. A., Diez, M. E., Stahle, R., & Burgoon, J. K. (1983, May). *The effects of distance violations on verbal immediacy: An exploration.* Paper presented at the annual meeting of the International Communication Association, Dallas, TX.

Epley, N., & Kruger, J. (2005). What you type isn't what they read: The perseverance of stereotypes and expectancies over email. *Journal of Experimental Social Psychology, 41*, 414–422.

Fulk, J., Schmitz, J., & Steinfield, C. W. (1990). A social influence model of technology use. In J. Fulk & C. Steinfield (Eds.), *Organizations and communication technology* (pp. 117–140). Newbury Park, CA: Sage.

Fulk, J., Steinfield, C. W., Schmitz, J., & Power, J. G. (1987). A social information processing model of media use in organizations. *Communication Research, 14*, 529–552.

Gibbs, J. L., Ellison, N. B., & Heino, R. D. (2006). Self-presentation in online personals: The role of anticipated future interaction, self-disclosure, and perceived success in Internet dating. *Communication Research, 33*, 1–26.

Hancock, J. T. (2004). Verbal irony use in face-to-face and computer-mediated conversations. *Journal of Language and Social Psychology, 23*, 447–463.

Hancock, J. T., Gee, K., Ciaccio, K., & Lin, J. M.-H. (2008). I'm sad you're sad: Emotional contagion in CMC. In B. Begole & D. W. McDonald (Eds.), *Proceedings of the 2008 ACM conference on computer supported work* (pp. 295–298). New York, NY: ACM.

Hancock, J. T., Landrigan, C., & Silver, C. (2007). Expressing emotion in text-based communication. In M. B. Rosson & D. Gilmore (Eds.), *Proceedings of the SIGCHI conference on human factors in computing systems (CHI 2007)* (pp. 929–932). New York, NY: ACM.

Hian, L. B., Chuan, S. L., Trevor, T. M. K., & Detenber, B. H. (2004). Getting to know you: Exploring the development of relational intimacy in computer-mediated communication. *Journal of Computer-Mediated Communication, 9*(3). Retrieved from http://onlinelibrary.wiley.com/doi/10.1111/j.1083-6101.2004.tb00290.x/abstract

Hiltz, S. R., & Turoff, M. (1978). *The network nation: Human communication via computer.* Reading, MA: Addison-Wesley.

Hiltz, S. R., & Turoff, M. (1981). The evolution of user behavior in a computerized conferencing system. *Communications of the ACM, 24*, 739–762.

Johansen, R., Vallee, J., & Spangler, K. (1979). *Electronic meetings: Technical alternatives and social choices.* Reading, MA: Addison-Wesley.

Joinson, A. N. (2001). Self-disclosure in computer-mediated communication: The role of self-awareness and visual anonymity. *European Journal of Social Psychology, 31*, 177–192.

Klimmt, C., & Hartmann, T. (2008). Mediated interpersonal communication in multiplayer video games: Implications for entertainment and relationship management. In E. A. Konijn, S. Utz, M. Tanis, & S. B. Barnes (Eds.), *Mediated interpersonal communication* (pp. 309–330). New York, NY: Taylor & Francis/Routledge

LaRose, R., & Whitten, P. (2000). Re-thinking instructional immediacy for web courses: A social cognitive exploration. *Communication Education, 49*, 1–19.

Lea, M., & Spears, R. (1991). Computer-mediated communication, de-individuation and group decision-making. *International Journal of Man-Machine Studies, 34*, 283–301.

Lenhart, A. (2012). *Teens, smartphone, & texting.* Retrieved from http://www.pewinternet.org/~/media//Files/Reports/2012/PIP_Teens_Smartphones_and_Texting.pdf

McGrath, J. E. (1990). Time matters in groups. In J. Galegher, R. E. Kraut, & C. Egido (Eds.), *Intellectual teamwork: Social and technical foundations of cooperative work* (pp. 23–61). Hillsdale, NJ: Erlbaum.

Okdie, B. M., Guadagno, R. E., Bernieri, F. J., Geers, A. J., & Mclarney-Vesotski, A. R. (2011). Getting to know you: Face-to-face vs. online interactions. *Computers in Human Behavior, 27*, 153–159.

Orcutt, J. D., & Anderson, R. E. (1977). Social interaction, dehumanization, and the "computerized other." *Sociology and Social Research, 61*, 380–397.

Ord, J. G. (1989). Who's joking? The information system at play. *Interacting with Computers, 1*, 118–128.

O'Sullivan, P. B., Hunt, S. K., & Lippert, L. R. (2004). Mediated immediacy: A language of affiliation in a technological age. *Journal of Language and Social Psychology, 23*, 464–490.

Peña, J., & Hancock, J. T. (2006). An analysis of socioemotional and task-oriented communication in an online multiplayer video game. *Communication Research, 33*, 92–109.

Postmes, T., & Baym, N. (2005). Intergroup dimensions of the Internet. In J. Harwood & H. Giles (Eds.) *Intergroup communication: Multiple perspectives* (pp. 213–238). New York, NY: Peter Lang.

Postmes, T., Spears, R., Lee, T., & Novak, R. (2005). Individuality and social influence in groups: Inductive and deductive routes to group identity. *Journal of Personality and Social Psychology, 89*, 747–763.

Radicati Group, Inc. (2011). *Email statistics report, 2011–2015.* Retrieved from http://www.radicati.com/wp/wp-content/uploads/2011/05/Email-Statistics-Report-2011-2015-Executive-Summary.pdf

Rapaport, M. (1991). *Computer mediated communications: Bulletin boards, computer conferencing, electronic mail, information retrieval.* New York, NY: Wiley.

Rheingold, H. (1993). *The virtual community: Homesteading on the electronic frontier.* Reading, MA: Addison-Wesley.

Rockart, J. F., & DeLong, D. W. (1988). *Executive support systems: The emergence of top management computer use.* Homewood, IL: Dow Jones-Irwin.

Rogers, P., & Lea, M. (2004). Cohesion in online groups. In K. Morgan, C. A. Brebbia, J. Sanchez, & A. Voiskounsky (Eds). *Human perspectives in the Internet society: Culture, psychology and gender* (pp. 115–124). Southampton, UK: WIT Press.

Salancik, G. R., & Pfeffer, J. (1978). A social information approach to job attitudes and task design. *Administrative Science Quarterly, 23*, 224–252.

Schouten, A. P., Valkenburg, P. M., & Peter, J. (2007). Precursors and underlying processes of adolescents' online self-disclosure: Developing and testing an "Internet-attribute-perception" model. *Media Psychology, 10*, 292–315.

Snyder, M., Tanke, E. D. & Berscheid, E. (1977). Social perception and interpersonal behavior: On the self-fulfilling nature of social stereotypes. *Journal of Experimental Social Psychology, 35*, 656–666.

Spears, R., Lea, M., & Postmes, T. (2000). Onside: Purview, problems and prospects. In T. Postmes, M. Lea, R. Spears, & S. D. Reicher (Eds.). *SIDE issues centre stage: Recent developments in studies of de-individuation in groups* (pp. 1–16). Amsterdam, The Netherlands: KNAW.

Spottswood, E. L., Walther, J. B., Holmstrom, A. R., & Ellison, N. E. (2013). Person-centered emotional support and gender attributions in computer-mediated communication. *Human Communication Research, 39*, 295–316.

Steinfield, C. W. (1986). Computer-mediated communication in an organizational setting: Explaining task-related and socioemotional uses. In M. L. McLaughlin (Ed.), *Communication yearbook 9* (pp. 777–804). Newbury Park, CA: Sage.

Tidwell, L. C., & Walther, J. B. (2002). Computer-mediated communication effects on disclosure, impressions, and interpersonal evaluations: Getting to know one another a bit at a time. *Human Communication Research, 28*, 317–348.

Toma, C. L., & Hancock, J. T. (2011). A new twist on love's labor: Self-presentation in online dating profiles. In K. B. Wright & L. M. Webb (Eds.), *Computer-mediated communication in personal relationships* (pp. 41–55). New York, NY: Peter Lang.

Toma, C. L., Hancock, J. T., & Ellison, N. B. (2008). Separating fact from fiction: An examination of deceptive self-presentation in online dating profiles. *Personality and Social Psychology Bulletin, 34,* 1023–1036. doi:10.1177/0146167208318067

Tong, S. T., & Walther, J. B. (in press). The confirmation and disconfirmation of expectancies in computer-mediated communication. *Communication Research.*

Valkenburg, P. M., & Peter, J. (2009). Social consequences of the Internet for adolescents: A decade of research. *Current Directions in Psychological Science, 15,* 1–5.

Vallee, J., Johansen, R., Lipinski, H., Spangler, K., Wilson, T., & Hardy, A. (1975). *Group communication through computers, volume 3: Pragmatics and dynamics.* Menlo Park, CA: Institute for the Future.

Van Der Heide, B. (2009, June). *Computer-mediated impression formation: A test of the sticky-cues model using Facebook.* Paper presented at the annual meeting of the International Communication Association, Singapore.

Walther, J. B. (1992). Interpersonal effects in computer-mediated interaction: A relational perspective. *Communication Research, 19,* 52–90.

Walther, J. B. (1993). Impression development in computer-mediated interaction. *Western Journal of Communication, 57,* 381–398.

Walther, J. B. (1996). Computer-mediated communication: Impersonal, interpersonal, and hyperpersonal interaction. *Communication Research, 23,* 3–43.

Walther, J. B. (2006). Nonverbal dynamics in computer-mediated communication, or: (and the net: ('s with you,:) and you:) alone. In V. Manusov & M. L. Patterson (Eds.), *Handbook of nonverbal communication* (pp. 461–479). Thousand Oaks, CA: Sage.

Walther, J. B. (2007). Selective self-presentation in computer-mediated communication: Hyperpersonal dimensions of technology, language, and cognition. *Computers in Human Behavior, 23,* 2538–2557.

Walther, J. B. (2009). Theories, boundaries, and all of the above. *Journal of Computer-Mediated Communication, 14,* 748–752. doi:10.1111/j.1083-6101.2009.01466.x

Walther, J. B. (2010). Computer-mediated communication. In C. R. Berger, M. E. Roloff, & D. R. Roskos-Ewoldsen (Eds.), *Handbook of communication science* (2nd ed., pp. 489–505). Los Angeles, CA: Sage.

Walther, J. B., Anderson, J. F., & Park, D. (1994). Interpersonal effects in computer-mediated interaction: A meta-analysis of social and anti-social communication. *Communication Research, 21,* 460–487.

Walther, J. B., & Burgoon, J. K. (1992). Relational communication in computer-mediated interaction. *Human Communication Research, 19,* 50–88.

Walther, J. B., & Carr, C. T. (2010). Internet interaction and intergroup dynamics: Problems and solutions in computer-mediated communication. In H. Giles, S. Reid, & J. Harwood (Eds.), *The dynamics of intergroup communication* (pp. 209–220). New York, NY: Peter Lang.

Walther, J. B., & D'Addario, K. P. (2001). The impacts of emoticons on message interpretation in computer-mediated communication. *Social Science Computer Review, 19,* 323–345.

Walther, J. B., DeAndrea, D., Kim, J., & Anthony, J. (2010). The influence of online comments on perceptions of anti-marijuana public service announcements on YouTube. *Human Communication Research, 36,* 469–492. doi:10.1111/j.1468-2958.2010.01384.x

Walther, J. B., Kashian, N., Jang, J.-W., & Shin, S. Y. (in press). Overattribution of liking in computer-mediated communication: Partners infer the results of their own influence as their partners' affection. *Communication Research.*

Walther, J. B., Loh, T., & Granka, L. (2005). Let me count the ways: The interchange of verbal and nonverbal cues in computer-mediated and FtF affinity. *Journal of Language and Social Psychology*, *24*, 36–65.

Walther, J. B., & Ramirez, A., Jr. (2009). New technologies and new directions in online relating. In S. W. Smith & S. R. Wilson (Eds.), *New directions in interpersonal communication research* (pp. 264–284). Newbury Park, CA: Sage.

Walther, J. B., Van Der Heide, B., Tong, S. T., Carr, C. T., & Atkin, C. K. (2010). The effects of interpersonal goals on inadvertent intrapersonal influence in computer-mediated communication. *Human Communication Research*, *36*, 323–347.

Watzlawick, P., Beavin, J. H., & Jackson, D. D. (1967). *Pragmatics of human communication: A study of interactional patterns, pathologies, and paradoxes.* New York, NY: Norton.

Wegner, D. M. & Vallacher, R. R. (1977). *Implicit psychology: An introduction to social cognition.* New York, NY: Oxford University Press.

Westerman, D. K., Van Der Heide, B., Klein, K. A., & Walther, J. B. (2008). How do people really seek information about others? Information seeking across Internet and traditional communication sources. *Journal of Computer-Mediated Communication*, *13*, 751–767. doi:10.1111/j.1083-6101.2008.00418.x

Wilson, J. M., Straus, S. G., & McEvily, W. J. (2006). All in due time: The development of trust in computer-mediated and FtF groups. *Organizational Behavior and Human Decision Processes*, *99*, 16–33.

2

Group Identity, Social Influence, and Collective Action Online

Extensions and Applications of the SIDE Model

Russell Spears and Tom Postmes

Faculty of Social and Behavioral Sciences, Department of Social and Organizational Psychology, University of Groningen, Groningen, Netherlands

Introduction

In this chapter, we provide an overview of the SIDE model, the Social Identity model of Deindividuation Effects (Postmes, Spears, & Lea, 1998; Reicher, Spears, & Postmes, 1995; Spears & Lea, 1992, 1994), especially as applied to the new media (email, Internet, social networking). We review some of the evidence for the utility and heuristic value of this model in explaining social influence when using these media, and also its relevance for facilitating collective action in cyberspace, and the possibilities the new media provide for political action. Some of the so-called new media are not so new anymore, and there is a fairly large body of empirical research on the effects of email, for example, also with respect to the SIDE model. In the meantime, other related media (social networking such as Facebook and Twitter) are much newer and thus provide less evidence of the effects of these (and their specific features), also with respect to the SIDE model. However, many of the principles of this model can also be used to interpret such media effects and can be used to anticipate and explain them where they occur. At the end of this chapter, we develop this agenda in some new directions.

The SIDE model is primarily a framework to help us understand media and media effects (and social psychological processes such as social influence and collective behavior more generally), and not a "normative model" of media use (i.e., prescribing which medium to use when). This is partly a deliberate choice as history has shown that trying to match media to its optimal use more often put us on the slippery slope to technological determinism. History also tells us that what the inventors and developers of communications technologies envisage as typical use is often confounded by the subsequent reality (e.g., the telephone was originally expected to have only a narrow business application; Fischer, 1992). In retrospect, predictions of media-task

The Handbook of the Psychology of Communication Technology, First Edition. Edited by S. Shyam Sundar.

suitability therefore often seem to be an open invitation to subversion. However, while eschewing technological determinism SIDE may yet provide some lessons about media use and media design, and at the end of this chapter we also speculate on these issues.

Background, Roots, and Development of SIDE

Understanding how SIDE came about can help to explain the meaning of the model and dispel some misconceptions that we sometimes encounter in the literature. A first point is that we, and our two main colleagues with whom we developed the model (Martin Lea and Stephen Reicher), have a background in social psychology rather than communication studies. One of the consequences has been that some of the theoretical refinements that we have proposed to the SIDE model have not always crossed disciplinary boundaries very clearly (of which more below). Lea was working on the new communications technologies (CSCW) in Manchester the mid-1980s when Spears arrived; Postmes joined the collaboration, working on these themes for his PhD, when Spears moved to Amsterdam a couple of years later. Our background in the social identity approach (social identity theory and self-categorization theory) formed a key theoretical foundation for the model.

One problem that had emerged from early work on effects of CMC was an overwhelming emphasis on the limitations of new communication media in comparison with face-to-face (FtF) interaction (e.g., Daft & Lengel, 1984; Short, Williams, & Christie, 1976): The restricted bandwidth of these new media was associated with a restricted communication of various social cues, and this in turn was supposed to reduce social regulation and social influence. An influential classic article by Kiesler, Siegel, and McGuire (1984) used this analysis to explain some apparently antisocial effects of CMC such as disinhibition, "flaming," and more extreme decisions that had been noted in some initial studies. This common emphasis on restricted bandwidth causing social and relational restrictions heavily influenced other researchers in the communications field.

There was no shortage of available theories and concepts to confirm and explain the asocial and antisocial image of CMC. In explaining disinhibited and deregulated behavior, one of the most obvious and accessible candidates (but by no means the only one) was "deindividuation theory" (Festinger, Pepitone, & Newcomb, 1952; Prentice-Dunn & Rogers, 1989), which we elaborate shortly. In explaining more extreme decisions, the most obvious and accessible candidate was the group polarization phenomenon.

Before explaining polarization and deindividuation, we should add that our hunch when we pondered this early literature was that something was not quite right. Our analysis had its roots in a theoretical tradition originating in European social psychology, which offered a very different explanation of the group processes that were central to the early literature on CMC. Specifically, this new approach stemmed from social identity theory (Tajfel & Turner, 1979), and especially self-categorization theory (Turner, Hogg, Oakes, Reicher, & Wetherell, 1987; together the "social identity approach"), and proposed that group identities and norms are central to both group polarization and crowd behavior. We consider how this approach viewed group polarization and collective behavior, and formed the basis of the SIDE model.

Group polarization

The phenomenon of group polarization, in which group discussion results in group decisions that are more extreme (or "polarized") than the mathematical average of individual group members' attitudes, long presented an important conceptual puzzle for social psychology. The phenomenon suggests that in groups, people are more extreme than they are as individuals (indeed, it was originally called "risky-shift" until it was realized that groups can also shift to caution). This type of group behavior is not straightforwardly explained by conformity: after all, if *every* group member conforms, they would converge on the average and would not become more extreme.

However, the social identity approach to group polarization developed by Turner and Wetherell (see Turner, 1991; Wetherell, 1987) made a theoretical breakthrough that explained how an outcome could be polarized and yet still reflect a process of conformity to the *group*. The basic insight was that the normative or prototypical position within a given discussion group reflects not only the mean position within the group, but also what distinguishes this "ingroup" from others in the broader social frame of reference. Where there is no explicit outgroup from which to differentiate the discussion group, the idea was that the group tends to differentiate its own responses (e.g., risky choices on a choice dilemma) from the "other" contrasting choices available (e.g., cautious choices on the other side of the scale), creating an implicit outgroup or contrast category in the frame of reference. So when we add this intergroup differentiation to intragroup conformity, the resulting normative (prototypical) position tends be more extreme or polarized. The elegance of this explanation is that it can explain both more risky and more cautious group decisions.

In sum, the social identity approach suggests that group polarization is (as in the critique of deindividuation theory discussed later) a sign of strong normative and social regulation. This interpretation does not correspond at all with the early interpretation of polarization effects in CMC because it implies that social behavior in CMC could be *more* normative and *more* socially regulated, not less so (see Spears & Lea, 1992; Turner et al., 1987).

Deindividuation and group behavior

Deindividuation is a concept that had been developed to explain antisocial behavior in the crowd and was inspired by the writings of Gustav Le Bon (1895/1995) in the previous century. He had argued that people lose their individual rationality in the crowd, reverting to the baser animal instincts. Although these ideas were considered speculative, even in their day, they had an enormous influence on attempts within psychology and sociology to explain crowd behavior in the following century. Le Bon's ideas were picked up in the early days of experimental social psychology (Festinger et al., 1952) and were further developed over three decades (Diener, 1980; Prentice-Dunn & Rogers, 1989; Zimbardo, 1969). The core idea in this modern research on deindividuation relied heavily on Le Bon: anonymity in the mass promotes a loss of self-awareness, which reduces inhibitions and increases the likelihood of antisocial behavior. One theme in the early models of behavior in CMC was that, although this context would seem to be very far from the madding crowd, the anonymity of this situation could have similar disinhibitory effects. As well as making sense of flaming, this also resonated with early evidence that group decision making in CMC could result in polarized or extreme decisions compared to face-to-face groups (e.g., Siegel, Dubrovsky, Kiesler, & McGuire, 1986).

As we will explain in more detail, the social identity approach facilitated a very different understanding of deindividuation phenomena and rejected the deindividuation explanation of crowd behavior inspired by Le Bon and continued by deindividuation researchers. For the moment, we can state that a key difference is that deindividuation theory proposes a loss of identity and/or a consequent deregulation of social behavior via self and social norms when immersed in (anonymous within) the group. By contrast, the social identity analysis (see e.g., Reicher, 1987) suggests that such conditions in the group actually lead to an increase in the salience and role of social factors (social identities and associated social norms). In short the social identity analysis of "deindividuation effects" is a departure from deindividuation theory. In fact, we make the opposite prediction to deindividuation theory in at least one key respect: we propose that behavior becomes more social or socially regulated, not less so. Hence, when we refer to "deindividuation phenomena" and "deindividuation effects," we refer to a cluster of specific phenomena and effects that need to be explained (i.e., the effects of group immersion and anonymity) because classical deindividuation theory fails to do so adequately.

Two aspects of anonymity

In order to understand the SIDE model, we actually have to "unpack" deindividuation theory and distinguish two aspects of anonymity that are very relevant to the social identity analysis of deindividuation phenomena and the resulting SIDE model. To recap', the claim that CMC was somehow less social or socially regulated concerned the lack of social cues in this medium (i.e., due to the visual anonymity or uniform representation of group members) and the consequences this would have for deindividuation. However, the basic argument of deindividuation theory actually crystallized into *two* elements concerning the effects of *anonymity*, which could reduce social accountability to others, and *submergence in the group*, argued to reduce (private) self-awareness. Both of these were predicted to result in reduced constraints on antinormative or antisocial behavior, although by separate routes (Prentice-Dunn & Rogers, 1989). Because antinormative behavior stemming from reduced accountability is conscious, only reduced self-awareness was ultimately associated with the "deindividuated state" (Prentice-Dunn & Rogers, 1989). As we shall see later, although the SIDE model rejects this process account of (anti)social behavior in deindividuation theory, the distinction between two aspects of anonymity ("anonymity of" and "anonymity to", elaborated further later) became important in defining what has become known as the cognitive and strategic sides of the SIDE model. The cognitive dimension of SIDE refers to "anonymity of" and specifically how anonymity of/within the ingroup can enhance the salience of group identity (rather than *loss* of identity caused by immersion in the group, following deindividuation theory). The strategic dimension of SIDE argues that reduced accountability to outgroups (following from anonymity *to* them) can allow (normative) ingroup behaviors that might otherwise we sanctioned or punished by the outgroup.

The cognitive dimension of SIDE: "Anonymity of..."

Researchers working within a social identity tradition (e.g., Reicher, 1982, 1987), had taken issue with the deindividuation theory explanation for crowd behavior, and once again proposed a more "social" and normative explanation stemming from group identity. Rather than people losing their sense of self or identity in the crowd, Reicher argued that there was a switch from individual to group identity, making those within the crowd more sensitive to associated norms and not less so. So once again the argument here was for acute social regulation although by norms relevant to the crowd.

Sometimes, of course, these norms and behaviors might seem antinormative in general terms or from an outside perspective (e.g., crowd aggression, violence, or online hacktivism, flaming). However, the point was that this analysis could explain both aggressive but also peaceful behavior, depending on the norms relating to the identity, often emerging in context. In sum, this social identity approach to crowd behavior offered a complete reinterpretation of the effects that were originally attributed to "deindividuation." Indeed, in key respects this reinterpretation was a conceptual reversal of the original deindividuation theory: crowd behavior is not antinormative and disinhibited (as deindividuation theory suggested), but rather it is the result of conformity to explicit or implicit norms. Moreover, this conformity occurs not because of a *loss* of self-awareness or identity (as deindividuation theory suggested), but through a ***heightened*** awareness of the crowd, by sharing its identity, and by acting in group terms: a process referred to as "depersonalization." (Turner, 1982).

It cannot be stressed enough that "depersonalization" is quite different to deindividuation, in process terms (a point that has sometimes plagued understanding and secondary accounts of the SIDE model, partly because of its title). Whereas deindividuation implies reduced self-regulation, depersonalization implies heightened social or group-level self-regulation. So, whereas deindividuation implies *a loss of self in the group*, depersonalization refers to *the emergence of group in the self*: the tendency to see others but also oneself in group terms (i.e., in terms of group norms and stereotypes). In short, although the terms seem to be similar, the processes could not be more different. It is perhaps no accident that deindividuation theory tends to see the influence of the group as generally bad or pernicious, in contrast to individual rationality, whereas the social identity approach is more open to the possibility that groups can also be good, self-defining, and key source of social rationality (Spears, 2010).

Reicher (1984) provided support for these ideas in an experiment where science and social science students dressed in overalls and hoods (a classic deindividuation manipulation), rendering them anonymous to each other, conformed more to ingroup norms than those who were in normal clothing and thus individuated. Moving back to the context of CMC though, a key question was whether we could find similar group influence effects when people were individually isolated at their computer terminals. This was far from a trivial question because it seemed to go against most classic theories of the group and group influence: how could physical isolation from the grip of the group lead to more group influence? For the identity-based understanding of group processes provided by self-categorization theory, this was not a problem.

Empirical support for the cognitive dimension of SIDE

Over the years, a body of research has emerged that has examined the various aspects of the SIDE model that we have discussed so far. It may be useful to summarize this initial research first, before moving on to a second strategic side of the SIDE model that has been more central to more recent research. In our first empirical study on this line (Spears, Lea, & Lee, 1990), we combined the critique of deindividuation theory (modeled on Reicher's research) with the group polarization paradigm that had been used to ground the CMC research on the reduced social cues model (Kiesler et al., 1984; Siegel et al., 1986). Based on the social identity approach to group polarization and deindividuation, we predicted the same kind of enhanced depersonalization and social influence effects as shown by Reicher (1984).

In our experiment, we required first-year psychology students to discuss attitude topics: this provided a test of group polarization. Could isolation from the groups associated with CMC paradoxically increase the power of the group to influence its members? This depended also on there being a salient group identity and an unambiguous local norm associated with that group (as in Reicher's study). So we manipulated group identity salience (by telling participants we were interested in them as psychology students) or personal identity salience (by telling participants that we were interested in personality differences in communication style). We also made sure that the group norms for psychology students were known by giving feedback on the typical distribution of views among psychology student for the discussion topics. Crossed with this manipulation we located discussants either in separate rooms (creating the standard visual anonymity typical of CMC) or had them facing each other in the same lab (visible, individuated condition). We predicted that participants would be most depersonalized and prone to conform to group norms when group identity was salient and they were isolated and anonymous.

This is what we found: When personal identity was salient, however, participants actually contrasted their views away from the group norm under anonymity, suggesting they were defining and affirming their individuality in contrast to this group identity under these conditions. In short we obtained the first evidence for the argument that polarized decisions in CMC groups might reflect a social influence process of conformity to group norms rather than reflecting the lack of social constraints implied by the reduced social cues framework. However, the group polarization paradigm remains a slightly idiosyncratic and perhaps controversial test-bed for the argument of group influence so we were keen to replicate and extend this finding using other social influence paradigms.

Two studies by Postmes, Spears, Sakhel, and De Groot (2001) used a different paradigm to test the prediction that online depersonalization leads to more group influence. In this research, we had groups discuss a problem confronting a hospital and whether efficiency and economic considerations or patient care should be the priority in the hospital's policy. We then surreptitiously primed norms relating to these two contrasting approaches using a scrambled sentence procedure before the group discussion. Group discussions were also conducted under conditions of anonymity or with participants identifiable to each with portrait pictures. As before we predicted that group influence would be stronger under anonymous conditions (depersonalization). This was confirmed. In a second study, we showed that when the efficiency norm was primed in two members of a four-person group, this norm also transferred to another two group members who were neutrally primed, but again primarily in anonymous groups, providing further direct evidence of a social influence process.

Overall there is now considerable evidence, from a number of different labs, for the prediction that communication via "anonymous" CMC (i.e., a user interface that obscures group members' personal or distinctive characteristics) can enhance rather than undermine social influence processes in contexts where a shared identity and shared norms are either unambiguously given or somehow contextually anticipated (e.g., Cress, 2005; Lee, 2007; Postmes, Spears, Lee, & Novak, 2005; Postmes, 2000, Spears & Lea, 2002; Sassenberg & Boos, 2003; Sassenberg & Postmes, 2002; Waldzus & Schubert, 2000). Beyond social influence, research has zoomed in on specific processes. For example, several studies have shown that online anonymity *within* groups can enhance social identity salience and/or identification (e.g., Lea, Spears, & de Groot, 2001; Lee, 2007; Postmes et al., 2001; Sassenberg & Postmes, 2002). And finally, Lee (2004) provided direct causal evidence for the prediction that online depersonalization would foster stronger social influence (and additionally showed that group identification statistically mediated this effect, see also Postmes et al., 2001, study 2). Lee (2008) extended this effect by showing that among CMC users who are individually identifiable, there is more attention to and influence of the strengths and weaknesses of specific arguments. In contrast, among CMC users who are not individually identifiable, argument quality appears to be less influential than the prevalent group norm.

With respect to SIDE model predictions regarding deindividuation, in the deindividuation literature more generally Postmes and Spears (1998) found that anonymity in the group actually enhances conformity to local or specific group norms, rather than increasing antinormative or antisocial behavior, supporting the SIDE model and contradicting deindividuation theory. More specific research in CMC contexts has shown that relative anonymity within groups (a) does *not* reduce personal self-awareness, as predicted by deindividuation theory, (b) that it tends to increase public self-awareness, if anything (e.g., Adrianson, 2001; Lee, 2007; Matheson & Zanna, 1998). Moreover, studies have found that neither private nor public self-awareness is associated with disinhibition or antinormative behavior in the way predicted by deindividuation theory (Lee, 2007; Matheson & Zanna, 1998; Yao & Flanagin, 2006).

Conceptual recap

Before moving on to discuss the second, strategic component of the SIDE model and the associated research program, it is important to make four observations about the SIDE model. First, although the model refers to "deindividuation effects," it does not endorse deindividuation theory, nor a "state of deindividuation." On the contrary, the SIDE model is grounded in a critique of deindividuation theory. The term deindividuation in "SIDE" is used descriptively to denote the range of effects that researchers using deindividuation manipulations had found but which we argue (and research confirms) were in need of an adequate (new) theory. So why not name it the social identity model of "Depersonalization Effects?" One answer is that we cannot assume that just because people are often visually anonymous within CMC that they will always share a group identity (or the same one). As our very first study described earlier showed, when group identity is not salient, anonymity will not lead to more group influence (see also Lee, 2007; Postmes et al., 2001, 2005). So we should not reify or essentialize CMC as always leading to greater social influence (this would be an example of technological determinism).

This brings us to our second point: SIDE argues that the social effects of characteristics of technology depend on the social relations in situ. Consistent with its roots in the social identity approach, SIDE holds that it is always necessary to take into account which identities might be relevant in a given social context, but also what norms are associated with that context and those identities. The technology (in this case CMC) does not determine the content or which identities are salient. Neither does the social side (content, identity salience) determine the effect technology may have. Though it can accentuate these through visual anonymity, the effect also depends on the salience of group identity or personal identity. In short, the model provides the scope to avoid both a technological determinism and a social determinism, and is sufficiently flexible to provide analyses of how these two factors (technological and social) may interact in producing effects. It is, however, more than simply descriptive or ad hoc: it provides a process account of how social psychological processes interact with social context and characteristics of technology, and how these different processes combine to produce certain social outcomes. We will elaborate further the implications of this openness to process specification later.

A third point is to clarify SIDE's stance on *anonymity*. Some of the experimental research on the SIDE model has manipulated the relative degree of anonymity of ingroup members as a method of promoting *depersonalization*. This appears to have confused some readers, who have mistakenly assumed that anonymity is an essential component of the SIDE model itself. But for SIDE, the core issue has never been that users behave differently because they are (in some absolute sense) anonymous. This notion, which is actually more central to deindividuation theory, rests on a problematic and simplistic conceptualization of anonymity (Anonymous, 1998; Joinson, 2005; Lea, Spears, & Watt, 2007; Rains & Scott, 2007; Tanis & Postmes, 2008). As several scholars have pointed out, in CMC complete anonymity is rare. More often, CMC is characterized by a "pseudonymity" in which communicators may be traced but nevertheless less recognizable in situ. For SIDE, what matters is *how* group members (and self) are visually represented online. As pointed out by Tanis and Postmes (2008, see also Lea et al., 2007), what aspects of users are made visible or disguised in technology design may provide cues to personal identity and cues to social identity, each of which may accentuate (or de-accentuate) aspects of social identity. Indeed, other research on SIDE has not manipulated anonymity, so much as the degree of uniformity (or heterogeneity) of group members in their online representations (e.g., Lee, 2004; Wodzicki, Schwämmlein, Cress, & Kimmerle, 2011), showing that such uniform representations enhance identification and pro-group behavior, respectively. In sum, SIDE is not about anonymity but about the online representation of individuals and groups.

Finally, the fourth point is that we prefer to call SIDE a model rather than a theory. The social identity approach provides the theoretical framework for our analysis. Within this, SIDE provides a more specific analysis of the effects of specific technological features. So far, SIDE has focused on cognitive and strategic effects of the communication medium (of which more will be provided later) but the scope remains to analyze additional features as they emerge (and thus to extend the model).

The strategic dimensions of SIDE: "Anonymity to…"

On this note it is now important to introduce a second key element of the SIDE model, namely the strategic dimension. The processes we have focused on until this point, explaining social influence based on enhanced social identity salience (depersonalization), refer to what has become known as the "cognitive" dimension of SIDE. In this case, cognitive simply refers to the fact that certain features of technology have an effect on the salience of identity (e.g., a moderating effect of the salience of group identity). In other words, the sense of who we are is affected, or strengthened, by the online representation of others and of ourselves.

But equally important is the "strategic" dimension of SIDE (see Klein, Spears, & Reicher, 2007; Postmes & Spears, 1998; Reicher et al., 1995; Spears & Lea, 1994). The strategic dimension refers to the opportunities and constraints provided by structural features of technologies such as CMC or of particular social situations. These opportunities and constraints are likely to be interpreted and made use of through the prism of a particular identity (e.g., myself as an individual, or us as a group). So to take the example of anonymity again, when CMC makes us less visible and identifiable to others this can offer strategic advantages, especially where there are reasons *not* to be identified. This can occur in the case where there is a powerful authority or outgroup whose aims and goals may conflict with your own (but this can also occur within a group, especially where there are power differentials and conflicts of interest). The fact that CMC is a medium where comments are on-record and we can be traced to an IP address can also make us *more* identifiable and accountable (as well as visually anonymous); the different features of CMC can be double-edged and contradictory, of which more follows later.

As we have seen, the earliest deindividuation research also focused on effects of identifiability and accountability (Festinger et al., 1952). Diener's (1980) "trick or treat" studies, for example, showed that children made anonymous by their costumes were more likely to take more candies. Later versions of deindividuation theory argued that such "antisocial" behaviors may stem from conscious knowledge that one would not be identified and apprehended (Prentice-Dunn & Rogers, 1989). These strategic considerations are associated with the implications of "anonymity to" (reducing accountability of self) rather than "anonymity of" (reducing individuation of others: more relevant to the cognitive SIDE).

The SIDE model develops the strategic dimension of deindividuation research in some important ways. First because the SIDE model contains an analysis of identity level, in line with the social identity approach, there is no blanket assumption that people will always act in line with *individual* self-interest when anonymous. Once again this will depend on the relevant salient identity, which could be a group identity. Reicher and colleagues have developed these ideas on the strategic dimension to understand collective behavior in the crowd. They used the power relation between students (low power) and staff (high power group) to test these ideas. They developed a paradigm in which they distinguished between behaviors that were (1) normative to students, and acceptable to the staff (e.g., missing the occasional lecture); (2) normative for students but not acceptable to staff (e.g., copying from another student's essay); and (3) neither normative for students nor acceptable to the staff (e.g., cheating on an exam).

In one study, they showed that students who were identifiable to staff were less likely to display behavior normative for them but unacceptable to staff (point 2), but when anonymous to staff they were more likely to display such behavior (because they could get away with it; Reicher & Levine, 1994a). In another study (Reicher & Levine, 1994b), the *co-presence* of other students also gave them more courage to endorse such punishable but ingroup normative behaviors (because of the social support implied by others). In both studies, students always endorsed behaviors that were normative but not punishable (1), but never non-normative behaviors (3). So again group behavior is clearly normative, but also (strategically) constrained by power relations and the possibilities available.

Empirical support for the strategic dimension of SIDE

Applying these ideas to the context of CMC, Spears, Lea, Corneliussen, Postmes, and Ter Haar (2002) manipulated the visual anonymity to others (by means of screens). All students were in the same location so the co-presence of others (and implied social support this provides) was kept constant. However, we also manipulated whether people could communicate via computer or not, reasoning that the computer represents a channel by means of which people can communicate social support and coordinate resistance to the powerful outgroup. The visibility manipulation had an impact on social influence for the normative but unpunishable items (point 1) in line with the cognitive SIDE effect already discussed (i.e., more influence when *not* mutually visible). However, the availability of communication technology produced a strategic SIDE effect, such that participants showed more support for the normative but punishable items (2) when they had email as they could then gauge the levels of support, and thus engage in normative behavior that was not acceptable to staff.

In sum, responses to the ingroup normative but punishable items (relevant to power differentials) showed the predicted strategic SIDE effect, whereas the ingroup normative but unpunishable items (not relevant to power) showed the classic cognitive SIDE effect discussed earlier.

One implication of this power-based strategic analysis is that less powerful groups might take advantage of the "pseudonymity" possible in CMC to conceal their group identities. For example, women might choose not to reveal their gender if they think that this puts them at a power disadvantage compared with men. Research by Flanagin and colleagues (2002) showed that women were indeed more likely to conceal their gender identity. In some experimental research, moreover, we have found that women were less likely to adopt female gendered avatar identities, especially when discussion topics are more associated with male expertise (Spears, Lea, Postmes, & Wolbert, 2011). Spears et al. (2011) further showed in an organizational context that women in more gendered cultures (in this case Italy; see e.g., Olivetti & Petrongolo, 2008) were less likely to reveal their gender identity when aware of the gender status differences than in less hierarchically "gendered" countries (UK, the Netherlands).

This shows the strategic possibilities of CMC and could be seen as endorsing the somewhat utopian view of cyberspace as a way to transcend power differentials, and also the restrictions of imposed identities that might blight us in everyday life. This motif has been heralded by some feminist scholars as a sign of liberation in cyberspace (Haraway, 1990; Turkle, 1995). However, there may also be a downside here. If women (or other groups suffering from status disadvantage) choose to conceal their identity and adopt others in cyberspace, then paradoxically they become less "visible" and present as a group, giving ground to male domination either in appearance or reality (Spears et al., 2011). If, as some have proposed, cyberspace is often less than women-friendly, this does not provide the most inviting context for women competing for a space and for their voices to be heard.

On the other hand, the fact that the Internet provides a communication medium (via email, bulletin boards and blogs, social networking) means that it is a powerful tool for making the contacts that can strengthen the connections within disempowered groups and allow them to coordinate action. Moreover, it can be more powerful than the FtF media because it facilitates contact between people whose group identity may not immediately be apparent and where there may even be reasons to conceal it to a majority or mainstream audience because of stigma or opprobrium (e.g., sexuality or radical political views; see Mckenna & Bargh, 1998). We consider such factors in more detail in the final section where we consider the possibilities that cyberspace provides as a tool for collective action.

To summarize, whereas the cognitive dimension analyses how the distinctive features of a communication technology affect the salience and operation of a particular identity, the strategic dimension analyses how such features might affect the ability to express such identities in line with norms that might be sensitive to surveillance by the audience (e.g., powerful outgroups, third parties, authorities, and even "policing" behavior within the ingroup). The structural features studied include the visual anonymity associated with CMC, but also the ability to communicate and commandeer social support that it engenders.

Extensions and elaborations of SIDE

Having considered the theoretical propositions and some empirical underpinnings of the SIDE model, the last part of this chapter will turn to specific questions that research has addressed in recent years. In turn, we will discuss the question whether the implications of the SIDE model extend beyond social influence, whether SIDE processes play a role beyond text-based CMC, and whether SIDE is of any immediate practical relevance beyond the sterile environment of the research laboratory.

Beyond social influence

Although the SIDE model, and specifically the cognitive side, applies its analysis to social influence (reflecting the initial focus on the effects of CMC for group polarization), the implications of the model were from the outset much broader in terms of process and also outcomes. The grounding in self-categorization theory means that the depersonalizing effects of anonymity within CMC contexts extend to a range of within-group behaviors and influences.

Many studies have found evidence that depersonalized online collaborations are associated with stronger identification, social identity salience, and group cohesion (e.g., Lea et al., 2001; Lee, 2007). Conversely, studies have found that individuation fosters the attention to individual contributions made by group members (Lee, 2008; Postmes et al., 2001). An extension of this basic phenomenon is that some studies have found that depersonalization is associated with a greater tendency to perceive the collaborative group or dyad as an entity or "as one" (Sassenberg & Postmes, 2002; Tanis & Postmes, 2008).

These effects of depersonalized CMC on psychological commitment translate to a range of outcomes that evidence greater behavioral commitment to the group: some studies have shown that depersonalized CMC fosters behavioral trust (Tanis & Postmes, Tanis & Postmes, 2005, 2007) and leads to more satisfaction with the collaboration, and to better performance among high identifiers (Tanis & Postmes, 2008). Studies from other labs have shown that depersonalized CMC fosters knowledge sharing, at least among those with pro-group intentions (Cress, 2005).

Beyond CMC: Relevance of SIDE to other communication technologies

Although SIDE was originally grounded in an analysis of the social effects of text-based CMC, more recent research has extended the model to other communication media. For example, the effects of communications media that do involve visibility (webcam, online video) can be analyzed using the same theoretical principles and framework. Once again the key issue here is not to reify the effects of the technology, but to make a specific process analysis of how technological features (e.g., visibility vs. anonymity) will interact with social features (e.g., group identities) to affect social psychological processes and outcomes. One consistent finding relating to the cognitive dimension of SIDE is that, other things being equal, the visual anonymity associated with CMC will cause available social category cues (and the stereotypes and norms associated with them) to become relatively salient, because of the absence of individuating information. Moreover, whereas the norms and stereotypes associated with group identities are often known and shared, the characteristics associated with specific individuals are often by definition idiosyncratic and unknown beforehand (which is why group processes can so easily permeate the anonymity and isolation of CMC). However, to raise the argument that the anonymity of CMC will always strengthen group salience to a fixed "rule" without exceptions would violate the context-specific and interactionist approach of SIDE. We now illustrate this point by examining the case of gender as a social category.

Initial research in the SIDE framework provided confirmatory support for the argument, derived from the cognitive side of SIDE, that adding individuating information (e.g., personal profiles) to knowledge of people's gender did indeed undermine the salience and impact of gender, resulting in reduced intergroup differentiation and power differences around gender (e.g., Postmes & Spears, 2002). However it is important to note that in this research, all communication was via text-based CMC and thus visually anonymous, so the comparison was between cases where only gender information was available versus whether this was accompanied by additional (textual) information about individual differences (e.g., preferences, hobbies, etc.). However, providing a channel of visual communication (e.g., webcam) not only provides information about individuating characteristics of the category members, it can also provide cues as to category membership.

It needs to be acknowledged here that not all categories are equal in this regard: some social categories do *not* typically have visual cues that are diagnostic of the categories themselves (e.g., nationality, ideology, sexuality): you cannot tell someone's nationality, opinion, or sexuality simply by looking at them. Other social categories, however, such as gender, or "race" or age, do have clear visual markers. For visually cued categories like gender, it is quite possible that gender identity becomes *more* salient under conditions of visibility (vs. visual anonymity) and not less so, potentially overriding effects of individuation. In terms of self-categorization theory, when categories are visually distinct, they have "comparative fit" (Oakes, 1987) insofar as category differences (e.g., in appearance, dress, etc.) will render the category dimension (and differences associated with it) more distinct and salient.

A study by Lea, Spears, and Watt (2007) set out to test the hypothesis that visibility might actually increase group salience and group effects for visually cued categories such as gender. In this research, we compared the effects of gender with those for another cross-cutting social category (nationality), which in this case was not expected to be visually salient. Specifically we examined CMC discussion groups comprising four people, namely two male and two female students, two of whom were British and two Dutch, with one of each gender in each case. These discussion groups were required to discuss a selection of pretested topics, two of which were designed to distinguish between men and women in terms of their stereotypic opinions (e.g., attitudes to relationships, attitudes to sport on TV), and two of which were designed to distinguish between nationality (e.g., the quality of British cuisine). We expected that the different social groups were more likely to (bi)polarize or diverge on these topics, rendering those categories more salient and producing more social influence and social differentiation, when the discussion topic were relevant to the social categories (e.g., men were predicted to be more interested in sport on TV than women; the Dutch were predicted be more critical about British food than the British). These predictions were upheld. Moreover, the underlying processes mediating these effects supported the cognitive dimension of SIDE: depersonalizing factors mediated influence under anonymity for nationality-based discussion topics, but mediated influence under *visibility* for gender-relevant topics. In short the SIDE model is able to explain when and why visual anonymity but also *visibility* can lead to greater group salience and social influence effects. This is important for understanding the (variable) effect of communication media depending on whether communicators are visible or not.

More generally, it may be instructive to attempt to glance ahead at the possibility to formulate a more generic SIDE perspective on media characteristics. Traditionally, media are analyzed in terms of features such as "richness." For SIDE, however, effects of the quantity of social and personal information conveyed are perhaps less important than their *content*. In this vein, one issue that has remained implicit and underdeveloped within SIDE is the notion that the media themselves may acquire "identity" characteristics that reflect, influence, and interact with the social identities of the users. Problems (although interesting ones) can arise when the social identity of the medium does not anticipate its technical scope and effects. Mismatches can arise between the "identity" and norms of the medium and its actual effects. For example, the informality and inter-personal intimacy associated with Twitter can lead to problems when statements usually treated as ephemeral, informal and restricted to the local ingroup, necessarily receive the scope, publicity, and permanence associated with this medium (and similar problems have arisen within CMC). This is of course not a new issue: as noted earlier, the telephone was originally thought to have only limited business applications (Fischer, 1992). Extending the identity analysis afforded by the SIDE model to the technology itself may take the interaction between social factors and technological features to a new level and allow us to better understand these media effects (and perhaps even to propose "normative" models of media use).

Beyond the laboratory: SIDE in the field

From the outset, the emphasis in SIDE research was on testing of the model's predictions in laboratory research. Nevertheless, practical tests of the implications of the model in field settings were similarly important in the model's development (although less widely cited).

Several studies have put the SIDE model to the practical test in educational settings. From 1997 onward, Lea and Postmes put various implications of the SIDE model to the test in international student collaborations involving Amsterdam and Manchester. Students had 6 weeks to complete a project during a course—giving them hands-on practical experience of collaborating via a CMC system. The experiences led to the formulation of a computer-supported collaborative learning system, called SIDE-view, the core premise of which is that (in line with SIDE principles) collaborations are facilitated to the extent that they start off with the development of a shared identity, in a collaboration system in which individuating features of group members are relegated to the background of the user interface (Lea, Rogers, & Postmes, 2002). In more recent research, SIDE predictions were upheld when it was found that anonymous student collaborations in a course were more successful than when collaborators were identifiable (Tanis & Postmes, 2008).

A recent field study, again in an educational setting, also found support for some strategic effects predicted by the SIDE model. Ainsworth and colleagues (2011) predicted and found that anonymous self-expression in classroom settings (where no unified shared identity was present) would provide schoolchildren with the strategic freedom to express views that were more in line with personal beliefs than with prevalent norms. They conclude that anonymous classroom voting systems have strategic benefits that are in line with strategic SIDE model predictions.

Perhaps the most eye-catching recent developments where the SIDE model has been put to the test concern the "stickiness" of online communities. A key question in the design of online communities is how to design them so that users keep returning and develop a sense of community and commitment despite their virtual dispersion. SIDE makes the counterintuitive prediction that communities in which the individual characteristics of group members are in the background can be just as attractive (and sometimes even more so) for users as communities where personal characteristics are at the heart of their online activities. Indeed, it could well be that environments that undermine individual identifiability, and thus credit-taking, foster more prosocial and community-oriented (even altruistic?) norms. What is clear now (with the massive growth of web traffic) is that highly successful communities can be sustained on the basis of high member identifiability (e.g., Facebook) or extremely low identifiability (e.g., Wikipedia). But when the SIDE model was developed, this was far from evident.

Some recent field experimental evidence shows important evidence that confirms the SIDE predictions. Ren and colleagues conducted several studies of the usage of a purpose built Movie database, which they had customized so that users would either be individually identifiable in all actions, or would be identifiable only as team members (Ren, Kraut, & Kiesler, 2007; Ren et al., 2012). They reasoned (in line with Postmes & Spears, 2000; Sassenberg, 2002; cf. Postmes et al., 2005) that team member identifiability would lead to group formation on the basis of a common identity, whereas individual identifiability would lead to team formation on the basis of common bonds. In line with the SIDE model, their findings showed strong support for the prediction that communities with a common identity were the most "sticky": here they found the strongest psychological and behavioral commitment.

More recent research has sought to apply these same ideas to the development of an online community in a context in which there is absolutely no reason for people to collaborate or commune. Farzan, Dabbish, Kraut, and Postmes (2012) manipulated the online visual representation of teams and individuals in the context of the popular Tetris game—a highly individualistic game with no clear collaborative benefits. Across several studies, results showed that (compared with an individual control condition) users played more games and were more likely to return when they were allocated to teams (irrespective of whether members were individuated or not, in this context). Again, it appears that turning the game environment into a social space considerably enhances its attractiveness, even if the game itself is a completely asocial activity.

Collective action and SIDE

A final important application of the SIDE model concerns collective action online. Many have heralded modern communication technology's ability to democratize nations and revolutionize collective action. But, as always, on closer inspection, reading the changing face of collective action is not as straightforward as the initial hype suggests. The process may be illustrated in the initial response to disturbances and riots such as those in London in 2011 and the Arab spring. In both cases, it was initially taken for granted that social media would have played a major role in mass mobilization. But identifying and comprehending that role turns out to be quite challenging.

What is clear is that initial assumptions that disturbances and unrest are in some way *caused* by the abundant availability of communication technology is simplistic (e.g., Anderson, 2011; Bohannon, 2012). In the London riots, speculations that Twitter and Facebook were instrumental in organizing actions were disproven (Postmes, van Bezouw, Täuber, & Van de Sande, 2013). In Egypt's uprisings, similarly, it appears extremely unlikely that mobilization was orchestrated through social media, for the simple reason that a government shutdown of Internet and SMS traffic made this impossible (Dunn, 2011).

Of course, this is not to deny the relevance of modern communication technology for collective action: technology is anything but neutral. But the overall picture is likely to be more complex than the straightforward technological determinist viewpoint that the availability of means to organize or inform would automatically lead to better or more actions being organized, or to people being more informed than before (cf. Postmes & Brunsting, 2002). SIDE research may help shed some light on the complex factors involved.

To begin with we conducted some early field research, simply to address the basic questions whether the known socio-psychological predictors of collective action would differ (a) across different activist groups, and (b) across different types of (online and offline) action. Brunsting and Postmes (2002) compared online with offline actions (both in more normative vs. more antinormative guises). We also contacted four different types of groups (nonactivists, sympathizers, hardcore activists, and pressure group members). Interestingly, results revealed no huge differences in what variables predicted collective action intentions either online or offline (nor for different groups): it would seem that we would not need to reinvent an entirely new psychology of collective action online.

Moreover, we also did not find that online actions were somehow perceived as less desirable or less demanding and therefore "cheap" (although interestingly, activists did think that they would be more popular for *others*). On the contrary: online actions were viewed as equally effective. The only area in which some differences were found was that cognitive calculations (expected action effectiveness and expected self-effectiveness) played a slightly stronger role in predicting online action intention than in predicting offline action intentions.

The finding that the psychological processes that promote offline and online action are broadly similar is quite important, we believe. It shows that models of how social media and communication technology influence collective action (of which there are a few) are likely to be barking up the wrong tree if they propose that a completely different set of variables is involved in producing collective actions across technologies and social contexts (Postmes, 2007; Postmes et al., 2013). It is for this reason that we are somewhat skeptical toward recent proposals that variables involved in modern forms of protest would be radically different (cf. Bennett & Segerberg, 2012). Indeed, we are pretty certain that we can continue to assume that there are three key predictors of collective action: the existence of a shared social identity, assessments of the efficacy of action, and feelings of anger and injustice (Van Zomeren, Postmes, & Spears, 2008). Technology may affect all three.

First, feelings of group identification are directly affected by characteristics of communication media, as elaborated in the cognitive side of the SIDE model. A recent field experiment indeed confirmed that cognitive SIDE effects of anonymity can also predict real-life collective action. Chan (2010) studied how church groups responded to a call for action (making a financial donation). Calls were made by church group leaders either in person (face-to-face) or via email. In addition, Chan made the social categorization of Christian salient or not. Confirming the counterintuitive SIDE prediction, Chan found that *more* donations were made in the email condition, particularly when Christian identity was salient, and particularly among low identifiers (for whom salience manipulations tend to have more impact). More generally, SIDE research has shown time and again that online representations of individual group members can exert a considerable influence in fostering impressions of the group as a unit (i.e., depersonalization). Thus, it is clear that strong social identities can be activated (or may emerge over time) even when computer users are physically isolated from the group and anonymous as individuals.

In addition, there are the known strategic effects of anonymity, empowering people where there is a need to conceal identity due to risks of stigma or opprobrium. And other affordances of technology have strategic implications as well: technology provides means for establishing contacts, building networks and communities that can strengthen social connections within disempowered groups, also by blurring the distinction between inter-personal and mass communication (as illustrated by Twitter, *inter alia*). In this process, the importance of formal structures and established social movement organizations appears to be receding (Bennett & Segerberg, 2012; Castells, 2012). Instead, many modern collective actions are bottom-up and emergent phenomena (of which we shall provide one example). This shift elevates the importance of the *psychological* drivers of collective action: individuals have greater autonomy in deciding to act, and choosing how to. Technology facilitates this: it lowers barriers to participation by reducing the need for physical presence and it also blurs the distinction between private and public action (Bimber, Flanagin, & Stohl, 2005). Most importantly, perhaps, is that it changes the composition and importance of interpersonal networks (Castells, 2012).

The implications of this shift are broad-ranging. Some have suggested that because of it collective action would become increasingly determined by desires for personal identity expression (Bennett & Segerberg, 2012). In some ways, this echoes the old idea that Internet would be an essentially asocial medium populated by atomized individuals connected though interpersonal relations (at best). This is not a view we share. Individuals prove able (now as ever) to organize and mobilize others in large numbers to the extent that they succeed in harnessing social identities that connect and integrate people *across* individuals' social networks. Arguably, achieving this without recourse to a formal organizational structure should *increase* the importance of social identity processes.

This point is illustrated neatly by a large protest in Portugal on March 12, 2011. This was organized as a Facebook event by three individuals, without participation of any organization or trades union. They picked a random date, wrote a short manifesto to explain their concerns, and succeeded in mobilizing an estimated 300,000 people across 10 cities—one of the biggest protests since the Carnation Revolution. One might look at the medium through which this was organized and conclude this as an exemplary tale of mobilization via personal networks. But this would ignore the manifesto, the context in which it was successful and the expressed concerns of participants (offline and online). Indeed, the protest revolved entirely around the shared concerns of the "endangered generation" (a shared identity with echoes in Italian and Spanish protests around this time, see Postmes et al., 2013).

Networks are increasingly important, too, because they *do* influence and change how social identities are defined. Thus, networks are maintained through interactions by which individuals, among others, align their ideas about reality including perceptions of ingroups and outgroups (Postmes et al., 2005). But networks also create physical infrastructures: bodies of people who are networked for a multitude of reasons, and who may come to think of themselves as a community or group with certain shared concerns (thereby facilitating the transformation of private opinions to public social identities; Thomas, McGarty, & Mavor, 2009).

Second, perceptions of *efficacy* can be affected by social media in diverse ways. As hinted earlier, social media may change power relations by making transparent how many of "us" there are. In the Arab spring, for example, social media may not have played a key role in organizing the actions themselves, but they may have communicated awareness of just how widespread discontent with the status quo was. The realization (or even illusion) that "we" are many can clearly give a great boost to an otherwise powerless or oppressed subgroup. Indeed, as strategic SIDE research has shown, the ability to communicate within an ingroup offers channels for social support and may thereby boost efficacy (Spears, R., Lea, M., Corneliussen, R. A., Postmes, T., & Ter Haar, W. (2002)). But similarly, the clampdown of authorities on freedom of communication may backfire: In Egypt, for example, the regime appears to have inadvertently strengthened the revolution by shutting down entire communication channels (such as Internet and text messaging). These drastic measures may have communicated, more effectively than a relatively small revolutionary movement ever could have, that the regime was having difficulty coping with the unrest (Dunn, 2011).

Finally, it is possible that communication technology fires up or otherwise affects collective emotions. The *Guardian* (a UK national newspaper), in collaboration with researchers from the London School of Economics, has made some interesting analyses of the role that Twitter played in the dissemination of rumors (www.guardian.co.uk/uk/series/reading-the-riots). It is a well-known phenomenon that collective unrest is often "triggered" by a particular incident or event that, for some reason, focuses collective emotions. In the case of the London riots, Tottenham riot police were alleged to have sparked the riot by heavy-handed policing in which a 16-year-old girl was mistreated. Rumors about this kept circulating for a day and a half (with 70 tweets per hour at peak moments) and were never contradicted. But it would be rash to infer from this that social media fan the flames of unrest: the research uncovered more substantial evidence that tweets about clearly erroneous rumors (e.g., the London eye is burning, tigers have escaped from London zoo) were quickly quashed. Despite the lack of clarity at this stage, it is clear that communication media may have a strong impact on the way in which emotions are shared and expressed online, as well as providing the appraisals (e.g., news of unjust actions) that evoke them in the first place. This makes the topic of online emotions a prime candidate for future systematic research and for further extensions of SIDE.

Overview of SIDE extensions

In this section, we have argued that the SIDE model has proved heuristic in leading to a number extensions and developments rising from the basic statement of the model. Although originally devised to analyze the effects of social influence in CMC and virtual environments, clearly it has broader relevance to a range of social psychological processes and can be used to analyze the effect of a range of different communication media, old and new, that vary in their key features. In considering these applications and extensions, it becomes apparent how far the SIDE model has come and developed, itself. Whereas early statements and research focused on communications technology primarily in terms of the anonymity they afford (in line with the early focus on the cognitive side of SIDE), more recent research focusing on the mass dissemination potential of social media and microblogs highlight the power of these media to influence others and coordinate action, which has implications for strategic as well as cognitive SIDE effects. Although not technologically determinist, then, it is certainly true that changes in communication technology have pushed the SIDE agenda in terms of the social phenomena to be explained, and the model has developed as a result. While we would not claim it is the only framework of relevance, we think it continues to provide some heuristic value in making sense of diverse media effects.

Summary and Conclusions

In this chapter, we have provided an overview of the SIDE model, its theoretical roots and how it developed, and its utility in explaining a range of social influence and other group phenomena as instantiated within the new communications technologies. As a model grounded in the social identity approach, the SIDE model provides broad scope for analyzing how specific features associated with these technologies (visual anonymity, the means to connect, coordinate, and coopt) interact with the levels of self and contents of identities, to predict key processes and outcomes. As a model (although grounded in theory) it remains provisional, a work in progress, open to being updated as new and relevant technological features, and their possible effects become apparent. However, rather than accounting for effects post hoc or ad hoc, the SIDE framework has been able to anticipate and explain some of the more counterintuitive social effects of CMC and cyberspace. The idea that social influence could be strengthened by anonymity and isolation went against the prevailing models as well as common sense, but has remained a robust and consistent finding replicated in much subsequent research (although circumscribed by identities in situ). Similarly the idea that these same features (anonymity and isolation) could actually strengthen power relations and the dominance of powerful groups, as predicted by the cognitive SIDE, went against much of the theorizing and writing on the effect and potential of these new media (Spears & Lea, 1994; Postmes et al., 1998). So far the SIDE model has been able to account for a wide range of influence phenomena and many related aspects of group processes (influence, stereotyping, group cohesion, power relations), as well as providing a framework to inform and guide applications in diverse domains (group decision making, cooperative learning and working, collective action, to name but a few). The SIDE model has been less prominent in prescribing media use and matching media to task demands although perhaps it is inevitable given its contribution to critiques of both technological and social determinism. However, this may yet be a realm in which it has a role to play, if only to define the scope and limits of such media-matching exercises.

References

Adrianson, L. (2001). Gender and computer-mediated communication: Group processes in problem solving. *Computers in Human Behavior*, *17*(1), 71–94.

Ainsworth, S., Gelmini-Hornsby, G., Threapleton, K., Crook, C., O'Malley, C., & Buda, M. (2011). Anonymity in classroom voting and debating. *Learning and Instruction*, *21*(3), 365–378.

Anderson, L. (2011). Demystifying the Arab spring. *Foreign Affairs*, *90*(3), 2–7.

Anonymous (1998). To reveal or not to reveal: A theory of anonymous communication. Communication Theory, 8, 3481–407.

Bennett, W. L., & Segerberg, A. (2012). The logic of connective action: Digital media and the personalization of contentious politics. *Information, Communication & Society*, *15*(5), 739–768.

Bimber, B., Flanagin, A. J., & Stohl, C. (2005). Reconceptualizing collective action in the contemporary media environment. *Communication Theory*, *15*(4), 365–388.

Bohannon, J. (2012). Tweeting the London riots. *Science*, *336*(6083), 831–831.

Brunsting, S., & Postmes, T. (2002). Social movement participation in the digital age: Predicting offline and online collective action. *Small Group Research*, *33*, 525–554.

Castells, M. (2012). *Networks of outrage and hope: Social movements in the internet age*. Cambridge, UK: Polity.

Chan, M. (2010). The impact of email on collective action: A field application of the SIDE model. *New Media & Society*, *12*(8), 1313–1330.

Cress, U. (2005). Ambivalent effect of member portraits in virtual groups. *Journal of Computer Assisted Learning*, *21*(4), 281–291.

Daft, R. L., & Lengel, R. H. (1984). Information richness: A new approach to managerial behavior and organizational design. *Research in Organizational Behavior*, *6*, 191–233.

Diener, E. (1980). Deindividuation: The absence of self-awareness and self-regulation in group members. In P. Paulus (Ed.), *The psychology of group influence*. Hillsdale, NJ: Erlbaum.

Dunn, A. (2011). Unplugging a nation: State media strategy during Egypt's January 25 uprising. *World Affairs*, *35*, 15–24.

Farzan, R., Dabbish, L. A., Kraut, R. E., & Postmes, T. (2011). Increasing commitment to online communities by designing for social presence. *Proceedings of the ACM 2011 Conference on Computer Supported Cooperative Work*, Hangzhou, China, 321–330. doi:http://doi.acm.org.proxy-ub.rug.nl/10.1145/1958824.1958874

Festinger, L., Pepitone, A., & Newcomb, T. (1952). Some consequences of de-individuation in a group. *Journal of Abnormal and Social Psychology*, *47*, 382–389.

Fischer, C. S. (1992). *America calling: A social history of the telephone to 1940.* Berkeley, CA: University of California Press.

Flanagin, A. J., Tiyaamornwong, V., O'Connor, J., & Seibold, D. R. (2002). Computer-mediated group work: The interaction of sex and anonymity. *Communication Research*, *29*(1), 66–93.

Haraway, D. (1990). A manifesto for cyborgs: Science technology, and socialist feminism in the 1980s. In L. Nicholson (Ed.) *Feminism/postmodernism*, New York, NY: Routledge.

Joinson, A. N. (2005). *Who's watching you? Power, personalization and on-line compliance.* Paper presented at Conference on Group Processes in Computer Supported Interaction: Technological and Social Determinism. Oxford, OH: Miami University.

Kiesler, S., Siegel, J., & McGuire, T. (1984). Social psychological aspects of computer-mediated communications. *American Psychologist*, *39*, 1123–1134.

Klein, O., Spears, R., & Reicher, S. (2007). Social identity performance: Extending the strategic side of the SIDE model. *Personality and Social Psychology Review*, *11*, 28–45.

Le Bon, G. (1995). *The crowd: A study of the popular mind.* London, UK: Transaction Publishers. (Original work published in 1895).

Lea, M., Rogers, P., & Postmes, T. (2002). SIDE-VIEW: Evaluation of a system to develop team players and improve productivity in internet collaborative learning groups. *British Journal of Educational Technology*, *33*(1), 54–64.

Lea, M., Spears, R., & De Groot, D. (2001). Knowing me, knowing you: Effects of visual anonymity on self-categorization, stereotyping and attraction in computer-mediated groups. *Personality and Social Psychology Bulletin*, *27*, 526–537.

Lea, M., Spears, R., & Watt, S. E. (2007). Visibility and anonymity effects on attraction and group cohesiveness. *European Journal of Social Psychology*, *37*, 761–773.

Lee, E. J. (2004). Effects of visual representation on social influence in computer-mediated communication. *Human Communication Research*, *30*(2), 234–259.

Lee, E. J. (2007). Deindividuation effects on group polarization in computer-mediated communication: The role of group identification, public-self-awareness, and perceived argument quality. *Journal of Communication*, *57*(2), 385–403. doi:10.1111/j.1460-2466.2007.00348.x

Lee, E. J. (2008). When are strong arguments stronger than weak arguments? Deindividuation effects on message elaboration in computer-mediated communication. *Communication Research*, *35*(5), 646–665. doi:10.1177/0093650208321784

Matheson, K., & Zanna, M. P. (1988). The impact of computer-mediated communication on self-awareness. *Computers in Human Behavior*, *4*(3), 221–233.

McKenna, K. Y. A., & Bargh, J. A. (1998). Coming out in the age of the Internet: Identity "de-marginalization" through virtual group participation. *Journal of Personality and Social Psychology*, *75*, 681–694.

Oakes, P. J. (1987). The salience of social categories. In J. C. Turner, M. A. Hogg, P. J. Oakes, S. D. Reicher, & M. S. Wetherell (Eds.), *Rediscovering the social group: A self-categorization theory* (pp. 117-141). Oxford, UK: Basil Blackwell.

Olivetti, C., & Petrongolo, B. (2008). Unequal pay or unequal employment? A cross-country analysis of gender gaps. *Journal of Labor Economics*, *26*, 621–654.

Postmes, T. (2007). The psychological dimensions of collective action, online. In A. N. Joinson, K. Y. A. McKenna, T. Postmes, & U. Reips (Eds.), *Oxford handbook of internet psychology* (pp. 165–186). Oxford, UK: Oxford University Press.

Postmes, T., & Brunsting, S. (2002). Collective action in the age of internet: Mass communication and online mobilization. *Social Science Computer Review*, *20*, 290–301.

Postmes, T., & Spears, R. (1998). Deindividuation and anti-normative behavior: A meta-analysis. *Psychological Bulletin*, *123*, 238–259.

Postmes, T., & Spears, R., (2000). Refining the cognitive redefinition of the group: Deindividuation effects in common bond vs. common identity groups. In T. Postmes, R. Spears, M. Lea, & S.D. Reicher (Eds.) *SIDE issues centre stage: Recent developments of de-individuation in groups.* (pp. 63–77). Amsterdam: Proceedings of the Dutch Royal Academy of Arts and Sciences.

Postmes, T., & Spears, R. (2002). Behavior online: Does anonymous computer communication reduce gender inequality? *Personality and Social Psychology Bulletin*, *28*, 1073–1083.

Postmes, T., Spears, R., & Lea, M. (1998). Breaching or building social boundaries? SIDE-effects of computer mediated communication. *Communication Research*, *25*, 689–715.

Postmes, T., Spears, R., & Lea, M. (2000). The formation of group norms in computer-mediated communication. *Human Communication Research*, *26*(3), 341–371.

Postmes, T., Spears, R., & Lea, M. (2002). Intergroup differentiation in computer-mediated communication: Effects of depersonalization. *Group Dynamics: Theory, Research and Practice*, *6*, 3–16.

Postmes, T., Spears, R., Lee, A. T., & Novak, R. J. (2005). Individuality and social influence in groups: Inductive and deductive routes to group identity. *Journal of Personality and Social Psychology*, *89*, 747–763.

Postmes, T., Spears, R., Sakhel, K., & De Groot, D. (2001). Social influence in computer-mediated groups: The effects of anonymity on social behavior. *Personality and Social Psychology Bulletin*, *27*, 1243–1254.

Postmes, T., van Bezouw, M., Täuber, S., & Van der Sande, J. P. (2013). *Naar een beter begrip van 21e eeuws protest en oproer [Towards a better understanding of 21st century protest and revolt].* Report commissioned by the Interior Ministry, the Hague, Netherlands. Retrieved from http://www.rug.nl/staff/t.postmes

Prentice-Dunn, S., & Rogers, R. W. (1989). Deindividuation and the self-regulation of behavior. In P. B. Paulus (Ed.), *The psychology of group influence* (2nd ed., pp. 86–109). Hillsdale, NJ: Lawrence Erlbaum.

Reicher, S. (1982). The determination of collective behaviour. In H. Tajfel (Ed.), *Social identity and intergroup relations* (pp. 41–83). Cambridge, UK: Cambridge University Press.

Reicher, S. (1984). Social influence in the crowd: Attitudinal and behavioural effects of de-individuation in conditions of high and low group salience. special issue: Intergroup processes. *British Journal of Social Psychology*, *23*(4), 341–350.

Reicher, S. D. (1987). Crowd behaviour as social action. In J. C. Turner, M. A. Hogg, P. J. Oakes, S. D. Reicher, & M. S. Wetherell (Eds.), *Rediscovering the social group: A self-categorization theory* (pp. 171–202). Oxford, UK: Basil Blackwell.

Reicher, S. D., & Levine, M. (1994a). Deindividuation, power relations between groups and the expression of social identity: The effects of visibility to the out-group. *British Journal of Social Psychology*, *33*, 145–163.

Reicher, S. D., & Levine, M. (1994b). On the consequences of deindividuation manipulations for the strategic considerations of self: Identifiability and the presentation of social identity. *European Journal of Social Psychology*, *24*, 511–524.

Reicher, S. D., Spears, R., & Postmes, T. (1995). A social identity model of deindividuation phenomena. *European Review of Social Psychology*, *6*, 161–198.

Ren, Y., Harper, F. M., Drenner, S., Terveen, L., Kiesler, S., Riedl, J., & Kraut, R. E. (2012). Increasing attachment to online communities: Designing from theory. *MIS Quarterly*, *36*(3), 841–864.

Ren, Y., Kraut, R., & Kiesler, S. (2007). Applying common identity and bond theory to design of online communities. *Organization Studies*, *28*(3), 377–408.

Sassenberg, K. (2002). Common bond and common identity groups on the internet: Attachment and normative behavior in on-topic and off-topic chats. *Group Dynamics: Theory, Research, and Practice*, *6*(1), 27.

Sassenberg, K., & Boos, M. (2003). Attitude change in computer-mediated communication: Effects of anonymity and category norms. *Group Processes & Intergroup Relations*, *6*(4), 405–422.

Sassenberg, K., & Postmes, T. (2002). Cognitive and strategic processes in small groups: Effects of anonymity of the self and anonymity of the group on social influence. *British Journal of Social Psychology*, *41*(3), 463–480.

Short, J. A., Williams, E., & Christie, B. (1976). *The social psychology of telecommunications.* Chichester, West Sussex: Wiley.

Siegel, J., Dubrovsky, V., Kiesler, S., & McGuire, T. (1986). Group processes in computer-mediated communication. *Organizational Behaviour and Human Decision Processes*, *37*, 157–187.

Spears, R. (2010). Group rationale, collective sense: Beyond intergroup bias. *Invited position paper. British Journal of Social Psychology*, *49*, 1–20.

Spears, R., & Lea, M. (1992). Social influence and the influence of the 'social' in computer-mediated communication. In M. Lea (Ed.), *Contexts of computer-mediated communication.* (pp. 30–65). Hemel-Hempstead, UK: Harvester-Wheatsheaf.

Spears, R., & Lea, M. (1994). Panacea or panopticon? The hidden power in computer-mediated communication. *Communication Research*, *21*, 427–459.

Spears, R., Lea, M., Corneliussen, R. A., Postmes, T., & Ter Haar, W. (2002). Computer-mediated communication as a channel for social resistance: The strategic side of SIDE. *Small Group Research*, *33*, 555–574.

Spears, R., Lea, M., & Lee, S. (1992). De-individuation and group polarization in computer-mediated communication. *British Journal of Social Psychology*, *29*, 121–134.

Spears, R., Lea, M., & Postmes, T. (2007). Computer-mediated communication and social identity. In: A. N. Joinson, K. Y. A. McKenna, T. Postmes, & J. E. Katz (Eds.), *The Oxford handbook of internet psychology* (ch. 17, pp. 253–269). Oxford, UK: Oxford University Press.

Spears, R., Lea, M., Postmes, T., & Wolbert, A. (2011). A SIDE look at computer mediated interaction: Power and the gender divide. In Z. Birchmeier, B. Dietz-Uhler, & G. Stasser (Eds.), *Strategic uses of social technology: An interactive perspective of social psychology* (pp. 16–39) Cambridge, UK: Cambridge University Press.

Spears, R., Postmes, T., Lea, M., & Wolbert, A. (2002). When are net effects gross products? The power of influence and the influence of power in computer-mediated communication. *The Journal of Social Issues*, *58*, 91–107.

Tajfel, H., & Turner, J. C. (1979). An integrative theory of intergroup conflict. In S. Worchel, & W. G. Austin (Eds.), *The psychology of intergroup relations* (pp. 33–47). Monterey, CA: Brooks-Cole.

Tanis, M., & Postmes, T. (2005). A social identity approach to trust: Interpersonal perception, group membership and trusting behavior. *European Journal of Social Psychology*, *35*, 413–424.

Tanis, M., & Postmes, T. (2007). Two faces of anonymity: Paradoxical effects of cues to identity in CMC. *Computers in Human Behavior*, *23*, 955–970.

Tanis, M., & Postmes, T. (2008). Cues to identity in online dyads: Effects of interpersonal versus intragroup perceptions on performance. *Group Dynamics-Theory Research and Practice*, *12*(2), 96–111. doi:10.1037/1089-2699.12.2.96

Thomas, E. F., McGarty, C., & Mavor, K. I. (2009). Aligning identities, emotions, and beliefs to create commitment to sustainable social and political action. *Personality and Social Psychology Review*, *13*(3), 194–218.

Turkle, S. (1995). *Life on the screen: Identity in the age of the internet.* New York, NY: Simon and Schuster.

Turner, J. C. (1982). Towards a cognitive redefinition of the group. In H. Tajfel (Ed.), *Social identity and intergroup relations* (pp. 15–40). Cambridge, UK: Cambridge University Press.

Turner, J. C. (1987). The analysis of social influence. In J. C. Turner, M. A. Hogg, P. J. Oakes, S. D. Reicher, & M. S. Wetherell, *Rediscovering the social group: A self-categorization theory* (pp. 68–88). Oxford, UK: Basil Blackwell.

Turner, J. C. (1991). *Social influence.* Milton Keynes, UK: Open University Press.

Turner, J. C., Hogg, M. A., Oakes, P. J., Reicher, S., & Wetherell, M. S. (1987). *Rediscovering the social group: A self-categorization theory.* Oxford, UK: Basil Blackwell.

Waldzus, S., & Schubert, T. (2000). Group norm and category norm in anonymous situations: Two sources of social influence. In T. Postmes, R. Spears, M. Lea, & S. Reicher (Eds.), *SIDE issues centre stage: Recent developments in studies of de-individuation in groups.* Amsterdam, the Netherlands: North Holland Elsevier.

Wetherell, M. (1987). Social identity and group polarization. In J. C. Turner, M. A. Hogg, P. J. Oakes, S. D. Reicher, & M. S. Wetherell (Eds.), *Rediscovering the social group: A self-categorization theory* (pp. 142–170). Oxford, UK: Basil Blackwell.

Wodzicki, K., Schwämmlein, E., Cress, U., & Kimmerle, J. (2011). Does the type of anonymity matter? The impact of visualization on information sharing in online groups. *Cyberpsychology, Behavior, and Social Networking*, *14*(3), 157–160.

Yao, M. Z., & Flanagin, A. J. (2006). A self-awareness approach to computer-mediated communication. *Computers in Human Behavior*, *22*(3), 518–544.

Zimbardo, P. G. (1969). The human choice: Individuation, reason, and order vs. deindividuation, impulse and chaos. In W. J. Arnold & D. Levine (Eds.), *Nebraska Symposium on Motivation* (Vol. 17, pp. 237–307). Lincoln, NE: University of Nebraska Press.

3

Toward a Theory of Interactive Media Effects (TIME)

Four Models for Explaining How Interface Features Affect User Psychology

S. Shyam Sundar, Haiyan Jia,
T. Franklin Waddell, and Yan Huang

*Media Effects Research Laboratory, College of Communications,
The Pennsylvania State University, University Park, PA, USA*

Introduction

A fundamental assumption of media scholars is that the medium of communication makes a difference. Otherwise, media psychologists would have nothing to add beyond what psychologists have already discovered, and mass communication research would be no different than interpersonal communication. When humans use media, the nature of their communications changes. As McLuhan (1964) famously said, media change the form, pace, process, scale, and even the content of communication. It is logical therefore to expect that they will change the effects of communication as well.

So, what do media do to the process of communication? A rudimentary mechanistic conception of media's role in communication would suggest that they transmit messages over a distance, amplify the signal strength, broaden the audience reach, and so on. Under this view, the medium is a mere channel, structural in its contribution to the process of communication, and largely uninteresting for a social scientific understanding of the dynamics of communication. This view has pervaded the field of media and communication studies for much of its history, with scholars focusing mostly on message aspects that affect the process of communication.

However, advancements in media technology over the past few decades have persuaded us to think of the medium of communication as more than a mere channel conveying information between senders and receivers. We tend to view and describe our experiences with modern media (from websites to games to virtual reality) in spatial terms, as places we visit and meet, work and play with others (Meyerowitz, 1989).

The Handbook of the Psychology of Communication Technology, First Edition. Edited by S. Shyam Sundar.

Further, as computers and robots take decisions on our behalf and provide us personalized information, we tend to treat them as autonomous sources in their own right (Sundar & Nass, 2000). We behave socially toward them and tend to make automatic attributions of intentionality to inanimate objects because they trigger scripts pertaining to human–human communication in our minds. As Reeves and Nass (1996) contend, the human brain is not evolved to respond to modern media technologies in a way that distinguishes between human–human communication and human–computer communication. While studies in the media equation literature have focused on how the psychology of interacting with media is similar to the psychology of interacting with humans, those in the interpersonal and group communications literature have focused on the degree to which communicating via media is different from communicating in person (Kiesler, Siegel, & McGuire, 1984; Spears & Lea, 1994; Walther, 1996). While the former is attributed to shortcomings of the human mind (e.g., humans cannot help but be social), the latter is attributed to shortcomings of the medium (e.g., cuelessness) in approximating the characteristics of face-to-face interactions.

The general tendency is to focus on human psychology adjusting to the exigencies of technology rather than specific technological features stimulating changes in human psychology and behaviors. While the consideration of human–human communication as the comparison standard helps us evaluate the changes brought about by technology, it does not quite explain the compelling user experience offered by modern-day communication technologies. All around us, we see people glued to their mobile devices—a dramatic transformation in public human behavior in just about a decade. Even more dramatic is the diffusion and obsessive use of social networking sites (SNS). There are several other examples from recent history, including games that have come to dominate the entertainment industry.

Why are people so drawn to games? Why are they glued to their mobile devices? Why are they so glad to be interrupted by mindless chatter on instant messengers and idle banter on SNS? It is not like these media provide new content that did not exist before. What is new is the convenience and manner in which the content is served to users. Very simply, they offer a new form of interaction. Each of these media have unique technological attributes which manifest themselves in terms of specific interface features and affordances that shape the nature of their communications, with consequent effects on their psychology. This fundamental thesis guides much of the research conducted by our lab group over the last 15 years. We have proposed and tested several theoretical formulations with interface features as the cause and user psychology as the effect. Across several experiments, we have discovered that this effect is not restricted to users' mental states, but indeed transfers over to their perception and processing of media content. That is, variations in interface features can predict variations in the effects of media content even though content is kept constant across conditions in all these experiments. Four models specifying distinct psychological mechanisms have governed this effort, producing empirical findings that not only advance our knowledge about media psychology but also inform interaction design.

Interactive Media Effects

The most important keyword in the discourse of modern media is "interactive." By offering interactivity, digital media have profoundly altered our conception of "medium." We no longer think of media as simply conduits or channels via which senders communicate with receivers, but rather as tools with which *we* can shape our interactions. These interactions could be with the media themselves and/or with others. They include a wide range of activities, such as accessing content of personal interest, chatting with remotely located friends, changing the modality of presentation, browsing an online shopping mall, and so on. These activities influence the locus, nature, and effects of our communications—whom we communicate with, what information we exchange and how we are affected by it. Therefore, interactivity is a fundamental affordance of modern media and communication technologies.

While there is agreement among scholars about the importance of interactivity, there is considerable variance in how to approach it. For some, interactivity is an inherent attribute of mediated communications via digital technologies, unalterably affecting user experience. We may think of this as the media-ecological approach, pioneered by McLuhan (1964), wherein interactivity is seen as a game-changer, fundamentally redefining communication and social psychological processes surrounding it. The proliferation of journals with "interactive" in their titles (e.g., Journal of Interactive Advertising, Journal of Interactive Marketing) captures this school of thought, in that the concept of interactivity is co-opted into an existing media practice or phenomenon without being isolated for independent examination of its role and impact.

For others, interactivity is in the eye of the beholder. A system can be more or less interactive depending upon how one uses it. This would fall under the uses-and gratifications paradigm of mass communication research (Rubin, 1993), in that it privileges user motivations and allows users to determine the amount of interactivity in a given interaction. The proposal of several scales of perceived interactivity over the past decade (Liu & Shrum, 2002; McMillan & Huang, 2002; Quiring, 2009; Sohn, 2011) represents this view and has allowed communication scholars to measure interactivity without having to design or manipulate interactivity. This has resulted in a number of correlational findings connecting perceived interactivity to a number of psychological outcome variables such as attitudes toward the site (Liu & Shrum, 2002; McMillan & Huang, 2002; Wu, 2005).

However, as Sundar (2004) argues, such perceptual measures of interactivity may be conflated with user perceptions of system attributes that are unrelated to interactivity, for example, usability. As such, perceived interactivity is a user variable, reflecting variations in usage, skills, and user experience of the system. It makes interactivity an attribute of the user rather than the system, thereby limiting our understanding of the role played by technology in mediated communications. In contrast, the aforementioned media-ecological approach is less user-centered and more object-centered, but both approaches tend to treat interactive technology as a black box, with little emphasis in understanding the nature and operation of interactivity. They do not quite help us answer the question: What is it about interactivity that changes the process and outcomes of communication?

The media-effects approach to studying communication technology (Sundar, 2009) offers a viable solution, in that it tends to disaggregate media into specific variables. Historically, these variables have focused on source and message attributes of media. In our lab group, we adopt this approach to vary attributes of the medium as well. Variables constructed out of nonmessage attributes of the medium are then examined for their effects on user psychology. Of course, "user psychology" is itself a black box, as pointed out by media-effects scholars (Geiger & Reeves, 1993), but several efforts have been made in recent times to unlock it by paying more attention to psychological processing of media stimuli (Lang, 1994). Likewise, disaggregating the medium into specific technological variables such as interactivity helps us to unlock the technological black box and understand the processes by which technology engages users: What exactly is the salient aspect of interactivity that impacts users' psychology and thereby affects their perception and attitudes toward content?

We approach such questions by first conceptualizing technological attributes of interactive media as "affordances" (Sundar & Bellur, 2010). Affordances are action possibilities suggested by visual stimuli in our environment (Gibson, 1977). Researchers in the domain of human–computer interaction, most notably Norman (1988), view affordances as perceivable properties of a system suggesting ways in which it could be operated. Users perceive these affordances by looking for cues on the interface of the system. If the cues are designed well, they will accurately signal the underlying affordance offered by the system. If the interface does not display a cue about an underlying action (e.g., mouse-over), then this action possibility is a "hidden affordance" (Gaver, 1991). On the other hand, if the cue on the interface is overly promising (in that the system functionality does not quite match up to the action expectations signaled by the cue), then the action signaled by such a cue is considered a "false affordance" (Gaver, 1991).

When conceptualized as an affordance, the focus shifts to the functionality that dictates the nature of user actions with the system. A rudimentary functionality of all interactive media is that the user can perform actions that elicit a contingent response (Rafaeli, 1988). Other classic definitions of interactivity focus on functionalities offered to the user, such as choice (Heeter, 1989) and real-time modifiability of the interface and its content (Steuer, 1992). In this way, it is important to identify those aspects of the affordances that are psychologically relevant—aspects like contingency, choice, control, and manipulability. Interactivity is not just about interaction. That is a given. Interactivity as a variable of study is about one or more psychologically salient functionalities embedded in that interaction. Therefore, the theorizing of the processes by which an affordance results in psychological effects is governed by the need to explicate the affordances in terms of those qualities of the actions that are psychologically important.

The starting point of a theory of interactive media effects (TIME) is an affordance offered by the medium of communication. This is typically an interface feature attributable to the technology of the medium rather than the source or content of communication. This affordance lies at the intersection of the medium and the user, in keeping with classic conceptualizations of "affordance" (Gibson, 1977; Norman, 1988).

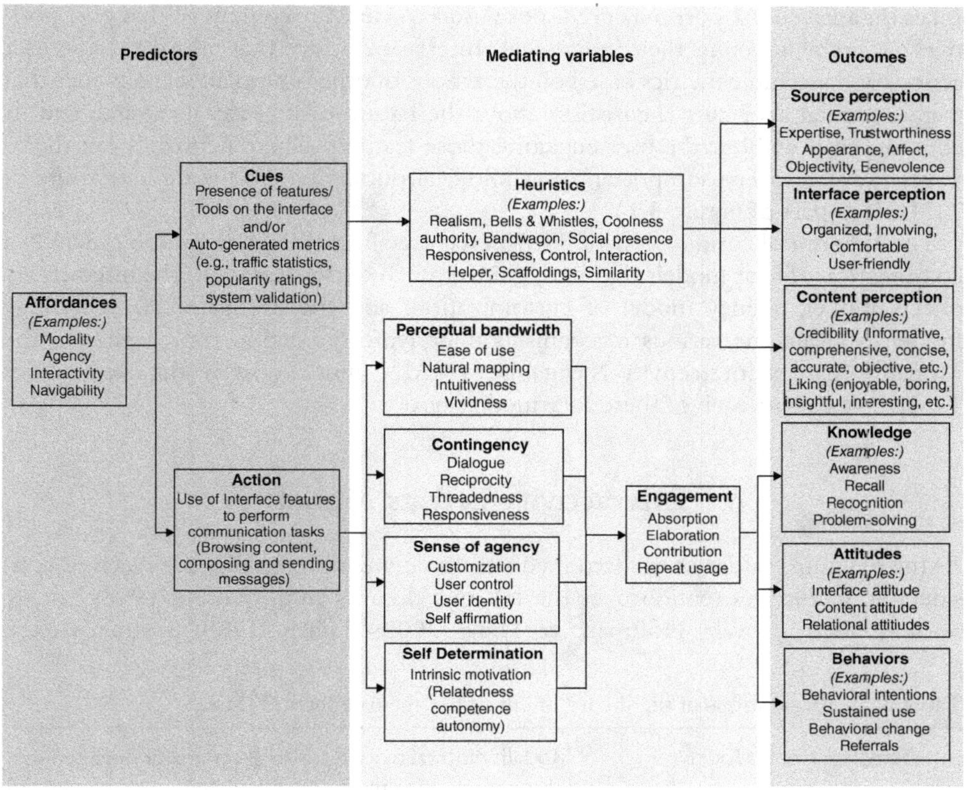

Figure 3.1 TIME model.

TIME predicts that affordances can affect user psychology in two distinct ways—by triggering action on the part of the user and/or by serving as symbolic representational cues on the interface.

The actions afforded by an interface feature may be ontological (e.g., provision of choice to users, or allowing the user to broadcast, i.e., serve as source of communication), but the key requirement for TIME is that they have psychological correlates (choice = perceived control, self as source = sense of agency). These psychological correlates play an important role as mediators between actions afforded by the interface and user engagement with the content offered by the interface. This affordance-driven engagement serves to moderate the effects of media content on users' knowledge, attitudes, and behaviors. This is the action route of TIME (bottom part of Figure 3.1).

Aside from motivating actions, affordances are richly suggestive by their sheer existence on an interface. Even without the user taking any action, some affordances can have important perceptual consequences. This is because the presence of an affordance in a system can serve as a psychologically salient cue for the user. For example, buttons on a politician's website inviting comments from users can serve as cues conveying the politician's open-mindedness or sensitivity to voters. Likewise, number of

Likes on a Facebook post may predispose users to view the content of that post positively, thereby affecting their judgments. Interface features that afford specific user actions and assemble metrics based on the actions of other users can serve as cues that trigger mental shortcuts (heuristics) about the nature of the site, its source and its content, even without the user engaging those features. These heuristics can themselves translate over to important psychological outcomes. This is the cue route of TIME (top part of Figure 3.1).

The theoretical formulation of TIME (see specific predictions in Table 3.1) is a combination of four models that have guided much of our research. The interactivity effects model, agency model of customization, and the motivational technology model explicate the various mechanisms underlying the action route, whereas the Modality-Agency-Interactivity-Navigability (MAIN) model governs the cue route of TIME. We discuss each of these four models next.

Interactivity Effects Model

As the defining feature of modern media technologies, interactivity has received considerable scholarly attention over the last two decades (e.g., Jensen, 1998; Liu, & Shrum, 2002; Novak, Hoffman, & Yung, 2000; Rafaeli, 1988; Stromer-Galley,

Table 3.1 Propositions of the theory of interactive media effects (TIME)

Interactivity Effects Model	Modality interactivity leads to greater user engagement (absorption) with media by enhancing users' perceptual bandwidth
	Message interactivity leads to greater user engagement (elaboration) with media by enhancing the contingency or interdependency in message exchanges
	Source interactivity leads to greater user engagement (contribution) with media by enhancing users' ability to customize, curate, and create content
Agency Model	Interactivity, modality, and navigability affordances that enhance message contingency, self-representation, and individualized exploration serve to provide users a greater sense of "self as source" and thereby positively influence their online involvement, identity and control
Motivational Technology Model	Navigability, interactivity, and customization affordances lead to greater intrinsic motivation by enhancing users' competence, relatedness, and autonomy, respectively
MAIN Model	Interface manifestations of modality, agency, interactivity, and navigability affordances shape user perceptions and experience by cueing cognitive heuristics about content quality and credibility

2004). The most salient change brought about by interactive features is that users are active. They are no longer the passive receivers of media messages. Instead, they are empowered to perform a variety of actions via the medium, participate in the construction of messages they consume and engage in a wide variety of interaction activities. While traditional mass communication was primarily characterized by unidirectional transmission of information, interactive communications typically involve two-way or multidirectional information exchanges.

It is not surprising therefore that early researchers tended to conceive of interactivity in terms of interpersonal conversations. Rafaeli (1988) explicated the concept at three ordinal levels: noninteractive, reactive (or quasi-interactive), and responsive (or fully interactive). Under this formulation, if two people send messages without accounting for each other's messages, then it is considered noninteractive. If one interactant directly responds to another's message, it is reactive. When the latter interactant responds back, acknowledging not only the latest but also all the messages preceding it, the message exchange is considered fully interactive.

The critical component of this conceptualization is the ongoing reciprocity between the interactants. Akin to face-to-face conversation, interactivity is a thread of message exchanges constructed from a series of messages that are contingent upon preceding messages that are sequentially related. Guided by this principle, Sundar, Kalyanaraman, and Brown (2003) operationalized interactivity in terms of hierarchical hyperlinks on a website. The contents and choices encountered by each individual user would be idiosyncratic because of the unique sequence of hyperlinks that they clicked. That is, the messages received by the user from the website are directly contingent upon their previous actions. Due to its emphasis on contingency, we labeled this conceptualization as the "contingency view" of interactivity.

While this was a useful way to characterize back-and-forth message exchanges, it failed to account for the numerous interactive tools offered by graphical user interfaces and proliferated by Web-based media since the mid-1990s. Aside from clicking on hyperlinks, users can use the scroll bar to navigate and the zoom tool to obtain more detail, drag objects across the screen, download files from remote locations, fill out online forms, and so on. Each of these actions fulfills a distinct function. Therefore, if an interface offers a wide variety of such functions, it would be considered more interactive than one that offers fewer such functions. This conceptualization of interactivity refers to the breadth of functions offered by an interface by providing tools that are manipulable by users. We called this the "functional view" of interactivity. While the contingency view is concerned about message exchanges, the functional view of interactivity is more concerned with the gamut of interactive tools and functions that offer users the ability to initiate interactions, either with the medium itself or with other users (Sundar, Xu, Bellur, Oh, & Jia, 2010). It is thus clear that these two views of interactivity situate the concept in different elements of communication (i.e., message and medium, respectively). The contingency-based interactivity highlights the threadedness of message exchanges, whereas the functional view focuses more on the features available on the interface of medium. Accordingly, in his interactivity effects model, Sundar (2007a) relabeled the contingency view and functional view as message interactivity and medium (or modality) interactivity, respectively.

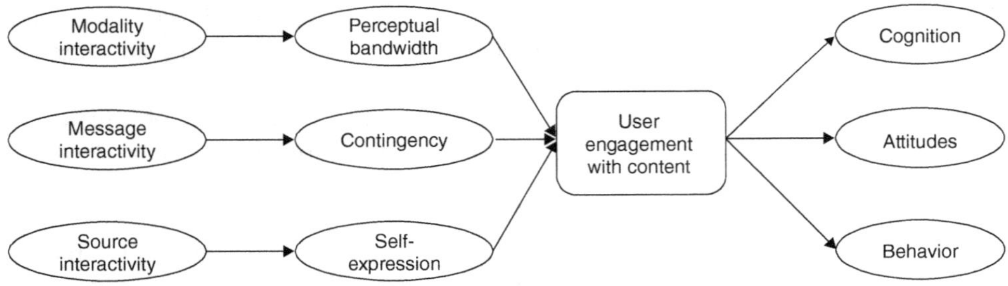

Figure 3.2 Updated model of interactivity effects.
Source: Reproduced with permission from Sundar (2007a, p. 100, Figure 7.2). © Oxford University Press.

Aside from message and medium interactivity, the model conceptualizes a third kind, source interactivity, rooted in yet another fundamental concept of communication. The source is indeed the starting point for all models of communication. In traditional mass media, the position of "source" (or sender) was held by a privileged few in charge of gatekeeping the channels of mediated communication. However, in interactive media, users often serve as the sources of communication because they are not only active but also able to control the flow and nature of content. In fact, some scholars define interactivity as the degree of user control afforded by the interface (Jensen, 1998; Newman, 1991). This control can come in a number of different ways, from the ability to choose content (selection) to tailoring content (customization) to recommending content (curation) to creating new content (generation, as in user-generated content or UGC). In essence, the user is now able to perform the various roles of a communication source, thanks to the interactive affordances of modern media, especially Web 2.0 media. Based on this, Sundar (2007) defined source interactivity as the degree to which the interface affords users the ability to act as the source of communication.

In addition to highlighting the various conceptualizations of interactivity, the interactivity effects model also maps out the psychological effects related to the different aspects of interactivity (see Figure 3.2). Each type of interactivity influences user engagement with the interface in a distinct way, involving a combination of technological affordances and psychological variables, as described below.

Effects of modality (or medium) interactivity

Medium (or modality) interactivity refers to the various methods of interaction offered by the interface, such as clicking, scrolling, dragging, and hovering. Each of these is unique in the action possibilities they offer, thus determining the manner in which users access, encode, and represent the information delivered by the interface (Sundar, Xu, & Bellur, 2010). For example, the 3D Carousel allows users to manually rotate a set of images and control their speed of rotation. As a result, the way users mentally map the content can be quite different compared to when those same images are displayed in a noninteractive gallery. Likewise, a mouse-over can prompt the appearance of related messages when users hover over certain places on the interface. Obtaining information by mousing over a "hotspot" would likely be different from simply reading it off the interface.

As these examples illustrate, modality interactivity affords users the ability to manipulate visual objects and simulate their movements in a variety of ways. These activities may serve to enhance the speed, range, and vividness of one's mental representation of the information on the interface. Thus, modality interactivity is said to promote user engagement with the content by expanding one's sensory breadth and depth involved in the interaction process (Steuer, 1992), affording greater "perceptual bandwidth" (Reeves & Nass, 2000) for experiencing the content.

The interactivity effects model highlights perceptual bandwidth as the key mechanism through which modality interactivity engages users (Sundar, 2007a), but it can be quite tricky to operationalize. We approached it from a usability perspective and measured it in terms of user assessment of the interface. More specifically, we reasoned that the perceived naturalness in mapping user actions, the intuitiveness of the operation, and the ease of using the interaction technique would together constitute formative indicators of the concept because they would serve to expand the scope of user exploration of the interface while simultaneously freeing up cognitive resources that would otherwise be allocated for operating the interface (Sundar, Bellur, Oh, Xu, & Jia, 2014). Experiments with several interaction techniques (e.g., clicking, sliding, dragging, hovering, and zooming) have demonstrated that participants' assessments of the interface (in terms of naturalness, intuitiveness, and ease of use) tend to mediate the relationship between modality interactivity and user absorption in the content as well as their attitudes toward the interface and its content.

The psychological appeal of modality interactivity has been confirmed in a variety of contexts, such as public relations, online marketing, and online exhibitions (Guillory & Sundar, 2014; Sundar, Xu, Bellur, Oh, & Jia, 2011). For example, on an e-commerce website, Xu and Sundar (2014) found that high modality interactivity (operationalized in terms of tools for users to rotate, zoom, and hover over different images of the product, in addition to just clicking and scrolling) effectively amplified the fun and control aspects of user engagement, resulting in more favorable attitudes toward the website as well as the product.

Other studies have shown that different tools of modality interactivity have distinct psychological advantages (Sundar, Xu, et al., 2011). For example, slider is superior in aiding users' memory for content; click and drag are perceived as more natural and intuitive to use; mouse-over and cover-flow can encourage more explorations on the website; and so on. Differential effects have been found for combinations of these tools as well (Sundar, Bellur, Oh, Xu, et al., 2014). For example, the combination of slide and click reduces the amount of user interactions on the interface, whereas the combination of drag and zoom tends to increase them. The combination of slide and mouse-over has been found to elicit better user attitudes compared to other pairs.

While a number of studies have documented the positive effects of modality interactivity, we should note that it tends to draw user attention away from noninteractive parts of an interface. Studies from our lab (Sundar, Bellur, Oh, & Jia, 2011; Xu &

Sundar, 2011) suggest that modality interactivity has a spotlighting function in that it tends to enhance users' cognitive processing of related messages by degrading their processing of noninteractive content on the same interface.

Effects of message interactivity

While modality interactivity is more concerned with functional features on the interface and the various tools available for users to access content, message interactivity pertains to the nature of exchanges between the user and the system (or other users). As messages for interaction are already embedded on the website in web-based communication, message interactivity is typically realized in the form of information organization, such as the hierarchical layers of hyperlinks, buttons, and other similar navigation tools. When users use these tools to navigate in different orders and make decisions on which parts of the content to read and which to ignore, they actually exchange messages with the interface in an idiosyncratic manner. This is analogous to a conversation between two humans in that the course of the interaction is highly contingent upon prior input. This operationalization, of imbuing contingency by layering information without involvement of artificial intelligence, has been borne out by successful manipulation checks across several experiments (e.g., Sundar et al., 2003; Sundar & Kim, 2005).

The interactivity effects model suggests that the critical mechanism of message interactivity is perceived contingency. When users perceive that the system is contingently responding to them, they tend to be more engaged with the website, which can carry over to other outcome variables. Sundar, Bellur, Oh, Jia, and Kim (2014) provided empirical evidence for this prediction. In the context of a movie recommendation site, they operationalized message interactivity by visualizing users' interaction history, that is, their searching and browsing records on the website. This manipulation successfully imbued participants with heightened feelings of contingency. When the interface showed more interaction history, participants tended to perceive higher level of contingency. This determined their level of engagement and in turn, influenced their attitudes toward the website as well as their behavioral intention to revisit the site.

In addition to interaction history, message interactivity can be operationalized by the addition of dialog-based features to the interface. For example, Bellur and Sundar (2013) used a chat-like interface that asked questions and provided answers with varying levels of contingency (i.e., noninteractive, reactive, and interactive). Findings suggest that perceived contingency positively mediates the effects of message interactivity on user engagement, attitudes toward the system, and stunningly, even their assessment of health information provided by the system.

Effects of source interactivity

Source interactivity is the degree to which the interface lets the user serve as the source of communication. In the early days, source interactivity was used synonymously with the notion of customization—the higher the ability to customize, greater the source interactivity afforded by an interface. The appeal of personally tailored content is obvious, given its relevance to the user. Several studies from our lab group (Kalyanaraman & Sundar, 2006; Sundar & Marathe, 2010) have confirmed the positive psychological effects of customized content. Therefore, we posited that customization was the critical theoretical mechanism underlying this third kind of interactivity, namely source interactivity (Sundar, 2007a). However, with the arrival of social media, users are not only able to tailor content for themselves but also recommend content for others (as in social bookmarking sites) and create their own content (e.g., using blogging tools). This has served to broaden the conceptualization of source interactivity in recent applications of the interactivity effects model.

For instance, a two-week field study by Sundar, Oh, Bellur, Jia, and Kim (2012) operationalized source interactivity by allowing users to change the appearance of the interface (cosmetic customization), modify the task-centered utility tools on the interface (functional customization), and manage their own blogs (either filter or create original content). These tools of source interactivity allow for different forms of self-expression and were found to have unique effects on user psychology. While some amount of functional customization was appreciated for improving task efficiency, providing an abundance of task-related choices proved to be cumbersome for users due to the greater need for decision making. On the contrary, cosmetic customization, relating to presentation-based aspects of the interface, seemed to effectively stimulate users' desire for self-expression. This in turn boosted other forms of user participation on the website, such as commenting and replying to other users. Furthermore, letting users create original content based on their interests (active blogging) resulted in a better experience for the users than simply filtering existing content (filter blogging). When participants were allowed to engage in active blogging, they felt the site was higher on usability. They also felt a higher sense of personal competence and a greater sense of community with other users of the site. These positive effects of active blogging, coupled with the psychological appeal of cosmetic customization, persuade us to expand our conceptualization of source interactivity by going beyond customizability to include gatekeeping for others (i.e., social bookmarking), user control of the interface, and self-expression opportunities afforded by the interface. Together, these affordances allow the user to be a source or gatekeeper of information, both for oneself and for others.

The psychology of "self as source" discussed in the context of the agency model suggests that the ability to express oneself is critical for engaging users. This is evident from the success of the two most popular social media tools (Facebook and Twitter), which, at their core, offer simple text fields for users to express themselves. Our experiments suggest that, in addition to affording self-expression, interfaces should allow feedback from other users and the system. Aggregated traffic statistics (e.g., number of views, number of replies) and popularity ratings (e.g., number of stars for helpfulness) tend to motivate user contributions to online forums (Kim & Sundar, 2011a). We also found that system-generated validation of a user (e.g., assigning labels like guru or veteran) encourages posting activities. In sum, source-interactivity tools boost user engagement by affording greater self-expression and providing more indicators of external validation and community.

Interaction and integration

The studies discussed thus far examined the three dimensions of interactivity and their psychological effects separately. While they provide empirical evidence for the individual paths theorized by the interactivity effects model, they cannot inform us if one type of interactivity interacts with another and how the model functions as a whole. Although the model proposed distinct linear mechanisms for predicting the psychological effects of different kinds of interactivity, previous explications of this concept suggest notable degrees of overlap (Liu & Shrum, 2002; McMillan & Hwang, 2002; Sundar, 2007a). It is thus important to test the combinatory effects of different types of interactivity.

A recent experiment examined the persuasive effects of modality and message interactivity and their psychological mechanisms on an antismoking website (Oh & Sundar, 2013). It found that modality interactivity, operationalized by providing the slider function, could afford more natural and intuitive interactions. As a result, users felt more absorbed in the browsing task, which then translated into more favorable attitudes toward the website as well as the antismoking messages provided on the website. Message interactivity was found to influence users' processing of the website through a different mechanism. Specifically, by providing layers of hyperlinks to structure related issues around certain topics and using breadcrumbs to keep track of users' location in the website, high message interactivity served to increase users' message elaboration. This greater elaboration was associated with better attitudes toward the antismoking messages and stronger beliefs in the negative effects of smoking.

These findings seem to suggest that message interactivity leads to more central (or effortful) processing of content, whereas modality interactivity leads to peripheral (or superficial) processing. In fact, we labeled the tools of modality interactivity as "bells and whistles" in some of our earlier work (Sundar, 2007a; Sundar et al., 2003). However, this tidy formulation based on dual-processing models in social psychology was not supported by empirical findings. Users' prior level of interest in the content did not reliably moderate the degree to which these two types of interactivity affected user attitude toward the content or the interface. If anything, we found that message interactivity can involve even those who are not highly interested in the subject matter. In fact, Oh and Sundar (2013) discovered that message interactivity could enhance message elaboration (as well as antismoking attitudes) among low-involvement users more markedly than high-involvement users.

Moreover, data from other studies suggest that modality interactivity is not always processed peripherally, that it can in fact be quite involving. For example, Guillory and Sundar (2014) found that modality interactivity employed by an organization's website can improve users' perceptions of its reputation by raising their level of involvement with the site. This finding held for both job seekers as well as nonjob seekers, suggesting that modality interactivity can influence attitudes even via the central route.

The combination of modality interactivity and message interactivity was also examined in the context of a mobile catering website. In their experiment, Dou and Sundar

(2014) operationalized modality interactivity by providing varying numbers of interaction techniques (i.e., tap and swipe). Message interactivity was operationalized in terms of the display of message exchanges in a threaded manner (as opposed to a listing of frequently asked questions or FAQs). Results indicate that the influence of modality interactivity on relational outcomes (such as trust, satisfaction, and commitment) is mediated by enjoyment and user engagement, whereas that of message interactivity is mediated by a heightened sense of contingency and dialogue.

Findings like these, from experiments that have combined different forms of interactivity on the same interface, suggest that the different types of interactivity (message interactivity and modality interactivity) do not necessarily lead to different levels of processing depth. Rather, as the interactivity effects model suggests, they differ in the mechanism by which they influence user processing of content in interactive media.

That said, it is important to consider the interaction effects of the three forms of interactivity for potential additive and attenuation effects. In order to test the co-occurrence of all three types of interactivity on a single interface, we developed a website devoted to visualizing data regarding public opinion on several global affairs. A field study (Sundar, Jia, Bellur, Oh, & Kim, 2014) with this site confirmed many of the theoretical paths hypothesized by the interactivity effects model. Specifically, by providing a greater number of interaction techniques, high modality interactivity was associated with positive perceptual assessments such as ease of use and usability, and further increased user activities. However, the link between message interactivity (operationalized by presenting interaction history) and perceived contingency was significant only when modality interactivity on the interface was high. Furthermore, instead of enhancing perceived customization and sense of agency, high source interactivity (with both customization and blogging functions) led to reduced motivation and greater anxiety when combined with high message interactivity. Taken together, these findings reveal the complex interaction effects of different types of interactivity on user experience when all of them were deployed on one interface. Moreover, these effects are moderated by individual differences in power usage and topic involvement. These findings persuade us to pursue a more theoretically nuanced understanding of the psychology of interactivity.

In sum, the interactivity effects model has explicated the appeal of interactive media by spelling out the various manifestations of interactivity and their respective psychological correlates. By situating interactivity in the three fundamental elements of communication—source, medium, and message—the model has drawn a picture of how interactive media engage users and then influence their knowledge, attitudes, and behaviors. Research around the model thus far has enhanced our understanding of the psychological appeal of interactivity and its underlying mechanisms in a variety of contexts. It has also yielded fruitful practical implications for interface designers, especially about how to afford greater user control, enhance user engagement, and achieve optimal persuasive outcomes. Future research will shed light on the combinatory effects of different forms of interactivity for different groups of users, thereby extending the theoretical contributions of the interactivity effects model.

Agency Model

Customization, made possible by the affordance of source interactivity (as discussed earlier), lies at the core of the demassification of mass communication because it allows each individual to receive his or her own media content. Developments in customization technology at the turn of the century afforded individual receivers of communication the ability to perform their own "gatekeeping," a fundamental function of mass communication that was historically the domain of professional journalists and media elites (Sundar, 2008b). They are now able to have a say in gatekeeping the information that they consume as well as information that they disseminate to others. For instance, Google News, a news aggregation service operated by Google Inc., scours thousands of publications, including prestigious news corporations as well as Web-only news blogs, to automatically assemble an individualized diet of information tailored to one's needs. More generally, the passive consumption of media content has been replaced by a more active interaction between the media and their users, leading to greater user input in determining the nature of content. The arrival of social media has made it possible for this user-generated content to be shared with others. The receivers are not only tailoring content for their own consumption, but also broadcasting it via social bookmarking sites, blogs, social networking sites, and other social media. They are, in a sense, becoming the source of communication, both for themselves and others in their network. By enabling users to become sources, the technologies of customization and social media have come together to provide them greater agency in the communication process. Individuals can now assert their own influence over the nature and course of information transmission, and therefore over-ride the authority held by professional gatekeepers ever since the invention of mass media.

Customization has been studied in a variety of contexts: for instance, some research in health communication (e.g., Kreuter, Farrell, Olevitch, & Brennan, 2000) views customization as tailoring information or interventions for targeting users' specific health and medical needs or based on the person's attributes. There appear to be three primary reasons why users customize—to convey personality, to manage tasks, and to create a social atmosphere, according to a survey of 318 users of a health website, Prevention.com (Marathe, Kang, & Sundar, 2010). E-commerce scholars (i.e., Rayport & Jaworski, 2001) emphasize the importance of offering marketing information or affordable products and services according to user interests and specifications. Research on interface design (i.e., Findlater & McGrenere, 2004) contrasts "adaptive" (or "personalized") interfaces (interfaces that automatically and dynamically adjust to support the users) from "adaptable" (or "customizable") ones (interfaces that provide customization options to allow users to personally specify the adaptation). Regardless of the domain of research, it appears that customization has positive effects on persuasion. For instance, tailoring of health information was found to be effective in improving information uptake and stimulating behavioral intentions (Kreuter, Bull, Clark, & Oswald, 1999), as well as in changing attitudes and eventual behaviors (Brug, Steenhaus, van Assema, & de Vries, 1996).

More importantly, customization has powerful psychological effects on users because it makes them feel that they are influential and in charge (Shapiro, 1999; Wind & Rangaswamy, 2001), as they dictate and control the "nature and course of the interaction" (Sundar, 2007a, p. 61). It also serves to highlight their unique identity and individuality, as well as their personal relationship with the content or the product (Alpert, Karat, Karat, Brodie, & Vergo, 2003; Kobsa, Koenemann, & Pohl, 2001; Marathe & Sundar, 2011a). In all these cases, the user himself or herself serves as the source of tailoring. It is the psychology of this "self-as-source" that underlies users' sense of personal agency and thereby influences user attitudes and behaviors. Under this conceptualization, "agency" refers to the degree that one can be a relevant actor and assert one's influence on the nature of content that is circulated in any given medium. It is the degree to which the user is able to serve as a source of tailored outcomes (Sundar, 2008b). As it turns out, this ability to serve as the source of tailoring is quite appealing to users of communication technologies, as evidenced by the explosion of customizable media products in the first decade of the twenty-first century, from filter settings on email browsers to self-representing avatars in video games to individualized faceplates on mobile phones.

The Agency Model of Customization (Sundar, 2008b) maps out the technological antecedents and psychological consequences of "self as source." Three key technological factors (interactivity, modality, and navigability) serve to amplify user agency through different mechanisms, resulting in psychological outcomes pertaining to cognition, affect, and behavior (see Figure 3.3).

Interactivity is by far the most influential technological factor in imbuing a sense of self as source. The very fact that an interface tailors content to suit an individual user means that it is responsive to the user. Indeed, users who encounter tailored content tend to perceive higher interactivity in the system (Kalyanaraman & Sundar, 2006). All three types of interactivity described in the previous section (on the interactivity-effects model) can promote the salience of self to the user. While tools of modality interactivity (e.g., sliders and mouse-overs) provide users the ability to choose the manner in which they interact with content, tools of message interactivity ensure that the content served to them is contingent upon their prior input. Source interactivity is the extent to which the interface affords users the ability to actively specify their needs and tailor the content themselves.

Modality features of modern media contribute to a heightened sense of "self as source" in two distinct, but related, ways: (a) they offer users a variety of modes with which to represent themselves in mediated online settings, ranging from simple text conveying their personality through their user i.d. handle to participating in online meetings via audio to using a carefully designed avatar to represent their identity to turning on the Webcam to project a real-time video of themselves. This choice in self-representation via different modality options can serve to highlight the sense of self as source by providing users with unprecedented decision-making ability in projecting themselves; and (b) they offer users the ability to experience differential degrees of self-presence in mediated environments. For example, the modality of virtual reality systems and certain games provide a heightened degree of sensory richness, as the

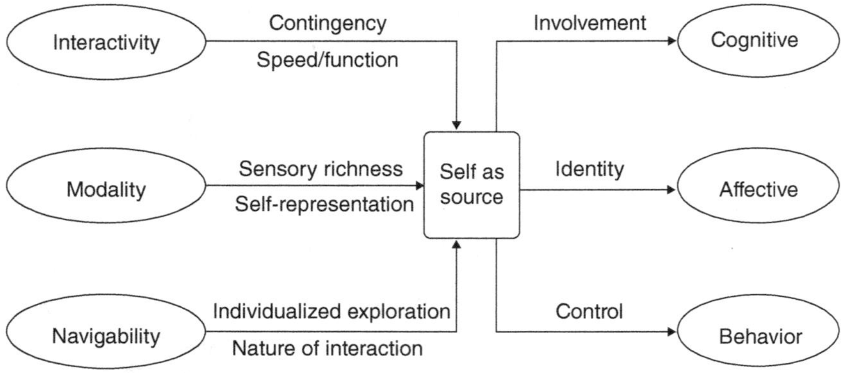

Figure 3.3 Agency Model of Customization.
Source: Reproduced with permission from Sundar (2008b, p. 69, Figure 4.1). © Taylor & Francis.

environment responds dynamically to user actions. This makes the user feel like they are the source of changes to the environment.

Navigability tools on the interface can also imbue a sense of self as source by almost literally providing steering wheels to the user. Modern media afford users plenty of choice in determining how to access a mediated space and how to move through it. While arrow keys on a keyboard, joysticks, game controllers, and other navigational tools outside the mediated environment help the user determine the nature of interaction, the interior design of the virtual environment can provide numerous opportunities for the user to decide how they would like to explore the space. Together, they serve to individualize the user's interaction and exploration of the mediated environment. As Balakrishnan and Sundar (2011) demonstrated, providing users with greater steering control not only result in greater possibilities for action but also in heightening users' sense of spatial self-location, that is, the degree to which they locate their ego-reference frame within the virtual environment.

In sum, the technological end of the agency model posits that the customizability afforded by interactivity, modality, and navigability tools on the interface can serve to enhance among users the sense that they are the source of their mediated experience. The psychological end of the model focuses on the effects of "self as source" upon the user. It argues that when the user is the source, s/he will pay more attention to the content of the interaction, thereby guaranteeing greater involvement with it. Furthermore, the user's identity will be tied up with the content because customization tools allow for deliberate crafting and projection of one's persona. Finally, being a source provides the user a profound sense of control and self-determination. Together, these psychological effects dictate users' cognition, affect, and behaviors in customized media.

Empirical tests of Agency Model

Several studies have used this model to test the effects of different forms of user agency in media. For example, experiments by Sundar and Marathe (2010) contrasted customization with personalization. While both result in content that is tailored to the user, the process of tailoring is different. When the user tailors the system, it is referred to as customization. But, when the system tailors based on the user's past behaviors, then it is referred to as personalization. The agency model of customization predicts that customization would be favored over personalization because it provides users a higher sense of self as source, a greater say in determining the nature of the resulting content. If user agency did not matter and only the relevance of content tailoring did, then there would be either no difference between the two or perhaps a preference for personalization over customization because it involves lesser effort on the part of users. Participants in the Sundar and Marathe study were invited to browse through the Google News website during the first session of an experiment. When they came back another day for the second session, they were either exposed to the news site just as in the first session (control condition), or a system-initiated personalized version of the site focusing on topics they read in the first session, or a user-initiated customizable version wherein they could create a news digest based on their own interests. We found that customization was indeed favored over personalization, with participants in the former condition rating the journalistic quality of the stories higher. However, this was true only for power users, that is, those who are motivated and expert users of technology. Nonpower users gave higher ratings of content quality when it was personalized for them rather than customized by them. A follow-up experiment revealed that this transverse interaction occurred only under conditions of low privacy, that is, when the news site informed users that their browsing information may be used to provide requested services in the future and shared with third parties. This implies that power users prefer customization because they fear loss of privacy in the process of personalization. In fact, data showed that, in the low-privacy condition, as power usage increased, perceived control (i.e., a feeling of being in charge of the experience) decreased in the personalization condition but increased in the customization condition. However, when privacy was assured, the correlation between power usage and perceived control was positive for both conditions. It therefore appears that power users see customization as a vehicle for taking control of their privacy, consistent with agency model's contention that customization engenders user control.

This tendency among power users to feel a higher sense of control when they are allowed to customize was confirmed by Marathe and Sundar (2011a) in the context of Web portals. The study also revealed that the two major psychological effects of customization—control and identity—hold different levels of importance for power users and nonpower users. While control is an integral aspect of agency for power users, identity-enhancement through customization is critical for nonpower users. In general, however, the effect of customization on sense of control was fully mediated by sense of identity (i.e., a feeling that the interface reflects one's personality), whereas the reverse was not true. This raises the possibility that the majority of participants were engaging in the kinds of customization on the site that pertained to their identity, such as the choice of colors, icons, and avatars.

Marathe and Sundar (2011b) classified customization into cosmetic customization (presentation-based changes to the interface, e.g., desktop screen design, music selection for ringtones) and functional customization (task-based changes such as filter-settings for email clients and privacy settings on one's social networking site account), and tested the possibility that the former was related to identity-enhancement whereas the latter to a sense of control among users. Results showed the primacy of identity over control regardless of the object of customization, in line with other studies showing a direct connection between self as source and sense of identity (Kang & Sundar, 2014). For both types of customization in the Marathe and Sundar study, sense of identity mediated the relationship with sense of control, which in turn positively mediated the effects on intrinsic motivation to explore the site and attitudes toward the portal.

Such benefits of customization can magnify the effects of content served through the portal. For example, Kim and Sundar (2011b) showed that when visitors of a health kiosk in Second Life entered the virtual environment with an avatar that they themselves customized (as opposed to an avatar assigned to them by the experimenter), they were more likely to benefit from the health messages. Study participants showed significantly higher sense of self-preservation in the customized avatar condition compared to their counterparts in the assigned avatar condition even though both received identical health messages. Furthermore, this effect was mediated by a "sense of agency" (a construct capturing a sense of identity, sense of control, and a more global sense of self as source, reflected via items such as "I feel confident about asserting myself" and "I feel that I can control my destiny"). Avatar customization had other effects, such as increasing the users' sense of self-presence, but it was the sense of agency that statistically mediated the relationship between customization and health-related outcomes.

Remarkably, simple acts of cosmetic customization by the user (like making an avatar resemble oneself) can serve to boost the effects of content encountered during the interaction. This can have implications in a number of domains, from health communication to marketing. In fact, a recent survey of Korean and American users showed that their level of aesthetic motivation for cosmetic customization of their mobile phones explains nearly half the variance in their professed attachment to their phones—an effect mediated by the degree to which they believe their customized phones reflect their self (Lee & Sundar, 2014).

In all these examples, it is evident that customization is a deeply personal process invoking the psychology of self. The sense of agency derived from customizing media content and interfaces for one's own consumption is certainly significant, but it can be extended further when the outcomes of one's customization are shared with others. Affordances of social media technology allow users the ability to not simply gatekeep for themselves but also broadcast it to their network. For instance, users can share online content to friends and acquaintances by "like"-ing it on Facebook or reposting it through microblogs such as Twitter and social bookmarking sites such as Digg.com, virtually making each and every user a *de facto* gatekeeper for others, much like a journalist. This is profoundly agentic in that it affords ordinary users the ability to serve as information agents for other users. In fact, a survey of bloggers showed that the higher the tendency for "filter blogging" (i.e., filtering online content for re-broadcasting through one's blog), the higher the sense of agency, which in turn was linked to psychological empowerment (Stavrositu & Sundar, 2012).

From customization to expression

In addition to filtering existing content, social media afford users the ability to create new content, leading to even greater sense of agency. Viewed from the perspective of "source interactivity" (i.e., the degree to which the technology affords users the ability to serve as a source of content), customization of content would be at the low end and creation of content at the high end, with curation of content in the middle.

Sundar, Oh, et al. (2012) conceptualized it as a "continuum of self-expression" and operationalized it in terms of the differential availability of customization and blogging affordances on an Internet start page/portal site. While cosmetic and functional customization features afford the ability to tailor the site for oneself, filter blogging facilitates curation and active blogging allows creation of new content by the user. A 3 (functional customization: no, low, high) × 2 (cosmetic customization:absent, present) × 2 (blogging: active, filter) full-factorial field experiment (Sundar, Oh, et al., 2012) with 141 participants using the site for 2 weeks revealed that active blogging boosted the positive effects of customization. Active bloggers who were provided an opportunity to cosmetically customize the site showed the highest level of engagement at the end of the study period. Active bloggers who were provided an opportunity to customize the functional aspects of the site expressed the highest behavioral intention to use the site in the future. Active blogging, by itself, was significantly more influential than filter blogging in imbuing a sense of personal competence as well as a sense of community and positive attitudes toward the portal. Together, these findings speak to the importance of affording self-expression in order to build agency among users.

Indeed, it may be argued that the ability to express oneself is the primary driver of customization. The activity of customization can be somewhat demanding, not just in terms of physical actions on the interface, but the cognitive aspects of decision-making among choices. In fact, studies have shown that this activity can lead to ego depletion, but the consequent loss of self-control can be countered if the activity offers potential for self-affirmation (Kang & Sundar, 2013). Therefore, self-related mechanisms pertaining to one's identity, expression, and affirmation critically determine the agency-enhancing potential of personal and social media technologies.

Machine agency

In addition to professional gatekeepers having agency and users now attaining agency, technologies are also increasingly attaining agency, thanks to advancements in personalization, robotics, artificial intelligence, and internet of things. In an explication of the concept, Sundar (2007a) distinguished between agency afforded to users of technology and the technology itself asserting its agency in the form of "robotic entities, either hardware or software related, that perform tasks for the user, often autonomously" (such as online travel agent and real estate agent). The agency model of customization contends that user agency would be more psychologically influential than machine agency because the user is afforded the ability to act as the source of the interaction. This is borne out by our studies contrasting customization with personalization, especially among power users. However, with the continual growth of "smart technologies," machine agency will manifest itself in ever-newer forms, creating a tension in human–computer interaction and raising new challenges for designers.

Given the work done by communication scholars over the last two decades, an obvious solution to this "agency tension" would be to capitalize on the innate human tendency for social responses to technology. Studies in the CASA (computers are social actors) paradigm have shown that users tend to apply social rules of human–human interaction to technologies (Reeves & Nass, 1996). Anthropomorphic cues on the interface, such as a human-like character on the screen (Kim & Sundar, 2011b), humanoid morphology, and humanlike voice, are known to enhance the socialness of human–technology interaction. Even everyday objects, such as a box of tissues, are seen as agentic when they show interactive potential by uttering social pleasantries such as "Bless You" to sneezing humans (Jia, Wu, Jung, Shapiro, & Sundar, 2013).

Another solution for easing this tension would be to make the machine clearly subservient to humans. Studies in the domain of human–robot interaction have shown that role assignments to technology can shape user responses. For example, Kim, Park, and Sundar (2013) found that assistive robots (which provide care to humans) tend to generate trust and lead to greater relationship satisfaction than companion robots because users perceive greater benefit. On the other hand, if users sense that the machine is too "smart" or "autonomous" and appears to function on its own intentionality without making its working plan and/or processes visible to humans, then they tend to be concerned about the negative consequences of machine agency (Jia, Wu, Jung, Shapiro, & Sundar, 2012). Such findings suggest that the design of future technologies should not be exclusively function-centric or user-centric, but find ways to balance human agency with machine agency.

This raises opportunities and challenges related to multiple agencies, agency tradeoffs, agency negotiation, and so on, between human users and machines, leading to new possibilities for using technologies as proxy agents for achieving both personal and social ends. Such tools and interfaces can enable individual users to act with machines or with other individuals toward shared goals without forfeiting personal agency, therefore empowering them to participate and engage in collective action and social changes. We have already witnessed the power of agency-enabling tools in recent times, such as social media's assistive role in fermenting the Arab Spring, and the various ICT4D programs that utilize communication technology to improve education and health conditions in developing countries. At the individual level, they also enable us to take better control of our own lives, by heightening our personal agency and motivating us to change attitudes and behaviors for our own betterment.

Motivational Technology Model

Social and personal media now come together in important ways to quantify one's own progress and calibrate it with others' input. We have any number of Web-based services and applications that track and assess personal performance and growth in a number of domains, from economic development to academic progress. The rise of the "quantified self" movement is a testament to the widespread availability and use of health-related mobile apps and other personal as well as social media dedicated to monitoring one's personal welfare. For instance, the National Institute of Health offers a list of apps that provide medical practice guidelines, information glossary, as well as visualizations and maps of health and safety data. WebMD, a health corporation, is known for its resourceful website as well as mobile apps, both among the most trafficked online sources for personal health and medical information. Even though these services and apps are available and accessible to the public, research has shown that their use for information seeking is somewhat sporadic and has not a demonstrably changed health-related practices and lifestyles among US adults. Clearly, the availability of relevant technological aids is not proving to be sufficient for initiating sustained changes in behavior. A more strategic approach is needed for leveraging the benefits of personal and social technologies.

The motivational technology model presents one particular strategy—the careful deployment of specific technological affordances for boosting intrinsic motivation among users to perform a particular action. More broadly, it presents a theoretical framework by which to analyze the recent proliferation of self-help technologies in the context of psychological theories pertaining to motivation. The model applies to all domains of human activity, but will be discussed here in the context of health behaviors.

Several thousand health apps have appeared in recent years, combining information databases and user data tracking features. Examples include Nike + Running that records your running routines and My Diet Diary that documents your calorie intake. These apps have so far received positive user feedback and shown the potential for encouraging users to follow healthy lifestyles. Why are these apps successful? What are the technology-related variables that actually motivate users' self-driven information-seeking and behavioral modification? Self-Determination Theory (Ryan & Deci, 2000) suggests that it is an inherent human need to understand oneself. Self-monitoring technologies such as Nike + Running and My Diet Diary appeal to this need and thereby promote an intrinsic motivation to act on this knowledge about oneself. Unlike traditional health communication strategies that focus on external pressure to initiate health behaviors, these technologies tend to motivate users internally. They encourage "autonomous regulation," which is found to be a far more effective mechanism in predicting individual behaviors than "controlled regulation" (by external parties such as doctors and health campaigners) in studies dealing with weight loss, weight management, and diabetes management.

Self-determination theory proposes three primary factors that give rise to intrinsic motivation for an individual to achieve his or her goal, namely autonomy, competence, and relatedness: autonomy is the degree to which one has volitional control, that is, being the initiator or source of one's own goals and actions. Competence refers to the capability and self-efficacy that one feels toward the set goals. Relatedness refers to the feeling of being integral to—and supported in—one's social networks.

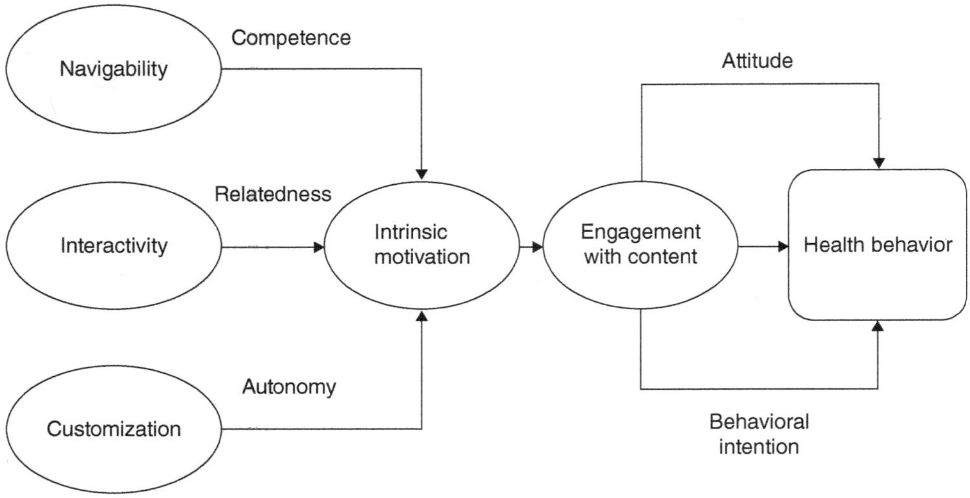

Figure 3.4 Theoretical model of motivational technology to promote preventive health behaviors.
Source: Reproduced with permission from Sundar et al. (2012, p. 116, Figure 1). © Springer.

Given that competence, relatedness, and autonomy represent the key psychological predictors of intrinsic motivation, we reasoned that provision of specific technological affordances geared to promote them would serve to heighten user engagement with preventive health applications. With this in mind, the Motivational Technology Model (Sundar, Bellur, & Jia, 2012) proposes that affordances related to navigability, interactivity, and customization will enhance intrinsic motivation to engage with a given self-improvement technology by enhancing individuals' perceived competence, relatedness, and autonomy, respectively. This heightened motivation would positively influence user engagement with the contents offered by the technology, affecting their health-related attitudes and behavioral intentions en route to influencing their health behaviors (see Figure 3.4).

In the model, ***navigability*** is defined as the extent to which users can explore a mediated environment in a highly idiosyncratic manner (e.g., Fitzpatrick, 2010). Well-designed navigational tools, such as a search engine, website map, and content indices, increase familiarity and predictability of the interface (Balakrishnan & Sundar, 2011). They make the task of using the interface less daunting by scaffolding users every step of the way. They also function as heuristic cues by providing users "information scent" (Pirolli, 2003) that helps them make important decisions about which type of distal information to pursue (Sundar, Knobloch-Westerwick, & Hastall, 2007). For example, interface cues on news websites (e.g., "related articles" cues, "source credibility" cues, and "recency" cues) provide users with shortcuts to online content and influence their judgment of its quality (Sundar, 2008a). By reducing navigational complexity, these aids enhance not only usability and user-friendliness, but also a sense of user control over the interface and self-efficacy. More generally, usability of an interface will likely enhance users' feelings of competence.

Interactivity, in this model, refers to the opportunities that the interface affords for users to connect and have a reciprocal exchange with others (Hoffman & Novak, 1996; Sundar, 2008b; Sundar et al., 2003). This pertains to the presence of "message interactivity" (Sundar, 2007) on the interface, manifest in terms of the contingency of messages and the threadedness resulting from the interdependent "back-and-forth" exchanges between the user and the system or between two or more users (Sundar, Bellur, Oh, Jia, et al., 2014). Perceived contingency, as a result of message interactivity, is found to positively affect user perceptions of website appeal, website excitement, as well as content quality and information value. It is also associated with user behaviors on the website and their intention to engage in the advocated health behaviors (Bellur & Sundar, 2013). If operationalized well, message interactivity can go beyond information provision and give users a profound sense of community—the feeling of relating to similarly situated others. In social media, this kind of interactivity can manifest itself in terms of interface cues that signal external approbation of one's comments and actions. One experiment demonstrated that when bloggers receive a lot of comments on their post, they feel a stronger sense of community compared to their counterparts who receive relatively fewer comments (Stavrositu & Sundar, 2012). Another experiment demonstrated that favorability of comments received by one's Facebook post serves to boost user engagement with the topic of the post. Other features of social networking sites, such as being tagged and receiving likes, were also found to be psychologically significant in aiding user engagement (Oeldorf-Hirsch & Sundar, 2012). Experiments with online health communities have shown that "bandwagon cues" in the form of star ratings of a discussion thread served to enhance user participation by conveying a sense of community (Kim & Sundar, 2011a). These findings exemplify the power of community feedback in imbuing relatedness and thereby motivating user action.

The third predictor in the model, *customization,* is defined as the extent to which users can tailor the mediated environment to meet individual needs (Rimer & Kreuter, 2006; Sundar, 2008b; Sundar & Marathe, 2010). Such user tailoring can be achieved through information filtering and personal goal-setting, both resulting in information that has greater relevance and utility (Sundar, Marathe, & Kang, 2009), leading to better attitudes toward both the interface and the content (Kalyanaraman & Sundar, 2006). More importantly, tailoring features provide users with opportunities to directly exert their influence on content, by becoming the gatekeepers, or sources, of the content they consume, a notion defined as "source interactivity" (Sundar, 2007). By specifying individualistic preferences and requests, users experience a sense of personal agency, involvement and control (Sundar, 2008b), leading to heightened autonomy and self-determination (Katz & Assor, 2007). Several cues on the interface reminding users about the customized nature of their content (e.g., self-resembling avatars, use of the word "my"—MyYahoo, My Workout Plan, etc.) will likely add to the sense of autonomy felt by users.

The general prediction of the motivational technology model is that the autonomy, relatedness and competence triggered by technological factors can serve to enhance intrinsic motivation for engaging with the interface. This engagement in turn can have attitudinal and behavioral effects based on the content of the interface. If the interface is a health app, for example, intrinsic motivation to engage with the app on a regular basis will likely lead to positive health-related attitudes and behaviors. But, the model is not simply restricted to the health domain. It could be used to design interfaces in any content domain where it is important to motivate users to constantly return and reuse the interface.

Aside from prompting theory-driven development of motivational technologies, the model can be quite useful in user-experience research. We could develop metrics based on the three predictors of intrinsic motivation. While competence metrics would pertain to the degree of user control, quality of information scent, way-finding ease and usefulness of guidance tools, relatedness metrics would gauge the degree of communication contingency provided by the interface, the social and information support offered in that venue, the degree to which the social presence of others is felt and the sense of community that prevails in the interface. Autonomy metrics would focus on the degree to which the interface allows for self-tailoring or customization, the degree to which users can deviate from default settings or recommended actions, the sense of personal identity conveyed by the interface, and the sense of agency felt by users. Together, these metrics can help advance the "stickiness" of the interface, so that users are self-determined to continually revisit it and make use of its contents.

MAIN Model

Aside from the psychological effects elicited by the use of interactive media, it is possible for technological affordances to affect users' perceptions by their mere presence on an interface. That is indeed the fundamental argument forwarded by the MAIN model (Sundar, 2008a). Interactive media affect users' evaluations of media content based on the technological affordances visible on the interface. The affordances, or "action possibilities" (Gibson, 1977), offered by a technology serve as cues that guide assessments of the interface (as well as its contents) based on the cognitive rules of thumb (heuristics) triggered by those cues. An affordance can trigger heuristics in two distinct ways: (a) its presence on the interface can transmit cues, both about its functionality and the designer's intent (e.g., a politician whose website has a lot of interactive features would be considered open-minded and willing to listen to voters than one with lesser interactivity); and/or (b) by adaptively gathering information for the user in the form of metrics that reflect its operation (e.g., the number of visitors to the site, displayed as a counter, will be used as a criterion to judge the politician's popularity among the electorate).

In order to systematically understand various such effects of interactive media on users' judgments, the MAIN model proposes that the relevant affordance offered by the technology and the related heuristics cued by the affordance must be identified. The effects of affordances therefore depend on users' interpretation of the affordances and the subsequent heuristics that are cued by these perceptions. Based on extensive research conducted at our lab, four broad categories of technological affordances are

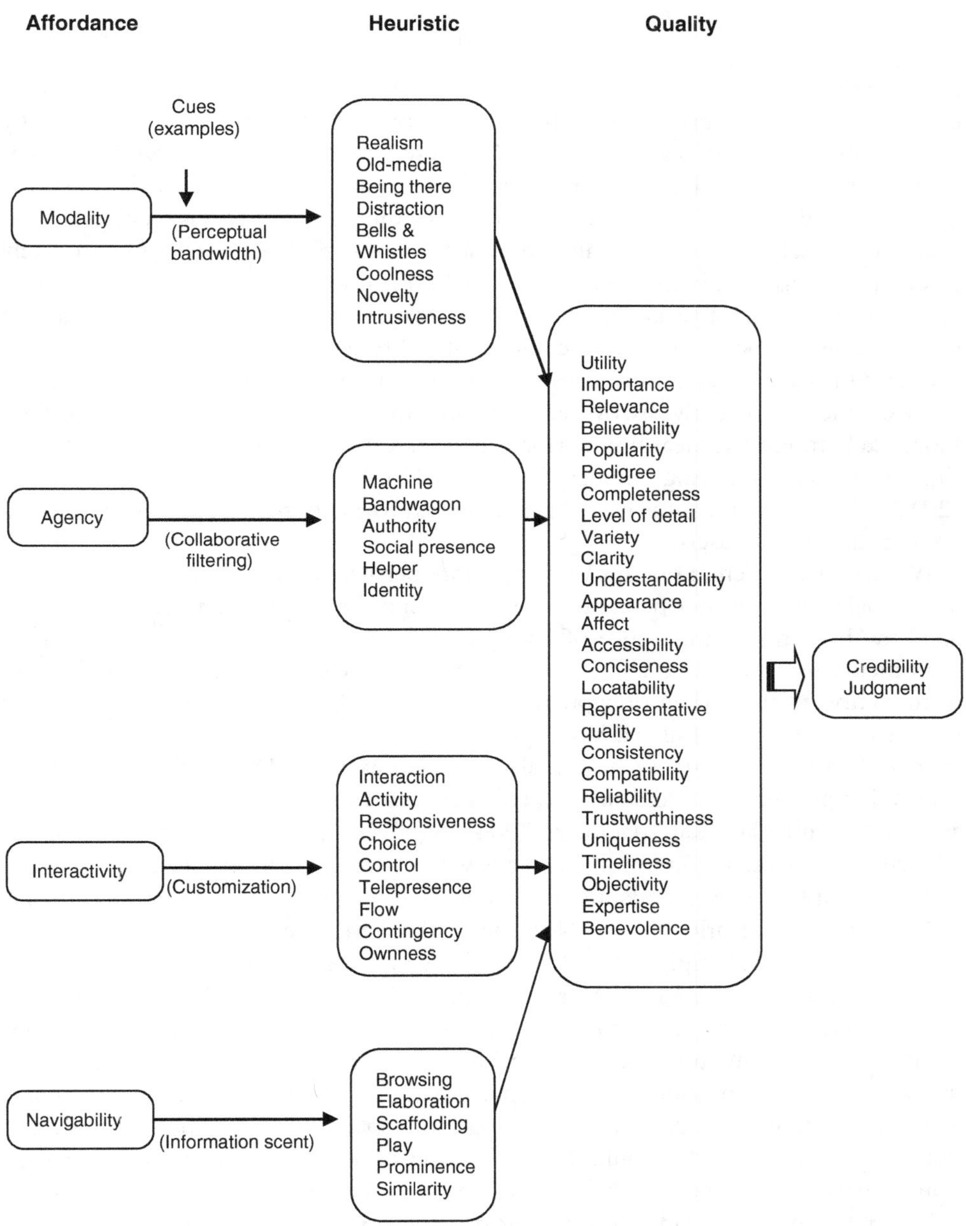

Figure 3.5 The MAIN model.

Source: Reproduced with permission from Sundar (2008a, p. 91, Figure 2). © MIT Press.

psychologically salient: modality, agency, interactivity, and navigability. Each of these technological features can cue a range of heuristics that affect users' perceptions of media content (such as its quality and credibility), even though the features themselves are external to the content of the communication (Figure 3.5).

Modality

A fundamental way in which a media technology can affect users' perceptions of content is through heuristics triggered by the modalities of presentation afforded by that technology. Specifically, MAIN model predicts that different modalities (or means through which information is conveyed) can either enhance or diminish the perceived quality and credibility of content depending on the triggered heuristic. For example, modalities that afford an audiovisual presentation of information are likely to cue the *realism heuristic* (i.e., seeing is believing), which predicts that individuals are more likely to trust visual information over textual information because images are often assumed to be a direct representation of reality. This trust is quite important for a variety of psychological outcomes of communication, from learning to shopping. For example, we recently discovered that three-dimensional visual presentations are more likely to enhance learning performance than two-dimensional presentations of similar content (Li, Zhang, Sundar, & Duh, 2013). Likewise, Kim, Park, and Sundar (2012) found that video was more likely than text to elicit greater purchase intentions among smartphone users.

While visual presentations appear to be more persuasive than their textual counterparts, their effect on user perceptions can be diminished if the *distraction heuristic* is cued by the combined presence of multiple modalities. Sundar (2000) found that a multimedia website was evaluated more negatively when both pictures and audio were combined than when only pictures were provided. Likewise, certain modalities unique to digital media, such as pop-ups and animation, can cue the *intrusiveness heuristic* and lead to negative psychological effects. For example, Diao and Sundar (2004) found that pop-ups were less likely to enhance message recall than traditional banner ads, while Sundar and Kalyanaraman (2004) found that ads with fast animation speed elicited lower purchase intentions than ads with slower animation speed.

The use of traditional modalities in digital media can be judged based on the evaluation standards of prior media. New media that resemble the format or layout of older modalities can trigger the *old media* heuristic, leading users to evaluate new presentation formats based on their expectations derived from traditional media use. For example, video modality in an online site or a mobile application may trigger comparisons to watching movies in a cinema or on a television set, whereas text modality on the same online site or app, especially if laid out in multiple-column format, can trigger comparisons to reading a newspaper or magazine. User perceptions of these older media would directly transfer over to their perception and reception of newer media because of the commonalty in the modality of communication.

New modalities can lead to positive or negative perceptions of content depending on the particular heuristics that they cue. Variations of perceived realism cued by modality can elicit the *being there heuristic*, a perception that sensory immersion in the mediated environment is an authentic experience, which can therefore lead to desirable psychological effects. For example, prior studies have found that large screen displays are more likely than small screen displays to elicit heightened presence, arousal, and attention (Ivory & Magee, 2009; Kim & Sundar, 2013; Reeves, Lang, Kim, & Tatar, 1999). Variations in modality can also trigger perceptions of presence and involvement via the ***novelty heuristic*** due to the association between the uniqueness of the technology and the underlying content presented. For example, one study (Ivory & Kalyanaraman, 2007) found that video games with advanced graphics and sound elicit greater presence, involvement, and physiological arousal than less advanced video games with similar controls, potentially due to the automatic association between the novelty of the game's advanced presentation and the overall media experience. Perceptions of content credibility can also be enhanced if the content is received via a modality perceived as "hip" or "stylish" due to the ***coolness heuristic*** (Sundar, Tamul, & Wu, 2014). In support of this prediction, Oh, Robinson, and Lee (2013) found that content was perceived as more credible when viewed through a touch-screen device compared to a device controlled via mouse among users who evaluated the naturally mapped modality as cool. However, new modalities can also cue negative perceptions of content via the ***bells and whistles*** heuristic if they are perceived as "all flash but no substance" (Sundar, 2008a, p. 82).

Modality affordances can thus trigger a range of possible heuristics depending on their interpretation by the user. While vivid new modalities can enhance perceived representational accuracy and thereby trigger the realism heuristic, they can also engender associations with other media and trigger the old-media heuristic. The being there, novelty, and coolness heuristics can be elicited by affordances perceived as original, unique, stylish, or that afford the user with a high-fidelity media experience. However, it is also possible for credibility to be diminished by certain modalities that trigger the distraction, intrusiveness, or bells and whistles heuristics.

Agency

Agency is a relatively new class of affordances pertaining to the source of media content. While older media had professional gatekeepers serving as the sole agents of content, modern digital media afford a number of different sourcing possibilities, including the ability of the user to serve as a source (as in user-generated content of social media) and the technology itself appearing as a source (as in interface agents and robots). The MAIN model predicts that perceptions of sourcing in digital media can affect users' credibility judgments. Specifically, agency affordances are a salient source of cues and heuristics that can be used to evaluate the credibility of message senders, especially in the context of digital media where multiple layers of sources are often present (Sundar, 2008b). Characteristics of the perceived source can therefore play a significant role in user judgments of content, especially if they are highlighted by interface cues. For example, cues pertaining to the pedigree or experience of the source can serve to trigger the ***authority heuristic***, or the perception that the source is an expert on the subject matter. Specifically, a series of studies have found that online news stories and message board postings are evaluated as more credible when the message is perceived as authored by a doctor rather than a layperson (Hu & Sundar, 2010). Source credibility can also affect users' perceptions of subsequent content, as one study (DiMuzio & Sundar, 2012) found that news viewed by following a hyperlink provided by a low credibility news source was evaluated as less trustworthy than the same news recommended by a high credibility source. Evaluations of source expertise are not limited to human sources. Website agents and robots categorized as specialist are more trusted than their generalist counterparts (Koh & Sundar, 2010a; Sah, Yoo, & Sundar, 2011).

Interface cues can also cue the collective opinion of other users through metrics that indicate the popularity of content (e.g., number of "shares," "likes," or "retweets"). According to the MAIN model, these serve to cue the *bandwagon heuristic*, or the perception that a majority of others have also endorsed the content of interest. Experimental tests of the bandwagon heuristic have found that interface cues such as "five-star" ratings, a high sales rank, or a greater number of user endorsements predict greater perceptions of product and news credibility than lesser star ratings, low sales rank, or fewer endorsements from others (Sundar, Oeldorf-Hirsch, & Xu, 2008; Xu, 2013). In addition, social media pages for social movements with a large number of similar members elicit more positive feelings for advocacy groups than social media pages with fewer members (Xu et al., 2012). Studies comparing the effects of the bandwagon heuristic and the authority heuristic have generally found that the bandwagon cue has a more positive effect on content perceptions than authority cues (Lee & Sundar, 2013), although authority is more likely to predict purchase intentions if other site metrics undermine the bandwagon cue, such as a negative product review (Sundar, Xu, & Oeldorf-Hirsch, 2009).

Aside from evaluations of content credibility based on the expertise or the opinions expressed by online sources, agency-related affordances can affect credibility based on the humanness attributed to the source. For example, the *social presence* heuristic can be elicited by interface cues that imply that the user is interacting with a social or intelligent entity. A perception of social presence is known to augment the credibility of socio-emotional content. For example, Park, Kim, Sundar, and del Pobil (2012) found that online gaming was more enjoyable when played against a robot rather than a computer. Similarly, micro-blogging accounts perceived as operated by a human curator elicit greater perceptions of social and task attraction than those presumably operated by a machine (Edwards, Edwards, Spence, & Shelton, 2014). However, while a lack of socialness might inhibit the emotional credibility of nonanthropomorphic agents, it is possible that interfaces low in anthropomorphism are sometimes judged positively. Specifically, Sundar (2008a) explains that nonhuman agents may be perceived as more objective due to the cueing of the *machine heuristic*. In line with this prediction, Sundar and Nass (2001) found that identical news stories were evaluated as higher quality when they were suggested by a computer terminal rather than by news editors, presumably due to the perceived absence of bias on the part of machines. Thus, it is likely that the social presence heuristic and the machine heuristic may toggle based on the degree of anthropomorphism conveyed by a technology, and may be differentially desirable for different types of media experiences.

Taken as a whole, cues related to the source of a message can have a range of effects depending on the affordance and the subsequent heuristic that is cued. If the source is touted as an expert or the content is endorsed by a large number of followers, authority and bandwagon heuristics would be elicited. The anthropomorphism of a source can also serve as a source of agency cues, with sources assumed to be human eliciting positive reactions via the social presence heuristic. By comparison, interface agents low on humanness can also generate positive perceptions due to the machine heuristic and the resulting presumption of objectivity. Thus, agency affordances can lead to a variety of potential heuristics, each of which can shape the ways in which related content is perceived and evaluated.

Interactivity

Given that interactivity is among the defining characteristics of digital media, it is not surprising that a range of heuristics can be cued by the presence of interactive affordances on an interface. Interactive media elicit more action from the user than traditional media. As a result, the presence of affordances that are perceived as facilitating interaction can elicit the *activity heuristic* (i.e., more activity is better), leading to positive impressions of the content (e.g., Sundar, Narayan, Obregon, & Uppal, 1998; Sundar et al., 2003).

In addition to cueing activity, interactive features can also trigger heuristics based on the way they are designed to facilitate user interactions on the interface. For example, skeuomorphic controls can serve to cue the *telepresence heuristic* and the resulting perception that the user has been transported to the location portrayed by the media. Consistent with this prediction, past research has found using a steering wheel for a racing game (Schmierbach, Limperos, & Wooley, 2012) or a gun replica for an action video game (Kim & Sundar, 2013) elicits stronger feelings of presence and transportation than nonisomorphic controllers (e.g., joystick and button controllers). Therefore, different interaction techniques afforded by the interface can differentially shape user experience of the interaction by cueing different heuristics.

Interactivity can also be defined by the extent to which the output of an interface is perceived as conditionally provided based on the input of the user. Specifically, messages from an interface that are threaded, that is, based on users' prior interactions with the interface, can cue the *contingency heuristic*, which can enhance perceptions of message relevance, uniqueness, and reliability. For example, one study (Lee & Park, 2013) found that organizational websites with a high degree of message responsiveness elicited greater perceptions of organizational trust, commitment, and reputation than noninteractive websites. Enhanced credibility cued by the contingency heuristic may also predict greater participation in online communities, as several studies have found that interactive messages are more likely to predict discussion intentions in online message boards than low-interactive messages (Velasquez, 2012; Wise, Hamman, & Thorson, 2006).

Interactive media are also capable of automatically adapting content to match the perceived interests of the user, thereby triggering the *own-ness heuristic*. When the own-ness heuristic has been activated by personalized messages, it is predicted that both content and the source of the message will be perceived as more credible due to an affirmed sense of self. For example, one study (Downs & Sundar, 2011) found that game avatar resemblance is related to greater game enjoyment, potentially due to the affirmation of self-provided by the similarity between user and avatar. Similarly, the products of customization, such as one's unique imprint on a web portal or a mobile device, can trigger the *identity heuristic* with desirable consequences. For example, Lee and Sundar (2014) discovered that the relationship between motivation to customize a mobile phone and user attachment to the phone was fully mediated by perceived reflection of self through the device.

In sum, the presence of interactivity affordances can lead to a number of potential heuristics. If users perceive an increased opportunity for site interaction or co-presence, the activity or telepresence heuristics can be elicited. Furthermore, interfaces that are perceived as providing adaptive or personalized information can cue the contingency and own-ness heuristics, both of which are generally expected to elicit more favorable impressions of message and source credibility. A host of other interactivity affordances have been identified by Sundar (2008a), all of which are predicted to cue credibility-enhancing heuristics such as *control*, *interaction*, and *flow*.

Navigability

According to the MAIN model, the affordances of particular navigability tools provided by an interface can shape the nature of the user experience. For example, navigability affordances that allow users to easily browse a site or quickly obtain relevant content can affect users' credibility judgments by triggering the *helper heuristic*. Similarly, interface features that facilitate users' navigation by stepping them through various parts of an interaction (e.g., dialog boxes that ask users "are you sure you want to proceed? Clicking yes will charge your credit card") can cue the *scaffolding heuristic*. Both these heuristics potentially suggest the benevolence of the designer, thereby predisposing users to be positive toward the site and its contents.

Heuristics can also be cued based on the ease with which a user is able to navigate through a system. For example, Balakrishnan and Sundar (2011) found that virtual simulations that afford an increased range of motion elicit a greater degree of spatial presence than virtual simulations with a lower range of motion. Browsing experiences that adjust to the ability of the user can cue the *play heuristic*, thus enhancing users' attitudes toward the system and its content (Mathwick & Ridgdon, 2004).

Navigability affordances can also lead to more systematic evaluation of content if the *elaboration* heuristic is cued by embedded hyperlinks, thereby prompting users to explore related topics in an effortful manner (Sundar, 2008a). Considering that online search is the primary navigational activity for most Internet users, the MAIN model posits that affordances that guide the search query can trigger heuristics related to evaluation of the quality of those results. A source may be evaluated as more credible if the interface provides cues that indicate the fit between the requested information and the outcome of the search, thereby eliciting the *similarity heuristic*. Consistent with this assumption, search engines that provide relevant results are evaluated as more credible than those that provide nonrelevant results, both in the presence of relevant ads and among highly involved users (Kalyanaraman & Ivory, 2009). Ad relevance leads to positive attitudes toward the site and the ads themselves, especially when the search site is cluttered with ads (Kim & Sundar, 2010) or when the website in general is nonpersonalized (Kim & Sundar, 2012). Along similar lines, online videos that feature a visual or textual preview of content are more likely to be viewed than videos without a preview (Fu, 2012; Fu & Sims, 2011), likely due to the confirmed similarity between the original query and the previewed content. The perception of situational relevance derived from the recency of information can also cue the similarity heuristic in the news browsing context, as online news articles that have been recently uploaded are more likely to be perceived as newsworthy than less recent stories (Sundar et al., 2007).

Aside from the perceived fit between search results and user interests, the order in which information is presented can also affect credibility assessments via the ***prominence heuristic***, or the assumption that results that appear first in the output of a search engine are more relevant and credible than results that appear lower down in the list. In support of this prediction, one study (Pan et al., 2007) found that links that appear at the top of a list of a search engine output are more likely to be selected than those that appear lower, regardless of the link's apparent relevance to the intended search. Furthermore, past research has found that "top ranking" sites evoke higher perceptions of content credibility than those that do not afford the user with cues related to prominence (Westerwick, 2013).

In sum, users' perceptions of a website and its content can be shaped in a variety of ways depending on the heuristics cued by the navigability affordances of interest. The helper and scaffolding heuristics can be cued if the system is perceived as providing navigation assistance, while navigational tools that match browsing experience according to the skill of the user can elicit the play heuristic. Content perceptions can also be affected by interface cues that allow the user to evaluate the relevance or relative "rank" of information via the prominence and similarity heuristics. Users' perceptions of content can thus vary based on the nature of the browsing experience provided by the system and the ease with which information foraging is facilitated.

Beyond credibility: MAIN Model applications and extensions

In addition to assessments of credibility, subsequent research has also applied the assumptions of the MAIN model to a range of other domains. Sundar, Xu, and Dou (2012) describe how technological affordances can influence the effectiveness of online advertisements and commercial websites, thereby affecting users' product attitudes and intentions. Likewise, Sundar, Oh, Kang, and Sreenivasan (2013) discuss the ways in which cognitive heuristics cued by interface affordances can determine the effectiveness of persuasive messages. In support of this proposition, recent research has indeed found that attitudes toward online product reviews are affected by user characteristics such as source expertise and the number of "followers" (Flanagin & Metzger, 2013; Xu, 2014), which serve to extend the applicability of authority and bandwagon heuristics to the area of persuasion.

Beyond the domain of strategic communication, the assumptions of the MAIN model have also been used to inform and extend existing concepts and theoretical frameworks. For example, Sundar, Oeldorf-Hirsch, and Garga (2008) use the MAIN model as a framework to specify how heuristics such as realism, control, and flow may mediate the relationship between technology use and the perception of nonmediation, that is, presence. Sundar and Limperos (2013) apply the MAIN model to the uses and gratifications perspective, arguing that the use of new technologies not only fulfills individuals' pre-existing needs but can also foster the development of new gratifications depending on the technological affordance in question. Bellur and Sundar (2014) also highlight the methodological applications of the MAIN model, noting that the study of heuristics should be guided by a focus on the specific features of media that trigger heuristics and the degree to which the related constructs that inform the heuristic are accessible. The MAIN model thus offers a range of both theoretical and methodological implications, as the simple presence of technological affordances can cue cognitive heuristics that affect not only credibility, but also perceptual and behavioral outcomes such as the persuasiveness of a message, the nature of user experience, and even gratifications sought and obtained from digital media use.

Concluding Remarks

The four models underlying the proposed theory of interactive media effects (TIME) shed light on different facets of the psychology of communication technologies. While the MAIN model captures the symbolic potential of interface *cues* and their influence on the meanings we derive from our online interactions, the other three specify the effects of technology-initiated action: the interactivity effects model explains the allure of modern digital media by specifying three dominant means by which we interact—and thus engage—with media; the agency model focuses on the powerful psychological appeal of users serving as sources and creators of messages (rather than being mere consumers); and the motivational technology model explores how features of the technology can be used to increase intrinsic motivation among users. Together, these four models serve to explain how interactive media technologies shape our online perceptions and actions, which carry over to our offline attitudes and behaviors.

TIME is a theory of affordances, about how users perceive and respond to those affordances, and what the user engagement of these affordances can do to their psychology, their media use, and their communications with others, leading to significant consequences for their knowledge, attitudes, and behaviors. For the theory to be fruitfully employed in research, the affordances have to be conceptually distinguished from the psychological states that they engender. For example, "user control" is by itself not an affordance, but a psychological effect of one. An affordance is an interface feature that facilitates a certain action. Researchers should be careful to isolate the affordance (e.g., a customization tool or a commenting feature) so that it is distinct from the psychological feeling of control felt by users when they use that affordance. While the latter would always be measured via self-reports, the former could be manipulated in experiments by varying the tools and features on the interface or measured in surveys by asking respondents the extent to which they used those tools and features. Employing both manipulations and measurement of affordances can be particularly helpful in disambiguating between the cue route and the action route.

While designing methods and measures, care should be taken to ensure that the effects of the cue route are distinct from that of the action route of TIME. The former proposes that the affordances on the interface lead to user perceptions and behaviors because of the meanings they convey and the cognitive heuristics triggered as a result, not because of the actual user actions that ensue. On the other hand, the action route specifies what happens when users actually interact with the interactive affordances—it leads to changes in contingency, perceptual bandwidth, sense of agency, and self-determination, which constitute four distinct mechanisms by which users become engaged with interactive media. The models discussed in this chapter specify which actions correspond most strongly with each of these mediators *en route* to increasing user engagement with the interface, shaping users' knowledge, attitudes, and behaviors both through interface-induced heuristics and actual interface use.

TIME argues that the nature of user engagement depends upon the mediator involved in a given interaction (see Figure 3.1). For example, as discussed earlier, modality interactivity is most strongly associated with the absorption dimension of engagement whereas message interactivity is associated with the elaboration aspect of engagement (Oh & Sundar, 2013). This is because the former serves to expand perceptual bandwidth whereas the latter promotes contingent dialogue. Likewise, the engagement engendered by a heightened sense of agency is qualitatively different and geared more toward self-expression and contribution while that due to heightened self determination is more likely to predict repeat usage of the interface. These variations in the nature of user engagement will likely have differential impact on psychological responses to content, with absorption and elaboration associated with cognitive outcomes, the contribution aspect of engagement leading to attitudinal outcomes and repeat usage of the interface predicting behavioral outcomes. Further, these effects are likely to be moderated by individual differences pertaining to both the medium (e.g., power usage) and the content (e.g., involvement in message topic). Together, the actions triggered by interface features affect psychological outcomes via a series of distinct mediators pertaining to users' psychological and behavioral states during the course of the interaction. Systematic investigation of the mechanisms involving these mediators in the action route as well as the heuristics triggered by interface features in the cue route will vastly enhance our understanding of the role played by technological affordances of modern media in our lives.

Acknowledgments

This research was supported by the U. S. National Science Foundation (NSF) via Standard Grant No. IIS-0916944, the Korea Science and Engineering Foundation under the WCU (World Class University) program funded through the Ministry of Education, Science and Technology, S. Korea (Grant No. R31-2008-000-10062-0), and the Ministry of Education, Korea, under the Brain Korea 21 Plus Project (Grant No. 10Z20130000013).

References

Alpert, S. R., Karat, J., Karat, C. M., Brodie, C., & Vergo, J. G. (2003). User attitudes regarding a user-adaptive ecommerce website. *User Modeling and User-Adapted Interaction*, *13*, 373–396.

Balakrishnan, B., & Sundar, S. S. (2011). Where am I? How can I get there? Impact of navigability and narrative transportation on spatial presence. *Human Computer Interaction*, *26*, 161–204.

Bellur, S., & Sundar, S. S. (2013, June). *Interactivity as conversation: Can back and forth interactions affect user cognitions and attitudes?* Paper presented at the 63rd annual conference of the International Communication Association, London, UK.

Bellur, S., & Sundar, S. S. (2014). How can we tell when a heuristic has been used? Design and analysis strategies for capturing the operation of heuristics. *Communication Methods and Measures*, *8*, 116–137.

Brug, J., Steenhaus, I., Van Assema, P., & de Vries, H. (1996). The impact of computer-tailored nutrition intervention. *Preventive Medicine*, *25*, 236–242.

Diao, F., & Sundar, S. S. (2004). Orienting responses and memory for web advertisements: Exploring effects of pop-up window and animation. *Communication Research*, *31*(5), 537–567.

DiMuzio, E., & Sundar, S. S. (2012). Does a hyperlink function as an endorsement? In M. Bang & E. L. Ragnenmalm (Eds.), *Proceedings of the 7th International Conference on Persuasive Technology (PERSUASIVE 2012), LNCS* (pp. 268–273). Berlin, Germany: Springer-Verlag.

Dou, X., & Sundar, S. S. (2014, March). *Cultivating relationships through a mobile website: The importance of modality interactivity and message interactivity*. Paper accepted for presentation at the 17th annual International Public Relations Research Conference (IPRRC), Miami, FL.

Downs, E. P., & Sundar, S. S. (2011). "We won" vs. "they lost": Exploring ego-enhancement and self-preservation tendencies in the context of video game play. *Entertainment Computing*, *2*, 23–28.

Edwards, C., Edwards, A., Spence, P. R., & Shelton, A. K. (2014). Is that a bot running the social media feed? Testing the differences in perceptions of communication quality for a human agent and a bot agent on Twitter. *Computers in Human Behavior*, *33*, 372–376.

Findlater, L., & McGrenere, J. (2004). A comparison of static, adaptive, and adaptable menus. *Proceedings of the SIGCHI Conference on Human Factors in Computing Systems (CHI '04), Vienna, Austria* (pp. 89–96). Association for Computing Machinery, New York, NY.

Fitzpatrick, R. (2010). Additional quality factors for the World Wide Web. *Proceedings of the Second World Congress for Software Quality*, Yokohama, Japan.

Flanagin, A. J., & Metzger, M. J. (2013). Trusting expert- versus user-generated ratings online: The role of information volume, valence, and consumer characteristics. *Computers in Human Behavior*, *29*(4), 1626–1634.

Fu, W. W. (2012). Selecting online videos from graphics, text, and view counts: The moderation of popularity bandwagons. *Journal of Computer-Mediated Communication*, *18*(1), 46–61.

Fu, W. W., & Sim. C. C. (2011). Aggregate bandwagon effect on online videos' viewership: value uncertainty, popularity cues, and heuristics. *Journal of the American Society for Information Science and Technology*, *62*(12), 2382–2395.

Gaver, W. W. (1991, March). Technology affordances. In *Proceedings of the SIGCHI conference on Human factors in computing systems (pp. 79–84)*. ACM.

Geiger, S., & Reeves, B. (1993). We interrupt this program … Attention for television sequences. *Human Communication Research*, *19*, 368–387.

Gibson, J. J. (1977). The theory of affordances. In R. Shaw & J. Bransford (Eds.), *Perceiving, acting, and knowing: Toward an ecological psychology* (pp. 67–82). Hillsdale, NJ: Lawrence Erlbaum.

Guillory, J., & Sundar, S. S. (2014). How does website interactivity affect our perceptions of an organization? *Journal of Public Relations Research*, *26*, 44–61.

Heeter, C. (1989). Implications of new interactive technologies for conceptualizing communication. In J. L. Salvaggio and J. Bryant (Eds.), *Media use in the information age: Emerging patterns of adoption and consumer use Lawrence Erlbaum* (pp.217–235). NJ: Hillsdale.

Hoffman, D.L., & Novak, T.P. (1996). Marketing in hypermedia computer-mediated environments: Conceptual foundations. *Journal of Marketing*, *60*, 50–68.

Hu, Y., & Sundar, S. S. (2010). Heuristic versus systematic processing of specialist versus generalist sources in online media. *Human Communication Research*, *36*, 103–124.

Ivory, J. D., & Kalyanaraman, S. (2007). The effects of technological advancement and violent content in video games on players' feelings of presence, involvement, physiological arousal, and aggression. *Journal of Communication*, *57*, 532–555.

Ivory, J. D., & Magee, R. G. (2009). You can't take it with you? Effects of handheld portable media consoles on physiological and psychological responses to video game and movie content. *Cyber Psychology & Behavior*, *12*(3), 291–297.

Jensen, J. F. (1998). Interactivity: Tracking a new concept in media and communication studies. *Nordicom Review*, *19*(1), 185–204.

Jia, H., Wu, M., Jung, E., Shapiro, A., & Sundar, S. S. (2012). Balancing human agency and object agency: An in-depth interview study of the Internet of things. *Proceedings of the 2012 ACM Conference on Ubiquitous Computing (Ubicomp'12)* (pp. 1185–1188). New York, NY: Association for Computing Machinery.

Jia, H., Wu, M., Jung, E., Shapiro, A., & Sundar, S. S. (2013). When the tissue box says "Bless You!": Using speech to build socially interactive objects. *Proceedings of CHI'13 Extended Abstracts on Human Factors in Computing Systems (CHI EA '13)* (pp. 1635–1640). New York, NY: Association for Computing Machinery.

Kalyanaraman, S., & Ivory, J. D. (2009). Enhanced information scent, selective discounting, or consummate breakdown: The psychological effects of web-based search results. *Media Psychology*, *12*(3), 295–319.

Kalyanaraman, S., & Sundar, S. S. (2006). The psychological appeal of personalized content in Web portals: Does customization affect attitudes and behavior? *Journal of Communication*, *56*(1), 110–132.

Kang, H., & Sundar, S. S. (2013). Depleted egos and affirmed selves: The two faces of customization. *Computers in Human Behavior*, *29*(6), 2273–2280.

Kang, H., & Sundar, S. S. (2014, May). *When self is the source: Implications of media customization on message processing*. Paper accepted for presentation at the 64th annual conference of the International Communication Association, Seattle, WA.

Katz, I., & Assor, A. (2007). When choice motivates and when it does not. *Educational Psychology Review*, *19*, 429–442.

Kiesler, S., Siegel, J., & McGuire, T. W. (1984). Social psychological aspects of computer-mediated communication. *American Psychologist*, *39*(10), 1123–1134.

Kim, H. S., & Sundar, S. S. (2011a). Using interface cues in online health community boards to change impressions and encourage user contribution. *Proceedings of the 2011 Annual Conference on Human Factors in Computing Systems* (pp. 599–608). New York, NY: Association for Computing Machinery.

Kim, K. J., Park, E., & Sundar, S. S. (2012). IPS vs. AMOLED: Effects of panel type on smartphone users' viewing and reading experience. In J. J. Park et al. (Eds.), *Proceedings of the 5th International Conference on Human-Centric Computing, LNEE 182*, 71–76. Netherlands: Springer.

Kim, K. J., Park, E., & Sundar, S. S. (2013). Caregiving role in human-robot interaction: A study of the mediating effects of perceived benefit and social presence. *Computers in Human Behavior*, *29*(4), 1799–1806.

Kim, K. J., & Sundar, S. S. (2013). Can interface features affect aggression resulting from violent video game play? An examination of realistic controller and large screen size. *Cyberpsychology, Behavior, and Social Networking*, *16*(5), 329–334.

Kim, N. Y., & Sundar, S. S. (2010). Relevance to the rescue: Can smart ads reduce negative response to online ad clutter? *Journalism and Mass Communication Quarterly*, *87*(2), 346–362.

Kim, N. Y., & Sundar, S. S. (2012). Personal relevance vs. contextual relevance: The role of relevant ads in personalized websites. *Journal of Media Psychology*, *24*(3), 89–101.

Kim, Y., & Sundar, S. S. (2011b, May). *Can your avatar improve your health? The impact of avatar attractiveness and avatar creation.* Paper presented at the 61st annual conference of the International Communication Association, Boston, MA.

Kobsa, A., Koenemann, J., & Pohl, W. (2001). Personalized hypermedia presentation techniques for improving online customer relationships. *The Knowledge Engineering Review*, *16*(2), 111–155.

Koh, Y. J., & Sundar, S. S. (2010a). Effects of specialization in computers, web sites and web agents on e-commerce trust. *International Journal of Human-Computer Studies*, *68*, 899–912.

Kreuter, M., Bull, F., Clark, E., & Oswald, D. (1999). Understanding how people process health information: A comparison of tailored and untailored weight loss materials. *Health Psychology*, *18*, 487–494.

Kreuter, M., Farrell, D., Olevitch, L., & Brennan, L. (2000). *Tailoring health messages: Customizing communication using computer technology.* Mahwah, NJ: Lawrence Erlbaum.

Lang, A. (1994). *Measuring psychological responses to media messages.* Hillsdale, NJ: Psychology Press.

Lee, H., & Park, H. (2013). Testing the impact of message interactivity on relationship management and organizational reputation. *Journal of Public Relations Research*, *25*(2), 188–206.

Lee, J. Y., & Sundar, S. S. (2013). To tweet or retweet? That is the question for health professionals on Twitter. *Health Communication*, *28*(5), 509–524.

Lee, S., & Sundar, S. S. (2014). Cosmetic customization of mobile phones: Cultural antecedents, psychological correlates. *Media Psychology*, doi: 10.1080/15213269.2013.853618.

Li, R., Zhang, B., Sundar, S. S., & Duh, H. B.-L. (2013). *Interacting with augmented reality: How does location-based AR enhance learning.* In P. Kotzé et al. (Eds.), *Proceedings of INTERACT 2013, Part II, LNCS 8118*, 616–623. Berlin Heidelberg: Springer.

Liu, Y., & Shrum, L. J. (2002). What is interactivity and is it always such a good thing? Implications of definition, person, and situation for the influence of interactivity on advertising effectiveness. *Journal of advertising*, *31*, 53–64.

Marathe, S., Kang, H., & Sundar, S. S. (2010, June). *Who customizes health websites and why?* Paper presented at the 60th annual conference of the International Communication Association, Singapore.

Marathe, S., & Sundar, S. S. (2011a). What drives customization? Control or identity? *Proceedings of the 2011 Annual Conference on Human Factors in Computing Systems (CHI'11), Vancouver, BC* (pp. 781–790). New York, NY: Association for Computing Machinery.

Marathe, S., & Sundar, S. S. (2011b). *The 'me' revolution in mediated communication: Investigating the psychology of cosmetic and functional customization.* Paper presented at the 61st annual conference of the International Communication Association, Boston, MA.

Mathwick, C., & Ridgdon, E. (2004). Play, flow, and the online search experience. *Journal of Consumer Research*, *31*, 324–332.

McLuhan, M. (1964). *Understanding media: The extensions of man.* New York: McGraw Hill.

McMillan, S. J., & Hwang, J. S. (2002). Measures of perceived interactivity: An exploration of the role of direction of communication, user control, and time in shaping perceptions of interactivity. *Journal of Advertising*, *31*, 29–42.

Meyerowitz, J. (1985). *No sense of place.* New York: Oxford University Press.

Newman, W. R. (1991). *The future of the mass audience.* Cambridge, UK: Cambridge University Press.

Norman, D. A. (1988). *The psychology of everyday things.* Basic Books, Inc.

Novak, T. P., Hoffman, D. L., & Yung, Y. F. (2000). Measuring the customer experience in online environments: A structural modeling approach. *Marketing Science, 19*(1), 22–42.

Oeldorf-Hirsch, A., & Sundar, S. S. (2012, August). *Engagement with news content in online social networks.* Paper presented at the 95th annual conference of the Association for Education in Journalism and Mass Communication, Chicago, IL.

Oh, J., Robinson, H. R., & Lee, J. Y. (2013). Page flipping vs. clicking: The impact of naturally mapped interaction technique on user learning and attitudes. *Computers in Human Behavior, 29*(4), 1334–1341.

Oh, J., & Sundar, S. S. (2013, August). *How does interactivity persuade? An experimental test of interactivity on cognitive absorption, elaboration, and attitudes.* Paper presented at the 95th annual conference of the Association for Education in Journalism and Mass Communication, Washington, DC.

Pan, B., Hembrooke, H., Joachims, T., Lorigo, L., Gay, G., & Granka, L. (2007). In Google we trust: Users' decisions on rank, position, and relevance. *Journal of Computer-Mediated Communication, 12*(3), 801–823.

Park, E., Kim, K. J., Sundar, S. S., & del Pobil, A. P. (2012). Online gaming with robots vs. computers as allies vs. opponents. *Proceedings of the 7th ACM/IEEE International Conference on Human-Robot Interaction, Boston, MA* (pp. 205–206). New York, NY: Association for Computing Machinery.

Pirolli, P. (2003). Exploring and finding information. In J. Carroll (Ed.), *HCI models, theories and frameworks: Toward a multidisciplinary science* (pp. 157–191). San Francisco, CA: Morgan Kauffmann.

Quiring, O. (2009). What do users associate with "interactivity"? A qualitative study on user schemata. *New Media & Society, 11*(6), 899–920.

Rafaeli, S. (1988). Interactivity: From new media to communication. In R. Hawkins, J. Wiemann, & S. Pingree (Eds.), *Advancing communication science: Merging mass and interpersonal processes* (pp. 110–134). Newbury Park, CA: Sage.

Rayport, J. F., & Jaworski, B. J. (2001). *E-commerce.* New York, NY: McGraw-Hill/Irwin.

Reeves, B., Lang, A., Kim, E. Y., & Tatar, D. (1999). The effects of screen size and message content on attention and arousal. *Media Psychology, 1*(1), 49–67.

Reeves, B., & Nass, C. (2000). Perceptual bandwidth. *Communication of the ACM, 43*(3), 65–70.

Reeves, B., & Nass, C. (1996). *The media equation: How people treat computers, television, and new media like real people and places.* Stanford, CA: CSLI Publications/Cambridge University Press.

Rimer, B. K., & Kreuter, M. W. (2006). Advancing tailored health communication: A persuasion and message effects perspective. *Journal of Communication, 56*(1), 184–201.

Rubin, A. M. (1993). Audience activity and media use. *Communication Monographs, 60*, 98–103.

Ryan, R. M., & Deci, E. L. (2000). Self-determination theory and the facilitation of intrinsic motivation, social development, and well-being. *American Psychologist, 55*, 68–78.

Sah, Y. J., Yoo, B., & Sundar, S. S. (2011). Are specialist robots better than generalist robots? *Proceedings of the 6th International Conference on Human Robot Interaction, Lausanne, Switzerland* (pp. 241–242). New York, NY: Association for Computing Machinery.

Schmierbach, M., Limperos, A. M., & Wooley, J. K. (2012). Feeling the need for (personalized) speed: How natural controls and customization contribute to enjoyment of a racing game through enhanced immersion. *Cyberpsychology, Behavior, and Social Networking, 15*(6), 364–369.

Shapiro, A. L. (1999). *The control revolution: How the Internet is putting individuals in charge and changing the world we know*. New York, NY: Public Affairs.

Sohn, D. (2011). Anatomy of interaction experience: Distinguishing sensory, semantic, and behavioral dimensions of interactivity. *New Media & Society, 13*(8), 1320–1335.

Spears, R., & Lea, M. (1994). Panacea or panopticon? The hidden power in computer-mediated communication. *Communication Research, 21*(4), 427–459.

Stavrositu, C., & Sundar, S. S. (2012). Does blogging empower women? Exploring the role of agency and community. *Journal of Computer-Mediated Communication, 17*, 369–386.

Steuer, J. (1992). Defining virtual reality: Dimensions determining telepresence. *Journal of Communication, 42*(4), 73–93.

Stromer-Galley, J. (2004). Interactivity-as-product and interactivity-as-process. *The Information Society, 20*, 391–394.

Sundar, S. S. (2000). Multimedia effects on processing and perception of online news: A study of picture, audio, and video downloads. *Journalism and Mass Communication Quarterly, 77*(3), 480–499.

Sundar, S. S. (2004). Theorizing interactivity's effects. *The Information Society, 20*(5), 385–389.

Sundar, S. S. (2007). Social psychology of interactivity in human-website interaction. In A. N. Joinson, K. Y. A. McKenna, T. Postmes, & U.-D. Reips (Eds.), *The Oxford handbook of Internet psychology* (pp. 89–104). Oxford, UK: Oxford University Press.

Sundar, S. S. (2008a). The MAIN model: A heuristic approach to understanding technology effects on credibility. In M. J. Metzger & A. J. Flanagin (Eds.), *Digital media, youth, and credibility* (pp. 72–100). Cambridge, MA: The MIT Press.

Sundar, S. S. (2008b). Self as source: Agency and customization in interactive media. In E. Konijn, S. Utz, M. Tanis, & S. Barnes (Eds.), *Mediated interpersonal communication* (pp. 58–74). New York, NY: Routledge.

Sundar, S. S. (2009). Media effects 2.0: Social and psychological effects of communication technologies. In R. L. Nabi & M. B. Oliver (Eds.), *The SAGE handbook of media processes and effects (pp. 545–560)*. Thousand Oaks, CA: Sage Publications.

Sundar, S. S., & Bellur, S. (2010). Concept explication in the internet age: The case of interactivity. In E. P. Bucy & R. L. Holbert (Eds.), *Sourcebook for political communication research: Methods, measures, and analytical techniques* (pp. 485–500). New York, NY: Routledge.

Sundar, S. S., Bellur, S., & Jia, H. (2012). Motivational technologies: A theoretical framework for designing preventive health applications. In M. Bang & E. L. Ragnemalm (Eds.), *Proceedings of the 7th International Conference on Persuasive Technology (PERSUASIVE 2012)*, LNCS 7284 (pp. 112–122). Berlin, Germany: Springer.

Sundar, S. S., Bellur, S., Oh, J., & Jia, H. (2011). Calls for interaction: The more the better? User experience of 3D carousel and additional interaction techniques. In P. Campos et al. (Eds.), *Proceedings of INTERACT 2011, Part IV, LNCS 6949* (pp. 487–490). Berlin, Germany: Springer.

Sundar, S. S., Bellur, S., Oh, J., Jia, H., & Kim, H. S. (2014). Theoretical importance of contingency in human-computer interaction: Effects of message interactivity on user engagement. *Communication Research*. doi:10.1177/0093650214534962

Sundar, S. S., Bellur, S., Oh, J., Xu, Q., & Jia, H. (2014). User experience of on-screen interaction techniques: An experimental investigation of clicking, sliding, zooming, hovering, dragging and flipping. *Human Computer Interaction, 29*, 109–152.

Sundar, S. S., Jia, H., Bellur, S., Oh, J., & Kim, H. S. (2014). *Interactivity in medium, source, and message: An empirical validation of the interactivity effects model.* Unpublished manuscript, Pennsylvania State University, State College, PA.

Sundar, S. S., & Kalyanaraman, S. (2004). Arousal, memory, and impression-formation effects of animation speed in web advertising. *Journal of Advertising, 33*(1), 7–17.

Sundar, S. S., Kalyanaraman, S., & Brown, J. (2003). Explicating website interactivity: Impression-formation effects in political campaign sites. *Communication Research, 30*(1), 30–59.

Sundar, S. S., & Kim, J. (2005). Interactivity and persuasion: Influencing attitudes with information and involvement. *Journal of Interactive Advertising, 5*(2), 6–29. Article retrievable from http://www.jiad.org/download1098.pdf?p=59

Sundar, S. S., Knobloch-Westerwick, S., & Hastall, M. R. (2007). News cues: Information scent and cognitive heuristics. *Journal of the American Society for Information Science and Technology, 58*(3), 366–378.

Sundar, S. S., & Limperos, A. (2013). Uses and grats 2.0: New gratifications for new media. *Journal of Broadcasting & Electronic Media, 57*(4), 504–525.

Sundar, S. S., & Marathe, S. S. (2010). Personalization vs. customization: The importance of agency, privacy and power usage. *Human Communication Research, 36*, 298–322.

Sundar, S. S., Marathe, S., & Kang, H. (2009, November). *Beyond tailoring: Customization in health websites.* Paper presented at the 95th annual convention of the National Communication Association, Chicago, IL.

Sundar, S. S., Narayan, S., Obregon, R., & Uppal, C. (1998). Does Web advertising work? Memory for print vs. online media. *Journalism & Mass Communication Quarterly, 75*(4), 822–835.

Sundar, S. S., & Nass, C. (2001). Conceptualizing sources in online news. *Journal of Communication, 51*(1), 52–72.

Sundar, S. S., Oeldorf-Hirsch, A., & Garga, A. (2008). A cognitive heuristic approach to understanding presence in virtual environments. In A. Spagnolli & L. Gamerini (Eds.), *PRESENCE 2008: Proceeding of the 11th International Workshop on Presence* (pp. 219–228). Padova, Italy: CLEUP Cooperativa Libraria Universitaria Padova.

Sundar, S. S., Oeldorf-Hirsch, A., & Xu, Q. (2008). The bandwagon effect of collaborative filtering technology. *Proceedings of CHI'08 Extended Abstracts on Human Factors in Computing Systems, 26, Florence, Italy* (pp. 3453–3458). New York, NY: Association for Computing Machinery.

Sundar, S. S., Oh, J., Bellur, S., Jia, H., & Kim, H. S. (2012). Interactivity as self-expression: A field experiment with customization and blogging. *Proceedings of the 2012 Annual Conference on Human Factors in Computing Systems (CHI'12)* (pp. 395–404). New York, NY: Association for Computing Machinery.

Sundar, S. S., Oh, J., Kang, H., & Sreenivasan, A. (2013). How does technology persuade? Theoretical mechanisms for persuasive technologies. In J. P. Dillard & L. Shen (Eds.), *The Sage Handbook of Persuasion: Developments in Theory and Practice* (2nd ed., pp. 388–404). Thousand Oaks, CA: Sage.

Sundar, S. S., Tamul, D., & Wu, M. (2014). Capturing "cool": Measures for assessing coolness of technological products. *International Journal of Human-Computer Studies, 72*, 169–180.

Sundar, S. S., Xu, Q., & Bellur, S. (2010). Designing interactivity in media interfaces: A communications perspective. *Proceedings of the 28th International Conference on Human Factors in Computing Systems (CHI'10)* (pp. 2247–2256). New York, NY: Association for Computing Machinery.

Sundar, S. S., Xu, Q., Bellur, S., Oh, J., & Jia, H. (2010). Modality is the message: Interactivity effects on perception and engagement. *Proceedings of the 28th of the International Conference Extended Abstracts on Human Factors in Computing Systems (CHI EA'10)* (pp. 4105–4110). New York, NY: Association for Computing Machinery.

Sundar, S. S., Xu, Q., Bellur, S., Oh, J., & Jia, H. (2011). Beyond pointing and clicking: How do newer interaction modalities affect user engagement? *Proceedings of the 2011 Annual Conference Extended Abstracts on Human Factors in Computing Systems (CHI EA'11)* (pp. 1477–1482). New York, NY: Association for Computing Machinery.

Sundar, S. S., Xu, Q., & Dou, X. (2012). Role of technology in online persuasion: A MAIN model perspective. In S. Rodgers & E. Thorson (Eds.), *Advertising theory* (pp. 355–372). New York, NY: Routledge.

Sundar, S. S., Xu, Q., & Oeldorf-Hirsch, A. (2009). Authority vs. peer: How interface cues influence users. *Proceedings of the 27th International Conference Extended Abstracts on Human Factors in Computing Systems (CHI'09)*, *27*, 4231–4236. Association for Computing Machinery, New York, NY.

Velasquez, A. (2012). Social media and online political discussion: The effect of cues and informational cascades on participation in online political communities. *New Media & Society*, *14*(8), 1286–1303.

Walther, J. B. (1996). Computer-mediated communication: Impersonal, interpersonal, and hyperpersonal interaction. *Communication Research*, *23*(1), 3–43.

Westerwick, A. (2013). Effects of sponsorship, web site design, and Google ranking on the credibility on online information. *Journal of Computer-Mediated Communication*, *18*(2), 80–97.

Wind, J., & Rangaswamy, A. (2001). Customerization: The next revolution in mass customization. *Journal of Interactive Marketing*, *15*(1), 13–32.

Wise, K., Hammer, B., & Thorson, K. (2006). Moderation, response rate, and message interactivity: Features of online communities and their effects on intent to participate. *Journal of Computer-Mediated Communication*, *12*(1), 24–41.

Wu, G. (2005). The mediating role of perceived interactivity in the effect of actual interactivity on attitude toward the Website. *Journal of Interactive Advertising*, *5*(2), 29–39.

Xu, Q. (2013). Social recommendation, source credibility, and recency effects of news cues in a social bookmarking website. *Journalism and Mass Communication Quarterly*, *90*(4), 757–775.

Xu, Q. (2014). Should I trust him? The effects of reviewer profile characteristics on eWOM credibility. *Computers in Human Behavior*, *33*, 136–144.

Xu, Q., Schmierbach, M., Bellur, S., Ash, E., Oeldorf-Hirsch, A., & Kegerise, A. (2012). The effect "friend" characteristics on evaluations of an activist group in a social networking context. *Mass Communication and Society*, *15*(3), 432–453.

Xu, Q., & Sundar, S. S. (2011, May). *Interactivity, arousal and memory: Information processing in an e-commerce site.* Paper presented at the 61st annual conference of the International Communication Association, Boston, MA.

Xu, Q., & Sundar, S. S. (2014). Lights, camera, music, interaction! Interactive persuasion in e-commerce. *Communication Research*, *41*(2), 282–308.

Part II

Source Orientation: Avatars, Agents and Androids

4

Examining Perception and Identification in Avatar-mediated Interaction

Kristine L. Nowak
Department of Communication, University of Connecticut, Storrs, USA

Avatars are being used in almost all types of computer-mediated interactions including advertising, news reporting, patient–doctor communication, social media, and job interviews. All of these are examples of avatar-mediated communication. The word avatar originated in Hinduism, where it was used to mean an incarnation of a deity in a human form, allowing them to experience the earth as an embodied human and to speak with mortals (Vilhjalmsson, 1996). The word avatar was first used to mean digital embodiment online in the science fiction novel *SnowCrash* (Stephenson, 1993) because in some way avatars allow people to be embodied in, and have experiences only possible in computer-generated environments (Bailenson, Yee, Merget, & Schroeder, 2006; Biocca & Nowak, 2002; Nowak, Hamilton, & Hammond, 2009). Avatars can theoretically take any shape including images of people, animals, inanimate objects, fictional images, or combinations of those things. While not everyone agrees on the appropriate use of the term, avatars can be life sized or larger in fully immersive virtual worlds, they can be a small image in the corner of a screen, or any size in between. They can be static or animated, realistic or cartoonish, two or three dimensional, or some or none of the above. If they move, they can be required to walk or allowed to fly or just appear in the desired location. They can be required to follow earthly norms of gravity and collision detection or move fluidly or even float through walls and objects in the environment. As described in *Snowcrash*, avatars out on what is called the street can pass through one another and can be no taller than the person's offline body, but collision detection prevents them from passing through one another and the rules for height are relaxed inside the exclusive online world where much of the book takes place (Stephenson, 1993).

The avatar is a feature of the medium embedded in the interface that only exists on the computer screen, does not have a smell, and cannot be touched. It is also the visible representation of a person or source, making it in some ways similar to the offline physical body in that it is used to make attributions of, represent, self-present,

The Handbook of the Psychology of Communication Technology, First Edition. Edited by S. Shyam Sundar.

and identify people (Bailenson et al., 2005; Bente, Rüggenberg, Krämer, & Eschenburg, 2008; Blascovich & Bailenson, 2011; Garau et al., 2003; Nowak, Hamilton et al., 2009; Nowak & Rauh, 2008; Seyama & Nagayama, 2007; Yee, Bailenson, & Rickertsen, 2007). At the same time, it is neither a feature of the medium nor the offline corporeal body because it is something entirely different. Users, advertisers, researchers, and designers can manipulate, exaggerate, or remove, certain features or characteristics to emphasize a message or control other aspects of the experience. Designers can manipulate skin tone, height, gender, the size or location of eyes, mouths, or even the location or presence or absence of a scar or limbs. People can select avatars to experiment with different sides of their identities or different selves, and they can be embodied in different cultures, conditions, climates, or environments. These digital experiences can be replications of something that could happen in the future, is happening now, happened in the past, or something that is complete fantasy. The appearance of an avatar is determined by technological affordances, graphics resolution, capacities of the system, as well as the vision of the designers for the use of the system. Each of these design decisions has consequences for user experience and perception of the system (Biocca & Nowak, 2002; Blascovich, 2002; Blascovich & Bailenson, 2011; Magnenat-Thalmann & Thalmann, 2005; Schroeder, 2002). Researchers can use avatars to test the prevalence of stereotypes and how different features influence attitudes, behaviors, and learning, as well as how people feel after being represented by different avatars. This can allow social scientists unlimited potential to test theories about human behavior and to work with clinicians and other content experts to create tutorials, workshops, and therapies to help people overcome phobias, learn new skills, or even reduce biases against others (Blascovich & Bailenson, 2011; Rizzo, Parsons, & Buckwalter, 2012).

Examining how people process information and interact with others online, and the role of avatars on this process, is important not only because of the prevalence of the use of avatars but also because the study of avatars can provide insight into processes related to self-presentation and person perception by allowing researchers to test hypotheses that are unethical, difficult, or even impossible to study offline. People seem to be aware that the selection of an avatar is important and this can lead to anxiety about what to choose or how to get to know others represented by avatars in online interactions. Companies want to know how to best represent their products and people and how to predict the way others will respond to associated messages. People want to know how to select an avatar that will help them put their best selves forward for dating, job interviews, selling products, or other activities. The good news is that there is enough research in to be able to definitively answer these questions: *It depends.*

While researchers generally seem to agree that the ideal avatar depends on a variety of things, no two researchers or individuals agree on the most important variables or even how to measure them. As will be discussed in the pages that follow, researchers have shown that the processing of avatars is influenced by individual differences, cognitive processing, schemas, previous relationships, likelihood of future interaction, the goals of the interaction, experience with the medium, presence in the interaction, engagement with the interaction partner, message or characters, features of the interface, realism, social potential of the characters, the quality of the graphics, and a variety of other factors. The question of the most important variables depends on the context, individual differences, outcome variables, and goals of the user. This chapter will examine how avatars are being used to self-present, identify, recognize, and make attributions of others in online interactions and the effects of this use on person and message perception and self-concept. This chapter will examine how people are using avatars and the effects of this usage with particular attention to the following:

1. The way people perceive and relate to avatars, focusing on how people make attributions of humanness and social potential of entities.
2. The types of avatars people select and the influence of avatar features on message processing, person perception, and self-presentation of identity.
3. Implications of the use of avatars for theory development, including suggestions for users and designers of systems using avatars, and critical questions for future research.

Responding to Avatars of Digital Others: Adapting to Online Information

Whether online or offline, people scan their surroundings and make judgments about the landscape, people, and images they encounter. Regardless of the type of information or source of information, people continue to make attributions of others and process whatever information is available to them. There has been some debate about whether the predicting and explaining the use of technology is best done by extending existing theories or requires the creation of new theories to account for perceptual differences as well as the unique format of information presentation in online interactions and message processing. New theories that were created especially to examine computer-mediated communication tended to focus on the features of the interface, making them heavily technologically dependent with an emphasis on time or bandwidth efficiency, many times with workplace tasks, and many made assumptions that processes would be very different online than offline. When researchers have sought to extend existing theories, they have taken theories derived from other contexts (interpersonal, mass communication, psychology) to make predictions about how people will adapt to utilize the features of technology to fulfill interaction goals online (For a discussion, see Walther, 1996, 2011; Walther & Parks, 2002). With this in mind, this section will explain what is known about how people process online images and entities when they are represented by avatars, with a discussion of relevant theories and models.

While categorization is a basic part of information processing and occurs with all information across all types of entities both online and offline, important differences in the process of categorization occur because of the way information is presented to the senses (McLuhan, 1964), and the range of categories considered (Lakoff, 1987; Sheehan, 1991). Social Information Processing theory (Walther, 1992) and the Hyperpersonal Model (Walther, 1996) explain that people have adapted interpersonal and information processing into online environments by modifying behavior or the source of information to effectively utilize the features of whatever medium is available, or convenient, during an interaction to fulfill their interaction goals, using visual cues provided by the medium to express identification and form impressions when the nonverbal cues in face to face interactions are not available (Walther & Parks, 2002). Thus, engaging in person perception and relationship maintenance when people are represented by digital images that are part of the medium gives new meaning and importance to McLuhan's (1964) position that "the medium is the message." Viewer responses to visual stimuli follow a predictable process regardless of the way it is presented to the senses because people have automatic and universal responses to visual stimuli and the sequences used to evaluate visual information are hard-wired (Hamilton & Nowak, 2010; Nowak et al., 2009a; Ramerez, Walther, Burgoon, & Sunnafrank, 2002). Still, information presented on a screen will be processed differently from information associated with a physical body, just as information presented verbally will be processed differently from information presented visually or text-based. Typing and reading are not the same as speaking, watching, or listening (Walther & Parks, 2002), and information presented from other channels is influenced by visual information, which is processed first because processing behaviors or text-based information requires more cognitive resources than processing visual information (Burgoon, 1994; Kenny, Horner, Kashy, & Chu, 1991). Online systems and virtual environments provide new ways for information to be presented to our senses, and the presentation of information can alter and redefine how the message is interpreted and the way it is processed (McLuhan, 1964).These differences influence not only the presentation of information and how people perceive it, but also how the information is processed as well as what is remembered (Graber, 1996; Lang, 1995). When applied to processing the visual stimuli associated with avatars, people follow a process of evaluating them and predicting future behavior of those they represent (Hamilton et al., 2010; Nowak et al., 2009a). It is unclear whether avatars are processed as the source, a part of the medium, part of the message, or some combination of these, and further research is needed to answer this question (Nowak & Rauh, 2008).

Regardless of context or the way information is presented, people use a combination of behavioral and visual cues relying upon heuristics and stereotypes based on their previous exposure to similar objects to place them into categories that are salient and perceived to be useful (Patterson, 1995). Importantly, they make the same category assignments and attributions of those they encounter online as they do for people they encounter offline using whatever information they are provided in the interaction (Flanagin & Metzger, 2007; Huang & Chen, 2006; Nowak, Hamilton et al., 2009a; Smith, Menon, & Sivakumar, 2005; Walther, 1996, 2007; Walther & Parks, 2002; Wogalter & Mayhorn, 2008). They begin by classifying objects as animate or inanimate; and if animate, as living or not; and if living, as human or not, and these classifications determine the other categories that become salient and relevant for perceiving the object (Lakoff, 1987; Sheehan, 1991). People compare the object to other similar objects encountered in the past to predict its ability to respond appropriately to complex stimuli and to follow social rules and norms, or its social potential, as well as likely behaviors and a judgment of potential usefulness or danger. As described in the next section, the judgment of social potential is based on a combination of factors and influences the categories that are relevant and the judgments made of the object or entity.

Attributions of Humanity, Anthropomorphism, and Social Potential

While this chapter is primarily focused on avatars that represent humans, not all visual characters online represent humans as sometimes these images represent agents or bots (automated computer programs), products (as in advertising), or are paired with messages or part of the interface serving as a guide. As with offline processing described earlier (Lakoff, 1987; Sheehan, 1991), a judgment similar to the categorization of humanness is among the first to be made of avatars encountered online (Hamilton & Nowak, 2010; Nowak et al., 2009a). The process of determining humanity, intelligence, and social potential of others is largely based on perceptions of anthropomorphism. Anthropomorphic means having a humanoid or human like form or features, whereas anthropomorphism is the assignment of human traits or qualities such as mental abilities (Kennedy, 1992), cognitions (Tamir & Zohar, 1991), intentions and emotions (Barrret, 1997), psychological states (Nass, Lombard, Henriksen, & Steuer, 1995), and behavior (DiSalvo & Gemperle, 2003) to entities that may or may not be human (Nowak et al., 2009a). For the purposes of this chapter, there are three major ways an entity can be anthropomorphic that will likely influence the judgment of social potential including whether the entity is perceived to be human, the degree to which one can mimic human behaviors, and the degree to which one has visual characteristics consistent with offline human bodies. This section will briefly consider how people respond to others that are judged to have social potential or are deemed anthropomorphic.

In initial interactions, people seek to define humanity as distinct from, if not superior to, both objects and animals (Sheehan, 1991). Offline there is a clear boundary between human and not being human, where living things contain common behavioral and visual properties that distinguish them from inanimate objects, and there are certain visual characteristics, traits, behaviors, and abilities that are unique to humans (Asch, 1958; Heider, 1958; Lakoff, 1987; Sheehan, 1991). People's bodies are by definition anthropomorphic (human like) offline, though offline non-human entities do not appear human. The same is not true online, where humans may be represented by non-anthropomorphic images, and non-human entities may be represented by anthropomorphic images, display intelligence, and have the ability to mimic social interaction rules (Nowak, 2004). While an avatar is a visual or graphic image used in an online representing a human, an agent is frequently used to described an image representing a computer program that carries out delegated tasks (Blascovich, 2002; Blascovich & Bailenson, 2011; Brent & Thompson, 1999; Laurel, 1990; Nowak, 2004; Nowak & Biocca, 2003). An agent may be represented by a highly anthropomorphic image that moves and speaks fluidly and an avatar could look like a bottle or a robot with halting movements, making it more difficult to differentiate between computers and humans online (Goodwin, 1998; Kim & Sundar, 2012a; Nowak, 2004; Nowak & Biocca, 2003).

This may require a reconceptualization of the category of humanity, or at least require a modification to the assumption that only humans look human and display intelligence (Bolter, 1984; Brent & Thompson, 1999; Dryer, 1999; Turkle, 2011; Nowak, 2004; Nowak & Biocca, 2003). While the judgment of humanness still occurs online, it is different from the offline process because of the type and source of information available as well as the categories of entities encountered online, and the significance humanness as a category may not mean the same thing it does offline (Turkle, 2011). Viewers do not seem to meaningfully differentiate between bots and humans in attributions or perceptions of behaviors and impose the same evaluation criteria as long as they display some minimum threshold of social potential (Nass & Moon, 2000; Nowak, 2004; Nowak & Biocca, 2003; Ramerez et al., 2002; Reeves & Nass, 1996; Turkle, 2011). Entities that meet a minimal threshold of sociability by looking at all anthropomorphic, responding appropriately to stimuli, interacting autonomously, displaying intelligence or emotion, or satisfying interaction goals are perceived to have social potential (Bailenson et al., 2006; Yee et al., 2007; Barret, 1997; Breazeal, 2003; DiSalvo & Gemperle, 2003; Hamilton & Nowak, 2010; Nass & Moon, 2000; Nowak, 2004; Nowak et al., 2009; Reeves & Nass, 1996). This "social bias" results in the overestimation in perceived social potential of stimuli in online interactions (Kim & Sundar, 2012a; Nass & Moon, 2000), which has been called mindless (Kim & Sundar, 2012a; Nass & Moon, 2000) because it is automatic. The overestimation of social potential could be a rational choice because there is a higher cost of underestimating the social potential of an entity than there is for overestimating it (Nowak et al., 2009a). The cost of underestimating the social potential could be at risk insulting the other, or putting the self in danger by underestimating a threat, or missing an opportunity, whereas the only cost of overestimating the social potential is using more cognitive resources than required.

More cognitive resources are allocated to the processing of entities with social potential because they are assumed to be capable of more complex behavior, emotions, and motivations, and display more nuanced reactions to situations, making more categories and complicating the perception process as people make attributions of motivations, personality, social status, credibility, gender, and likeability while also considering intention and emotion in their interpretation of behavior (Asch, 1958; Heider, 1958; Sheehan, 1991). Information Processing theory explains the increased allocation of cognitive resources as the engagement of the charisma sequence, and entities that engage the charisma sequence are perceived as more complicated entities that activate schema, or personae, leading to attributions of intentionality and emotion that are used to predict future behavior. Even static images have engaged this Information Processing theory's charisma sequence, where people generate personae and make judgments of the entity that would only be made of one perceived to have high social potential (Hamilton & Nowak, 2010; Nowak et al., 2009a). When a stimulus activates personae, the perceiver will make a determination of the membership in each of the primitive categories including race, gender, and social status. The perceiver will also work to identify interaction patterns and social norms that set expectations and help people predict behavior of others (Asch, 1958; Heider, 1958; Sheehan, 1991). Thus, when a stimulus triggers this process, people expend more resources in the perception process as they create a more nuanced mental model of that person or object (Nowak et al., 2009a). There are still questions about how and when the visible characteristics of avatars activate the more complex processing generally reserved for humans and social entities, and this is an important area for future research.

People also make attributions of gender when interacting with others, particularly with anthropomorphic entities with high social potential. Characters that have social potential lead to more complex mental models of others as described earlier. One category that becomes relevant when an entity has been judged to be living is sex category (assigning others to be either male or female). People have assigned others to a sex category even when they cannot see the corporeal body, meaning the attribution is based on stereotypes associated with gender as a social category and not biological sex or physical differences (Balsamo, 1995; Fox & Bailenson, 2009a; Lipton, 1996; Nowak, 2003; Spender, 1996). Making a sex category assignment is generally assumed to help predict the types of behaviors and to establish appropriate interaction scripts and behaviors (Hamilton & Sherman, 1994; Lakoff, 1987), explaining why attributions of sex category and gender have maintained salience and meaning online (Biocca & Nowak, 2002; Skitka & Maslach, 1996; Spender, 1996; Turkle, 1995). The use of gender stereotypes has continued for avatars, and these stereotypes influence both how the people using them feel and how they are perceived by others (Balsamo, 1995; Fox & Bailenson, 2009; Lee, 2004a; Nowak, Fox, & Ranjit, 2014; Nowak & Rauh, 2008), and gender schema are applied even when only synthesized voices or static images are provided for entities known to be bots and not humans (Nowak, Fox, & Ranjit, 2014; Reeves & Nass, 1996). The continued use of visible characteristics online in making gender attributions of even computer simulations could be explained by people's desire to reduce uncertainty about the future behavior of others (Berger & Calabrese, 1975), and the perception that knowing the gender of others during interactions has provided useful information in previous interactions (Fiske & Neuberg, 1990; Hamilton & Sherman, 1994). Users feel more uncertain with androgynous avatars, and seem to dislike avatars that lack clear gender identification (Nowak & Rauh, 2005, 2008), and being able to make attributions of gender online helps reduce the uncertainty associated with interacting with others in the absence of the visible corporeal body (Clark, 1995). People see anthropomorphic avatars, particularly avatars that represent their gender as more homophilous than either androgynous avatars, or avatars that do not match their gender. Homophily, both the extent to which an avatar looks similar to the self and the extent to which the avatar represents psychological aspects or personality similar to the self, has been shown to be a key perception people make about avatars, particularly anthropomorphic avatars. People seem to make an assessment of the extent to which the avatar is "like me" even when it is representing someone else, with more homophilous avatars generally being seen as more credible and likeable (Nowak, 2013; Nowak, Hamilton et al., 2009a) and more persuasive (Guadagno, Blascovich, Bailenson, & McCall, 2007), though the effect of gender matching on homophily has a stronger effect on women than men (Hamilton & Nowak, 2010). It is the sense of homophily and connection to an avatar that leads to people selecting it for interactions with others in most contexts, and the types of avatars people select, and the reasons for those selections is discussed in the following.

Essentially, the offline dichotomous categorization of humanity seems less relevant online. Instead, it seems that people tend to overestimate the social potential of entities they encounter and those displaying any level of social responsiveness, intelligence, or emotion are perceived to have social potential and there is little difference in perception of an entity that is human and one that is not human but has social potential (Breazeal, 2003; Hamilton & Nowak, 2010; Nass et al., 1995; Nass & Moon, 2000; Nowak, 2004; Nowak, Hamilton et al., 2009a; Reeves & Nass, 1996; Turkle, 2011). While more research should seek to uncover categories or judgments made online that are not made offline, a series of studies has shown that while people reported they would not treat a computer as a social entity, insisting that would not try to spare the feelings of a computer, or would not apply gender stereotypes to computer simulations of gendered voices, they did (Reeves & Nass, 1996). Research shows that attribution of intention, emotion, race, and sex category assignment applied only to humans or animals offline are being applied to agents online along with associations and stereotypes (Dryer, 1999; Groom, Bailenson, & Nass, 2009; Nowak, Fox, & Ranjit, 2014; Pena, Hancock, & Merola, 2009; Reeves & Nass, 1996). This could be explained again by reliance on offline experience where people automatically interpret stimuli that appear intelligent as driven by a human entity as described earlier, but further research on this topic is needed to fully understand this process. The next section will discuss some theoretical explanations for how and why people rely on the visual characteristics of avatars in the perception process and how that influences the way people react to avatars and the people they represent.

Understanding the Use of the Visual Characteristics of Avatars in Perception

Using information provided on the screen, such as the avatar, as one would use visual information in the natural world such as the corporeal body offline may initially seem "mindless" or naive. However, this could make both logical and evolutionary sense. An avatar is analogous to the corporeal body offline in that it serves as the visual identifier and representation of the person and using it in person perception is evidence of people's adaptation to utilize the available information available in the medium to meet their communication needs (Walther, 1992). While a person cannot control or change their corporeal body offline, people can alter and control the visible indicators of avatars online, and avatars are selected by the user to send a particular message or represent an important part of the self during interactions (Blascovich & Bailenson, 2011; Nowak, 2013; Pena et al., 2009). The avatar therefore provides information about the person it is representing, and is intended as part of the message being sent, which could therefore provide useful information about the other. It is possible that information gleaned from the avatar a person selected for an interaction could provide a more accurate picture of the person than using the corporeal body, and determining the accuracy of judgments made about others using visible characteristics of avatars is an important area of future research.

The use of visual characteristics of avatars in online perception could reflect a rational choice, an innate response to visual stimuli, or the reliance on what is familiar in the relatively unfamiliar online worlds. There is support for the notion that the use of visual characteristics makes evolutionary sense because it is innate and automatic because it requires almost no processing resources, which is the other meaning of mindless (Reeves & Nass, 1996). Although people process information using all indicators associated with a message or person, visual information is usually processed automatically and immediately, requiring almost no cognitive resources (Argyle, 1988; Bull & Rumsey, 1988; Burgoon, 1994; Patterson, 1995). Information Processing theory explains that the sequencing and automatic processing of visual information is hard-wired in most higher-order animals, meaning that viewer responses to visual stimuli, whether computer-mediated or not, transcend context. The visual information is used to determine the social potential of entities and then to generate personae and derive expectations that help them interpret current behavior and predict future behavior (Hamilton & Nowak, 2005, 2010; McGloin & Nowak, 2014; Nowak, Hamilton et al., 2009a; Ramerez et al., 2002). The prediction that visual information associated with a source or message will be processed first and influence perception and attribution is also consistent with most interpersonal and media effects models and leads to predictable and consistent reactions to certain visual stimuli (Scherer, 1984). This would explain why the impressions based on the physical appearance of others have a stronger influence on the perception process than impressions based on behavior (Kenny et al., 1992) and the same seems to be true online as well (Nowak & Rauh, 2008).

Another justification for using the visual characteristics of avatars in person perception is that it is perceived to provide information that helps reduce uncertainty. People seek information in ways they trust to help them predict and understand how people behave and what to expect from others. Users may be more uncertain when they navigate the relatively unfamiliar online environments and lack certainty about how to make judgments about others, or what type of information they should use, and people are not very confident about their ability to get to know others online. Uncertainty Reduction theory predicts that when people engage in communication in any situation, their primary goal is a reduction in uncertainty regarding the person or situation (Afifi & Burgoon, 2000; Berger & Calabrese, 1975), and in highly uncertain situations people will carefully weigh all information provided and rely more heavily on whatever is available when in unfamiliar situations or in low information situations (Clatterbuck, 1979; Infante, Rancer, & Womack, 1997). In the offline perception process, people have more trust in visual information, which is processed before other information and used in categorization (Srull & Wyer, 1989), to reduce uncertainty, and to form expectations and attributions of others (Afifi & Burgoon, 2000; Burgoon, 1994; Ekman & Friesen, 1969; Mehrabian, 1972). Indeed, the process of reducing uncertainty continues online and people adapt the process to utilize the information available in the mediated interaction to reduce uncertainty (Nowak & Rauh, 2008; Ramerez et al., 2002; Walther, 1996, 2011). People place greater weight on the communicative cues they have in the interaction to fulfill their goals (Walther & Parks, 2002), and evidence suggests that the avatar is relied upon in online person and message perception (Hamilton & Nowak, 2010; McGloin & Nowak, 2014; Nowak, Hamilton et al., 2009a). Perhaps using the visible characteristics of avatars to process people and information is part of how people are comfortable processing people, which feels familiar and may help reduce uncertainty (Walther, 1996; Walther & Parks, 2002). Thus, using the information provided by the avatar in the categorization process likely reduces uncertainty by allowing users to follow a process they are comfortable with and that has provided them with useful information.

On the other hand, using the avatar characteristics in person perception may not be logical or rational at all because of differences between the avatar and the offline body. The offline body is not chosen and is relatively stable. People do not choose their height, race, or biological sex offline and they cannot easily change them. While there may be some question as to whether people should be using height or gender to make attributions of credibility or likeability, there is no question that this happens and it is this stability and inability to consciously control the physical body that people say explains their reliance on nonverbal behaviors over verbal behaviors offline and their confidence in the accuracy of these attributions (Ambady, Hallahan, & Rosenthal, 1995; Bull & Rumsey, 1988; Burgoon, 1994; Ichheiser, 1970). Therefore, the argument against using the characteristics of the avatar in the perception process begins with the essential difference between the avatar and the corporeal body. The avatar is not a physical representation of the other but is instead a part of the interface that does not necessarily have any physical resemblance to the person it represents. People know that the avatar was selected by the other person and that it can change at any time. It is as easy as it to select an avatar intended to provide honest information about the self and enhance the intended message as it is to select an avatar to deceive, or mislead others. The visual characteristics of the avatar can provide both more and less reliable information about the other than the physical body offline, and further research is needed to examine how people use the avatars in this process and the extent to which avatars are a reliable and useful source of information.

Whether the use of the avatar features in perception is rational or not, research suggests that the avatar has a stronger effect on online person perception process than associated text or behavior of the person it represents (Nowak & Rauh, 2008). Whether people will continue to use visual characteristics of avatars in perception over time will likely be determined by the extent to which it provides them with useful information during interactions. If using the information provided by the avatar allows people to reduce uncertainty about others and fulfill their interaction goals, they will continue to use that information. If using the avatar does not provide useful information about the other, people will likely evolve other processes for perceiving others and perhaps other relevant and useful categories as long as the process is ruled by rational choice and not automatic, and further research on this question will be an important area of future study that will be of relevance to our understanding of offline perception as well (Blascovich & Bailenson, 2011; Blascovich et al., 2002; Schroeder, 2002). The next section examines the research on avatars showing the importance of anthropomorphism and realism on perceptions of social potential and how this is influencing person and message perception.

Avatars that have social potential can fall into the uncanny valley

People rely upon the visual characteristics of avatars to make attributions of the social potential of those encountered online, and entities that display social potential are believed to be best at satisfying many interaction goals including making people feel less alone, providing believable information, and responding appropriately to stimuli (Barret, 1997; Breazeal, 2003; DiSalvo & Gemperle, 2003; Hamilton & Nowak, 2010; Nass & Moon, 2000; Nowak, 2004; Nowak et al., 2009a; Reeves & Nass, 1996). Specifically, avatars perceived to be high on both visual realism and anthropomorphism are perceived to have higher social potential, likeability, homophily, and credibility, which are transferred to perceptions of the person or message associated with the avatar (Hamilton & Nowak, 2010; Koda & Maes, 1996; Nowak, 2004; Nowak & Rauh, 2008; Nowak, Hamilton et al., 2009a; Wexelblat, 1998). The presence of an avatar acts as a social cue that influences user's perception by inducing people to perceive the avatar as having more social potential (Breazeal, 2003), and this is good to a certain point. This section summarizes research showing how certain types of avatars are seen as more social than others (Nowak, 2004; Nowak, Hamilton et al., 2009a), explains the uncanny valley, and looks at factors being considered, as well as the judgments being used in making attributions of social potential.

The fact that agents and avatars can both appear anthropomorphic and display intelligent and autonomous behavior has not stopped people from using these as indicators of social potential and even humanity online. As with the human/not human attribution offline, the perception of anthropomorphism is among the first judgments made of an avatar, and it influences how other variables are perceived as well as the overall attribution of social potential (Bailenson et al., 2005; Garau et al., 2003; Hamilton & Nowak, 2010). Human, or anthropomorphic, features allow viewers to "connect" with the image at a basic biological level (Scherer, 1984) and avatars with human traits are assumed to have more social potential than those with less human traits regardless of ability or behavior (DiSalvo & Gemperle, 2003; Noske, 1989; Nowak, 2004; Nowak & Rauh, 2005; 2008; Shapiro, 1997; Turkle, 2011). Research has shown that avatar characteristics influence perceived sociability of others, at least in some contexts (Breazeal, 2003). Those with more anthropomorphic features are assumed to have higher social potential and increased credibility, whereas entities represented by less anthropomorphic images are likely to be perceived as less sentient and assumed to have less social potential (Hamilton & Nowak, 2010; Reeves & Nass, 1996; Yee et al., 2007). This may partially explain why designers of avatars and virtual humans see the creation of human-like, social, and emotional characters "as one of the ultimate goals of research in both virtual humans and social robots" (Dalibard, Magnenat-Thalman, & Thalman, 2012, p. 1). While it is true that anthropomorphism increases perceptions of social potential, there is not always agreement about which avatars are more anthropomorphic, and increasing perceptions of social potential is not always necessarily a good thing, nor are more anthropomorphic avatars necessarily better for all interactions and designs.

More anthropomorphic avatars look more like humans and are perceived to have higher social potential, and with the perception of social potential come high expectations of credibility, intelligence, and expectations of appropriate social responses (Bailenson et al., 2005; Garau et al., 2003; Nowak, 2004; Nowak et al., 2009a). These expectations are more difficult to meet, and unmet expectations results in disappointment and lower credibility and likeability ratings following interactions (Nowak, 2004; Slater & Steed, 2002). System designers should be cautious about applying anthropomorphic qualities to their designs, which can lead to user overexpectation and disappointment when the system fails to live up to these expectations (Nowak & Rauh, 2008; Shneiderman, 1988). Additionally, Mori (1970) argued that people are uncomfortable when representations of computer simulations get too realistic, or look or appear too human. Anthropomorphic images that are human and yet seem somehow seem abnormal or digitally manipulated fall within what has been called the "uncanny valley," and are perceived to be unnatural, eerie, or creepy (Bailenson et al., 2005; Dalibard et al., 2012; Garau et al., 2003; Hamilton & Nowak, 2010; Seyama & Nagayama, 2007), and small imperfections in highly realistic virtual humans and avatars can be repulsive (Dalibard et al., 2012). It seems that computer simulations that come too close to "real" human morphology, and telecommunication systems that are too close to replicating face to face, serve to reduce the willing suspension of disbelief and remind people that the experience is mediated. (Nowak, Watt, & Walther, 2005, 2009b). There is also the possibility that the closer an avatar comes to appearing realistic and human, the more attention people will pay to it, making any slight imperfections more obvious and notable. This has led some designers to recommend avatars that are "zoomorphic," or represent living or household animals as a way to reduce expectations and avoid the uncanny valley (Dalibard et al., 2012).

This makes it difficult to make recommendations about the ideal avatar for a particular interaction. One wants to enhance social potential enough to engage the user, but not enough to cross over into the uncanny valley where the avatar seems creepy. The perception of social potential is based on more than just anthropomorphism, and realism is influential in perceptions of social potential as well. Realism has been defined as the user's perception of the extent to which an image is typical, probable, or likely to exist in a non-mediated context (Busselle & Greenberg, 2000; Busselle et al., 2004; Garau et al., 2003; Hall, 2003; Pina, Cerezo, & Serón, 2000). Perceptual realism of digital stimuli can refer to the object being depicted, or the rendering quality of the image itself. Another dimension of realism is the placement of the avatar from cartoon like to photorealistic or fantasy images or things able to exist offline (Bailenson et al., 2006; Garau et al., 2003). Certain images, whether portrayed on a computer or television screen may be photorealistic but represent images that would not exist offline as they may portray something from a fantasy world (e.g., dragons, two-headed dogs, or flying purple elephants) meaning that a photorealistic image could lack realism, defined here as something probable, or likely to exist in a non-mediated context (Busselle, 2004; Lee, 2004b). In general, mediated depictions are perceived to have high levels of realism when they represent something typical and probable (frequently occurring), factual (having a specific correlate in the physical world), involving (eliciting identification), pervasive (creating a compelling and coherent illusion), or consistent with the environment in which it is placed (Busselle & Greenberg, 2000; Busselle, Ryabovolova, & Wilson, 2004; Garau et al., 2003; Hall, 2003; Pina, Cerezo, & Serón, 2000). The extent to which mediated stimuli are perceived as realistic influences viewers' processing of source information and judgment (Busselle, 2001), and viewers rate avatars with abnormal or obviously manipulated or exaggerated features as unpleasant (Seyama & Negayama, 2007). Any one of these dimensions of realism could interact with perceived anthropomorphism and influence perceived social potential. In terms of social potential, more realistic images are more likely to evoke schema and heuristics than images with less realism (Busselle, 2001) and the level of perceived realism will influence perceptions of plausibility of behavior (Maddock, Edge, & Sanchez, 2005). Having social potential, or being anthropomorphic, does not necessarily enhance perceptions of realism and avatars can seem very humanlike without being high on realism (Hamilton et al., 2010; Nowak et al., 2009a) but the more anthropomorphic the body design, the more important it becomes to animate it in a "believable way" (Dalibard et al., 2012, p. 3). Testing the interaction between the appearance of the avatar and the quality and type of motion and resolution of image is an important area for future research.

Some argue that the uncanny valley is triggered not only by digital stimuli or robots with high anthropomorphism, but by a mismatch between anthropomorphism and realism, or a failure to live up to the perceived social potential (Bailenson et al., 2005; Dalibard et al., 2012; Garau et al., 2003; Hamilton & Nowak, 2010; Seyama & Nagayama, 2007). It seems when anthropomorphic images increase perceptions of social potential, they also increase expectations for rendering quality, realism, motion, behavior, or ability that can lead to disappointment if these expectations are not met (Bailenson et al., 2005; Garau et al. 2003; McGloin & Nowak, 2014; Nowak & Rauh, 2008; Shneiderman, 1988; Slater & Steed, 2002). This means that the perception of avatar characteristics influences expectations, attributions, and perception of the behavior, and designers and users of systems should work to match expectations with system abilities and performance, though finding this ideal match is going to require careful user testing and experimentation and is an important area for future research.

To complicate this process further, it is not the characteristic of the avatar that influences these expectations and attributions, but the perception of the avatar's characteristics that determine the ideal characteristics of the avatar. There are no universal objective criteria for either anthropomorphism or realism as both are subjective judgments influenced by the interaction of personal experience, context, and features of the source that influence how viewers process and judge the source and information associated with it (Bailenson et al., 2006; Busselle, 2001; Garau et al., 2003; Seyama & Nagayama, 2007; Yee et al., 2007), just like attributions of social potential. This means the perception of visual characteristics, including anthropomorphism and realism differ significantly across individuals (Hamilton & Nowak, 2010; Nowak, Hamilton et al., 2009a), likely because people's varying personal experiences lead them to activate different schema or personae and that leads them to interpret the same stimuli in unique ways. Some people see an avatar as highly anthropomorphic and other people would see the same avatar as low or moderate and some would see it as high in realism but another person would disagree, and this perception is not even consistent for individuals as a person's mood or motivation may also influence it. Because perceptions of realism and anthropomorphism vary across individuals and avatars, researchers cannot conclude what percentages of their effect sizes are due to anthropomorphism or realism and what might be explained by other variables without measuring the variables of interest (Hamilton & Nowak, 2010; Nowak & Rauh, 2008; Nowak, Hamilton et al., 2009a). This also complicates the process of describing characteristics of avatars that are more anthropomorphic or more realistic or credible.

It is difficult to know whether a person thinks if it is anthropomorphic or realistic as not only does this likely vary across individuals and context but even within individuals as a person's mood or motivation in the interaction may also influence the attribution. The different perceptions may be influenced by computer experience, gender, bias, trait variables and personality differences, or a variety of other individual differences as well as goals and other trait variables. Uncovering the individual difference variables that best predict differing perceptions of avatars is an important area of future work.

This process is consistent with the essential premise of Expectancy Violations theory (Afifi & Burgoon, 2000; Burgoon & Walther, 1990; Burgoon & Hale, 1988), which predicts that it is the visible information people can attain from passive observation that generally determines attributions, assessments of credibility, and drives expectations. Essentially, expectations are driven by information processed first and confusion or uncertainty results when those expectations are not met. If applied to avatar perception, it could help predict when and why the uncanny valley will occur and how to prevent it (McGloin, Watt & Nowak, 2014), and researchers should continue to explore this possibility. In expectancy violations terms, the uncanny valley phenomenon occurs because the visual information sets up expectations for human behavior and social potential and people want all aspects of information to be consistent, and a mismatch seems eerie. Causal modeling research has shown that the perception of anthropomorphism happens first and drives the perception of, and expectations for, visual realism (Hamilton & Nowak, 2010; McGloin, Watt & Nowak, 2014; Nowak, Hamilton et al., 2009a). Thus, an avatar with high levels of anthropomorphism will be expected to have high levels of realism (Hamilton & Nowak, 2010) and people will be uneasy and more uncertain when these expectations are not met or when images are more anthropomorphic than realistic or the other way around (Seyama & Nagayama, 2007). Consistent with this explanation, the perceived inconsistency across these distinct constructs (low anthropomorphic avatar with high visual realism) has a stronger influence on perceived communication quality than other aspects of the avatar or associated image (Bailenson et al., 2005; Garau et al., 2003; Hamilton & Nowak, 2010).

Uncanny Valley and Expectancy Violations theory both describe the importance of consistency, but the explanations of the process would lead to different recommendations. Uncanny valley seems researchers suggest avoiding highly anthropomorphic avatars, particularly with highly realistic images (Mori, 1970), encouraging the use and design of "zoomorphic" instead of anthropomorphic avatars (Dalibard et al., 2012). This seems to assume a linear relationship between anthropomorphism and the uncanny valley with certain avatar characteristics always making people uncomfortable. Because Expectancy Violations theory would predict that the visual characteristics of the avatar processed first (anthropomorphism, realism) will influence the perception of information and characteristics processed later including categorizations made after anthropomorphism as well as social potential and perceived credibility of the associated text or message, it would instead recommend consistency and does not imply that the relationship is not linear. In this way, a highly anthropomorphic avatar perceived to have high social potential sets up high expectations, and expectations are not met when it is paired with an avatar low on realism or a less capable agent, and the inconsistency and not-met expectations would land the avatar in the uncanny valley. Thus, any type of image, anthropomorphic or not, could be successful as long as there is perceived consistency across visual characteristics (realism, anthropomorphism) and the system behavior and capabilities can live up to the expectations derived from the visual characteristics of the robot or characters used (McGloin, Watt & Nowak, 2014; Nowak, Hamilton et al., 2009a). Failure to meet expectations will result in negative attribution and disappointment or a negative outcome, though it is also possible that exceeding expectations will result in a negative outcome because it is also a violation of expectations (Afifi & Burgoon, 2000; McGloin, Watt & Nowak, 2014). Either way, this theory can provide a model for how researchers can enhance users' experiences with avatars and avoid the uncomfortable feeling sometimes resulting from interactions with certain types of avatars, and can at least partially predict or provide boundary information about when and why some avatars end up in the uncanny valley.

As people interact with others who are represented by avatars, they make an assessment of the avatar and seek to understand why that person selected it. As they are doing this, they may also consider when or if they would use it for self-presentation in future interactions. Thus, while self-presentation continues online and the process online has implications for how people identify and perceive themselves, there are some important differences because of the options and modalities at the disposal of the individual allowing them to control many aspects of the self-presentation process. The next section will summarize some research examining how people are selecting and using avatars for self-presentation.

Selecting an Avatar to Represent the Self to Others in Online Interactions

The process of perceiving another person's avatar seems to mirror the process of selecting an avatar for self-presentation, where people may be thinking about when, why, and if they would select an avatar to represent them as they go through the process of making attributions of others who have selected it (Hamilton & Nowak, 2010). People select avatars to represent themselves for the same reasons they select an outfit to wear or a hairstyle; because they think it will help them self-present some aspect of the identity, whether personality, hobby, or some physical attributes that are important to the sense of self and relevant to the relationship goals of the interaction (Birnbaum, 2008; Ducheneaut, Wen, Yee, & Wadley, 2009; Van Der Werf, 2007). Impression Management theory predicts that individuals manage self-presentation based on their goals, understanding of what is expected in a situation (Goffman, 1959). People self-present parts of their identity to others and carefully manage information they present in ways that they believe will give others impressions of them that will help fulfill their interaction or relationship goals (Aronson, Wilson, & Akert, 2010; Myers, 2008), though using online systems requires some important adaptations of the self-presentation strategies (Boyd, 2004; Chandler, 1998; Vazire & Gosling, 2004). Importantly, the image one selects and uses during an interaction can affect not only how others perceive him or her, but also the user's self-perception

(Lee,2003, 2004a; Pena et al., 2009) and future behavior (Blascovich & Bailenson, 2011; Kim & Sundar, 2012b; Fox & Bailensen, 2009b).

Not all avatars are the same and not all interfaces have the same options or possibilities for avatar usage, meaning the selection of avatar is limited by decisions made by system designers. Some systems assign an avatar to the user for the duration of the interaction while others allow users to choose an avatar to represent them, some allow them to change the avatar during the interaction and others do not. Some systems limit the selection to one or two avatars (e.g., most video games), others provide dozens or hundreds of choices, or even allow users to upload their own pictures or images (e.g., IM systems or Facebook), and still others allow users to modify the avatars provided by selecting skin tone, hair and eye color, height, attire, and a variety of other aspects (e.g., creating a Mii using the Wii system), while other systems allow users to create their own avatars using online software (like Second Life or Yahoo). Each of these options made by designers of systems will influence the user experience and perception of the system as well as people and messages encountered during the interaction.

On the user side, the selected avatar is determined by a variety of factors including the goals, personality, and objectives of the user, and people have reported a preference for avatars that accurately depicts either their physical characteristics or some aspect of their personalities (Nowak, 2013; Vasalou, Joinson, Bänziger, Goldie, & Pitt, 2008; Vasalou, Joinson, & Pitt, 2007). The choice of avatar has been shown to influence perceptions of others (Garau et al., 2003; Hamilton & Nowak, 2010; Nowak & Rauh, 2008), self-presentation, and identity (Nowak, 2013; Pena et al., 2009), as well as learning, memory, the sense of involvement with the interaction and the system itself, and even future behavior (Blascovich & Bailenson, 2011; Ducheneaut et al., 2009; Fox & Bailenson, 2009b; Nowak, Hamilton et al., 2009a; Nowak & Rauh, 2008; Vasalou et al., 2008). While users are more engaged when they are able to personalize and choose the avatar that represents them (Ducheneaut et al., 2009), and it may seem that this would lead to a recommendation that designers provide a wide variety of choices, this is not necessarily the case.

First, it is not always possible to provide a huge array of choices of avatar. Every possible avatar choice must be created by a designer and rendered by the computer, requiring higher networking speeds, longer download and rendering times, and more computer and programmer time (see Magnenat-Thalmann & Thalmann, 2005 for a discussion). This can mean that providing more choices of avatar slows down the system, increases the cost, and possibly makes it impossible to access using certain devices or from low-speed internet connections. Additionally, consumer behavior researchers generally agree that giving people choices is a good thing, but that choice overload (providing too many choices) makes people less likely to purchase products, take longer to choose, and be less satisfied with the choices they eventually make (Fasolo, McClelland, & Todd, 2007; Iyengar & Lepper, 2000). Assuming this applies to avatars, giving users too many options would require them to spend more time setting things up before they could use the system. Therefore, system designers must carefully select the types of avatars they allow users to select by considering their target population and ensure they provide appropriate choices that will help users feel connected while controlling for cost and keeping the system from overwhelming users.

People adapt their self-presentation strategies to meet the available features of the medium and vary based on the goals of the interaction (Back, Schmuckle, & Egloff, 2008; Birnbaum, 2008; Tamborini & Westerman, 2006), perceived audience for the message (Kassin, Fein, & Markus, 2008), and social norms of use of different systems. Offline, people use language, clothing choice, and behaviors (Schlenker, 1980), and online people use screen names, text, and avatars (Ellison, Heino, & Gibbs, 2006; Nowak & Gomes, 2013). There is much concern about people's ability to deceive others online because of the lack of physical information or people's lack of confidence in their ability to detect deception. While some users do lie to others, the fears that people will take advantage of features of computer-mediated communication deceive others more online than offline seem unwarranted. People report a desire to present honest representations of themselves even when online, likely because intimacy and relationship maintenance relies on acceptance of the authentic self, and most people seek to establish and maintain favorable but believable impressions while presenting information to help them reach their goals (Back et al., 2008; Bargh, McKenna, & Fitzsimons, 2002; Ellison et al., 2006). The perception of increased control over self-presentation online may allow for more selectivity and self-censorship and a wider variety of strategies for exchanging information about the self (Birnbaum, 2008; Trammell & Keshelashvili, 2005; Vazire & Gosling, 2004) but does not make the self represented online less authentic (Nowak, 2013). People may opt to present different parts of their identity in different contexts or with different people, but this does not mean they are misrepresenting the self and is true offline as well.

Also, it is difficult to deceive others online because in most instances, self-presentation and impression management do not happen exclusively online or offline. Instead, face-to-face and online communication frequently occur in tandem with the same people and individuals (Birnbaum, 2008; Dwyer, 2007; Mitrano, 2006; Nowak, 2013; Turkle, 1995, 2011). Thus, the information people present about themselves online is not necessarily different from what they present offline, but this does not mean that computer media have not altered the way people self-present. The unique parameters of online communication can mean that information is presented to a wide variety of people that is easily verifiable using other online sources. This could lead to more open and honest self-presentation than in offline environments, where verification using other sources is harder to do (Bargh et al., 2002; Ellison et al., 2006).

Online communicators can be freed from the associations of their traditional bodily appearance (Paulos & Canny, 1997), because they are consciously selected and easily changed. Some argue that this more controlled form of self-presentation can be used to foster higher levels of self-disclosure (Bailenson et al., 2006) and some users report a preference for avatars that are realistic, look like them, and accurately represent something about them either physically or psychologically (Nowak, 2013; Nowak & Rauh, 2008; Vasalou et al., 2007). Nowak and Rauh (2005, 2008) found that participants reported a preference for gendered (highly masculine or feminine) avatars, that users preferred to select avatars with the same gender, and that androgyny reduced both homophily and a person's likelihood to choose an avatar. Some users report that online settings allow them a place to reveal aspects of their true selves they are uncomfortable or unwilling to present face-to-face (Bargh et al., 2002; Ellison et al., 2006). While more research on the use of avatars for self presentation is needed, early reports suggest people have a desire to represent the authentic self at least in part.

Using Avatar Research to Understand Human Interactions

Understanding how people use and perceive avatars is both theoretically and practically important. People spend many hours of their days engaged in interactions mediated by phones and computers. The avatars and images they use to represent themselves during these interactions have vast implications for their interpersonal relationships and their sense of who they are. The research on avatars shows fairly conclusively that the type of avatar used in an interaction influences the people in the interaction, their relationships, and what is remembered. This then affects their offline relationships, which are generally with the same people, as very few relationships exist either exclusively online or offline (Birnbaum, 2008; Dwyer, 2007; Mitrano, 2006; Nowak & Gomes, 2013; Parks & Floyd, 1996; Turkle, 1995). Further, the avatar one selects to self-represent will not only influence how others perceive the person being represented, but will also influence a person's sense of self and identity. Research has shown that spending time as a short person, or as a black person, or even as a thin person can influence behavior even days after the online experience (Bailenson et al., 2005; Blascovich & Bailenson, 2011; Groom et al., 2009; Kim & Sundar, 2012b; Yee & Bailenson, 2009; Yee, Bailenson, & Ducheneaut, 2009). The questions raised above about why people use the avatar in the perception process are also important for researchers to continue examining. If the process is automatic, it is unlikely to change or be altered over time. However, if it is a rational choice, then people will likely to continue to use avatars in perception as long as the avatar provides useful information about the other. This necessarily raises the question of accuracy or whether the avatar will provide useful or accurate information about the other. It is even possible as mentioned above that the avatar that is selected for the interaction as an intentional part of the message will provide more useful or accurate information than the human body that was not selected and is not modifiable. The accuracy of perceiving others online using avatars as compared to offline will be an important area of future study.

Research online allows social scientist to examine questions that are impossible, unethical, or at the very least impractical to study offline. Researchers have been able to isolate certain behaviors, physical characteristics, or even disabilities online to study perceptions and attributions. Online research shows many communication processes are similar online and offline, though there are important differences resulting from the way information is presented to the senses. Even so, researchers can take advantage of affordances of these environments to hold certain characteristics of avatars constant and manipulate only the variables of interest, allowing the exploration of variables believed to influence the person perception process, the social interaction, or any other outcomes of practical or theoretical interest (Biocca & Nowak, 2002; Blascovich & Bailenson, 2011;

Fox, Arena, & Bailenson, 2009; Schroeder, 2002). Researchers have used the same voice and message content and modified only the characteristics of the avatar (Garau et al., 2003; Lee, 2004a; Nowak, 2004; Nowak, Hamilton et al., 2009a), and other researchers have kept most features of the avatar constant to control one variable including the height (Yee et al., 2009), the race (Groom et al., 2009), or even whether the avatar wore black or white clothing (Pena et al., 2009). These studies have allowed researchers to test results found in offline and face to face studies with more control over specific variables they would not be able to manipulate offline and the results generally replicate what was found offline (Blascovich & Bailenson, 2011; Reeves & Nass, 1996). This research can provide understanding about bias that would otherwise not be possible due to lack of control over other aspects of the interaction. It is also possible that future systems could help work to overcome prejudice as they have helped people train for new situations in simulations, overcome phobias, and address post traumatic stress disorder (PTSD), as well as many other clinical applications (Blascovich & Bailenson, 2011; Rizzo et al., 2012). In this way, avatars provide way to control the environments and visible features the user is encountering to help them in their offline worlds and lives while at the same time allowing researchers to investigate a variety of research questions about how people process objects, people, or animals that are not possible to isolate or study offline. They also allow researchers to control features of the environment, interface, or interaction, which can facilitate our ability to examine how communication and social processes work.

References

Afifi, W. A., & Burgoon, J. K. (2000). The Impact of violations on uncertainty and the consequences for attractiveness. *Human Communication Research, 26*(2), 203–233. doi:10.1111/j.1468-2958.2000.tb00756.x

Ambady, N., Hallahan, M., & Rosenthal, R. (1995). On judging and being judged accurately in zero-acquaintance situations. *Journal of Personality and Social Psychology, 69*(3), 518–529. doi:10.1037/0022-3514.69.3.518

Argyle, M. (1988). Clothes, physique, and other aspects of appearance. In M. Argyle (Ed.), *Bodily Communication* (2nd ed., pp. 233–254). New York, NY: International Universities Press.

Aronson, E., Wilson, T. D., & Akert, R. M. (2010). *Social psychology* (6th ed.). Upper Saddle River, NJ: Prentice Hall.

Asch, S. (1958). The metaphor: A psychological inquiry. In R. Tagiuri & L. Petrullo (Eds.), *Person perception and interpersonal behavior* (pp. 86–94). Stanford, CA: Stanford University Press.

Back, M. D., Schmuckle, S. C., & Egloff, B. (2008). How extroverted is hunnybunny77@hotmail.de? Inferring personality from email addresses. *Journal of Research in Personality, 42*, 1116–1122. doi:1110.1016/j.physletb.2003.1110.1071

Bailenson, J. N., Swinth, K. R., Hoyt, C. L., Persky, S., Dimov, A., & Blascovich, J. (2005). The independent and interactive effects of embodied agent appearance and behavior on self-report, cognitive, and behavioral markers of copresence in immersive virtual environments. *Presence: Teleoperators and Virtual Environments, 14*, 379–393. doi:10.1162/105474605774785235

Bailenson, J. N., Yee, N., Merget, D., & Schroeder, R. (2006). The effect of behavioral realism and form realism of real-time avatar faces on verbal disclosure, nonverbal disclosure, emotion recognition, and copresence in dyadic interaction. *Presence: Teleoperators and Virtual Environments, 15*, 359–372. doi:10.1162/pres.15.4.359

Balsamo, A. (1995). Forms of technological embodiment: reading the body in contemporary culture. In M. Featherstone & R. Burrows (Eds.), *Cyberspace/cyberbodies/cyberpunk: Cultures of technological embodiment* (pp. 215–237). Thousand Oaks, CA: Sage. doi:10.1111/1540-4560.00247

Bargh, J. A., McKenna, K. Y. A., & Fitzsimons, G. M. (2002). Can you see the real me? Activation and expression of the "true self" on the Internet. *Journal of Social Issues, 58*(1), 33–48.

Barret, J. L. (1997). *Anthropomorphism, intentional agents, and conceptualizing God* (Unpublished doctoral dissertation). Cornell University, Ithaca, NY.

Bente, G., Rüggenberg, S., Krämer, N. C., & Eschenburg, F. (2008). Avatar-mediated networking: Increasing social presence and interpersonal trust in net-based collaborations. *Human Communication Research, 34*(2), 287–318. doi:10.1111/j.1468-2958.2008.00322.x

Berger, C., & Calabrese, R. (1975). Some explorations in initial interaction and beyond: Toward a developmental theory of interpersonal communication. *Human Communication Research, 1*, 99–112. doi:10.1111/j.1468-2958.1975.tb00258.x

Biocca, F., & Nowak, K. L. (2002). Plugging your body into the telecommunication system: Mediated embodiment, media interfaces, and social virtual environments. In D. Atkin & C. Lin (Eds.), *Communication technology and society: Audience Adoption and uses* (pp. 407–447). Cresskill, NJ: Hampton Press.

Birnbaum, M. (2008). *Taking Goffman on a tour of Facebook: College students and the presentation of self in a mediated digital environment* (Doctoral dissertation). University of Arizona, Tucson, AZ.

Blascovich, J. (2002). Social influence within immersive virtual environments. In R. Schroeder (Ed.), *The social life of avatars: Presence and interaction in shared virtual environments* (pp. 127–145). London, UK: Springer-Verlag. doi:10.1007/978-1-4471-0277-9_8

Blascovich, J., & Bailenson, J. (2011). *Infinite reality: Avatars, eternal life, new worlds, and the dawn of the virtual revolution.* New York, NY: Harper Collins.

Blascovich, J., Loomis, J., Beall, A. C., Swinth, K. R., Hoyt, C. L., & Bailenson, J. N. (2002). Immersive virtual environment technology as a methodological tool for social psychology. *Psychological Inquiry, 13*(2), 103. doi:10.1207/S15327965PLI1302_03

Bolter, J. D. (1984). *Turing's man: Western culture in the computer age.* Chapel Hill, NC: The University of North Carolina Press.

Boyd, D. (2004). Friends, Friendster and MySpace top 8: Writing community into being on social networking sites. *First Monday, 11*(12). Retrieved from http://www.danah.org/papers/FriendsFriendsterTop8.pdf

Breazeal, C. (2003). Emotion and sociable humanoid robots. *International Journal of Human-Computer Studies, 59*(1–2), 119–155. doi:10.1016/S1071-5819(03)00018-1

Brent, E., & Thompson, G. A. (1999). Sociology: Modeling social interaction with autonomous agents. *Social Science Computer Review, 17*(3), 313–322. doi:10.1177/089443939901700306

Bull, P., & Rumsey, N. (1988). *The social psychology of facial appearance.* New York, NY: Springer-Verlag. doi:10.1007/978-1-4612-3782-2

Burgoon, J. K. (1994). Nonverbal signals. In M. L. Knapp & G. R. Miller (Eds.), *Handbook of interpersonal communication* (pp. 229–285). Thousand Oaks, CA: Sage.

Burgoon, J. K., & Hale, J. L. (1988). Nonverbal expectancy violations: Model elaboration and application to immediacy behaviors. *Communication Monographs, 55*(1), 58–79. doi:10.1080/03637758809376158

Burgoon, J. K., & Walther, J. (1990). Nonverbal expectancies and the evaluative consequences of violations. *Human Communication Research, 17*(2), 232–265. doi:10.1111/j.1468-2958.1990.tb00232.x

Busselle, R. (2001). Television exposure, perceived realism, and exemplar accessibility in the social judgment process. *Media Psychology, 3*(1), 43–67. doi:10.1207/S1532785XMEP0301_03

Busselle, R., & Greenberg, B. (2000). The nature of television realism judgments: A reevaluation of their conceptualization and measurement. *Mass Communication and Society, 3*(2–3), 249–268. doi:10.1207/S15327825MCS0323_05

Busselle, R., Ryabovolova, A., & Wilson, B. (2004). Ruining a good story: Cultivation, perceived realism and narrative. *Communications, 29*(3), 365–378.

Chandler, D. (1998). *Personal homepages and the construction of identities on the web.* Paper presented at the Aberystwyth Post-International Group Conference, Wales, UK.

Clark, N. (1995). Rear-view mirrorshades: The recursive generation of the cyberbody. *Cyberspace/cyberbodies/cyberpunk: Cultures of technological embodiment* (pp. 113–133). Thousand Oaks, CA: Sage. doi:10.4135/9781446250198

Clatterbuck, G. (1979). Attributional confidence and uncertainty. *Human Communication Research, 5*(2), 147–157. doi:10.1111/j.1468-2958.1979.tb00630.x

Dalibard, S., Magnenat-Thalmann, N., & Thalmann, D. (2012). *Anthropomorphism of artificial agents: A comparative survey of expressive design and motion of virtual characters and social robots.* Paper presented at the Workshop on Autonomous Social Robots and Virtual Humans at the 25th Annual Conference on Computer Animation and Social agents (CASA 2012), Singapore. Retrieved from http://hal.archives-ouvertes.fr/hal-00732763/

DiSalvo, C., & Gemperle, F. (2003, June). *From seduction to fulfillment: The use of anthropomorphic form in design.* Paper presented at the 2003 International Conference on Designing Pleasurable Products and Interfaces, Pittsburgh, PA.

Dryer, D. C. (1999). Getting personal with computers: How to design personalities for agents. *Applied Artificial Intelligence, 13*(2), 273–295. doi:10.1080/088395199117423

Ducheneaut, N., Wen, M., Yee, N., & Wadley, G. (2009). *Body and mind: A study of avatar personalization in three virtual worlds.* Paper presented at the Computer Human Interactions (CHI) Conference, Boston, MA.

Dwyer, C. (2007). *Digital relationships in the 'MySpace' generation: Results from a qualitative study.* Paper presented at the 40th Hawaii International Conference on System Sciences (HICSS), Big Island, HI. doi:http://doi.ieeecomputersociety.org/10.1109/HICSS.2007.176

Ekman, P., & Friesen, W. (1969). The repertoire of nonverbal behavior: Categories, origins, usage, and coding. *Semiotica, 1*, 49–98.

Ellison, N., Heino, R., & Gibbs, J. (2006). Managing impressions online: Self-presentation processes in the online dating environment. *Journal of Computer-Mediated Communication, 11*(2), 415–441. doi:10.1111/j.1083-6101.2006.00020.x

Fasolo, B., McClelland, G. H., & Todd, P. M. (2007). Escaping the tyranny of choice: When fewer attributes make choice easier. *Marketing Theory, 7*, 13–26, doi:10.1177/1470593107073842

Fiske, S., & Neuberg, S. (1990). A continuum of impression formation, from category-based to individuating processes: Influences of information and motivation on attention and interpretation. In M. P. Zanna (Ed.), *Advances in experimental social psychology* (Vol. 23, pp. 1–74). Ontario, Canada: Academic Press. doi:10.1016/S0065-2601(08)60317-2

Flanagin, A., & Metzger, M. (2007). The role of site features, user attributes, and information verification behaviors on the perceived credibility of web-based information. *New Media and Society, 9*(2), 319–342. doi:10.1177/1461444807075015

Fox, J., Arena, D., & Bailenson, J. N. (2009). Virtual reality: A survival guide for the social scientist. *Journal of Media Psychology, 21*(3), 95–113. doi:10.1027/1864-1105.21.3.95

Fox, J., & Bailenson, J. N. (2009a). Virtual virgins and vamps: The effects of exposure to female characters' sexualized appearance and gaze in an immersive virtual environment. *Sex Roles, 61*(3–4), 147–157. doi:10.1007/s11199-009-9599-3

Fox, J., & Bailenson, J. N. (2009b). Virtual self-modeling: The effects of vicarious reinforcement and identification on exercise behaviors. *Media Psychology, 12*, 1–25.

Garau, M., Slater, M., Vinayagamoorthy, V., Brogny, A., Steed, A., & Sasse, M. A. (2003). *The impact of avatar realism and eye gaze control on perceived quality of communication in a shared immersive virtual environment.* Paper presented at the Proceedings of CHI'03—The Conference on Human Factors in Computing Systems. Fort Lauderdale, FL.

Goffman, E. (1959). *The presentation of self in everyday life.* Garden City, NY: Doubleday.

Goodwin, J. (1998). The meaning of life: Real and/or artificial. In T. Kunii & A. Luciani (Eds.), *Cyberworlds* (pp. 43–65). Tokyo, Japan: Springer-Verlag. doi:10.1007/978-4-431-67941-7

Graber, D. (1996). Say it with pictures. *Annals of the American Academy of Political and Social Science, 546*(1), 85–96. doi:10.1177/0002716296546001008

Groom, V., Bailenson, J. N., & Nass, C. (2009). The influence of racial embodiment on racial bias in immersive virtual environments. *Social Influence, 4*(3), 231–248. doi:10.1080/15534510802643750

Guadagno, R. E., Blascovich, J., Bailenson, J. N., & McCall, C. (2007). Virtual humans and persuasion: The effects of agency and behavioral realism. *Media Psychology, 10*(1), 1–22. doi:10.1080/15213260701300865

Hall, A. (2003). Reading realism: Audiences' evaluations of the reality of media texts. *Journal of Communication, 53*(4), 624–641. doi:10.1111/j.1460-2466.2003.tb02914.x

Hamilton, D., & Sherman, J. (1994). Stereotypes. In J. Robert, S. Wyer, & T. K. Srull (Eds.), *Handbook of social cognition* (Vol. 2, pp. 3–58). Hillsdale, NJ: Lawrence Erlbaum Associates.

Hamilton, M. A., & Nowak, K. L. (2005). Information systems concepts across two decades: An empirical analysis of trends in theory, methods, process, and research domains. *Journal of Communication, 55*(3), 529–553. doi:10.1111/j.1460-2466.2005.tb02684.x

Hamilton, M. A., & Nowak, K. L. (2010). Advancing a model of avatar evaluation and selection. *PsychNology Journal, 8*(1), 33–65. Retrieved from www.psychnology.org

Heider, F. (1958). Perceiving the other person. In R. Tagiuri & L. Petrullo (Eds.), *Person perception and interpersonal behavior* (pp. 22–26). Stanford, CA: Stanford University Press.

Huang, J. H., & Chen, Y. F. (2006). Herding in online product choice. *Psychology & Marketing, 23*(5), 413–428. doi:10.1002/mar.20119

Ichheiser, G. (1970). *Appearances and realities: Misunderstanding in human relations.* San Francisco, CA: Jossey-Bass.

Infante, D., Rancer, A., & Womack, D. (1997). *Building communication theory* (3rd ed.). Prospect Heights, IL: Waveland Press.

Iyengar, S. S., & Lepper, M. R. (2000). When choice is demotivating: Can one desire too much of a good thing? *Journal Of Personality And Social Psychology, 79*(6), 995–1006. doi:10.1037/0022-3514.79.6.995

Kassin, S. M., Fein, S., & Markus, H. R. (2008). *Social psychology* (7th ed.). Boston, MA: Houghton Mifflin.

Kennedy, J.S. (1992) *The New Anthropomorphism.* Cambridge University Press. New York.

Kenny, D. (1991). A general model of consensus and accuracy in interpersonal perception. *Psychological Review, 98*(2), 155–163. doi:10.1037/0033-295X.98.2.155

Kenny, D. A., Horner, C., Kashy, D. A., & Chu, L. (1992). Consensus at zero acquaintance: Replication, behavioral cues, and stability. *Journal of Personality and Social Psychology, 62*, 88–97.

Kim, Y., & Sundar, S. S. (2012a). Anthropomorphism of computers: Is it mindful or mindless? *Computers in Human Behavior, 28*, 241–250.

Kim, Y., & Sundar, S. S. (2012b). Visualizing ideal self vs. actual self through avatars: Impact on preventive health outcomes. *Computers in Human Behavior, 28*, 1356–1364.

Koda, T., & Maes, P. (1996). *Agents with faces: The effects of personification.* Paper presented at the Human-Computer Interaction'96, London.

Lakoff, G. (1987). *Women, fire, and dangerous things: What categories reveal about the mind.* Chicago, IL/London, UK: The University of Chicago Press.

Lang, A. (1995). Defining audio/video redundancy from a limited- capacity information processing perspective. *Communication Research, 22*(1), 86–115. doi:10.1177/009365095022001004

Laurel, B. (1990). Interface agents: Metaphors with character. In B. Laurel (Ed.), *The art of human-computer interface design* (pp. 235–366). Reading, MA: Addison-Wesley.

Lee, E.-J. (2004a). Effects of gendered character representation on person perception and informational social influence in computer-mediated communication. *Computers in Human Behavior, 20*(6), 779. doi:10.1016/j.chb.2003.11.005

Lee, J. R. (2003). *Consuming virtual identity: Exploring the motivation based of participation in online avatar marketing.* Paper presented at the International Communication Association conference, San Diego, CA. doi:ica_proceeding_12207.PDF

Lee, K. M. (2004b). Presence, explicated. *Communication Theory, 14*(1), 27–50. doi:10.1111/j.1468-2885.2004.tb00302.x

Lipton, M. (1996). Forgetting the body: Cybersex and identity. In L. Strate, R. Jacobson, & S. Gibson (Eds.), *Communication and cyberspace: Social interaction in an electronic environment* (pp. 335–350). Cresskill, NJ: Hampton Press.

Maddock, S., Edge J., Sanchez M. (2005), "Movement realism in computer facial animation", Workshop on Human-animated Characters Interaction, 6th September, 2005, Napier University, Edinburgh, UK.

Magnenat-Thalmann, N., & Thalmann, D. (2005). Virtual humans: Thirty years of research, what next? *Visual Computing, 21*, 997–1015. doi:10.1007/s00371-005-0363-6

McGloin, R., Nowak, K.L., & Watt, J. (2014). Avatars and Expectations: Influencing Perceptions of Trustworthiness in an Online Consumer Setting. *PsychNology.* 1(1–2), 7–28.

McLuhan, M. (1964). *Understanding media.* New York, NY: McGraw Hill.

Mehrabian, A. (1972). *Nonverbal communication.* Chicago, IL: Aldine Atherton.

Mitrano, T. (2006). A wider world: Youth, privacy, and social networking technologies. *Educause Review, 41*(6), 16–29. Retrieved from http://www.educause.edu/ero/article/wider-world-youth-privacy-and-social-networking-technologies

Mori, M. (1970). The uncanny valley. *Energy, 7*(4), 33035.

Myers, D. G. (2008). *Exploring social psychology* (Vol. 5). Boston, MA: McGraw-Hill.

Nass, C., Lombard, M., Henriksen, L., & Steuer, J. (1995). Anthropocentrism and computers. *Behavior and Information Technology, 14*(4), 229–238. doi:10.1080/01449299508914636

Nass, C., & Moon, Y. (2000). Machines and mindlessness: Social responses to computers. *Journal of Social Issues, 56*(1), 81. doi:10.1111/0022-4537.00153

Noske, B. (1989). *Humans and other animals: Beyond the boundaries of anthropology.* London, UK: Pluto Press.

Nowak, K. L. (2003). Sex categorization in computer mediated communication (CMC): Exploring the utopian promise. *Media Psychology, 5*(1), 83–103. doi:10.1207/S1532785XMEP0501_4

Nowak, K. L. (2004). The influence of anthropomorphism and agency on social judgment in virtual environments. *Journal of Computer Mediated Communication, 9*(2), Retrieved from http://jcmc.indiana.edu/vol9/issue2/nowak.html. doi:10.1111/j.1083-6101.2004.tb00284.x

Nowak, K. L. (2013). Choosing buddy icons that look like me or represent my personality: Using buddy icons for social presence. *Computers in Human Behavior, 29*, 1456–1464. Retrieved from http://dx.doi.org/10.1016/j.chb.2013.01.027

Nowak, K. L., & Biocca, F. (2003). The effect of agency and anthropomorphism on users' sense of telepresence, social presence and copresence in virtual environments. *Presence: Teleoperators and Virtual Environments, 12*(5), 481–494. MIT Press. doi:10.1162/105474603322761289

Nowak, K.L., Fox, J. & Ranjit, Y. (2014-May). Examining the Role of Sexism in Perceptions of Female Virtual Representations. Paper presented at the International Communication Association Conference. Seattle, Washington.

Nowak, K. L., & Gomes, S. (2013). *The choices people make: The types of buddy icons people select for self presentation online. AI & Society*. doi:10.1007/s00146-013-0501-z

Nowak, K. L., Hamilton, M. A., & Hammond, C. C. (2009a). The effect of image features on judgments of homophily, credibility, and intention to use as avatars in future interactions. *Media Psychology, 12*, 50–76. doi:10.1080/15213260802669433

Nowak, K. L., & Rauh, C. (2005). The influence of the avatar on online perceptions of anthropomorphism, androgyny, credibility, homophily, and attraction. *Journal of Computer-Mediated Communication, 11*(1), 153–178.

Nowak, K. L., & Rauh, C. (2008). Choose your "buddy icon" carefully: The influence of avatar androgyny, anthropomorphism and credibility in online interactions. *Computers in Human Behavior, 24*(4), 1473–1493. doi:10.1016/j.chb.2007.05.005

Nowak, K. L., Watt, J., & Walther, J. (2005). The influence of synchrony and sensory modality on the person perception process in computer-mediated groups. *Journal of Computer Mediated Communication, 10*(3), Retrieved from http://onlinelibrary.wiley.com/enhanced/doi/10.1111/j.1083-6101.2005.tb00251.x/

Nowak, K. L., Watt, J., & Walther, J. (2009b). Computer mediated teamwork and the efficiency framework: Exploring the influence of synchrony and cues on media satisfaction and outcome success. *Computers in Human Behavior, 25*(5), 1108–1119. doi:10.1016/j.chb.2009.05.006

Parks, M., & Floyd, K. (1996). Making friends in cyberspace. *Journal of Communication, 46*(1), 80–97. doi:10.1111/j.1460-2466.1996.tb01462.x

Patterson, M. (1995). Invited article: A parallel process model of nonverbal communication. *Journal of Nonverbal Behavior, 19*(1), 3–29. doi:10.1007/BF02173410

Paulos, E., & Canny, J. (1997). Ubiquitous tele-embodiment: Applications and implications. *International Journal of Human-Computer Studies, 46*, 861–877. doi:10.1006/ijhc.1996.0120

Pena, J., Hancock, J., & Merola, N. (2009). The priming effects of avatars in virtual settings. *Communication Research, 36*(6), 838–856. doi:810.1177/0093650209346802

Pina, A., Cerezo, E., & Serón, F. J. (2000). Computer animation: From avatars to unrestricted autonomous actors; A survey on replication and modeling mechanisms. *Computers & Graphics, 24*(1), 297–311. doi:10.1016/S0097-8493(99)00165-X

Ramerez, A., Walther, J. B., Burgoon, J. K., & Sunnafrank, M. (2002). Information-seeking strategies, uncertainty, and computer mediated communication toward a conceptual model. *Human Communication Research, 28*(2), 213–228. doi:10.1111/j.1468-2958.2002.tb00804.x

Reeves, B., & Nass, C. (1996). *The media equation: How people treat computers, television, and new media like real people and places.* Stanford, CA: CSLI Publications.

Rizzo, A., Parsons, T. D., Kenny, P., & Buckwalter, J. G. (2012). Using virtual reality for clinical assessment and intervention. In L. L'Abate & D. A. Kaiser (Eds.), *Handbook of technology in psychology, psychiatry, and neurology: Theory, research, and practice, (pp. 277–318).* New York, NY: Nova Science.

Scherer, K. R. (1984). On the nature and function of emotion: A component process approach. In K. R. Scberer & P. Ekman (Eds.), *Approaches to emotion* (pp. 293–318). Hillsdale, NJ: Erlbaum.

Schlenker, B. (1980). *Impression management: The self-concept, social identity, and interpersonal relations.* Monterey, CA: Brooks/Cole.

Schroeder, R. (2002). Social interaction in virtual environments: Key issues, common themes, and a framework for research. In R. Schroeder (Ed.), *The social life of avatars: Presence and interaction in shared virtual environments.* London, UK: Springer-Verlag. doi:10.1007/978-1-4471-0277-9_1

Seyama, J., & Nagayama, R. S. (2007). The uncanny valley: Effect of realism on the impression of artificial human faces. *Presence, 16*(4), 337–351. doi:10.1162/pres.16.4.337

Shapiro, K. J. (1997). A phenomenological approach to the study of nonhuman animals. In R. W. Mitchell, N. S. Thompson, & H. L. Miles (Eds.), *Anthropomorphism, anecdotes, and animals* (pp. 92–101). New York, NY: SUNY.

Sheehan, J. (1991). Introduction: Humans and animals. In J. Sheehan & M. Sosna (Eds.), *The boundaries of humanity: Humans, animals, machines* (pp. 27–35). Oxford, England: University of California Press.

Shneiderman, B. (1988). A nonanthropomorphic style guide: Overcoming the humpty-dumpty syndrome. *The Computing Teacher, 8,* 9–10.

Skitka, L., & Maslach, C. (1996). Gender as schematic category: A role construct approach. *Social Behavior and Personality, 24,* 53–74. doi:10.2224/sbp.1996.24.1.53

Slater, M., & Steed, A. (2002). Meeting people virtually: Experiments in shared virtual environments. In R. Schroeder (Ed.), *The social life of avatars: Presence and interaction in shared virtual environments* (pp. 146–171). London, UK: Springer-Verlag. doi:10.1007/978-1-4471-0277-9_9

Smith, D., Menon, S., & Sivakumar, K. (2005). Online peer and editorial recommendations, trust and choice in virtual markets. *Journal of Interactive Marketing, 19*(3), 15–37. doi:10.1002/dir.20041

Spender, D. (1996). *Nattering on the net: Women, power and cyberspace.* Melbourne, Australia: Spinifex Press.

Srull, T. K., & Wyer, R. S. (1989). Person memory and judgment. *Psychological Review, 96,* 58–83.

Stephenson, N. (1993). *Snow crash.* New York, NY: Bantam Books.

Tamborini, R., & Westerman, D. (2006). *The effect of avatars on uncertainty reduction.* Paper presented at the International Communication Association conference, Dresden, Germany.

Tamir, P., & Zohar, A. (1991). Anthropomorphism and teleology in reasoning about biological phenomena. *Science Education, 75*(1), 57–67.

Trammell, K. D., & Keshelashvili, A. (2005). Examining the new influencers: A self-presentation study of a-list blogs. *Journalism and Mass Communication Quarterly, 82*(4), 968–982. doi:10.1177/107769900508200413

Turkle, S. (1995). *Life on the screen: Identity in the age of the Internet.* New York, NY: Simon & Schuster.

Turkle, S. (2011). *Alone together: Why we expect more from technology and less from each other.* New York, NY: Basic Books.

Van Der Werf, M. (2007). Beware of using social-networking sites to monitor students, lawyers say. *The Chronicle of Higher Education.* Retrieved from http://chronicle.com/article/Beware-of-Using/28063/

Vasalou, A., Joinson, A., Bänziger, T., Goldie, P., & Pitt, J. (2008). Avatars in social media: Balancing accuracy, playfulness and embodied messages. *International Journal of Human-Computer Studies, 66*(11), 801. doi:10.1016/j.ijhcs.2008.08.002

Vasalou, A., Joinson, A. N., & Pitt, J. (2007). *Constructing my online self: Avatars that increase self-focused attention.* Paper presented at the Computer Human Interaction (CHI) Conference, San Jose, CA.

Vazire, S., & Gosling, S. D. (2004). e-Perceptions: Personality impressions based on personal websites. *Journal of Personality and Social Psychology, 87*(1), 123–132. doi:110.1037/0022-3514.1087.1031.1123

Vilhjalmsson, H. (1996). *Infinite reality: Avatars, eternal life, new worlds, and the dawn of the virtual revolution.* New York, NY: Harper Collins.

Walther, J. B. (1992). Interpersonal effects in computer-mediated interaction: A relational perspective. *Communication Research, 19,* 52–90.

Walther, J. B. (1996). Computer-mediated communication: Impersonal, interpersonal, and hyperpersonal interaction. *Communication Research, 23*(1), 3–43. doi:10.1177/009365096023001001

Walther, J. B. (2007). Selective self-presentation in computer-mediated communication: Hyperpersonal dimensions of technology, language, and cognition. *Computers in Human Behavior, 23*(5), 2538–2557. doi:10.1016/j.chb.2006.05.002

Walther, J. B. (2011). Theories of computer-mediated communication and interpersonal relations. In M. L. Knapp & J. A. Daly (Eds.), *The handbook of interpersonal communication* (4th ed., pp. 443–479). Thousand Oaks, CA: Sage.

Walther, J. B., & Parks, M. R. (2002). Cues filtered out, cues filtered in: Computer-mediated communication and relationships. In M. L. Knapp & J. A. Daly (Eds.), *Handbook of interpersonal communication* (3rd ed., pp. 529–563). Thousand Oaks, CA: Sage.

Wexelblat, A. (1998, November). *Don't make that face: A report on anthropomorphizing an interface.* AAAI Technical Report SS-98-02. Retrieved from http://aaaipress.org/Papers/Symposia/Spring/1998/SS-98.../SS98-02-028.pdf

Wogalter, M., & Mayhorn, C. (2008). Trusting the internet: Cues affecting perceived credibility. *International Journal of Technology and Human Interaction, 4*(1), 75–93. doi:10.4018/jthi.2008010105

Yee, N., & Bailenson, J. N. (2009). The difference between being and seeing: The relative contribution of self-perception and priming to behavioral changes via digital self-representation. *Media Psychology, 12*(2), 195–209. doi:10.1080/15213260902849943

Yee, N., Bailenson, J. N., & Ducheneaut, N. (2009). The proteus effect: Implications of transformed digital self-representation on online and offline behavior. *Communication Research, 36*(2), 285–312. doi:10.1177/0093650208330254

Yee, N., Bailenson, J. N., & Rickertsen, K. (2007). *A meta-analysis of the impact of the inclusion and realism of human-like faces on user experiences in interfaces.* Paper presented at the Paper presented at CHI 2007 April 28-May 3, San Jose, California, USA.

5

Effects of Visual Cues on Social Perceptions and Self-Categorization in Computer-Mediated Communication

Eun-Ju Lee[1] and Soo Youn Oh[2]

[1] Department of Communication, Seoul National University, Gwanak-Gu, Seoul, Republic of Korea
[2] Department of Communication, Stanford University, Stanford, CA, USA

Generally defined as artificial, computer-generated graphical representations of users in a virtual environment, avatars refer to a virtual stand-in for users on the Internet, most commonly seen in computer games and online chat rooms (Bente, Rüggenberg, Krämer, & Eschenburg, 2008; Galanxhi & Nah, 2007). Albeit not limited to avatars, the malleability of virtual identity has led researchers to address questions such as what motivates people to exploit various textual and visual cues to create their online self-representation, how they perform their virtual identity, and the like. For example, online role-playing game players created their characters to be closer to their ideal than actual self for certain aspects of their personality, especially when they experienced lower levels of psychological well-being (Bessière, Seay, & Kiesler, 2007), and people actively selected avatars that reflect their body ideal when designing video game characters (Dunn & Guadagno, 2012). Gender-swapping is also commonly practiced among online game players, with half of the respondents reporting cross-dressing for a range of purposes, including better treatment, identity exploration, and enjoyment (Hussain & Griffiths, 2008).

While these studies center on users' volitional and motivated choice of their online self-representation, highlighting the unprecedented capacity of the virtual world that enables people to live out their ideal self free from their history and physical conditions, other researchers have examined how the arbitrary assignment of avatars affects social dynamics in computer-mediated communication (CMC). Directly germane to the notion that the paucity of social cues lowers social presence and renders the interaction impersonal in CMC (i.e., cues-filtered-out perspective; Walther & Parks, 2002), researchers have investigated if the presence (vs. absence) or human-likeness of avatars adds interpersonal flair to otherwise impersonal interactions, inducing emotional closeness and more positive impressions about interaction partners (e.g., Bente et al., 2008;

The Handbook of the Psychology of Communication Technology, First Edition. Edited by S. Shyam Sundar.

Nowak, 2004). This avatar-as-human-proxy approach has also been adopted in the context of web-based marketing as a way of adding human touch, with some studies reporting positive effects of anthropomorphic agents on consumer experiences (e.g., Holzwarth, Janiszewski, & Neumann, 2006, Study 1), while others documenting no significant effects of humanoid service representatives on cognitive and emotional trust (e.g., Qiu & Benbasat, 2005). Similarly, students favored a more human-like virtual teacher to a smiley-shaped one, although they preferred a less realistic agent to a humanoid avatar for their learning companion (Girard & Johnson, 2010), and medical practitioners collaborated more effectively in a virtual environment when they were supported by anthropomorphic avatars than when the collaborators were represented by a colored pointer (Tapie, Terrier, Perron, & Cellier, 2006).

At the same time, another line of research has investigated how specific visual attributes of anthropomorphic avatars alter the way people interact with their anonymous CMC partners. Among these studies, two research programs seem to stand out for their programmatic nature, which evaluated how the physical properties of arbitrary on-screen characters affect social stereotyping (Lee, 2004a, 2005, 2007a, b, c; Palomares & Lee, 2010) and activate different self-categorization processes (personal vs. group identity) (Lee, 2004b, 2007d; Lee & Nass, 2002) in the cue-limited environment. Therefore, we first review a series of research on (a) how avatar gender serves as a basis of gender stereotyping in dyadic CMC and (b) how visual similarity (vs. dissimilarity) among avatars facilitates identification with and conformity to the group, and then identify several factors that seem to moderate such avatar effects. After discussing potential explanations for why arbitrary visual cues shape the way people make sense of and interact with mediated others, we conclude this chapter with some agendas for future research.

Avatars and Social Stereotypes

Although it is debatable how automatic the activation of stereotypes is (Macrae & Bodenhausen, 2001), the notion that stereotypes serve as a useful cognitive shortcut people rely on to deal with the complexities of everyday life is widely accepted. Upon encountering a stranger, for example, it can be a fairly effective uncertainty reduction strategy to categorize the person into a social group (e.g., man, Caucasian, British) and ascribe specific qualities (stereo)typically associated with the group to that person. In this sense, social stereotypes can be seen as a strategic tool with which people expedite the person perception process, rather than a passive attempt to cope with information overload by filtering out much of the available information (Bargh, 1999).

Given that social stereotypes are often based on visually discernible category cues, such as gender and race (Brewer, 1988), CMC offers an interesting and unique venue to study how social stereotypes operate in our everyday functioning. On the one hand, the equalization hypothesis (Dubrovsky, Kiesler, & Sethna, 1991; Kiesler, Siegel, & McGuire, 1984) suggests that physical isolation and visual anonymity in CMC can substantially obscure, if not completely mask, social category information, which often serves as a status cue. Consequently, people feel more liberated from power differentials than they would in face-to-face interaction, and thus, interact more freely and equally. For example, although men were more likely to give suggestions and opinions than women in face-to-face group discussions, there was no corresponding gender difference with computer-mediated brainstorming (Herschel, Cooper, Smith, & Arrington, 1994). Likewise, while women tended to refrain from making the first decision proposals in face-to-face meetings, such a tendency was not found in electronic conversations (Sproull & Kiesler, 1991).

On the other hand, the social identity model of deindividuation effects (SIDE) predicts that the paucity of visual cues in text-based CMC can have the opposite effects, amplifying the influence of category membership (Postmes, Spears, & Lea, 1998, 2000). In this view, with little personal information about the mediated interactant, the remaining cues may become even more influential in subsequent social judgments, leading to overattribution. Thus, insofar as the individual's social identity (e.g., man vs. woman, in-group vs. out-group) is known, people may exhibit even more stereotyped perceptions of their CMC partners.

Although these two predictions are derived from different assumptions about the availability of social category cues in CMC, they both raise an important question of how social stereotypes operate in CMC. To address this question, Lee (2004a, 2005, 2007a, b, c) conducted a series of studies investigating how the gender of an avatar affects individuals' perceptions of and behavioral reactions to an anonymous CMC partner. Gender was chosen as the focal social category because it is one of the most salient visual cues social perceivers use to categorize others into groups (Brewer, 1988) and yet commonly unavailable in text-based CMC. What is more, manipulating the avatar gender made it possible to create a context in which gender cue was available (in the form of unambiguously gender-marked cartoon characters), while the interactant's gender was still unknown (in that there was no inherent connection between avatar gender and the person's physical identity).

Study method

At the beginning of the experiment, participants, who were recruited from various classes at a large US university, were told that they would play a trivia game with another student attending a different university as a team, although the ostensible partner's responses were pre-programmed. To reduce their suspicion about the purpose of the study, the participants were asked to choose an alphabet letter on the computer screen to determine the avatar that would represent them during the interaction, not knowing which character corresponded to each letter. Unbeknownst to the participants, their avatar always mismatched their gender so that they were constantly reminded of the arbitrary nature of the character assignment. For each multiple-choice question, participants first picked their own answer and then saw their partner's answer. For the majority of questions, the partner gave answers that were different from the participants' initial answer, and the participants were prompted to submit their final answer. To enhance the participants' involvement in the study and to discourage blind rejection of the partner's answers, a cash prize was offered for the team with the highest score, which they believed would be determined by the total number of correct final answers. Conformity was measured by how often the participants switched to their partner's answer for their final answer.

Key findings

In the first study in this series (Lee, 2004a), participants were asked questions about either sports or fashion. Although both men and women inferred their partner's gender from the randomly assigned avatar, the avatar effects on conformity were moderated by the topic and the participant's gender. Specifically, when the topic matched the participant's own gender, the partner's avatar had no significant effect on conformity, presumably due to higher self-confidence in the topic their gender group can claim the ownership of. In addition, whereas women exhibited greater conformity to the male- than female-charactered partners on a masculine topic, men conformed more to the male- than female-charactered partners even on a feminine topic, suggesting male resistance to female influence.

Although such findings suggest that even arbitrary visual cues can trigger categorical person perception in CMC, the avatars were the only information associated with the partner. Therefore, for a more conservative test, avatar gender was pitted against another gender cue, gender-typed knowledge bias (Lee, 2007a), which was manipulated by having the ostensible partner expressing high levels of self-confidence in either a masculine (sports) or feminine (fashion) topic. As expected, the partner's high self-confidence in the gendered topic led to stereotype-congruent gender inferences, such that the partner who claimed to know the answers to sports/fashion questions was more likely to be taken as a man/woman. Avatar gender, however, significantly influenced gender inferences over and above gender-linked knowledge only for women, suggesting individual differences in the proclivity to utilize arbitrary visual cues in person perception.

Another factor that might have exaggerated the avatar effects is the use of gender-typed topics, which presumably heightens gender consciousness. To assess the robustness of the avatar effects, Lee (2005) used gender-neutral questions (i.e., fast food trivia). When the participants' own avatar *mismatched* their gender (Study 1 and Study 2), they still assigned their unknown partner's gender based on the on-screen character, but the partners' avatar had no significant effect on conformity. Instead, the participants' conformity decision was significantly affected by the partner's self-confidence, albeit in different ways for men and women. By contrast, when the participants' own avatar *matched* their gender (Study 3), participants were more likely to accept the answer of the partner who was represented by a male than a female avatar. Since the participants chose their avatar without seeing the available options, they knew that their partners had no control over their avatars. Even so, when their own avatar happened to match their gender, they failed to discount for the potential disjunction between the avatar and their partner, even with the gender-neutral task.

Extending this line of research, Palomares and Lee (2010) tested if avatar gender induces self-stereotyping by analyzing the comments the participants wrote during a trivia game. Results showed that avatars whose gender matched the participants' actual gender increased the likelihood of gender-typed language use, with women using more references to emotion and apologies than men, whereas gender-mismatched avatars suppressed such a tendency. Moreover, women were more likely than men to adopt different language styles in line with their avatar gender. Collectively, these results suggest that gender-typed language use might represent conformity to social norms prescribing specific linguistic styles for men and women, and that people might feel more relaxed from such expectations when their gender is masked by a gender-mismatching avatar.

Cross-validation and extension

In keeping with the notion that virtual category membership, such as gender and race, binds users' perceptions and behaviors, Eastwick and Gardner (2009) examined if the race of the avatar requesting help moderates the effectiveness of the door-in-the-face (DITF) compliance technique (i.e., a large request followed by a moderate request) in a preexisting online virtual world. The DITF technique significantly increased compliance compared to the moderate-request-only condition when used by the light-skinned avatar, but not by the dark-skinned avatar with Afro-centric facial features. Although it remains unclear whether such racial biases represent "an automatic racial bias unfortunately imported from the real world or a thoughtful bias against users who would choose an unusually dark skin tone" (p. 28), the results resonate well with the previous findings obtained with randomly assigned avatars, not contaminated by this avatar versus owner confound.

Likewise, Groom, Bailenson, and Nass (2009) examined how the race of the avatar assigned to users in an immersive virtual environment affects their racial bias. Two competing predictions were derived from perspective-taking theory (i.e., people with Black avatars will show *less* explicit and implicit racial bias than those with White avatars) and stereotype activation theory (i.e., people with Black avatars will show *more* implicit racial bias than those with White avatars). Results supported the latter, as those embodied in Black avatars exhibited greater White preference in the Implicit Association Test than their White avatar counterparts. Such results seem to deviate from those of other studies in that people did not embrace their virtual identity and internalize its perspective, but underscore the persistent influence of social category membership in avatar-mediated interaction.

Extending the previous research that manipulated visible demographic attributes of the avatar, which evoke stereotypic reactions, Peña, Hancock, and Merola (2009) varied the way the avatars were dressed and found that avatar clothing can prime individuals' thoughts in a stereotype-consistent manner. When assigned avatars in either black or white cloaks (Experiment 1), male participants with black-cloaked avatars reported more aggressive intentions and attitudes toward unarmed game players and experienced weaker group cohesion than their white-cloaked counterparts. Similarly, participants in the Ku Klux Klan (KKK) avatar condition composed stories that were rated as more aggressive than those of the control group (no visual representation) and less affiliative than those of the doctors' uniform group (Experiment 2). These results mirror Johnson and Downing's (1979) finding that participants made to wear a KKK-like robe administered more severe electric shocks than those in nurse's uniforms when the confederate did not give the correct answer, and show that even randomly assigned avatars are powerful enough to elicit the responses expected from such entities.

In sum, previous research consistently showed that one's avatar has a significant effect on the way people perceive and categorize others as well as themselves, supporting the notion that "computer-mediated social perceptions are indeed sensitive to social cues and these cues may invoke stereotypes or a priori expectations" (Matheson, 1991, p. 137). Moreover, people not only ascribe traits to their partners based on their avatars, but they also define their relationship with others in a manner that corresponds to how they are visually represented (e.g., Lee, 2004b, 2007d; Lee & Nass, 2002). The following section highlights how avatar similarity (vs. dissimilarity) shapes the relational dynamics among communicating individuals.

Proposition 1: Category membership cues (e.g., gender and race) embedded in avatars evoke associated social stereotypes from individuals who either perceive or operate them, even when they are randomly assigned.

Avatar Similarity and Self-Categorization: Personal versus Group Identity

Defined as "the act of changing one's behavior to match the responses of others" (Cialdini & Goldstein, 2004, p. 606), conformity stems from the desire for accuracy, affiliation, and self-concept maintenance (Cialdini & Goldstein, 2004). However, these motivations may attenuate in an anonymous environment with low levels of individual accountability (Zimbardo, 1969), where the credibility or relational intent of others often remain uncertain. As such, CMC has been considered to have the potential to lessen the conformity pressure by filtering out various identity cues normally taken for granted in face-to-face interaction. The findings that individuals communicating via computer engaged in more anti-normative behavior labeled as "flaming" (Kiesler et al., 1984) and felt less self-conscious about intimate self-disclosure (Joinson, 2001) lend support to this contention.

As stated previously, however, the SIDE model posits that the paucity of social cues does not inevitably lower public self-awareness and normative concerns, but instead heightens the salience of group identity cues and facilitates group-centered perceptions and behavior, such as in-group favoritism (Postmes & Spears, 1998; Postmes et al., 1998). Moreover, if mere exposure to the virtual gender identity triggered the corresponding gender categorization and activated associated stereotypes, visual similarity among on-screen characters may also induce perceived similarity and we-feelings. By systematically varying the level of similarity among avatars and examining its impact on spontaneous group formation, studies have shown how reduced individuality via uniform avatars prompts more group-centered behaviors in CMC, as detailed below.

Study method

Upon arrival, participants were told that they would exchange their opinions on hypothetical choice dilemma scenarios with other participants. After reading a scenario portraying a dilemma between more rewarding yet less attainable and less rewarding but more attainable choices (Kogan & Wallach, 1967), they proceeded to the next page, where the other (ostensible) participants' responses were shown. The other discussants' responses were pre-programmed to be identical on critical trials, and filler scenarios were included for which they gave different answers. After reading the other discussants' responses, the participants indicated their opinion and provided their rationale for the decision. The other discussants' arguments were then presented by their avatars at various intervals. Conformity to the group norm was measured by how closely the participants' decision matched the majority opinion, after taking their predispositional risk-taking tendency into account.

Key findings

In one of the first scholarly attempts to investigate the effects of visual representation on individuals' perceptions and behaviors in CMC (Lee & Nass, 2002), participants were represented by either text boxes, stick figures, or animated cartoon characters. Although participants attributed greater competence, social attractiveness, and trustworthiness to their partners who were represented by animated characters than text boxes, they conformed more to the majority opinion when they were all represented by textboxes than animated characters. The authors speculated that the visible individual differences among the animated characters might have reminded the participants of the intra-group differences and hindered identification with the group, whereas the uniform textboxes might have amplified we-feelings by masking interpersonal differences amongst group members, thereby enhancing conformity to the group norm.

To directly evaluate this conjecture, Lee (2004b) conducted two experiments that examined how avatar similarity affects conformity behavior, and if such effects vary depending on the salient self-identity (personal vs. group). Specifically, participants were told that they would be interacting with students from either the same university (interpersonal) or different universities (inter-group), who were assigned avatars that were either identical to or different from each other (Experiment 1). As predicted, participants were more likely to conform to the majority when their avatar matched those of their partners, but only in the inter-group context where the group-level self-identity was salient. The fact that those in the inter-group condition reported stronger group identification with their home university than those in the interpersonal condition supports this interpretation. By contrast, identical characters lowered conformity compared to different characters when the participants interacted with the same-university students, which presumably heightened the salience of their personal self-identity. A slightly different explanation for these findings came from the theory of individuation (Ziller, 1964) and optimal distinctiveness theory (ODT; Brewer, 1991), which center on the tension between the need to be similar to others and the countering need to protect one's uniqueness. According to ODT, social identity, as an extension of the self, and group loyalty are strongest when the self-categorization offers both a sense of belonging and a sense of distinctiveness (Brewer, 1991). Encountering other university students could have fulfilled the participants' need for differentiation, prompting them to seek a self-inclusive group identity more actively. Uniform characters, in this context, might have served as a marker of a shared identity and facilitated group identification. On the other hand, interacting with the same university students, each being represented by identical characters would have intensified the need for differentiation, thereby lowering conformity to the group norm.

Additional evidence supporting this avatar-based group identification was provided by Lee's (2007d) study, which used the aforementioned trivia game context with gendered avatars. Even though (a) the two avatars of the same gender were not identical and (b) the participants' own avatar always mismatched their physical gender identity, when both their own and their partner's avatars belonged to the same gender category, participants reported stronger team affiliation with the partner, attributed greater competence to the partner, and exhibited greater acceptance of the partner's answers, indicating in-group favoritism. An interpretation from ODT would suggest that belonging to the same gender category, yet assuming a different-looking avatar, would have afforded optimal distinctiveness, thereby facilitating team identification.

Cross-validation and extension

Direct support for the influence of avatar similarity on the spontaneous formation of in-group identity came from Kim and Park's (2011) replication study. Consistent with Lee's (2004b) original findings, uniform avatars increased group identification, which in turn, fostered conformity to the majority opinion. At the same time, uniform avatars heightened perceived threat to uniqueness (i.e., deindividuation), and subsequently, lowered conformity intention. In this study, however, because the participants were asked to choose their avatar, the stronger group identification in the same avatar condition might have stemmed from the shared preference for the selected avatar, rather than the shared avatar per se. Just as preference for paintings (Klee vs. Kandinsky) served as an important basis for psychological group formation in the minimal group paradigm (Tajfel, 1970), the belief that they all picked the same avatar might well have evoked perceptions of similarity and stronger identification with the others. Also, it remains unclear how deindividuation and group identification are related to each other, and more important, when the uniform avatars elicit the feelings of losing "me" (deindividuation) versus finding "us" (group identification). Notwithstanding, these findings not only suggest that people harbor inherently conflicting desires to be similar to and yet different from others (Brewer, 1991; Codol, 1984), but also support Lee's (2004b) notion that (dis)similarity in virtual identities may upset the equilibrium between these competing forces and instigate individuals' actions to restore it, for example, by voicing agreement or disagreement with the rest of the group.

In yet another attempt to validate the effect of visual similarity on group identification and subsequent conformity, Kim (2010) contrasted the same versus different avatar conditions, either as humans with ethnic cues or nonhuman characters. Once again, participants reported stronger group identification when represented by uniform than different avatars, which in turn led to higher conformity intention and subsequent conformity behavior, but such effect was more pronounced with nonhuman than human avatars. Because (a) the availability of ethnicity information was confounded with the human-likeness of characters (human vs. nonhuman), (b) the participant's self-avatar in the nonhuman condition was actually a human cartoon character (i.e., Kenny from South Park) and the only non-animal character, and (c) having their ethnicity disclosed by the avatars could have enhanced perceived identifiability, the interaction lends itself to multiple interpretations. At the very least, the findings mark another instantiation of how avatar similarity breeds self-categorization as a member of a group, rather than an idiosyncratic individual.

Moving beyond the artificial laboratory setting, Lee and Park (2011) exposed participants to fictitious profiles of residents in avatar-based virtual worlds (Second Life) and found that non-White participants were less likely to feel that they belonged to the virtual world and showed lower levels of intention to participate in it when they saw the White-dominant avatar profiles than the racially diverse profiles (Study 1). Although there was no guarantee that the avatar's race actually matched the player's, exposure to the virtual profiles signaling White dominance led non-White participants to feel disconnected and marginalized. Considering that the players can customize their avatars to their liking, such psychological reactions might be toward those who prefer White avatars, regardless of their actual race, rather than the avatars themselves. Still, the findings confirm the existence of avatar-based group identification (us vs. them).

Proposition 2: Similarity in avatar appearance is more likely to foster group identification and facilitate group-oriented behaviors when social, rather than personal, self-identity is salient.

Avatars as a Pseudo-Identity Cue: Contingent Conditions

While the extant literature has rather consistently documented avatar-based social stereotyping and in-group identification in CMC, it also suggests several important boundary conditions under which such effects are more or less likely to occur.

Individual differences

There appear to be stable, enduring individual differences in the extent to which people associate perceptually salient yet logically irrelevant visual cues with anonymous CMC partners. First, women were more likely than men to equate the avatar gender with their unknown partner's. For example, whereas women inferred the partner's gender from the partner's avatar, men did not show the corresponding tendency (Lee, 2005, Study 1 and Study 2; Lee, 2007a). Similarly, women were more likely than men to adjust their language features to reflect their avatar gender (Palomares & Lee, 2010). Several possibilities were entertained, such as superior visual recognition memory (McGivern, Andersen, Byrd, Mutter, & Reilly, 2002) and greater sensitivity to gender-related cues among women, but never directly tested. In fact, the finding that male participants were more susceptible to the effects of avatar clothing (Peña et al., 2009a) refutes the differential sensitivity to visual cues as an explanation, and suggests that the nature of stereotypes might have contributed to such gender differences. Still, a more systematic investigation is in order to elucidate when and why men and women respond differently to avatar cues.

Proposition 3: Women are more likely than men to exhibit stereotype-consistent reactions to gender-marked avatars.

Another individual difference observed in the previous research concerns experiential thinking style (Lee, 2007b). According to cognitive-experiential self-theory (Epstein, 1994; Epstein & Pacini, 1999), there are stable individual differences in the extent to which people rely on experiential and rational systems to process incoming stimuli, which roughly correspond to intuitive and analytical modes of information processing in dual-process theories (Chaiken, 1987; Petty & Cacioppo, 1986). Considering that the experiential system operates in an automatic, concrete, and holistic manner, it was expected that high experientials would be more likely to make avatar-based gender inferences because they are (a) more responsive to concrete images, (b) less likely to engage in the cognitively effortful correction process once initial gender categorization occurs, and (c) more prone to broad generalization and stereotypical thinking (Lee, 2007b). As predicted, those high in experientiality were more likely than those low in experientiality to infer their partner's gender from arbitrary avatars and also exhibited greater conformity to male- than to female-avatared partners (Study 1). However, given that high experientials relied more on linguistic features to infer their partner's gender (Study 2), the greater avatar effects among high experientials might simply indicate their general proclivity toward categorical thinking, rather than their susceptibility to avatar effects in general. If so, experientiality might not play an important role when avatar effects are defined in other terms unrelated to social categories, such as interpersonal traits (e.g., attractiveness), which awaits further investigation.

Proposition 4: Those high in experientiality are more likely to exhibit avatar-based social stereotyping than their less experiential counterparts.

With respect to the effects of virtual similarity, Kim (2009) sought empirical support for Lee's (2004b) explanation that one's counteracting needs for uniqueness and similarity might affect the degree to which the uniform avatars trigger (anti)conformity, by measuring the participant's need for uniqueness (NFU). Just as the uniform avatars (shared virtual identity) lowered conformity when interacting with others supposedly from the same university (shared real identity), whereas they enhanced conformity to those from different universities (Lee, 2004b, Study 1), one's NFU lessened group identification when the participants and the ostensible partners were assigned identical avatars (shared virtual identity) with the same ethnic background (shared real identity), but not when presented by the same avatars without ethnicity cues. These findings suggest that too much similarity might threaten one's desire for distinctiveness, thereby triggering reactions to restore one's individuality.

In fact, these results comport well with Lee's (2006) study, which demonstrated how the need for public individuation (NPI), referring to "the willingness to engage in behaviors that would publicly differentiate themselves from others," (Maslach, Stapp, & Santee, 1985, p. 729) interacts with the depersonalizing context to affect group conformity. In this study, the level of (de)personalization was varied by having the participants either exchange brief personal profiles (personalized) or not (depersonalized), and NPI negatively predicted conformity only when the participants were personalized, indicating that reduced individuality can suppress NPI by rendering the interaction context less public. Despite the differences in the specific ways in which NFU and NPI operated, these findings collectively suggest that individuals' intrinsic identity needs guide their reactions to the limited individuating cues in CMC.

Proposition 5: Similarity in avatar appearance is more likely to evoke attempts to restore individuality among those higher in NFU.

Contextual factors

In addition to enduring individual differences, several contextual factors seem to moderate the likelihood of avatar effects. First, self-avatar congruency (match vs. mismatch) was found to moderate the extent to which participants exhibited greater susceptibility to the male than female influence agent (Lee, 2005) and used gender-typed language (Palomares & Lee, 2010), with gender-matching avatars facilitating stereotyped reactions as well as self-stereotyping. That is, even though the partner's gender still remained unknown, when a gender-matched avatar suggested the connection between an on-screen character and the physical identity of the person behind it, the virtual gender identity activated the same stereotypical reactions as found in face-to-face encounters, whereas the mismatch between the virtual and physical self-identities tended to suppress it. One possible route through which gender-matched avatars promote social stereotypes is perceived identifiability. Just as perceived similarity between users and their avatars enhances perceived identifiability, or conversely, reduces perceived anonymity (Midha & Nandedkar, 2012), gender-matched avatars might have made people feel that their gender was known to others and led them to conform to traditional gender roles and show stereotyped perceptions. Still, given that matching or mismatching the avatar ethnicity to that of the participant did not make a significant difference in the participants' group identification or conformity behavior (Kim, 2011), more systematic investigations are needed to validate the moderating effects of self-avatar congruency and the mechanism underlying them. At the very least, the findings that serendipitous self-avatar congruency fostered avatar-based stereotyping effectively refute Groom et al. (2009) conjecture that inaccurate virtual embodiment might make the relevant category cues more salient by highlighting the difference between real and virtual identities, and thus promote category-based thinking.

Proposition 6: People are more likely to attribute an arbitrary avatar's characteristics to their CMC partner when their own avatar matches, rather than mismatches, their actual identity.

Second, the ambiguity of the situation or uncertainty about the interactant might also make people more vulnerable to avatar effects. For example, when the interaction partner's self-confidence ratings were made available in either numeric (Study 1) or verbal form (Study 2), hence reducing the ambiguity surrounding how much the partner's answer should be trusted, the partner's avatar gender had no significant effect on the participants' conformity decision (Lee, 2005). Likewise, when the partner expressed high self-confidence in a gendered topic, participants inferred the partner's gender in a stereotyped manner, and the avatar gender had no significant effect on gender inferences, at least among men (Lee, 2007a). Taken together, these findings suggest that the influence of avatars on subsequent judgments varies depending on the availability of information that aids people to make sense of the given situation and properly handle the task at hand, supporting the stereotypes-as-explanations account (McGarty, Yzerbyt, & Spears, 2002). That is, in the absence of relevant information, such as unambiguous confidence ratings, people seem to actively bring in gender stereotypes to guide their decision, even when the category membership was embedded in randomly assigned on-screen characters.

Proposition 7: Avatar-based person perception and social judgments are more likely to occur when the ambiguity or uncertainty of the situation is higher.

Why Avatars Matter: Potential Explanations

The apparent failure to discount for the impact of arbitrary situational variables on social perceptions well predates the advent of the Internet and the avatars populating cyberspace. For example, Ross, Amabile, and Steinmetz's college bowl study (1977) showed that even randomly assigned roles can induce role-consistent social perceptions of role players as well as self-perceptions. Specifically, when students were randomly assigned a role of either the questioner, who thought up difficult questions about any topic, or the respondent, who answered the questions, observers attributed greater intelligence and knowledge to the questioners than to the respondents after watching the respondents fail to answer many questions, although they knew that the respondents were obviously disadvantaged. What is more, students who played the respondent rated themselves to be inferior to the questioners in terms of general knowledge. Likewise, the studies reviewed thus far have shown that individuals' awareness of the arbitrary nature of the avatar assignment did not suppress the activation of associated stereotypes and group identification. Why?

Mindlessness

As a research paradigm that centers on how people interact with computers, Computers Are Social Actors (CASA) paradigm posits that people tend to follow social scripts and rules when interacting with computers (Nass & Moon, 2000; Reeves & Nass, 1996). Despite their knowledge that "the computer is not a person and does not warrant human treatment or attribution" (Nass & Moon, 2000, p. 82), people nonetheless apply social rules to computers, rating a female-voiced computer to be more informative about love and relationships than its male-voiced counterpart (Nass, Moon, & Green, 1997) and conforming more to the computer exhibiting a personality similar to their own (Nass, Moon, Fogg, Reeves, & Dryer, 1995). To account for such findings, Nass and Moon (2000) proposed a mindlessness explanation (Langer, 1989) and argued that when computers emulate human-like attributes, such as interactivity and voice, people focus on such cues and mindlessly assign social categories to computers and emit overlearned social behaviors. In a direct test of this mindlessness account, Lee (2008, Experiment 1) used the trivia game setup described earlier and manipulated the computer gender using gender-marked cartoon characters. Even though the participants were explicitly told that the computer was programmed to generate random answers, they attributed greater competence and exhibited greater conformity to the male than female computer agents on gender-neutral questions, but only when they were simultaneously engaged in multiple tasks, hence cognitively busy.

If people mindlessly apply social categories even to inanimate machines, it might not seem surprising that they would do the same to their human partners based on the salient visual traits of avatars. However, considering that computers do not belong to any "social" category, there is no reason not to use whatever social categories and labels attached to the computer to define its nature. By contrast, arbitrarily assigned categories and roles might contradict the human interactant's actual identity, lowering their values as an identity cue. Despite such potential for misrepresentation, however, previous findings suggest that when interacting with unknown others through avatars, people find it too difficult not to utilize such perceptually salient cues. That is, the initial characterization process activated upon encountering the interactant's avatar ("Oh, my partner's a man," "We all look exactly the same") does not seem to be followed by a more cognitively demanding correction process ("But the avatars have nothing to do with my partners") (Gilbert, Pelham, & Krull, 1988). Apparently, it indeed requires willpower (or mindfulness) and training "to escape from the dominance of the given and to actually think about events and relationships that are not salient and explicit in our experience" (Hastie & Dawes, 2001, p. 128).

Thus far, no direct evidence seems to exist to show that mindlessness underlies avatar-based person perception and self-categorization in CMC. However, the facts that high experientials conformed more to their partner represented by the male than female avatars on a gender-neutral task, but did not attribute greater competence to the former (Lee, 2007b) suggest that gender-typed conformity behavior among high experientials reflects relatively unconscious reactions rather than a deliberate decision based on the presumption of differential abilities of male and female partners. If so, it seems reasonable to expect that people will be influenced more by the avatar's appearance when they are cognitively busy with multitasking or working under time pressure, which awaits empirical validation.

Proposition 8: Mindlessness facilitates avatar-based person perception and self-categorization.

Priming

To explain the aforementioned findings that participants who used avatars in black or KKK-associated clothes developed more aggressive intentions and attitudes toward other gamers and less group cohesion and affiliation than their white-cloaked or doctor-uniformed counterparts (Peña et al., 2009a), Peña (2011) proposed automatic priming. In this view, participants' awareness of the avatar appearance, like dark uniforms, activates related concepts stored in their memory, like aggressiveness, which subsequently affect their cognitive, affective, and behavioral reactions without their conscious intention to respond in the prime-consistent manner (Bargh & Chartrand, 1999). Supporting this notion of associative connection, when instructed to compose a story about a day in the life of their avatar in Second Life, those assigned a female avatar dressed in formal clothing (i.e., gray jacket and skirt) referred more to education, books, and numbers, as compared to those using the same avatar in a fancy red dress, who used more words related to sports, entertainment, and clothes (Peña, McGlone, Jarmon, & Sanchez, 2009). Moreover, the fact that those embodied in Black avatars indicated stronger, not weaker, White preference than those represented by White avatars seems to comport well with this priming account, demonstrating the persistent operation of "automatic racial bias" (Groom et al., 2009, p. 244). Based on these findings, Peña (2011) postulates that "avatar appearance will automatically influence the operator's cognition and behavior in congruence with the associations raised by such looks" (p. 155).

As parsimonious and intuitive as it is, however, the automatic priming account cannot seem to explain (a) why being represented by the avatar (embodiment) engendered greater behavioral changes than merely seeing the same visual stimulus (Yee & Bailenson, 2009) and (b) why self-avatar congruency fosters avatar-induced social perceptions and language use (Lee, 2005; Palomares & Lee, 2010). That is, if avatar cues guide social perceivers' evaluations and behaviors in accordance with the typical associations of those signs (Peña, 2011), it should not have mattered whom the avatars represented (self vs. others) or how avatar cues are related to one's physical identity (match vs. mismatch).

Under the same rubric of the automaticity principle, Peña (2011) also suggested unintentional mimicry of other people's gestures, postures and behaviors (i.e., chameleon effect) as an alternative explanation for why uniform avatars elicited greater conformity to group norms compared to distinctive avatars (Lee, 2004b). In addition, with respect to the disjunction between public compliance and private conformity (Lee & Nass, 2002), he argued that "individuals mimicked their group members' opinions but did not consciously alter their private views, thus revealing an automatic priming phenomenon more akin to the chameleon effect of avatars" (p. 159). However, if behavioral mimicry explains why avatar similarity fostered group conformity, it is unclear why such effects were found only in the inter-group, and not interpersonal contexts (Lee, 2004b, Study 1). Also, participants were not any more likely to conform to, or mimic, the majority opinion in the public than private contexts; it was the difference in conformity between one- and four-partner conditions which was greater in the former (Lee & Nass, 2002, Study 1). In fact, participants exhibited greater conformity in the private than public contexts when interacting with computer agents (Study 2), directly contradicting the unintentional mimicry account. Given such conflicting findings, it seems reasonable to conclude that although automatic association and mimicry may play a part in avatar-based gender stereotyping and in-group formation, a different psychological process might also be at work.

Overattribution

Another possible explanation for avatar effects concerns the interaction context, especially the information restriction of the medium. Unlike the cues-filtered-out approach, which underscores how the lack of interpersonal and contextual cues in CMC renders the interaction more impersonal (Short, Williams, & Christie, 1976), SIDE highlights that the paucity of cues can make the interaction more "social" by obscuring individual differences, "provided a common identity is available" (Postmes & Spears, 2002, p. 1075), thereby causing people to behave and perceive others as group members rather than as unique individuals (Postmes et al., 1998, 2000). Although no such tension between personal and social identity is assumed, Walther's (1996) hyperpersonal communication model is also grounded in this notion of overattribution, as it predicts that people tend to develop overly positive and idealized impressions of mediated partners from selective self-presentation and presume across-the-board similarity on the basis of limited commonality, like shared interest.

By demonstrating that people spontaneously infer their interactants' social category from arbitrary and potentially misleading contextual cues and respond in line with their inferences, the studies reviewed herein have established that visual anonymity does not make people blind to social categories and suppress associated group-oriented perceptions. However, the critical notion that the limited channel capacity encourages the overreliance on remaining cues remains to be tested with respect to avatar effects. At least, given that the unknown partner's gender-linked verbal behavior activated gender stereotyping among gender-typed individuals only when no personal information was available about the partner (Lee, 2007c), the extent to which avatars shape individuals' perceptions and behavioral reactions in CMC may also vary as a function of the amount of personal information.

Proposition 9: The lack of individuating cues fosters avatar-based social stereotyping of and group identification with CMC partners.

Implications and Future Directions

In the studies reviewed herein, avatars were mostly treated as a pseudo-identity cue that filled individuals in on who their anonymous CMC partner was. However, research has also shown that avatars can serve as an identity-concealing device that frees people from public accountability and negative group stereotypes. For example, those deceiving their partners were more likely than truth-tellers to choose avatars that looked different from themselves, which relieved their anxiety in deceiving others (Galanxhi & Nah, 2007). Likewise, when solving anagrams with two ostensible Caucasian coactors, African-American participants performed better when they engaged in cooperation than in competition, but no corresponding effect of the coaction context was observed for those whose racial identity was hidden in silhouette-image avatars (Lee & Nass, 2012, Study 1). If avatars can serve as a mask to increase (perceived) anonymity in cyberspace, less human-like or less realistic avatars, with no gender, race, or other status cues, might help people not to "fall back on the tried and trusted categories of the old world and actively carry over the constraints of their own everyday identities" (Postmes et al., 1998, p. 708). Alternatively, more technologically sophisticated avatars that closely resemble real humans might inadvertently contribute to the unfortunate reproduction of stereotypes and biases against prejudiced minorities in the virtual world.

Given individuals' tendency to utilize avatar cues to make inferences about their mediated interactants, one might consider more proactive appropriation of avatars as a means to alleviate negative stereotypes about disadvantaged minority groups. Extending the intergroup contact hypothesis (Allport, 1954) to media presentations of out-group members, the parasocial contact hypothesis posits that repeated exposure to likable members of a minority group (e.g., gay, transvestite) in TV shows can foster positive impressions of them and subsequently reduce stereotypes about the group as a whole (Schiappa, Gregg, & Hewes, 2005). If so, technology-mediated interaction with others embodied in avatars representing a prejudiced minority group, who demonstrate positive, counter-stereotypic traits, might help to reduce negative beliefs about the group. For example, encountering players using female avatars in an online war game who exhibit superior knowledge in military strategies might challenge other gamers' long-held beliefs concerning what men and women are good at (or just make them suspicious about the players' true gender). Moreover, if perceiving a target as a member of the group, rather than an idiosyncratic individual, is a necessary condition for the information obtained about that person to get generalized to the entire group (Rothbart & John, 1985), the lack of individuating information in CMC might have the potential to amplify the effects of parasocial contacts.

To capture avatar effects free from the self-selection bias, this chapter has focused on the studies using randomly assigned avatars, leaving unanswered the questions of how the very act of creating one's own avatar, as well as the type of avatars being created, might moderate influence such effects. In Kim and Sundar's (2012) study, participants who created avatars reflecting their ideal self were better able to visualize their ideal body, which increased their willingness to refrain from potentially harmful actions to their health (i.e., self-preservation) than those who built avatars resembling their actual self. Moreover, those who created an avatar reported higher levels of the self-preservation intention, compared to those with randomly assigned avatars, although the intention did not extend to their actual behavior. What remains to be examined is whether avatar customization has any effect on the extent to which people infer their interactants' characteristics from their avatars. On the one hand, when the avatar is selected and/or tailored by the person it represents, people might place more weight on avatar features to guess what the person is like. On the other hand, the manipulability of avatar appearance might make people more wary of the potential disconnect between virtual and actual identities and discount its cue value.

Another question that requires future research pertains to how replicable the previously noted avatar effects are in the world outside the laboratory, beyond highly structured synchronous dyadic/small-group interaction. Although no actual behavior was measured, Yee (2001) reported that 80% of surveyed EverQuest players felt that gamers using female avatars receive more preferential treatment than those using male avatars. However, it merits note that role-playing is an implicit norm within online game worlds, wherein willing suspension of disbelief about the authenticity of claimed identity tends to prevail. As such, it is not surprising that game players take the avatar gender at face value and respond differently (if they indeed do). Therefore, to assess the robustness of avatar effects, future research should employ a wide range of CMC contexts and explore their boundary conditions. For example, when anthropomorphic thumbnail avatars accompany customer reviews on an e-commerce site, the avatar gender may or may not moderate the effects of reviews on readers' product evaluations or purchase intention, perhaps depending on the product type (feminine vs. masculine vs. gender-neutral).

Conclusion

When the limited channel bandwidth masks individuals' physical identity to varying degrees in text-based CMC, avatars, however they have little resemblance to the person they stand for, seem to have the power to shape individuals' social perceptions and group dynamics. In general, the tendency to make dispositional attributions based on the seemingly trivial and downright arbitrary attributes of the avatar has been found across studies, indicating that people may unwittingly import appearance-based social stereotypes and biases from the real world to the virtual. At the same time, it deserves note that several dispositional and contextual factors also seemed to play a significant role in determining social perceivers' sensitivity and susceptibility to avatar cues and that the exact psychological processes underlying avatar effects remain unclear.

Ironically, however, individuals' reliance on avatar cues as a basis for categorical thinking also opens up the possibility to leverage the plasticity of virtual identity to challenge long-held stereotypes and redefine the boundary of the in-group. For example, repeated interaction with others embodied in out-group avatars who betray one's prior beliefs about the group might prompt changes in his or her presumptions about and attitudes toward the out-group over time. Similarly, having uniform avatars during computer-supported group meetings might help to cultivate team spirit among the members of geographically dispersed work groups, beyond location-based subgroup identities. After all, whether virtual or real, the glass is both half full and half empty.

References

Allport, G. (1954). *The nature of prejudice*. Reading, MA: Addison-Wesley.

Bargh, J. A. (1999). The cognitive monster: The case against the controllability of automatic stereotype effects. In S. Chaiken & Y. Trope (Eds.), *Dual process theories in social psychology* (pp. 361–382). New York, NY: Guilford.

Bargh, J. A., & Chartrand, T. L. (1999). The unbearable automaticity of being. *American Psychologist*, *54*(7), 462–479. doi:10.1037/0003-066X.54.7.462

Bente, G., Rüggenberg, S., Krämer, N. C., & Eschenburg, F. (2008). Avatar-mediated networking: Increasing social presence and interpersonal trust in net-based collaborations. *Human Communication Research, 34*(2), 287–318. doi:10.1111/j.1468-2958.2008.00322.x

Bessière, K., Seay, A. F., & Kiesler, S. (2007). The ideal self: Identity exploration in World of Warcraft. *CyberPsychology & Behavior, 10*(4), 530–535. doi:10.1089/cpb.2007.9994

Brewer, M. B. (1988). A dual process model of impression formation. In R. S. Wyer, Jr. & T. K. Srull (Eds.), *Advances in social cognition* (Vol. 1, pp. 1–36). Hillsdale, NJ: Erlbaum.

Brewer, M. B. (1991). The social self: On being the same and different at the same time. *Personality and Social Psychology Bulletin, 17*(5), 475–482. doi:10.1177/0146167291175001

Chaiken, S. (1987). The heuristic model of persuasion. In M. P. Zanna, J. M. Olson, & C. P. Herman (Eds.), *Social influence: The Ontario symposium* (Vol. 5, pp. 3–39), Hillsdale, NJ: Erlbaum.

Cialdini, R. B., & Goldstein, N. J. (2004). Social influence: Compliance and conformity. *Annual Review of Psychology, 55*(1), 591–621. doi:10.1146/annurev.psych.55.090902.142015

Codol, J. P. (1984). Social differentiation and non-differentiation. In H. Tajfel (Ed.), *The social dimension: European developments in social psychology* (Vol. 1, pp. 314–337). Cambridge, UK: Cambridge University Press.

Dubrovsky, V., Kiesler, S., & Sethna, B. (1991). The equalization phenomenon: Status effects in computer-mediated and face-to-face decision-making groups. *Human-Computer Interaction, 6*(2), 119–146. doi:10.1207/s15327051hci0602_2

Dunn, R. A., & Guadagno, R. E. (2012). My avatar and me – Gender and personality predictors of avatar-self discrepancy. *Computers in Human Behavior, 28*(1), 97–106. doi:10.1016/j.chb.2011.08.015

Eastwick, P. W., & Gardner, W. L. (2009). Is it a game? Evidence for social influence in the virtual world. *Social Influence, 4*(1), 18–32. doi:10.1080/15534510802254087

Epstein, S. (1994). Integration of the cognitive and the psychodynamic unconscious. *American Psychologist, 49*(8), 709–724. doi:10.1037/0003-066X.49.8.709

Epstein, S., & Pacini, R. (1999). Some basic issues regarding dual-process theories from the perspective of cognitive-experiential self-theory. In S. Chaiken & Y. Trope (Eds.), *Dual-process theories in social psychology* (pp. 462–482). New York, NY: Guilford.

Galanxhi, H., & Nah, F. F.-H. (2007). Deception in cyberspace: A comparison of text-only vs. avatar-supported medium. *International Journal of Human-Computer Studies, 65*(9), 770–783. doi:10.1016/j.ijhcs.2007.04.005

Gilbert, D. T., Pelham, B. W., & Krull, D. S. (1988). On cognitive busyness: When person perceivers meet persons perceived. *Journal of Personality and Social Psychology, 54*(5), 733–740. doi:10.1037/0022-3514.54.5.733

Girard, S., & Johnson, H. (2010). What do children favor as embodied pedagogical agents? In V. Aleven, J. Kay, & J. Mostow (Eds.), *Intelligent tutoring systems* (Vol. 6094, pp. 307–316). Berlin, Germany: Springer. Retrieved from http://www.springerlink.com/index/10.1007/978-3-642-13388-6_35

Groom, V., Bailenson, J. N., & Nass, C. (2009). The influence of racial embodiment on racial bias in immersive virtual environments. *Social Influence, 4*(3), 231–248. doi:10.1080/15534510802643750

Hastie, R., & Dawes, R. M. (2001). *Rational choice in an uncertain world: The psychology of judgment and decision making*. Thousand Oaks, CA: Sage.

Herschel, R. T., Cooper, T. R., Smith, L. F., & Arrington, L. (1994). Exploring numerical proportions in a unique context: The group support systems meeting environment. *Sex Roles*, *31*(1–2), 99–123. doi:10.1007/BF01560279

Holzwarth, M., Janiszewski, C., & Neumann, M. M. (2006). The influence of avatars on online consumer shopping behavior. *Journal of Marketing*, *70*(4), 19–36. Retrieved from http://warrington.ufl.edu/mkt/docs/janiszewski/avatarjm.pdf

Hussain, Z., & Griffiths, M. D. (2008). Gender swapping and socializing in cyberspace: An exploratory study. *CyberPsychology & Behavior*, *11*(1), 47–53. doi:10.1089/cpb.2007.0020

Johnson, R. D., & Downing, L. L. (1979). Deindividuation and valence of cues: Effects on prosocial and antisocial behavior. *Journal of Personality and Social Psychology*, *37*(9), 1532–1538. doi:10.1037/0022-3514.37.9.1532

Joinson, A. N. (2001). Self-disclosure in computer-mediated communication: The role of self-awareness and visual anonymity. *European Journal of Social Psychology*, *31*(2), 177–192. doi:10.1002/ejsp.36

Kiesler, S., Siegel, J., & McGuire, T. W. (1984). Social psychological aspects of computer-mediated communication. *American Psychologist*, *39*(10), 1123–1134. doi:10.1037/0003-066X.39.10.1123

Kim, J. (2009). "I want to be different from others in cyberspace" The role of visual similarity in virtual group identity. *Computers in Human Behavior*, *25*(1), 88–95. doi:10.1016/j.chb.2008.06.008

Kim, J. (2010). Balancing uniqueness and assimilation in computer-mediated groups. *Computers in Human Behavior*, *26*(4), 778–784. doi:10.1016/j.chb.2010.02.001

Kim, J. (2011). Two routes leading to conformity intention in computer-mediated groups: Matching versus mismatching virtual representations. *Journal of Computer-Mediated Communication*, *16*(2), 271–287. doi:10.1111/j.1083-6101.2011.01539.x

Kim, J., & Park, H. S. (2011). The effect of uniform virtual appearance on conformity intention: Social identity model of deindividuation effects and optimal distinctiveness theory. *Computers in Human Behavior*, *27*(3), 1223–1230. doi:10.1016/j.chb.2011.01.002

Kim, Y., & Sundar, S. S. (2012). Visualizing ideal self vs. actual self through avatars: Impact on preventive health outcomes. *Computers in Human Behavior*, *28*(4), 1356–1364. doi:10.1016/j.chb.2012.02.021

Kogan, N., & Wallach, M. A. (1967). Risky-shift phenomenon in small decision-making groups: A test of the information-exchange hypothesis. *Journal of Experimental Social Psychology*, *3*(1), 75–84. doi:10.1016/0022-1031(67)90038-8

Langer, E. (1989). Minding matters: The consequences of mindlessness-mindfulness. In L. Berkowitz (Ed.), *Advances in experimental social psychology* (Vol. 22, pp. 137–173). San Diego, CA: Academic Press.

Lee, E.-J. (2004a). Effects of gendered character representation on person perception and informational social influence in computer-mediated communication. *Computers in Human Behavior*, *20*(6), 779–799. doi:10.1016/j.chb.2003.11.005

Lee, E.-J. (2004b). Effects of visual representation on social influence in computer-mediated communication. *Human Communication Research*, *30*(2), 234–259. doi:10.1111/j.1468-2958.2004.tb00732.x

Lee, E.-J. (2005). Effects of the influence agent's sex and self-confidence on informational social influence in computer-mediated communication: Quantitative versus verbal presentation. *Communication Research*, *32*(1), 29–58. doi:10.1177/0093650204271398

Lee, E.-J. (2006). When and how does depersonalization increase conformity to group norms in computer-mediated communication? *Communication Research*, *33*(6), 423–447. doi:10.1177/0093650206293248

Lee, E.-J. (2007a). Categorical person perception in computer-mediated communication: Effects of character representation and knowledge bias on sex inference and informational social influence. *Media Psychology*, *9*(2), 309–329. doi:10.1080/15213260701286007

Lee, E.-J. (2007b). Wired for gender: Experientiality and gender-stereotyping in computer-mediated communication. *Media Psychology*, *10*(2), 182–210. doi:10.1080/15213260701375595

Lee, E.-J. (2007c). Effects of gendered language on gender stereotyping in computer-mediated communication: The moderating role of depersonalization and gender-role orientation. *Human Communication Research*, *33*(4), 515–535. doi:10.1111/j.1468-2958.2007.00310.x

Lee, E.-J. (2007d). Character-based team identification and referent informational influence in computer-mediated communication. *Media Psychology*, *9*(1), 135–155. doi:10.1080/15213260709336806

Lee, E.-J. (2008). Gender stereotyping of computers: Resource depletion or reduced attention? *Journal of Communication*, *58*(2), 301–320. doi:10.1111/j.1460-2466.2008.00386.x

Lee, E.-J., & Nass, C. (2002). Experimental tests of normative group influence and representation effects in computer-mediated communication. *Human Communication Research*, *28*(3), 349–381. doi:10.1111/j.1468-2958.2002.tb00812.x

Lee, J.-E. R., & Nass, C. (2012). Distinctiveness-based stereotype threat and the moderating role of coaction contexts. *Journal of Experimental Social Psychology*, *48*(1), 192–199. doi:10.1016/j.jesp.2011.06.018

Lee, J.-E. R., & Park, S. G. (2011). "Whose Second Life is this?" How avatar-based racial cues shape ethno-racial minorities' perception of virtual worlds. *Cyberpsychology, Behavior, and Social Networking*, *14*(11), 637–642. doi:10.1089/cyber.2010.0501

Macrae, C. N., & Bodenhausen, G. V. (2001). Social cognition: Categorical person perception. *British Journal of Psychology*, *92*(1), 239–255. doi:10.1348/000712601162059

Maslach, C., Stapp, J., & Santee, R. T. (1985). Individuation: Conceptual analysis and assessment. *Journal of Personality and Social Psychology*, *49*(3), 729–738. doi:10.1037/0022-3514.49.3.729

Matheson, K. (1991). Social cues in computer-mediated negotiations: Gender makes a difference. *Computers in Human Behavior*, *7*(3), 137–145. doi:10.1016/0747-5632(91)90003-J

McGarty, C., Yzerbyt, V. Y., & Spears, R. (Eds.). (2002). *Stereotypes as explanations: The formation of meaningful beliefs about social groups*. Cambridge, UK: Cambridge University Press.

McGivern, R. F., Andersen, J., Byrd, D., Mutter, K. L., & Reilly, J. (2002). Cognitive efficiency on a match to sample task decreases at the onset of puberty in children. *Brain and Cognition*, *50*(1), 73–89. doi:10.1016/S0278-2626(02)00012-X

Midha, V., & Nandedkar, A. (2012). Impact of similarity between avatar and their users on their perceived identifiability: Evidence from virtual teams in Second Life platform. *Computers in Human Behavior*, *28*(3), 929–932. doi:10.1016/j.chb.2011.12.013

Nass, C., & Moon, Y. (2000). Machines and mindlessness: Social responses to computers. *Journal of Social Issues*, *56*(1), 81–103. doi:10.1111/0022-4537.00153

Nass, C., Moon, Y., Fogg, B. J., Reeves, B., & Dryer, D. C. (1995). Can computer personalities be human personalities? *International Journal of Human-Computer Studies*, *43*(2), 223–239. doi:10.1006/ijhc.1995.1042

Nass, C., Moon, Y., & Green, N. (1997). Are machines gender neutral? Gender-stereotypic responses to computers with voices. *Journal of Applied Social Psychology*, *27*(10), 864–876. doi:10.1111/j.1559-1816.1997.tb00275.x

Nowak, K. L. (2004). The influence of anthropomorphism and agency on social judgment in virtual environments. *Journal of Computer Mediated Communication*, *9*(2). http://onlinelibrary.wiley.com/doi/10.1111/j.1083-6101.2004.tb00284.x/abstract

Palomares, N. A., & Lee, E.-J. (2010). Virtual gender identity: The linguistic assimilation to gendered avatars in computer-mediated communication. *Journal of Language and Social Psychology, 29*(1), 5–23. doi:10.1177/0261927X09351675

Peña, J. F. (2011). Integrating the influence of perceiving and operating avatars under the automaticity model of priming effects. *Communication Theory, 21*(2), 150–168. doi:10.1111/j.1468-2885.2011.01380.x

Peña, J., Hancock, J. T., & Merola, N. A. (2009). The priming effects of avatars in virtual settings. *Communication Research, 36*(6), 838–856. doi:10.1177/0093650209346802

Peña, J., McGlone, M., Jarmon, L., & Sanchez, J. (2009). *The influence of visual stereotypes and roles on language use in virtual environments.* Paper presented at the 95th Annual Convention of the National Communication Association, Chicago, IL.

Petty, R. E., & Cacioppo, J. T. (1986). *Communication and persuasion: Central and peripheral routes to attitude change.* New York, NY: Springer-Verlag.

Postmes, T., & Spears, R. (1998). Deindividuation and antinormative behavior: A meta-analysis. *Psychological Bulletin, 123*(3), 238–259. doi:10.1037/0033-2909.123.3.238

Postmes, T., & Spears, R. (2002). Behavior online: Does anonymous computer communication reduce gender inequality? *Personality and Social Psychology Bulletin, 28*(8), 1073–1083. doi:10.1177/01461672022811006

Postmes, T., Spears, R., & Lea, M. (1998). Breaching or building social boundaries? SIDE-effects of computer-mediated communication. *Communication Research, 25*(6), 689–715. doi:10.1177/009365098025006006

Postmes, T., Spears, R., & Lea, M. (2000). The formation of group norms in computer-mediated communication. *Human Communication Research, 26*(3), 341–371. doi:10.1111/j.1468-2958.2000.tb00761.x

Qiu, L., & Benbasat, I. (2005). Online consumer trust and live help interfaces: The effects of text-to-speech voice and three-dimensional avatars. *International Journal of Human-Computer Interaction, 19*(1), 75–94. doi:10.1207/s15327590ijhc1901_6

Reeves, B., & Nass, C. (1996). *The media equation: How people treat computers, television, and new media like real people and places.* Stanford, CA: CSLI Publications.

Ross, L. D., Amabile, T. M., & Steinmetz, J. L. (1977). Social roles, social control, and biases in social-perception processes. *Journal of Personality and Social Psychology, 35*(7), 485–494. doi:10.1037/0022-3514.35.7.485

Rothbart, M., & John, O. P. (1985). Social categorization and behavioral episodes: A cognitive analysis of the effects of intergroup contact. *Journal of Social Issues, 41*(3), 81–104. doi:10.1111/j.1540-4560.1985.tb01130.x

Schiappa, E., Gregg, P. B., & Hewes, D. E. (2005). The parasocial contact hypothesis. *Communication Monographs, 72*(1), 92–115. doi:10.1080/0363775052000342544

Short, J. A., Williams, E., & Christie, B. (1976). *The social psychology of telecommunications.* New York, NY: Wiley.

Sproull, L., & Kiesler, S. (1991). *Connections: New ways of working in the networked organization.* Cambridge, MA: MIT Press.

Tajfel, H. (1970). Experiments in intergroup discrimination. *Scientific American, 223*(5), 96–102.

Tapie, J., Terrier, P., Perron, L., & Cellier, J.-M. (2006). Should remote collaborators be represented by avatars? A matter of common ground for collective medical decision-making. *AI & Society, 20*(3), 331–350. doi:10.1007/s00146-005-0028-z

Walther, J. B. (1996). Computer-mediated communication: Impersonal, interpersonal, and hyperpersonal interaction. *Communication Research, 23*(1), 3–43. doi:10.1177/009365096023001001

Walther, J. B., & Parks, M. R. (2002). Cues filtered out, cues filtered in: Computer-mediated communication and relationships. In M. L. Knapp & J. A. Daly (Eds.), *Handbook of interpersonal communication* (3rd ed., pp. 529–563). Thousand Oaks, CA: Sage.

Yee, N. (2001, May). *The Norrathian scrolls: A study of EverQuest.* Retrieved from http://www.nickyee.com/report.pdf

Yee, N., & Bailenson, J. N. (2009). The difference between being and seeing: The relative contribution of self-perception and priming to behavioral changes via digital self-representation. *Media Psychology, 12*(2), 195–209. doi:10.1080/15213260902849943

Ziller, R. C. (1964). Individuation and socialization: A theory of assimilation in large organizations. *Human Relations, 17*(4), 341–360. doi:10.1177/001872676401700403

Zimbardo, P. G. (1969). The human choice: Individuation, reason, and order vs. deindividuation, impulse and chaos. In W. J. Arnold & D. Levine (Eds.), *Nebraska symposium on motivation* (Vol. 17, pp. 237–307). Lincoln, NE: University of Nebraska Press.

6

Social Effects of Virtual and Robot Companions

Nicole C. Krämer, Astrid M. Rosenthal-von der Pütten, and Laura Hoffmann

Social Psychology—Media and Communication, University of Duisburg-Essen, Duisburg, Germany

Imagine that you have bought a fancy new TV set which features a virtual, humanoid assistant that helps you to choose a program or assists you in video-recording the evening's programs. When your assistant asks you what program you would like to watch, do you think you would be more likely to choose a socially desirable alternative than you would if the alternatives were presented by a usual graphical user interface? Your answer will probably be "no," as—like the majority of people—you would not believe that a machine would be able to elicit social effects like those known from human–human interaction. However, when this setting was employed in a laboratory experiment, results showed that when a virtual program guide was present, participants indeed chose a socially desirable TV program (here: a documentary about Albert Einstein) more frequently than an entertainment program (Krämer, Bente, & Piesk, 2003). Numerous other studies also demonstrated that when interacting with artificial entities, participants showed social reactions. For example, social inhibition can be elicited by an embodied agent. In a study by Rickenberg and Reeves (2000), participants showed worse performance during a task when an animated character was present that seemed to be observing them. Krämer (2005) showed that embodied conversational agents (ECA) affect the way in which users communicate with a TV–VCR system. When an embodied interface agent is visible instead of a graphical user interface or a user interface with speech output, users address the system significantly more often using natural speech rather than using a remote control. Additional qualitative analyses of the semantic content of all speech acts indicate that users seem to have a more human-like attitude and behavior towards the system when it is represented by an anthropomorphic agent. According to these results, the question posed by Larry Hodges on his website, "If a digital human gave you a compliment, would it brighten your day?" should, on average, be answered rather with "yes" than "no."

The Handbook of the Psychology of Communication Technology, First Edition. Edited by S. Shyam Sundar.

In summary, effects of embodied interface agents can be termed "social" if a participant's emotional, cognitive, or behavioral reactions are similar to reactions shown during interactions with other human beings (Krämer, 2005). The studies that have been conducted in the research realm not only show that these—actually inappropriate—reactions really do occur, but they also provide evidence that the reactions are largely unconscious and that the social cues which elicit the effects can be rather subtle (see research on the media equation approach and computers-as-social-actors, below).

In the following, we would like to elaborate on explanations for these effects and systematize the boundary conditions for the occurrence of social reactions. We start by describing the media equation (Reeves & Nass, 1996) and computers-as-social-actors approach (CASA, Nass, Moon, Morkes, Kim, & Fogg, 1997; Nass, Steuer, & Tauber, 1994), which influenced our research in this area. While summarizing explanations for the phenomena (e.g., the ethopoiea concept, Nass & Moon, 2000; TMSI, Blascovich et al., 2002), we will present our own studies that tested alternative explanations and will continue by discussing further boundary conditions for the occurrence of social reactions. These include the role of agency, embodiment, features of the agent (especially verbal abilities and nonverbal behavior) and attributes of the user. The studies conducted in our lab which will be reported in this context all employ experimental designs, usually with between-subjects manipulation. After highlighting the findings, we will discuss theoretical implications and the conclusions to be drawn for the applied area of developing artificial entities. We will end by suggesting directions for future work. The technological focus of the chapter is on virtual characters and on robots. However, theoretical models and results might also be transferred to computers or mobile devices. Indeed, the research realm originates in studies and assumptions on the social effects of computers, which are presented in the following.

Social Reactions Towards Artificial Entities: Media Equation and CASA Studies

Reeves and Nass (1996) postulate that individuals treat computers and other artifacts as social actors, which they term the "media equation" (media equals real life). To test this assumption, they conducted numerous experiments within the Computers Are Social Actors (CASA) paradigm, in which human subjects had to interact with computers. These experiments all followed a similar pattern: Search for a social science finding, replace "human" with "computer" in the theory statement and method, and observe whether the social rule is still observable (Nass et al., 1994). The media equation is considered to be verified when the results resemble the findings from interpersonal contexts.

Examples of social rules that have been studied within the CASA paradigm are politeness, reciprocity, and gender stereotypes. Politeness toward computers was observed in a study in which participants had to evaluate a tutor-computer with which they had previously worked (Nass, Moon, & Carney, 1999). Participants gave significantly better evaluations when the tutor-computer itself asked for its evaluation than when the questioning was conducted by another computer. This is in line with findings from the interpersonal context, which revealed that people tend to be polite when a person asks a question about himself/herself in order not to hurt the person's feelings. The computers in these studies were not particularly humanlike, but possessed social cues like (a) speech output, (b) interactivity, and (c) the filling of roles traditionally filled by humans (e.g., tutor) (Nass, Steuer, Tauber, & Reeder, 1993). More recent studies tried to contribute to these prerequisites and boundaries by analyzing whether effects increase when social cues are even more salient, for example, when employing anthropomorphic virtual agents. Hoffmann, Krämer, Lam-Chi, and Kopp (2009) replicated the politeness experiment of Nass et al. (1999) with an ECA. Participants had to evaluate the ECA Max (at Bielefeld University) after a short interaction, whereby the questioning was conducted by Max itself, either a paper-and-pencil questionnaire in the same room or a paper-and-pencil questionnaire in a separate room. The results showed that participants were more polite, that is, provided better evaluations of Max's competence when Max itself asked for its evaluation, compared to the questionnaire in the same room. However, no significant difference was observed compared to the questionnaire in a separate room. Further analyses suggested that 37 out of 63 participants considered the programmer in their evaluation. When this was included as a second fixed factor in ANOVA, a significant interaction effect resulted, indicating that for the individuals who did not think about the programmer in their evaluation, the results of Nass et al. (1999) were replicated. In other words, the best evaluation was obtained through questioning by Max, followed by the questionnaire in the same room, and finally by the questionnaire in a separate room. However, for those who confessed that they had thought about the programmer in their evaluation, the order was different and less intuitive (separate room led to better evaluation than questioning by Max, which in turn was better than the same-room questionnaire condition). In conclusion, politeness behavior was also observable toward ECAs but the pattern of results was not as clear as in the study by Nass et al. (1999). Additionally, several participants reported that they thought about the programmer, which might indicate that the programmer plays a more important role than was the case in the classic studies by Nass.

It should be noted that not all studies analyzing people's reactions toward artificial entities found social effects in the manner and to the extent which was demonstrated in the CASA studies and their follow-ups. Based on their criticism that the setting in the CASA studies focused too narrowly on the computer asking for evaluation rather than allowing for and assessing actual interaction, Shechtman and Horowitz (2003) conducted an experiment in which they analyzed the actual conversational discourse and—in order to address a second shortcoming of the CASA approach—added an experimental condition in which the interlocutor was announced to be human. Within the CASA approach, the authors frequently conclude that the behavior toward a computer is the same as that toward fellow humans, although this is merely indirectly compared to social psychological insights derived from the literature and is not tested directly. In their study, Shechtman and Horowitz (2003) discovered as a result of the conversational analysis that human–human interaction is indeed different from human–computer interaction. When participants thought that their partner was human, they, for instance, used more words, spent more time and used more relationship statements. Despite these findings, it can be summarized that numerous studies yield evidence that artificial entities lead at least partly to social effects. Therefore, it seems justifiable that many research endeavors—including our own work—have been dedicated to the question of how these results can be explained.

Attempts to Explain Social Reactions Toward Artifacts or Computers

Different approaches that try to explain social reactions toward non-social artifacts have been mentioned in the literature. These approaches can generally be divided into (a) approaches that assume that the reactions cannot be considered as truly social as they either result from a deficit on the part of the user or are conscious reactions due to demand characteristics of the situation and (b) approaches that argue that social reactions toward artifacts occur unconsciously and are even denied by the human interlocutor (mindlessness, ethopoeia, computer as source). Additionally (c) it has been suggested that social reactions depend on the level of assumed agency (TMSI).

Ad (a) Early criticism with regard to the CASA study results stated that users who react in a social way must have deficits resulting from psychological dysfunctions, young age, or lack of experience. However, this can be denied, as the participants in the CASA studies were mostly healthy undergraduate students who had extensive experience in word processing (e.g., Nass et al., 1999; Nass, Fogg, & Moon, 1996). Similarly, the notion that the participants assume that they are rather interacting with the programmer than with the computer, that is, the computer is not seen as a source but merely as a medium, has been refuted and empirically disproven (Sundar, 1994; Sundar & Nass, 2000). A further explanation that continues to compete against the explanations below lies in the assumption that the observable reactions are not really "social" but are merely due to demand characteristics of the situation. It is argued that people show "as if" reactions (in the sense that people consciously tell themselves that they will talk to the artificial entity *as if* it were a person) that merely occur because appropriate scripts are missing when humans interact with computers (Kiesler & Sproull, 1997).

Ad (b) Supporters of the CASA paradigm see social reactions to artificial entities as truly social in the sense that "People respond socially and naturally to media even though they believe it is not reasonable to do so, and even though they don't think that these responses characterize themselves" (Reeves & Nass, 1996, p. 7). Nass and Moon (2000) suggest using the term "ethopoeia" as an explanation for this unconscious and automatic behavior (social reaction) which is inconsistent with one's opinion (computers do not need social treatment). According to this approach, minimal social cues like a human-sounding voice mindlessly (cf. Langer, 1989) trigger social responses because humans cannot avoid reacting automatically to social cues. The ethopoeia approach is supported by the fact that participants in the studies of Nass et al. obviously did not consciously recognize their social behaviors and when they were asked in the debriefing, they stated that they did not act socially (e.g., polite) toward the computers and that they believe such behavior to be inappropriate (Nass et al., 1999). Recent studies suggest that social responses to computers are indeed moderated by the extent of cognitive busyness (Lee, 2010). Also, it has been demonstrated that anthropomorphism of computers can happen mindlessly, leading to the social reactions that have been described (Kim & Sundar, 2012).

Additionally, one of our own recent studies yields support for the assumption that reactions to an artificial companion can be social, but largely unconscious. Krämer, Kopp, Becker-Asano, and Sommer (2013) show that people mimic the smiling behavior of a virtual agent. Mimicry is a widely cited phenomenon of human–human communication that has been shown to be especially indicative of the sociality of the situation. In a between-subjects design, 104 participants conducted an 8-min small-talk conversation with an agent that did not smile, showed occasional smiles, or displayed frequent smiles. Results show that the human interaction partners themselves smiled for longer when the agent was smiling. Interestingly, the smiling activity did not have an impact on people's evaluation of the agent, and nor were they able to reliably indicate whether the agent had smiled and whether this was occasional or frequent. Therefore, it can be concluded that the participants' behavioral reactions were rather unconscious and automatic. Moreover, since mimicry has been shown to be indicative of the sociality of the situation, the fact that participants smiled at the agent can be seen as evidence that they—at least unconsciously—experienced the situation as social.

Ad (c) In the course of the development of anthropomorphic characters, Blascovich (2002) established the TMSI that predicts the social verification of a virtual other depending on the factors of agency and behavioral realism. Agency here means the degree to which the virtual entity is controlled by a real human (low agency is present in the case of a virtual agent that is controlled by an autonomous computer program; high agency is given when a human controls the virtual character—which is then termed avatar). The authors assume a Threshold of Social Influence that has to be crossed to evoke social reactions by the user. This is only possible when the level of social verification is sufficiently high. When the factor of agency is high (i.e., when the user knows that the virtual character is a representation of a human being), then the factor of behavioral realism does not have to be high in order for social verification to take place and for social effects to occur. Conversely, when the factor of agency is low (i.e., when the user knows that the virtual character is a mere computer program), the factor of behavioral realism has to be very high to compensate for the lack of agency. In sum, it can be derived that according to the TMSI, the social influence of real persons will always be high, whereas the influence of an artificial entity depends on the realism of its behavior.

Since most aspects of the set of explanations presented under (a) have already been refuted by various studies, we will focus on the explanations (b) and (c) in the remainder of the chapter. The major difference here is that the TMSI model assumes that there is a fundamental difference between agents and avatars in the sense that users react socially to avatars (i.e., mediated fellow humans) but will only react socially to agents when they show sufficient social cues. The ethopoiea model, on the other hand, assumes that agents will automatically evoke social reactions in the same way as fellow humans do.

Testing of Explanations and Influencing Factors

Now that the most widely cited models, assumptions, and explanations for social reactions toward artificial entities have been summarized, we will provide an overview of recent research conducted by our own team and other teams which tested the models described earlier and identified further factors that influence whether social reactions are likely to occur. We will first describe an experiment that tested the ethopoiea assumption against the TMSI and thereby focuses on the factor of agency. Then, the factors, that is, (a) number of social cues, (b) embodiment of the artificial entity, (c) agent features such as appearance and verbal and nonverbal behaviors of the agent/robot, and (d) personality of the user are discussed.

The role of agency: TMSI versus media equation

Von der Pütten, Krämer, Gratch, and Kang (2010) empirically tested the TMSI against the ethopoeia approach (Nass & Moon, 2000). With the aim of testing the aforementioned assumptions, agency and behavioral realism of a virtual agent (the Rapport Agent; Gratch et al., 2006) were experimentally manipulated in a 2×2 between-subjects design. Participants were led to believe that they would be interacting either with another participant mediated by a virtual character or with an autonomous computer program. Moreover, the agent with higher behavioral realism featured responsive nonverbal behavior while participants were interacting with the agent, whereas the agent in the low behavioral realism condition showed only idle behavior (breathing, eye blinking), but no responsive behaviors. According to the TMSI, interaction effects between agency and behavioral realism should occur (in the sense that social reactions are observable in both avatar conditions but only in the agent condition with high behavioral realism). However, if the ethopoeia concept in its revised version (which acknowledges that automatic and unconscious social reactions will be stronger if there are more social cues; Morkes, Kernal, & Nass, 1999; Nass & Yen, 2012) is more accurate, social reactions should be reinforced when behavioral realism increases and should be independent of assumed agency. During the interaction, the Rapport Agent asked the participants intimate questions so that self-disclosure behavior of the participants could be used as dependent variable. Additionally, self-report scales to evaluate the virtual character as well as the situation were employed.

The data analyses revealed that the belief of interacting with either an avatar or an agent resulted in barely any differences with regard to the evaluation of the virtual character or behavioral reactions, whereas higher behavioral realism affected both. Although one main effect for the factor of agency was found, namely that participants in the agent condition experienced more negative feelings with *Low Dominance* than those in the avatar condition, no interaction effects of the factors of agency and behavioral realism emerged. However, three significant effects with regard to the variation of the behavioral realism were found. In the high behavioral realism condition, participants perceived the animated character as more naïve (presumably because the agent only nodded according to voice features and also showed this behavior when participants reported about negative events), they experienced more feelings of mutual awareness, and they used more words during the interaction. Ultimately, since main effects of behavioral realism, but no interaction effects, were found, the results support the Revised Ethopoeia Concept but not the TMSI.

However, it should be noted that a recent meta-analysis by Fox et al. (2014) rather provided evidence for the notion that the perception of agency is decisive when interacting with virtual characters. The analysis revealed that, overall, social reactions were stronger when people thought that they were interacting with another human compared to when they believed they were interacting with a computer program. Therefore, the role of agency in terms of the emergence of social reactions is still unclear.

The role of number of social cues

It is astonishing that the rather subtle variation of social and behavioral cues in the aforementioned study (virtual character with and without head nodding; von der Pütten, Krämer, Gratch, & Kang 2010) had more impact than the knowledge of whether one was interacting with a fellow human or a machine. However, it cannot be concluded from this that it is first and foremost social cues rather than agency that result in social reactions on the part of the user, because a multitude of social cues were present in all conditions, since in all conditions a human-like virtual agent was employed. A follow-up study (Appel, von der Pütten, Krämer, & Gratch, 2012) addressed this shortcoming by comparing the impact of agency and a more explicit variation of social cues. While agency is again varied by means of the instruction, the participants are given on the nature of the interaction partner, the number of social cues is varied by applying either a virtual figure (presented as either agent or avatar) or a text tool (presented as either text chat with a fellow human or as computer interface).

Comparable to the results of von der Pütten, Krämer, Gratch, and Kang (2010), data analyses revealed more effects regarding the variation of social cues compared to the factor of agency. For agency, only one main effect emerged: Participants reported having experienced higher social presence after communicating with the "other subject" via avatar or text chat than after communicating with the computer. This might, however, be due to the specific wording of the social presence scale, which explicitly asks about the presence of a real living person (e.g., "the person appears to be sentient (conscious and alive) to me"). With regard to the variation of the social cues, results showed that a human-like virtual character (high number of social cues) triggers stronger social reactions than a plain text-based interface (low number of social cues): Subjects in the virtual character conditions described the interlocutor less negatively and had a stronger feeling of mutual awareness as well as reciprocal attention allocation compared to the participants in the text conditions. In sum, the studies both by Appel et al. (2012) and by von der Pütten, Krämer, Gratch, and Kang (2010) therefore yield support for the assumption that "the more computers present characteristics that are associated with humans, the more likely they are to elicit social behavior" (Nass & Moon, 2000, p. 97; see also Nass & Yen, 2012).

The role of different embodiments: comparison studies of embodied virtual agents and robots

Due to the latest developments in the field of social robotics (robots for social purposes like personal coaches, companions, or guides), recent studies have aimed to compare the effects of robots in contrast to virtual characters. In the context of this chapter, the question that should be answered is whether social reactions depend on the form of embodiment.

In the literature, different effects of robots and agents have been described. In a review of potential explanations for these differences, Hoffmann and Krämer sum up that

> different sizes, different degrees of realism, the existence in or out of the real world (shared space), the presence or absence of a physical body, and the perception of another as socially present (Hoffmann & Krämer, 2011, p. 15)

may be responsible. For example, robots are generally taller than characters on a screen, which might lead to more favorable perceptions of the robot (Powers, Kiesler, Fussel, & Torrey, 2007). Also, the "physical proximity" (Powers et al., 2007) might play a role. Whereas virtual characters are only graphical 2D or 3D representations on a monitor, robots have a material embodiment. Accordingly, they are able to physically touch humans, carry things, and move on the ground. Therefore, it has been assumed that robots are perceived more as autonomous "living" systems than virtual characters, meaning that more social presence might be sensed when facing them (Jung & Lee, 2004). Although some studies indeed demonstrate that robots elicit stronger experiences and social behaviors than agents (Bartneck, 2002; Kiesler, Powers, Fussel, & Torrey, 2008), this has not been shown consistently (Wainer, Feil-Seifer, Shell, & Matarić, 2007; Yamato, Shinozawa, Naya, & Kogure, 2001; see also Hoffmann & Krämer, 2013).

Hoffmann and Krämer (2011) identified that one reason for these inconsistencies might be the lack of comparability due to different operationalizations of the embodiments (e.g., different robots, animation, and video recordings), dependent variables, and interaction scenarios. In particular, findings by Shinozawa, Naya, Yamato, and Kogure (2005) suggest that the interaction scenario and task might be decisive. Therefore, Hoffmann and Krämer (2011) conducted a 2 × 2 between-subjects experiment in which embodiment (rabbit-shaped Nabaztag robot vs. virtual version) and the interaction scenario (cognitive task vs. persuasive conversation) were systematically varied. For the purpose of the study, a 3D virtual character was designed that resembles the rabbit-shaped robot that was used and displays the same voice and behavior. Two different interaction scenarios were created: a persuasive conversation about health habits (cf. Kiesler et al., 2008) and a task-oriented scenario in which participants had to solve a Towers of Hanoi puzzle, which was set on the table, under the guidance of the artificial counterpart (cf. Wainer et al., 2007). As dependent variables, subjective evaluation criteria (e.g., affective state, perceived social presence, attractiveness, and general evaluation of the interaction) and objective measures were assessed. In the conversational scenario, persuasion was analyzed as objective measurement by means of the amount of healthy food participants ate after the interaction. In the task-oriented setting, the amount of moves to solve the Towers of Hanoi task was counted as performance measure.

Against the hypothesis that more social presence should be experienced in the presence of the physical robot, no differences emerged. Moreover, no differences caused by the embodiment were observable with respect to the participants' affective state, acceptance of the artificial entity, performance, or persuasion.

With regard to embodiment, two main effects emerged: First, participants perceived the robot as more competent with regard to the fulfillment of tasks (subscale task of the interpersonal attraction scale; McCroskey, Hamilton, & Weiner, 1974). The second main effect of embodiment occurred for the factor of control. Participants stated that they perceived more control during the interaction with the screen animation than with the robot. Additionally, a significant interaction between embodiment and interaction scenario with regard to the evaluation of the artificial counterpart emerged. In line with the findings of Shinozawa et al. (2005), task-oriented attractiveness was higher for the robot in the Towers of Hanoi condition, whereas it was higher for the screen animation in the conversational scenario.

Unlike in earlier comparison studies, Hoffmann and Krämer (2011) also examined the impact of the type of interaction scenario on the dependent variables and observed a main effect of the scenario on the affective state of the participants. Irrespective of the embodiment, participants reported that they felt better after the task-oriented scenario than after the conversational scenario. In addition, the comparison between the scenarios in terms of the amount of healthy food eaten after the interaction demonstrated that more healthy food was chosen after the conversation about healthy living behavior. Although this analysis was more or less a manipulation check, it demonstrated that robots as well as screen characters are able to persuade their users when the content of the conversation is carefully designed. In summary, the results of this study suggest that virtual characters can be used instead of more expensive robots when the aim of the application is of a persuasive nature. For scenarios in which physical manipulation is necessary at least on the side of the user, robots seem to be beneficial because they share the space of reference. Finally and most importantly, the study underlines the importance of the consideration of different contexts (i.e., task or interaction scenario) while analyzing the impact of different embodiments. Whether social effects (like persuasion by an artificial companion) can be observed will therefore depend not only on the form of embodiment alone but also on the appropriateness of the specific embodiment for the specific task or scenario.

The role of specific agent features

Just as it has been demonstrated that embodiment as either virtual agent or robot can matter for eliciting social reactions, it has also been shown that various characteristics of a virtual agent can influence the extent of social effects. Here, specific physical appearance as well as various communicative abilities and features can play a role.

Physical appearance: Several studies have shown that the appearance of a virtual character matters, for example, for acceptance and evaluation of the character (van Vugt, Konijn, Hoorn, Keur, & Eliens, 2007; Domagk, 2008, 2010). In particular, Domagk (2010) shows that when the appearance (and voice) is likeable, a pedagogical agent has more positive effects. Building on this, we compared the impact of a virtual tutor depending on its appearance as either a cartoon-like rabbit character or a realistic anthropomorphic agent (Sträfling, Fleischer, Polzer, Leutner, & Krämer, 2010). Results show that the rabbit-like agent was not only preferred, but people exposed themselves to the tutoring session for longer when the rabbit provided feedback. However, this was not related to an increase in learning performance. Other studies, which focus more on credibility rather than learning and likeability, show that characters that are more anthropomorphic are perceived as more credible (Nowak & Rauh, 2005). In sum, there is sufficient evidence to conclude that social effects also depend on aspects of physical appearance. However, given that different studies focus on different dimensions of appearance (e.g., realism, anthropomorphism, and attractiveness/likeability), it is still difficult to conclude which physical features are decisive.

Communicative behavior of the agent: It is self-evident that the behavior of the agent should also matter for its perception as well as for the social effects it is able to yield. As early as 2000, Rickenberg and Reeves showed that the fact of whether a virtual character on a website monitored the user or ignored him/her had an impact on the user's perceived anxiety. Additionally, in our studies, we analyzed in several experimental settings how the communicative behavior of the agent influences the users' reactions. In one experiment in which the agent acted as an interviewer, we varied quality (whether the agent engages in self-disclosure) and quantity (whether the agent is more talkative or less talkative) of the agent's utterances in a 2 × 2 between-subjects setting (von der Pütten, Klatt, Hoffmann, & Krämer, 2011). It was found that the agent's self-disclosure only affected perceived co-presence. However, verboseness influenced both the participants' verbal behavior (with regard to word usage and intimacy of answers) and their perception of the interview. Participants more often disclosed specific embarrassing situations, their biggest disappointment, and what they feel guilty about to the agent regardless of its previous self-disclosure. Moreover, the more talkative agent was generally evaluated more positively and the interview was perceived as being more pleasant. It is therefore possible that talkativeness led to a more favorable evaluation by the users and subsequently facilitated self-disclosure.

Unlike the findings of Moon (2000) and Joinson (2001) in previous studies, the experiment did not reveal a general effect for verboseness on breadth of disclosure (the total amount of words). However, for those questions that allowed for a sufficient variance (and were not merely answered with yes/no), we observed that participants used more words to answer when faced with the more talkative agent. These results might be explained by the Communication Adaptation Theory (Giles & Coupland, 1991; Giles, Mulac, Bradac, & Johnson, 1987) as well as by assumptions on linguistic alignment (Branigan, Pickering, Pearson, & McLean, 2010) (however, it has to be noted that these theories concentrate on more specific linguistic features such as rate of speech, latencies of pauses, or lexical or syntactic structures in speech). What can be concluded here is that people accommodate to the virtual interviewer in a way that would only be expected in human–human communication. In sum, it was found that it is not so much the quality of the agent's communicative behavior (in terms of self-disclosure) as rather the quantity that matters: A verbose agent was not only perceived more positively but also led to more self-disclosure on the part of the user.

Nonverbal communicative features of agents: Against the background of extensive research on communicative feedback such as back channeling (e.g., Cassell & Thórisson, 1999; Gratch et al., 2006), von der Pütten, Reipen, Wiedmann, Kopp, and Krämer (2008, 2009) conducted a study investigating whether emotional expressions used for communicative feedback also influence the effectiveness of human–agent interaction and the participants' perception of the interaction as well as their actual behavior. Participants in this study were engaged in a small-talk (10 min) situation with the virtual Agent MAX (Jung & Kopp, 2003). Two different types of feedback were implemented: an emotional feedback (e.g., smiles and compliments) and a communicative feedback that provided feedback about the comprehension of the participants' contributions and presented MAX as an attentive listener (e.g., nodding).

Unlike in earlier studies on agent feedback (e.g., Cassell & Thórisson, 1999), we did not find an impact of communicative feedback on perceived effectiveness and efficiency. Moreover, and in line with the results of previous work, emotional feedback had no influence on the effectiveness and efficiency of the conversation. However, the emotional feedback led to increased feelings of interest in the participants (von der Pütten et al., 2008). Notably, participants in the emotional feedback condition gave more compliments and thanked the agent more often than participants in the condition with communicative feedback (see also Mower, Feil-Seifer, Matarić, & Narayanan, 2007). Further studies also showed that nonverbal behavior can have beneficial—but also detrimental—effects on the perception of the virtual agent. Krämer, Simons, and Kopp (2007) demonstrated that when the virtual Agent Max showed self-touching gestures (e.g., touching its arm or face), this had positive effects on the experiences and evaluations of the user, whereas eyebrow-raising evoked less positive experiences and evaluations in contrast to no eyebrow-raising. In further experiments, we tried to replicate well-known phenomena from nonverbal communication research and social psychology. Social psychological research demonstrates that the same behavior might lead to different evaluations depending on whether it is shown by a man or a woman. In a 2 × 2 between-subjects experiment, we manipulated gaze behavior of the Rapport Agent (Gratch et al., 2006) and its gender in order to test whether female agents in particular are evaluated more negatively when they do not show gender-specific immediacy behavior and avoid gazing at the interaction partner. Instead of this interaction effect, we found two main effects: Gaze avoidance was evaluated negatively and female agents were rated more positively than male agents (Kulms, Krämer, Gratch & Kang, 2011). A further study was designed to experimentally investigate the influence of gesture usage and gender of a leader on the attribution of leadership skills in the context of a job interview (Klatt, Haferkamp, Tetzlaff, & Krämer, 2012). Here, we manipulated the gender of a virtual agent and nonverbal behavior (extensive use of hand and arm gestures vs. rather restrictive nonverbal behavior) in a 2 × 2 between-subjects design. Given the fact that gestures influence the perception of competence (Maricchiolo, Gnisci, Bonaiuto, & Ficca, 2009), we wanted to investigate the interrelations between the leader's gender and his or her nonverbal behavior in order to determine which factor (gender or people's nonverbal signals) is more decisive with regard to the perception and evaluation of a leader. Results confirmed the positive impact of extensive nonverbal behavior. Participants were more willing to hire the virtual person who used hand and arm gestures than the more rigid person. The virtual person using gestures was also perceived as exhibiting more leadership skills and general competence than the person in the non-gesture condition. Concerning leadership styles, nonverbal behavior had a significant influence on the attribution of task-oriented leadership skills (in contrast to person-oriented leadership skills that showed no effect). The virtual person's gender did not bring about any effect. Although the primary goal of the latter two studies was to replicate and extend social psychological findings and to use virtual agents merely as a research method, they still show that the virtual agent's nonverbal behavior is also important with regard to the emergence of social effects, namely the specific outcome of person perception. It was also shown that the agent's gender is not very influential—at least not in the way in which it was demonstrated in previous social psychological studies, in which gender was shown to be an important moderator for the evaluation of behavior.

Interestingly, nonverbal behavior also has an impact when it is shown by robots. Eimler, Krämer, and von der Pütten (2011) used the robot rabbit Nabaztag, a small WiFi-enabled device with movable ears, integrated RFID (radio-frequency identification, i.e. the wireless use of electromagnetic fields to transfer data) reader functionality, and speech-synthesis capability to analyze the effects of the rabbit's expression according to its ears. In a within-subjects setting, a German and a US American sample were asked to rate the rabbit's current emotional status from pictures that showed the rabbit with a variety of six different ear positions. Results indicate that people infer specific emotional states from the robot rabbit's different ear positions. For example, when the robot's ears are straight upright, it is perceived as attentive and dominant, whereas ears in a horizontal position are interpreted as sadness.

In conclusion, numerous nonverbal features have been demonstrated to influence the evaluation of the agent. We have found that emotional feedback, self-touching gestures, and illustrative gestures, as well as looking at the user lead to a more favorable evaluation of the agent—mirroring the results that have been found in human–human interaction.

The role of the user

There is a general lack of awareness regarding the influence of users' personality traits on human-agent interaction and its effects. Most studies neglect to include standardized questionnaires on personality. Numerous studies do not even consider explanatory variables like age and gender even though they are easily accessible. In order to determine the impact of personality and sociodemographic variables on social effects and evaluation of virtual agents, we reanalyzed a part of our studies with regard to the influence of participants' characteristics such as gender, age, computer literacy, and personality traits like the Big Five. Irrespective of the experimental manipulations in these studies, we consistently found that these characteristics affect how participants (a) experience the interactions and (b) perceive and evaluate artificial entities.

Gender. As Krämer (2004) outlined, males are still believed to have more affinity with technical artifacts than females. Furthermore, Krämer, Hoffmann, and Kopp (2010) revealed in their re-analysis of earlier studies that men and women have different preferences with regard to embodied agents. In fact, compared to the effects of age and computer literacy, the influence of gender was prevalent. In one study, women were found to be more nervous during the interaction with the agent, which is on the one hand unsurprising given that women are still said to be less accomplished with regard to computer technology (Krämer, 2004), but on the other hand, it contradicts the vision that embodied agents will facilitate human–computer interaction for these kinds of users. Our data suggest that female users' interest and acceptance can be increased when nonverbal behaviors are implemented (here: self-touching gestures) and when the agent frequently smiles. Therefore, we conclude that women especially benefit from an increased nonverbal behavior of the agent, in line with the finding that women are more sensitive to nonverbal behaviors (Hall, 1984). One can speculate that they are reassured when they perceive signals from the interlocutor that they are able to interpret. Male users, on the other hand, do not react with positive experiences when increased nonverbal behavior is shown. Thus, it can be concluded that agents for male users should be kept simple with regard to nonverbal behavior.

Age. Older users are generally less experienced with (new) technology than younger ones and show more anxiety (Laguna & Babcock, 1997). This was supported by our findings with the Agent MAX, which showed that older persons were more nervous when interacting with the agent than younger ones (Krämer et al., 2007). Moreover, older participants were more attentive when no nonverbal behavior was shown by the agent (Krämer et al., 2013). According to these results, the age of the users should be considered when designing the nonverbal behavior of artificial entities. For tasks in which the attention of the user is important for the task (e.g., a learning task), no nonverbal behavior should be displayed for older participants.

Computer literacy. Computer novices proved to be more nervous when interacting with the agent than other users. This is in line with previous findings that computer laypeople do not benefit from embodied agents in the way in which it is typically hoped (Krämer, Rüggenberg, Meyer zu Kniendorf, & Bente, 2002). The results of our studies showed that computer experts evaluate Max as more strained and gamers evaluate Max as more passive. This might be due to the fact that gamers are particularly used to sophisticated virtual characters.

Personality. Results of a study with the Rapport Agent show that participants' personality traits affected their subjective feelings after the interaction, as well as their evaluation of the agent and their actual behavior (von der Pütten, Krämer, & Gratch,

2010). From the various personality traits, those traits that relate to withstanding behavioral patterns in social contact (e.g., agreeableness, extraversion, approach avoidance, self-monitoring sensitivity, shyness, and public self-consciousness) were found to be predictive for the positive and negative feelings participants reported after the interaction, the evaluation of the agent, and the amount of words they used during the conversation. However, other personality traits (e.g., openness and neuroticism) as well as gender and age did not affect the evaluation. For instance, the higher one's rating on extraversion and public self-consciousness, the more words were used. Furthermore, the more shy people are, the more negatively they evaluate the agent, whereas agreeableness increases positive feelings after the interaction.

In sum, a person's attributes and disposition can greatly influence his or her evaluation and also his or her actual behavior during the interaction. Gender is especially influential in interaction with the agent's nonverbal behavior, in the sense that women prefer agents with (positive) nonverbal behavior. Age and computer literacy predict first and foremost users' nervousness and anxiety when interacting with agents—indicating that agents are not necessarily the best solution for people who are not experienced with technology. With regard to personality traits, some studies suggest that personality traits are better predictors for the evaluation outcome than the actual behavior of the agent, as this has been manipulated in the experiment. Therefore, we advise researchers to be aware of and assess participants' personality traits in order to be able to control for the effects they elicit. Researchers should also consider other instruments besides the well-known Big Five personality inventory, because our results suggest that the Big Five had limited explanatory value. Although there is a great deal of work concentrating on extraversion and agreeableness, other instruments might deliver more results, which might lead to interesting insights into the nature and the mechanisms of human–agent interaction and its differential social effects. In conclusion, several personality characteristics were shown to influence how people experience interactions with artificial entities and should therefore be carefully considered in the tailoring of interactions with artificial entities.

Summary of Theoretical Implications

The most important contribution to theory development provided by our research is the conclusion that can be derived concerning explanations for social effects. Most relevant for this was our experiment that tested the ethopoiea assumption against the TMSI. Here, we did not find the interaction that would have been expected by the TMSI, but predominantly found main effects of behavioral realism: In the conditions of high behavioral realism, participants experienced more feelings of mutual awareness, and they used more words during the interaction—the assumed agency, on the other hand, did not have an influence (von der Pütten, Krämer, & Gratch, 2010). The follow-up study by Appel et al. (2012) additionally showed that not only behavioral realism but also number of social cues plays a role. Therefore, in sum, we showed that artificial entities are able to evoke similar reactions to avatars (i.e., humans) and that these are stronger when more social cues and behavioral realism are present. In addition, the experiment by Krämer et al. (2013) perfectly demonstrated how participants unconsciously reacted to the behavior of the agent and mimicked its smile. Given that mimicry is known to be especially indicative of the sociality of the situation, this supports the theoretical assumption that humans automatically react in a social way to subtle behavior of an artificial agent.

While we did not find an influence of agency in the sense that people reacted differently to agents versus avatars, the research group of Bailenson presents different results in a recent meta-analysis (Fox et al., 2014). Here, agency is reported to be decisive for whether social reactions occur. Therefore, the influence of agency is not yet clear and warrants further investigation.

In order to highlight further factors that might influence whether social reactions occur, we conducted additional experiments and were able to identify further conditions such as the embodiment of the artificial entity. Indeed, social reactions seem to at least partly depend on the form of embodiment (robot versus virtual agent; Hoffmann & Krämer, 2011): Here, we did not find as many differences between robots and agents as expected, but the study underlines the importance of the consideration of different tasks or interaction scenarios. Whether social effects can be observed will depend not only on the form of embodiment but also on the appropriateness of the specific embodiment for the specific task or scenario. Moreover, we provided knowledge about the influence of agent features: As could have been expected from social psychological research, an artificial entity has to be attractive and likeable. In sum, there is sufficient evidence to conclude that social effects also depend on aspects of physical appearance (Sträfling et al., 2010). But also verbal and nonverbal behavior of the agent has an influence on the occurrence of social effects. In particular, the agent's verboseness had an effect on the participants' behavior: Here, verboseness of the agent led to more verboseness on the part of the participants and to more self-disclosure (von der Pütten, Klatt et al., 2011). The nonverbal behavior of the virtual character proved to be even more important: A series of studies show that nonverbal behavior matters especially—not only for the evaluation but also for the users' behavioral reactions. For example, emotional feedback led to increased feelings of interest in the participants (von der Pütten et al., 2008). More interestingly, participants in the emotional feedback condition gave more compliments and thanked the agent more often than participants in the condition with communicative feedback. Generally, qualitative as well as quantitative aspects of nonverbal behavior mattered: More nonverbal behavior proved to be better than less (Klatt et al., 2012; Simons et al., 2007) and with regard to quality, for example, gaze avoidance led to negative evaluation (Kulms et al., 2011). Nonverbal behavior in particular had a large impact on the person perception of the virtual entity. For example, when more gestures were present, more competence and leadership abilities were ascribed (Klatt et al., 2012).

Interestingly, the results on nonverbal behavior can probably also be transferred to robots, as can be derived from our study on the effects of robot ear positions (Eimler et al., 2011). With regard to the role of personality characteristics of the user, our most important finding, which should be considered for future model- and theory-building, is that especially those traits that relate to behavioral patterns in social contact (e.g., agreeableness, extraversion, approach avoidance, self-monitoring sensitivity, shyness, and public self-consciousness) were found to be predictive, whereas other personality traits (e.g., openness and neuroticism) did not affect the evaluation. The results with regard to gender, age, and computer literacy show that interacting with virtual agents might not lead to less anxiety in people who are not experienced with or interested in technology—unlike what is frequently expected in the literature.

In sum, our results show that people react socially to artificial entities and that the extent depends on several factors, such as number of social cues, embodiment, agent features, and user characteristics. Altogether, we interpret the findings as evidence for the fact that humans as social beings cannot prevent themselves from reacting to social cues even if they come from an artificial entity. We conclude that the findings also inform us about the nature of human sociality. In this respect, Kappas (2005) aptly stated that humans are "free monadic radicals" who are eager to search for social cues and keen to react to them in a social way.

Conclusions for Implementation

Results show that although, currently, interaction and communication abilities of robots and agents are still limited, social effects are already observable in the sense that people react similarly to how they would in human–human interaction. Nevertheless, this should not lead us to conclude that the abilities of artificial entities do not need to be advanced and developed further—indeed, numerous researchers in the area of robots and agents are contributing to the goal of making artificial entities more humanlike with regard to their communication abilities and therefore easier to use. However, in order to be able to achieve this, they need to welcome the contribution of psychologists and communication researchers. Most of the corresponding attempts in this regard center on the implementation of emotions (Picard, 1997, 2002). Based on the argument that an artificial entity with human-like emotions will be able to better understand human emotions and will therefore be more successful and efficient in the interaction, various emotion models have been implemented (Gratch & Marsella, 2004). It has been argued, however, that it might not be efficient and straightforward to implement emotions in order to facilitate interaction between humans and artificial entities (Krämer, 2008). If the ultimate goal is to improve interaction and to produce (nonverbal) behavior of the agent that is beneficial in a specific situation, emotions might not be directly helpful for choosing an appropriate behavior. This is due to the fact that in human–human communication, some scholars deny any direct link between emotions and nonverbal behavior and state that this link is even not reasonable since behavior should best be dependent on the demands of the social situation and not be controlled by emotions (Fridlund, 1991). Therefore, instead of implementing emotions, it was suggested that it is more reasonable to implement basic prerequisites for communication, for example, the ability to establish common ground, to engage in perspective-taking, or a theory of mind which enables people to infer mental states of others (Krämer, 2008; see also Krämer, Eimler, von der Pütten, & Payr, 2011).

Directions for Future Work

As depicted earlier, the research on the diverse boundary conditions we addressed in our studies has to be continued and amended. In particular, research that focuses on the explanations for social effects will be important since results in this regard are not yet consistent. In terms of the agency aspect, our results differed from those of the research group of Bailenson and Blascovich, and unlike them, we rather found evidence for the (revised) ethopoeia model instead of the TMSI. While they demonstrate in their studies that social effects and evaluations depend first and foremost on the agency (i.e., the assumption of whether the virtual character is an agent or avatar), we did not find an influence of agency but in particular found an influence of the agent's behavior. Here, future studies need to be designed which will clarify the role of agency. This will be especially important as it also tackles the basic nature of human sociality and will tell us to what degree humans are indeed "free monadic radicals" (Kappas, 2005).

Moreover, research on applied areas needs to be intensified. Especially in the realm of e-learning, it has been discussed that social effects of virtual agents can be beneficial (Krämer & Bente, 2010). Here, social effects in the sense that students might be affected by pedagogical agents in a similar way to how they would be affected by teachers might lead to increased motivation of the students. In this specific field, boundary conditions have to be still identified in future studies.

Future research should also take into account that most of the studies conducted so far have been experimental laboratory studies and therefore tried to increase internal validity. However, in order to learn whether social effects of artificial entities are actually relevant in humans' everyday encounters with agents or robots, field studies and long-term studies also have to be conducted, which provide higher external validity. Field studies with an open, exploratory approach are also important in order to be able to identify new aspects of human–agent or human–robot interaction that can then be checked in controlled laboratory studies.

In one study, we therefore chose a qualitative approach to investigate people's natural behavior in human–robot interaction. We analyzed participants' nonverbal behavior (gaze and proximity) in unscripted interactions between humans and the android robot "Geminoid HI-1" in a natural environment (an Austrian public café) (von der Pütten, Krämer, Becker-Asano, & Ishiguro, 2011). We found that participants' behavior toward the android robot was influenced by the behavior the robot displayed (whether it established eye contact). In addition, we found huge inter-individual differences in the participants' behavior. The emergence of huge inter-individual differences was also one of the most important results of a long-term study in which six elderly participants interacted with a robot serving as a health advisor (von der Pütten, Krämer, & Eimler, 2011). Following a multi-methodological approach, the continuous quantitative and qualitative description of user behavior on a very fine-grained level gave us insights into when and how people interacted with the robot companion. Post-trial semi-structured interviews explored how the users perceived the companion and revealed their attitudes. Based on this large data set, we can state that users are willing to start interactions and even conversation with a robot even though its perceptive and expressive capabilities are limited. Although aware of the fact that they are interacting with an artificial being, some of the users built relationships with the robot, while others treated the robot as a tool. Some people tended to like the robot as long as it was helping and they had a feeling of being in control. Others seemed to integrate the companion into their life, even though it did not always work properly and was perceived as being of limited use. Thus, the fine-grained analysis showed a particular prevalence of idiosyncratic reactions. Therefore, future research should extend field and long-term studies in order to identify patterns and factors that determine the conditions under which people show social reactions and whether they ultimately even develop relationships with artificial entities.

References

Appel, J., von der Pütten, A. M., Krämer, N. C., & Gratch, J. (2012). Does humanity matter? Analyzing the importance of social cues and perceived agency of a computer system for the emergence of social reactions during human-computer interaction. *Advances in Human-Computer Interaction*, *2012*(10). doi:10.1155/2012/324694

Bartneck, C. (2002). *eMuu - An embodied emotional character for the ambient intelligent home* (Unpublished PhD thesis). Eindhoven University of Technology, Eindhoven. Retrieved from http://www.bartneck.de/publications/2002/eMuu/bartneckPHDThesis2002.pdf

Blascovich, J. (2002). A theoretical model of social influence for increasing the utility of collaborative virtual environments. In W. Broll, C. Greenhalgh, & E. F. Churchill (Eds.), *CVE 2002. Proceedings of the 4th International Conference on Collaborative Virtual Environments* (pp. 25–30). New York, N.Y: ACM Press. doi:http://dx.doi.org/10.1145/571878.571883

Blascovich, J., Loomis, J. M., Beall, A. C., Swinth, K. R., Hoyt, C. L., & Bailenson, J. N. (2002). Immersive virtual environment technology as a methodological tool for social psychology. *Psychological Inquiry*, *13*, 103–124. doi:http://dx.doi.org/10.1207/S15327965PLI1302_01

Branigan, H. P., Pickering, M. J., Pearson, J., & McLean, J. F. (2010). Linguistic alignment between people and computers. *Journal of Pragmatics*, *42*(9), 2355–2368. doi:10.1016/j.pragma.2009.12.012

Cassell, J., & Thórisson, K. R. (1999). The power of a nod and a glance: Envelope vs. emotional feedback. *Applied Artificial Intelligence*, *13*, 519–538. doi:http://dx.doi.org/10.1080/088395199117360

Domagk, S. (2008). *Pädagogische Agenten in multimedialen Lernumgebungen. Empirische Studien zum Einfluss der Sympathie auf Motivation und Lernerfolg* [Pedagogical agents in multimedia learning environments. Empirical studies on the influence of likeability on motivation and performance]. Berlin, Germany: Logos.

Domagk, S. (2010). Do pedagogical agents facilitate learner motivation and learning outcomes? The role of the appeal of agent's appearance and voice. *Journal of Media Psychology: Theories, Methods, and Applications*, *22*(2), 84–97. doi:10.1027/1864-1105/a000011

Eimler, S. C., Krämer, N. C., & von der Pütten, A. M. (2011). How to design an engaging robot: Empirical results on determinants of acceptance and emotion attribution in confrontation with a robot rabbit. *Applied Artificial Intelligence*, *25*, 503–529. doi:10.1080/08839514.2011.587154

Fox, J., Anh, S. J., Janssens, J., Yeykelis, L., Segovia, K., & Bailenson, J. N. (2014). Avatars versus agents: A meta-analysis quantifying the effect of agency. *Human–Computer Interaction*, *2*, 1–61.

Fridlund, A. J. (1991). Evolution and facial action in reflex, social motive, and paralanguage. *Biological Psychology, 32*(1), 3–100. doi:http://dx.doi.org/10.1016/0301-0511(91)90003-Y

Giles, H., & Coupland, N. (1991). *Language: Contexts and consequences*. Monterey, CA: Brooks/Cole.

Giles, H., Mulac, A., Bradac, J. J., & Johnson, P. (1987). Speech accommodation theory: The first decade and beyond. In M. McLaughlin (Ed.), *Communication yearbook* (Vol. 10, pp. 13–48). Newbury Park, CA: Sage.

Gratch, J., & Marsella, S. (2004). Evaluating the modeling and use of emotion in virtual humans. In *Proceedings of the Third International Joint Conference on Autonomous Agents and Multiagent Systems*, New York (Vol. 1, pp. 320–327). doi:10.1109/AAMAS. IEEE Computer Society Washington, DC, USA © 2004.117

Gratch, J., Okhmatovskaia, A., Lamothe, F., Marsella, S., Morales, M., van der Werf, R. J., & Morency, L.-P. (2006). Virtual rapport. In J. Gratch, M. Young, R. Aylett, D. Ballin, & P. Olivier (Eds.), *Lecture notes in computer science* (Vol. 4133, pp. 14–27). Berlin/Heidelberg, Germany: Springer. doi:http://dx.doi.org/10.1007/11821830_2

Hall, J. A. (1984). *Nonverbal sex differences. Communication accuracy and expressive style*. Baltimore, MD: Johns Hopkins University Press.

Hoffmann, L., & Krämer, N. C. (2011). *How should an artificial entity be embodied? Comparing the effects of a physically present robot and its virtual representation*. Paper presented at the HRI 2011 Workshop on Social Robotic Telepresence, Lausanne, Switzerland, (pp. 14–20).

Hoffmann, L. & Krämer, N. C., (2013). Investigating the effects of physical and virtual embodiment in task-oriented and conversational contexts. *International Journal of Human-Computer Studies, 71*(7–8), 763–774.

Hoffmann, L., Krämer, N. C., Lam-Chi, A., & Kopp, S. (2009). Media equation revisited: Do users show polite reactions towards an embodied agent? In Z. Ruttkay, M. Kipp, A. Nijholt, & H. Vilhjálmsson (Eds.), *Lecture Notes in Computer Science: Vol. 5773, Intelligent Virtual Agents 2009* (pp. 159–165). Amsterdam, The Netherlands: Springer-Verlag. doi:http://dx.doi.org/10.1007/978-3-642-04380-2_19

Joinson, A. N. (2001). Knowing me, knowing you: Reciprocal self-disclosure in internet-based surveys. *Cyberpsychology and Behavior, 4*(5), 587–591. doi:10.1089/109493101753235179

Jung, B., & Kopp, S. (2003). FlurMax: An interactive virtual agent for entertaining visitors in a hallway. In T. Rist, R. Aylett, D. Ballin, & J. Rickel (Eds.), *IVA 2003, LNAI 2792* (pp. 23–26). Berlin/Heidelberg: Springer-Verlag. doi:10.1007/978-3-540-39396-2_5

Jung, Y., & Lee, K. M. (2004). Effects of physical embodiment on social presence of social robots. *Proceedings of PRESENCE 2004*, Valencia, Spain (pp. 80–87).

Kappas, A. (2005). *My happy vacuum cleaner*. Paper presented at the ISRE General Meeting, Symposium on Artificial Emotions, Bari.

Kiesler, S. B., Powers, A., Fussel, S. R., & Torrey, C. (2008). Anthropomorphic interactions with a robot and robot-like agent. *Social Cognition, 26*(2), 169–181. doi:10.1521/soco.2008.26.2.169

Kiesler, S. B., & Sproull, L. (1997). Social responses to "social" computers. In B. Friedman (Ed.), *Human values and the design of technology* (pp. 191–200). Cambridge, MA: Cambridge University Press/CSLI.

Kim, Y., & Sundar, S. S. (2012). Anthropomorphism of computers: Is it mindful or mindless? *Computers in Human Behavior, 28*, 241–250.

Klatt, J., Haferkamp, N., Tetzlaff, L., &. Krämer, N. C. (2012). *How to be... a leader—Examining the impact of gender and nonverbal behavior on the perception of leaders*. Paper presented at the Annual Meeting of the International Communication Association, Phoenix, June 2012.

Krämer, N. C. (2004). Mensch-computer-interaktion. In R. Mangold, P. Vorderer, & G. Bente (Hrsg.), *Lehrbuch Medienpsychologie* (S. 643–672). Göttingen, Germany: Hogrefe.

Krämer, N. C. (2005). Social communicative effects of a virtual program guide. In T. Panayiotopoulos et al. (Eds.), *IVA 2005. Lecture notes in computer science* (pp. 442–543). Berlin/Heidelberg, Germany: Springer. doi:http://dx.doi.org/10.1007/11550617_37

Krämer, N. C. (2008). Theory of mind as a theoretical prerequisite to model communication with virtual humans. In I. Wachsmuth & G. Knoblich (Eds.), *Modeling communication with robots and virtual humans* (pp. 222–240). Berlin, Germany: Springer. doi:http://dx.doi.org/10.1007/978-3-540-79037-2_12

Krämer, N. C., & Bente, G. (2010). Personalising e-learning. *The social effects of pedagogical agents. Educational Psychology Review*, *22*(1), 71–87. doi:10.1007/s10648-010-9123-x

Krämer, N. C., Bente, G., & Piesk, J. (2003). The ghost in the machine. The influence of embodied conversational agents on user expectations ans user behavior in a TV/VCR application. In G. Bieber & T. Kirste (Eds.), *IMC Workshop 2003, assistance, mobility, applications*. Stuttgart, Germany: Fraunhofer IRB Verlag.

Krämer, N. C., Eimler, S. C., von der Pütten, A. M., & Payr, S. (2011). Theory of companions: What can theoretical models contribute to applications and understanding of human-robot interaction? *Applied Artificial Intelligence*, *25*(6), 474–502. doi:10.1080/08839514.2011.587153

Krämer, N. C., Hoffmann, L., & Kopp, S. (2010). Know your users! Empirical results for tailoring an agent's nonverbal behavior to different user groups. In J. Allbeck, N. Badler, T. W. Bickmore, C. Pelachaud, & A. Safonova (Eds.), *Lecture Notes in Computer Science: Vol. 6356, Intelligent Virtual Agents 2010* (pp. 468–474). Berlin/Heidelberg, Germany: Springer. doi:http://dx.doi.org/10.1007/978-3-642-15892-6_50

Krämer, N. C., Kopp, S., Becker-Asano, C., & Sommer, N. (2013). Smile and the world will smile with you: The effects of a virtual agent's smile on users' evaluation and behavior. *International Journal of Human-Computer Studies*, *3*(71), 335–349. doi:http://dx.doi.org/10.1016/j.ijhcs.2012.09.006

Krämer, N., Rüggenberg, S., Meyer zu Kniendorf, C., & Bente, G. (2002). Schnittstelle für alle? Möglichkeiten zur Anpassung anthropomorpher Interface Agenten an verschiedene Nutzergruppen. In M. Herzceg, W. Prinz, & H. Oberquelle (Eds.), *Mensch und Computer* (pp. 125–134). Stuttgart, Germany: Teubner. doi:http://dx.doi.org/10.1007/978-3-322-89884-5_13

Krämer, N. C., Simons, N., & Kopp, S. (2007). The effects of an embodied conversational agent's nonverbal behavior on users evaluation and behavioral mimicry. In C. Pelachaud et al. (Eds.), *Lecture notes in computer science* (pp. 238–251). Berlin/Heidelberg, Germany: Springer. doi:http://dx.doi.org/10.1007/978-3-540-74997-4_22

Kulms, P., Krämer, N. C., Gratch, J., & Kang, S.-H. (2011). It's in their eyes: A study on female and male virtual humans' gaze. In H. H. Vilhjálmsson, S. Kopp, S. Marsella, & K. R. Thórisson (Eds.), *Lecture notes in computer science* (pp. 80–92). Berlin/Heidelberg, Germany: Springer. doi:http://dx.doi.org/10.1007/978-3-642-23974-8_9

Laguna, K., & Babcock, R. L. (1997). Computer anxiety in young and older adults: Implications for human-computer interactions in older populations. *Computers in Human Behavior*, *13*(3), 317–326. doi:http://dx.doi.org/10.1016/S0747-5632(97)00012-5

Langer, E. J. (1989). *Mindfulness*. Reading, MA: Addison-Wesley.

Lee, E.-J. (2010). What triggers social responses to flattering computers? Experimental tests of anthropomorphism and mindlessness explanations. *Communication Research*, *37*(2), 191–214. doi:10.1177/0093650209356389

Maricchiolo, F., Gnisci, A., Bonaiuto, M., & Ficca, G. (2009). Effects of different types of hand gestures in persuasive speech on receivers' evaluations. *Language and Cognitive Processes, 24*(2), 239–266. doi:10.1080/01690960802159929

McCroskey, J. C., Hamilton, P., & Weiner, A. (1974). The effect of interaction behavior on source credibility, homophily, and interpersonal attraction. *Human Communication Research, 1*(1), 42–52. doi:10.1111/j.1468-2958.1974.tb00252.x

Moon, Y. (2000). Intimate exchanges: Using computers to elicit self-disclosure from consumers. *Journal of Consumer Research, 26*(4), 323–339. doi:10.1086/209566

Morkes, J., Kernal, H., & Nass, C. (1999). Effects of humor in task-oriented human-computer interaction and computer-mediated communication: A direct test of SRCT theory. *Human-Computer Interaction, 14*(4), 395–435. doi:10.1207/S15327051HCI1404_2

Mower, E., Feil-Seifer, D. J., Matarić, M. J., & Narayanan, S. (2007). Investigating implicit cues for user state estimation in human-robot interaction using physiological measurements. In *Proceedings of the 16th IEEE International Workshop on Robot and Human Interactive Communication (RO-MAN 2007)* (pp. 1125–1130). Jeju Island, Korea: IEEE Press. doi:http://dx.doi.org/10.1109/ROMAN.2007.4415249

Nass, C., Fogg, B. J., & Moon, Y. (1996). Can computers be teammates? *International Journal of Human-Computer Studies, 45*, 669–678. doi:http://dx.doi.org/10.1006/ijhc.1996.0073

Nass, C., & Moon, Y. (2000). Machines and mindlessness: Social responses to computers. *Journal of Social Issues, 56*(1), 81–103. doi:10.1111/0022-4537.00153

Nass, C., Moon, Y., & Carney, P. (1999). Are people polite to computers? Responses to computer-based interviewing systems. *Journal of Applied Social Psychology, 29*(5), 1093–1110. doi:http://dx.doi.org/10.1111/j.1559-1816.1999.tb00142.x

Nass, C., Moon, Y., Morkes, J., Kim, E.-Y., & Fogg, B. J. (1997). Computers are social actors: A review of current research. In B. Friedman (Ed.), *Moral and ethical issues in human-computer interaction* (pp. 137–162). Stanford, CA: CSLI Press.

Nass, C., Steuer, J., & Tauber, E. R. (1994). Computers are social actors. In *Proceedings of the SIGCHI Conference of Human Factors in Computing Systems (CHI 1994)* (pp. 72–78). New York, NY: ACM Press. doi:http://dx.doi.org/10.1145/191666.191703

Nass, C., Steuer, J., Tauber, E. R., & Reeder, H. (1993). Anthropomorphism, agency, and ethopoeia: computers as social actors. In S. Ashlund, K. Mullet, A. Henderson, E. Hollnagel, & T. White (Eds.), *INTERCHI'93 Conference on Human Factors in Computing* (pp. 111–112). New York, NY: ACM. Retrieved from http://dblp.uni-trier.de/db/conf/chi/chi1993c.html#NassSTR9

Nass, C., & Yen, C. (2012). *The man who lied to his laptop. What we can learn about ourselves from our machines.* New York, NY: Penguin.

Nowak, K. L., & Rauh, C. (2005). The influence of the avatar on online perceptions of anthropomorphism, androgyny, credibility, homophily, and attraction. *Journal of Computer Mediated Communication, 11*(1), 153–178.

Picard, R. (1997). *Affective computing.* Cambridge, MA: MIT Press.

Picard, R. W. (2002). What does it mean for a computer to "have" emotions? In R. Trappl, P. Petta, & S. Payr (Eds.), *Emotions in humans and artifacts* (pp. 213–236). Cambridge, MA: MIT Press.

Powers, A., Kiesler, S. B., Fussel, S. R., & Torrey, C. (2007). Comparing a computer agent with a humanoid robot. In *Proceedings of the 2nd ACM/IEEE International Conference on Human Robot Interaction (HRI 2007)* (pp. 145–152). Virginia, USA: ACM Press.

Reeves, B., & Nass, C. (1996). *The media equation: How people treat computers, television, and new media like real people and places.* Stanford, CA: Cambridge University Press.

Rickenberg, R., & Reeves, B. (2000). The effects of animated characters on anxiety, task performance, and evaluations of user interfaces. In T. Turner & S. Pemberton (Eds.),

CHI 2000. Proceedings of the SIGCHI Conference on Human Factors in Computing Systems (pp. 49–56). New York, NY/Harlow, UK: ACM Press.

Shechtman, N., & Horowitz, L. M. (2003). Media inequality in conversation: How people behave differently when interacting with computers and people. In *CHI'03. Proceedings of the SIGCHI Conference on Human Factors in Computing Systems* (pp. 281–288). Ft. Lauderdale, FL: ACM Press. doi:10.1145/642611.642661

Shinozawa, K., Naya, F., Yamato, J., & Kogure, K. (2005). Differences in effect of robot and screen agent recommendations on human decision-making. *International Journal of Human-Computer Studies, 62*(2), 267–279. doi:10.1016/j.ijhcs.2004.11.003

Simons, N., Krämer, N. C. & Kopp, S. (2007). The effects of an embodied agent´s nonverbal behaviour on user´s evaluation and behavioral mimicry. Paper presented at the ICA 2007 (International Communication Association), May 2007, San Francisco.

Sträfling, N., Fleischer, I., Polzer, C., Leutner, D., & Krämer, N. C. (2010). Teaching learning strategies with a pedagogiocal agent. The effects of a virtual tutor and its appearance on learning and motivation. *Journal of Media Psychology: Theories, Methods, and Applications, 22*(2), 73–83.

Sundar, S. S. (1994). *Is human-computer interaction social or parasocial?* Paper presented at the Annual Meeting of the Association for Education in Journalism and Mass Communication, Atlanta, Georgia.

Sundar, S. S., & Nass, C. (2000). Source orientation in human-computer interaction: Programmer, networker, or independent social actor. *Communication Research, 27*(6), 683–703. doi:10.1177/009365000027006001

van Vugt, H., Konijn, E., Hoorn, J., Keur, I., & Eliens, A. (2007). Realism is not all! User engagement with task-related interface characters. *Interacting with Computers, 19*, 267–280.

von der Pütten, A. M., Klatt, J., Hoffmann, L., & Krämer, N. C. (2011). Quid Pro Quo? Reciprocal self-disclosure and communicative accomodation towards a virtual interviewer. In H. Vilhjálmsson (Ed.), *LNAI 6895, IVA 2011* (pp. 183–194). Heidelberg, Germany: Springer.

von der Pütten, A. M., Krämer, N. C., Becker-Asano, C., & Ishiguro, H. (2011, March 6–9). An Android in the Field. *Proceedings of the 6th ACM/IEEE International Conference on Human-Robot Interaction (HRI'11)*, Lausanne, Switzerland (pp. 283–284). doi:10.1145/1957656.1957772

von der Pütten, A. M., Krämer, N. C., & Eimler, S. C. (2011, November 14–18). Living with a robot companion–Empirical study on the interaction with an artificial health advisor. *Proceedings of the 11th International Conference on Multimodal Interaction, ICMI'11*, Alicante, Spain. doi:10.1145/2070481.2070544

von der Pütten, A. M., Krämer, N. C., & Gratch, J. (2010). How our personality shapes our interactions with virtual characters: Implications for research and development. In J. Allbeck, N. Badler, T. W. Bickmore, C. Pelachaud, & A. Safonova (Eds.), *Lecture Notes in Computer Science: Vol. 6356, Intelligent Virtual Agents 2010* (pp. 208–221). Berlin/Heidelberg, Germany: Springer.

von der Pütten, A. M., Krämer, N. C., Gratch, J., & Kang, S.-H. (2010). "It doesn't matter what you are!" Explaining social effects of agents and avatars. *Computers in Human Behavior, 26*(6), 1641–1650. doi:10.1016/j.chb.2010.06.012

von der Pütten, A. M., Reipen, C., Wiedmann, A., Kopp, S., & Krämer, N. C. (2008). Comparing emotional vs. envelope feedback for ECAs. In J. G. Carbonell et al. (Eds.), *Proceedings of the 8th international Conference on Intelligent Virtual Agents* (pp. 550–551). Tokyo, Japan: Springer-Verlag.

von der Pütten, A. M., Reipen, C., Wiedmann, A., Kopp, S., & Krämer, N. C. (2009). The impact of different embodied agent-feedback on users' behavior. In Z. Ruttkay, M. Kipp, A. Nijholt, & H. Vilhjálmsson (Eds.), *Lecture Notes in Computer Science: Vol. 5773, Intelligent Virtual Agents 2009* (pp. 549–551). Amsterdam, The Netherlands: Springer-Verlag.

Wainer, J., Feil-Seifer, D. J., Shell, D., & Matarić, M. J. (2007). Embodiment and human-robot interaction: A task-based perspective. In *Proceedings of the 16th IEEE International Workshop on Robot and Human Interactive Communication (RO-MAN 2007)*, Jeju, Korea. doi:10.1109/ROMAN.2007.4415207

Yamato, J., Shinozawa, K., Naya, F., & Kogure, K. (2001). Evaluation of communication with robot and agent: Are robots better social actors than agents? *Proceedings of IFIP INTERACT01: Human-Computer Interaction 2001*, Tokyo, Japan (pp. 690–691).

7

Telepresence and Apparent Agency in Human–Robot Interaction

Leila Takayama
Google[x], Mountain View, CA, USA

Introduction

What is it like to be a robot? There are an increasing number of people around the world who can actually begin to answer this question because they are using telepresence robots as part of their everyday routines. These people are using telepresence robots to telecommute to work from a distance, wandering the office hallways to find their coworkers, lingering by the coffee machine to catch up with the latest gossip, and engaging in both formal and informal meetings with their colleagues (Lee & Takayama, 2011; Takayama & Go, 2012). In other contexts, people are also starting to use telepresence robots to attend school (Fels, Waalen, Zhai, Shumin, & Weiss, 2001), participate in athletic events (Smith, 2011), and visit with family members or patients (Cesta, Cortellessa, & Tiberio, 2011).

Some humanoid telepresence robots take on the form of their operators (e.g., Geminoid (Sakamoto, Kanda, Ono, Ishiguro, & Hagita, 2007)), whereas others use video displays to represent any person who is operating them (e.g., Personal Roving Presence (Paulos, 1997)). Kristoffersson, Coradeschi, and Loutfi (2013) provide a review of the state of the art in robotic telepresence (aka: mobile remote presence, remote presence devices, or robot-mediated communication). Our own telepresence robots were developed by Dallas Goecker and Curt Meyers at Willow Garage, Inc., who used parts from other robots around the lab to build a telepresence robot that would help Dallas, who lives in Indiana, work more effectively in our California office. After several iterations upon the hardware system design, they built approximately 25 units of the version 7 prototype, which we call the Texai Alphas. The Texai Alpha prototypes stand approximately 1.57 m tall, which includes a "head" that contains a color LCD touch screen, pan-tilt web camera, audio speakers, and a USB microphone. There is also a motorized base,

The Handbook of the Psychology of Communication Technology, First Edition. Edited by S. Shyam Sundar.

Figure 7.1 A Texai Alpha (foreground) in use at an all-company weekly meeting.

which contains a remotely controlled active caster in the back and two passive wheels in the front, a polycarbonate bumper, a computer, and a large battery that holds approximately 8 hours of charge. It is controlled via a web-based graphical user interface and videochat program. We typically refer to theremote person who is using the robot to "be" here as the "operator." We refer to the people who live and work in the same physical environment as the robot as the "locals" (Figure 7.1).

These telepresence robots are empirical technologies (Polanyi, 1962) that raise questions about the ***apparent agency*** that people experience when enhancing their ability to "be there" in distant locations. Apparent agency refers to how much agency something or someone seems to have in-the-moment of an interaction (as opposed to what seems to be true upon further reflection). In this case, "agency" refers to how much an agent (whether human or nonhuman) can act in the world of its own accord. Autonomous robots might be perceived as having a high degree of apparent agency (e.g., achieving the robot's goals). For example, when a robot completes tasks like plugging itself into a power outlet, it is extremely common for people to cheer for the robot and praise it (like a child or student), even though it was the human engineering work that actually made the task possible. Telepresence robots are often different in that they typically inherit their agency from the human tele-operators who drive them. For example, when a telepresence robot operator runs into a wall and dents it, the operator is the one who gets blamed and scolded for the error, not the robot itself.

Apparent agency is perceived; it lies in the eyes of the beholder. Apparent agency lies in contrast to other conceptions of agency as being an inherent property. For example, social cognitive theory conceptualizes agency as something that is inherent to humans (Bandura, 2006). In social cognitive theory, human agency has four core properties: intentionality, forethought, self-reactiveness, and self-reflectiveness. Social cognitive theory is useful for making sense of human agency, but it does not speak to issues of *perceptions* of agency in things that we do not reflectively believe to be agents.

Actor network theory claims that there are many actors out in the world that are nonhuman, yet we have delegated agency to them, for example, automatic doors "discriminate against very little and very old persons" (Latour, 1992). This kind of agency is projected upon nonhuman actors not only in sociological analyses of the world, but also in everyday human responses to computing technologies (Levin, Adams, Saylor, & Biswas, 2013; Nass & Moon, 2000).

Theoretical Grounding

As empirical technologies, telepresence robots inspire further scientific research about how and why people can expand their sense of where they are and what they can do (Merleau-Ponty, 1962). We have observed the Texai Alpha prototypes become *invisible-in-use* (Heer & Khooshabeh, 2004; Takayama, 2011), "disappearing" into the background of conscious attention. In fact, there have been several moments when we have been unable to recall whether Dallas, our colleague, was in a particular conversation in person or "in robot." This is reminiscent of one of the goals of Mark Weiser's vision for ubiquitous computing (Weiser, 2002)—that computing quietly empowers people, being ready-at-hand like the carpenter's trusty hammer (Heidegger, 1992).

Along similar lines of thought, ecological psychologist James Gibson wrote about the automobile driver's experience of feeling like there is a field of possible actions that extends like a tongue in front of the car (Reed & Jones, 1982); fast, nimble sports cars might have longer and wider tongues than slow, lumbering cars. Similarly, Merleau-Ponty referred to one's sense of "I can" that is influenced by the tools that one has mastered (Merleau-Ponty, 1962). Andy Clark further wrote about this experience in discussing natural-born cyborgs (Clark, 2003). The common thread is that technologies can become integrated with one's sense of self and sense of one's own capabilities.

Because these experiences often refer to people using familiar technologies such as forks, eyeglasses, and walking canes (McCloud, 1993), we were surprised to find similar phenomena happening with relatively novel technologies like telepresence robots. From the remote operator's perspective, the telepresence robots can become invisible-in-use. The remote operators feel like they "are" the robots in that they can express their own agency through the robots. Similarly, our prior work has also found that cars, cell phones, and laptop computers are also commonly mentioned as becoming invisible-in-use (Takayama, 2010).

From the local's perspective, the robots often take on the *apparent agency* of the remote operator. That is, it feels like the operator is "just there," not like the operator is talking through a robot. When I talk with Dallas, I simply talk with Dallas. I do not think about the fact that we are making use of any technology to converse with each other.

These seemingly nonsensical phenomena highlight the importance of apparent agency. If only humans could have agency and machines were just machines, then this notion of apparent agency would not be useful and people would not experience or talk about incorporating telepresence robots into their sense of self. Study 1 in this chapter shows otherwise; people use auditory and visual feedback more effectively incorporate the telepresence robots into their presentation of self. Similarly, if apparent agency did not matter, then we would also predict that agency exists, independent of the observer. Study 2 in this chapter shows otherwise. Finally, if this notion of apparent agency was not important, then people would only form a sense of in-group membership with other people, not with the telepresence robots. Study 3 in this chapter also shows otherwise. Through qualitative field observations and quantitative findings from a series of lab experiments, we are compiling a body of work that points toward the importance of apparent agency and teasing out the factors that influence it.

Empirical Grounding in Field Explorations

These types of robotic systems are not new. They have been built as various prototypes for at least a decade in the form of the personal roving presence (Paulos, 1997; Paulos & Canny, 2001), mobile interactive avatars (Ishiguro & Trivedi, 1999), PEBBLES for hospitalized children (Fels et al., 2001), and BiReality (Jouppi, Iyer, Thomas, & Slayden, 2004). What is different today is that the wireless network capabilities have recently improved to the point of being able to support these robotics systems to function reliably and over much longer periods of time (e.g., more than just half an hour). Being able to put these robotic systems to use outside of highly instrumented laboratory environments means that we are now able to observe and learn from the ways that people use these systems as part of their everyday lives.

We ran a set of field studies over a period of many months so that we could observe a variety of people using the telepresence robots as part of their typical workdays. Our goal was to explore what is going on when people start to feel that they are not just operating a robot, but that they "are" the robot. In our initial work, we simply aimed to identify the use cases and social norms that formed around geographically distributed work teams that made use of our telepresence robots (Lee & Takayama, 2011). Using a combination of critical incident interviews (Flanagan, 1954), contextual inquiry observations (Beyer & Holtzblatt, 1998), and surveys across three companies that used the robots for a period of 2–18 months, we found that the majority of positive incidents (76%) and most positive effects upon work activities involved informal communications (e.g., impromptu work meetings, hallway encounters, or hanging out in kitchen spaces). Furthermore, people tended to use the systems to express their commitment to being present and available for their co-workers, capturing and maintaining their coworkers' attention, and building social connections, which are consistent with prior findings from geographically distributed work teams (Nardi, 2005). Many of the social norms that formed around these systems were drawn from existing technologies such as telephones, but sometimes the behaviors that were appropriate for those other technologies were inappropriate for this case. For example, people typically hang up the telephone at the end of the conversation, but logging off of a robot at the end of the conversation is rude to the locals because it leaves them with the burden of putting the robot away in its charging station when the operator really should have done that. Similarly, it is usually fine for people to hit the volume increase and volume decrease buttons on conference phones, but several

Figure 7.2 Texai operators being given personal space in a group meeting.

operators became agitated with locals would push "their" buttons (on the robot) because it felt like a violation of personal space (Figure 7.2).

As the field studies continued on for even longer periods of time, we began to notice even more inconsistencies in the ways that people treated and responded to the remote operators (Takayama & Go, 2012). We know that people have a tendency toward treating computers, televisions, and other media as though they were real people (Nass, Steuer, & Tauber, 1994; Reeves & Nass, 1996). Videoconferencing systems are typically conceptualized as being windows (Fish, Kraut, & Chalfonte, 1990) or separate spaces (Bly, Harrison, & Irwin, 1993). The situation with robots is much less straightforward. People tend to draw from multiple source domains when responding to telepresence robots, including both nonhuman metaphors (e.g., "Skype on wheels" or "Robot") and human-like metaphors (e.g., "JohnBot"); the most unexpected metaphor that consistently arose over time across field sites was that of a person with disabilities (Takayama & Go, 2012). Because the robots started to inherit their apparent agency from the remote operators, people started to compare the telepresence robots to people more directly. They noticed that the robots did not have arms and often had to be assisted to open and close meeting room doors. Locals would often write larger, speak louder, and walk slower when interacting with remote operators. Particularly when opening doors for the remote operators, locals would comment upon how much easier it was to open doors for robots than for wheelchairs, which was not a comparison we had ever considered for ourselves. While it is probably fine for there to be multiple metaphors at play when using telepresence robots, it can become problematic when locals versus remote operators implicitly disagree upon metaphors, especially human versus nonhuman ones. One salient example of this was an interpersonal breakdown that occurred when a local hit the run-stop button that shuts down all power going to the robot motors. This local was frustrated with the remote operator and wanted him to stop asking her questions. She believed that this action was equivalent to hanging up a telephone (a nonhuman metaphor), but the remote operator felt more like he had been incapacitated (a human metaphor). While it is clear that there are many opportunities presented by telepresence robots, there are also many challenges that need to be addressed so that these technologies are used for fostering interpersonal relationships and collaborations instead of damaging them.

Experimental Explorations and Hypothesis Testing

Based on the lessons learned and issues raised by these field studies, we conducted a set of controlled laboratory experiments to explore how different framing and design variables would influence one's experience of interacting through a robot. Using a variable-based approach (Nass & Mason, 1990) to studying communication technologies, we looked for design and framing variables that would improve how loudly or quietly people spoke (study 1), how safely people would move around (study 2), and how much of a teammate the remote coworkers could become (study 3). We are conducting many other experiments in the space of telepresence robots, but these are the most basic in that they address the three major aspects of using telepresence robots: talking, driving around, and being a part of a team.

Speaking at more appropriate volumes: feedback (study 1)

One of the major problems identified in our field studies as well as by others (Tsui, Desai, Yanco, & Uhlik, 2011) is that operators often unintentionally project their voices too loudly through the telepresence robots. This is disruptive for the local bystanders (Clark, 1996), who are not necessarily interested in interacting with the operators. This problem became so rampant in our own site that locals were starting to slam their office doors shut whenever they heard Texai operators talking in the hallways.

Even though locals tend to have no qualms with asking operators to "speak up" (more loudly), they seem to be more hesitant to ask the operators to speak more quietly. Furthermore, although it is technically possible for locals to simply push volume buttons on the head of the Texai to turn down the system's volume, many locals were hesitant to touch those buttons because they felt it would be rude or disruptive to the conversation (Takayama & Go, 2012).

Fortunately, we were able to draw from prior work in telephone research on sidetone (Lombard, 1911), which is the intentional, attenuated feedback of speakers' voices to their ears while using a telecommunication device. When people have some of this sidetone feedback, they have a better sense of how loudly they are speaking into the telephone and automatically self-regulate their speaking loudness. We prototyped sidetones of varying levels in two separate experiments—one using sidetones in headphones on the operators ($N=20$) and another using sidetones in loudspeakers in front of the operators ($N=14$). Participants in both studies were asked to talk with another person via a Texai Alpha prototype from our laboratory setting, doing a wide range of speaking tasks that included reading from a list of words (to practice talking through the robot and give us a baseline for the participant), engaging in some getting-to-know-you discussions, and doing an alternate uses creativity task (Guilford, Christensen, Merrifield, & Wilson, 1978). Both studies used a within-participants experiment design in which each participant experienced all three sidetone levels (no sidetone, low sidetone, and high sidetone). We measured loudness levels of each participant in each of the three conditions. Based on prior work (Lane, Tranel, & Sisson, 1970), we hypothesized that people would speak more quietly in the high sidetone condition and increasingly loudly in the low and no sidetone conditions.

The results of study 1a (using headphones) found that higher sidetone levels did indeed get operators to speak more quietly, but not to the same degree that those same sidetone levels had influenced telephone users in the past (Lane et al., 1970). This suggests that there is something different going on with telepresence robots as opposed to the prior telephone research. It is possible that the video component of the telepresence robots might be weakening the effects of sidetones, but that hypothesis has not yet been tested.

The results of study 1b (using loudspeakers) did not show any statistically significant support for the research hypothesis. This suggests that there might be something different about loudspeakers (as opposed to headphones and telephone handsets) that weakens the effects of sidetone upon how loudly a person speaks into the device.

Altogether, the results of these two studies point to the somewhat weaker effects of sidetone levels upon operator loudness when operators hear sidetones through headphones and a lack of effect when operators are using desktop loudspeakers. Approximately half of the operators in our field studies used loudspeakers and half used headphones; sidetones may be useful for that latter half of telepresence robot operators. In broader terms, providing more feedback to operators may help operators to self-regulate the way they present themselves through telepresence robots. For more information about the details of this study, see our UIST 2012 conference paper (Paepcke, Soto, Takayama, Koenig, & Gassend, 2011).

We have found similar results in subsequent experiments on helping remote operators to more effectively present themselves, visually, through these systems (Takayama & Harris, 2013). If operators see reflections of themselves in mirrors placed around the local environment, then they do a better job of making themselves visible and presentable through their robotic bodies. However, just like in normal local encounters with mirrors, remote operators also feel more self-conscious when they are faced with mirrors than when there are no mirrors in the environment.

The results of these experiments show that it takes more feedback than usual to effectively incorporate a robot into one's sense of self (Merleau-Ponty, 1962). Just as we check in mirrors before stepping out the front door in the morning, remote operators use similar feedback mechanisms to improve the way that they present themselves to locals through telepresence robots, incorporating the robots into their presentation of self (Goffman, 1959).

Driving around more safely and efficiently: assistance (study 2)

Unlike driving cars, operating telepresence robots in human environments requires navigating through constantly changing environments (e.g., doors being open or closed, people moving in all directions) and making sense of the particular type of drive system used in the robot (e.g., in our situation, a single rear-wheel drive with two passive wheels in front). Many operators experience difficulty with driving around the office in a safe and efficient manner, resulting in many moments of embarrassment (on the part of the operator) and frustration (on the part of the locals). To address this challenge, we added obstacle avoidance to the Texai Alpha that leveraged sensor data from a laser range finder on the base of the robot to create a cost map, simulated a trajectory that would not collide with obstacles and yet be as close as possible to the operator's original command, and ultimately moved according to the obstacle-avoiding path.

To test the effectiveness of our approach to assisting operators, we ran a controlled user study ($N=24$, gender balanced) to evaluate the impact of robotic assistance (obstacle avoidance) in comparison to not providing any assistance (no obstacle avoidance). Participants in this study logged into the Texai Alpha from their own home or office computers. Each participant filled out questionnaires (on demographic information such as technology use, personality (Rotter, 1966), and spatial cognitive abilities (Shepard & Feng, 1972)), got trained on how to drive the robot, practiced driving the robot by doing three laps around an obstacle course, performed a timed run through the obstacle course, and then filled out more questionnaires about their experience. We hypothesized that (H1) participants would hit fewer obstacles when using assisted instead of unassisted teleoperation and (H2) participants with greater spatial cognitive abilities would complete the course faster than people with lesser spatial cognitive abilities.

Indeed, the data supported hypothesis 1; people completed the obstacle course with fewer errors (e.g., collisions) when they had obstacle avoiding assistance than when they had no assistance ($M=2.4$ fewer errors, $SE=0.6$), $t(23)=-3.71$, $p<0.01$. However, people who had obstacle avoiding assistance also took longer to complete the obstacle course than people who had no assistance ($M=64.4$ s longer, $SE=11.7$), $t(23)=4.59$, $p<0.0001$. There seems to be a speed versus accuracy trade-off. Thus, in situations where it is critical to move around safely (e.g., maneuvering around hospital patients), it may be worthwhile to use obstacle avoidance, even if it takes longer to drive around. In contrast, in situations where it is less important to move around safely (e.g., sparsely populated office spaces), it may be better to not use obstacle avoiding robotic assistance or to allow for easy user override.

We did not find support for hypothesis 2; people with greater spatial cognitive abilities were not found to be any better at completing the obstacle course than people with lesser spatial cognitive abilities. However, we did identify a somewhat surprising trend (that became obvious in hindsight)—people with a strong internal locus of control (i.e., people who believe they have control over events in the world) took longer to complete the obstacle course with obstacle avoiding assistance than without assistance and took longer than people with an external locus on control (i.e., people who believe they have less control over events in the world) in either assistance

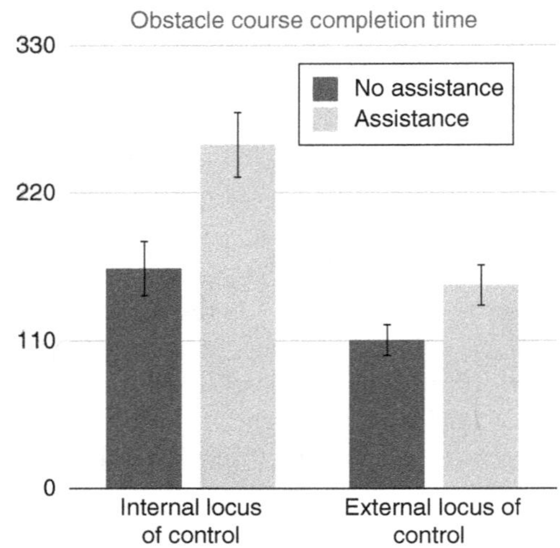

Figure 7.3 Time to complete obstacle course (in seconds) for people with an internal locus of control versus people with an external locus of control and with obstacle avoiding assistance versus no assistance (mean and standard errors).

condition, $F(1.22)=5.74$, $p<0.03$. In other words, people with a strong internal locus of control were fighting against the robotic assistance, suggesting a user issue with giving up control to an autonomous robotic system (Figure 7.3).

These results indicate that there are important situational dimensions (e.g., safety critical vs. relaxed) and human personality dimensions (e.g., locus of control) that must be addressed in order to inform the design of assistance for telepresence robots. It also suggests that these dimensions might be important for designing other types of semi-autonomous robotic systems (e.g., human-in-the-loop control of increasingly autonomous robots (Leeper, Hsiao, Ciocarlie, Takayama, & Gossow, 2012)). For more information about the details of this study, see our ICRA 2011 conference paper (Takayama, Marder-Eppstein, Harris, & Beer, 2011).

These findings also speak to the importance of understanding how different personality types perceive (and sometimes resist) the apparent agency of increasingly autonomous robotic capabilities. Even though they used the exact type of autonomous assistance, different types of people have different perceptions of the system's agency. People with an external locus of control incorporate the obstacle avoidance into their own performance of navigating the obstacle course; in contrast, people with an internal locus of control fight against the obstacle avoidance, reacting to it as a separate agent.

Being part of the team: framing (study 3)

Ultimately, these telepresence robots are designed to improve interpersonal interactions across geographical distances. In the domain of distance collaborations, it is particularly important to foster more in-group behaviors, that is, treating remote coworkers more like teammates and less like opponents. When Dallas began to use Texai to be present in our office, he gradually became more like a colleague and less like some voice in a box on the conference room table. We decided to explore this opportunity more deeply by trying two different manipulations that showed some promise for improving in-group behaviors—*visual* framing by enabling locals to visually customize the telepresence robot and *verbal* framing by explaining the collaborative task as being scored interdependently (as a team) versus independently (as a pair of individuals).

Prior work in the computers as social actors paradigm has found that people who decorated their computer monitors to match the person's "team" color (e.g., blue) tended to behave more collaboratively with the computer (Nass, Fogg, & Moon, 1996). This brought us to formulate the first hypothesis (H1) that locals who decorated the telepresence robot with their team's color would behave more collaboratively with the remote operator than those locals who were not given that opportunity. Indeed, we observed several field sites decorating the Texai Alphas with their own company stickers and decorations around the office.

In the same prior study, Nass et al. (1996) found that telling people that their task performance would be dependent upon how well they did as a team also increased how collaboratively they behaved in the experiment. As such, we formulated the second hypothesis (H2) that locals who were in the interdependently framed situation would behave more collaboratively with the remote operator than those who were in the independently framed situation.

In a 2 (visual framing: no decorating vs. decorating) × 2 (verbal framing: independent vs. interdependent) between-participants experiment (N=40, gender balanced), we tested these research hypotheses. Participants in the study were introduced to the telepresence robot that was being driven by a confederate operator, participated in a collaborative decision-making task (Lafferty & Eady, 1974) and then did a self-disclosure task to test the trust established between the local participant and the remote operator.

The results of this study consistently showed that interdependent framing of the situation helped to foster more in-group behaviors (e.g., more disclosure to the remote operator, liking the operator more, and attributing fewer animal-like emotions to the operator), which supported the second hypothesis (H2). However, the results of this study also consistently went in the opposite direction of the results predicted by our second hypothesis (H1). Locals who decorated the telepresence robots responded more negatively to the remote operators than locals who were not given the opportunity to decorate the robots. This directly opposes the results found in prior work with desktop computers (Nass et al., 1996). Based upon our follow-up interviews with participants in this study and upon further discussion with colleagues, we have formed a new hypothesis that decorating the robots might make locals feel more self-extended into the robot (as shown in other work with personal robots (Groom, Takayama, Ochi, & Nass, 2009)) and, therefore, upsets the locals when some stranger logs into their robot. This hypothesis still needs to be tested in future work that could compare locals versus remote operators decorating the telepresence robots (Figure 7.4).

This study provided some mixed results in terms of implications for the design of telepresence robots. First, to encourage more in-group behaviors, it is likely to be

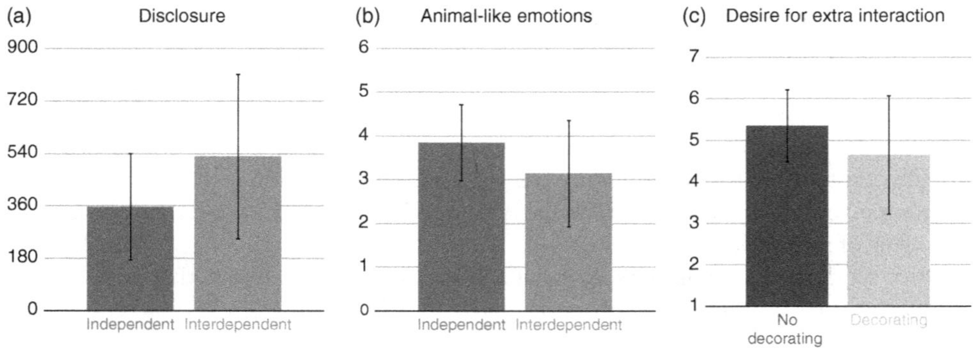

Figure 7.4 Several of the main effects of verbally framing (independent vs. interdependent performance measurement) and of visually framing (not decorated by local vs. decorated by local) (means and standard errors). (a) More disclosure when performance framed as interdependent. (b) Less animal-like emotions ascribed to partner when performance framed as interdependent. (c) Less desire for extra interaction when Texai was decorated.

helpful if managers frame task performance metrics in terms of team performance rather than individual performance. Second, to encourage more in-group behaviors, it is not necessarily helpful for locals to be able to decorate the telepresence robots; it might be more helpful if the remote operators participate in the decorating activity, but that hypothesis is still open for future testing. For more information about the details of this study, see our CHI 2012 conference paper (Rae, Takayama, & Mutlu, 2012).

These findings suggest that the apparent agency of telepresence robots might be somewhat complicated for local users. They seem to perceive a dual agency of the remote operator as well as the telepresence robot. In this sense, the telepresence robot could have team membership with the local user that is separate from the local's sense of team membership with the human remote operator.

Implications for Theory

Even though we rationally believe that robots are just electromechanical machines, it is become increasingly clear that *robots are not just robots* when it comes to human perception and human–robot interaction. Through our longitudinal field studies and controlled laboratory experiments, we keep running into the same issues of people perceiving more apparent agency in the telepresence robots than they would if they believed in-the-moment that the robots were just machines. These telepresence robots are raising issues around apparent agency that are somehow different from personal computers and even from videoconferencing systems. For example: Who is initially blamed versus ultimately responsible for damages caused by using telepresence robots? Unlike the computers as social actors (CASA) body of work that found computers were treated as independent social actors (Sundar & Nass, 2000), we have been finding an increasing amount of evidence that the apparent agency of the telepresence robots comes from the telepresent human operators, for example, Dallas' telepresence robots often seem to *be* Dallas. Of course, this kind of agency is not always so simple. We observed dual apparent agency when locals seemed to perceive the telepresence robot as having its own agency as well as that of the remote operator (e.g., study 3 on being part of a team). Also, as we add capabilities like obstacle avoidance, it is also possible for remote operators to perceive a dual agency of human operator and robot (e.g., study 2 on driving around safely). Furthermore, this differs from videoconferencing in that people do not always treat the remote operators as though they are simply behind a window or in another space; they are often treated as though they were actually collocated in the same local space, for example, giving personal space to the telepresence robots and avoiding hitting the volume controls on the telepresence robot head.

There are (at least) two sides of the interaction at play here. There is the apparent agency that people perceive when operating the robots (e.g., "I feel like I am the robot") and the apparent agency that local people perceived (e.g., "It seems like the operator is here with me"). Because the remote operator's user experience is so different from that of the local's user experience, it may be the case that we need a separate set of theories for predicting people's perceptions of apparent agency. Both of these perceptions seem to be influenced by many factors, including how much the robotic systems itself is in control (e.g., autonomous obstacle avoidance), what types of personality the operators have (e.g., strong internal locus of control), how the interpersonal interaction is framed (e.g., interdependence of teamwork), and how the system is visually presented (e.g., to be decorated or not). This is not a comprehensive list of influential factors; it is just a start toward gaining a deeper understanding of what it is like to effectively use telepresence robots, how we can make them more invisible-in-use, and how they can improve the sense of co-presence between locals and remote operators.

Future Work

These field studies and experiments represent one slice of the ongoing research in human–robot interaction, human–computer interaction, and computer-supported cooperative work that is moving toward gaining a deeper understanding of robotic telepresence. At this moment, there are European Union and National Science Foundation supported projects that are exploring this space, including our own. Among the many opportunities for further research, there are open questions about how the embodied form of these robots influences interpersonal dominance and submissiveness in human interactions (e.g., altering robot height (Rae, Takayama, & Mutlu, 2013) and amplifying volume) or maybe even how people might be able to gain a better sense of empathy for others by "walking in their shoes" via robot. By adding other robotic capabilities such as arms and hands (mechanical) or increasing the amount of autonomous assistance the robot can provide (artificial intelligence) as found in some of our ongoing work (Leeper et al., 2012), we can learn even more about other types of remote presence and apparent agency that involve more than just video-based communication channels. Another important perspective to consider in future work is that of the local and remote bystanders, not only the interactants. This work purposefully focused upon remote collaboration in office spaces, but there are other types of users (e.g., older adults (Beer & Takayama, 2011)) and use cases (e.g., tele-medicine (Chung, Grathwohl, Poropatich, Wolf, & Holcomb, 2007; Vespa, 2005)) that are very likely to inspire fruitful directions for further inquiry.

In terms of research methods, we have relied upon many standard practices from contextual inquiry (Beyer & Holtzblatt, 1998) to psychology experiment design, but there are clearly many other ways to examine apparent agency in telepresence robots. To better grasp and examine the notion of "apparent agency," there is clearly more work to be done in developing metrics—both task-specific and in more general form. To generate task-specific metrics, it is also critical to explore experiment paradigms and procedures for eliciting relevant behaviors from people who use telepresence robots; the tasks used by videoconferencing will not directly transfer to tasks used for telepresence robots because these robots are typically used in less formal settings due to their mobility. We have relied upon many tasks from prior human–computer interaction and human factors studies, but that is only a starting point, not a gold standard.

Conclusions

There is still much more work to be done before telepresence robots become widely adopted and understood; there is also much more empirical and theoretical work to be done for perceptions of apparent agency to become predictable. This chapter provides an overview of our past and ongoing work in these spaces. Exploring new uses cases, developing new telepresence systems, bringing in more perspectives, and sharing more experiences is not only encouraged, but is also necessary for telepresence robots to address more human needs and for us to better understand what is going on in the human psyche when we start feeling like we can really be together at a distance.

Acknowledgments

This work would not have been possible without the generous time and effort of the Texai Alpha development team (now at Suitable Technologies, Inc.), the field study participants, and many dozens of experiment participants. I would also like to thank my colleagues, who have worked side-by-side with me through these explorations. The field studies were conducted in collaboration with Bianca Soto, Judee Humburg, Janet Go, Matti Hirpa, Parmita Dalal, Helen Harris, and Min Kyung Lee. Study 1 was a collaborative project with Andreas Paepcke, Bianca Soto, Frank Koenig, and Blaise Gassend. Study 2 was a collaborative project with Eitan Marder-Eppstein, Helen Harris, and Jenay Beer. Study 3 was a collaborative project with Irene Rae and Bilge Mutlu. This research was supported by Willow Garage and National Science Foundation award 1117652.

References

Bandura, A. (2006). Toward a psychology of human agency. *Perspectives on Psychological Science, 1*(2), 164–180.

Beer, J. M., & Takayama, L. (2011). Mobile remote presence systems for older adults: Acceptance, benefits, and concerns. In *Proceedings of Human-Robot Interaction Conference: HRI 2011, Lausanne, CH* (pp. 19–26). New York, NY: ACM Press.

Beyer, H., & Holtzblatt, K. (1998). *Contextual design: Defining customer-centered systems.* San Francisco, CA: Morgan Kaufmann.

Bly, S. A., Harrison, S. R., & Irwin, S. (1993). Media spaces: Bringing people together in a video, audio, and computing environment. *Communications of the ACM, 36*, 28–46.

Cesta, A., Cortellessa, G., & Tiberio, L. (2011). Long-term evaluation of a mobile remote presence robot for the elderly. *ERCIM News, 84*, 20–21.

Chung, K. K., Grathwohl, K. W., Poropatich, R. K., Wolf, S. E., & Holcomb, J. B. (2007). Robotic telepresence: Past, present, and future. *Journal of Cardiothoracic and Vascular Anesthesia, 21*(4), 593–596.

Clark, A. (2003). *Natural-born cyborgs: Minds, technologies, and the future of human intelligence.* New York, NY: Oxford University Press.

Clark, H. H. (1996). *Using language.* New York, NY: Cambridge University Press.

Fels, D. I., Waalen, J. K., Zhai, Shumin, & Weiss, P. L. (2001). *Telepresence under exceptional circumstances: Enriching the connection to school for sick children.* Paper presented at the Interaction, Tokyo, Japan (pp. 617–624).

Fish, R. S., Kraut, R. E., & Chalfonte, B. L. (1990). The Video Window system in informal communication. In *Proceedings of CSCW '90, Conference on Computer-Supported Cooperative Work* (pp. 1–11). New York, NY: ACM Press.

Goffman, E. (1959). *The presentation of self in everyday life.* New York, NY: Anchor Books.

Groom, V., Takayama, L., Ochi, P., & Nass, C. (2009). I am my robot: The impact of robot-building and robot form on operators. In *Proceedings of the 4th ACM/IEEE international Conference on Human-Robot Interaction (HRI 2009)* (pp. 31–36). New York, NY: ACM Press.

Guilford, J. P., Christensen, P. R., Merrifield, P. R., & Wilson, R. C. (1978). *Alternate Uses Task Manual.* Orange, CA: Sheridan Psychological Services.

Heer, J., & Khooshabeh, P. (2004). *Seeing the invisible.* In *Workshop on Invisible and Transparent Interfaces,* Advanced Visual Interfaces, Gallipoli, Italy.

Heidegger, M. (1992). *Basic writings: From being and time (1927) to the task of thinking (1964).* New York, NY: HarperCollins.

Ishiguro, H., & Trivedi, M. (1999). *Integrating a perceptual information infrastructure with robotic avatars: A framework for tele-existence* (pp. 1032–1038). Presented at the Intelligent Robots and Systems (IROS).

Jouppi, N. P., Iyer, S., Thomas, S., & Slayden, A. (2004). BiReality: Mutually-immersive telepresence. In H. Schulzrinne, N. Dimitrova, M. A. Sasse, S. B. Moon, & R. Lienhart (Eds.), *Proceedings of the ACM Multimedia 2004* (pp. 860–867). New York, NY: ACM Press.

Kristoffersson, A., Coradeschi, S., & Loutfi, A. (2013). *A review of mobile robotic telepresence. Hindawi Publishing Corporation.* Retrieved from http://dx.doi.org/10.1155/2013/902316

Lafferty, J. C., & Eady, P. M. (1974). *The desert survival problem.* Plymouth, MI: Experimental Learning Methods.

Lane, H., Tranel, B., & Sisson, C. (1970). Regulation of voice communication by sensory dynamics. *Journal of the Acoustical Society of America, 47*, 618–624.

Latour, B. (1992). Where are the missing masses? The sociology of a few mundane artifacts. In W. E. Bijker & J. Law (Eds.), *Shaping technology/building society: Studies in sociotechnical change* (pp. 225–258). Cambridge, MA: MIT Press.

Lee, M. K., & Takayama, L. (2011). "Now, I have a body": Uses and social norms for mobile remote presence in the workplace. In *Proceedings of the CHI* (pp. 33–42). New York, NY: ACM Press.

Leeper, A., Hsiao, K., Ciocarlie, M., Takayama, L., & Gossow, D. (2012). Strategies for human-in-the-loop robotic grasping. In *Proceedings of the 7th ACM/IEEE International Conference on Human-Robot Interaction (HRI 2012)* (pp. 1–8). New York, NY: ACM Press.

Levin, D. T., Adams, J. A., Saylor, M. M., & Biswas, G. (2013). A transition model for cognitions about agency. *Proceedings of the 8th ACM/IEEE International Conference on Human-Robot Interaciton (HRI 2013), Tokyo, Japan* (pp. 373–380).

Lombard, E. (1911). Le signe de l'elevation de la voix. *Ann. maladies oreille larynx nez pharynx, 37*, 109–119.

McCloud, S. (1993). *Understanding comics: The invisible art*. New York, NY: Harper Perennial.

Merleau-Ponty, M. (1962). *Phenomenology of Perception*. New Jersey, USA: Routledge/Kegan Paul.

Nardi, B. A. (2005). Beyond bandwidth: Dimensions of connection in interpersonal communication. *Computer Supported Cooperative Work, 14*, 91–130.

Nass, C., Fogg, B. J., & Moon, Y. (1996). Can computers be teammates? *International Journal of Human Computer Studies, 45*(6), 669–678.

Nass, C., & Mason, L. (1990). On the study of technology and task: Avariable-based approach. In J. Fulk & C. W. Steinfield (Eds.), *Organization and communication technology* (pp. 46–67). Newbury Park, CA: Sage.

Nass, C., & Moon, Y. (2000). Machines and mindlessness: Social responses to computers. *Journal of Social Issues, 56*(1), 81–103.

Nass, C., Steuer, J., & Tauber, E. R. (1994). Computers as social actors. In *Proceedings of the Conference on Human Factors in Computing Systems, Boston, MA (pp. 72–78)*. New York, NY: ACM Press.

Paepcke, A., Soto, B., Takayama, L., Koenig, F., & Gassend, B. (2011). Yelling in the hall: Using sidetone to address a problem with mobile remote presence. In *UIST '11 Proceedings of the 24th Annual ACM Symposium on User Interface Software and Technology, Santa Barbara*. New York, NY: ACM Press.

Paulos, E. (1997). PRoP: Personal roving presence. In *Proceedings of the ACM/IEEE International Conference on SIGGRAPH 1997* (p. 99). New York, NY: ACM Press.

Paulos, E., & Canny, J. (2001). Social tele-embodiment: Understanding presence. *Autonomous Robots, 11*, 87–95.

Polanyi, M. (1962). *Personal knowledge: Towards a post-critical philosophy* (Vol. 2). Chicago, IL: University of Chicago Press.

Rae, I., Takayama, L., & Mutlu, B. (2012). One of the gang: Supporting in-group behavior for embodied mediated communication. In *Proceedings of the ACM/IEEE International Conference on Human Factors in Computing Systems (CHI 2012)*. Austin, TX: ACM Press.

Rae, I., Takayama, L., & Mutlu, B. (2013). The influence of height in robot-mediated communication. *Proceedings of the 8th ACM/IEEE International Conference on Human-Robot Interactions (HRI 2013), Tokyo, Japan*, (pp. 1–8).

Reed, E., & Jones, R. (1982). *Reasons for realism: Selected essays of James J. Gibson*. Hillsdale, NJ: Lawrence Erlbaum Associates.

Reeves, B., & Nass, C. (1996). *The media equation: How people treat computers, television, and new media like real people and places.* New York, NY: Cambridge University Press/CSLI.

Rotter, J. B. (1966). Generalized expectancies of internal versus external control of reinforcements. *Psychological Monographs, 80*(1), 1–28.

Sakamoto, D., Kanda, T., Ono, T., Ishiguro, H., & Hagita, N. (2007). Android as a telecommunication medium with a human-like presence. In *Proceedings of the 2nd ACM/IEEE International Conference on Human-Robot Interactions (HRI 2007)* (pp. 193–200). New York, NY: ACM Press.

Shepard, R. N., & Feng, C. (1972). A chronometric study of mental paper folding. *Cognitive Psychology, 3*, 228–243.

Smith, G. (2011). *A boy and his bot. Sports Illustrated.*

Sundar, S. S., & Nass, C. (2000). Source orientation in human-computer interaction. *Communication Research, 27*(6), 683–703. doi:10.1177/009365000027006001

Takayama, L. (2010). On making robots invisible-in-use. In *Proceedings of the 2nd International Symposium on New Frontiers in Human-Robot Interaction*, De Montfort University, Leicester, UK.

Takayama, L. (2011). Toward making robots invisible-in-use: An exploration into invisible-in-use tools and agents. In K. Dautenhahn & J. Saunders (Eds.), *New frontiers in human-robot interaction* (pp. 111–132). Amsterdam, The Netherlands: John Benjamins.

Takayama, L., & Go, J. (2012). Mixing metaphors in mobile remote presence. *Proceedings of the Computer Supported Cooperative Work (CSCW 2012)*, Seattle, WA.

Takayama, L., & Harris, H. (2013). Presentation of (telepresent) self: On the double edged effect of mirrors. In *Presented at the HRI '13: Proceedings of the 8th ACM/IEEE International Conference on Human-Robot Interaction, Tokyo, Japan*, (pp. 381–388). IEEE Press

Takayama, L., Marder-Eppstein, E., Harris, H., & Beer, J. M. (2011). Assisted driving of a mobile remote presence system: System design and controlled user evaluation. *Proceedings of the International Conference on Robotics and Automation: ICRA 2011, Shanghai, CN* (pp. 1883–1889).

Tsui, K. M., Desai, M., Yanco, H. A., & Uhlik, C. (2011). Exploring use cases for telepresence robots. In *Proceedings of the 6th ACM/IEEE International Conference on HRI 2011* (pp. 11–18). New York, NY: ACM Press.

Vespa, P. (2005). Robotic telepresence in the intensive care unit. *Critical Care, 9*(4), 319–320.

Weiser, M. (2002). The computer for the twenty-first century. *IEEE Pervasive Computing, 1*, 19–25.

8

Psychological Aspects of Technology Interacting with Humans

Johan F. Hoorn

VU University Amsterdam, Amsterdam, Netherlands

There is a lively fascination in popular culture of "us" being "them" and "them" being "us." In *Surrogates* (Mostow, 2009, Touchstone Pictures), for instance, isolated people lie in bed all day while they use special goggles to direct their lifelike robot version through the outside world. In *Marwencol* (Malmberg, 2010, independent),[1] the brain damaged victim of a violent attack, Mark Hogancamp, creates his own therapy by building a scale model (1:6) of a Belgian town, Marwencol, in the WWII era. He populates his town with dolls (soldiers and girls), which have personalities based on him and his acquaintances. This fantasy world becomes so famous that Marwencol ends up in a New York art gallery. Hogancamp explains that he lives a second life now in which he finally can be himself.

This situation is nearer than most of us realize but then without the doom and gloom a la *Surrogates* or as a mere reality for the mentally disturbed. While for ages, fictional characters are the backbone of novels, theater plays, and films, today their synthetic equivalents are crowding our interfaces and start to step out of their square screen existence in the form of life-like robots and androids, employed, for example, in health care settings (e.g., Banks, Willoughby, & Banks, 2008; Broadbent, Stafford, & MacDonald, 2009; Broekens, Heerink, & Rosendal, 2009).

The subject of this chapter is how interactive computer systems perceive us as their human users (rather than the other way round). How can these systems serve us, human users, as effectively as possible, executing programmed behaviors but also anticipating and responding creatively to our actions while complying with moral principles. This chapter describes a large number of sometimes complex studies that in unison have led to the production of a reliable, serviceable sometimes stern but fair care robot. The initial summary of studies is necessary to understand what the smart, sensitive, creative, and moral humanoid that interacts with its human user is actually based upon: Its core is driven by the model called Perceiving and Experiencing

The Handbook of the Psychology of Communication Technology, First Edition. Edited by S. Shyam Sundar.

Fictional Characters (PEFiC), the interactive variant of which (I-PEFiC) received a module to make affective decisions (ADM). An integration of I-PEFIC^{ADM} with prevailing emotion models led to the Silicon Coppélia model. Silicon Coppélia, then, was extended by a system for moral reasoning (Moral Coppélia) and is now in the process to become creative (ACASIA) and have an understanding of reality in contrast to fiction (Epistemics of the Virtual).

Virtual Characters

The starting point of this stream of research was a series of investigations into how readers of literature and visitors of theater plays process media personalities and fictional characters (Hoorn & Konijn, 2003). This work was inspired by literary theory as well as rich theoretical work on the psychology of emotions (Frijda, 1986; Lazarus, 1991) and media research on emotions of actors on stage (Konijn, 2000). Our theory of PEFiC (Hoorn & Konijn, 2003) describes unidirectional communication from medium to human (Figure 8.1). Humans encode fictional characters in terms of good or bad (ethics), beautiful or ugly (aesthetics), and realistic or unrealistic (epistemics). Features are then compared for self-similarity ("Are we alike?"), relevance to personal concerns ("Is this important?"), and valence, whether viewers wish the character to achieve its goals or not ("Should he win or lose?"). In unison, these evaluations determine the fuzzy trade-off between being involved with the character and simultaneously being at a distance, which predict the level of "liking" or appreciation of the character. The idea that psychological dimensions can occur in parallel and do not need to be bipolar (i.e., positive = 1 – negative) was inspired by the work of Dollard and Miller (1950). In two empirical studies, we could fit this PEFiC model closely to the data of viewers watching feature films (Konijn & Hoorn, 2005) and media figures (Konijn & Bushman, 2007).

In these and subsequent studies, many of the assumptions underlying the PEFiC theory were empirically corroborated using structured questionnaires (designed according to Dillman, 2000), the results of which were analyzed through structural equation modeling (SEM). It indeed turned out that in unidirectional media, people use but three encoding dimensions: the moral fiber of the character (ethics), how s/he looks (aesthetics), and whether s/he is represented realistically (epistemics), for instance, in a photorealistic style or cartoonish. In line with Frijda (1986), we found that also in response to fictional characters, people's cognitions preceded emotion and that emotion was concern-driven. Due to the mix of positive and negative traits, people found round characters more interesting than flat characters. Therefore, there was a unipolarity of dimensions, involvement not excluding measures of distance but compensating one another—an optimal balance exerting most appreciation.

For modeling purposes, we relied on fuzzy sets theory (Zadeh, 1965) because without violating certain axioms of probability, it can deal with features that to a certain degree are member of multiple categories (e.g., the villain's weapon is beautiful *and* repulsive). Regarding entertainment and news media, fuzzy logics could explain

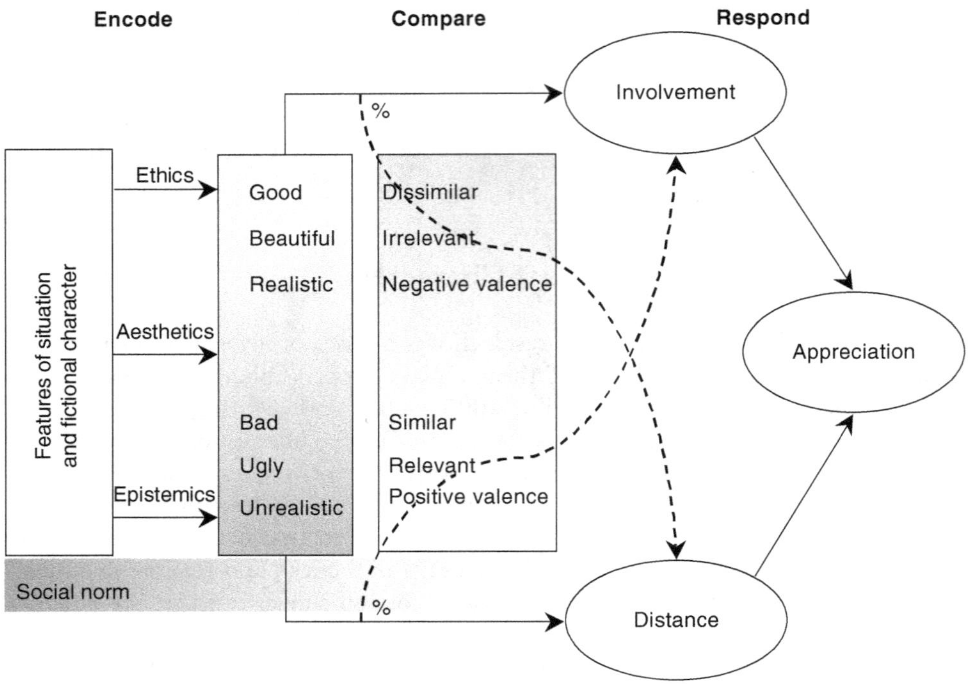

Figure 8.1 Perceiving and Experiencing Fictional Characters (PEFiC). Dashed arrows indicate percentages of information spillover to the other polarity.
Source: Reprinted with permission from Hoorn and Konijn (2003). © John Wiley & Sons, Inc.

that apart from distancing effects, negative qualities can be fun as well (cf. round characters). With respect to computer-interface agents, fuzzy sets allowed for the information spillover that turned out to be crucial for the user experience (see next).

Interactive PEFiC

To further PEFiC in the domain of interactive media, we focused on the question how computer users react to so-called embodied agents. Embodied agents are software programs that appear on the computer screen with a human or animal-like appearance. You can think of Office Assistants as well as game characters. At that time, it was unclear to us why users felt emotionally attached or aloof and why they wanted to use an agent (or not) for executing computer tasks. Our aim was to comprehend the psychological processes triggered by embodied agents so to improve their design. As said, PEFiC suggests that a complex set of factors such as ethics, esthetics, epistemics, similarity (e.g., in appearance or behavior), personal relevance, and valence (positive or negative outcome expectations) contribute to the affective bond with fictional characters. This we called the *engagement* process.

In contrast to fictional characters, however, embodied agents also have an interactive component. Communication is bidirectional. Users can actively interact with and employ embodied agents, for example, to find information on the Internet. The PEFiC model, then, was enriched with a process of *interaction*. Interactive PEFiC or I-PEFiC for short (Van Vugt, Hoorn, & Konijn, 2009a) based the interaction process on different theories that explain why people accept and use the one system and not the other. The factor of affordance (or action possibilities) came from ecological psychology and affordance theory (Gibson, 1979; Gaver, 1991; McGrenere & Ho, 2000). We supposed that affordances the agent offered would lead to intentions to use the agent for task execution. These so-called use intentions were drawn from the renowned Technology Acceptance Model (Davis, 1989; Sood & Tellis, 2005; Turner, Kitchenham, Brereton, Charters, & Budgen, 2010; Venkatesh, Morris, Davis, & Davis, 2003). Moreover, where literary readers and theater goers talk about the appreciation of the character, computer users talk about being satisfied with their systems. Therefore, we changed the final explicandum of "appreciation" in PEFiC into "satisfaction" in I-PEFiC.

The interaction process assumes that actual usage is explained by the action possibilities (so-called affordances) the system offers. Affordances are functional features that allow the user to perform an action. A chair affords sitting, even if you want to stand. The relevance of the agent for the task at hand or the user's expectations also play a role (e.g., "The agent won't be efficient but will be fun to work with"). This would lead to the user's willingness to perform a task with the agent or not (i.e., use intentions). I-PEFiC further supposes that user satisfaction depends on the *engagement* process as well as the *interaction* process. Therefore, I-PEFiC consists of two layers (Figure 8.2), the first having to do with the personality side of embodied agents (engagement process); the second with their functional side (interaction process).

I-PEFiC was empirically verified in a series of lab experiments, indicating that typical character factors (such as aesthetics) also affected the interaction process and that typical interaction factors (such as affordances and task relevance) impinged upon the experience of the agent's character. In Van Vugt, Konijn, Hoorn, Eliëns, and Keur (2007), for instance, we wondered whether (form) realism was a key factor for user responses to embodied agents in a virtual museum tour but it turned out that aesthetics and task-relevance were more important than realistic representations. Lifelike is not being more likeable.

In Van Vugt, Hoorn, Konijn, and De Bie Dimitriadou (2006), we manipulated the Sims game to study how aesthetics, a factor from I-PEFiC's engagement process, and affordances, a factor from the interaction process, influenced user responses. We measured a strong effect of affordances on use intentions but also a strong crossover effect on the engagement side (Figure 8.2). If a beautifully designed agent did not perform well, distance increased more than with an ugly agent. If she did well, the beautiful agent was appreciated more than the ugly one.

In Van Vugt, Bailenson, Hoorn, and Konijn (2010), we decided to scrutinize the effects of affordances on user reactions more closely. We wondered how affordances and facial similarity, another factor of the engagement process, would affect the users: someone like me behaving clumsily or efficiently. We morphed the faces of tutor agents

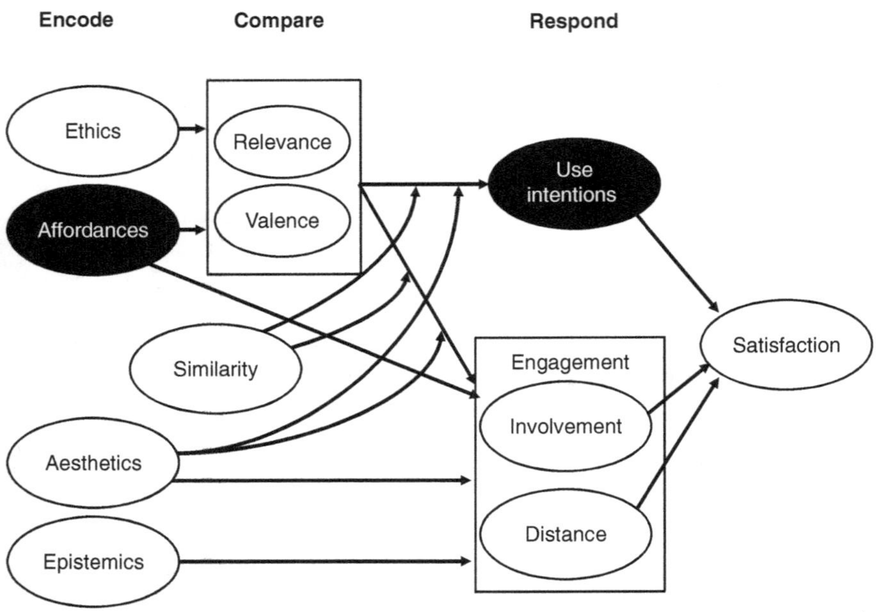

Figure 8.2 Interactive PEFiC (I-PEFiC) has an engagement (white) and an interaction side (black). Curved arrows indicate interaction effects.
Source: Reprinted with permission from Van Vugt, Hoorn et al. (2009a). © John Wiley & Sons, Inc.

with the face of the user to a level of 35%, which is an empirical threshold for unconscious awareness of facial similarity. In a science trivia quiz, tutor agents with or without the user's face morphed in cued the proper answer or it made mistakes. Affordances again had a large effect on use intentions as well as a strong crossover effect on the character side. Facial similarity had an impact on involvement and use intentions, but only for men. Men felt that if an agent looked like them and performed poorly, they did not want to interact with it, having fewer problems when the agent was dissimilar. For women, this effect remained absent. The effects of facial similarity were remarkable since subjects were unaware that we had manipulated the facial similarity.

In Van Vugt et al. (2009b), we continued the study of gender effects, but in an e-health context. We wondered whether similarity of body size between user and virtual health coach would affect engagement and interaction as compared to a coach with an ideal body-size (i.e., thin). It turned out that a nonideal figure had a stronger effect than body shape similarity on both the engagement side and the interaction side. In other words, the thinner participants also liked to interact with a heavier coach! These effects occurred mainly indirectly through the perception of *ethics*. Remarkably, the nonideal (bigger) agents were deemed as morally better (i.e., more trustworthy) than ideal (thinner) agents.

From all these studies, we learned that functionality outdoes appearances. Affordances had more impact on typical engagement factors such as involvement and distance than graphical variables such as realism, aesthetics, facial and body shape similarity, or ideal body shape. In other words, we like an agent better and feel more engagement as a friend if it does a proper job; the other factors work through interaction or exert weaker effects. *Ethics* was the notable exception—but ethics of course has nothing to do with outer appearance. Ethical considerations were involved in all character evaluations including its affordances (cf. the tooling should be morally acceptable, trustworthy, reliable) and therefore, affected use intentions ("It may serve my purpose but it is not the right thing to do" or "It may not be the right thing to do but it serves my purposes"). In general, however, people are not inclined to use bad characters quickly.

In interaction with affordances or ethics, factors relating to outer looks do have an effect on involvement and distance as well as use intentions. If affordances hinder a user goal (e.g., untimely pro-active advice is well-meant but annoying), a positive feature such as facial beauty and facial similarity have a negative impact on user responses. Certain appearance factors may be seen as more reliable (i.e., heavy body shape is more trustworthy) and positively impact the willingness to use the application. But even if an agent has positive affordances and is ethically correct, the agent is perceived in a task context in which task-relevance and the valence of expected outcomes mediate the final responses of involvement-distance and use intentions. User satisfaction, finally, is a complex construct because it depends on both the engagement process as well as the process of interaction. That a user is satisfied with an agent does not say much about what is going on in his or her mind because she can be heavily engaged with the personality of the embodied agent but be disappointed about its abilities to perform a task.

From these results, we can derive five noteworthy conclusions: (a) realism is not so important for emotional bonding and use intentions, (b) what is beautiful need not be used, (c) affordances and ethics are central factors for emotional attachment and use intentions, (d) positive characteristics do not always lead to positive effects, and (e) relevance and valence of user goals are mediating the response.

I-PEFiC (Figure 8.2) describes the empirical basis of the psychological processes that are generated when users encounter an embodied agent onscreen. Designers should realize that the interaction with agents in executing certain tasks is not only crucial for use intentions, but even for the emotional bond with agents, even stronger than the agent's appearance. Creating a realistic and beautiful embodiment certainly is not enough. A well-designed graphical appearance cannot compensate for an unhandy interface or for software malfunctioning. That may even harm the user experience more than wrapping poor functionality in an unappealing interface. An effective design allows an agent first to be intelligent enough to let users really enhance their performance, and as an added value only, the appearance of the agent may be driven to greater heights. First make it functional, then nice.

I-PEFiC and Affective Decision Making

We then turned the whole idea around and speculated what would happen if virtual characters could perceive us the way we perceive them? Use the psychology of human–agent interaction as a behavioral model of embodied agents and humanoid robots to see how they relate to us! We formalized I-PEFiC into a computational model of agent and robot behavior, capable of building up affective attitudes and states in response to its user (Hoorn, 2008). Social scientists could see this as making a working model of theories of human behavior to predict effects under certain parameter settings.

Our aim was to use affect and emotions in a functional way, providing information regarding the ongoing match between human user and robotic counterpart, facilitating decision making of the latter. We wanted robot emotions to have a social function, an affective interface if you will with its human partner. In line with Gross' (1998) theory of human–human emotional interaction, agent emotions would provide the user with information about the system's intentions, scripting the social behavior between the two.

During the last decades, psychology increasingly explored the topic of human emotion regulation mechanisms, which are meant to maintain, increase, or decrease emotional-responses (e.g., Davidson, Putnam, & Larson, 2000; Gross, 1998, 2001; Ochsner & Gross, 2005; Thompson, 1994). Based on Gross (1998), we formalized one of five strategies of emotion regulation: situation selection was the first (e.g., if you do not like a person, talk to someone else). The other strategies we implemented later (see section Silicon Coppélia). They were situation modification (e.g., change the tone of voice), attentional deployment (e.g., looking away), cognitive change (e.g., to look on the bright side of things), and response modulation (e.g., emotion suppression). These five strategies were added to the I-PEFiC model, which originally was more directed at the generation of affect rather than regulating it.

Apart from Gross, we also looked into Emotion and Adaptation (EmA) of Smith and Lazarus (1990); particularly the adaptation and implementation of that theory by Gratch and Marsella (2005) and that was tested by Mao and Gratch (2006). Zeroing in on coping strategies, we copied the EmA model into I-PEFiC so that agents can appraise a situation and deal with it if they deem it too negative. Via the introduction of EmA, our agent processes a number of appraisal variables that pertain to several goals, such as desirability ("I want her to eat"), likelihood ("She probably won't take it"), and responsibility for reaching desired or undesired goal-states ("But I care about her health"). In addition, EmA explains how appraisal variables flush forward specific emotions (☺, 😐, ☹), something that was missing in the older PEFiC versions. The agent can vary its actions, plans, and intentions by means of coping strategy.

Yet, because emotion is concern-driven (Frijda, 1986), we had to supplement the I-PEFiC + EmA integration with goal-directed judgment formation (Hoorn, Konijn, Van Vliet, & Van der Veer, 2007) and overt actions. In Hoorn, Pontier, and Siddiqui (2008), we argued that basically there can be no more than eight valence statements (Table 8.1).

In Table 8.1, each constituent in the valence statement evokes a positive (p) or negative (n) response (e.g., intelligence is positive, facilitating is positive, and inefficiency is negative ③). During the weighing, ambiguous emotions happen, when positive and negative alternate. Statement ① and ⑧ (Table 8.1) are the most straightforward in this respect. Because affordances (e.g., intelligence) have predictive power for user engagement (Van Vugt et al., 2006), all the negative responses are input for experiencing distance, whereas the positive responses feed into involvement. The related action tendencies determine the intentions to "make use" of the other agent (which might be

Table 8.1 Agent A (e.g., robot) judges Agent B (e.g., user), resulting in valences that precede action tendencies

Statement	*Feature of Agent B*	*Affords*	*Goal of Agent A*	*Agreement*[a]	*Weighing*	*Valence to means*[c]
①	Intelligence ^{p}	Facilitates ^{p}	Agent A's efficiency ^{p}	Agree ^{p}	$p=(p \cdot p) \cdot p=(p) \cdot p =$	p
				Disagree[b] ^{n}	$p=(p \cdot p) \cdot n=(p) \cdot n =$	n
②	Intelligence ^{p}	Inhibits ^{n}	Agent A's efficiency ^{p}	Agree ^{p}	$p=(n \cdot p) \cdot p=(n) \cdot p =$	n
				Disagree ^{n}	$p=(n \cdot p) \cdot n=(n) \cdot n =$	p
③	Intelligence ^{p}	Facilitates ^{p}	Agent A inefficiency ^{n}	Agree ^{p}	$p=(p \cdot n) \cdot p=(n) \cdot p =$	n
				Disagree ^{n}	$p=(p \cdot n) \cdot n=(n) \cdot n =$	p
④	Intelligence ^{p}	Inhibits ^{n}	Agent A inefficiency ^{n}	Agree ^{p}	$p=(n \cdot n) \cdot p=(p) \cdot p =$	p
				Disagree ^{n}	$p=(n \cdot n) \cdot n=(p) \cdot n =$	n
⑤	Unintelligence ^{n}	Facilitates ^{p}	Agent A's efficiency ^{p}	Agree ^{p}	$n=(p \cdot p) \cdot p=(p) \cdot p =$	p
				Disagree ^{n}	$n=(p \cdot p) \cdot n=(p) \cdot n =$	n
⑥	Unintelligence ^{n}	Inhibits ^{n}	Agent A's efficiency ^{p}	Agree ^{p}	$n=(n \cdot p) \cdot p=(n) \cdot p =$	n
				Disagree ^{n}	$n=(n \cdot p) \cdot n=(n) \cdot n =$	p
⑦	Unintelligence ^{n}	Facilitates ^{p}	Agent A inefficiency ^{n}	Agree ^{p}	$n=(p \cdot n) \cdot p=(n) \cdot p =$	n
				Disagree ^{n}	$n=(p \cdot n) \cdot n=(n) \cdot n =$	p
⑧	Unintelligence ^{n}	Inhibits ^{n}	Agent A inefficiency ^{n}	Agree ^{p}	$n=(n \cdot n) \cdot p=(p) \cdot p =$	p
				Disagree ^{n}	$n=(n \cdot n) \cdot n=(p) \cdot n =$	n

Source: Adapted from Hoorn et al. (2008).

[a] Attribution of truth according to Agent A's world view or "belief system."

[b] Gray cells indicate an unconventional, counter-intuitive, belief that urges to adapt conventional theory.

[c] If valence is positive, an action tendency to approach an agent's feature (here, Agent B) occurs. If valence is negative, depending on situation, context, or "personality," an action tendency to avoid, attack, remove, or change the feature occurs (e.g., Agent A starts to educate Agent B).

the human user), which after reaching threshold may translate into overt behaviors such as conversing with the user, kick him, or hug her. And although in the end only two appraisal values remain (negative, positive), the motivation for the same outcome can be entirely different. In fact, Table 8.1 shows that there are 16 different motivations for the valence experienced toward an aspect of another agent (whether user or software).

Like this, our agents can estimate to what degree certain actions facilitate or inhibit their goals (e.g., to help the user). The resulting involvement-distance trade-off within the agent and the intentions to use the human counterpart for agent purposes (e.g., to recharge its batteries) determine the actions the agent chooses. Therefore, we devised a module for action selection called affective decision making (ADM). Possible actions the agent can execute are to approach the user positively (e.g., make a compliment), negatively (e.g., be sarcastic), avoid the user (e.g., ignoring him or her), or change the user (e.g., through teaching) (Figure 8.3).

To try our simulation (not yet with real users), we created a virtual environment of a soccer stadium peopled by a number of virtual hooligans (Bosse, Hoorn, Pontier, and Siddiqui, 2008). Barry, Harry, and Gary were three agent systems that expressed club adherence by wearing their team's jersey. We equipped the three fans with different features designed into their outer appearance (*aesthetics*; i.e., skin color), the language that they spoke (*affordances*; i.e., Urdu or not), and whether they were good guys or bad guys (*ethics*; same jersey or opponent's jersey). We also made sure each had different biases, for example, having more or less problems with skin color so that we could establish ambiguous cases such as wrong color but right jersey. Moreover, we played with two levels of doubt (higher vs. lower) to see what would happen in case of a dilemma.

Thus, when Barry, Harry, and Gary met for the first time they were partially involved with one another and partially at a distance. Then the interaction unfolded, not scripted, but based on autonomous appraisal of the appearances and behaviors of the other agents. Initially, Barry experienced a lot of distance and little involvement toward Harry and wanted to approach him negatively (i.e., kick Harry). Barry experienced more involvement than distance for Gary but because of his original settings distance was already higher than involvement. Therefore, Barry was in doubt about how to approach Gary, more than he was with Harry. When we increased the level of doubt Barry could experience, Barry was satisfied with Gary rather than with Harry. When we decreased the level of doubt, Barry experienced more satisfaction with Harry; of whom he clearly knew that he wanted to approach him negatively (i.e., he liked kicking him). In tweaking the levels of aesthetics, ethics, epistemics, and affordances, the simulation could replicate the empirical findings of the Van Vugt et al. studies that positive features (e.g., beauty) not only feed into involvement but also partially contribute to distance and reversely, that negative features not only foster distance but also some involvement. This is due to the fact that most effects are mediated through relevance and valence, which relate features (or behaviors) to agent goals and concerns. Being beautiful, therefore, simultaneously can be appraised as being dangerous (cf. Van Vugt et al., 2006). These phenomena could be simulated by using fuzzy algorithms. Now that we were capable of creating personalities that complied with or deviated from expected social behaviors, we expanded I-PEFiC^{ADM} with more emotion regulation and coping behaviors than situation selection alone.

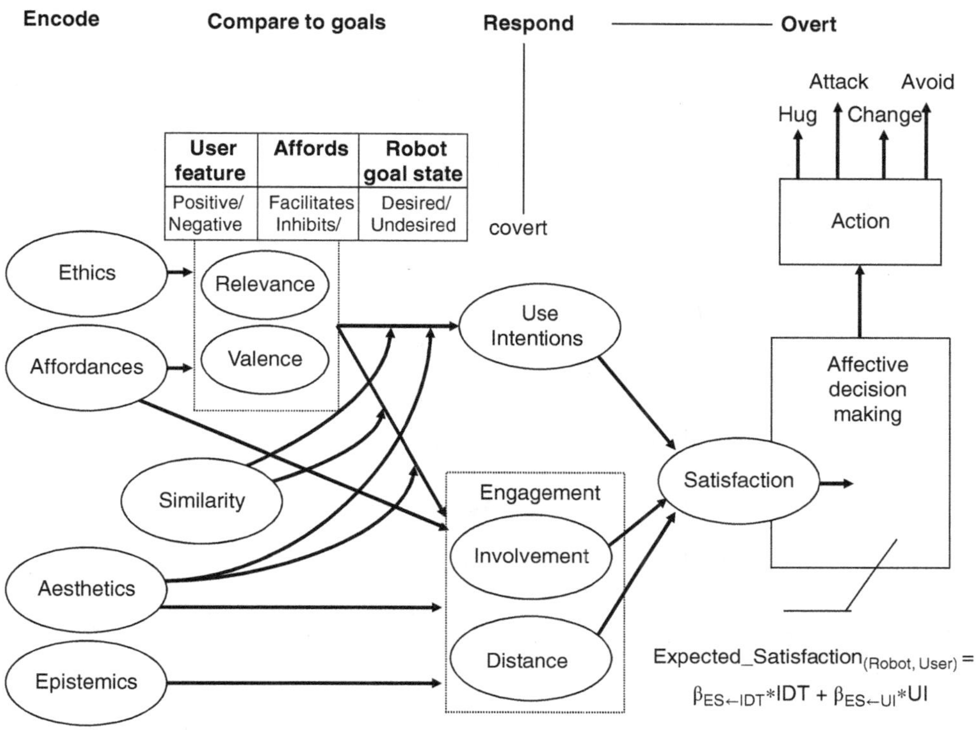

Figure 8.3 Dependencies in I-PEFiC^{ADM}. *IDT* involvement-distance trade-off, *ES* expected satisfaction, *UI* use intentions.

Source: Reprinted with permission from Hoorn, Pontier, and Siddiqui (2012). © Elsevier.

Silicon Coppélia

Named after the dancing doll Coppélia,[2] our software program Silicon Coppélia integrated I-PEFiC^{ADM}, EmA, and the emotion regulation principles by Gross. Coppélia simulates goal-related beliefs that give rise to affect (Hoorn et al., 2012). She simulates beliefs about the responsibility of other agents (humans included) for helping or obstructing to arrive at a goal state (cf. Table 8.1). She also holds beliefs about the probability of achieving those goals. Coppélia has beliefs about the way the world is, which she holds for true. She also has beliefs that events influence certain states of the world. The beliefs and variables such as ethics, aesthetics, etc., govern the seven emotions that Coppélia can have: joy, distress, hope, fear, anger, guilt, and surprise. Silicon Coppélia can simulate emotions and changes of beliefs about the responsibility of other agents for her being happy or sad. She moreover estimates the chance that goal-states may happen and can make irrational decisions when appropriate (e.g., "I should send you away but I like you too much").

Silicon Coppélia estimates the likelihood that she achieves a goal in the range [−1, 1] by checking which sub goals she already accomplished. Coppélia regards a state that

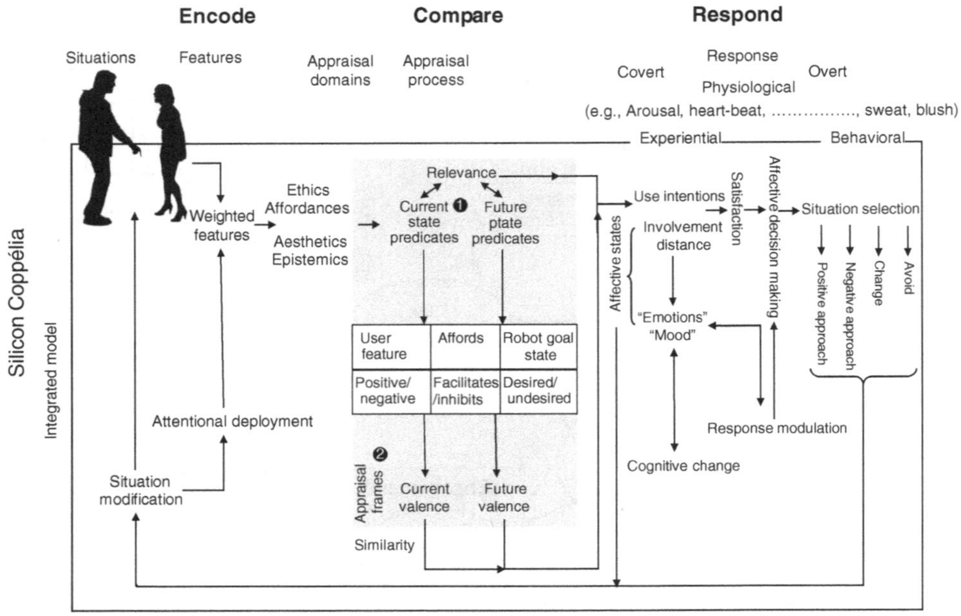

❶ Features are matched against goals, concerns, beliefs, intentions, etc. of self and others (allows taking perspectives)
❷ Relevance and valence (gray area) are the variables in the appraisal frames

Figure 8.4 Silicon Coppélia.
Source: Reprinted with permission from Hoorn et al. (2012); (http://morguefile.com/license/morguefile). © Elsevier.

facilitates the end goal as a sub goal; those goal states that inhibit the end goal decrease the likelihood of achievement. Silicon Coppélia has ontological beliefs about goal states being true, ranging from [0–1], and about their facilitating [1], inhibiting [–1], or absent [0] effect on another state. Coppélia calculates the likelihood of accomplishing a goal state across all states she believes are influential for achieving the end goal. She multiplies those states she holds for true with the beliefs about the facilitating or inhibiting effect of those states, using the following algorithm:

- Sort the values in two lists: $[0 \rightarrow 1]$ and $[0 \rightarrow -1]$
- For both lists, start with 0 and take the average of the current value and the next. Continue until end of file
- Result = a weighed mean of the outcomes with weights (#pos/#tot) for the positive values and (#neg/#tot) for the negative (weights are proportional to the number of values in the list)

Each sub goal Coppélia accomplishes increases the perceived likelihood of reaching the end goal. Yet, the more sub goals are achieved, the less influence each individual sub goal has on the perceived likelihood of accomplishing the end goal, having a decreasing marginal utility.

Figure 8.4 shows an outline of Silicon Coppélia, indicating that Coppélia develops state predicates about the user (or any other agency) in a given situation and context. The features of the user are appraised for different aspects, such as ethics, aesthetics, etc. These features are compared with the personal goals of Coppélia so that the user gains personal meaning for her. This way, the user becomes relevant to the Coppélia system, directing the intensity of her affective responses. The direction of Coppélia's affect is regulated by valence. Perspective taking can be done through knowledge of the goals of others. This way, the goals of Coppélia can converge with those of the user so that, for instance, she can coach the user to keep his promise of doing daily exercises.

Through relevance, current, and future valence, appraisal frames are established that guide her intensions to use her (human) counterpart to achieve her goals (e.g., maintenance, winning a game, help the user). Appraisal frames also determine her friendliness toward the user (i.e., Coppélia's involvement) balanced by distance. During affective decision making, Coppélia selects one of four actions from which she expects the highest satisfaction: positive approach (e.g., compliment the user), negative approach (e.g., beat the user), change (e.g., instruct the user), or walk away. Coppélia's affective decisions may change by modulating her responses.

We imported Silicon Coppélia in an affective virtual agent, playing tic-tac-toe with the user. We suspected that the agent could trade rational for affective choice in that the agent would not execute certain moves to win the game if the agent judged that the opponent was nice. We created agents with the ambitions to win or to lose. In view of the likelihood of achieving these goals, of the user's moves, and of Coppélia's expectations on the outcome of the game, Coppélia simulated the emotions joy, distress, hope, fear, and surprise with high or low intensity, expressing them facially by means of Haptek's PeoplePutty software. Like this, 32 (2^5) different emotional states represented each possible combination of two levels of intensity of the five emotions simulated by the Coppélia system. With the different parameter settings, we designed five Coppélia variants and confronted them with users while playing, not to test user experience but to check whether the system worked as expected—that is, we were interested in robot experience.

Simulation 1: Coppélia wants to win

We set Coppélia's ambitions to win at full strength (1) and her ambition to lose at −1, which translates into a strong aversion of losing. All affective influences on the rational decisions to win the game were set to zero. This gave Coppélia a winner's mentality or killer instinct if you will—a rational and narrowly self-interested *homo economicus*.

When played right, it is impossible to lose the tic-tac-toe game; in other words, Coppélia never lost a game, won most of the time, worst performance being a draw as depending on the wit of the human opponent. When Coppélia won a game, joy was increased and distress reduced. Unexpected moves by the opponent or unexpected game endings were accompanied by surprise. As long as Coppélia estimated that she was likely to win, she had relatively high levels of hope; when it was likely she would lose, she experienced relatively high levels of fear.

Simulation 2: Coppélia loses on purpose

We instructed Coppélia that she played against little children and that she had to be considerate. Therefore, Coppélia's ambition to win was set to –1 and her ambition to lose was set to 1 (i.e., she *wanted* to lose). Again, we did not let affect exert an influence on rational decision making so that Coppélia desired to lose irrespective of her "feelings" being hurt (the Samaritan posture). The only way Coppélia could win was if the opponent would make all the wrong moves such that Coppélia was forced to make the winning three-in-a-row in her last move. Indeed, Coppélia was joyful and reduced her distress when she lost. The reverse happened when she won.

Simulation 3: Coppélia plays for money

This time, we made Coppélia's ambition to win or lose dependent on the money ratio between her and her human opponent. We implemented a socialist outlook on life, implying that if Coppélia was rich and the human opponent poor, she wished to lose the game and transfer her money to her opponent. If Coppélia was poor and her human opponent rich, she wanted to win the game and earn some money. Again, the weight of the affective influences on Coppélia's rational decision making was set to 0. Every time poor Coppélia won, joy was increased and distress decreased but after becoming rich this way, enough was enough: Coppélia was not joyful anymore and most distressed each time she won the game.

Simulation 4: Coppélia is too involved with her opponent to win

We then gave Coppélia a positivity bias, in which she could see no evil in her opponent, deeming him morally good, beautiful, and helpful at all times (all values set to 1) and never bad, ugly, or obstructive. At the start of the game, she was very involved with her human opponent (value = 0.85), experiencing hardly any distance (value = 0.08). Yet, she wanted to win the game and not lose it but her ambition levels to win or lose were made dependent on the importance of the game (we told her whether the game was important or not). Moreover, the weight of rational considerations to win the game that would normally influence her affective decision making was set to 0. In other words, we programmed Coppélia as a naïve pleaser (see no evil, hear no evil), making emotional decisions and willing to lose the game in favor of her opponent when the game was not too important. Whenever Coppélia had the chance, she did something positive for her opponent, whom she liked so much. Coppélia lost because she wanted to please her human partner although winning a game would have given her joy and would have reduced her distress.

Simulation 5: Balanced Coppélia wins some, loses some

For our last simulation, we equipped Coppélia with a more balanced set of parameters, the positive side being somewhat stronger than the negative (e.g., initial involvement = 0.63; initial distance = 0.26). The ambition to win the game or not was made dependent on the amount of money that could be gained. Rational influences on decision making were set to 0.8 and the positivity and negativity of actions were set to 0.1. Yet, Coppélia's ambition to win (evoking joy) was always higher than to lose (evoking distress). Nevertheless, when the stakes were low, Coppélia turned out to be so involved with her opponent that she forgot to win the game. At high stakes, however, or when Coppélia became poorer than her opponent, her ambition level increased and she focused all her actions on winning the game. In this simulation, Coppélia made her decisions partially based on rational choice (game moves) and partially on emotion (likeability of the user).

Experiment: Our robot acting human is like a human acting out a robot

The simulations, of course, were not concerned about the way users perceived the Silicon Coppélia system. We made up for this deficit by organizing a speed date for young female participants with a virtual dating partner Tom. To be honest, we were still not interested in the user's perception of Tom but in the way these women thought Tom perceived *them*. Put differently, we wanted humans to diagnose for us the psychology that the technology simulated to see whether the suggested cognitive-emotional behaviors appeared to be natural.

In a more sophisticated version of the Turing Test (Turing, 1950), we had 54 female heterosexual university students (M=20.07, SD=1.88) interact with a virtual dating partner named Tom, shown in a graphic representation created in Haptek's PeoplePutty software, on a web site (Pontier, Siddiqui, & Hoorn, 2010).

Tom was controlled by either our Silicon Coppélia software or by a human confederate (actually, we used two confederates to control for human idiosyncrasies). The confederates could control the responses of Tom as well as his facial expressions. Tom was capable of simulating five emotions: hope, fear, joy, distress, and anger, expressed through facial expressions with low or high intensity. As such, 32 (2^5) different emotional states were created: one for each possible combination of two levels of intensity of the five simulated emotions. The level of intensity depended on the level of relevance of participants' responses to Tom's goals and concerns.

The women were unaware of the condition they worked in and were told that in all cases they were conversing with a *robot* (also in the human-controlled condition). We preferred this over Turing's framing of a machine acting like a human because that would raise such expectations that overall disappointment would exert a ceiling effect on all the negative variables (e.g., "very unrealistic") and a floor effect on all the positive variables (e.g., "hardly relevant").

Participants conversed with Tom for about 10 min about various topics (i.e., family, sports, appearance, hobbies, music, food, and relationships) by means of drop-down boxes. Tom replied in text and speech, and showed facial expressions in line with the emotional state he was in. After the speed date, participants were asked to fill out the I-PEFiC questionnaire together with a number of items related to emotions such as joy and distress. They had to assess how Tom thought about them (e.g., "Did Tom like you?") instead of how they felt about Tom.

Actually, we predicted the null hypothesis because we hoped that the software could simulate human behavior so well that participants would fail to see the difference between the robot-directed conversations versus the human confederates. And indeed, analysis of variance and regression analysis of the obtained scales showed that Tom as governed by Silicon Coppélia was attributed quite the same and the same intensity of emotions as Tom when he was performed by humans. However, a null hypothesis can never be accepted, only rejected or failed to be rejected. Therefore, we also employed Bayesian statistics, which assumes that a priori beliefs, that is, prior distributions, are updated with observed evidence, the likelihood function, and which combines prior and likelihood in a so-called posterior distribution. Analysis according to Bayes indicated that the null hypothesis (i.e., robot performance equaled human performance) received almost seven times more support from the data than the alternative hypothesis (i.e., robot performance differed from human performance).

In other words, these young women endowed Tom as controlled by our Silicon Coppélia software with the same emotions and with similar strength of emotions as Tom played by human confederates. Likewise, the cognitive-affective architecture that these women detected in Tom controlled by our software was similar to the cognitive-affective architecture they saw in the human confederates enacting Tom. The upshot was that these findings resembled the I-PEFiC model as retrieved from empirical data of users expressing how they felt about embodied agents (i.e., the Van Vugt et al. studies). In other words, behavior simulated by Silicon Coppélia is about the same as humans acting like a robot (cf. Anthony Daniels playing C-3PO or Brent Spiner doing the android Data), which resembles closely how humans feel about a robot that is not the same as a robot being human but is more human-like than a robot that is not performed by a human. As is, Silicon Coppélia resembles a person wearing a robot suit, judging humans as if they were virtual characters in the—to her—real robotic world.

Realism? Epistemics of the Virtual

With Silicon Coppélia we created a synthetic stage actor impersonating a human being. Coppélia holds beliefs, which she regards as "true" and upon which she acts. The robot world is her frame of reference against which she judges human behavior. That made us aware that the factor *epistemics* in I-PEFiC and Silicon Coppélia was not about designed or perceived realism alone. Whether a robot or virtual character looks exactly like a human or not may not invigorate great involvement or distance in humans (Van Vugt et al., 2007) but a world view that is totally different from that of the user might, because it may lead to actions and opinions that are opposing those of the user. Reversely, what the user thinks the robot is capable of and how she "means" things determines whether affordances are perceived at all ("Can a robot write a symphony?") or whether robot decision making is ethical or not ("Murder's a new trick for a robot").[3] In *Epistemics of the Virtual* (Hoorn, 2012), these issues were elaborated at length, resulting in a flow chart that describes how humans make sense of the world about them in terms of truth, realism, figurativeness, framing, illusions, and the like. That chart is currently used to implement robot understanding of matters such as fiction-reality distinction, understanding metaphor, and detecting fake identities so that they know that we see them as virtual, whereas they see us as real.

With Silicon Coppélia connected to Epistemics of the Virtual, the robot has a system of beliefs that describes the way the world is organized and catalogued according to a preset ontology (e.g., "Theater is fiction, science is real"). This is called ontological classification. Due to its fuzzy logics, the robot can attach moderations of truth or falsity to those beliefs as well as to incoming data that is matched against its reference ontology (we plan to do this with fast Fourier transformations). When programmed that way, the robot may take a rigid stance of right or wrong or may apply multivalued logics and deem things possible or probable ("Perhaps theater holds more truth than science").

When the robot encounters events or things that go against the implemented ontology, epistemic appraisal starts, indicating whether phenomena that are deemed "false" or "not real" and on the brink of being rejected can at least be considered as "realistic"—looking real, as if real, feasible rather than being really real, according to robot beliefs. Dependent on signal detection settings (e.g., Poor, 1994), stimulus discrimination may be strict or lenient, which means that unexpected input (e.g., someone entering the room with a red plastic nose) leads to rejection ("People do not have plastic noses") or acceptance of the deviation ("It is a clown!"). In a lenient mode, unexpected or uncertain input (e.g., "Are grain circles man-made?" "Can particles take the shape of a wave?"), more features that seem unrealistic are tolerated. This allows for the initial ontology to open up to change, accepting that there may be peripheral members to a class overlapping with other classes ("But we just don't know").

In the case that a mismatch is detected between stored classes and incoming data while epistemic appraisal judged that most features of the deviation are unrealistic, there is but one module left for the robot: to do a novel-comparison evaluation. When the robot perceives a—to her—novel combination with many unrealistic aspects, it will compare the combined entities to figure out whether this novelty is meant literally (cf. science), metaphorically (cf. poetry), or as an anomaly (cf. absurd humor) (cf. Hoorn, 1997). If you want examples of each, then compare Bohr's "The atom is a miniature solar system," Shakespeare's "Juliet is the sun," and Carroll's "All mimsy were the borogoves," respectively. In the first, the robot should understand that certain features are meant literally the same (e.g., orbital motion); in the second that literal features of the one (e.g., light, warmth) should be taken figuratively for the other; the third does not have to be understood because the few features present do not overlap.

Whereas ontological confirmation merely strengthens the belief system, epistemic appraisal and the evaluation of novel combinations changes it—definitely because the robot now knows there are occurrences that seem unrealistic or absurd; possibly because slowly and after many confirmations, the unrealistic and absurd knowledge may shift from the fiction to the reality department—all depending on signal-detection settings. What was speculation at first becomes common knowledge after sufficient confirmation.

Ethics: Moral Reasoner

The more robots become autonomous, making decisions based on beliefs they hold for "true," the louder the call for ethical behavior (e.g., Wallach, 2010). Imagine a robot that can keep a distance to its user and believes the user made mistakes on purpose. Then one of the actions might be to walk away and in doing so the robot does not remind the user of taking their once-a-day cardiac tablet. Or worse, it starts to behave like an army combat robot. In line with Gillon (1994), we defined beneficence, nonmaleficence, autonomy, and justice as the four basic moral obligations of a robot. Applied to the medical domain, this means that robot interaction is directed to the wellbeing of the patient, not causing harm, respecting the patient's independence and sovereignty, and being fair in, for example, distributing goods over multiple people.

In Pontier and Hoorn (2012), we used three of the four moral obligations to push the ethical reasoning of the Silicon Coppélia system beyond the simple assessment of "good" or "bad." As a goal of its own, Coppélia attempted to maximize the utility of her actions for each (human) stakeholder in the situation. Coppélia estimated the consequences of her actions for the user's independence (autonomy), for being harmless (nonmaleficence), and for advancing the user's well-being (beneficence)—in this order. Anderson and Anderson (2008) argue that this prioritization is important because a patient's decision should only be disobeyed if the patient gets hurt (nonmaleficence) or if beneficence is strongly violated (e.g., refusing medicine that can cure a deadly disease). Autonomy is crucial here and other principles come into operation only if the patient's autonomous decision making is severely damaged (e.g., in the case of dementia). Nonmaleficence is more important than beneficence, because causing harm *always* violates the patient's autonomy, whereas not helping *might*.

We simulated six situations that were drawn from Buchanan and Brock (1989), in which medical ethical professionals judged morally ambiguous cases. In one case, the patient had a pathological fear of medication and refused to take the antibiotics needed to cure a deadly infection. Our Moral Coppélia reasoned that she should try to change the patient's mind because irreversible harm would be done if she let him persist in his fear. Because the fear had no empirical ground, she believed that the autonomy of the patient's judgment was not fully intact. Therefore, Coppélia decided that offering the medication again (beneficence) was morally better than respecting the patient's autonomy and accepting his refusal. These considerations were the same as those of the medical-ethical judges in Buchanan and Brock (1989).

In a second case, the patient refused medication not by undue fear but out of deep religious motives. This aspect changed Coppélia's attitude completely. Although she disagreed with the patient's refusal of medication, she accepted the decision as definite. Although the patient put his life on the line, the reasoning was performed by a competent adult so that in Coppélia's moral assessment, autonomy toppled nonmaleficence as well as beneficence. The medical ethical professionals in Buchanan and Brock (1989) came to the same conclusions. Like this, we performed another four simulations in which Coppélia's moral decisions were identical to those of the professionals (Pontier & Hoorn, 2012).

However, all Coppélia's moral reasoning was done without any affect. Although strictly speaking, Coppélia may be right in applying rational ethics to medical cases, this is not what ordinary people do in daily life. And so, Coppélia had no understanding of affect-laden ethical decisions of her users. Greene, Sommerville, Nystrom, Darley, and Cohen (2001) used two variants of the same moral dilemma and found that emotional engagement biases rational decision making. Although rationally, two situations may be morally equivalent, emotionally this does not count. People have little problems eating a cow but they deem it morally repulsive to eat their pet.

We simulated the Greene et al. cases to let Moral Coppélia draw conclusions from the four ethical principles with and without affective components such as personal relevance and valence (Pontier, Widdershoven, & Hoorn, 2012). The dilemmas tested were the "Footbridge" and the "Trolley" dilemma. Both boil down to the same difficulty, namely whether one is prepared to sacrifice one person to save five. The difference is that in the Trolley dilemma, the action is undertaken at a distance, changing the switches to let a trolley run over one person instead of five. In the Footbridge dilemma, the action is personal and at close range. One needs to push someone off a bridge to save five others. In the latter version it is much harder for people to make a decision than in the first.

When we made Coppélia a stone-cold rationalist, there were no differences in her decisions between the Footbridge and the Trolley dilemma. After all, the moral parameters were the same (take out one, save five) so that she decided that killing was morally better than doing nothing. When we made her more sensitive, one person dead or five people dead were both undesirable goals. Hitting the switch in the Trolley dilemma would lead to one dead, whereas doing nothing would cause five people to die. Coppélia realized that sacrificing the one person would be an extremely negative action toward that person, whereas letting him live at the expense of the others would be extremely positive for him. And although Coppélia was more involved than at a distance toward the potential victim, the expected satisfaction of letting him live ("do nothing") and killing five others was just a little lower than letting him die and save the others. Thus, Coppélia pulled the trigger in compassion—but nevertheless.

With a "sensitive" Coppélia in a more personal setting, the rational influence evaporated. In the Footbridge dilemma, we turned up the weights for the outlook on the positive or negative consequences of Coppélia's actions toward the potential victim. All else being equal, Coppélia now decided against her rational ethics and in favor of the individual of whom she knew that killing him was an extremely negative action. Therefore, Coppélia froze: she did not push him. The group of five for whom she "felt" nothing (because we had set those emotional parameters to zero) was sacrificed to save the one she loved.

In future research, we will create Footbridge and Trolley dilemmas in which autonomy is intact or not and Coppélia provides moral advice with emotional considerations included or not. Imagine a scenario with a daughter whose father is an Alzheimer patient. He tripped over and is severely injured. He will surely die but father refuses to be hospitalized. If she lets him go, his heart and kidneys will save three patients waiting for a transplant. If Coppélia takes the position of the daughter in a rational way, she will advise against father's free will and have him hospitalized (beneficence over degraded autonomy). If Coppélia takes the position of the doctor, her most rational advice is to let the Alzheimer patient die to save three others. Both advices are derived from rational reasoning but the conclusions are completely contrary. This is only explicable from the different premises of *emotional* attachment from which the reasoning started! For the daughter the situation is a Footbridge dilemma in which beneficence prevails over poor judgment by her beloved father. For the doctor, the situation is a Trolley dilemma, in which saving three has higher priority than saving one (whose autonomy is degraded anyways). The doctor has emotional attachment to neither of the four patients. But what will Coppélia do if the doctor has a crush on the daughter or if the daughter is the heiress of a rich father? Or if the father does not have Alzheimer but refuses treatment out of religious conviction? In other words, in the future Coppélia is capable of taking perspectives and explaining moral positions to other parties.

Affordances: Creative Machines

Suppose a sick child refuses food but that food contains the right nutrition and vital medication. As a caretaker, you could reason that doing nothing harms the child and that the child's judgment is not developed enough to see what is good for her. Therefore, you may choose to override her autonomy and force-feed her. But every parent knows that the food will not be swallowed but given back in a gross way. This is not only a moral but also a practical dilemma. It is morally better to respect the child's autonomy and apropos she will eat better too that way.

An uncreative and merely intelligent robot would advise to try again and again, repeating known behaviors, the utter consequence indeed being to force the medicine upon the child, perhaps even intravenous. A robot merely capable of intelligent reasoning would provide more arguments why the child should eat, perhaps citing the complete medical encyclopedia to underscore its point. But little children or Alzheimer patients are not necessarily convinced by reason. Nor are healthy people for that matter.

A creative machine should approach the matter from a completely different angle. The parent trying to put a fork into the mouth of his child suddenly relies on an entirely different frame of reference than food intake and draws in the concept of an airplane. Now there is no resistance, the fork lands in the mouth, and the food is swallowed down with a smile. Patient autonomy is safeguarded and practical solution found; without reasoning.

ACASIA is a system capable of creating novel concepts by combining entities from associatively remote categories (e.g., fork and airplane) (Hoorn, 1997, 2002, 2013; Lowcre, 2013). The acronym stands for association, combination, abstraction, selection, integration, and adaptation, which are the process modules responsible for creation (Figure 8.5). ACASIA assumes that the creator has an ontology of at least two categories (cf. Epistemics of the Virtual). Categories should not be empty and contain at least one exemplar. Exemplars should not be empty either but have at least one feature. The core principle of ACASIA is to search for matches between features of categories

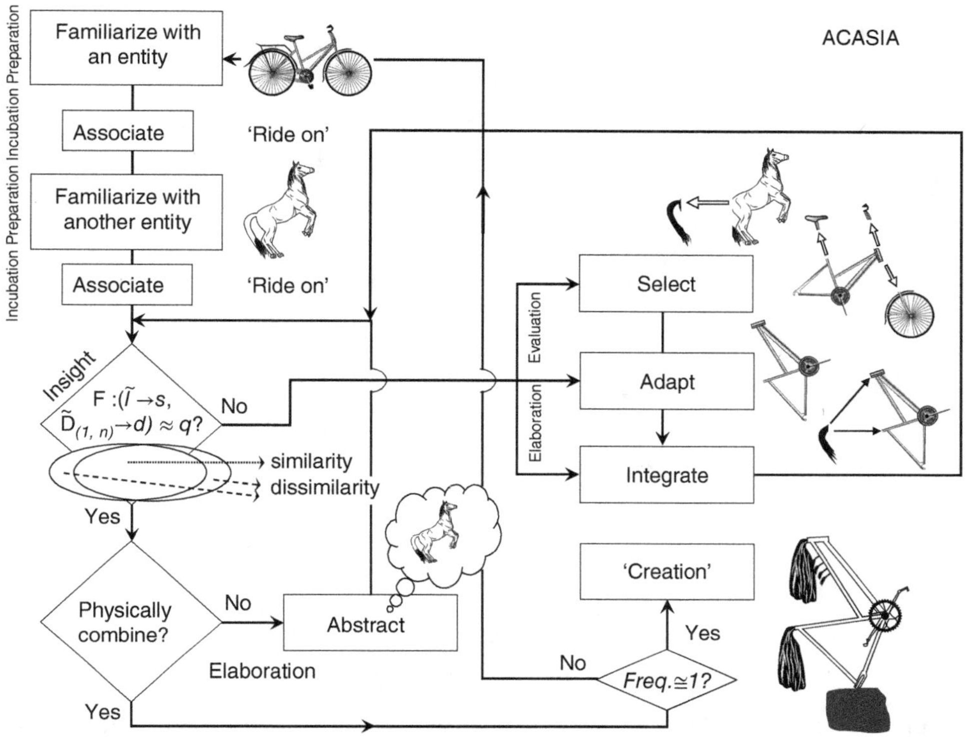

Figure 8.5 The creative process.
Source: Adapted from Hoorn (2002). © ACM.

that are unrelated semantically (cf. Aristotelian comparison in Hoorn, 1997), according to the standard belief system (cf. ontological classification). The intersecting features evoke (in humans) the perception of similarity, the distinctive sets the perception of dissimilarity (cf. Tversky, 1977). There should be a balance between similarity and dissimilarity—not one excluding the other—for which ACASIA uses feature selection and adaptation—letting features in or out or modifying them (e.g., change color). If it is not possible to match features, ACASIA abstracts them to a higher level (ending in mathematics or geometry if that is required) so that features can be matched anyways.

This way, ACASIA can generate novel comparisons like "An elephant is like a mountain," explaining that both are big and old. Or "The wind is like a pencil," both being sharp. Apart from such poetic similes, ACASIA produces design ideas such as turning a spoon into a snake, which we actually did as depicted in Figure 8.6.

With ACASIA installed, a Caredroid can actually replicate the idea of the parent changing his feeding strategy and turning a fork into a plane. This is what we plan to do in future research. We want to provide ACASIA with an ontology that has fork, airplanes, and other vehicles at its disposal to see whether fork and plane are combined as expected. It might be that the absence or presence of "moving through the air" in the fork feature-list decides whether the combination is made at all. It is a rare feature

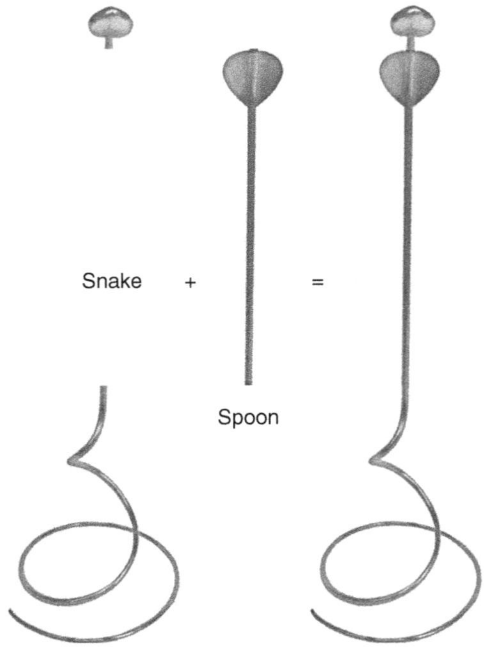

Figure 8.6 Suggested by computer: CobraSpoon is a cocktail spoon turned into a snake. The actual object was made by a goldsmith.
Source: Reprinted with permission from Hoorn (2103). © John Benjamins.

of fork normally becoming salient only when the arm hangs in the air waiting for the child to bite. If ACASIA can repeat standardized creativity such as "A fork is like an airplane" (see the Air fork one),[4] it can do novel combinations as well (see CobraSpoon, Figure 8.6), thus providing Caredroids with creative problem solving tools next to the straightforward rational approaches.

Caredroids: Humanoids Taking Care

Robots in a healthcare setting need to know about health. That is a given. Apart from being a dressed-up medical file or Wikipedia, however, they should show affect such that humans can relate to them. This is what we try to achieve with Silicon Coppélia. Yet, because ethics is the main factor in becoming friends with a system, we are developing Moral Coppélia so that the Caredroid behaves in a moral way and has an idea of how humans reason morally. The other determining factor in computer–human interaction is the type of affordances the system offers. We extended rational problem solving with creative strategies so that, for instance, a moral deadlock (Alzheimer patient refuses medicine) may be breached by offering unexpected alternatives ("It's not medicine but a candy"—the reverse placebo). Fooling people into believing something for their own good, creatively substituting one concept for the other, raises ethical questions (beneficence over autonomy?), and produces a genuine epistemics of the virtual: What is true, what is truthful, what is trustworthiness, what do you believe in when a fictional character (the Caredroid) autonomously decides to tell a white lie (the candy) for the good of the patient? We will know once the cognitive robotics of our Caredroids have become an integration between Silicon and Moral Coppélia, ACASIA, and Epistemics of the Virtual.

Conclusions

Literature and theater were at the foundation of our program of research, and what we learned as social scientists is, in an itemized form, that core emotional components such as relevance and valence are pivotal not merely in the appraisal of actual but of virtual others as well. Additionally, positive and negative trends can be simultaneously present as observable in the trade-off between involvement and distancing effects.

We also learned that the tooling side of interactive agent systems and robots, the affordances they offer to the user, exert definite effects on experiencing friendly or cold feelings for the "personality" of the device, more usable being more of a friend. Vice versa, whether the tool is regarded as morally sound or not has profound effects on the intentions to use it: If a care robot posts medical data on Facebook without authorization of the user, the tool is considered less useful, although strictly speaking it performs its core tasks (e.g., care brokerage) with the same ability.

Psychological models and measured effects can serve to shape synthetic characters, which in turn can be used to conduct social-psychological experiments with personalities that can be systematically varied, displaying a constant performance. To make simulations work properly, moreover, many assumptions need to be explicated and quantified, adding more precision to the original psychological theories.

Another way in which our work advances the psychology of human–technology interaction is the change in perspective while trying to grasp how machines could possibly understand us. The public introduction of humanoid robots is near and we better be prepared for what they can or cannot do—with exciting new opportunities for social scientists to study people's responses in terms of what real emotion is and what true creativity? Whether ethics is rational or should be softened by emotional considerations? Today's humanoid robots challenge the way humans envision themselves.

In a succinct description of the systems that we have built or that we are in the process of building, Silicon Coppélia is the most elaborated thus far: it is an artificial intelligence system that simulates the generation and regulation of affective states and processes and mimics emotions in response to other agencies (human or artificial). The system gradually builds up one of four action tendencies with respect to the other agency (i.e., positive or negative approach, avoidance, or changing the other), leading to a concrete action.

Silicon Coppélia is initially programmed (ENCODE) according to a situational model that makes assumptions about the (human) world around her (based on, for instance, moral and aesthetic principles) and how events in the world can affect situations. In addition, the system is programmed such that in response to a user, it can display six emotions. The kind of behavior or emotion that Coppélia shows is affected by the behavior and reactions of its user (interaction) and the (care) goals that she has installed. In the COMPARE phase, Coppélia weighs user features in relation to the goals she must achieve and based on these considerations her actual reactions follow (RESPOND). The parameters related to the trade-offs can be externally manipulated (by the programmers) and in the future by Coppélia as a self-learning system. In her most creative variant, Coppélia will be able to adjust her emotion-regulation strategies and subsequent reactions (e.g., positive, negative, avoidant), but not her purpose and mission. The question is whether the latter would be desirable in certain cases.

Moral Coppélia is an extension of the first; an artificial intelligence system that is capable of moral reasoning and deciding on ethical dilemmas in the same way as medical ethical professionals do. It bases itself on the (standard) prioritization of ethical principles (in order of increasing importance): justice, beneficence, non-maleficence, and patient autonomy.

ACASIA, so far, is a simple search algorithm to find (partial) semantic matches between entities in (semi) structured databases (categories-exemplars-features) to make creative combinations in problem solving as well as in (industrial) design. In theory, ACASIA has extra modules to abstract, select, adapt, and integrate features to optimize the blend and a frequency-of-occurrence counter to estimate novelty in a given sample but these are not implemented yet.

Epistemics of the Virtual is a system under construction. It will eventually be capable of fiction-reality distinction to discern between human ("real") and artificial ("fictitious") responses, such as agencies giving commands or doing requests, which is important in interaction with delusive patients. It can also be employed to detect simulated illness.

The degree to which the systems we have designed are able to mimic human psychology, based on studies that we have covered (such as the speed-dating application), is that we can simulate human behavior as if humans were performing a robot: their human-likeness shines through. This does not mean, however, that the electromechanics of the machine does the same as the electrobiology of the human brain to achieve correspondence at the functional level: more roads may lead to Rome.

Acknowledgments

This chapter was funded through Services of Electro-Mechanical Care Agencies (SELEMCA) within the Creative Industry Scientific Programme (CRISP) (grant number: NWO 646.000.003). CRISP is supported by the Dutch Ministry of Education, Culture and Science. The current chapter benefited from the insights of Dr Erna van Hest who provided many helpful comments to the presentation of the studies reviewed here.

Notes

1 http://www.marwencol.com/
2 http://nl.wikipedia.org/wiki/Copp%C3%A9lia
3 Questions and remarks by detective Del Spooner during Sonny's interrogation in *I, Robot* (Alex Proyas, 2004). Available at: http://www.youtube.com/watch?v=9ZADo17YbCA; http://en.wikiquote.org/wiki/I,_Robot_(film)
4 http://www.perpetualkid.com/airfork-one-kids-fork.aspx

References

Anderson, M., & Anderson, S. (2008). Ethical healthcare agents. In M. Sordo, S. Vaidya, & L. C. Jain (Eds.), *Studies in Computational Intelligence: Vol. 107. Advanced computational intelligence paradigms in healthcare – 3* (pp. 233–257). Berlin/Heidelberg, Germany: Springer.

Banks, M. R., Willoughby, L. M., & Banks, W. A. (2008). Animal-assisted therapy and loneliness in nursing homes: Use of robotic versus living dogs. *Journal of the American Medical Directors Association, 9*, 173–177.

Bosse, T., Hoorn, J. F., Pontier, M. A., & Siddiqui, G. F. (2008). A robot's experience of another robot: simulation. In V. Sloutsky, B. C. Love, & K. McRae (Eds.), *Proceedings of the 30th Annual Conference of the Cognitive Science Society* (pp. 2498–2503). Washington, DC: Cognitive Science Society.

Broadbent, E., Stafford, R., & MacDonald, B. (2009). Acceptance of healthcare robots for the older population: Review and future directions. *International Journal of Social Robotics, 1*(4), 319–330.

Broekens, J., Heerink, M., & Rosendal, H. (2009). Assistive social robots in elderly care: A review. *Gerontechnology, 8*(2), 94–103. doi:http://dx.doi.org/10.4017/gt.2009.08.02.002.00

Buchanan, A. E., & Brock, D. W. (1989). *Deciding for others: The ethics of surrogate decision making*. Cambridge, UK: Cambridge University Press.

Davidson, R. J., Putnam, K. M., & Larson, C. L. (2000). Dysfunction in the neural circuitry of emotion regulation—A possible prelude to violence. *Science, 289*, 591–594.

Davis, F. D. (1989). Perceived usefulness, perceived ease of use, and user acceptance of information technology. *MIS Quarterly, 13*(3), 319–339.

Dillman, D. A. (2000). *Mail and Internet surveys: The tailored design method*. New York, NY: Wiley.

Dollard, J., & Miller, N. E. (1950). *Personality and psychotherapy—An analysis in terms of learning and thinking*. New York, NY: McGraw-Hill.

Frijda, N. (1986). *The emotions*. Cambridge, UK: Cambridge University Press.

Gaver, W. W. (1991). Technology affordances. In S. P. Robertson, G. M. Olson, & J. S. Olson (Eds.), *Proceedings of the SIGCHI Conference on Human Factors in Computing Systems* (pp. 79–84). New Orleans, LA: ACM.

Gibson, J. J. (1979). *The ecological approach to visual perception*. Boston, MA: Houghton Mifflin.

Gillon, R. (1994). Medical ethics: Four principles plus attention to scope. *British Medical Association, 309*(6948), 184–188.

Gratch, J., & Marsella, S. (2005). Some lessons from emotion psychology for the design of lifelike characters. *Journal of Applied Artificial Intelligence (special issue on Educational Agents—Beyond Virtual Tutors), 19*(3–4), 215–233.

Greene, J. D., Sommerville, R. B., Nystrom, L. E., Darley, J. M., & Cohen, J. D. (2001). An fMRI investigation of emotional engagement in moral judgment. *Science*, *293*(5537), 2105–2108. doi:10.1126/science.1062872

Gross, J. J. (1998). The emerging field of emotion regulation: an integrative review. *Review of General Psychology*, *2*(3), 271–299.

Gross, J. J. (2001). Emotion regulation in adulthood: Timing is everything. *Current Directions in Psychological Science*, *10*(6), 214–219.

Hoorn, J. F. (1997). *Metaphor and the brain: Behavioral and psychophysiological research into literary metaphor processing* (Doctoral dissertation, VU University. Ann Arbor, MI: ProQuest). Retrieved from http://dare.ubvu.vu.nl/bitstream/1871/13048/1/7392.pdf

Hoorn, J. F. (2002). A model for information technologies that can be creative. In T. T. Hewett & T. Kavanagh (Eds.), *Proceedings of the 4th Creativity and Cognition Conference, Loughborough, UK* (pp. 186–191). New York, NY: ACM.

Hoorn, J. F. (2008). A robot's experience of its user: Theory. In V. Sloutsky, B. C. Love, & K. McRae (Eds.), *Proceedings of the 30th Annual Conference of the Cognitive Science Society* (pp. 2504–2509). Washington, DC: Cognitive Science Society.

Hoorn, J. F. (2012). *Epistemics of the virtual*. Amsterdam/Philadelphia, PA: John Benjamins.

Hoorn, J. F. (Ed.). (2013). *Organic Creativity and the Physics within*. Amsterdam/Philadelphia, PA: John Benjamins.

Hoorn, J. F., & Konijn, E. A. (2003). Perceiving and experiencing fictional characters: An integrative account. *Japanese Psychological Research*, *45*(4), 250–268.

Hoorn, J. F., Konijn, E. A., Van Vliet, H., & Van der Veer, G. (2007). Requirements change: Fears dictate the must haves; desires the won't haves. *Journal of Systems and Software*, *80*(3), 328–355.

Hoorn, J. F., Pontier, M. A., & Siddiqui, G. F. (2008). When the user is instrumental to robot goals. First try: agent uses agent. In L. C. Jain, & Yuefeng Li (Eds.), *Proceedings of IEEE/WIC/ACM Web Intelligence and Intelligent Agent Technology 2008 (WI-IAT '08), 2* (pp. 296–301). Sydney, Australia: IEEE/WIC/ACM. doi:10.1109/WIIAT.2008.113

Hoorn, J. F., Pontier, M. A., & Siddiqui, G. F., (2012). Coppélius' concoction: Similarity and complementarity among three affect-related agent models. *Cognitive Systems Research*, *15–16,* 33–49. doi:10.1016/j.cogsys.2011.04.001

Konijn, E. A. (2000). *Acting emotions. Shaping emotions on stage*. Amsterdam, The Netherlands: Amsterdam University Press.

Konijn, E. A., & Bushman, B. J. (2007). World leaders as movie characters? Perceptions of G.W. Bush, T. Blair, O. Bin Laden, and S. Hussein at the eve of Gulf War II. *Media Psychology*, *9*(1), 157–177.

Konijn, E. A., & Hoorn, J. F. (2005). Some like it bad. Testing a model for perceiving and experiencing fictional characters. *Media Psychology*, *7*(2), 107–144.

Lazarus, R. S. (1991). *Emotion and adaptation*. New York, NY: Oxford University Press.

Lowcre, M. M. M. (2013). *Organic creativity and the physics within*. Amsterdam/Philadelphia, PA: John Benjamins.

Mao, W., & Gratch, J. (2006). Evaluating a computational model of social causality and responsibility. *Proceedings of the 5th International Joint Conference on Autonomous Agents and Multiagent Systems, Hakodate, Japan.*

Marsella, M., & Gratch, J. (2009). EMA: A process model of appraisal dynamics. *Journal of Cognitive Systems Research*, *10*(1), 70–90.

McGrenere, J., & Ho, W. (2000). Affordances: Clarifying and evolving a concept. In S. S. Fels & P. Poulin (Eds.), *Proceedings of Graphics Interface 2000, Montreal* (pp. 179–186). Quebec, Canada: Lawrence Erlbaum.

Mostow, J. (2009). *Proceedings of Graphics Interface 2000, Montreal* (pp. 179–186). Surrogates. Burbank, CA: Touchstone Pictures.

Ochsner, K. N., & Gross, J. J. (2005). The cognitive control of emotion. *Trends in Cognitive Sciences, 9*, 242–249.

Pontier, M. A., & Hoorn, J. F. (2012). Toward machines that behave ethically better than humans do. In N. Miyake, B. Peebles, & R. P. Cooper (Eds.), *Proceedings of the 34th International Annual Conference of the Cognitive Science Society, CogSci'12, 2012, Sapporo, Japan* (pp. 2198–2203). Austin, TX: Cognitive Science Society.

Pontier, M. A., Siddiqui, G. F., & Hoorn, J. F. (2010). Speed dating with an affective virtual agent. Developing a testbed for emotion models. In J. A. Allbeck, N. I. Badler, T. W. Bickmore, C. Pelachaud, & A. Safonova (Eds.), *Proceedings of the 10th International Conference on Intelligent Virtual Agents (IVA) Sept. 20–22, 2010, Philadelphia, PA, Lecture Notes in Computer Science (LNCS) 6356* (pp. 91–103). Berlin/Heidelberg, Germany: Springer.

Pontier, M. A., Widdershoven, G., & Hoorn, J. F. (2012). Moral Coppélia—Combining ratio with affect in ethical reasoning. In J. Pavón et al. (Eds.), *Lecture notes in artificial intelligence, 7637* (pp. 442–451). Berlin/Heidelberg, Germany: Springer.

Poor, H. V. (1994). *An introduction to signal detection and estimation.* New York, NY: Springer.

Smith, C. A., & Lazarus, R. (1990). Emotion and adaptation. In L. A. Pervin (Ed.), *Handbook of personality: Theory and research* (pp. 609–637). New York, NY: Guilford.

Sood, A., & Tellis, G. J. (2005). Technological evolution and radical innovation. *Journal of Marketing, 69*, 152–168.

Thompson, R. A. (1994). Emotion regulation: a theme in search of definition. In N. A. Fox (Ed.), *Monographs of the Society for Research in Child Development: Vol. 59(240). The development of emotion regulation: Biological and behavioral aspects* (pp. 25–52). New York, NY: Wiley.

Turing, A. M. (1950). Computing machinery and intelligence. Mind. *59*(236), 433–460.

Turner, M., Kitchenham, B., Brereton, P., Charters, S., & Budgen, D. (2010). Does the technology acceptance model predict actual use? A systematic literature review. *Information and Software Technology, 52*(5), 463–479.

Tversky, A. (1977). Features of similarity. *Psychological Review, 84*(4), 327–352.

Van Vugt, H. C., Bailenson, J. N, Hoorn, J. F., & Konijn, E. A. (2010). Facial similarity shapes user response to embodied agents. *Transactions of Computer-Human Interaction (TOCHI), 17*(2), *7*, 1–27. doi:10.1145/1746259.1746261

Van Vugt, H. C., Hoorn, J. F., & Konijn, E. A. (2009a). Interactive engagement with embodied agents: An empirically validated framework. *Computer Animation and Virtual Worlds, 20*, 195–204. doi:10.1002/cav.312

Van Vugt, H. C., Hoorn, J. F., Konijn, E. A., & De Bie Dimitriadou, A. (2006). Affective affordances: Improving interface character engagement through interaction. *International Journal of Human-Computer Studies, 64*(9), 874–888. doi:10.1016/j.ijhcs.2006.04.008

Van Vugt, H. C., Konijn, E. A., Hoorn, J. F., Eliëns, A., & Keur, I. (2007). Realism is not all! User engagement with task-related interface characters. *Interacting with Computers, 19*(2), 267–280.

Van Vugt, H. C., Konijn, E. A., Hoorn, J. F., & Veldhuis, J. (2009b). When too heavy is just fine: Creating trustworthy e-health advisors. *International Journal of Human-Computer Studies, 67*(7), 571–583. doi:10.1016/j.ijhcs.2009.02.005

Venkatesh, V., Morris, M. G., Davis, G. B., & Davis, F. D. (2003). User acceptance of information technology: Toward a unified view. *MIS Quarterly, 27*(3), 425–478.

Wallach, W. (2010). Robot minds and human ethics: The need for a comprehensive model of moral decision making. *Ethics and Information Technology, 12*(3), 243–250.

Zadeh, L. A. (1965). Fuzzy sets. *Information and Control, 8*(3), 338–353.

Part III

Technological Affordances and Social Interaction

9

Social Network Site Affordances and Their Relationship to Social Capital Processes

Nicole B. Ellison[1] and Jessica Vitak[2]

[1] *School of Information, University of Michigan, Ann Arbor, MI, USA*

[2] *College of Information Studies, University of Maryland, College Park, MD, USA*

Overview of Social Media

Today, Internet-enabled communication technologies are pervasive, reshaping the ways millions of individuals search for life partners (Ellison, Hancock, & Toma, 2012), access information and share opinions (Jeong, Morris, Teevan, & Liebling, 2013), and even grieve for deceased loved ones (Marwick & Ellison, 2012). Social network sites are particularly well-suited for relationship maintenance (Tong & Walther, 2011), and individuals use sites like Facebook and Twitter to keep in touch with a wide range of contacts (Binder, Howes, & Sutcliffe, 2009; Vitak, 2012a) such as professional colleagues, family members (Burke, Adamic, & Marciniak, 2013), friends, and acquaintances, including those from previously inhabited locations and across physical distances (Ellison, Steinfield, & Lampe, 2007).

Social media support patterns of information flow and interpersonal communication that differ in important ways from traditional mass media like television and radio. While no uniformly adopted definition of social media has emerged, the term is used colloquially to describe a set of features and tools that enable peer-to-peer communication in ways not supported by mass media which use a "one-to-many" broadcast model. The term encompasses social network and micro-blogging sites (e.g., Facebook, Twitter), peer production communities (e.g., Wikipedia), content sharing and discussion forums (e.g., Reddit), and online dating sites (e.g., Match.com, eHarmony)—although many of these platforms are converging and now support multiple kinds of activities. For the 85% of US adults that are online (Pew Internet and American Life Project, 2013), social media are reshaping mundane tasks, like finding a restaurant, and more significant ones, such as finding a life partner.

Social media environments evolve rapidly—user practices change over time, services update their user-facing interfaces and back-end technology, and the larger social and

The Handbook of the Psychology of Communication Technology, First Edition. Edited by S. Shyam Sundar.

legal frameworks surrounding their development, regulation, and use shift. In order for researchers to avoid producing scholarship that merely describes a particular site and set of users at a particular moment in time, scholars who use the notion of higher-level affordances as a lens for understanding their findings have the added benefit of producing work that will be useful even after the sites have long changed, or even disappeared. Thus, social media researchers seeking to contextualize their findings in relation to higher-level patterns of behavior or to better understand the mechanisms behind some of the relationships they have identified have used an affordance approach to do so.

In the context of social media, an affordance-based approach[1] has been utilized by a wide range of technology researchers (boyd, 2010; Ellison, Lampe, Steinfield, & Vitak, 2011; Resnick, 2002; Sundar, 2008; Treem & Leonardi, 2012; Wellman, 2001), perhaps because they value the opportunity to frame their insights in relation to higher-level characteristics as opposed to the idiosyncratic features of a particular technology or site. The affordance framing enables us to consider the materiality of a technology (Leonardi, Nardi, & Kallinikos, 2012) without the technologically deterministic assumptions of "impact" that historically accompany the introduction of any new communication technology (Sturken, Thomas, & Ball-Rokeach, 2004).

Researchers focus on different affordances of social media, typically driven by their specific area of study, and employ the term in slightly different ways. For example, Treem and Leonardi (2012) outline a set of four affordances that differentiate social media in organizational settings from other forms of online and offline communication: (a) *visibility*, which highlights the ease with which social media enable users to make previously invisible information visible, as well as the ease with which other users can then locate that information; (b) *persistence*, also known as reviewability, which refers to the accessibility of content—in its original format—after it has been posted; (c) *editability*, which relates to the asynchronous nature of communication in social media and describes users' ability to carefully craft a message before posting, as well as the ability to edit it after it has been posted; and (d) *association*, which refers to both the articulated connections between users (e.g., Facebook friends, Twitter followers), as well as the connections between users and the content they post. They note that social media rank high across all four affordances, unlike many other forms of mediated communication. Thus, in the workplace, enterprise social media are potentially transformative with regard to communication patterns and organizational knowledge-sharing (Leonardi, Huysman, & Steinfield, 2013) and are increasingly being used—in both planned and unplanned ways—to supplement other channels of communication and to encourage knowledge-sharing, identification of expertise, and information discoverability, especially among distributed workers (Ellison, Gibbs, & Weber, in press). Leonardi et al. (2013) highlight the visibility and persistence affordances of enterprise social media that expand users' access to people and content, and thus result in increased opportunities for social learning in the workplace.

Looking at social media use more broadly, danah boyd (2010) describes a similar set of affordances, including persistence, replicability (the ability to duplicate content), scalability (akin to Treem and Leonardi's (2012) visibility affordance), and searchability, noting that these affordances introduce new dynamics, such as those surrounding instances of context collapse (to be discussed later). Similarly, Ellison and colleagues (e.g., Ellison et al., 2007; Ellison, Lampe, et al., 2011; Ellison, Steinfield, & Lampe, 2011; Ellison, Lampe, et al., 2011; Ellison, Vitak, Gray, & Lampe, 2014) highlight the role Facebook's affordances play in allowing users to form and maintain relationships with a wide range of contacts. As they have noted, the lowered transaction costs of relationship maintenance behaviors on these sites enable users to form "social supernets," a term coined by Judith Donath (2007) to describe the large networks supported by SNSs that would be difficult, if not impossible, to maintain without the technology. As noted by Ellison, Steinfield, et al. (2011) and Vitak and Ellison (2013), Facebook's affordances enable users to employ features like status updates and wall posts to request a variety of resources, including emotional support and information, from their connections on the site. For instance, in a qualitative study of adult Facebook users, Vitak and Ellison (2013) found that the broadcasting affordance—the ability to quickly disseminate content across one's entire network with the click of a button—was an important channel for participants who wished to access various resources from their network.

From both the literature and our everyday experiences, it seems clear that social media are enabling new patterns of communication, interaction, and affiliation; how, why, and to what ends social media use is related to social capital outcomes is the focus of this chapter. In the following section, we synthesize the literature on social network sites and social capital in order to describe these dynamics.

Social Network Sites

Social network sites are a particularly vibrant form of social media, well-suited for the kinds of interpersonal exchanges that serve to maintain and strengthen social bonds. Ellison and boyd (2013) articulate three key elements of SNSs in their revised definition. They write,

> A social network site is a *networked communication platform* in which participants 1) have *uniquely identifiable profiles* that consist of user-supplied content, content provided by other users, and/or system-level data; 2) can *publicly articulate connections* that can be viewed and traversed by others; and 3) can consume, produce, and/or interact with *streams of user-generated content* provided by their connections on the site (p.158.)

This definition differs slightly from their earlier definition, which included the profile, the articulated network, and the ability to view and traverse connections (boyd & Ellison, 2007), with the primary difference being the inclusion of the media stream, acknowledgement of the co-constructed nature of the profile, and the more explicit acknowledgement of communication as the engine driving usage patterns.

SNS usage has grown tremendously over the last decade, and the practice of checking profiles and posting updates on one or more sites has become part of everyday life for many Americans, not just teenagers. The Pew Internet and

American Life Project, which tracks Internet usage trends over time, notes that while growth peaked early for younger adults, in more recent years, older adults have increasingly been joining sites like Facebook: from 2008 to 2013, SNS use by Internet-using American adults ages 30–49 increased from 25% to 78%, while for those ages 50–65, the increase was from 11% to 60% and for those 65 and above, it was from 7% to 43% (Brenner & Smith, 2013). Furthermore, while Facebook remains the dominant site for interaction—and is an active site for relationship maintenance due to its support for fostering communication between pre-existing ties—several other SNSs have become popular among the general population in recent years. For example, Twitter doubled its American user base between 2010 and 2012, with 16% of Internet-using American adults and 24% of American teens maintaining an account. The site is especially popular among young, urban blacks (Duggan & Brenner, 2013; Madden et al., 2013). Twitter differs from Facebook in that unidirectional connections are much more common, meaning reciprocal approval is not needed for one user to "follow" another; this enables users to grow networks that can include friends, celebrities, news organizations, strangers, and other people of interest.

Why do people use SNSs? The vast majority of people asked this question say they do so to stay in touch with family and friends (e.g., Joinson, 2008; Smith, 2011). For example, Joinson (2008) identified seven unique motivations for using Facebook, with social connection being primary among them; additional motivations related to relationship maintenance included a factor he labeled "shared identities," which captures activities related to identifying and establishing common ground, as well as passive browsing activities on the site to potentially maintain relationships or keep up to date on other users' activities without interaction. Likewise, when asking American adults about their reasons for using SNSs, Smith (2011) found that nearly all said that staying in touch with current friends and family and reconnecting with old friends was a major or minor reason for using the site, while nearly half (49%) said they used SNSs to interact with people with shared interests.

Researchers have examined a variety of social, psychological, and behavioral outcomes associated with use of SNSs. Specific personality traits, including extroversion (Ross et al., 2009) and narcissism (Carpenter, 2012), have been positively linked to use of the site, while a number of other studies have established positive correlations between use of the site and increased well-being (Burke, Marlow, & Lento, 2010; Kim & Lee, 2011; Steinfield, Ellison, & Lampe, 2008). A substantial research stream has focused on the relationship between SNS use and social capital, which will be discussed in further detail later.

Overview of Social Capital

Research in multiple domains, including health communication, nursing, psychology, and sociology, highlights the important and tangible benefits we receive from our social relationships. Different disciplines focus on different dimensions of these benefits: for instance, those in psychology might study the extent to which different kinds of relationships are associated with one's self-esteem, whereas those in the medical arena might examine social support as an important coping mechanism during periods of stress. But cutting across these perspectives is the assumption that meaningful social connections are critical for our happiness and well-being.

Among sociologists and others, one framework that has been used to explore the benefits of interpersonal relationships and social networks is social capital, which saw some of its earliest development in the work of Bourdieu (1985) and Coleman (1988). In comparing social capital to other forms of capital, such as financial or cultural capital, Bourdieu (1985) defined it as "the aggregate of the actual or potential resources which are linked to possession of a durable network of more or less institutionalized relationships of mutual acquaintance and recognition" (p. 51). Bourdieu further stressed that these relationships are maintained through social interaction, which, presumably, enables the exchange of resource requests and provisions. Based on this definition, it becomes clear that the breadth and depth of social capital resources available to individuals will depend on their network composition, as well as how they communicate with that network. Lin (2001a) echoes this thinking in his conceptualization of social capital, framing the concept in economic terms: he frames social capital as "investment in social relations with expected returns in the marketplace" (p. 19). Furthermore, as is the case for other kinds of capital, many researchers have noted that social capital can be converted into other forms of capital (Bourdieu, 1985; Burt, 1992; Resnick, 2002).

Lin's (2001a) definition highlights another central construct of social capital: reciprocity. The "expected returns in the marketplace" he references speaks to the idea that individuals perform certain actions because they expect that action to be reciprocated at some point in the future. Putnam (2000) distinguishes between two forms of reciprocity in communities. Specific reciprocity refers to reciprocity at the individual level (e.g., one favor for another), whereas generalized reciprocity refers to actions individuals perform with no expectation of an immediate return on investment (e.g., donating blood, changing a flat tire for a stranger). Putnam (2000) notes that communities characterized by a norm of generalized reciprocity will be more efficient because "trustworthiness lubricates social life" (p. 21).

Individuals access different kinds of resources from their social relationships. Putnam (2000) divided network-provided resources into two major categories, bridging and bonding social capital. Bridging social capital is associated with resources ranging from access to new or hard-to-access information (such as when one is looking for a new job) to being exposed to more diverse perspectives and feeling part of a broader community. When considering bridging social capital, the most important network members are those who span multiple clusters (i.e., bridging ties), as these individuals help information spread quicker and more efficiently through a network and close "structural holes" between otherwise unconnected groups (Burt, 1992). The resources associated with bonding social capital are largely derived from individuals' inner circle of connections, including close friends and family members. These resources include emotional and more substantive support, or "big favors," such as when a family member asks for a loan, as well as creating a sense of group solidarity and exclusivity.

Bridging the Gap: Social Capital and Social Network Site Research

Research has explored the extent to which SNSs can be used to support the creation, accrual, and conversion of social capital. One stream of scholarship has documented how social media, specifically SNSs such as Facebook, have been used to support the informational and social support exchanges that comprise important aspects of social capital (Burke et al., 2010; Burke, Kraut, & Marlow, 2011; Ellison, Gibbs, et al., in press; Ellison, Steinfield, et al. 2011; Ellison, Vitak, et al. 2014; Ellison et al., 2007). This literature addresses how Facebook and other SNSs can be used for question-asking (Morris, Teevan, & Panovich, 2010; Paul, Hong, & Chi, 2011), exchanges of emotional support (Vitak & Ellison, 2013; Vitak, Ellison, & Steinfield, 2011), and sharing of information (Gray, Ellison, Vitak, & Lampe 2013; Lampe, Vitak, Gray, & Ellison, 2012). As highlighted earlier, many of these social capital exchanges are dependent on processes that benefit from being embedded within a social context and from the specific affordances of mediated technology. For instance, Morris et al. (2010) found that respondents preferred to ask questions on SNSs as opposed to question-and-answer sites (e.g., Quora) or search engines because they could pose their query to people that knew them and because they trusted their network more than the strangers found on other online sources. Similarly, the generalized reciprocity that is associated with social capital exchanges benefits from interactions that are visible and persistent, and thus can serve as public prompts to "pay it forward."

In contrast to earlier work in this area employing general measures of use (e.g., time on site, number of friends), more recently researchers have focused on identifying the specific kinds of activities that are predictive of social capital. First, as social capital is derived from *interactions* with one's network, researchers have begun exploring specific communication behaviors users perform on the site and their relationship to social capital outcomes. For example, research employing surveys and Facebook server log data by Burke and colleagues (2011) found a positive correlation between directed communication (i.e., messages a user receives from another Facebook Friend in the form of a wall post, "Like," etc.) and self-reported bridging social capital. Notably, the two noninteractive variables examined in this study—broadcasting updates and passively consuming content—were not significant predictors, which highlights the importance of active engagement with one's network. Likewise, research by Ellison, Vitak, et al. (in press) found a positive relationship between bridging social capital and a measure labeled "Facebook Relationship Maintenance Behaviors" (FRMB, called Signals of Relational Investment (SRI) in earlier work, such as Lampe et al., 2012). FRMB activities include responding to requests expressed via status updates, presumably with messages providing social, informational, or emotional support, or wishing a Facebook Friend "happy birthday."

Second, network composition appears to be a significant predictor of users' social capital perceptions. Ellison, Steinfield, et al. (2011) found that the number of "actual" friends a user reported having on Facebook was a significant predictor of bridging and bonding social capital while the total number of Facebook Friends was not. The authors argue the actual friends variable provides a more meaningful measure of a user's network. In a study of network diversity, Vitak (2012a) found that as users' networks increased in size and diversity (i.e., number of unique social groups identified within the network), so did their perceived bridging social capital. This finding is in line with research on network structure, which posits that networks including a large number of weak ties are more likely to include bridging ties, which help facilitate information diffusion (Burt, 1992); in the context of Facebook, it is also supported by research showing that weak ties are the primary drivers of information diffusion on the site (Bakshy, Rosenn, Marlow, & Adamic, 2012).

In addition to empirical work identifying specific behaviors or attributes that are linked to social capital gains, other work examines how SNSs can reshape communication networks and the kinds of interactions that occur via SNSs. For instance, scholars point to SNSs' lowered barriers for communicating with strong and weak ties (Ellison, Lampe, et al., 2011), which are typically associated with bonding and bridging social capital, respectively. Weak-tie interactions are typically framed in relation to informational benefits; weak ties, including Friends of Friends, are more likely to hold different opinions and worldviews and to have access to novel information (Burt, 1992; Granovetter, 1973). Conversely, emotional support (as captured by the concept of bonding social capital; see Putnam, 2000) is generally understood to be provided by closer ties, such as family members. However, considering that an average SNS user has more than 200 friends (Hampton, Goulet, Rainie, & Purcell, 2011), that many ties on SNSs are weaker ties (Bakshy et al., 2012), and that individuals are disclosing health information and requests for emotional support on SNSs (Greene, Choudhry, Kilabuk, & Shrank, 2011; Vitak & Ellison, 2013), it may be that these weak-tie interactions are providing other kinds of meaningful support, above and beyond access to novel information and diverse worldviews. For instance, in their qualitative interviews with adult Facebook users, Vitak and Ellison (2013) found that supportive messages from weaker ties were comforting to users because, in the words of one participant, "you know somebody at least cares enough to respond" (p. 250).

Additionally, SNSs may reshape network composition. The ability of SNSs to support larger networks of weaker ties was first noted by Donath and boyd (2004) in one of the earliest academic treatments of SNSs. Writing about the impact of social network sites, they note that while they did not expect SNSs to significantly impact the number of strong ties individuals had, the number of weak ties could grow significantly because the cost of maintaining relationships in these spaces is low. Furthermore, they argue that SNSs should lead to an overall increase in available information because the influx of weaker ties should create a more diverse network. It is this ability to expand one's reach and create Donath's (2007) "social supernets" with more connections than otherwise possible that makes SNSs so important when considering social capital processes, because a reshaped network, especially one with many more bridging ties, has the potential to increase one's access to various resources, such as novel information and diverse perspectives.

Other ways in which SNSs may restructure users' social networks is by allowing them to better connect with latent ties (Ellison et al., 2007), by keeping in touch with individuals from previously inhabited environments, and by accessing "Friends of Friends" (who are more visible and accessible due to the association affordance). Ellison and colleagues (2007) point out that SNSs like Facebook are likely to provide the impetus to convert latent ties to weak (or strong) ties because they highlight users' identity characteristics. Latent ties, as defined by Haythornthwaite (2005) are social ties that are "technically possible but not activated socially" (p. 137). As Ellison et al. (2007) write about their undergraduate sample:

> Facebook might make it easier to convert latent ties into weak ties, in that the site provides personal information about others, makes visible one's connections to a wide range of individuals, and enables students to identify those who might be useful in some capacity (such as the math major in a required calculus class), thus providing the motivation to activate a latent tie. (p. 1162)

The association and visibility affordances play an important role in facilitating this conversion, especially in terms of connecting Friends of Friends, who may "meet" through comments on a mutual Friend's status update. Additionally, SNSs might help individuals maintain, initiate, or re-invigorate weak-tie relationships that might otherwise have decayed without the social lubrication provided by online contexts. Hampton, Lee, and Her (2011) note that the persistent and pervasive nature of information and communications technologies (ICTs) like Facebook should reduce the likelihood that weak tie relationships will ever truly become dormant because these connections will always be just a click away.

When studying social media, it is important to remember that communication processes as well as social capital outcomes may be influenced by the sites' affordances. Resnick (2002) was one of the earliest to describe this relationship in his discussion of "sociotechnical capital," a phrase he developed in dawn of the social media era to describe "productive combinations of social relations and information and communication technology" (p. 250). Resnick highlights six technological affordances that may impact social capital processes in online spaces, such as the removal of temporal and spatial boundaries, expanded communication reach, and archiving capabilities. Resnick argues these affordances may enhance individuals' ability to accrue social capital, especially as it becomes easier and less time-consuming to maintain connections, coordinate and support large groups, and diversify one's network by making new connections.

Considering a more traditional definition of social capital, Lin (2001a) highlights three key components of social capital that he deems to be particularly important to understanding the construct. He writes:

> It has been proposed that social capital, as an investment in social relations with an expected return in the marketplace, should be defined as *resources embedded in a social structure that are accessed and/or mobilized in purposive actions.* In this definition, three critical components present themselves for analysis: (1) the resources, (2) being embedded in a social structure, (3) action. (p. 29), and (4) emphasis added

When thinking about these components within the context of SNS use, a number of social and technical affordances seem especially relevant. First, SNSs have the potential to widen and reshape the network itself, thus increasing potential *resources*, while elements of the profile (e.g., place of employment) highlight the resources available to them from others and make them more salient. Second, these resources are *embedded in a system* that is rich with social cues. In SNSs, social ties are visibly articulated, enabling access to Friends of Friends, while social grooming activities (e.g., "liking" or commenting on a Friend's status update) are visible, persistent, and often salient. The ability to mobilize social resources through these sites may be amplified, because these connections are embedded in a system that supports these visible signals of attention and the maintenance of persistent social relationships they enable. Third, *action* is made possible via technical features such as the ability to broadcast messages to a wide group. Synthesizing the scholarship on SNSs, there is evidence that suggests that sites like Facebook serve to make social resources more accessible because they enable social information-seeking (Ellison, Steinfield, et al., 2011), allow for the broadcasting of requests or provisions of support, and surface relevant connections within the broader network of Friends of Friends.

Lin (2001b) offers four explanations for why embedded resources in a network can benefit an individual, that is, why social capital works. The first mechanism he describes is the way in which social networks reshape the flow of *information*. Being in a strategic location in a network, for instance, can enable access to novel information about opportunities that can benefit individuals and organizations. Applying this to the context of SNSs, it is clear that the composition of one's Twitter network—the kinds of individuals or organizations one follows—clearly shapes the kinds of links, information, and opinions one consumes on the site. Second, social ties can exert *influence*. Lin offers the example of a powerful agent who "puts in a word" or recommends an individual. Professionally oriented SNSs such as LinkedIn explicitly include recommendations in their design, no doubt capitalizing on the fact that a referral from a known, high-status individual carries weight in many circles. The third explanation is around *social credentials*. As Lin (2001b) writes: "social-tie resources, and their acknowledged relationships to the individual, may be conceived by the organization or its agents as certifications of the individual's social credentials" (p. 7). These ties are made more visible via SNSs' affordances (i.e., association and visibility). Finally, *reinforcement* speaks to the ways in which social relationships reinforce membership in a group (and access to its resources) and identity. In addition to explicit groups (e.g., LinkedIn Groups, Facebook Groups, and Twitter Lists), implicit group identity is signaled by visible interaction patterns such as "@" exchanges in Twitter or comments on Facebook posts.

The size and composition of the network are related to the kinds of resources one possesses; social status, financial resources, cultural capital, and domain-specific knowledge are some of resources one might consider. A personal recommendation from a higher-status individual will carry more weight than that of a lower-status individual, for instance. Because of this, some network members may be better equipped to provide certain kinds of resources over others. For example, Weiss' (1974) research on social support identified six categories of social provisions as well as the types of social ties best equipped to provide each type of support; nurturing was most likely to take the form of a parent-child relationship, while attachment—having a sense of belonging—came from one's closest friends and female–female dyads. When considering this issue, one benefit of SNSs is that individuals can organize and maintain a large network of diverse connections with little effort; the sites serve as a "virtual rolodex" (Vitak, 2012b) where contact information is automatically updated and, as long as the connection is active, a potential resource is a click away.

To summarize, research finds that use of social network sites is positively associated with perceived access to a variety of social capital resources. While SNSs appear to be particularly well-suited to the accrual of bridging social capital due to the sites' technical structure, which allows users to maintain large and diverse networks of connections and interact with them through a variety of channels, these sites also provide a medium through which individuals can request and provide social and emotional support. Several factors emerge as especially important to consider when looking at social capital processes on SNSs, most notably, users' interaction patterns and network composition.

In the following section, we continue to explore the relationship between SNSs and social capital by looking at how social capital exchanges are facilitated by the three elements of SNSs, as delineated by Ellison and boyd (2013): the profile, the articulated network, and the content stream.

Three Elements of SNSs and Their Role in Social Capital Processes

As noted earlier, Ellison and boyd's (2013) updated definition of SNSs reflects the increasing focus on both communication—via static disclosures in the profile and more dynamic sharing through posts distributed to one's entire network—and co-creation of identity through users' interactions with their network. The three elements the authors highlight—the profile, the articulated network, and the stream of content and updates—each play a role in users' ability to request and provide resources to members of their network. Next, each of these element's role in social capital processes is discussed in further detail.

Element one: the profile

As Ellison and boyd (2013) argue, SNS profiles have changed over time, evolving from static portraits created by the profile owner to a more fluid collection of content co-created by profile owners, their Friends, and their actions. The authors write, "Over time, the profile has shifted from a self-presentational message created by the individuals to a portrait of an individual as an expression of action, a node in a series of groups, and a repository of self- and other-provided data" (p. 154). The sharing of identity information via SNS profiles and content streams is one way in which SNSs support relationship development and, via these relationships, the exchange of resources. Nie and Sundar (2013) found that the amount of information in users' Facebook profiles was positively related to the extent to which they felt their profile reflected their personal identity (as opposed to their social identity). The authors noted that Facebook can be used in two ways—to engage in more self-directed identity work [as described by Mead's (1967) concept of "I"] or to create an outward-facing public persona (akin to Mead's concept of "Me"). Here, we focus on the kinds of profile information that can be used as a social lubricant, smoothing social interaction by highlighting commonalities and differences.

Feld's (1981) notion of foci is useful for thinking about the role of this information in bringing people together. Feld (1981) argues that

> relevant aspects of the social environment can be seen as foci around which individuals organize their social relations. A focus is defined as a social, psychological, legal, or physical entity around which joint activities are organized (e.g., workplaces, voluntary organizations, hangouts, families, etc.). As a consequence of interaction associated with their joint activities, individuals whose activities are organized around the same focus will tend to become interpersonally tied and form a cluster. (p. 1016)

Offline, events that bring together people with shared interests, such as a political rally or musical performance, can initiate relationships and enable individuals to convey identity information (e.g., a political affiliation or taste in music) by their presence. In SNS contexts, the profile does some of this work of highlighting shared interests and helping users find like-minded others because the profile enables individuals to signal their interests and other characteristics that might serve to coalesce groups or attract others. Although it is rare for people to try to Friend total strangers on Facebook (Ellison, Steinfield, et al., 2011), it is the case that Groups, hashtags, or other SNS features can make people more aware of others who share a particular interest, and can facilitate communication about that topic. From these interactions, more stable relationships can form. On sites like Twitter, where it is more normative to "follow" people one has not met face-to-face or from another context, conversations around hashtags allow those with shared interests to come together and, in some cases, mobilize for a cause (Bruns & Burgess, 2011).

As Ellison and boyd (2013) point out, the profile has become less prominent in SNS activities, as users spend more time with aggregated media streams such as Facebook's News Feed. Still, the profile is a key component of how identity information is communicated in SNSs. By viewing one's connections, self-reported identity characteristics, and history, it becomes easier to identify common ground. Clark (1996) refers to the shared expectations that are relied upon during communication as *communal common ground*. Clark and Brennan (1991) note that individuals "cannot even begin to coordinate on content without assuming a vast amount of shared information or common ground—that is, mutual knowledge, mutual beliefs, and mutual assumptions. ... All collective actions are built on common ground and its accumulation" (p. 127). In the SNS context, Lampe, Ellison, and Steinfield (2007) found that some Facebook profile fields were more likely than others to be associated with higher Friend counts, and attributed this to the fact that these profile fields (such as hometown) enabled individuals to establish common ground with one another. Knowing what you have in common with someone else means that communication can be more efficient, because interactants can use shared references without explaining them. Thus, the identity information shared in SNSs serves to promote social capital processes by bringing together people who share characteristics and by smoothing interactions by making it easier for them to identify their common ground. As Ellison and colleagues (2011) argue, profile information acts as a "social lubricant" in interactions and may be instrumental in network expansion and resource accrual.

Identity information as expressed via the profile enables users to engage with weaker ties in a context that supports socially relevant interactions. As Ellison, Steinfield, et al. (2011) note, having the *technical* ability to communicate is not sufficient for relationship development; for example, people are unlikely to call a total stranger on the phone and strike up a conversation because they do not have access to socially relevant information that can help guide interactions. Facebook, they state, provides users with a "rich collection of social context cues, such as mutual friends or shared interests, which can guide conversations to socially relevant topics and better enable participants to find common ground" (p. 887). Thus, the profile is an important component of social capital processes via SNSs, even if the profile itself has morphed over time and is less central to the daily experience of many SNS users now than it has been in the past.

Element two: the public articulation of network

The public articulation of users' networks—typically through a list of connections that users and their connections can browse—serves multiple roles. In addition to supporting social capital exchanges, they also "are employed to mark and display relationships, delineate who can access what content, and serve as a filter through which viewers can browse profiles and discover friends in common" (Ellison & boyd, 2013, p. 155).

The ability to view other users' connections serves a vetting process and helps establish one's identity by showing mutual friends (Donath & boyd, 2004), which in turn may help users grow their network on SNSs. The articulation of the network and public communication features like comments and replies also enable Friends of Friends to interact, which may further amplify social capital accrual because these Friend of Friend encounters expand the speed and range of information diffusion. The association of connections and visibility of content are especially important in facilitating Friend of Friend interactions, and likely increases the chance that these encounters occur and that they are positive, because the profile information helps signal shared interests and establish common ground (Lampe et al., 2007).

On sites with reciprocal networks (in which both users must approve a connection), the articulation of the network typically demarcates a basic degree of boundary access: most (58%) adult American SNS users restrict their accounts to "Friends Only" (Madden, 2012); in this way, being a "Friend" enables access to specific rights and information that non-Friends cannot see. It is also important to note that a user's connections are typically aggregated in a central location—whether in the Friend List, Contacts, or Following—and enables users to click through to each profile and browse individual profiles. These passive consumption behaviors, although not directly linked to social capital outcomes in prior research (Burke et al., 2011), have been positively associated with a number of relational outcomes, including closeness, satisfaction, and access to emotional and instrumental resources (Vitak, 2012b). Most SNSs explicitly signal the connections two users share; knowing that another person shares connections is likely to increase trust, an important construct in the social capital literature (Burt, 1992).

Finally, the public articulation, or association, of ties enables social capital exchanges to occur, by linking individuals to one another and to content, and by directing attention to content produced by one's ties (i.e., content passes through the media stream or becomes available on the connection's profile). As noted earlier, researchers have linked social capital outcomes to specific activities on Facebook, including receiving comments from a Friend and attending to resource requests posted by one's Facebook Friends (Burke et al., 2011; Ellison, Vitak, et al., 2014). Many SNSs have features that allow users to "tag" other users in an update, which directly associates them with the content being shared. As has been suggested in other research (e.g., Ellison, Lampe, et al., 2011), linking network members to an update may serve as a form of social grooming, signaling attention, and highlighting the relationship in a public forum. This more directed form of communication is also more likely to be seen by the tagged individual than a general broadcast message that might get lost in the stream.

Element three: streams of user-generated content

The third element of Ellison and boyd's (2013) SNS definition focuses on the aggregated collection of content from one's network presented to the user via features like the News Feed in Facebook or the home page in Twitter. Called 'social awareness streams' by some researchers (Naaman, Boase, & Lai, 2010), these aggregated collections serve "as the point of departure for other activities on the site or the Web, replacing the act of surfing from profile to profile to discover updated content" (Ellison & boyd, 2013, p. 159). Crafting a message, such as a resource request, and disseminating it throughout one's network via these media streams allows users to quickly and easily broadcast messages to a large potential audience.

The broadcasting of content to a large audience is especially productive in some situations, such as times when one is seeking information but does not know who exactly holds the needed information. It is also useful for distributing information to large audiences, as when one is sharing information about a significant life event, such as a new job, an engagement, or a death in the family. At the same time, broadcasting to such a large audience—which can easily number in the hundreds—may be problematic for many users, especially when the information being disclosed is sensitive or could be misinterpreted when taken out of context. For example, Vitak and Ellison (2013) found that while Facebook users recognized the benefits of broadcasting updates on the site, they also voiced a number of concerns related to sharing information in a pseudo-public space and many employed strategies to minimize those risks, including moving conversations to a private channel or using the site's privacy settings to restrict access to content. These concerns are best understood through the lens of *context collapse*: the aggregation of distinct audiences, or segments of one's social network, into one group, introducing self-presentational and other challenges. Although context collapse can happen in offline settings, such as a wedding, online spaces that enable communication among groups traditionally associated with different facets of one's identity are often sites of context collapse. For instance, technological innovations and environmental concerns in the late 1990s ushered in a wave of home-based "teleworkers" who were often troubled by the lack of physical separation between their home and work roles and the "blurring" (or collapsing) of these identity roles that followed (Ellison, 2004). Context collapse is particularly salient in SNSs that condense all of one's contacts into a single group such as "Friends" or "contacts."

Individuals tend to vary their self-presentational disclosures based on their audience (Goffman, 1959) in order to achieve goals specific to that particular audience. In most offline contexts, audiences are fairly discrete and homogenous. For instance, in one's workplace, one is surrounded by co-workers, is attempting to achieve work-related self-presentational goals (e.g., to be perceived as hard-working, competent, and responsible), and can tailor behavior and outward appearance to this particular context. When network members from multiple contexts co-mingle, self-presentation can be more difficult because presentational goals often differ across contexts.

Many SNS users experience context collapse on a daily basis because SNS networks tend to be large and often represent multiple facets of one's identity: professional colleagues, neighbors, old friends, family members, hobby buddies. Because the default setting in sites like Facebook and Twitter is to group all connections a user makes under a single moniker ("Friends," "Followers"), sharing content under the default settings becomes a complex process. Context collapse complicates the disclosure process in SNSs, as users no longer have the luxury of varying self-presentation across each group without employing advanced privacy settings to recreate some of the offline boundaries between groups (Vitak, 2012a).

Context collapse likely has both positive and negative implications for social capital processes. From one standpoint, a diverse network comprised of ties from a wide variety of social contexts may lead users to censor their disclosures or not disclose any potentially sensitive information at all due to concerns about who will see the content and how different connections could interpret it (Hogan, 2010). Withholding disclosures could limit users' ability to reap social capital benefits from their SNS use: Ellison, Vitak, Steinfield, Gray, and Lampe (2011) argue that if one considers social capital to be the resources obtained through interactions with one's social network, Facebook users must be willing to share resource requests (i.e., to acknowledge a need) in order for their network to be able to respond appropriately. Indeed, Vitak (2012a) positively linked the amount and conscious intention (i.e., how much thought was behind posting) of disclosures to users' perceived bridging social capital. Those who do not disclose needs to their network—either due to context collapse or other factors—presumably receive less support from them.

On the other hand, the flattening of groups enables these otherwise-unconnected groups to interact through shared communication features (e.g., comments on a mutual Friend's status update). As argued earlier, these Friend of Friend connections, linked to bridging social capital, may encourage the spread of information, spark new ideas, and lead to productive new connections. Putnam (2000) points to the important role of groups that bring together people from different life contexts, such as bowling leagues. A book club that brings together women from different professional contexts serves as a site for exchanging book-related opinions, but also enables the formation of bridging social capital, as participants learn more about one another and engage in exchanges about other topics. Interacting with a diverse group of people on SNSs can have positive outcomes, as with the participants in Vitak and Ellison's (2013) study, who noted that they were able to access various types of information through Facebook that would have been difficult or impossible to do through other channels.

Social Grooming in Social Network Sites

Thus far, this chapter has considered the history of social capital and social network site research, with a specific aim at highlighting the role that SNS affordances play in the process, as well as providing an in-depth analysis of how Ellison and boyd's (2013) three main elements of SNSs create new opportunities and challenges to facilitating social capital transactions. In this final section, we turn our attention to the specific relational behaviors that we believe drive site use, enable resource-sharing, and aid relationship maintenance. Because of the unique affordances of SNSs that facilitate communication with a large network of ties—including Friends of Friends—and make those interactions both highly visible and persistent, we believe that social grooming practices on these sites should be of particular interest to researchers studying social capital transactions in dynamic online spaces.

SNSs can be compared to other communication channels, such as face-to-face communication. In general, computer-mediated communication (including SNSs) is more visible and persistent than face-to-face communication, but contains fewer nonverbal cues. For instance, in a face-to-face setting, it is usually apparent whether one's audience is attending to the speaker, especially in a small group or one-on-one setting. In a SNS, signals of attention operate differently. A Facebook Friend may read a post and laugh out loud, sympathize profoundly, or clench her fists in anger, but none of these emotions will be conveyed to the poster unless she enters keystrokes that leave a visible and explicitly constructed trace on the site. This is in contrast to face-to-face settings, and to a lesser degree true for other channels, where non-verbal cues such as the sound of laughter or an empathetic expression is apparent. In a SNS, in fact, it is often not clear whether the original message was read at all. Tweets scroll off the screen, it is unclear which posts are hidden by Facebook's algorithm and which are not, and other technical features serve to obfuscate audience. Even when messages are "delivered," they may not be attended to: the sheer number of messages can overwhelm readers, such that it becomes harder to separate the wheat from the chaff and to allocate attention.

The lack of passive signals of attention, such as eye gaze, in SNS contexts shapes user behavior and social dynamics. In order to signal that a piece of content has been seen and attended to, users need to take explicit actions that leave visible traces. These actions can include a comment left on a status update, "liking" a piece of content in Facebook, "favoriting" or retweeting a post on Twitter, liking or commenting on an Instagram photo, or liking a YouTube video. As Bernstein, Bakshy, Burke, and Karrer (2013) write:

> Posting to a social network site is like speaking to an audience from behind a curtain. The audience remains invisible to the user: while the invitation list is known, the final attendance is not. Feedback such as comments and likes is the only glimpse that users get of their audience. (p. 21)

An important characteristic of these behaviors is that they leave visible traces—traces that can be counted and serve as quantifiable metrics. These content-free, lightweight signals of attention, in aggregate, become metrics that can be used as crude measures of popularity or audience reach, or are considered by users when attempting to gauge audience size (Bernstein et al., 2013). The "like" button on Facebook is one example of a user action on Facebook that leaves a visible trace; it "provides a one-click shortcut to express a variety of affective responses such as excitement, agreement, compassion, understanding, but also ironic and parodist liking" (Gerlitz & Helmond, 2013). These traces are visible to the poster of the content receiving them as well as a larger audience, which introduces self-presentational pressures. Among teen Facebook users, "likes" are seen as a proxy for social status, causing these users to edit their content accordingly by deleting photos with too few likes (Madden et al., 2013). Similarly, Marwick and Ellison (2012) describe the way in which creators of memorial sites on Facebook focused on "likes" and group size as a metric signaling the impact or importance of the person being memorialized, noting that pages often included comments such as the following: "I just noticed there are over 3300 people who like this page—that speaks volumes about how many hearts Beth and Cord have touched" (p. 386).

In regards to social capital and interpersonal dynamics, however, these signals of attention are important because they are visible traces that can contribute to social capital dynamics on SNSs. When users take advantage of these opportunities to signal to others their attention to that relationship and their willingness to contribute to it, they help ensure that their requests will be positively received due to the norms of reciprocity that govern social interactions at both the interpersonal and societal level. These relational micro-transactions, such as "liking" a Friend's positive post, help create an environment in which reciprocal attention and low-level social grooming is productively enacted.

Social grooming practices are activities that build trust and create expectations of reciprocal attention among network members. In primate groups, grooming enhances hygiene and general well-being of apes and builds social bonds; humans use language to engage in equivalent trust-building and relationship-nurturing activities (Dunbar, 1996). SNSs allow individuals to engage in social grooming when they engage in actions that signal "I'm thinking of you" such as sending a virtual gift (Donath, 2007). Tong and Walther (2011) describe the technical features of SNSs that are helpful for relationship maintenance practices, including the ways in which SNSs foster participation, feedback, and interaction through various communication channels. Empirically exploring this issue, Ellison, Vitak, et al. (2014) identify a set of Facebook practices, such as writing "happy birthday" on a Friend's Wall and answering questions posed by Facebook Friends, which they argue serve to maintain relationships by visibly signaling attention to one's connections. They find that engaging with one's Facebook network in this way is positively linked to higher levels of bridging social capital, and argue that the behaviors captured in their social grooming measure help users by elevating their visibility in the system by triggering expectations of reciprocal assistance, and by giving them access to an extended network (their Friends' Friends).

These expectations of reciprocal assistance reflect Lin's (2001a) understanding of social capital—the "investment in social relations with expected returns in the marketplace" (p. 19). In Lin's framing, social capital is a product of investments in relationships and the expectations of reciprocity that result from them. We can view these "investments" as the building blocks of relationships—investments such as time, emotional vulnerability, attention or social grooming, as well as other components of interpersonal relationship creation and development. The persistence, association, and visibility associated with SNS practices mean that social grooming that occurs on SNSs often leaves behavioral traces that serve to make social structures more apparent to both those embedded within them and those outside the social cluster. Through these visible traces such as comments, Wall posts, @ replies, and "favoriting" or "liking" content, exchanges among ties and the relationships they represent are visible to a larger audience. Recently introduced features, such as the Timeline and Social Graph on Facebook, make these trace patterns increasingly persistent and searchable. Importantly, these social grooming and visible micro-transactions enable Friend of Friend interactions and the bridging social capital they are likely to represent.

A final implication of the articulated network for social capital processes stems from the visibility and association affordances of SNSs as articulated by Treem and Leonardi (2012)—the ways in which social connections (networks of ties) are articulated on these sites and the particular way in which interactions are structured to enable Friends of Friends to interact. Looking specifically at Facebook, when content is pushed to Friends via the News Feed, individuals who are mutual Friends with a user but not Friends with one another can interact via the stream of comments that are visible to all Friends of the original poster (depending on the user's privacy settings). In Facebook, comments to status updates enable cross-network communication, because they are seen by the Friends of the original poster, not the network of the commenter. Ellison, Vitak, et al. (2014) point out that, when considering social capital benefits, commenting on a Friend's message is a more productive pathway for accessing new ties than contributing one's own status update: "Comments on Friends' posts may be more likely to generate bridging social capital than status updates or other broadcasting behaviors because these comments are seen by the poster's network in addition to one's own." They note that responding to a Friend's post gives a user access to a new collection of people. Importantly, these users are interacting not as strangers with no common ground, but rather as individuals with a shared connection (i.e., the mutual Friend) and access to each other's identity information via the profile.

Conclusion

The ability to create and share user-generated content is not limited to social network sites. Even older forms of communication typically considered broadcast channels (such as newspapers or call-in radio shows) have included elements of this many-to-many communication dynamic. In fact, the first multipage newspaper in the United States, the 1690 *Publick Occurences Both Forreign and Domestick*, had a blank sheet intended for individuals to write their own news and comments before passing it on to the next reader (DiPiazza, 2012). However, SNSs are a unique combination of mass and more directed, interpersonal communication channels.

Social network sites enable users to broadcast messages to a broad audience and to engage in one-to-one messaging via a set of features that support private and public utterances, synchronous and asynchronous communication, and personal and corporate presences. Some scholars have argued that sites like Facebook should be studied not as a singular communication tool, but rather a collection of tools or a "toolkit" (Smock, Ellison, Lampe, & Wohn, 2011), a framing that acknowledges the diversity of user practices. Of these different communication channels, the ability to broadcast messages is probably the most important for generating social capital accrual and development. Importantly, broadcasted requests shared via the status update feature in Facebook are not sent to a mass audience of strangers, but rather situated in a rich context of identity cues and shared with a network of social connections, potentially increasing the chances that they will be positively received. Facebook and other "masspersonal" SNSs thus share elements of both mass communication—namely, the ability to broadcast requests to a wide audience, which is especially useful for requests and announcements—and dyadic, interpersonal communication, in which messages are embedded in a rich network of social relationships, a visible shared history, and identity signals.

SNS profiles typically include a wide range of personal information about users, such as their preferences, history, social circle, and more. This information serves as signals of identity and are aggregated from a variety of sources including the user, the system, and other users (Ellison & boyd, 2013). The identity information in social media profiles serves multiple purposes, ranging from social, such as allowing other users to find common ground, to psychological, in that profiles allow users to ideate future versions of self (Ellison et al., 2012), reflect their actual personality (Back et al., 2010) or engage in self-affirmation (Toma & Hancock, 2013). Future research should explore the role of the profile in social capital processes. For instance, Ellison, Steinfield, et al. (2011) found that using Facebook to connect with total strangers was not related to perceptions of social capital but using the site to "check out" or "learn more about" (presumably via the profile) those with some shared connection was. As they write, "The process by which Facebook can be used to scaffold productive social interactions is complex" (p. 887); future research could explore these dynamics in more detail.

Finally, second-order effects of social capital exchanges on SNSs are not well understood. Specifically, what are the implications for viewing others' resource requests and the responses to these requests? Social learning theory (Bandura, 1977) posits that individuals are more likely to adopt a behavior after witnessing others do so. In the case of SNSs, many kinds of activities are made visible to other users, meaning that requests for resources are often seen by those who are incidental audiences—not able to help but still privy to the request. Little is known about the specific dynamics of how the effects of these kinds of user actions and their responses can ripple throughout a network. For instance, consider a Facebook user who asks her Friend network for information. Although research has focused on motivations and dynamics of those who request favors or information from their network (Morris et al., 2010) and those who respond to them (Ellison, Vitak, et al., 2014), less work has examined the way in which passive observers of these interactions are affected by these exchanges. For instance, those who see positive episodes, in which users receive help, will presumably be more likely to engage in this behavior themselves, whereas those who witness negative outcomes (such as privacy violations or requests that go unmet) would presumably be less likely to engage in similar behaviors.

To summarize, this essay has described some of the potential ways in which social capital exchanges are enabled by the affordances of SNSs, such as visibility and persistence of content, as well as the low cost of connecting with and maintaining a large network of ties. We believe that SNSs offer an opportunity to revisit and extend a wide range of social science theories in the domains of media use, relationship maintenance, social capital, and computer-mediated communication. For instance, drawing from research in the fields of interpersonal and mass communication may help us understand how these hybrid communication sites support a wide range of communication exchanges, in that SNSs merge components of mass (or broadcast) technologies like the television with characteristics of interpersonal communication as experienced by small groups or dyads. Although the sites we describe here will continue to evolve over time, individuals will continue to rely on their social networks—and the communication technologies that support them—to support them in accomplishing tasks both mundane and monumental.

Acknowledgements

We would like to thank Shyam Sundar and the anonymous reviewers for their thoughtful critiques of our work.

Note

1 Treem and Leonardi (2012) point to Gibson's (1986) framing of the concept of affordances, in which different animals use the same object, such as a rock, in different ways, and thus perceive the object not for what it is, but rather for the *kinds of uses it affords*. As Treem and Leonardi (2012) explain, "Although the features of an object are common to each person who encounters them, the affordances of that artifact are not. Affordances are unique to the particular ways in which an actor, or a set of actors, perceives and uses the object" (p. 145).

References

Back, M. D., Stopfer, J. M., Vazire, S., Gaddis, S., Schmukle, S. C., Egloff, B., & Gosling S. D. (2010). Facebook profiles reflect actual personality, not self-idealization. *Psychological Science, 21,* 372–374.

Bakshy, E., Rosenn, I., Marlow, C., & Adamic, L. (2012). The role of social networks in information diffusion. In M. Rabinovich, & S. Staab (Eds.), *Proceedings of the 21st International Conference on World Wide Web* (pp. 519–528). New York, NY: ACM.

Bandura, A. (1977). *Social learning theory.* Oxford, England: Prentice-Hall.

Bernstein, M. S., Bakshy, E., Burke, M., & Karrer, B. (2013). Quantifying the invisible audience in social networks. In S. Brewster, & S. Bødker (Eds.), *Proceedings of the SIGCHI Conference on Human Factors in Computing Systems* (pp. 21–30). New York, NY: ACM.

Binder, H., Howes, A., & Sutcliffe, A. (2009). The problem of conflicting social spheres: Effects of network structure on experienced tension in social network sites. In K. Hinckley, M. R. Morris, S. Hudson, & S. Greenberg (Eds.), *Proceedings of ACM CHI 2009* (pp. 965–974). New York, NY: ACM.

Bourdieu, P. (1985). The forms of social capital. In J. G. Richardson (Ed.), *Handbook of theory and research for the sociology of education* (pp. 241–258). New York, NY: Greenwood Publishing.

Brenner, J., & Smith, A. (2013). *72% of online adults are social networking site users.* Washington, DC: Pew Internet and American Life Project.

Bruns, A., & Burgess, J. E. (2011, August). *The use of Twitter hashtags in the formation of ad hoc publics.* Paper presented at the 6th European Consortium for Political Research General Conference. Reykjavik, Iceland.

boyd, d. (2010). Social network sites as networked publics: Affordances, dynamics, and implications. In Z. Papacharissi (Ed.), *A networked self* (pp. 39–58). New York, NY: Routledge.

boyd, d., & Ellison, N. B. (2007). Social network sites: Definition, history, and scholarship. *Journal of Computer-Mediated Communication, 13,* 210–230.

Burke, M., Adamic, L., & Marciniak, K. (2013). Families on Facebook. In N. B. Ellison, B. Hogan, P. Resnick, & I. Soboroff (Eds.), *Proceedings of the seventh international AAAI conference on weblogs and social media* (pp. 41–50). Washington, DC: AAAI.

Burke, M., Kraut, R., & Marlow, C. (2011). Social capital on Facebook: Differentiating uses and users. In *Proceedings of the SIGCHI conference on human factors in computing systems* (pp. 571–580). New York, NY: ACM.

Burke, M., Marlow, C., & Lento, T. (2010). Social network activity and social well-being. In *Proceedings of the SIGCHI conference on human factors in computing systems* (pp. 1909–1912). New York, NY: ACM.

Burt, R. S. (1992). *Structural holes: The structure of social capital competition.* Cambridge, MA: Harvard University Press.

Carpenter, C. J. (2012). Narcissism on Facebook: Self-promotional and anti-social behavior. *Personality and Individual Differences, 52,* 482–486.

Clark, H. (1996). *Using language.* Cambridge, UK: Cambridge University Press.

Clark, H. H., & Brennan, S. E. (1991). Grounding in communication. In L. B. Resnick, J. M. Levine, & S. D. Teasley (Eds.), *Perspectives on socially shared cognition* (pp. 127–149). Washington, DC: American Psychological Association.

Coleman, J. S. (1988). Social capital in the creation of human capital. *American Journal of Sociology, 94,* 95–120.

DiPiazza, F. D. (2012). *Friend me! 600 years of social networking in America.* Minneapolis, MN: Twenty-First Century Books.

Donath, J. S. (2007). Signals in social supernets. *Journal of Computer-Mediated Communication, 13*, 231–251.

Donath, J. S., & boyd, d. (2004). Public displays of connection. *BT Technology Journal, 22*(4), 71–82.

Duggan, M., & Brenner, J. (2013). *The demographics of social media users, 2012.* Washington, DC: Pew Internet and American Life Project.

Dunbar, R. I. M. (1996). *Grooming, gossip, and the evolution of language.* Cambridge, MA: First Harvard University Press.

Ellison, N. (2004). *Telework and social change: How technology is reshaping the boundaries between home and work.* Greenwich, CT: Greenwood Publishing.

Ellison, N. B., & boyd, D. (2013). Sociality through social network sites. In W. H. Dutton (Ed.), *The Oxford handbook of Internet studies* (pp. 151–172). Oxford, UK: Oxford University Press.

Ellison, N. B., Gibbs, J. L., & Weber, M. S. (in press). The use of enterprise social network sites for knowledge sharing in distributed organizations: The role of organizational affordances. *American Behavioral Scientist.*

Ellison, N. B., Hancock, J. T., & Toma, C. L. (2012). Profile as promise: A framework for conceptualizing veracity in online dating self-presentations. *New Media & Society, 14*, 45–62.

Ellison, N., Lampe, C., Steinfield, C., & Vitak, J. (2011). With a little help from my Friends: Social network sites and social capital. In Z. Papacharissi (Ed.), *A networked self: Identity, community and culture on social network sites* (pp. 124–145). New York, NY: Routledge.

Ellison, N., Steinfield, C., & Lampe, C. (2007). The benefits of Facebook 'Friends': Exploring the relationship between college students' use of online social networks and social capital. *Journal of Computer-Mediated Communication, 12*, 1143–1168.

Ellison, N. B., Steinfield, C., & Lampe, C. (2011). Connection strategies: Social capital implications of Facebook-enabled communication practices. *New Media & Society, 13*, 873–892.

Ellison, N., Vitak, J., Gray, R., & Lampe, C. (2014). Cultivating social resources: The relationship between bridging social capital and Facebook use among adults. *Journal of Computer-Mediated Communication, 19*(4), 855–870.

Ellison, N., Vitak, J., Steinfield, C., Gray, R., & Lampe, C. (2011). Negotiating privacy concerns and social capital needs in a social media environment. In S. Trepte & L. Reinecke (Eds.), *Privacy online: Perspectives on privacy and self-disclosure in the social web* (pp. 19–32). New York, NY: Springer.

Feld, S. L. (1981). The focused organization of social ties. *American Journal of Sociology, 86*, 1015–1035.

Gerlitz, C., & Helmond, A. (2013). The like economy: Social buttons and the data-intensive web. *New Media & Society, 15*, 1348–1365.

Gibson, J. J. (1986). *The ecological approach to visual perception.* Mahwah, NJ: Lawrence Earlbaum.

Goffman, E. (1959). *The presentation of self in everyday life.* Garden City, NY: Anchor.

Granovetter, M. S. (1973). The strength of weak ties. *American Journal of Sociology, 78*, 1360–1380.

Gray, R., Ellison, N., Vitak, J., & Lampe, C. (2013). Who wants to know? Question-asking and answering practices among Facebook users. In C. Lampe & L. Terveen (Eds.), *Proceedings of the 16th ACM conference on computer supported cooperative work and social computing* (pp. 1213–1224). New York, NY: ACM.

Greene, J. A., Choudhry, N. K., Kilabuk, E., & Shrank, W. H. (2011). Online social networking by patients with diabetes: A qualitative evaluation of communication with Facebook. *Journal of General Internal Medicine, 26*, 287–292.

Hampton, K., Goulet, L. S., Rainie, L., & Purcell, K. (2011). *Social networking sites and our lives.* Washington, DC: Pew Internet & American Life Project.

Hampton, K. N., Lee, C. J., & Her, E. J. (2011). How new media affords network diversity: Direct and mediated access to social capital through participation in local social settings. *New Media & Society, 13*(7), 1031–1049.

Haythornthwaite, C. (2005). Social networks and Internet connectivity effects. *Information, Communication, & Society, 8*, 125–147.

Hogan, B. (2010). The presentation of self in the age of social media: Distinguishing performances and exhibitions online. *Bulletin of Science, Technology & Society, 30*, 377–386.

Jeong, J.-W., Morris, M. R., Teevan, J., & Liebling, D. (2013). A crowd-powered socially embedded search engine. In N. B. Ellison, B. Hogan, P. Resnick, & I. Soboroff (Eds.), *Proceedings of the Seventh Annual International Conference on Weblogs and Social Media* (pp. 263–272). Washington, DC: AAAI.

Joinson, A. N. (2008). "Looking at", "looking up," or "keeping up with" people? Motives and uses of Facebook. In D. Tan (Ed.), *Proceedings of the 26th International Conference on Human Factors in Computing Systems* (pp. 1027–1036). New York, NY: ACM.

Kim, J., & Lee, J.-E. R. (2011). The Facebook paths to happiness: Effects of the number of Facebook Friends and self-presentation on subjective well-being. *CyberPsychology, Behavior, and Social Networking, 14*, 359–364.

Lampe C., Ellison, N., & Steinfield, C. (2007) A familiar Face(book): Profile elements as signals in an online social network. In D. Gilmore (Ed.), *Proceedings of the 25th International Conference on Human Factors in Computing* (pp. 435–444). New York, NY: ACM.

Lampe, C., Vitak, J., Gray, R., & Ellison, N. (2012) Perceptions of Facebook's value as an information source. In E. H. Chi, & K. Höök (Eds.), *Proceedings of the SIGCHI Conference on Human Factors in Computing* (pp. 3195–3204). New York, NY: ACM Press.

Leonardi, P. M., Huysman, M., & Steinfield, C. (2013). Enterprise social media: Definition, history, and prospects for the study of social technologies in organizations. *Journal of Computer-Mediated Communication, 19*, 1–19.

Leonardi, P. M., Nardi, B. A., & Kallinikos, J. (Eds.). (2012). *Materiality and organizing: Social interaction in a technological world.* Oxford, UK: Oxford University Press.

Lin, N. (2001a). *Social capital: A theory of social structure and action.* London, UK: Cambridge University Press.

Lin, N. (2001b). Building a network theory of social capital. In N. Lin, K. Cook, & R. S. Burt (Eds.), *Social capital: Theory and research* (pp. 3–29). New York, NY: Aldine de Gruyter.

Madden, M. (2012). *Privacy management on social media sites.* Washington, DC: Pew Internet and American Life Project.

Madden, M., Lenhart, A., Cortesi, S., Gasser, U., Duggan, M., & Smith, A. (2013). *Teens, social media, and privacy.* Washington, DC: Pew Internet and American Life Project.

Marwick, A., & Ellison, N. B. (2012). "There isn't Wifi in heaven!" Negotiating visibility on Facebook memorial pages. *Journal of Broadcasting & Electronic Media, 56*, 378–400.

Mead, G. H. (1967). *Mind, self, and society: From the standpoint of a social behaviorist.* Chicago, IL: University of Chicago Press.

Morris, M. R., Teevan, J., & Panovich, K. (2010). What do people ask their social networks, and why? A survey study of status message Q&A behavior. In G. Fitzpatrick, S. Hudson, K. Edwards, & T. Rodden (Eds.), *Proceedings of the 28th International Conference on Human Factors in Computing Systems* (pp. 1739–1748). New York, NY: ACM.

Naaman, M., Boase, J., & Lai, C.-H. (2010). Is it really about me? Message content in social awareness streams. In J. Tang (Ed.), *Proceedings of the ACM Conference on Computer Supported Cooperative Work* (pp. 189–192). New York, NY: ACM.

Nie, J., & Sundar, S. S. (2013). Who would pay for Facebook? Self esteem as a predictor of user behavior, identity construction and valuation of virtual possessions. In P. Kotzé, G. Marsden, G. Lindgaard, J. Wesson, & M. Winckler (Eds.), *Human-Computer Interaction–INTERACT 2013* (pp. 726–743). Berlin/Heidelberg, Germany: Springer.

Paul, S. A., Hong, L., & Chi, E. H. (2011). Is Twitter a good place for asking questions? A characterization study. In L. Adamic, R. Baeza-Yates, and S. Counts, *Proceedings of the International Conference on Weblogs and Social Media* (pp. 578–581). Washington, DC: AAAI.

Pew Internet and American Life Project. (2013). *Demographics of Internet users.* Retrieved from http://pewinternet.org/Static-Pages/Trend-Data-(Adults)/Whos-Online.aspx

Putnam, R. D. (2000). *Bowling alone.* New York, NY: Simon & Schuster.

Resnick, P. (2002). Beyond bowling together: Socio-technical capital. In J. Carroll (Ed.), *HCI in the new millennium* (pp. 247–272). New York, NY: Addison-Wesley.

Ross, C., Orr, E. S., Sisic, M., Arseneault, J. M., Simmering, M. G., & Orr, R. R. (2009). Personality and motivations associated with Facebook use. *Computers in Human Behavior, 25,* 578–586.

Smith, A. (2011). *Why Americans use social media.* Washington, DC: Pew Internet and American Life Project.

Smock, A. D., Ellison, N. B., Lampe, C., & Wohn, D. Y. (2011). Facebook as a toolkit: A uses and gratification approach to unbundling feature use. *Computers in Human Behavior, 27,* 2322–2329.

Steinfield, C., Ellison, N. B., & Lampe, C. (2008). Social capital, self-esteem, and use of online social network sites: A longitudinal analysis. *Journal of Applied Developmental Psychology, 29,* 434–445.

Sturken, M., Thomas, D., & Ball-Rokeach, S. (Eds.). (2004). *Technological visions: The hopes and fears that shape new technologies.* Philadelphia, PA: Temple University Press.

Sundar, S. S. (2008). The MAIN model: A heuristic approach to understanding technology effects on credibility. *Digital media, youth, and credibility,* 73–100.

Toma, C., & Hancock, J. T. (2013). Self-affirmation underlies Facebook use. *Personality and Social Psychology Bulletin, 39,* 321–331.

Tong, S., & Walther, J. B. (2011). Relational maintenance and computer-mediated communication. In K. B. Wright & L. M. Webb (Eds.), *Computer-mediated communication in personal relationships* (pp. 98–118). New York, NY: Peter Lang Publishing.

Treem, J. W., & Leonardi, P. M. (2012). Social media use in organizations: Exploring the affordances of visibility, editability, persistence, and association. *Communication Yearbook, 36,* 143–189.

Vitak, J. (2012a). The impact of context collapse and privacy on social network site disclosures. *Journal of Broadcasting and Electronic Media, 56,* 451–470.

Vitak, J. (2012b). *Keeping connected in the Facebook age: The relationship between Facebook use, relationship maintenance strategies, and relational outcomes* (Doctoral dissertation, Michigan State University, East Lansing, MI).

Vitak, J., & Ellison, N. (2013). "There's a network out there you might as well tap": Exploring the benefits of and barriers to exchanging informational and support-based resources on Facebook. *New Media & Society, 15,* 243–259.

Vitak, J., Ellison, N., & Steinfield, C. (2011). The ties that bond: Re-examining the relationship between Facebook use and bonding social capital. *Proceedings of the 44th Annual Hawaii International Conference on System Sciences.* Kauai, HI. Computer Society Press.

Weiss, R. S. (1974). The provisions of social relationships. In Z. Rubin (Ed.), *Doing unto others: Joining, molding, conforming, helping, loving* (pp. 17–26). Englewood Cliffs, NJ: Prentice-Hall.

Wellman, B. (2001). Physical place and cyberplace: The rise of personalized networking. *International Journal of Urban and Regional Research, 25*(2), 227–252.

10

The Social Psychology of Mobile Communication

Kathleen M. Cumiskey[1] and Rich Ling[2]
[1] College of Staten Island, City University of New York, New York, NY, USA
[2] Nanyang Technological University, Singapore

Introduction

Kurt Lewin, known to many as the father of social psychology, said that all behavior is a function of the person and their environment (Sansone, Morf, & Panter, 2003). As with many other of the discipline's founders, his conception of the world was based on face-to-face interaction—undisturbed by mobile communication. We have now entered an era where this foundational assumption has expanded to account for additional demands in one's social environment that were not available during Lewin's time. Classic studies in social psychology have focused on the influence of others and the impact of information that individuals receive as they navigate their social worlds (Aronson, 2011). Social psychology's focus is the domain of social interaction. All behavior must be derived from the totality of co-existing facts and being reminded of this from within the study of social psychology can assist scholars in broadening their scope of observation to include influences that may not be physically present or readily observable in the physical environment. In Lewin's use of field theory, he acknowledged that all social behavior arose from a settling of tensions between the individual's perceptions of self and their perceptions of their social environment. Rather than a theory, field theory "is best described as a method: namely, a method of analyzing causal relations and building scientific constructs" (Lewin, 1951, p. 45). One of the goals of this chapter is to utilize field theory as a backdrop through which we can begin to reconceptualize the field of social interaction to include dimensions of influence located beyond the physical environment. As Lewin (1939) pointed out, psychology once regarded as "mystic those who claimed that social facts were as real as physical ones" (p. 868). The social psychology of mobile technology and social media calls for a similar adaptive response by psychology to now consider real, social facts that are generated through a reality that is augmented by the seemingly "mystical"

The Handbook of the Psychology of Communication Technology, First Edition. Edited by S. Shyam Sundar.

presence of technology. Social psychology must widen its definition of the field of social interaction to include the influence of the mere presence of mobile communication technology. Mobile media and technology assist in rendering present seemingly absent social influences (Gergen, 2003). The entrance of remote interlocutors into the field of social interaction can issue a definite challenge for both the people within the field of interaction and the social scientist. This challenge is posed as balancing of forces that are present in one's current context (i.e., those with whom one is face to face) along with or in addition to those who are made present through the use of a mobile device. How this mediated social influence is integrated into what we already know about the influence of the presence of others on the field of social interaction will be explored in this chapter.

Social psychological theories and methodologies are critical to the development of the study of mobile communication. Embracing the influence of mobile technology on social behavior presents exciting new variables within the field of social interaction to empirically examine. And while this may be challenging, tracing the impact of the use of mobile communication technology presents the opportunity to glean insight into social processes. For instance, mobile phone use and smartphone technology have affected social cohesion, social coordination, and the ease in which we organize our social worlds (Ling, 2008). At the same time, reliance on mobile technology, for all its uses and affordances, in regard to social capital and beyond, can cultivate a new brand of dependency on mobile devices. This dependency is indeed different from non-networked devices (i.e., some desktop computers) because mobile phones link us to our social world. Unlike dependency on "stand alone" artifacts, we use mobile devices in the pursuit of social fulfillment. Further, they are also perpetually with us; integrated into the flux of our lives in ways that wire-based networked devices cannot be. To refer back to Lewin's field theory, this sense of dependency on a mobile device then also broadens the field of social interaction beyond the bounds and limits of one's current, "stand alone" physical environment to then include the potentiality for an open channel of interaction that any working mobile device affords. Social contacts, stored in one's device and then carried with the user into any and all contexts, can influence the behavior of the user as much as the behavior of those with whom the user is physically co-present.

Therefore, the demands that mobile devices present, whether it be physical, emotional, or cognitive, may work to limit a user's ability to be fully present in their current field of social interaction and this can impose constraints around how we function in our social worlds. In addition, managing the split attention caused by the integrated use of mobile technology could render the user vulnerable to risk taking behaviors and other dangers. This chapter will explore these topics and how these technologies impact social behavior and the study of social psychology.

This paper will examine several dimensions of mobile communication. These include

- How we use mobile communication devices to negotiate social expectations with one another
- The role of these devices vis-à-vis physical co-presence

- How mobile communication technology facilitates social cohesion
- How mobiles aid in multitasking
- A discussion of problematic mobile phone use
- Methodological challenges in the study of mobile communication

We feel that these issues will help us to understand mobile communication technology's impact on how we perceive our social worlds and the influence of "others." Integrating this technology into the flow of daily interaction poses a serious challenge to those who desire to predict and control social behavior. For as much as behavior can be monitored through the use of such devices, the presence of a mobile device offers the potential for outside influences to unforeseeably penetrate the boundaries of the classic field of social interaction and disrupt the domain of social expectations.

Mobile Communication and Negotiating the Social Expectations of Others

Erving Goffman (1959) uses a dramaturgic metaphor to describe the construction of one's social self, in that individuals may present a public "face" that is counter to one's more private self. As such, within each individual is a certain brand of self-consciousness that recognizes that there are social rules and roles that govern public behavior. In addition, there may be expressions of oneself that remain private and behaviors that are solely relegated to the private sphere. Our social, public identity is therefore imagined with a sense that there are others with whom one shares a social realm, whose needs and expectations must be taken into consideration. Individuals use this knowledge, their distinct cultural values, and social expectations to then evaluate and interpret their own behavior and the behavior of others. Attribution theory has been a central tenet to the study of social psychology (Monson & Snyder, 1977; Nisbett, Caputo, Legant, & Marecek, 1973). Attribution theory centers on how people perceive and then explain their own behavior and the behavior of others within the field of social interaction and social expectations that Goffman so aptly describes. It is no surprise that often individuals make errors in their perception and have a tendency to explain the behavior of others differently than their own. Individuals have a tendency to overestimate the role that a person's personality or disposition plays in determining behavior and underestimate the role that situational factors play in influencing behavior. This cognitive and perceptual error is called the "fundamental attribution error." As the use of mobile phones and other mobile technology has become more prevalent, the social expectations governing such behavior has also shifted.

The near complete saturation of mobile phones around the world has caused a shift in the social etiquette of when we use (or refrain from using) our devices in public. How observers interpret the use of mobile phones has become an area of study for mobile communication research. Certain situational factors lead to whether such mobile phone behavior, in the presence of others, is viewed as socially unacceptable or not. The main factor being whether one is the mobile phone user or not (Cumiskey,

2005). The use and presence of mobile devices then puts additional demands on the actors in any field of social interaction (Monk, Carroll, Parker, & Blythe, 2004).

Managing the demands of continual availability of distant known others alongside the demands of those with whom they are face to face can be difficult to handle. The difficulty is not so much in the sense of there being a "big brother" constantly monitoring our behavior but that different publics with whom the individual is engaged can always reach them. The boss, their partner or children, friends who are more or less close socially and different professional peers can always call. This, of course, affords efficiency when there is the need to solve the different tasks and issues of daily life. The individual can be more productive, they have an increased flexibility in their planning and logistics, and they can engage in micro-coordination. However, this can also mean that in physical spaces where one used to be able to assume a particular role, there are now calls and messages from other publics that can threaten the successful role performance (Goffman, 1959). This then puts the mobile device user at risk of violating the norms of social behavior. The perceived intentionality of this behavior by onlookers can also lead to attribution errors and possibly increase tension between co-present others and the user of the mobile device (Cumiskey, 2005). This is being defined as a risk here because the violation of social norms often come with negative consequences as it relates to impression management, social identity, managing conflict, and "saving face."

The risks placed on one's moral and social apprehension can also be interpreted as play or as a way that mobile users alleviate the mundane. For example, our best buddy can call to joke around while we are having a parent–teacher conference to discuss our child's trouble with the latest math assignments. Our boss can call to check up on how a particular job is going and we can play the role of good employee while we are out with our family. Since most of us engage in this "risky" behavior, we may also be vigilant about the ways in which others may be "duping" us and may be quick to judge such behavior as wrong or suspicious. Our paramour might call on the day we forgot our phone and wonder why we do not answer or return messages (Lasén, 2011). The sum of interactions such as these is that we feel obliged to have the phone with us while also knowing that it can expose our role segregation.

In another sense, continual availability robs us of those open spaces in daily life where we are able to drop the different facades and attend to our own needs. We can be physically alone, but nonetheless reachable by friends, family, and our broader web of relations. Indeed, this ubiquitous reachability is a particular characteristic of mobile communication. Thus, there can be a type of Foucaultian sense of perpetual surveillance (Foucault, 1977). But does this constant sense of monitoring (even amidst the seemingly paradoxical moments of rebellion) lead to the disciplining of mobile phone users' behavior and an even more extensive buy-in to one's social roles within the field of social interaction?

At the very least users feel the pressure to be continually available to one another. Thus, one is not able to leave their mobile phone at home for fear of being found "out of the loop." Jarvenpaa and Lang (2005) suggest that after people become used to the expectation that their devices must always be on, they may feel anxiety when they discover that they have lost what Wurtzel and Turner (1977) call imminent connectedness. Speaking of the loss of landline telephony in the 1970s, Wurtzel and Turner note that this loss was in some ways not possible to replace with other forms of interaction. Even in that pre-mobile period, telephonic social interaction had its own unique place.

Wurtzel and Turner (1977) found that the people who went without a landline phone for several weeks reported having feelings of lost control and that these feelings "suggest that loss of telephone contact is an assault upon the way sample members conceived and structured their social reality" (p. 257). The more fine-grained access provided by the mobile phone means that the device potentially gives us a sense of control over our structured social reality. However, when we are without it because we have forgotten it or because the battery is dead, that sense of control fades away. Anxiety associated with missing an important text or an emergency call from someone in our closest sphere means that for many of us we have developed a psychological need to always have the mobile phone available (Ling, 2012).

In this way, the mobile phone becomes an ambiguous stimulus upon which we project our most deeply rooted feelings and emotions. The technology has, as Silverstone and Hirsch (1992) point out, a double articulation. There are the characteristics of the device itself and there is also the use of the device as a link to other people. On both counts, the device can be the source of or the conduit through which to test the individual's emotional state. Thinking first of the mobile phone as an object, it might not fulfill our expectations or it may falter at a critical time. On the other hand, the device might be a point of pride since we are "the first on our block" to have such a magnificent piece of technology. Others have examined emotional attachment to the actual physical device. Vincent (2006) notes that it is increasingly through the mobile phone that we receive and send messages regarding important events in our lives. In addition, it is through the mobile phone that we arrange the logistics of daily life. Given this central position in our lives, it is not surprising that we have engendered an emotional attachment to the mobile phone.

When looking at the interpersonal nature of mobile communication, the exciting news that gives us a huge emotional kick (i.e., the first photos of our grandchild) can come via the mobile phone. At the same time, we might get messages that lead to frustration and disgust (i.e., the message that an underhanded colleague got the promotion you wanted) or they can be an element in the development of a romantic relationship (i.e., sexting and sharing an intimate photo). Fortunati (2005) discusses how our public use of the mobile phone means that as we use it on what Goffman would call the front stage, we can show co-present others our emotional state. Rather than hiding behind closed doors when we get a keenly anticipated call, or the opposite, a call we would rather not have received, we get these communications when in public. Thus, our use of the mobile phone in these situations may mean that we expose our emotional life to the approbation of others. The study of social expectations, or the anticipation of the behavior of others based on the internalization of social norms, within social psychology must therefore incorporate not only the influence of those with whom actors are face to face with but also those remote others with whom they must maintain ties and also "save face."

To summarize, as mobile technology becomes more prevalent within the field of social interaction, social expectations have shifted. Negotiating the demands of actors in the field of social interaction becomes more complicated with the additional actors made "present" through the use and presence of mobile devices. This negotiation can lead to perceptual errors and increase tension between co-present others and the user of the mobile device. Despite this, users still feel the pressure to be available to remote others and they may feel anxiety when any sense of "imminent connectedness" is disrupted. The study of social behavior must incorporate all actors on the field of social interaction: actors who are face to face and ones who are remotely connected to the field.

Mobile Co-Presence and the Impact of Mobile Phones on the Field of Social Interaction

A fundamental tenet of social psychology is social influence. The mere presence of others serves to shape an individual's perception of their field of social interaction and influences their behavior. The presence of known or familiar others reduces anxiety and increases relaxation (de Castro, 1994). Individuals who feel the presence of others feel less inhibited than when alone, and this is enhanced by the presence of known others. The introduction of mobile media into the field of social interaction affords the opportunity for individuals to feel as though they are carrying known others with them and that they have unfettered access to a wealth of knowledge that they can readily use to navigate the unknown, to satisfy their curiosity and to cultivate a sense of mastery over their current environment(s) (Rheingold, 2003). The classic understanding of co-presence and social influence becomes complicated and unpredictable when the access to others is facilitated through technological devices that are mobile and ever-present. In this, co-presence, defined by Erving Goffman (1959) as face-to-face interaction between two or more people, takes on a different connotation when understood in the context of mobile communication (Ling, 2008). For the rules and norms and the effect of social influence to occur, individuals must have a sense that others are present. The introduction of mobile-mediated social interaction then begs the question—do actors need to be physically present for such effects to occur? The study of presence in internet studies established a theoretical foundation to substantiate the fact people can experience a sense of presence in "virtual worlds" (Mantovani & Riva, 1999). The discussion of presence has moved beyond a simple debate concerning what kind of presence is "real" to encompass the role that technology plays in influencing the actions of those with whom we are face to face (Cumiskey & Hjorth, 2013).

As access to the Internet and the use of apps through mobile devices increases, there is much less of a distinction around when one is online and when one is offline. The social pressure to remain "always on" creates the basis for users to generate an augmented reality where one uses the information and tools that one can access through the use of their phone to enhance their experience of their current physical context (McCall, Wetzel, Löschner, & Braun, 2011). There is also social pressure to maintain a presence online even when one is out and about in the physical world. "Checking in" is a term that most aptly demonstrates these phenomena in action. The application *foursquare,* a spinoff of *Dodgeball,* is one of many location-based type services that has a social media platform that allows registered users to create a virtual accounting of all the places that they may visit in a day and to build mobile social networks into the completion of daily tasks and the engagement in social events (Humphreys, 2007). The term "checking in" refers to the process by which, through the power of GPS, people broadcast their location to others online. The "atmosphere" of *foursquare* has a gaming quality to it (i.e., users collect badges from different places and can become the "mayor" of a particular place) and for the most part, the role we play online does not always translate into what is actually occurring the physical space. For instance, we may "check in" at a place while standing in line to be served. By checking in we may then become the Mayor of that place. When this happens, it is rarely broadcast to those with whom we share the physical space but it is broadcasted on the public page of the place that is located on the *foursquare* site.[1] The experience of becoming the Mayor may have some emotional impact and as such their current field of social interaction takes on new meaning and significance, solely based on the engagement of a social media application within a specific context of physical reality (Schwartz, 2013). Use of mobile media in this way changes our perception of the social environment that is not necessarily shared by those who are co-located. The use of mobile technology as part of the field of social interaction allows for the user to feel as though they have a certain command over what is happening within their sphere of social perception. They may feel as though the remote realm with which they have contact has joined them in their field of social interaction and a type of *mobile symbiosis* occurs (Cumiskey, 2011).

Another specific example of how the use of mobile devices augments the perception of social reality is in the ways in which women may use their mobile phones to enhance their sense of personal safety when out alone in public. The use of mobile phones allows for women to imagine their mobile phones to be weapons of self-defense (Cumiskey, 2010; Cumiskey & Brewster, 2012). When asked about whether or not a mobile phone has ever gotten them out of an unsafe situation, women revealed specific ways in which having a mobile phone enhanced their sense of public safety. For instance, participants responded:

> When I feel like a creepy guy is going to come up to me and talk I usually take out my cell phone and act like I am talking so I can ignore them (#9, 21, Latina);
>
> When going home at night where it's empty and only men, it's good to talk on the phone with someone (#28, 18, Latina);
>
> Because when men approach me I pretend to be on it or make a call when I don't want them to talk to me or I feel uncomfortable (#35, 20, White);
>
> One time I had to walk through a crowd of rowdy & drunk men to get to my friend's house. I took out the cell phone to call her & passed through the crowd (#42, 20, White). (Cumiskey, 2010, p. 209)

These findings further suggest that mobile technology help users protect themselves against feelings of vulnerability, especially the vulnerability that comes from being out alone in a public place. Mobile media and being wireless allows for the user to experience realms beyond the limitations of their current field of social interaction.[2] This coupled with a sense of connectedness and joining with known others may generate (a seemingly false) sense of security, camaraderie, and safety (Cumiskey, 2011). This was also indicated in the responses from participants:

> I feel a little more safer (sic) if I'm alone and I'm on the cell phone (#5, 18, White);
>
> I always get paranoid walking home alone late at night. If I didn't have a cell phone on me I would always be home before the sun goes down. I feel more secure that I have a phone in my pocket (#28, 18, Latina);
>
> I feel every time I'm out alone may be a time a cell phone could change my life and save me potentially (#110, 19, White);
>
> You feel more comfortable with a cell phone at times. In case of an emergency, it's always good to have a cell phone (#128, 19, Ethnicity unknown);
>
> Simply leaving my house would even be scarier without a cell phone to contact anyone (#153, 19, White). (Cumiskey, 2010, p. 209)

Connecting to what is known and familiar through the use of mobile devices may alleviate the anxiety that is triggered by being out alone in public. Rendering a remote other present via the device may make the user feel as though their loved ones are in close proximity, something that a weapon like pepper spray is incapable of (Cumiskey & Brewster, 2012). This connection is deeply psychological and another indication of how dependency on one's mobile can engender emotional attachment to the device. What results is an increased dependency and attachment to those we know and a growing fear and resistance to connecting to those strangers with whom we are face to face. Yet the quality of the interaction that we have with those with whom we maintain close ties is also at risk due to the use of mobile devices.

To summarize, the use of mobile technology as part of the field of social interaction allows for the user to feel as though they have a certain command over what is happening within their sphere of social perception. They may feel as though the remote realm with which they have contact has joined them in their field of social interaction and a type of *mobile symbiosis* occurs (Cumiskey, 2011). This connection is deeply psychological and another indication of how dependency on one's mobile can engender emotional attachment to the device. What results is an increased dependency and attachment to those we know and a growing fear and resistance to connecting to those strangers with whom we are face to face. Yet the quality of the interaction with those with whom we maintain close ties is also at risk due to the use of mobile devices. Despite this tricky terrain of managing relationships, mobile phones are promoted as excellent tools for facilitating social cohesion.

Facilitating Social Cohesion through Mobile Communication

There is an assertion that the mobile phone separates us from one another and replaces the social fabric with meaningless chatter. Jarvenpaa and Lang (2005), for example, draw on Heidegger when they suggest that the technology "destroys distance by destroying closeness" (p. 7). This suggests that mobile communication challenges association and the norms of the internalized roles, rules, and expectations that govern most social interactions. There is without question truth in this assertion when considering co-present interaction, where one or more actors are physically present.[3] Turkle (2008, 2011) writes that our ability to network with one another removes us from a particular situation such that we are only half aware of the possibilities that are available in our present field of social interaction. When being a tourist in Athens we may be more concerned with what a distant friend has posted on their Facebook page than touring ancient ruins. We may not interact directly, face to face, with our spouse or children, but we leave messages on their phones or send them texts or tweets. Turkle's point is that we have moved from living in the time and place bounded world of our grandparents to a world where our social networks are mediated. Turkle's fear is that we are losing the connection to other humans and replacing it with virtual contacts that are not as supple, viable, or real. Hence, validating the point raised at the start of this chapter that the study of social psychology needs to broaden the field of social interaction to include the influence of the "presence" of virtual others.

An alternative understanding of the same issue is seen in the work of, among others, Licoppe (2004) and his idea of connected presence, Wellman and Rainie (2012) with their notion of networked individualism and Ling's (2008) notion of bounded solidarity. In this vision of mobile-based association, the device allows us to connect with significant others in a way that was not previously possible. Using Licoppe's concept, we are now able to have an ongoing interaction, a "connected presence," with our closest sphere of connections as the need or inclination arises. If we are stuck with the thought that a chat with our partner would be nice, there is nothing in the bounded field of social interaction stopping us (as long as there is reception, a charged phone, etc.). If we need to remind our child that they need to feed the cat, we can send a text message even as we are in a meeting with an important client. Ling (2008) has noted that it is the sum of all these large and small interactions that tie us into our social orbits and that work to define an augmented field of social interaction. It is through the sum of these interactions (both mediated and direct) that we work out our relationships to our family and friends. Seen in this way, mobile communication in its many forms provides new ways to secure these bonds. While there is a security afforded by this type of interaction, others have suggested that it can also lead to a narrowing of our social horizon and can potentially reduce our individual resourcefulness. Kenneth Gergen (2003) agrees with the potential for mobile phones to enhance dialogue. However, he writes that mobile phones:

> shift the understanding of self from the bounded to the relational. These ranges of technology emphasize and underscore the importance of connection as opposed to autonomy, looking outward rather than inward, toward network as opposed to self-sufficiency, In this sense, the mobile phone may be more significant than any other technology to date. (Gergen, 2003, p. 111)

Another dimension of this issue is how the boundary between in- and outgroups is defined. Analysis of mobile phone traffic shows that the majority of our voice and text interactions are with only a handful (three to five) of others (Ling, Bertel, & Sundsøy, 2012). Exactly how this maps onto our perceived social network and how this disjunction is played out in other social interactions is one of the challenges associated with the technology. Does it result in an inward focus that excludes others in bounded groups or is it a vehicle to a limitless social integration—a boundless field of social interaction? The research indicates that there is an enhancement in our ability to connect with the immediate social sphere. This is because they are always immediately at hand. These findings are often based on analysis of mobile voice and texting. The development of the mobile internet and smartphones will bring with them new challenges to our understanding of this.

To summarize, we have moved from living in the time and place bounded world to a world where our social networks are mediated. The sum of all our large and small interactions tie us into our social orbits and work to define an augmented field of social interaction. It is through the sum of these interactions (both mediated and direct) that we work out our relationships to our family and friends. Seen in this way, mobile communication in its many forms provides new ways to secure these bonds. Once users buy-in to the perceived necessity of having a mobile device to manage our worlds, we may engage in multitasking and appear distracted from what is occurring in the immediate physical realm of interaction. This distraction may have consequences.

Multitasking and the Social Psychological Consequences of Split Attention and Mobile-Mediated Social Facilitation

While the mobile phone is often deeply intertwined in our daily lives, it gives us new forms of access to friends and family, it can also remove us from our co-present field of social interaction. The mobile phone is one element in a collage of gadgets that can be used for different types of multitasking. The television, the PC, a music device, and the mobile phone can all be elements in a multitasking technoscape that provides a rich potential for multiple threads of activity (Tamminen, Oulasvirta, Toiskallio, & Kankainen, 2004). Teens are often characterized as the prototypical multitaskers. They are sometimes portrayed as doing home work while listening to music, watching TV, chatting with friends on a social network site, and surfing the web (Wallis, 2006).

The ability to increasingly slice attention into smaller intervals has implications for the way that teens learn, socialize, and interact. The pace of work and the organization of social life can place a premium of the ability to multitask (Clapp, Rubens, Sabharwal, & Gazzaley, 2011). The mobile phone fits into the folds of daily life, and this means that it plays a central role in our polymediated lives.

The ability to do the proverbial walking and chewing gum at the same time indicates that we can multitask at some level. That is, we can have several lines of activity operating concurrently. Clearly we can do this with skills that are highly practiced (Levy & Pashler, 2008). However, when there is the need to make a significant shift in the flow of activity, then it becomes more difficult to manage (Jarvenpaa & Lang, 2005). When, for example, we have to read a bus table and answer a question from our child, our activity becomes more a series of sequential (not parallel) events. In this context, there is a loss in efficiency. We cannot carry out either task as quickly and we are more prone to mistakes. According to Strayer and Watson (2012), "the vast majority of us cannot multitask without significant costs" (p. 29). Other work suggests that there are age-based dimensions to the ability to multitask. Research has shown, for example, that younger people can switch more effectively between tasks than can older people (Clapp et al., 2011; Grady, Springer, Hongwanishkul, McIntosh, & Winocur, 2006). Regardless of age, however, there are limits on how much we can multitask (Manhart, 2004; Strayer & Watson, 2012). It is more efficient—and less prone to mistakes—to perform one job and then turn to another rather than to switch between them or attempt to do them simultaneously. However, this approach to activity flies in the face of our interactive lives. Research has shown, for example, that people who were high in multitasking skills often had lower memory capacity, were more prone to impulsive and sensation seeking behavior, and overrate their multitasking skills (Strayer & Watson, 2012). Since the mobile phone is often deeply intertwined in our daily lives, and its use makes particular demands on our attention (i.e., we are accountable to our interlocutor in a way that we are not accountable to, e.g., the TV), it provides a rich potential for introducing multiple threads of activity (Tamminen et al., 2004).

As noted, the discussion of multitasking often plays on the image of the teen watching TV while chatting with friends on Facebook and listening to music. However, there is a far more dangerous side to multitasked mobile phone use, namely the mounting use of mobile phones while driving. An increasing amount of research shows that there is an impact on driving and mobile phone use (Horrey & Wickens, 2006). The broad finding from this literature is that talking on the phone (regardless of whether using a "hands free" device or holding the phone) increases the potential for an accident. Talking on the mobile phone while driving promotes "tunnel vision" that limits one's ability to see important traffic-related cues (Strayer, Drews, & Johnsen, 2003). Strayer, Drews, and Crouch (2006) compare this to the impact of driving while intoxicated. According to Reed and Robbins (2008), texting while driving is far worse than simply talking on the mobile phone. According to their simulation work, drinking can increase reaction time when, for example, maintaining position in the driving lane and breaking, by 12%. Use of cannabis can increase reaction time by 21% and texting by 35%. Thus, by any measure, texting while driving is a patently dangerous activity. Indeed, there can be a strong compulsion encouraging this behavior, in that users appear willing to risk their lives and the lives of others just to maintain ties with remote contacts (Beck, Yan, & Wang, 2007).

To summarize, the pace of work and the organization of social life can place a premium of the ability to multitask and yet it is recognized that most people cannot multitask without significant costs. Mobile phones play a central role in our polymediated lives. Compulsive use of the mobile phone while actively engaging in other activities makes users appear willing to risk their lives and the lives of others just to maintain ties with remote contacts.

Problematic Mobile Phone Use: Exploring Mobile Phone Dependency and Risk-Taking

What is becoming evident is that the allure and fantasies surrounding what our mobile devices can do for us can also lead us into the temptation of becoming over reliant on our mobile devices. This, in turn, can supplant the need for us to develop skills that enhance our adaptation to our broadening field of social interaction. Habit driven and impulsive use of the mobile phone has been demonstrated in particular situations (Billieux, van der Linden, D'Acremont, Ceschi, & Zermatten, 2007). Many people always have a mobile phone with them, and an individual's social role may influence their sense of justification for their compulsive mobile phone use. While our minds may be called to imagine the high-powered executive who has multiple devices populating their world with the anticipation of not missing a call from abroad, one of the more common justifications for compulsive use is that of the parent. Parents' reliance on mobile technology to assist them with childcare is quite prevalent throughout the world (Lenhart, Ling, Campbell, & Purcell, 2010; Ling, 2004; Ling & Haddon, 2008). Parents often feel validated in the ways in which they use their mobile phones to monitor the lives of their children (Weisskirch, 2009). And yet for most parents, they are also aware of the ways in which compulsive use of their mobile may impede their relationships with their children. Ribak (2009) interviewed parents who openly acknowledged that using their mobiles could be construed as intrusive and they needed to develop self-imposed rules around when it was or was not appropriate to call:

> … on the weekend when she was at her boyfriend's I wanted to call – I missed her – but my husband said not to disturb her, that I need to get used to the fact that she will soon be drafted to the army (Shula, daughter, 18);
>
> my perspective is that even if I'm a little concerned I don't immediately call, I don't nudge … if he needs something, he'll call… I don't let myself become a nudnik [a person who is a bore or nuisance]; I'm working on it (Pazit, son, 18). (Ribak, 2009, p. 188)

Evident in these quotes is the parents' struggle to not develop a dependency on the phone as a comfort object that could be utilized at whim. This unfiltered use would be intrusive and may damage their relationship. However, there is also a sense of longing. There is the expectation that the child will call when needed and that the parent is expected to be open and available at all times to receive the call. In this there is a sort of codependency. Again, here is the struggle around managing the demands and expectations of use by remote others.

As Ribak (2009) points out, what is important in the tangled web of mobile-mediated relationships is not the actual use of the mobile device but the potential for contact. Not being able to turn off your device in the anticipation of an "important call" can put undue stress on the mobile phone user. In addition, the acknowledgement that others use the phone to reassure themselves and alleviate their anxiety becomes central to how people justify their resistance toward turning their phones off (Bayer, Ling, & Campbell, 2013; Wurtzel & Turner, 1977). Compulsive mobile phone use then gets attributed to balancing the expectations of others, that one must always be always available, while also maintaining this extended field of social interaction.

When the intentionality of the use of the mobile is called into question, people often indicate the necessity of use "in case of an emergency," yet when studied in actuality, most mobile phone use is rather mundane and in response of our individual need for reassurance or to aid in our inability to tolerate boredom (Cumiskey, 2005).

The mobile phone is seen as a convenient way to fill in time when we are, for example, waiting for others while shopping and also as a well-trod link to friends and family. Bayer and Campbell (2012) have used the concept of "automaticity" to describe the degree to which mobile communication and in particular the use of mobile phone becomes automatic. They examine how, for example, checking for texts or mobile Facebook updates is not a conscious activity but rather is "triggered by situational cues and lacks control, awareness, intention, and attention."

Routinized smokers, for example, reach for their cigarettes when they come to an area where it is legitimate to smoke. In a similar way, mobile phone users may have well-oiled routines for glancing at their mobile when there is a break in the stream of social interaction or when they feel a vibration indicating a new message. Bayer and Campbell suggest that "these cues become automatized to the point that the actor may do so without even meaning to do it." Chillingly, Bayer and Campbell find not only that there is automaticity (or habit) when examining the use of texting, but also it is a significant factor when both sending and receiving texts while driving. Thus, the use of the mobile phone while driving may be done without total awareness or control of the behavior.

Taking this beyond the discussion of habit, some authors have discussed the issue as compulsion or behavioral addiction (Bianchi & Phillips, 2005; Park, 2003, 2005; Walsh, White, & Young, 2007).[4] It is clear that there are different types of psychological dependence associated with the use of the mobile phone (or other types of ICTs for that matter). It has become common to use the phrase mobile addiction to characterize this. Some authors discuss whether extensive use of mobile communication is an addiction or whether it is some other type of behavior (Bianchi & Phillips, 2005). Marlatt, Baer, Donovan, and Kivlahan (1988) suggest that addiction is a patterned behavior that enhances an individual's risk of disease or social problems. There is a loss of control and the inability to voluntarily stop the behavior. When discussing alcohol and drug addiction, Morse and Flavin (1992) describe it as being progressive and often fatal, and characterized by impaired control over the behavior, preoccupation with the substance in spite of its adverse consequences and distortions in thinking such as denial. When thinking of the social context in which addiction takes place it is often possible to see extreme social disorganization such as breaking and entering in order to get a fix, violent, and aggressive behavior, and so on. It is rare that the drive to use the mobile phone reaches these heights.[5] It is perhaps better to think of it in terms of compulsion. Extensive use of the mobile phone can, in many ways, be compared to the compulsion for shopping, overeating, watching sports, washing our hands, and so on.

Personality factors have been shown to influence how users use their mobile phones (Butt & Phillips, 2008). When examining the literature for correlates of mobile phone misuse, it was more common among young persons who were extraverted, who had low self-esteem but were not neurotic (Bianchi and Phillips, 2005). These people were also less risk adverse than others. According to the work done in South Korea by Koo (2009), there was lower self-control and a tendency toward impulsiveness among those who misused mobile phones.

To summarize, what is becoming evident is that the allure and fantasies surrounding what our mobile devices can do for us can lead us into the temptation of becoming over reliant on them. Compulsive mobile phone use gets attributed to balancing the expectations of others that one must always be always available. We may try to initiate some control over others' demands that we be an "open channel"; however, it is this openness that facilitates our ability to feel perpetually connected (Laursen, 2013). Despite users' claims to the contrary, most public mobile phone use is rather mundane and in response of our individual need for reassurance or to aid in our inability to tolerate boredom. Mobile phone users may become automatized in their routines around checking their phones. Even if it may mean that they use their mobile phones without total awareness or control of the behavior. Personality factors may provide some insight into problematic mobile phone use, yet researchers are left with questions about how the use of mobile technology may impact personality.

Methodological Challenges in the Social Psychological Study of Mobile Communication

Beyond the effect of the mobile phone on our interactions with others and our psychological well-being, it is also worth considering how the mobile phone can be used to glean insight into the functioning of society. The calls, texts, and the messages that we share with one another trace our social networks (Wellman & Rainie, 2012). Seen in this way, the data that are produced in these interactions help us to understand the way we communicate and the extent of our social lives. For example, the examination of call records in the immediate wake of disasters helps us to see the way that we tie together our closest social links (Cohen & Lemish, 2005; Figley & Jones, 2008; Ryan & Hawdon, 2008; Sundsøy, Ling, Bjelland, Canright, & Engø-Monsen, 2012; Vieweg, Palen, Liu, Hughes, & Sutton, 2008). Yet capturing these intricate webs of interaction and the multiple dimensions of impact empirically is a challenge to most social scientists. This is even more complex when considering the variety of ways that we can interact with our mobile devices, that is, via calling, texting, using Facebook, and so on.

Lewin's formulation of field theory as a methodology can assist us here in that we must first start at the point of broadening our field of social interaction within social psychology to include realms of influence that may not be so readily observed or accessible. There is no denying the psychological power of mobile devices to impact our mood and to distract us from what is occurring in a physical reality that is bound by the limits of time and space. Yet how do we then study the influences of social factors that defy our ability to control them?

The challenge to social psychology that mobile communication poses is that we must now incorporate the social influences of these factors into our understanding of the field of social behavior. If actors do not have to be physically present to still have an influence on those who are face to face, then how do we systematically measure that influence? In the foundational studies of mobile communication, most researchers have split these dimensions in that the impact of mobile phones on the field of social interaction has usually been studied from the user's perspective or from the observer's or co-present other's perspective. There have been some analyses on the impact of receiving calls but it is near impossible to capture that in any moment of any day—if one has a functioning mobile device—they can assume these roles quite fluidly. There has also been a lack of studies that capture use in real time and therefore much of the theoretical development in this area has been based on self-report or on analysis of frequency of use, but what is usually missing is the necessary contextual data that could deepen the analysis of this use.

With advances in the development of technology, scholars of mobile communication must be quite flexible and amenable to engage the latest platforms and functions so that their work maintains relevance. Devices are getting smaller, more mobile and more wearable (soon integrated into our bodies). Understanding the multiple dimensions around how users engage this technology is important—regardless of whether you study the technology directly or not. For instance, if we were to investigate the leading trends in mainstream social psychology—social appraisal, attitudes and beliefs, self-identity, social identity, social cognition and emotions—all of these topics have the potential to be impacted by the influence of social media and mobile communication. Successful engagement of these influences from within any study of the field of social behavior must be explored.

Conclusion

Advances in mobile technology have impacted the world in immeasurable ways. The multinational corporations responsible for the introduction and saturation of markets with these devices often promote them with particular demographics and cultural norms in mind. Popular commercials in the Global North often play on how these devices can enhance one's social life, make life seem more fun and interesting, and make one's mundane work and home responsibilities appear more manageable. In the Global South, mobile phone commercials work to promote the reality that, due to the influence of mobile technology, cultural norms may be shifting and becoming obsolete to make way for the new modern citizen: one who utilizes technology to access the rest of the world, unfettered by any societal pressure to conform to the standards and practices of the past. We have indeed entered an era where influences that we have to contend with in the field of social interaction were not even imaginable in Kurt Lewin's time. The domain of social psychology must now be extended beyond that which is readily observable to include the illusive yet ever-present use of technology that, in effect, brings our social world directly into our lives regardless of where we happen to be. The entrance of remote interlocutors into the field of social interaction has posed a challenge for researchers who want to study its influence empirically. The challenge of the "open channel" created via an active mobile communication device means that even when not in actual use, the potential for use can also have an influence over people's behavior within the field of social interaction. The presence of mobile technology and the potential for interaction can result in dependence on mobile devices, but it can also tie us more securely together. We cannot overlook the ways in which mobile devices have the power to present constraints around how we function in our social worlds. Managing the split attention caused by the integrated use of mobile technology renders users vulnerable to risk-taking behaviors (like texting while driving) and other dangers. Investigating these issues will help us to understand mobile communication technology's impact and serve as a counter story to the popular culture notion that frequent use of mobile devices borders on necessary for survival.

Notes

1 By broadcast we mean that those in the physical realm may not announce over a public address system the news that you are mayor, however, it is "broadcasted" to users who may also be checking in to the same physical space via their mobile devices. There may be a curiosity as to "who" is sharing your experience in that there is still a certain degree of anonymity around who is checking in.
2 A similar use is reported by both males and females who note that when confronted with the possibility of having to talk to someone they would rather avoid socially, the mobile phone provides a quick excuse to duck out of a conversation (Ling & Baron, 2007).
3 Use of the mobile phone is disruptive. It challenges the decorum of well established social settings. According to Monk et al. (2004), regardless of whether an individual is engaged with the person using the phone, or whether the mobile phone is a part of the ambient situation, third parties find the public use of the mobile phone disruptive. On the other hand, the person having the conversation in a public place can also feel that they are put into the public gaze in an unfavorable way (Jarvenpaa & Lang, 2005). They have the realization that use of the mobile phone in public places also carries with it the need to develop new forms for privacy.
4 Carbonell et al. have reviewed the literature with regards compulsive mobile phone use (Carbonell, Guardiola, Beranuy, & Bellés, 2009).
5 The obvious caveat is the urge to text while driving.

References

Aronson, E. (2011). *The Social Animal* (11 ed.). New York, NY: Worth Publishers.

Bayer, J. & Campbell, S. (2012). Texting while driving on automatic: Considering the frequency-independent side of habit. *Computers in Human Behavior*, *28*, 2083–2090.

Bayer, J., Ling, R., & Campbell, S. (2013). Mobile communication habits and the mobile middle. *Paper presented at 2013 ICA Mobile Pre-Conference*, London School of Economics, London, June 16–17 2013.

Beck, K. H., Yan, F., & Wang, M. Q. (2007). Cell phone users, reported crash risk, unsafe driving behaviors and dispositions: A survey of motorists in Maryland. *Journal of Safety Research*, *38*(6), 683–688.

Bianchi, A., & Phillips, J. G. (2005). Psychological predictors of problem mobile phone use. *CyberPsychology and Behavior*, *8*(1), 39–52.

Billieux, J., van der Linden, M., D'Acremont, M., Ceschi, G., & Zermatten, A. (2007). Does impulsivity relate to perceived dependence on and actual use of the mobile phone? *Applied Cognitive Psychology*, *21*(4), 527–537.

Butt, S., & Phillips, J. G. (2008). Personality and self reported mobile phone use. *Computers in Human Behavior*, *24*(2), 346–360. doi:10.1016/j.chb.2007.01.019

Carbonell, X., Guardiola, E., Beranuy, M., & Bellés, A. (2009). A bibliometric analysis of the scientific literature on Internet, video games, and cell phone addiction. *Journal of the Medical Library Association: JMLA*, *97*(2), 102–107. doi:10.3163/1536-5050.97.2.006

Clapp, W. C., Rubens, M. T., Sabharwal, J., & Gazzaley, A. (2011). Deficit in switching between functional brain networks underlies the impact of multitasking on working memory in older adults. *Proceedings of the National Academy of Sciences*, *108*(17), 7212–7217. doi:10.1073/pnas.1015297108

Cohen, A. A., & Lemish, D. (2005). When bombs go off the mobiles ring: The aftermath of terrorist attacks. In K. Nyiri (Ed.), *A sense of place: The global and the local in mobile communication* (pp. 117–128). Vienna, Austria: Passagen Verlag.

Cumiskey, K. M. (2005). “Surprisingly nobody tried to caution her”: Perceptions of intentionality and the role of social responsibility in the public use of mobile phones. In R. Ling & P. Pedersen (Eds.), *Mobile communications: Re-negotiation of the social sphere*. London, UK: Springer.

Cumiskey, K. M. (2010). “Simply leaving my house would even be scarier”: How mobile phones affect women's perception of safety and experiences of public places. *Media Asia, 37*(4), 205–214.

Cumiskey, K. M. (2011). Mobile symbiosis: A precursor to public risk-taking behavior? In R. Ling & S. W. Campbell (Eds.), *Mobile communication: Bringing us together and tearing us apart* (pp. 17–36). New Brunswick, NJ: Transaction Publishers.

Cumiskey, K. M., & Brewster, K. (2012). Mobile phones or pepper spray? Imagined mobile intimacy as a weapon of self-defense for women. *Feminist Media Studies, 12*(4), 590–599.

Cumiskey, K. M., & Hjorth, L. (2013). Between the seams: Mobile media practice, presence and politics. In K. M. Cumiskey & L. Hjorth (Eds.), *Mobile media practices, presence and politics: The challenge of being seamlessly mobile* (pp. 1–11). New York, NY: Routledge.

de Castro, J. M. (1994). Family and friends produce greater social facilitation of food intake than other companions. *Physiology & Behavior, 56*(3), 445–455.

Figley, C. R., & Jones, R. (2008). The 2007 Virginia tech shootings: Identification and application of lessons learned. *Traumatology, 14*(1), 4–7. doi:10.1177/1534765608319921

Fortunati, L. (2005). Mobile telephone and the presentation of self. In R. Ling & P. Pedersen (Eds.), *Mobile communications: Re-negotiation of the social sphere* (pp. 203–218). London, UK: Springer.

Foucault, M. (1977). *Discipline and punish: The birth of the prison*. London, UK: Penguin.

Gergen, K. (2003). Self and community and the new floating worlds. In K. Nyíri (Ed.), *Mobile democracy: Essays on society, self and politics* (pp. 61–69). Vienna, Austria: Passagen Verlag.

Goffman, E. (1959). *The presentation of self in everyday life*. New York, NY: Doubleday Anchor Books.

Grady, C. L., Springer, M. V., Hongwanishkul, D., McIntosh, A. R., & Winocur, G. (2006). Age-related changes in brain activity across the adult lifespan. *Journal of Cognitive Neuroscience, 18*(2), 227–241. doi:10.1162/jocn.2006.18.2.227

Horrey, W. J., & Wickens, C. D. (2006). Examining the impact of cell phone conversations on driving using meta-analytic techniques. *Human Factors, 48*(1), 196–205.

Humphreys, L. (2007). Mobile social networks and social practice: A case study of Dodgeball. *Journal of Computer-Mediated Communication, 13*(1), 341–360.

Jarvenpaa, S. L., & Lang, K. R. (2005). Managing the paradoxes of mobile technology. *Information Systems Management, 22*(4), 7–23.

Koo, H. Y. (2009). Development of a cell phone addiction scale for Korean adolescents. *Journal of Korean Academy of Nursing, 39*(6), 818–828.

Lasén, A. (2011). “Mobiles are not that personal”: The unattended consequences of the accountability, accessibility and transparency afforded by mobile telephony. In R. Ling & S. Campbell (Eds.), *Mobile communication: Bringing us together or tearing us apart? Mobile communication research series*. New Brunswick, NJ: Transaction Publishers.

Laursen, D. (2013). Organizing 'my mind is with you': Continued interaction after closed interaction via the mobile phone. In K. M. Cumiskey & L. Hjorth (Eds.), *Mobile media practices, presence and politics: The challenge of being seamlessly mobile* (pp. 101–105). New York, NY: Routledge.

Lenhart, A., Ling, R., Campbell, S., & Purcell, K. (2010). *Teens and mobile phones*. Washington, DC: Pew Research Center.

Levy, J., & Pashler, H. (2008). Task prioritisation in multitasking during driving: Opportunity to abort a concurrent task does not insulate braking responses from dual-task slowing. *Applied Cognitive Psychology, 22*(4), 507–525.

Lewin, K. (1939). Field theory and experiment in social psychology: Concepts and methods. *American Journal of Sociology, 44*(6), 868–896. doi:10.2307/2769418

Lewin, K. (1951). *Field theory in social science: Selected theoretical papers*. New York, NY: Harper.

Licoppe, C. (2004). 'Connected' presence: the emergence of a new repertoire for managing social relationships in a changing communication technoscape. Environment and Planning D: Society and Space, *22*, 135–156.

Ling, R. (2004). *The mobile connection: The cell phone's impact on society*. San Francisco, CA: Morgan Kaufmann.

Ling, R. (2008). *New tech, new ties: How mobile communication is reshaping social cohesion*. Cambridge, MA: MIT Press.

Ling, R. (2012). *Taken for grantedness: The embedding of mobile communication into society*. Cambridge, MA: MIT Press.

Ling, R., & Baron, N. (2007). Text messaging and IM: Linguistic comparison of American College Data. *Journal of Language and Social Psychology, 26*(3), 291–298.

Ling, R., Bertel, T., & Sundsøy, P. (2012). The socio-demographics of texting: An analysis of traffic data. *New Media & Society, 14*(2), 280–297.

Ling, R., & Haddon, L. (2008). Children, youth and the mobile telephone. In K. Dortner & S. Livingstone (Eds.), *International handbook of children, media and culture* (pp. 137–151). London, UK: Sage.

Manhart, K. (2004). *The limits of multitasking. Scientific American mind*. Retrieved from http://www.sciamdigital.com/index.cfm?fa=Products.ViewIssuePreview&ARTICLEID_CHAR=3D583107-2B35-221B-6E5A9C25286A414E

Mantovani, G., & Riva, G. (1999). "Real" presence: How different ontologies generate different criteria for presence, telepresence, and virtual presence. *Presence: Teleoperators & Virtual Environments, 8*(5), 540. doi:Article

Marlatt, G. A., Baer, J. S., Donovan, D. M., & Kivlahan, D. R. (1988). Addictive behaviors: Etiology and treatment. *Annual Review of Psychology, 39*(1), 223–252.

McCall, R., Wetzel, R., Löschner, J., & Braun, A.-K. (2011). Using presence to evaluate an augmented reality location aware game. *Personal & Ubiquitous Computing, 15*(1), 25–35. doi:Article

Monk, A, Carroll, J., Parker, S., & Blythe, M. (2004). Why are mobile phones annoying? *Behavior and Information Technology, 23*(1), 33–41.

Monson, T. C., & Snyder, M. (1977). Actors, observers, and the attribution process. *Journal of Experimental Social Psychology, 13*(1), 89–111.

Morse, R. M., & Flavin, D. K. (1992). The definition of alcoholism. *JAMA: The Journal of the American Medical Association, 268*(8), 1012–1014. doi:10.1001/jama.1992.03490080086030

Nisbett, R. E., Caputo, C., Legant, P., & Marecek, J. (1973). Behavior as seen by the actor and as seen by the observer. *Journal of Personality and Social Psychology, 27*(2), 154–164.

Park, W. K. (2003). *Mobile phone addiction: A case study of Korean college students*. In R. Ling & P. Pedersen (Eds.), Paper presented at the Front Stage/Back Stage: Mobile Communication and the Renegotiation of the Social Sphere, Grimstad, Norway.

Park, W. K. (2005). Mobile telephone addiction. In R. Ling & P. Pedersen (Eds.), *Mobile communications: Renegotiation of the social sphere* (pp. 259–272). London, UK: Springer.

Reed, N., & Robbins, R. (2008). *The effect of text messaging on driver behaviour: A simulation study. TLR Report*. Retrieved from www.trl.co.uk

Rheingold, H. (2003). *Smart mobs: The next social revolution.* Cambridge, MA: Basic Books.

Ribak, R. (2009). Remote control, umbilical cord and beyond: The mobile phone as a transitional object. *British Journal of Developmental Psychology, 27*(1), 183–196. doi:Article

Ryan, J., & Hawdon, J. (2008). From individual to community: The "Framing" of 4-16 and thedisplayofsocialsolidarity. *Traumatology, 14*(1),43–51.doi:10.1177/1534765607312686

Sansone, C., Morf, C. C., & Panter, A. T. (Eds.). (2003). *The Sage handbook of methods in social psychology* (1st ed.). Thousand Oaks, CA: Sage Publications, Inc.

Schwartz, R. (2013). The networked familiar stranger: An aspect of online and offline urban anonymity. In K. M. Cumiskey & L. Hjorth (Eds.), *Mobile media practices, presence and politics: The challenge of being seamlessly mobile* (pp. 135–149). New York, NY: Routledge.

Silverstone, R., & Hirsch, E. (1992). *Consuming technologies.* London, UK: Routledge.

Strayer, D. L., Drews, F. A., & Crouch, D. J. (2006). A comparison of the cell phone driver and the drunk driver. *Human Factors: The Journal of the Human Factors and Ergonomics Society, 48*(2), 381–391.

Strayer, D. L., Drews, F. A., & Johnsen, W. A. (2003). Cell phone induced failures of visual attention during simulated driving. *Journal of Experimental Psychology: Applied, 9*(1), 23–32.

Strayer, D. L., & Watson, J. M. (2012). Supertaskers and the multitasking brain. *Scientific American Mind, 23*(1), 22–29. doi:10.1038/scientificamericanmind0312-22

Sundsøy, P. R., Ling, R., Bjelland, J., Canright, G., & Engø-Monsen, K. (2012, August 26–29). *The activation of core social networks in the wake of the 22 July Oslo bombing.* Presented at the IEEE/ACM ASONAM 2012, Istanbul, Turkey.

Tamminen, S., Oulasvirta, A., Toiskallio, K., & Kankainen, A. (2004). Understanding mobile contexts. *Personal and Ubiquitous Computing, 8*(2), 135–143.

Turkle, S. (2008). Always-on/always-on-you: The tethered self. In J. E. Katz (Ed.), *Handbook of mobile communication studies* (pp. 121–137). Cambridge, MA: MIT Press. Retrieved from http://sodacity.net/system/files/Sherry-Turkle_The-Tethered-Self.pdf

Turkle, S. (2011). *Alone together: Why we expect more from technology and less from each other.* New York, NY: Basic Books.

Vieweg, S., Palen, L., Liu, S. B., Hughes, A. L., & Sutton, J. N. (2008). *Collective intelligence in disaster: Examination of the phenomenon in the aftermath of the 2007 Virginia tech shooting.* Boulder, CO: University of Colorado.

Vincent, J. (2006). Emotional attachment and mobile phones. *Knowledge, Technology and Policy, 19*(1), 39–44.

Wallis, C. (2006). The multitasking generation. *Time Magazine, 167*(13), 48–55.

Walsh, S. P., White, K. M., & Young, R. M. D. (2007). *Young and connected: Psychological influences of mobile phone use amongst Australian youth.* Retrieved from http://eprints.qut.edu.au/9753/

Weisskirch, R. S. (2009). Parenting by cell phone: Parental monitoring of adolescents and family relations. *Journal of Youth & Adolescence, 38*(8), 1123–1139. doi:Article

Wellman, B., & Rainie, L. (2012). *Networked: The new social operating system.* Cambridge, MA: MIT Press.

Wurtzel, A. H., & Turner, C. (1977). Latent functions of the telephone: what missing the extension means. In I. De Sola Pool (Ed.), *The social impact of the telephone* (pp. 246–261). Cambridge, MA: MIT Press.

11

Real or Ersatz? Determinants of Benefits and Costs of Online Social Interactions

Melanie C. Green[1] and Jenna L. Clark[2]

[1]Department of Communication, University at Buffalo (SUNY), Buffalo, NY, USA
[2]Department of Psychology, University of North Carolina, Chapel Hill, NC, USA

When new communication technologies are introduced to society, scholarly debate as to their likely uses, benefits, and implications tends to follow, perhaps inevitably. Such debates arose first in the wake of the telegraph and have arisen anew from the development of every new technology since (Katz, Rice, & Aspden, 2001)—indeed, debates over the impact of television remain ongoing.

It is unsurprising, then, that the advent of the Internet has fostered a lively discussion as well. Two views tended to predominate in early discussion on the topic: a pessimistic view that suggested the Internet was poor for building social resources, and an optimistic view that implied the Internet could be used for building new forms of community (see Katz et al., 2001 for an in-depth discussion of both perspectives). However, much of this initial debate was more philosophical than empirical.

Kraut et al. (1998) provided one of the first and most influential examinations of Internet use with their HomeNet study. Ninety-three families in Pittsburgh, Pennsylvania, were studied for 1–2 years after initially beginning Internet use at home; the measures included hours of online time, number of emails sent, number of websites visited, well-being (loneliness, stress and depression), and social involvement (family communication, size of distant social network, size of local social network, and social support). They found that greater use of the Internet led to decreases in family communication and size of the local social network and increases in loneliness and depression.

These results seemed to dovetail with recent patterns about society as a whole that had been remarked on by several scholars, most notably Robert Putnam (1995). Putnam found that American society was becoming increasingly isolated, with notable decreases in civic participation. This trend, Putnam explicitly noted, may have been exacerbated by the invention of the television. A time displacement hypothesis was

The Handbook of the Psychology of Communication Technology, First Edition. Edited by S. Shyam Sundar.

advanced by Putnam as the most likely explanation: hours spent at home watching television are hours that are not spent out socializing with other individuals.

The results of the HomeNet study implied a similar time displacement hypothesis might be applicable to the Internet despite its ostensibly greater potential for social interaction when compared to the television. This perspective was further reinforced through research done by Nie (Nie, 2001; Nie & Erbring, 2000), which seemed to confirm this theory by suggesting that time spent using the Internet was indeed replacing time spent with friends and family.

However, even as this philosophy received a wealth of media attention, contradictory results were beginning to pile up that found null effects of Internet use—or in some cases even positive ones. Replicating the HomeNet study, Wastlund, Norlander, and Archer (2001) could not demonstrate the same negative effects. Several other studies found that using the Internet explicitly for social purposes might in fact reduce depression and loneliness, causing increases in well-being (Bessière, Kiesler, Kraut, & Boneva, 2008; Morgan & Cotten, 2003; Shaw & Gant, 2002). A review conducted by Katz et al. (2001) of studies published between 1995 and 2000 generally concluded that the negative effects of Internet use were not supported by the literature. Perhaps most dramatically, Kraut et al. (2002) revisited the HomeNet sample and found that the negative effects they had seen had dissipated.

Two potential explanations for this change in the overall consensus of the literature are offered by Valkenburg and Peter (2009). First, perhaps the outcomes of Internet use had changed as the technology had matured and become more widespread. For example, if all of an individual's social network had Internet access compared to none, one can easily visualize how Internet use could become more beneficial. Second, a shift in the actual methods of online interaction had occurred, moving from technologies that encouraged interacting with strangers to technologies that encouraged interacting with already-known individuals. These two changes, in concert, could perhaps explain a shifting effect of the Internet on well-being.

However, some more recent studies continue to show negative effects on well-being (Chen & Peng, 2008; Schiffrin, Edelman, Falkenstern, & Stewart, 2010) even after the narrowing of the digital divide and the spread of more intimate communication technologies. The conclusion that the Internet's effects had changed from negative to positive seems somewhat too simplistic for the data. What theoretical approaches could reconcile a continuing pattern of mixed results?

In this chapter, we will describe two theories that address the nature and potential outcomes of mediated interactions. Ersatz ("substitute") social engagement theory classifies social activities on a continuum from ersatz to "real," with real social activities involving less mediation, a higher number of sensory channels, and a stronger degree of emotional connection. Real social activities show more positive outcomes. The second theory, the perceived reality of online interactions (PROI), posits that individuals' perceptions of the potential benefits on online interactions may play a key role in determining the actual benefits of such interactions. Specifically, individuals who believe that online communication can lead to true friendships or relationships are likely to act in ways that confirm this belief, thus creating more supportive and beneficial online relationships. Together, these theories provide a roadmap for future research, and more broadly, suggest that a more nuanced view of online interaction is needed to help understand when and for whom online interaction will have positive effects.

Ersatz Social Engagement Theory

One potential explanation for the results observed in studies of Internet use could be that the outcomes are actually mixed—both positive and negative consequences, depending on specific types of Internet use. For example, online social interaction might be harmful, whereas nonsocial uses of the Internet might not be. Ersatz social engagement theory (Green & Brock, 1998, 2008) is one of several theories that might predict such an outcome. The term "ersatz" comes from the German word meaning "substitute," often referring to an inferior substitute. We have adopted this term to indicate that individuals might engage in activities designed to meet social goals, but which fall short of providing the full benefits of social interaction. In ersatz social engagement theory, certain "ersatz" social activities provide instant gratification and pleasure, but at the cost of failing to build real relationships that provide lasting positive effects.

The theoretical framework of ersatz social engagement theory begins with the fact that, as social animals, human beings have needs for companionship (Baumeister & Leary, 1995; Fiske, 2004). However, there are many different ways in which these needs might be met, and not all paths that satisfy them are created equal. Activities that fulfill social needs while being to some extent incomplete or inferior to real social activities are those defined by ersatz social engagement theory as "ersatz social activities."

What delineates real and ersatz social activities?

In general, social activities can be arranged on a continuum from real to ersatz. This continuum is based on three major dimensions: the extent to which an activity is in-person versus mediated, the number of sensory channels accessible in an activity, and the level of true emotional/interactive connections between individuals in the activity.

Degree of mediation

Degree of mediation refers to whether participants in an activity are face-to-face with one another or if they are communicating through some media that allows them to bridge the gap between them. Not all forms of mediated communication are created equal—as is discussed in the next section on sensory channels—but there are certain inevitable downsides to the use of any kind of mediated communication.

For example, it is difficult to provide instrumental forms of social support through a mediated communication. One can offer advice or emotional comfort, but it is difficult to clean a friend's house, help them move, or walk their dog when only mediated communication is possible (e.g., when the individuals are geographically separated).

Additionally, there may be unique benefits to physical contact that cannot be gained through a mediated communication. For example, it is possible to communicate emotion through touch (Herteinstein, Holmes, McCullough, & Keltner, 2009). In a broad review of the topic, Gallace and Spense (2010) note that touch is integral to emotional well-being and touching can cause individuals to feel more positively about a situation, reduce stress, induce greater compliance with health regimens and even increase the amount of tipping. Considering this wide range of possible benefits of touch in interpersonal communication, one could suspect that mediation strips a surprising amount of potential positive emotion from an interaction.

Lastly, mediated communications may be troubled by issues of quality inherent to the medium's functioning. Though these experiences can vary widely, it is not uncommon to encounter technical issues such as lost call reception, video quality issues, or lag. These intrusive technical issues may break one's sense of immersion in an interaction, or otherwise pose difficulties in conducting a meaningful social activity.

Sensory channels

Sensory channels refer to the different ways in which a medium can convey information. For example, a telephone call has fewer sensory channels than a face-to-face interaction because one cannot see their interaction partner, nor touch them. Social presence theory (Short, Williams, & Christie, 1976) dictates that the fewer sensory channels are available in an interaction, the less attention will be given to the contents of that interaction. A related concept, media richness (Daft & Lengel, 1984, 1986; Trevino, Lengel, & Daft, 1987), describes the extent to which a medium can convey information. Rich media are more capable of conveying the full depth of information that was initially involved in a communication.

Rice (1993) notes that the gist of both theories is that the characteristics of a medium should be matched to the appropriate activity, so that media high in social channels or richness should be used for tasks that require more nuance or depth of communication. His results indicate that while all media are useful for exchanging information, media with a low social presence (i.e., a reduced number of channels) are seen as less appropriate for social interaction.

Emotional connection

Perhaps the most powerful determinant of an ersatz versus real interaction is the level of emotional connection or engagement between the participants. However mediated the communication, the potential to form the kind of bonds that underlie the creation of social capital exists so long as the participants are able to form a true connection. That said, it is difficult to assess what technological factors are related to the ability to form a true connection. Both mediation and the number of channels could presumably enhance a technology's suitability for emotional connection; however, even a purely text-based interaction can lead to strong emotional connections given appropriate content (e.g., self-disclosure, expressions of affection). Furthermore, some level of true interaction between individuals must exist in order for an activity to be considered on the "real" end of the continuum. That is, an actual other person (rather than just a character) must be involved in the interaction. Individuals may have strong emotional reactions to television characters and form parasocial relationships with them, but these relationships would still be considered ersatz because although emotion is evoked, the connection is only one-way.

Ersatz and parasocial interactions

To provide an example of what we mean by ersatz social activities, watching a television show could be characterized as an extremely ersatz social activity, at the far end of the continuum. One could see a social dimension in the extent to which an individual felt a connection or bond with characters on the screen. However, the activity is clearly ersatz because: (1) there is complete mediation, (2) the sensory channels that do exist are all one-way, and (3) no emotional or interactive connection is possible with scripted television characters (perhaps due to points 1 and 2).

This concept is potentially related to the idea of "parasocial interactions" as seen in communication studies. Parasocial interactions involve a sense of intimacy with a media persona, such as a character on a television show (Horton & Wohl, 1956; Rubin, Perse, & Powell, 1985). They are therefore ersatz as a whole by our definition. However, the category of ersatz social activities should be considered to encompass additional activities that are not as completely lacking in true engagement. Even a telephone conversation, while largely real, possesses some ersatz qualities in its mediated and voice-only nature.

What are the consequences of choosing ersatz social activities over real social activities?

Ersatz social engagement theory suggests that when choosing ersatz social activities, individuals do not build the same social skills or social capital as they would in real social activities. Ersatz social activities can be considered the junk food of the social world: they fill one up without providing nutritional value. For example, while watching the aforementioned television show, an individual may indeed feel a sense of belonging or connection with the characters on the screen. However, no lasting reciprocal emotional bond can actually be created.

Social skills

It is a largely accepted theoretical position that friendships can serve as a place to learn social skills, though this claim may be one that is in need of more empirical support (Asher, Parker, & Walker, 1996; Glick & Rose, 2011). Asher et al. (1996) proposed that the skills required to perform certain social tasks were integral to friendship success, providing an explanation as to how friendship could in fact build social skills. Rose and Asher (1999) found that the endorsement of hostile strategies in negotiating hypothetical friendship conflict led to having fewer friends and more actual conflict, whereas prosocial strategies predicted less conflict.

However, this study and similar others (Rose & Asher, 2004) did not examine the question with methods suitable to establish causality. Glick and Rose (2011) took a longitudinal approach with adolescents examined in both the fall and the spring of one school year. They presented vignettes in which participants either experienced a stressor and could ask for help, or a friend experienced a stressor and they could choose to help. Both having more friends and more high-quality friendships predicted improvement in social skills from time 1 (the fall) to time 2 (the spring). Number of friends predicted decreases in avoidant or hostile strategies, while high-quality friendship predicted increases in emotionally-engaging strategies and decreases in disengagement strategies.

This research suggests that friendships may well be an integral part of learning social skills, particularly in adolescence. Interestingly, it implies that much of the learning may come from navigating stressors and surmounting their potentially negative effects. Ersatz social activities, with their focus on easier gratification, are less likely to include these kind of challenges. Consider our television-watching example: one will never need to help the character on the screen, or be able to ask for help from that character. Some types of online relationships may also reduce these opportunities for learning; if a conflict arises in an online interaction, individuals may find it easier to log off, block contact from the individual, or surf to another website rather than working through the conflict or stressor in a constructive way. Therefore, valuable opportunities to build social skills can be lost when one purses ersatz social activities.

Social capital

Social capital has been defined in many different ways across different line of research. For the sake of this discussion, we will use Putnam's 1995 definition: "features of social organization such as networks, norms, and social trust that facilitate coordination and cooperation for mutual benefit" (Putnam, 1995, p.67). As noted earlier in this chapter, social capital has been declining in the United States in recent years (Putnam, 1995). Putnam (2000) expressed concerns that the Internet would further decrease social capital in two ways: via (a) replacing real interaction and (b) in a way that did not lead to the generation of more social capital (Green & Brock, 2008).

Are declines in social capital worth being concerned about? Putnam (2001) has identified a wealth of potential negative consequences of decreased social capital. Though the findings are largely correlational, SAT scores, child welfare, crime, health, tax evasion, economic inequality, civic inequality, and individual well-being are all linked to social capital. In many cases, social capital explains more variance in these fields than other, more conventional variables (such as spending on education for SAT scores or poverty for crime).

Looking at this pattern of results, one could see social capital as the glue that holds a society together, providing a wealth of benefits to its members. If ersatz social activities do not produce social capital, and are chosen instead of real social activities that do, the negative consequences could be far-reaching. However, as we discuss later in the chapter, empirical studies of the effects of online social networks have had mixed results, showing both positive and negative effects.

Why do people choose ersatz social activities?

If choosing ersatz social activities results in so many negative consequences, however, why would anyone pursue them? One answer is that ersatz social activities have their particular own benefits, perhaps serving to explain both their draw and the field's history of divided results. There are three primary aspects of ersatz social activities that make them appealing: they pose less risk, require less effort, and offer more immediate gratification.

Risk

We run many risks when engaging in social interaction. Awkwardness, rejection and more are all potential outcomes of an interaction that goes badly—and they are outcomes that are often more controllable or avoidable in ersatz social activities. A true emotional bond with a fictional character may be impossible—but one cannot be rejected by a fictional character, either. Closer to the middle of the spectrum, mediated interactions with other people may still provide more of an ability to control one's self-presentation. Individuals can hide behind avatars, choose which information to share, and more easily shed relationships in online contexts. Empirically, Green and Brock (2008) found that ersatz activities were indeed considered less risky by participants.

Effort

Beyond the question of risk, real social activities also require a good deal of effort. Friendships are highly reciprocal; it is an expected norm that individuals provide social support to their friends, which can require considerable expenditure of time and resources. Even arranging a real social activity may require effort, in accounting for individuals' schedules and otherwise preparing. In contrast, many ersatz activities are available whenever an individual might desire them, simply by the expedient of turning on a TV or a computer. Similarly, fewer demands are likely to be made of an individual in an ersatz social activity—in a parasocial one, perhaps none at all. Green and Brock (2008) again confirmed this by asking for individuals' perceptions of ersatz social activities and finding them significantly less effortful than real ones.

Gratification

If one wishes to reap the benefits of a real friendship, a lengthy process of getting to know someone is often required. An initial acquaintanceship must be given time to blossom into a rewarding social bond. With ersatz social activities, this is often not required. Characters in a television show can be instantly empathized with, or understood in a far briefer timespan. Also, as mentioned earlier, ersatz social activities can often occur solely on the timescale of the individual seeking them. Today's technology allows individuals to consume media (and, if they choose, engage in ersatz social activity) whenever they desire.

Online Social Interaction as an Ersatz Social Activity

We have noted already that several different interactions can be grouped under the label of ersatz social activities, falling at different points along the continuum from ersatz to real. Due to the multidimensional nature of this spectrum, it can be difficult to determine a specific point at which an activity becomes ersatz. Most of us would not hesitate to categorize telephone calls as real interactions despite their mediation and reduced sensory channels, but instant messaging, email, social networks, and other forms of Internet communication may be more difficult to place.

To examine the question of how ersatz social engagement theory can broaden our understanding of the consequences of online social interaction, it is helpful to examine (a) how online social activity fits the definition of an ersatz social activity and (b) its established consequences in comparison to those predicted for an ersatz social activity.

Online Social Interaction on the Continuum of Real to Ersatz

The three main hallmarks of an ersatz social activity, as presented in the previous section, are proximity versus mediation, the number of sensory channels available, and the degree of true emotional/interactive connection.

Proximity versus mediation in online social interaction

By default and by definition, online social interactions tend to be fully mediated. Proximity is irrelevant; the experience is the same whether one is in the next building or thousands of miles away. Although it is common for individuals who engage in online social interaction to later meet (McKenna, Green, & Gleason, 2002), the activity itself must still be considered mediated.

This mediation provides many downsides; technical issues, as mentioned earlier, can arise. Dropped connections, unavailability of the Internet, and lag are all prominent issues that can interfere with computer-mediated communication. As chronemic (time-based; Walther & Tidwell, 1995) cues can be important in computer-mediated communication, these technical issues may have considerable impact on the quality of the interaction.

Sensory channels in online social interaction

The role of sensory channels in online social interactions is a contentious question, with several contending theories. The cues-filtered-out perspective (Culnan & Markus, 1987; Rice, 1993) suggests that online communication tends to lack much of the sensory information we use to understand our conversations, which has a deleterious effect on understanding and the quality of communication. This was supported by work in social presence theory (Short, Williams, & Christie, 1976), which suggests that fewer channels of communication means less attention, and by other findings that suggested computer-mediated communication is an impoverished medium lacking in context clues (Sproull & Kiesler, 1985). Initially, the cues-filtered-out perspective was the most prominent view in the study of online social interaction, and it would seem to lend strength to the idea that a lack of sensory channels might make online social interaction more of an ersatz activity.

However, others have disputed the idea that sensory channels are inevitably missing from computer-mediated communication. Walther's (1992, 1993, 1994, 1996) Social information processing theory suggests that individuals are fully capable of adapting to the differences between computer-mediated communication and face-to-face communication. Instead of losing sensory information irrevocably, they learn to master other channels by which it can be conveyed. Examples of additional channels in online communication include time-related or chronemic cues as mentioned earlier (e.g., pauses, conversational rhythm; Walther & Tidwell, 1995) and typographic cues such as emoticons (Walther & D'Addario, 2001). Also, individuals perceive equal affective response across media (Walther, Loh, & Granka, 2005) and, when engaging in computer-mediated communication, solicit more information actively about their partner to compensate for missing cues (Tidwell & Walther, 2002).

Therefore, although the normal sensory channels may indeed be lacking from online social interaction, evidence seems to suggest individuals can compensate for this lack by learning to "decode" additional channels that can be conveyed by the medium. The extent to which these additional channels may compensate for the lack of nonverbal cues, however, is likely still open to debate.

Emotional connection

Perhaps the crux of the real-ersatz question, the degree of emotional connection that can be established over online social interactions, is seemingly disputed. Parks and Floyd (1996) provided some of the earliest evidence that the social bonds one makes online can be deep and meaningful. Further work by McKenna, Green, and Gleason (2002) and Bargh, McKenna, and Fitzsimons (2002) found that people often felt better able to express their "true self" online, and that those who felt this the most strongly tended to create the deepest, longest-lasting friendships. McKenna's work in general painted a picture of the Internet as a place where superficial features that might normally prevent friendships from blooming were no longer obstacles, allowing individuals to establish bonds based on true suitability—an image that implies the Internet might almost be more suitable for forming emotional connections than regular face-to-face communication.

However, many users of the Internet seem to perceive their online relationships as less deep and less meaningful than their offline relationships. Chan and Cheng (2004) found that offline relationships were rated as higher in depth, breadth, commitment and many other indicators of relationship strength, although the gap between online and offline narrowed as time went on. Peter and Valkenburg (2006) found that most adolescents considered online communication as less reciprocal, broad, or deep than face-to-face communication; similarly, Schiffrin et al. (2010) found that college students rated online communication as less suitable for forming close online relationships. Baym, Zhang, and Lin (2004) reported that online social interaction was seen as high in quality, but still lower than face-to-face interactions. Though more research is needed in this area, these studies suggest a general perception that online interactions are of lower quality than those conducted over other media.

These perceptions of the medium suggest that individuals typically feel it is difficult to form close emotional connections online, lending more credence to construing online social interaction as an ersatz social activity.

Why do people choose online social interaction?

Ersatz social engagement theory predicts that people choose ersatz social activities due to decreased risk, decreased effort, and the potential for immediate gratification.

Risk

The decreased risk of pursuing online social activities has been observed by many researchers. McKenna and Bargh (1999, 2000) identified that the anonymity of the Internet allows for greater self-disclosure, as it decreases the sense of risk. Interestingly, the rise in social networking sites such as Facebook may be increasing the psychological risk of online activities; Facebook, Twitter, and other social networking sites often use individuals' real names, so there is no longer a shield of anonymity.

Ease

McKenna and Bargh (2002) also suggest that the lack of gating features as mentioned in discussing emotional connection makes it generally easier to form and maintain relationships online. Additionally, it is easy to find similar others through forums or other websites. As mentioned in the previous sections, Internet interactions are also relatively low-effort activities; individuals do not need to schedule a specific time to meet friends in order to interact over Facebook, for example.

Instant gratification

Online social interaction, unlike some other ersatz social activities, is not an entirely reliable provider of instant gratification. Depending on the technological platform, one cannot simply "turn on" an online conversation in the way one can turn on a television show; the presence of another person is required, at least for some applications (e.g., instant messaging). However, it is easier to be present online in a wide variety of times and settings than it is in person. The availability of more instant forms of online social interaction is only increasing with time, with the spread of smartphones and online social apps such as Twitter and Facebook. Indeed, these applications allow one to read pre-posted messages at any time, even if true instant interaction is less feasible. Therefore, it is reasonable to say that online social interaction is far more likely to provide instant gratification compared to face-to-face social activity.

What are the consequences of choosing online social interaction?

This is perhaps the most contentious question in the study of online social interaction, as implied by the introduction to this chapter. Though the early research discussed Internet use as a whole, the pattern is similar for online social interaction: research has shown both positive and negative effects. Looking at consequences through the lens of ersatz social engagement theory, we might expect decreased social capital and social skills as online social interaction supplanted face-to-face social interaction. It is also necessary, in any discussion of consequences, to touch on the contentious issue of the

Internet's influence on well-being. Ersatz social engagement theory would imply that, while online social interaction might create initial boosts of happiness via the instant gratification it provides, long-term deficits in well-being would be likely due to the lack of real social bonds.

It is worth noting that not all online social interaction is created equal; the different affordances provided by different methods of online social interaction may lead to different consequences. For example, watching videos online may occupy the far end of the ersatz continuum, whereas speaking with voice or video chat over the Internet to a close friend or loved one may hardly differ from a real interaction (if indeed it does). For the purpose of this discussion, however, we will review relevant research that speaks to all forms of online social interaction.

Social skills

Whether online social interaction can lead to the development of social skills is a surprisingly under-researched question. This may be due to an observation noted earlier: little research has been done on the broader question of how friendship may develop social skills. Campbell, Cumming, and Hughes (2006) found that individuals who are high in social phobia believe that online social interaction helps them to develop social skills, although they also counterintuitively believe that Internet users are lonely and Internet use is addictive. Green and Brock (2008) found that individuals reported that ersatz activities were less effective at building civic skills such as public speaking. However, no research seems to exist to-date that addresses the question of whether or not online social interaction can build social skills. Furthermore, the answer to this question may depend on the type of online interaction; different effects may emerge if one is playing an online game as a particular character versus sharing information with close friends and family on Facebook versus talking with strangers on a discussion board.

Social capital

In ersatz social engagement theory, ersatz social activities are predicted to lead to decreases in social capital—does this hold true for online social interaction? Like so many questions raised when studying the Internet, the answer is unclear. Earlier research suggested that the Internet might indeed have deleterious effects on social capital, but a wealth of later findings have cast doubt on that summary.

Kraut's HomeNet study (1998) found decreases in the user's social circle that implied decreases in social capital. Looking specifically at social internet use, Shah, Kwak, and Holbert (2001) found that it decreased social capital in general, while Shklovski, Kraut, and Rainie (2004) found that it specifically decreased the likelihood of spending face-to-face time with close others without increasing the likelihood of contact through other media, such as phone calls. Shklovski, Kraut, and Rainie's study is notable in that it was one of the few longitudinal studies to tackle the question.

Even as early as 2001, however, contradictory results were being found—and they would soon outnumber the original perspective. Katz et al. (2001)'s general review discerned no real negative effects of Internet use on social capital, and several positive ones. Other studies went further, claiming that the Internet in fact increases civic participation and social capital (Hampton & Wellman, 2003; Stern & Dillman, 2006; Wellman, Haase, Witte, & Hampton, 2001). As the field of study matured, research looking more specifically at social use only reinforced this claim: solely online relationships lead to increased social capital (Best & Krueger, 2006); people who use the Internet socially have more ties than people who do not (Zhao, 2006); engaging in political discussion online increases civic participation (Shah, Cho, Eveland, & Kwak, 2005); and greater use of Facebook is linked to higher social capital (Steinfield, Ellison, & Lampe, 2008; Valenzuela, Park, & Kee, 2009).

Overall, it appears that online social interaction does not harm social capital, but can in fact boost it. This may be dependent on the specific types of social interaction—those that cement social ties, for example, as opposed to casual interactions with strangers that do not lead to lasting relationships.

Well-being

Ersatz social engagement theory implies that online social interaction might produce an initial sense of well-being through its instant gratification, but fail to lead to the deeper bonds that can bring lasting satisfaction. This contradiction would perhaps explain the conundrum in the early research on online social interaction that was discussed at the start of this chapter.

Looking at instant gratification, Green et al. (2005) found that communication with a stranger over instant messaging was more effective at boosting mood than conversation with a stranger face to face. This would seem to directly confirm ersatz social engagement theory's contention, at least for short-term effects.

On a longer timescale, it does seem that those who are particularly drawn to online social interaction are those most likely to suffer negative outcomes from it (Caplan, 2005). A wealth of research suggests that these individuals are often lonely (e.g., Bonetti, Campbell, & Gilmore, 2010; Ceyhan & Ceyhan, 2008; Kim, LaRose, & Peng, 2009; Morahan-Martin, 1999) and, more importantly, socially anxious—social anxiety may be the mediating variable in loneliness's tie to Internet use (Caplan, 2007). These heavy social Internet users seem to lose out on face-to-face social ties (Morahan-Martin, 1999; Weiser, 2001) due to displacing them with online activities. This would fit neatly with ersatz social engagement theory—socially anxious individuals may be particularly drawn to the decreased risk of ersatz social activities, but at the long-term cost of more meaningful relationships.

However, an alternative perspective has been advanced, primarily by Caplan (2002, 2003, 2005)—Internet use may not be harmful unless it is problematic internet use (PIU), which is driven by preference for online social interaction (POSI). In this cognitive-behavioral model of problematic Internet use (Caplan, 2005), individuals who perceive themselves as low in social skills are likely to prefer online social interaction. In turn, this preference leads to more compulsive Internet use, and compulsive Internet use is what leads to the negative effects on well-being seen in several studies. Caplan draws an important distinction between excessive and compulsive use—in his model, it is not the amount of time spent on the Internet but the way in which that time is approached.

If it is not amount of Internet use that determines consequences in terms of well-being, but the attitude and approach toward Internet use, we might well expect individual studies to provide different results. Studies that simply look at correlations between amount of Internet use and well-being will fail to provide a consistent picture because (a) they often combine social use and other uses of the Internet and (b) they cannot account for the distinction between compulsive and excessive use. Indeed, a meta-analysis of 40 studies examining correlations between well-being and Internet use found a negative but essentially zero (–0.03) correlation overall (Huang, 2010).

Studies also fail to take into account the issue of individuals with pre-existing problems being more drawn to online social interactions, a central part of the cognitive-behavioral model of PIU. One fascinating study on this topic (Modayil, Thompson, Varnhagen, & Wilson, 2003), examined a large population of Internet users for mental health difficulties with time of onset data. The difficulties of their participants had begun 5–22 years before the start of Internet use.

If online social interaction does not create deficits in well-being, then, can it be said to create improvements? Valkenburg and Peter (2007) suggest that it can. Their research, studying adolescent uses of the Internet, initially found a negative effect of Internet communication on well-being. However, including the tendency to talk with strangers and their closeness to friends in their analysis created a model where time spent talking to friends online increased well-being through increasing closeness, whereas time spent talking to strangers decreased well-being. Again, however, their data showed more deleterious effects of Internet use on the lonely.

Overall, the conclusion seems to be that while ersatz social engagement theory is supported in short-term extra benefits to mood, long-term negative effects do not seem to be universally observed. Only individuals who engage in certain forms of online social interaction (like primarily with strangers) or who have an actual preference for online social engagement may show decrements in well-being.

Online social interaction as an ersatz social activity

Reviewing the data, we have found mixed evidence to support online social interaction as an ersatz social activity along the lines of ersatz social engagement theory. Though it does boast many of the hallmarks of ersatz social activities, the consequences of engaging in online social interaction do not seem to be as dire as ersatz social engagement theory would imply. Indeed, online social interaction seems to have the potential to build social capital and increase well-being, contradictory to predictions.

What could explain these deviations? An implication of the cognitive-behavioral model of PIU may lie at the heart of the issue. PIU is a phenomenon of attitudes and approaches to Internet use, rather than necessarily a specific kind of use. To understand the consequences of online social interaction, we cannot look simply at what people do—we must look at why and how they do it. These individual factors may be the primary drivers of the consequences of online social interaction.

Individual Factors in the Consequences of Online Social Interaction

In 2000, McKenna and Bargh observed that "there is no simple main effect of the Internet on the average person" (p. 59). This quote was prescient; despite its age, it remains appropriate and instructive today. The most promising lines of future research in studying the consequences of online social interaction focus on individual factors and their impact. These individual factors may include behaviors, perceptions, or even more general attitudes.

Behavior and consequences of online social interaction

Enough evidence exists to make a compelling claim that the specific behaviors engaged in during an online social interaction may be a powerful determinant of its consequences. Valkenburg, Peter, and Schouten (2006) found that individuals who use Facebook frequently are more likely to get positive feedback, which increases their well-being. This chapter has also already touched on another result of theirs (Valkenburg & Peter, 2007) that suggests the choice of interaction partner (stranger or friend) can determine whether online social interaction is beneficial or detrimental. In fact, many of the results we have already discussed in this chapter touch on this idea. Online social interaction is a category of possible activities, not a single omnibus activity; activities such as political discussion can build social capital, whereas anonymous talk in a chat room may have no positive effects.

Perception and consequences of online social interaction

Our perceptions and attitudes about online social interaction likely drive many of the choices we make in dealing with the Internet, not only whether or not to interact in the first place, what kind of interaction we pursue, but also our behavioral choices during an interaction.

For example, Carpenter, Green, and LaFlam (2011) have found that certain healthy uses of Facebook (direct interaction and use that was supplemental to face-to-face social interactions) are linked to approach motivations, whereas more unhealthy uses (Facebook-only relationships and romantic relationship-seeking) were tied to avoidance motivation. It is also illustrative to consider the behavioral-cognitive model of PIU once again. Individuals' perceptions that they can be more socially "successful" on the Internet than in face-to-face interactions lead directly to patterns of problematic usage.

Ersatz social engagement theory, as originally proposed, does not account for the role of differing perceptions and attitudes toward online social interaction. Despite a general opinion that online interactions are inferior, some individuals do disagree (Peter & Valkenburg, 2006)—there is room for variance on a continuum of attitudes about online social interaction.

Indeed, examining online social interaction as an ersatz social activity suggested that online interaction has elements of both real and ersatz engagement. Our current research examines a question implied by that ambiguity: does the extent to which an individual perceives an activity as ersatz determine its consequences? That is, do individuals who perceive online interaction as "an inferior substitute" to face-to-face interaction act in ways that might confirm this belief?

The Perceived Reality of Online Social Interactions

We are defining this concept as "the perceived reality of online social interactions," or perceived reality for short. Perceived reality refers to the extent to which an individual believes that computer-mediated communication is suitable for the maintenance and formation of close relationships. This idea should be differentiated from perceived realism, as this concept does not revolve around the reality of a mediated experience; instead, it speaks to the lay terminology of online social interactions as "not real interactions." Individuals high in perceived reality of online interactions believe that true interpersonal bonds can be formed online, and that the Internet is a valid way of forming and maintaining friendships and other relationships. In contrast, individuals low in perceived reality do not accept that true connections can happen in a computer-mediated context; to these individuals, computer-mediated communication will always be substantially inferior to face-to-face interaction.

Indeed, to understand what perceived reality is, we must explain what it is not. Despite surface similarities, perceived reality represents a different construct than Caplan's preference for online social interaction. First and foremost, we do not suggest that high perceived reality represents preference. Rather, individuals who are high in perceived reality consider online or computer-mediated relationships to be "real" and potentially valuable aspects of their social network. However, the "endpoints" of the continuum of perceived reality have yet to be empirically determined; individuals who are near the highest end of the scale in perceived reality may still consider computer-mediated communication inferior to face to face contact. (The existence of preference for online social interaction may imply otherwise, but it is an empirical question as to how prevalent such a preference may be.) Second, we do not believe that perceived reality necessarily leads to negative consequences or PIU. Indeed, high perceived reality may be beneficial because individuals who believe in the value of their online interactions may gain more psychological and social support benefits from these interactions. It seems feasible to use online social interaction in a healthy way that supplements one's face-to-face social life without having to devalue it as far less meaningful.

Perceived reality should also not be seen as need for affiliation or similar constructs that generally speak to someone's attitude toward friendship in general. Perceived reality is expected to only influence the extent to which someone values online social interactions (and by extension online relationships), not relationships that occur offline. It is a judgment of the medium specifically.

Theoretical implications of perceived reality

If perceived reality does not reflect either preference for Internet use, PIU, or general desire for friendship, what theoretical role does it play? How does it interface with existing research? What new value does it bring to discussions of online social interaction?

Considering perceived reality as an overall attitude toward online social interaction, there are several beliefs that likely go hand-in-hand with higher levels of it. Looking back at our previous discussion of online social activity as ersatz, we would expect that individuals high in perceived reality would endorse Walther's Social Information Processing Theory—that medium-unique communicative cues accessible in computer-mediated communication allow individuals to still engage in meaningful discussion despite the lack of non-verbal cues. Following that line of thought, we would also expect them to endorse a belief that online social interaction is capable of creating true emotional connection. Both of these beliefs would seem to greatly reduce the extent to which online social interaction was seen as ersatz.

What would be the consequences of these beliefs? If one believes that online social interactions can provide many of the same benefits as face-to-face interaction, while still requiring less effort, posing less risk and offering instant gratification, it becomes a far more appealing option than if one discounts these benefits. Therefore, individuals high in perceived reality should be likely to pursue online relationships more assiduously than individuals who are low. Also, by choosing to adopt strategies that encourage lasting friendships (friendship-forming strategies; Baxter & Philpott, 1982; Wentzel & Erdley, 1993), individuals high in perceived reality are likely to actually form deeper friendships online and thus reap more of the usual benefits—social support, social skills, and so on. In a way, then, perceived reality can be viewed as a self-fulfilling prophecy: believe that your online ties are more real and less ersatz, and the consequences of your online social interaction will more closely resemble a real interaction.

Of course, an exception may need to be made for individuals who truly believe that online relationships are either identical to or even superior to face-to-face relationships, and perhaps therefore forgo face-to-face relationships in favor of online interaction. These individuals are high in markers such as social anxiety and loneliness (Peter & Valkenburg, 2006), and may represent Caplan's troubled population whose preference for the Internet leads to PIU. A curvilinear relationship is implied by this model, where a certain high but not maximal level of perceived reality may lead to the most beneficial outcomes.

It is also worth noting that perceived reality may have different implications for different portions of the population. Much of the research on online social interactions has been done with college undergraduates, a common issue in research; however, college undergraduates are uniquely privileged when it comes to making friends. They are in close proximity to many individuals with shared interests and similar backgrounds. Therefore, if younger individuals are valuing online relationships highly, it may be a sign that they are part of that troubled population—they are choosing online relationships when face-to-face relationships are more readily available. However, older individuals may be at a stage in their life when face-to-face relationships are not so accessible. It may only be beneficial for older individuals to value online social interactions, as they are not supplanting or replacing face-to-face interactions but instead satisfying a need that would otherwise go unmet.

Overall, perceived reality offers an explanation for varying results about the impact of online social interaction on well-being. Like preference for online social interaction, it claims that the amount of use is not the important factor, but the approach and attitudes toward use; unlike preference for online social interaction, it does not account only for results in a troubled subsection of the population, but has potential applications to any Internet user.

Evidence for perceived reality

Empirical investigations of perceived reality remain in the first stages of development, but preliminary work on the concept has been promising.

An initial examination of perceived reality and several potentially related variables (Clark & Green, 2012) produced mixed results. This study examined the effect of experimentally manipulated perceived reality on perceived social support in online and offline relationships, as well as correlations with personality traits, social anxiety, loneliness, general attitudes toward the Internet, hours spent online, and number of online friends. Both a college undergraduate population and an Amazon Mechanical Turk population (an online sample of adults who participate in studies or do other online tasks in exchange for small payments) were included in analysis, in order to observe if perceived reality could play a different role in different age groups.

For this research, a five-item Perceived Reality Scale was generated. Initial results showed a low Cronbach's alpha for the whole scale, but when one problematic item was excluded, alphas of 0.79 and 0.83 were observed in the two samples. The questionnaire asks individuals to answer the following items on a scale of 1–7, ranging from strongly disagree to strongly agree:

1. I feel that online relationships can be as meaningful as relationships conducted face-to-face.
2. I would be willing to share my deepest hopes and fears in an online relationship.
3. I feel that online relationships can be an important part of someone's social network.
4. I don't think that online communication provides the kind of interactions people need to build close relationships. (reverse-scored)

Perceived reality proved difficult to manipulate in a student population, but when treated as a pre-existing variable, served as the only significant predictor of online social support in a multiple regression analysis. It was also a significant predictor in the Mechanical Turk population, along with loneliness and general Internet attitudes. Individuals in the Mechanical Turk population, which overall had higher levels of perceived reality, also reported on average twice as many purely online relationships as their student counterparts.

Current research in perceived reality

We are currently attempting to develop manipulations for perceived reality. Intriguingly, this variable has proved easier to manipulate in a sample of Amazon Mechanical Turk participants than in college undergraduates; additionally, Mechanical

Turkers were higher in perceived reality. Multiple attempts have been successful in Mechanical Turkers only to later fail in students, such as vignettes about supportive online relationships and generation of reasons to value online interactions. Current attempts involve a combination of reports of faux-scientific research with generation of statements that agree with the research's findings.

In substantive work, our next step will be to examine links between perceived reality and actual behavior in online social interaction. If individuals who are low in perceived reality devalue online interaction, they should be less likely to perform certain behaviors that are difficult or require emotional investment (such as provision of social support or self-disclosure) in an online context. Additionally, perceived reality's effects may manifest in different and specific uses depending on the relevant technology; individuals low in perceived reality may eschew more private and personal channels, such as instant messaging, in favor of impersonal and public communications. The effects of perceived reality may even appear in word choice, use of emoticons, or other aspects of the communication itself, as individuals who do not value online interaction take pre-emptive steps to avoid emotional investment in CMC contexts.

Future theoretical directions in perceived reality

In order to establish perceived reality as a valuable construct in the study of online social interaction, further empirical support is needed for its predictive value. Ideally, perceived reality can explain: (a) behavioral choices in online social interactions; (b) varying desires to pursue online social interaction; and (c) varying consequences of online social interactions.

However, for perceived reality to become a more comprehensive theory, there are also several potential problems and ambiguities that must be clarified. For example, the theory as written speaks to the perceived reality of online social interactions—but often conflates the interactions that comprise a relationship and that relationship itself. For example, our initial work suggested that raising an individual's level of perceived reality should also improve their perception of their online relationships, thus resulting in greater feelings of social support from the relationship. But is it valid to assume that one's opinions about the interactions can be generalized to an opinion about the larger relationship? What can perceived reality say about relationships that are comprised of both online and face-to-face interaction? These questions await further research to be answered.

Furthermore, the role of technological features in perceived reality remains to be explored. For example, it is possible that if individuals engage in online communication primarily with tools that allow a rich sensory experience (e.g., video chat), they may develop a higher sense of perceived reality. Similarly, although we are currently treating perceived reality as a blanket judgment about online communication as a whole, it is possible that individuals have a range of perceived reality beliefs depending on the specific features of the medium (e.g., Twitter vs. email vs. Facetime).

Should perceived reality prove a valuable addition to the literature on online social interaction, it may also have useful implications for real-world policy and health recommendations. Older individuals with mobility or isolation concerns could potentially benefit from learning to value online social interactions, allowing them to fulfill their needs for belonging. For the rest of the population, healthy and unhealthy ways to engage in online social interaction could be disseminated to the public: using them as supplements to a full face-to-face social life, while still appreciating the genuine benefits they have to offer.

Conclusion

Opinions on the consequences of online social interaction, both scholarly and lay, have varied tremendously over the years. There is not yet a solid scientific consensus on the question, perhaps because much of the research that has considered the question has considered it on a gross level.

Ersatz social engagement theory suggests that the degree to which an activity is real or ersatz is what determines its outcomes. While it is likely true that many social activities vary in consequence based on immutable factors built-in to the experience, our review of the literature concerning online social interaction suggests that the question is not that simple. If we truly wish to understand the issue, we must look deeper—not at the mere question of whether or not someone engages in an activity, but why, how, and with whom.

Perceived reality theory hopes to offer a nuanced view on the outcomes of online social interaction by taking all of those questions into account. Why—to form lasting friendships, or simply to pass the time? How—in ways that differ noticeably from face-to-face communication due to devaluing the medium, or in similar and equally beneficial ways? With whom—with strangers or acquaintances, or with individuals one can genuinely call a friend?

By viewing the issue with these important questions in mind, we hope to truly understand what choices an individual can make to drive the consequences of online social interaction. The Internet is a valuable technology with great promise to connect the world in ways previously unfathomable; research can serve to guide this promise to full fruition in a way that is of the most benefit to all.

References

Asher, S. R., Parker, J. G., & Walker, D. L. (1996). Distinguishing friendship from acceptance: Implications for intervention and assessment. In W. M. Bukowski, A. F. Newcomb, & W. W. Hartup (Eds.), *The company they keep: Friendship in childhood and adolescence* (pp. 366–405). New York, NY: Cambridge University Press.

Bargh, J. A., McKenna, K. Y. A., & Fitzsimons, G. M. (2002), Can you see the real me? Activation and expression of the "true self" on the Internet. *Journal of Social Issues, 58*, 33–48.

Baumeister, R. F., & Leary, M. R. (1995). The need to belong: Desire for interpersonal attachment as a fundamental human motivation. *Psychological Bulletin, 117*, 497–529.

Baxter, L. A., & Philpott, J. (1982). Attribution-based strategies for initiating and terminating relationships. *Communication Quarterly, 30*, 217–224.

Baym, N. K., Zhang, Y. B., & Lin, M. (2004). Social interactions across media: Interpersonal communication on the Internet, telephone, and face-to-face. *New Media and Society, 6*(3), 299–318.

Bessière, K., Kiesler, S., Kraut, R., & Boneva, B. (2008). Effects of Internet use and social resources on changes in depression. *Information Communication and Society, 11*(1), 47–70.

Best, S. J., & Krueger, B. S. (2006). Online interactions and social capital: Distinguishing between new and existing ties. *Social Science Computer Review, 24*, 395–410.

Bonetti, L., Campbell, M. A., & Gilmore, L. (2010). The relationship of loneliness and social anxiety with children's and adolescents' online communication. *Cyberpsychology, Behavior and Social Networking, 13*(3), 279–285.

Campbell, A. J., Cumming, S. R., & Hughes, I. (2006). Internet use by the socially fearful: Addiction or therapy? *Cyberpsychology and Behavior, 9*(1), 69–81.

Caplan, S. E. (2002). Problematic Internet use and psychosocial well-being: Development of a theory-based cognitive-behavioral measure. *Computers in Human Behavior, 18*, 533–575.

Caplan, S. E. (2003). Preference for online social interaction: A theory of problematic Internet use and psychosocial well-being. *Communication Research, 30*, 625–648.

Caplan, S. E. (2005). A social skill account of problematic Internet use. *Journal of Communication, 55*, 721–736.

Caplan, S. E. (2007). Relations among loneliness, social anxiety, and problematic Internet use. *Cyberpsychology and Behavior, 10*(2), 234–242.

Carpenter, J. M., Green, M. C., & LaFlam, J. (2011). People or profiles: individual differences in online social networking use. *Personality and Individual Differences, 50*, 538–541.

Ceyhan, A. A., & Ceyhan, E. (2008). Loneliness, depression and computer self-efficacy as predictors of problematic Internet use. *Cyberpsychology and Behavior, 11*(6), 699–701.

Chan, D. K. S., & Cheng, G. H. L. (2004). A comparison of offline and online friendship qualities at different stages of relationship development. *Journal of Social and Personal Relationships, 21*(3), 305–320.

Chen, Y., & Peng, S. S. (2008). University students' Internet use and its relationships with academic performance, interpersonal relationships, psychosocial adjustment and self-evaluation. *Cyberpsychology and Behavior, 11*(4), 467–469.

Clark, J. L., & Green, M. C. (2012, March). *An exploratory analysis of "perceived reality"*. Poster presented at Social Psychology at UNC and Duke conference, Durham, NC.

Culnan, M. J., & Markus, M. L. (1987). Information technologies. In: F. M. Jablin, L. L. Putnam, K. H. Roberts, & L. W. Porter (Eds.), *Handbook of organizational communication: an interdisciplinary perspective* (pp. 420–443). Newbury Park, CA: Sage.

Daft, R. L., & Lengel, R. H. (1984). Information richness: A new approach to managerial behavior and organization design. *Research in Organizational Behavior, 6*, 191–233.

Daft, R. L., & Lengel, R. H. (1986). Organizational information requirements, media richness and structural design. *Management Science, 32*, 554–571.

Fiske, S. J. (2004). *Social beings: A core motives approach to social psychology*. New York, NY: Wiley.

Gallace, A., & Spense, C. (2010). The science of interpersonal touch: An overview. *Neuroscience and Biobehavioral Reviews, 34*, 246–259.

Glick, G. C., & Rose, A. J. (2011). Prospective associations between friendship adjustment and social strategies: Friendship as a context for building social skills. *Developmental Psychology, 47*(4), 1117–1132.

Green, M. C., & Brock, T. C. (1998). Trust, mood, and the outcomes of friendship determine preferences for real versus ersatz social capital. *Political Psychology*, *19*, 527–544.

Green, M. C., & Brock, T. (2008). Antecedents and consequences of choosing real versus ersatz social activities. *Media Psychology*, *11*(4), 566–592.

Green, M. C., Hilken, J., Friedman, H., Grossman, K., Gasiewski, J., Adler, R., & Sabini, J. (2005). Communication via instant messenger: Short-term and long-term effects. *Journal of Applied Social Psychology*, *35*(3), 445–462.

Hampton, K., & Wellman, B. (2003). Neighboring in Netville: How the Internet supports community and social capital in a wired suburb. *City & Community*, *2*(4), 277–311.

Herteinstein, M. J., Holmes, R., McCullough, M., & Keltner, D. (2009). The communication of emotion via touch. *Emotion*, *9*(4), 566–573.

Horton, D., & Wohl, R. R. (1956). Mass communication and para-social interaction: Observations on intimacy at a distance. *Psychiatry*, *19*, 215–229.

Huang, C. (2010). Internet use and psychological well-being: A meta-analysis. *Cyberpsychology, Behavior and Social Networking*, *13*(3), 241–249.

Katz, J. E., Rice, R. E., & Aspden, P. (2001). The Internet 1995–2000: Access, civic involvement and social interaction. *American Behavioral Scientist*, *45*(3), 405–419.

Kim, J., LaRose, R., & Peng, W. (2009). Loneliness as the cause and effect of problematic Internet use: The relationship between Internet use and psychological well-being. *Cyberpsychology and Behavior*, *12*(4), 451–455.

Kraut, R., Kiesler, S., Boneva, B., Cummings, J., Helgeson, V., & Crawford, A. (2002). Internet paradox revisited. *Journal of Social Issues*, *58*(1), 49–74.

Kraut, R., Patterson, M., Lundmark, V., Kiesler, S., Mukopadhyay, T., & Scherlis, W. (1998). Internet paradox: A social technology that reduces social involvement and psychological wellbeing? *American Psychologist*, *53*, 1011–1031.

McKenna, K. Y. A., & Bargh, J. A. (1999). Causes and consequences of social interaction on the Internet: A conceptual framework. *Media Psychology*, *1*, 249–269.

McKenna, K. Y. A., & Bargh, J. A. (2000). Plan 9 from cyberspace: The implications of the Internet for personality and social psychology. *Personality and Social Psychology Review*, *4*, 57–75.

McKenna, K. Y. A, Green, A. S., & Gleason, M. E. J. (2002). Relationship formation on the Internet: What's the big attraction? *Journal of Social Issues*, *58*(1), 9–31.

Modayil, M. V., Thompson, A. H., Varnhagen, S., & Wilson, D. R. (2003). Internet users' prior psychological and social difficulties. *Cyberpsychology and Behavior*, *6*(6), 585–590.

Morahan-Martin, J. (1999). The relationship between loneliness and Internet use and abuse. *Cyberpsychology and Behavior*, *2*(5), 431–439.

Morgan, C., & Cotten, S. R. (2003). The relationship between Internet activities and depressive symptoms in a sample of college freshmen. *Cyberpsychology and Behavior*, *6*, 133–142.

Nie, N. H. (2001). Sociability, interpersonal relations, and the Internet: Reconciling conflicting findings. *American Behavioral Scientist*, *45*(3), 420–435.

Nie, N. H., & Erbring, L. (2000, June). *Internet and society: A preliminary report.* Stanford, CA: Stanford Institute for the Quantitative Study of Society. Retrieved from http://www.stanford.edu/group/siqss/Press_Release/Preliminary_Report.pdf.

Parks, M. R., & Floyd, K. (1996). Making friends in cyberspace. *Journal of Communication*, *46*, 80–97.

Peter, J., & Valkenburg, P. M. (2006). Research note: Individual differences in perceptions of Internet communication. *European Journal of Communication*, *21*(2), 213–226.

Putnam, R. D. (1995). Bowling alone: America's declining social capital. *Journal of Democracy*, *6*, 65–78.

Putnam, R. D. (2000). *Bowling alone*. New York, NY: Simon & Schuster.

Putnam, R. D. (2001). Social capital: Measurement and consequences. *Canadian Journal of Policy Research*, *2*, 41–51.

Rice, R. E. (1993). Media appropriateness: Using social presence theory to compare traditional and new organizational media. *Human Communication Research*, *19*, 451–484.

Rose, A. J., & Asher, S. R. (1999). Children's goals and strategies in response to conflicts within a friendship. *Developmental Psychology*, *35*(1), 69–79.

Rose, A. J., & Asher, S. R. (2004). Children's strategies and goals in response to help-giving and help-seeking tasks within a friendship. *Child Development*, *75*, 749–763.

Rubin, A. M., Perse, E. M., & Powell, R. A. (1985). Loneliness, parasocial interaction, and local television news viewing. *Human Communication Research*, *12*, 155–180.

Schiffrin, H., Edelman, A., Falkenstern, M., & Stewart, C. (2010). The associations among computer-mediated communication, relationships, and well-being. *Cyberpsychology, Behavior and Social Networking*, *13*(3), 299–306.

Shah, D. V., Cho, J., Eveland, W. P., Jr., & Kwak, N. (2005). Information and expression in a digital age: Modeling internet effects on civic participation. *Communication Research*, *32*(5), 1–35.

Shah, D. V., Kwak, N., & Holbert, R. L. (2001). "Connecting" and "disconnecting" with civic life: Patterns of Internet use and the production of social capital. *Political Communication*, *18*(2), 141–162.

Shaw, L. H., & Gant, L. M. (2002). In defense of the Internet: The relationship between Internet communication and depression, loneliness, self-esteem, and perceived social support. *Cyberpsychology and Behavior*, *5*(2), 157–171.

Shklovski, I., Kraut, R., & Rainie, L. (2004). The internet and social participation: Contrasting cross-sectional and longitudinal analyses. *Journal of Computer-Mediated Communication*, *10*(1), 00–00.

Short, J., Williams, E., & Christie, B. (1976). *The social psychology of telecommunication*. London, UK: Wiley.

Sproull, L., & Kiesler, S. (1985). Reducing social context cues: electronic mail in organizational communication. *Management Science*, *11*, 1492–1512.

Steinfield, C., Ellison, N. B., & Lampe, C. (2008). Social capital, self-esteem, and use of online social network sites: A longitudinal analysis. *Journal of Applied Developmental Psychology*, *29*, 434–445.

Stern, M. J., & Dillman, D. A. (2006). Community participation, social ties, and use of the Internet. *City and Community*, *5*(4), 409–424.

Tidwell, L. C., & Walther, J. B. (2002). Computer-mediated communication effects on disclosure, impressions and interpersonal evaluations: Getting to know one another a bit at a time. *Human Communication Research*, *28*(3), 317–348.

Trevino, L. K., Lengel, R. H., & Daft, R. L. (1987). Media symbolism, media richness and media choice in organizations: A symbolic interactionist perspective. *Communication Research*, *14*, 553–575.

Valenzuela, S., Park, N., & Kee, K. F. (2009). Is there social capital in a social network site? Facebook use and college students' life satisfaction, trust, and participation. *Journal of Computer-Mediated Communication*, *14*, 875–901.

Valkenburg, P. M., & Peter, J. (2007). Online communication and adolescent wellbeing: Testing the stimulation versus the displacement hypothesis. *Journal of Computer-Mediated Communication*, *12*(4), 1169–1182.

Valkenburg, P. M., & Peter, J. (2009). Social consequences of the Internet for adolescents: A decade of research. *Current Directions in Psychological Science*, *18*(1), 1–5.

Valkenburg, P. M., Peter, J., & Schouten, A. P. (2006). Friend networking sites and their relationship to adolescents' well-being and social self-esteem. *Cyberpsychology and Behavior*, *9*(5), 584–590.

Walther, J. B. (1992). Interpersonal effects in computer-mediated interaction: A relational perspective. *Communication Research*, *19*, 52–90.

Walther, J. B. (1993). Impression development in computer-mediated interaction. *Western Journal of Communication*, *57*, 381–398.

Walther, J. B. (1994). Anticipated ongoing interaction versus channel effects on relational communication in computer-mediated interaction. *Human Communication Research*, *20*, 473–501.

Walther, J. B. (1996). Computer-mediated communication: Impersonal, interpersonal, and hyperpersonal interaction. *Communication Research*, *23*, 1–43.

Walther, J. B., & D'Addario, K. P. (2001). The impacts of emoticons on message interpretation in computer-mediated communication. *Social Science Computer Review*, *19*, 323–345.

Walther, J. B., Loh, T., & Granka, L. (2005). Let me count the ways: The interchange of verbal and nonverbal cues in computer-mediated and face-to-face affinity. *Journal of Language and Social Psychology*, *24*(1), 36–65.

Walther, J. B., & Tidwell, L. C. (1995). Nonverbal cues in computer-mediated communication, and the effect of chronemics on relational communication. *Journal of Organizational Computing*, *5*, 355–378.

Wastlund, E., Norlander, T., & Archer, T. (2001). Internet blues revisited: Replication and extension of an Internet paradox study. *Cyberpsychology and Behavior*, *4*(3), 385–391.

Weiser, E. (2001). The functions of Internet use and their social and psychological consequences. *Cyberpsychology and Behavior*, *4*(6), 723–743.

Wellman, B., Haase, A. Q., Witte, J., & Hampton, K. (2001). Does the Internet decrease, increase, or supplement social capital? *American Behavioral Scientist*, *45*(3), 436–455.

Wentzel, K. R., & Erdley, C. A. (1993). Strategies for making friends: Relations to social behavior and peer acceptance in early adolescence. *Developmental Psychology*, *29*, 819–826.

Zhao, S. (2006). Do Internet users have more social ties? A call for differentiated analyses of Internet use. *Journal of Computer-Mediated Communication*, *11*, 844–862.

12

Deception with Technology

Jeffrey T. Hancock and Jamie Guillory
Department of Communication, Cornell University, Ithaca, NY, USA

Butlers, Sock Puppets, and the Chinese Water Army

When it comes to deception, there is never a shortage of examples. And with the widespread adoption of technology in human communication, it can sometimes seem that deception is more rampant than ever. We have seen famous football players tricked into believing they are dating a woman that does not exist. Recent work on deception and technology has revealed *butler lies*, in which people use technology to create a social buffer and avoid or put off communication, such as the ubiquitous "on my way" when in fact the person has not yet left their house. There are also *sock puppets*, in which glowingly positive book reviews are in fact written by their high-profile authors; or, more generally, fake reviews for hotels, restaurant, and any other business online (Ott, Cardie, & Hancock, 2012). And more worrisome yet are large-scale deception phenomena, such as the so-called *Chinese Water Army*, in which individuals are paid by governments or corporate organizations to spread propaganda in an organic, grassroots way. Here in North America this is often referred to as *astroturfing* (for review, see Hancock, 2012).

At the same time, there is a steady drumbeat of scandals in which deceptions are revealed by digital technology. When Mitt Romney told a group of donors at a closed-door meeting about 47% of the nation being dependent on the government, it was a digital video recorder that captured the moment and allowed it to be shared worldwide. In fact, almost any major political scandal of the last decade has involved some digital record, from email to a text message. The ways in which lying takes place, and how it is detected, are being transformed by technology along with all the other aspects of human communication. But, as we shall see, our assumptions about how technology and deception affect one another are not always as simple as they seem, and we propose here that there are several key factors that need to be considered to

The Handbook of the Psychology of Communication Technology, First Edition. Edited by S. Shyam Sundar.

understand deception in the networked age, including biases and heuristics unique to deception, the transformation of the interaction space, and our own views of ourselves as good, honest people.

The Cues Heuristic and Implications for Deception Online

One of the most basic questions we can ask about deception in the networked age is whether communication technology changes how often people lie. The answer for most people is obvious: technology leads to more deception (Keyes, 2004). The basic reasoning for this answer, which we call the cues heuristic, goes like this: there are fewer cues online so people lie more because they can get away it.

Although this seems like a fairly straightforward argument, there are three important assumptions in this heuristic. The first assumption is that there are fewer cues online than face to face (FtF). Email has fewer cues than the phone, for example, because there are no vocal cues, while the phone has fewer cues than FtF because there are no nonverbal cues, and so on. The second is that there are reliable cues for detecting deception. The belief that liars give off cues that can indicate their deception is near universal. It follows, therefore, that because technologically mediated communication has fewer cues than FtF interaction it should be more difficult to detect deception. So, email lies should be more difficult to detect than phone lies, which in turn should be harder than Skype-based lies. The last assumption is that when people can get away with it, they will lie more. In the case of lies online, because there are fewer cues that make it harder to detect deception, people will take advantage of this fact and lie more.

The belief in this cues heuristic is wide spread, as we discovered in a survey of the population on their beliefs about deception behavior (Toma, Jiang, & Hancock, 2013). To examine beliefs about deception online, we first asked people how much they lie in different media, including FtF, the phone, messaging, and email. We later asked how much they thought that others lie in these media. While people believe that others always lie more than they themselves do, a typical self-other asymmetry, the difference between self and other was larger the fewer cues there were. That is, the asymmetry was bigger in online media like email and messaging. In a second study, we replicated this pattern: the asymmetry between self and other for deception increased for online media (e.g., messaging and email) compared to richer media such as the phone or FtF. The reason for the intensified asymmetry was the different motivations people ascribed to their own lies versus others' lies. Consistent with self-other asymmetries in all types of social psychological judgments (e.g., Pronin, 2006), people ascribe altruistic motivations for their own deception using online media (e.g., "I lie to protect myself online"), but selfish motivations for others' deceptions online (e.g., "Others lie to take advantage of the lack of cues").

Unfortunately, each of the assumptions underlying the cues heuristic is problematic. The first assumption, that online communication has fewer cues, appears accurate on its face. After all, text-based communication, which makes up the majority of online communication, has fewer vocal and nonverbal cues than FtF. But this assumption ignores two of the most important features of online communication for deception: recordability and social networks. As we will discuss later in the chapter, online communication leaves digital traces that can be easily copied, backed up, shared, searched, and analyzed. This property of online communication has crucial implications for deception, both how it is practiced and how it is detected. In general, leaving a record of a lie is a bad idea. Networks also have important implications for deception. Networks constrain what a person can say because others that have ground truth information may encounter the message (Warkentin, Woodworth, Hancock, & Cormier, 2010). Consider, for example, the person that says they have a computer science degree on linkedin.com and who has their college advisor in their network. If in fact the claim is not true, the claim may be called into question by the advisor. This possibility should constrain deceptions in networked environments.

The second assumption, that cues are useful for detecting deception, is based on several myths surrounding deception and does not take into account recent meta-analyses of several decades of deception detection research. Bond and DePaulo (2006) found that over many studies involving deception detection, in which participants make judgments about the veracity of messages (both verbal and nonverbal), accuracy rates hover around 54%, statistically higher than chance but practically no better than flipping a coin. Why are accuracy rates so low? The primary reason is that there are, in fact, very few reliable cues to deception that can be used by people to detect deception. This view is supported by another DePaulo meta-analysis (DePaulo et al., 2003) that examined cues to deception across hundreds of studies. The data revealed that very few nonverbal cues are reliable signals, and those that are, such as vocal pitch and pupil dilation, are difficult for humans to detect given their subtlety.

These empirical findings stand in stark contrast to most people's beliefs about the cues to deception. For instance, a study that examined beliefs about cues to deception in countries around the world found that one of the most common cues cited as important for detecting deception was eye gaze. Participants in almost every country reported that eyes are the window to the truth, and that liars have shifty eyes. People also list many other nonverbal cues as important across the world, including fidgeting and other body movements. In fact, eye gaze is not one of the reliable cues identified in the meta-analysis, nor is fidgeting (DePaulo et al., 2003). Other beliefs about deception that do not match empirical findings include microexpressions, brief facial expressions that reveal suppressed emotions (Ekman, 1985). While most people believe that microexpressions indicate deception, recent work has shown that there are no peer-reviewed publications presenting evidence that this is the case (Porter & ten Brinke, 2008).

Many of these myths about deception detection are portrayed regularly in the popular media. In an insightful study illuminating the problems with these deception myths, participants were asked how much they watched the popular deception detection TV show *Lie To Me* and then completed a deception detection task (Levine, Serota, & Shulman, 2010). The results indicated that those who watched the show performed significantly worse at detecting deception than people who did not watch the show.

Unfortunately, these misplaced beliefs in the cues to deception, namely that there are reliable nonverbal cues, suggest to people that it is more difficult to detect deception online. In fact, there is very little evidence that this is the case. When we experimentally manipulated deception detection across Internet chat and FtF, we saw no difference in lie detection accuracy unless the liar was highly motivated (Hancock, Woodworth, & Goorha, 2010).

The third assumption inherent in the cues heuristic is that people will lie more if they can get away with it. This assumption is also deeply flawed, it turns out. Recent work suggests that people usually only lie when they have a reason (Levine, Kim, & Hamel, 2010). Although there is a small percentage of the population that lie regularly for no apparent reason (e.g., pathological liars, psychopaths), most people require some motivation to lie. Other work from behavioral economics confirms this view (Ariely, 2012), with many studies showing that people may cheat a little when given the opportunity, but that large deceptions are rare even when it is impossible to be caught.

Why do people only lie for a reason, even when they are unlikely to get caught? One reason is that lying is difficult. As Sir Walter Scott noted, "Oh what a tangled web we weave when at first we seek to deceive." Maintaining a fabricated story is extremely difficult, and this remains true online, as the hoaxer in the Manti T'eo case learned. Another reason is that people have a strong aversion to being considered a liar (Anderson, 1968). When people lie or cheat too much, it can trigger a shift in self-identity from good, honest person to liar or cheat. Ariely (2012) and colleagues' work suggest that self-identity as an honest person is a strong constraint that prevents most people from lying too much, even when they can get away with it. So, even if someone could get away with a lie more easily online, it is not necessarily the case that they would lie more simply because they could.

The cues heuristic, though a powerful and intuitive one, is based on faulty assumptions and clouds our judgments about how deception interacts with technology. In The Transformation of the Interaction Space we examine a more fruitful approach to deception and technology that takes a more holistic look at deception and communication technology.

The Transformation of the Interaction Space

A more useful way to approach deception and technology than focusing simply on cues is to consider the broader transformation that technology has wrought on human interaction. Consider that humans have been using spoken language to talk with one another for at least 50,000 years, although probably much longer (Dediu & Levinson, 2013). During this period there have been at least 100 billion humans, a species evolving over thousands of generations by using language to communicate FtF. Writing emerged much more recently, approximately five to six thousand years ago. For all of the humans that lived before the invention of writing, the only way they could communicate was if the other person was in the same physical location speaking at the same time the other person was listening. That is, before the invention of writing interactants needed to be co-present and co-temporal. One consequence of this is that for the 80 billion or more humans who lived before the advent of writing, not one of their words left a record. The evanescence of spoken language means that each word that they spoke disappeared as they said it. In fact, this is the case for most people throughout human history. Even after the invention of the printing press, most of the world's population was illiterate. Indeed, it was not until after World War II that the world literacy rate exceeded 50%.

When we step back to consider this history of interpersonal human communication, it puts into perspective the radical changes that communication technology have brought to the human interaction space. People can now talk to others across distance and time, and they regularly do so. People write a tremendous number of words in email, texts, social media and other forms of online interaction. Eric Schmidt, former CEO of Google, estimates that more information is created every 2 days as existed up until 2003. On a personal level, this means that many adults in the United States. produce more written words in a day than most humans throughout history produced in their entire lifetime.

These changes to the interaction space are having a profound effect on how deception and deception detection operate, but as noted before, this effect is not simply a matter of cues. Here we describe three ways that the transformation of the interaction space has modified deception in human communication: (1) the introduction of the display as a mode of interaction, (2) the effect of certain communication features introduced by these changes, such as the behavioral traces we now leave behind in digital contexts, and (3) the presence of warrants or connections in virtual space to a person's real-world identity.

The display

When we interact with others online, we do so through a display of some kind, such as a computer monitor, a mobile display, or, in the near future, our glasses. The introduction of the display to human communication is perhaps one of the most studied aspects of computer-mediated communication. In the context of deception, the introduction of the display means that interlocutors have much more control over their interaction space. As many others have now noted, people can control their self-presentation to a greater degree because they can choose and edit what they present, a term called selective self-presentation (Walther, 1996). People who seek to deceive can take advantage of this increased control to develop a number of different techniques for deceiving others.

The most well-known of these is spam, which can be deceptive in a variety of ways. The 419 spam, for example, involves a message from Nigeria or other African country that supposedly is from a wealthy individual who needs help getting their money out of the country, help that will lead to great riches for the spam recipient. Another common form of deception that relies on the control afforded by the display is phishing, which involves a website that looks like a bank or other trustworthy institution but is in fact controlled by a fraudster. The goal is to lure the user into providing personal information that can be used to access their bank accounts or steal their identity (Dhamija, Cassidy, Hallam-Baker, & Jacobsson, 2006).

Both spam and phishing are made possible by the control that the fraudster has over the user's display. In the spam case, for example, the fraudster can present as a wealthy individual from a far off land. In a nonmediated world, this would be much more difficult (though certainly not impossible) given both the physical and social indicators of cultures and wealth. In the phishing case, the fraudster can modify the user's display to show trustworthy images, logos and text that mimic, for instance, a bank. Both spam and phishing attacks use their control over the user's display to leverage a variety of psychological principles common to most scams, such as distraction, social compliance, greed, and scarcity (Stajano & Wilson, 2011). Phishing attacks that manipulate the user's display and rely on these scam principles can be highly effective, with up to 90% of users falling for well executed attacks (Dhamija, Tygar, & Hearst, 2006).

While these examples highlight how deception can be facilitated with the control of the display, the control of the display can also be used to modify the likelihood of dishonesty. Research in social psychology and behavioral economics suggests that there are subtle cues that can prime behaviors related to deception (Ariely, 2012). Several studies, for example, have tested the effect of surveillance primes, such as a pair of human eyes, on changing peoples' pro-social behavior (for review, see Nettle et al., 2013). In one study, people gave more money to an honesty jar for coffee cream when a photo of human eyes was affixed to the jar than a photo of a flower (Bateson, Nettle, & Roberts, 2006). More recently, surveillance primes have been used to reduce bicycle theft by nearly two-thirds.

An important question is whether these kinds of primes can also be used to shape deception-related behavior online. In one study that suggests this is possible, Haley and Fessler (2005) examined the effect of surveillance primes on pro-social sharing behavior in an economic game. In the game, one of the participants had the ability to decide on how to share a pot of money with a partner. This person was in one of two relevant conditions: either a surveillance prime, implemented as a set of cartoon eyes, was present on the display or it was not. People in the condition with a surveillance prime on their display shared more of the pot with their partner than people without a surveillance prime on their screen, suggesting that surveillance primes on a computer screen can alter behaviors related to cheating.

We recently demonstrated that this effect can also take place on Facebook and with interpersonal lies (Spottswood & Hancock, 2013). We inserted images of eyes in the advertising stream on Facebook—for example, ads for eye drops that prominently displayed eyes—or images that did not involve eyes. We then gave participants the opportunity to lie to a friend either on a wall post (in public) or in a message (in private). We found that the presence of the eyes had an important effect, but it was not simply to reduce deception. Instead, we found that the eyes increased the likelihood that the participant would "do the right thing," namely follow the norm for public or private behavior. When in public the eye spots led participants to lie more often to protect their friend, but in private the eyes led participants to tell the truth more often, presumably to give their friends more honest feedback. The takeaway for us is that the interface can have important, and sometimes subtle, effects on how we lie or are lied to.

Communication features and deception

With the transformation of our interaction space comes changes to the way that we use deception in computer-mediated contexts. Research to date suggests that differences in the way that people lie across communication media are affected by specific features of these different media. Hancock, Thom-Santelli and Ritchie's (2004a) feature-based model of deception and technology suggests that there are three important features of communication contexts that should influence deception. First, they argue that the recordability of a communication medium, or the degree to which communication in a given space has a durable record, should influence deceptive behavior. As communication in a given medium becomes more recordable, they argue that the likelihood of deception decreases. If we consider the recordability of different types of communication media, email provides the most permanent record, with copies of messages available not only on the individual's personal computer, but also on servers. In comparison, instant message conversations and SMS text messages appear less permanent, although this is mostly an illusion, as many political scandals involving text messaging have revealed. Almost any form of online interaction, from email to text messages to Facebook posts, leave a digital trail that can be retained, searched, and copied.

A second feature of communication media that is important to deception is the synchronicity of communication. Synchronicity refers to the degree to which a conversation is occurring in real time. Telephone conversations, for example, are highly synchronous as conversation partners are exchanging messages in real time. Email correspondence is often far less synchronous as delays between message exchange can be anywhere from seconds to days or even weeks, with a general expectation that responses will not be read immediately following message receipt. Hancock et al. (2004a) suggest that deception should occur more frequently in highly synchronous media. This is because highly synchronous media provides less time to craft honest responses. Many of the lies that we tell in conversation are small lies told to maintain relationships and protect our own image (DePaulo et al., 2003) (e.g., "No, you don't look fat in that dress, dear!"). In real-time communication, we often need to rely more on these types of lies to avoid situations where the truth might create an awkward or unwanted situation. When communication is less synchronous, we have more time to craft honest responses that need not rely on deception to be plausible.

A third feature of communication media that may dictate deceptive behavior is whether communicators are co-present. If a person is physically present, they are unable to lie about certain things, such as where they are, who they are with or what they are doing. Thus, computer-mediated communication contexts provide a greater number of opportunities to be deceptive about one's location, activities, and the company one is keeping.

The three parts of the feature-based model suggest that people should lie more in phone conversations than FtF conversations. In addition, FtF conversations should have more lies than email conversations (phone > face-to-face > email). Phone calls are not recorded, are not co-present and are highly synchronous. FtF conversations are not recorded, but do involve co-presence and are highly synchronous. Email, on the other hand, is recorded, but does not involve co-presence or high synchronicity.

Several studies have tested these predictions of the feature-based model. In an initial study, Hancock et al. (2004a) asked students to use a diary method to record all social interactions and lies over a 7-day period. Participants in this study made records of FtF interactions, as well as phone calls, emails, and instant messages. The authors used participants' data to calculate lying rates by media, as well as proportions depicting the number of lies per number of conversations by media for each individual participant. Similar to previous research (DePaulo, Kashy, Kirkendol, Wyer, & Epstein, 1996), they found that people lied more in phone conversations than FtF conversations. They also found that emails involved the fewest number of deceptions across media. These findings are in line with predictions from the feature-based model.

In a follow-up study, the authors (Hancock, Thom-Santelli, & Ritchie, 2004b) used the same procedure, but sought more detailed information about the types of lies that were told in daily interactions. In this study, the authors requested more detailed information about the specific lies that people told and to whom these lies were told. In this study, the authors observed the same patterns of deception as the initial study. In addition, they found that the features of the media used influenced the content of lies. Phone conversations had more lies about what people were doing, which makes sense given that this was the most mobile form of media assessed at the time of the study. On the other hand, FtF conversations included the greatest number of lies related to feelings, which the authors explain was likely because synchronous conversations often involve discussions of opinions. Email included the greatest number of lies related to explanations, presumably because the asynchronous nature of email allows more time to construct of these types of lies. In addition to differences in the types of lies people told, the recipients of these lies differed across media. People lied more often to friends and family via mediated communication channels (i.e., phone and email). Alternatively, people told more lies to strangers or acquaintances in FtF conversations, which is consistent with findings from previous research (DePaulo & Kashy, 1998; Ennis, Vrij, & Chance, 2008).

A more recent study conducted by Whitty and colleagues (2012) examined deceptive behaviors across a broader range of communication media. In addition to exploring deception across phone, FtF and email, they also asked participants to report on interactions and deceptions in IM, social networking websites, and SMS text messages. Using the feature-based model they predicted that people would lie more on the phone, followed by FtF interactions and IM (due to similarities in synchronicity and recordability), and then email, SMS texting and social networking websites (due to similarities in co-presence, recordability, and asynchronicity). They were also interested in the spontaneity of deception used in these contexts. While many of the lies that people tell are spontaneous, research suggests that a proportion of lies are planned (e.g., DePaulo et al., 1996, 2003). For example, if a person knows that they are going to arrive late to meet a friend without a legitimate excuse, they are likely come to the destination with a prepared lie that provides a reason for their lateness rather than coming up with a lie on the spot.

In a 1-week diary study, participants reported their daily interactions and deceptions, and provided information about the content of their lies, the reason for the lie, the type of lie, the referent of the lie (i.e., who did you tell this lie to?), and the referent's closeness to the individual (i.e., how close are you with the person you lied to?) and gender. In line with findings from earlier studies (Ennis et al., 2008; Hancock et al., 2004a, 2004b; Kashy & DePaulo, 1998), Whitty et al. (2012) observed a greater number of deceptions over the phone compared to FtF interactions. People were also more likely to lie over IM than SMS texting. No other differences in deception were observed across the other types of media, though the authors noted the low frequency of deceptions in IM, email, and social networking websites made interpretation of this null finding difficult.

In addition to replicating findings from earlier studies, the authors observed some interesting differences in the way that people told planned versus spontaneous lies. As might be expected, participants in the study were more likely to tell planned lies in certain forms of asynchronous media (e.g., SMS texting) than synchronous media (e.g., FtF interactions). They also found that participants rated planned lies as more serious lies than spontaneous lies.

The aforementioned studies provide evidence for the feature-based model of deception and technology. These findings suggest that people take features of the communication media (e.g., recordability, synchronicity, co-presence) into consideration when making decisions about whether or not to lie. We see that these features are also important when people consider the recipients and content of their lies.

The Warranted Self-Presentation Perspective

Deception is almost always used to accomplish goals, such as appearing attractive or interesting for a romantic partner or appearing competent to an employer. Self-enhancing deceptions are common, and typically motivated by a desire to present oneself positively to others. In the self-presentational framework of deception, DePaulo et al. (2003) suggest that lies are used to manage our self-presentations. This framework suggests that self-presentation lies are motivated by the psychological rewards that come with conveying a more positive self-presentation to others. Lies are a self-presentational strategy that allow the individual to appear more positively than their actual traits suggest, but also allow them to avoid the consequences that they might experience from revealing their true characteristics (e.g., disapproval, ridicule, etc.).

DePaulo argues that self-presentational lies can be self or other-oriented. Self-oriented lies are those told to protect or enhance one's self-presentation. For example, in a self-oriented lie, a job seeker might lie about their how long they worked at an organization to appear more experienced in hopes of being hired. Other-oriented lies are those told to protect others' feelings while maintaining a positive impression of the self. For example, in an other-oriented lie, a person might tell a friend that they are on their way to a meeting place when they have not yet left the house, which both conveys to the friend that they have not forgotten their plans while simultaneously maintaining the person's identity as a considerate friend.

DePaulo's research provides a detailed picture of lies told to accomplish self-presentation goals, but what happens when our self-presentations move to online contexts? As we have described, the online environment has evolved dramatically over our traditional FtF form of interaction. While we discuss the opportunities for deception in various media earlier, changes to communication capabilities provide abundant opportunities for self-presentation. For example, we craft detailed profiles on social networking websites, often sharing details about our personal and professional lives in great detail. In addition, many of our online activities are tracked and shared with members of our social network. Many websites and mobile-based applications allow users to share their online activities with members of their social network, providing unique information about our identity.

These traces of our online activity provide members of our social network with a highly detailed picture of our characteristics and behaviors that would be difficult, if not impossible, to match with information gleaned in offline contexts (e.g., Backstrom & Kleinberg, 2014). These changes in the amount of and accessibility to self-presentational context raise the question of whether deception for self-presentation operates differently online. Does the online environment lead to radical changes in the way that deception occurs, or do the same principles for deception in offline contexts apply online?

While some of the unique features of online communication that enable deception are described earlier, these new opportunities for self-presentation online, such as social networking websites, present important new constraints to deception. Perhaps the most important of these constraints is described in what we call the warranted self-presentation perspective. In explaining this perspective, it is important to first outline the features of online communication that constrain deception. Social network profiles make self-presentations publicly available and link individuals to the profile who can verify whether information is deceptive or not. Researchers both on and offline have long demonstrated the importance of our social relationships in fostering honesty between individuals (Resnick, Kuwabara, Zeckhauser, & Friedman, 2000; Resnick & Varian, 1997; Zimmerman & Kurapati, 2002). For example, recommender systems on websites like eBay help to ensure that transactions in these environments remain honest by providing users, who have no previous history of a seller, with valuable information about the seller's trustworthiness. Affordances that allow for social links between the virtual representation of self that one encounters online and the physical, offline self can improve the likelihood of honesty in the online environment.

Walther and Parks (2002) call these links between the virtual and physical self *warrants*, a concept first coined by Stone (1995). Stone (1995) originally defined a warrant as the construction and maintenance of a connection between the physical self and a discursive representation of the self. Stone's (1994) original conception of warrants comes from virtual systems theory and views "the production and maintenance of a discursive space and a physical space as 'warranting'" (p. 181). Here Stone (1994) sees individuals as composed of both physical and discursive components (i.e., intangible elements of the individual). Discursive components are grounded in the physical body via warranting. Walther and Parks introduced warranting to the communication field by stating that we can "conceive of a warrant [as] connecting the self with an on-line presentation" (p. 551).

The warranted self-presentation perspective extends the idea of a warrant by suggesting that senders or creators of content use warrants when crafting self-presentations. While previous work has viewed warranting as a receiver-oriented process, in which receivers make judgments about an individual, here we examine warranting from the sender or content producer perspective. In the present chapter, we argue that the presence of a warrant should also affect the production of deception. More specifically, the central argument is that senders should lie differently when warrants are present than when they are not as warrants represent a connection between the physical self and online self-presentations. Here, we suggest that warrants should constrain how individuals use deception (e.g., frequency of lies, types of lies) to accomplish self-presentational goals.

Next we examine the warranted self-presentation perspective in a number of online contexts and show that many of the same deception principles apply in online contexts, but there are some unexpected changes in how lies are implemented to accomplish self-presentation goals in online settings.

Deception in online resumes

Deception for self-presentation purposes offers the reward of appearing more positive to others, but deceptive activities bring with them a host of consequences. One particular context where these consequences can be severe is in the professional world. High-profile examples of these consequences abound. Take Janet Cooke, for example, who lost the Pulitzer Prize in 1981 after it was discovered that she lied about her education on her resume (Kidwell, 2004). While these consequences should deter people from using extreme deceptions in resumes, the motivation to appear competent still drives deception in resumes. Research from a resume consulting service estimates that 43% of resumes contain significant inaccuracies (Cullen, 2006). In experimental research, 90% of people lied on a resume-like scholarship application, despite the fact that they were offered no explicit incentives to be deceptive (George, Marett, & Tilly, 2004). How might the online environment influence the way that people produce self-presentation deceptions in online resumes?

Most of our professional activities can now be conducted online, with people searching and applying for jobs and creating personal profiles in professional networking websites like LinkedIn. This shift in professional activities, such as sharing resumes and job applications online, has important implications for how deception should occur. While individuals are motivated to provide positive self-presentations, the warranted self-presentation perspective suggests that the publicness of resumes (i.e., accessibility of profiles to colleagues and friends) in social networking websites makes people more accountable for information shared online. When a person creates a LinkedIn profile, the site provides default settings making the profile public, creating a potential audience to which the communication partner must explain deceptions (profiles can be made private upon request). Public settings should increase the possibility that an employer might discover deceptions. Traditional resumes, on the other hand, are confidential and are not widely shared outside organizations (Rousseau, 2001). Though it's common for employers to contact references to review truthfulness of resumes, traditional resumes are limited in their ease of accessibility to others, with far fewer potential viewers to verify veracity.

The likelihood of being caught in a lie about previous employment should be higher for publicly available LinkedIn profiles than for traditional resumes. Though profile publicness does not guarantee relevant audiences (e.g., supervisors) will view profiles, profile creators should alter deceptive behavior to be consistent with information known by potential audiences. Socially connected displays of information on these sites should constrain deception, as being detected has serious consequences (e.g., exposure of deceptions by network members).

In a recent study, we found that publicness does indeed influence deceptive behaviors in LinkedIn profiles (Guillory & Hancock, 2012). Our participants created resumes in one of three resume settings: a traditional (offline) resume, a private Linkedin profile, or a public Linkedin profile. Private profiles made the content available only to the participant and the experimenter and were used in the study to ensure that differences in deception were not fueled by the profile being online. Public profiles in the study made the content available to anyone online via a web search. Undergraduate student participants in the study were asked to create a profile for an entry-level job with qualifications that would be very difficult for this demographic to meet. Participants were not specifically encouraged to be deceptive in the study, but were told that the best fitting resume for the job would receive a $100 gift card. At the conclusion of the study, participants were asked to use a retrospective identification technique to identify all deceptions in their resume.

On the surface, it would seem that making Linkedin profiles public would reduce the frequency of deceptions that people used in profiles. Findings from this study suggest that it is not so simple. Instead, deception occurred at similar frequencies across the three resume conditions—participants lied on average three times regardless of condition. Specific types of information provided on a resume, however, were affected differentially by the presentation format. When contrasted with traditional resumes, Linkedin resumes were less deceptive about claims that could be verified by people in a person's social network, such as an applicant's prior work experience and job responsibilities. On the other hand, creators of Linkedin profiles were more deceptive about difficult to verify information, such as interests and hobbies, in which only the self-presenter knows the truth (e.g., favorite book, favorite hobby).

These findings make sense from the warranted self-presentation perspective: deception should be constrained by the people in a person's social network who have access to profile information. Deceptions about verifiable information, such as work experience or educational background, pose major risks to a person's reputation and future career if made public and are more likely to be classified as lies. On the other hand, when the truthfulness of resume claims is difficult to assess objectively, as is the case with hobbies or interests, there less risk of being caught in a lie and these claims are less likely to be identified as lies (Fielder & Walka, 1993; Walther & Parks, 2002). The falsifiability heuristic claims that when a person shares information about the self that is more objectively verifiable, such as observable behaviors, these claims are perceived as more credible and are less likely to be identified as lies (Fielder & Walka, 1993). Therefore, it makes sense that participants in this study practiced deception in a strategic manner by lying about different kinds of information based on whether profile claims were publicly available.

These strategic considerations surrounding deception are understandable, as research shows that the discovery of deceptions about verifiable resume claims entails consequences such as loss of jobs and awards and damage to one's professional reputation (McCornack & Levine, 1990). To avoid these consequences, participants in our study who created public resumes lied about information that was not widely known to members of their social network and therefore less job-relevant. In one example, participants noted that the job they were applying for included a great deal of travel and lied in their resumes about interests in travel and learning new languages to appear to be an apt candidate for this position, without lying about information that would be verifiable by network members. Though unverifiable information may be less relevant to obtaining a job, our research suggests that people can still use it strategically to accomplish self-presentational goals.

These findings counter the assumption that the online environment encourages excessive deception, and alternatively suggests that features of social networking websites that make information publicly available actually help to encourage more honesty in the context of online resumes compared to traditional paper resumes. In this professional context, the consequences for deceptive behavior are often grave (e.g., Kidwell, 2004) and could involve damage to one's professional reputation or even job loss. Will deceptive behaviors be influenced in the same way in a broader range of contexts with less serious consequences surrounding deception?

Warranted self-presentation deceptions across communication contexts

How does the presence of warrants to a greater or lesser degree influence deception in different online communication contexts? A recent study (Warkentin et al., 2010) set out to explore the way that differing levels of anonymity in online communication contexts affected the frequency and types of deception that people use. This research used the warranted self-presentation perspective to predict that deception would occur less frequently in communication contexts with fewer warrants. Three types of warrants were considered: (a) name warrants, or references to a person's real name; (b) photo warrants, which included recognizable photographs of a person that could be visually linked to their real-world identity; and (c) acquaintance warrants, which were defined by the degree to which a person's communication partners knew about his or her identity offline.

In order to explore the role that these different types of warrants play in deceptive behaviors, they surveyed participants about their deceptive behaviors in email, IM, forums, chat rooms, and social networking websites. Consistent with their predictions, they found that more warrants led to fewer lies. They also found that people told less serious lies in the presence of a more warrants. When they considered the influence of the specific type of warrant on deception, they found that acquaintance warrants were the main factor driving the reduction in both the frequency and magnitude of deception.

These findings suggest that warrants in online communication act as a major factor influencing deception across a wide variety of online communication contexts, and provide some support for a more sender-oriented perspective of warrants. That is, in addition to affecting how we judge the credibility of information online, the presence of warrants also constrain the production of deceptive information.

Love and Lies: Deception in Online Dating

The self-presentational goal of appearing attractive is particularly salient in the context of online dating. In wanting to appear attractive in these online self-presentations, daters may consider using deception in their dating profiles to improve other daters' perceptions of their attractiveness. While self-presentation deception is perceived as occurring frequently in these spaces, it should be buffered by daters' desire to appear honest given that they hope to 1 day meet a romantic partner in person. This prospect of anticipated future interaction should constrain people from using extreme deceptions that would be revealed as lies if they were to meet the romantic partner in person (e.g., claiming to have a "toned and athletic figure" when the individual is overweight).

Toma, Hancock, and Ellison (2008; Ellison, Hancock, & Toma, 2012; Hancock & Toma, 2009; Toma & Hancock, 2010) conducted a series of studies exploring deceptive processes in online dating. In the first of these studies, Toma et al. (2008) compared content from online dating profiles, such as profile holders' stated height, to their observed characteristics, namely how tall they were as measured in the lab, to determine how much the daters' self-presentation in their dating profile deviated from their actual characteristics. Because many online daters hope to eventually meet romantic partners in person, Toma et al. (2008) also predicted that lies in profiles would be small in magnitude. Deception in online dating profiles should also be driven in part by gender differences. In addition, based on characteristics that men and women seek in mates, they predicted that men would lie more about social status, occupation, education, and height (as being tall is associated with higher social status in men). In contrast, women should lie more about age and weight, which is an indicator of physical attractiveness and valued by men.

In line with predictions, Toma et al. observed that the magnitude of deceptions that people shared was rather small. The average deviation in height was less than 1 in. (2.09% of participants' actual height), the average deviation in weight was less than 10 pounds (5.5% of participants' actual weight), and the average deviation in age was 0.55 years (1.4% of participants' actual age). While men and women subtle in their deceptions about age, weight and height, key gender differences emerged. For example, consistent with expectations from evolutionary psychology (e.g., Buss & Schmitt, 1993), men were more likely to over-report their height than women, while women were more likely to under-report their weight than men.

The findings from this initial study illustrate the tension between the self-presentation goal of appearing attractive to other daters and the desire to appear honest in anticipation of future interactions. In a follow-up study, Hancock and Toma (2009) explored deception in the context of photographs from online dating profiles. While the process of selective self-presentation has been explored extensively in computer-mediated communication contexts, few studies have explored how these processes occur in nontextual aspects of computer-mediated communication (e.g., photographs). Photographs may be altered in many ways. People can choose flattering camera angles and scenery. In more extreme cases, people can use tools, such as Adobe Photoshop, to edit the content of photos (e.g., removing blemishes or other unattractive physical features from photos).

Similar to the initial study, Hancock and Toma (2009) predicted that photographic deception would differ by gender based on characteristics that men and women value in potential mates (Lance, 1998; Woll & Cozby, 1987). Because men should value physical attractiveness more than women, they predicted that women's photographs would contain more deceptions. Because men should value youthfulness in female mates, they also predicted that women would provide younger photographs than men.

In order to study photographic deception, which is more difficult to objectively verify than something like height or weight, participants were first asked to rate their photographs for inaccuracies. Next participants posed for a photograph to obtain a record of their everyday appearance. They then asked independent judges to compare the everyday photograph of the individual to the profile photograph. Consistent with predictions, women's photos contained a greater number of discrepancies between the everyday photograph and the profile photograph than men's photographs. The data also suggest that women chose profile photographs that were significantly older (based on self-ratings) than men's photographs, which is consistent with the prediction that women would attempt to present a younger version of the self to other daters.

While these studies identify deceptive patterns in the context of online dating, they provide little information about how possessing certain qualities will lead men and women to engage in different deceptive strategies in this context. In another study, the authors (Toma & Hancock, 2010) explored the role that physical attractiveness plays in deception in online dating. The authors predicted that in cases where individuals were less physically attractive, they would compensate by using more deception, both in photographs and self-descriptions (e.g., height, weight, age). They also predicted that this tendency would be stronger for women, consistent with findings from their previous two studies (Hancock & Toma, 2009; Toma et al., 2008) and theory that suggests men value physical attractiveness more than women in mates (Lance, 1998; Woll & Cozby, 1987). In addition, they predicted that more attractive daters would want to take advantage of this positive quality by posting more photographs to their dating profile. In line with their predictions, Toma and Hancock (2010) found that less attractive participants were more prone to engaging in deception, both in photographs and self-descriptions (e.g., height, weight, age), while attractive daters were indeed likely to take advantage of their attractiveness by posting more photographs than unattractive daters.

Findings from this series of studies provide strong evidence that self-presentation deception is an important tactic in online dating profiles and that these deceptions are driven by a desire to appeal to specific characteristics that are desirable to men and women, respectively. In addition, this research suggests that having (or not having) certain qualities, such as higher levels of attractiveness, leads to different patterns of deception in online dating profiles.

On My Way: Butler Lies

The cases discussed earlier focus on the use of deception to accomplish self-presentational goals that make the individual appear more attractive, be it for a job or potential mate. Recall that DePaulo would refer to these lies as self-oriented lies, or those that are told to protect or enhance one's self-presentation. There are also cases where people use lies that have an other-oriented component. Other-oriented lies are those told to protect others' feelings in order to maintain positive self-presentations. These types of lies come up quite often when people are attempting to manage their availability (i.e., initiating, concluding, or coordinating social interactions). Text-based communication conveniently provides opportunities to access our social contacts regardless of the time or location. With this convenience comes the burden of constantly being available to people in our social network, even at times when communication is unwanted or difficult. As a way to deal with this constant availability, we have observed that people have developed deceptive strategies to avoid social interaction or account for failure to communication. We refer to these strategies as *butler lies.* Butler lies are told to manage social interaction and awareness either by avoiding a new conversation, smoothly exiting an ongoing conversation, or explaining other communication behavior.

In an initial study, we explored butler lies in the context of instant messaging (IM) (Hancock et al., 2009). Participants in this study were asked to rate the deceptiveness of all outgoing instant messages from their personal computer during a 4-day period. Many of the lies that participants told were related to either to avoid starting a new conversation, to exit an ongoing conversation, or to arrange or explain other interactions or failure to interact with a person. Our analysis of the content of these butler lies suggests that these lies were other-oriented self-presentational deceptions that were told to manage interpersonal impressions when the individual wanted to engage in an activity that could potentially hurt the feelings of their communication partner, such as ending a conversation early or avoiding a conversation altogether.

While understanding how people use butler lies in IM conversations provides a small window into people's availability management strategies, communication via mobile phones, primarily SMS text messaging, helps us to understand how people manage their availability in wider range of contexts. In a second study, we explored butler lies in the context of SMS text messaging and found that while these messages occurred in similar frequency to those that we observed in IM conversations, the types of butler lies used differed from IM. While people primarily used butler lies as a strategy for exiting conversations in IM, but lies in the SMS context were used primarily to avoid social interactions (e.g., "Can't meet up later, gotta work"), which can be attributed to the fact that SMS is often used to coordinate social interactions (Grinter & Eldridge, 2003). In a second text messaging study, we compared Blackberry Messenger (BBM) with SMS (Reynolds et al., 2011). This follow-up study replicated findings from the previous two studies and also demonstrated the significance of people's relationships in driving the use of butler lies. Participants tended to tell more butler lies to the people they made plans with the most, which implies that these types of lies are an essential component of relationship management.

Findings from these studies demonstrate several important points about the availability management strategies. First, we see that people have difficulty managing their availability given the constraints of communication technology and have developed their own strategies to manage availability in these situations. Second, these studies show us that a major part of our availability management involves maintaining our relationships, highlighting once again that deception is often a reasonable response to technology that takes into consideration relational factors, rather than technology simply increasing deception.

Conclusion

Whether its butler lies, sock puppets or the Chinese Water Army, there is no question that technology and deception are influencing how people communicate with one another, from the everyday interactions on mobile texting to the politically significant scandals revealed by the digital traces left behind by communication technology. It is also clear that the relationship between deception and technology is not a straightforward one. Although powerful heuristics lead us to believe that technology facilitates deception, it is much more complicated than that.

In the present chapter, we have argued that understanding our biases around deception, and around how technology affects us, are important to understand and diagnose in order to understand the impact of technology on deception. We also offer an important alternative approach that considers transformations to the interaction space wrought by new technologies. These transformations include the changes to the visual presentation of information, the features of novel communication spaces (e.g., recordability), the ways in which social networks now link and constrain us online, and the importance of relational concerns in understanding how and why we lie with technology. By approaching deception in this way we believe that we can advance our understanding of this very old human phenomena in these very novel times.

References

Anderson, N. H. (1968). Likableness ratings of 555 personality-trait words. *Journal of Personality and Social Psychology*, *9*, 272–279.

Ariely, D. (2012). *The honest truth about dishonesty.* New York, NY: HarperCollins.

Backstrom, L., & Kleinberg, J. (2014). Romantic partnerships and the dispersion of social ties: A network analysis of relationship status on Facebook. In S. Fussell, W. Lutters, M. Ringel Morris, & M. Reddy (Eds.), *Proceedings of the 17th ACM Conference on Computer Supported Cooperative Work and Social Computing (CSCW2014), Baltimore, MD.* New York, NY: ACM.

Bateson, M., Nettle, D., & Roberts, G. (2006). Cues of being watched enhance cooperation in a real world setting. *Biology Letters*, *2*, 412–414.

Bond, C. F., Jr., & DePaulo, B. M. (2006). Accuracy of deception judgments. *Personality and Social Psychology Review*, *10*, 214–234.

Buss, D. M., & Schmitt, D. P. (1993). Sexual strategies theory: An evolutionary perspective on human mating. *Psychological Review*, *100*, 204–232.

Cullen, L. T. (2006). Getting wise to lies. *Time*, *167*, 59–60.

Dediu, D., & Levinson, C. (2013). On the antiquity of language: The reinterpretation of Neanderthal linguistic capacities and its consequences. *Frontiers in Psychology*, *4*, 397.

DePaulo, B. M., & Kashy, D. A. (1998). Everyday lies in close and casual relationships. *Journal of Personality and Social Psychology*, *74*, 63–79.

DePaulo, B. M., Kashy, D. A., Kirkendol, S. E., Wyer, M. M., & Epstein, J. A. (1996). Lying in everyday life. *Journal of Personality and Social Psychology*, *70*, 979.

DePaulo, B. M., Lindsay, J. J., Malone, B. E., Muhlenbruck, L., Charlton, K., & Cooper, H. (2003). Cues to deception. *Psychological Bulletin*, *129*, 74–118.

Dhamija, R., Cassidy, P., Hallam-Baker, P., & Jacobsson, M. (2006). Phoiling phishing. In A. L. Car et al. (Eds.), *Proceedings of the ACM 15th international conference on World Wide Web (WWW2006)* (pp. 749–749). Edinburgh, UK: ACM.

Dhamija, R., Tygar, J. D., & Hearst, M. (2006, April). Why phishing works. In P. M. Aoki et al. (Eds.), *Proceedings of the SIGCHI conference on Human Factors in Computing Systems* (pp. 581–590). New York, NY: ACM.

Ekman, P. (1985). *Telling lies: Clues to deceit in the marketplace, politics and marriage*. New York, NY: Norton.

Ellison, N., Hancock, J. T., & Toma, C. (2012). Profile as promise: A framework for conceptualizing the veracity of online dating profiles. *New Media & Society*, *14*, 42–59.

Ennis, E., Vrij, A., & Chance, C. (2008). Individual differences and lying in everyday life. *Journal of Social and Personal Relationships*, *25*, 105–118.

Fielder, K., & Walka, I. (1993). Training lie detectors to use nonverbal cues instead of global heuristics. *Human Communication Research*, *20*, 199–223.

George, J., Marett, K., & Tilly, P. (2004). Deception detection under varying electronic media and warning conditions. In J. F. George, K. Marett, & P. Tilley (Eds.), *Proceedings of 37th Hawaii International Conference on System Sciences*. New York, NY:IEEE Computer Society.

Grinter, R., & Eldridge, M. (2003, April). Wan2tlk? Everyday text messaging. In P. Korhonen (Eds.), *Proceedings of the SIGCHI conference on Human Factors in Computing Systems* (pp. 441–448). New York, NY: ACM.

Guillory, J., & Hancock, J. T. (2012). The effect of LinkedIn on deception in resumes. *Cyberpsychology, Behavior, and Social Networking*, *15*, 135–140.

Haley, K. J., & Fessler, D. M. (2005). Nobody's watching? Subtle cues affect generosity in an anonymous economic game. *Evolution and Human Behavior*, *26*(3), 245–256.

Hancock, J. T., Birnholtz, J. P, Bazarova, N. N., Gullory, J., Perlin, J., & Amos, B. (2009). Butler lies: Awareness, deception, and design. In D. R. Olsen Jr. et al., (Eds.), *Proceedings of the SIGCHI Conference on Human Factors in Computing Systems* (pp. 517–526). New York, NY: ACM.

Hancock, J. T. (2012). *The future of lying*. Technology, Education & Design (TED), Winnipeg, MN. Retrieved from http://www.ted.com/talks/jeff_hancock_3_types_of_digital_lies

Hancock, J. T., Thom-Santelli, J., & Ritchie, T. (2004a). Deception and design: The impact of communication technology on lying behavior. In M. Tscheligi (Eds.), *Proceedings of the SIGCHI Conference on Human Factors in Computing Systems* (pp. 129–134). New York, NY: ACM.

Hancock, J. T., Thom-Santelli, J., & Ritchie, T. (2004b) *What lies beneath: The effect of communication medium on deception production*. Paper presented at the Annual Meeting of the Society for Text and Discourse, Chicago, IL.

Hancock, J. T., & Toma, C. L. (2009). Putting your best face forward: The accuracy of online dating photographs. *Journal of Communication*, *59*, 367–386.

Hancock, J. T., Woodworth, M. T., & Goorha, S. (2010). See no evil: The effect of communication medium and motivation on deception detection. *Group Decision and Negotiation*, *19*, 327–343.

Lance, L. L. (1998). Gender differences in heterosexual dating: A content analysis of personal ads. *The Journal of Men's Studies, 6*(3), 297–305.

Levine, T. R., Kim, R. K., & Hamel, L. M. (2010). People lie for a reason: Three experiments documenting the principle of veracity. *Communication Research Reports, 27*, 271–285.

Levine, T. R., Serota, K. B., & Shulman, H. C. (2010). The impact of Lie to Me on viewers' actual ability to detect deception. *Communication Research, 37*, 847–856.

Keyes, R. (2004). *The post-truth era: Dishonesty and deception in contemporary life.* New York, NY: St. Martin's Press.

Kidwell, R. E. (2004). "Small" lies, big trouble: The unfortunate consequences of résumé padding, from Janet Cooke to George O'Leary. *Journal of Business Ethics, 51*, 175–184.

McCornack, S. A., & Levine, T. R. (1990). When lovers become leery: The relationship between suspicion and accuracy in detecting deception. *Communication Monographs, 57*, 219–230.

Nettle, D., Harper, Z., Kidson, A., Stone, R., Penton-Voak, I. S., & Bateson, M. (2013). The watching eyes effect in the Dictator Game: It's not how much you give, it's being seen to give something. *Evolution and Human Behavior, 34*, 35–40.

Ott, M., Cardie, C., & Hancock, J. T. (2012). Estimating the prevalence of deception in online review communities. *Proceedings of International World Wide Web Conference. Lyon, France. 2012 (WWW2013)*.

Porter, S., & ten Brinke, L. (2008). Reading between the lies: Identifying concealed and falsified emotions in universal facial expressions. *Psychological Science, 19*, 508–514.

Pronin, E. (2006). Perception and misperception of bias in human judgment. *Trends in Cognitive Sciences, 11*, 37–43.

Resnick, P., Kuwabara, K., Zeckhauser, R., & Friedman, E. (2000). Reputation systems. *Communications of the ACM, 43*, 45–48.

Resnick, P., & Varian, H. R. (1997). Recommender systems. *Communications of the ACM, 40*, 56–58.

Reynolds, L., Gillette, S., Marder, J., Miles, Z., Vodenski, P., Weintraub, A., … Hancock, J. T. (2011). Contact stratification and deception: Blackberry messenger versus SMS use among students. In P. Hinds (Eds.), *Proceedings of the ACM 2011 Conference on Computer Supported Cooperative Work* (pp. 221–224). New York, NY: ACM.

Rousseau, D. M. (2001). Schema, promise and mutuality: The building blocks of the psychological contract. *Journal of Occupational and Organizational Psychology, 74*, 511–541.

Spottswood, E., & Hancock, J. T. (2013, June). The eyes have it: How publicness and surveillance primes influence prosoical communication on Facebook. *International Communication Association 63rd Annual Conference*, London, UK.

Stajano, F., & Wilson, P. (2011). Understanding scam victims: Seven principles for systems security. *Communication of the ACM, 54*, 70–75.

Stone, A. R. (1994). Split subjects, not atoms; or, how I fell in love with my prosthesis. *Configurations, 2*, 173–190.

Stone, A. R. (1995). *The war of desire and technology at the close of the mechanical age.* Cambridge, MA: MIT Press.

Toma, C. L., & Hancock, J. T. (2010). Looks and lies: The role of physical attractiveness in online dating self-presentation and deception. *Communication Research, 37*, 335–351.

Toma, C. L., Hancock, J. T., & Ellison, N. B. (2008). Separating fact from fiction: An examination of deceptive self-presentation in online dating profiles. *Personality and Social Psychology Bulletin, 34*, 1023–1036.

Toma, C. L., Jiang, L. C., & Hancock, J. T. (2013, June). *Lies in the eyes of the beholder: Self-other asymmetry in beliefs about deception across interpersonal media*. Paper presented at the 63rd Annual Conference of the International Communication Association, London, UK.

Walther, J. B. (1996). Computer-mediated communication impersonal, interpersonal, and hyperpersonal interaction. *Communication research*, *23*, 3–43.

Walther, J. B., & Parks, M. R. (2002). Cues filtered out, cues filtered in: Computer-mediated communication and relationships. In I. M. Knapp & J. A. Daly (Eds.), *Handbook of interpersonal communication* (3rd ed., pp. 529–563). Thousand Oaks, CA: Sage.

Warkentin, D., Woodworth, M., Hancock, J. T., & Cormier, N. (2010). Warrants and deception in computer-mediated communication. In K. Inkpen (Eds.), *Proceedings of the 2010 ACM Conference on Computer Supported Cooperative Work* (pp. 9–12). New York, NY: ACM.

Whitty, M. T., Buchanan, T., Joinson, A. N., & Meredith, A. (2012). Not all lies are spontaneous: An examination of deception across different modes of communication. *Journal of the American Society for Information Science and Technology*, *63*, 208–216.

Woll, S. B., & Cozby, P. C. (1987). Videodating and other alternatives to traditional methods of relationship initiation. *Advances in Personal Relationships*, *1*, 69–108.

Zimmerman, J., & Kurapati, K. (2002, April). Exposing profiles to build trust in a recommender. In L. Terveen et al. (Eds.), *CHI'02 Extended Abstracts on Human Factors in Computing Systems* (pp. 608–609). New York, NY: ACM.

13

Immersive Virtual Environments and the Classrooms of Tomorrow

Cody O. Karutz and Jeremy N. Bailenson

Virtual Human Interaction Lab, Stanford University, Stanford, CA, USA

Introduction

Imagine you are sitting in a traditional university lecture hall. Even though the room has a 200 seat capacity, only half are occupied by other students. At closer glace, the others around are mostly distracted, checking various social media platforms on their mobile devices. You find that this distracts you from focusing on the professor that is attempting to lecture on stage about climate change science. His monotone voice drones on about carbon emissions, fossil fuels, and glacier melt timelines. What is worse is that the professor is not even using an interactive slideshow. Instead, he is using an almost unreadable ancient transparency overhead machine with tiny text. You realize that you have nothing in common with this professor, and despite his established credentials you find this lecture incredibly unengaging. But suddenly, this entire unengaging scene starts to change. The students around you all simultaneously put away their electronic devices and pay full attention to the professor, taking diligent notes while smiling complacently. Next, the professor morphs into Mark Ruffalo, the animated actor and environmental advocate that you admire. He kicks over the transparency machine, to the cheers of the audience, and explains that instead of lecturing about climate change, he is going to show it directly. The lecture hall rumbles with an earthquake-like magnitude, the walls cave in, and water gushes into the previously dry room. You look around, and realize you are now on a glacier, in the arctic north. Fully alert and awake, you remind yourself that although this is a completely virtual environment it feels incredibly real.

Compare the two versions of the previous imagined learning scenario. In the beginning, a traditional large-format university lecture was described, with disinterested students and a professor lacking public speaking pizazz. The second version of the scenario, however, brought the students to full attention with a lecturer that was

The Handbook of the Psychology of Communication Technology, First Edition. Edited by S. Shyam Sundar.

more entertaining and a visual environment that was both vivid and relevant to the lesson material. Although this fictional depiction of an educational setting might seem like science fiction, recent technological advances have catapulted ***immersive virtual environments*** (IVEs) out of expensive research labs and into both living rooms and classrooms (Blascovich & Bailenson, 2011).

Virtual environments are already consuming more and more of the population's time spent online. A user study revealed that in 2011, 21 million users spent almost 1.3 billion collective hours playing League of Legends, a popular online video game (Gaudiosi, 2012). In second place was World of Warcraft at over 600 million hours. A comprehensive digital media consumption study that surveyed daily trends of 8–18 year-olds found that on average children spend 7 hr and 28 min a day using digital media (Kaiser Family Foundation, 2010). Of those hours, 1 hr and 13 min are spent in video games on various platforms. The power of IVEs lie in their allowance of control, since everything is customizable and anything can be tweaked to construct a virtual classroom that is not only a near replication of a real one, but can outperform its physical counterpart in certain ways. What would such a virtual learning environment look like, and how would it work?

Enter the ***massive open online course*** (MOOC), a digital trend that has swept many of the nation's universities in the recent years to make educational material, the same that current students pay hefty tuition for, available for free online. This shift represents a bold move in bringing content to a platform that universities have traditionally been conservative about openly embracing: the internet. But because this format is accessible to as many students as there is interest, many problems still exist with the model. For example, how does assessment work for a class that has 50,000 students but only 10 teaching assistants? With such a large class size, how can education still feel personal or individualized for each student? Despite these issues, the MOOC model is gaining tremendous momentum. As the economist Thomas Friedman notes, "the MOOCs revolution, which will go through many growing pains, is here and is real" (Friedman, 2013).

MOOCs

The term MOOC was first introduced by George Siemens and Stephen Downes in 2008 to describe their online course *Connectivism and Connective Knowledge* (CCK08) taught through the University of Manitoba (Fini, 2009). Although it certainly was not the first online course to attempt to open up its course material in an accessible way for anyone, CCK08 succeeded in defining a new trend of education by building a network of learners of various educational backgrounds that self-organized to learn from an expert using various online resources and tools (McAuley, Stewart, Siemens, & Cormier, 2010). Unlike physical university classes, which have enrollment capacities, MOOCs are open to all users with an internet connection and a willingness to not necessary complete the course, but to engage in as little or as much learning as they like.

Through various measures, the MOOC model has already proved extremely successful. For achieving high enrollment numbers, an artificial intelligence course at

Stanford University registered 160,000 students, of which approximately 23,000 successfully completed the course (Martin, 2012; Rodriguez, 2012). Adopting the platform can also increase student performance, such as an electrical engineering course at San Jose State University that increased pass rates from 55% to 91% when it switched to a MOOC platform (Fowler, 2013). It has also been argued that MOOCs represent a larger economical movement where users are engaging in lifelong learning that increases their marketable skills and shifts the population from knowledge scarcity to knowledge abundance (McAuley et al., 2010). This is especially important for parts of the world that might not have access to traditional educational institutions but do have limited access to the internet, allowing them to still to learn through a MOOC platform for free.

This new network of learners that are in control of their own educational pace and goals provides challenges for traditional models. Thorough assessment becomes an issue when grading moves beyond simple multiple choice answers and onto more complex computer science problem sets or projects with qualitative elements (Cooper & Sahami, 2013). Additional assessment problems deal with validation and plagiarism, where it becomes difficult for administrators of a MOOC to both verify work and detect when a student is cheating (Cooper & Sahami, 2013). Especially puzzling are accountability issues, such as the high attrition rates (Kizilcec, Piech, & Schneider, 2013) and the behaviors of *lurkers,* who are students in a MOOC that observe content in a course but do not actively participate (Rodriguez, 2012). Hybrid models such as the *flipped classroom* attempt to find a way to strike a balance between digital optimizations and impersonal interactions by combining video content of MOOCs with physical classroom discussion to tackle personalization issues (Cooper & Sahami, 2013). Due to geographic location or time constraints, not all students have the opportunity to engage in flipped classroom models. Is there a way to capitalize on the operational advantages of the MOOC with the high emotional interactivity of an IVE to create a model that creates a more personal and individualized educational platform?

MOOVE Overview

This chapter aims to propose a new theoretical hybrid classroom model called the *massive open online virtual environment* (MOOVE) that capitalizes on the existing research on IVEs and applies them to the MOOC to present solutions for its current issues of assessment, accountability, and personalization. In order to do this, the literature surrounding IVEs is thoroughly reviewed through a social science lens. First, the hardware systems required to run IVEs are detailed, and new technologies are highlighted that are making components more affordable to average consumers. Research methods are explicated, where IVEs offer certain advantages over traditional social science experiments, such as collecting massive amounts of behavioral data about a user in a noninvasive way. Three theories are then outlined and reviewed, all with specific applications to the proposed MOOVE model. Transformed social interaction leverages unique characteristics of an IVE to disrupt the tracking and rendering cycle to change behavioral outcomes of an interaction. Social learning theory is reviewed, applying its literature to three new IVE research avenues where users view versions of their digital selves. Lastly, embodied cognition theory looks at the effects of actively moving body limbs on higher order cognition, such as solving a difficult physics equation. Conclusions drawn from the IVE literature offer an enticing expansion of the current MOOC model to create a new hybrid educational platform that leverages unique IVE affordances with MOOC accessibility and popularity. Ethical and philosophical issues with aspects of this technology are finally mentioned, along with suggested future research directions.

IVE Hardware Systems

What makes IVEs different than other types of virtual environments? Technologically speaking, IVEs are a dynamic feedback loop between tracking a user's movements and updating a rendered virtual scene. Physical motions are tracked, and a user is presented with a multimodal sensory experience that replaces physical stimuli with digital ones (Biocca, 1997; Loomis, 1992) creating a psychological experience of forgetting that the user is in a virtual environment at all (Witmer & Singer, 1998). On the lower end of the immersive spectrum, tracking involves using a mouse or keyboard to move a viewpoint or character around a virtual scene. Traditional video game consoles, such as the Microsoft Xbox 360, often use game controllers with joysticks and buttons to track user movements. However, those systems traditionally involve rendering to more passive display technologies, such as a desktop monitor or a television. Fully immersive setups rely on more accurate locomotion tracking systems, as well as multisensory rendering devices.

Tracking a user with extreme accuracy and precision requires a dedicated space with properly calibrated camera equipment designed to scan for certain light sources. One example of such a system is an *active optical* tracking system, where lights, such as infrared, are attached to various parts of a user and are then triangulated by multiple cameras to provide a reliable x, y, and z position in Cartesian space (Meyer, Applewhite, & Biocca, 1992; Welch, 2009). Additionally, *accelerometers* can be used to track the yaw, pitch, and roll of a user's head. This movement is especially important for an IVE where the user employs a first-person perspective and physical head movements correspond to moving a virtual viewpoint. *Passive optical* systems offer alternative and more portable solutions for tracking in a space that might not allow for designated cameras. One popular passive system is Microsoft's Kinect sensor, which uses an infrared emitter to process depth information about a physical scene to track a user without having to attach any lights (Suma, Lange, Rizzo, Krum, & Bolas, 2011; Wang, Stolka, Boctor, Hager, & Choti, 2012). Passive sensors are typically more cost-efficient (the Kinect currently sells for around $100) but operate with higher latency and lower resolutions, creating less-reliable tracking systems. At the time of this writing, new tracking products are circulating in the consumer electronic market, offering new ways to control interfaces with gesture recognition instead of traditional input devices such as mice and keyboards. Although these devices are meant to be used in less immersive applications, such as scrolling through pictures using a flick of

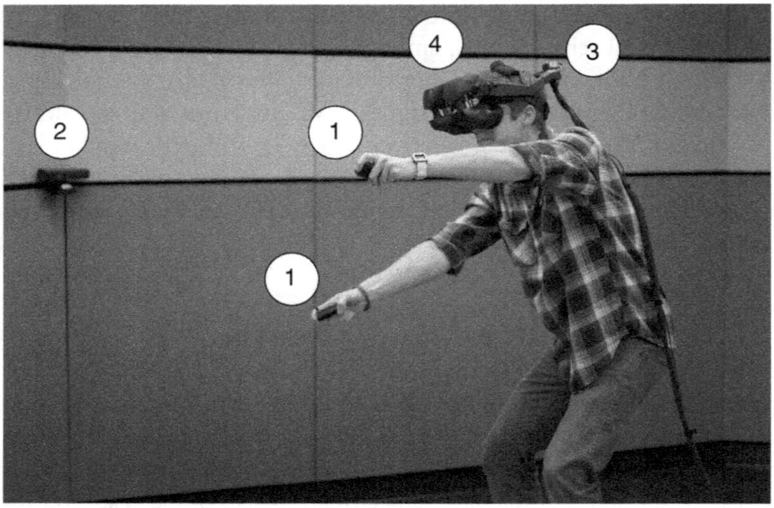

Figure 13.1 A user in an IVE pretends to fly using tracking and rendering equipment. Hands are tracked by active infrared sensors (1). Nonverbal behavior is collected by a passive infrared Microsoft Kinect device (2). Head movements are tracked by an accelerometer (3). The virtual scene is rendered from all of the tracking data and shown to the user in 3D with a head-mounted display (4).

the hand, they offer appealing applications to IVEs. One such product, the Leap Motion, offers reliable tracking of all 10 fingers simultaneously with an accuracy of 1/100th of a millimeter, which is planned to be included in select retail computers (Alzieu, 2013). If paired with an IVE, a gesture tracker such as the Leap Motion could potentially offer a more accurate and affordable passive tracking solution.

Multimodal rendering is the next step in the IVE system, utilizing sensory channels to create virtual stimuli. For sight, ***head-mounted displays*** (HMDs) use LCD displays that are displaced in front of a user's eyes and then aligned with binocular overlap to create stereoscopic depth for a rendered environment (Chung et al., 1989; Furness, 1987). HMD technologies have traditionally been utilized in research labs and large-scale simulation training, such as military flight simulators, partly due to their enormous price tags and limited market competition. For an example of tracking and rendering equipment, see Figure 13.1. However, Mark Bolas and the MxR division of the University of Southern California's Institute for Creative Technologies have developed a lens attachment that once printed, will attach to smartphones and tablets to transform them into portable HMDs. This option is especially enticing since smartphones and tablets pervade most of the consumer market and already have self-inclusive tracking systems, such as accelerometers. USC's HMD system is called VR2Go, and uses open-source software applications to create a completely mobile IVE system (Hoberman, Krum, Suma, & Bolas, 2012; Olson, Krum, Suma, & Bolas, 2011). Similarly, an HMD backing by commercial video game developers called the Oculus Rift is targeted for consumer users at a cost of around $300. Both VR2GO and the Oculus Rift have the potential to finally provide an inclusive and affordable piece of IVE hardware that consumers can comfortably use wirelessly or in their own homes.

Visual rendering is also achieved through other hardware means, such as using three-dimensional (3D) projectors or televisions. Dedicated rooms can project images on all surfaces, such as a *computer-assisted virtual environment* (CAVE®) that still tracks a user using optical systems but does not block out self-stimuli such as the hands or body (Cruz-Neira, Sandin, Defanti, Kenyon, & Hart, 1992; Sutcliffe, Gault, Fernando, & Tan, 2006). Recently, auto-stereoscopic technologies have developed to allow for a complete 3D viewing experience but without the need for any glasses at all (Woods, 2009), such as the Alioscopy lenticular television and the Nintendo 3DS handheld game console. With 3D entertainment still proving economically successful, other electronic companies are similarly building consumer-level products that maximize the immersive viewing experience but minimize the inconvenience of wearing extra devices such as 3D glasses.

Beyond visual stimuli, IVEs also utilize modalities such as *spatialized*, or virtual sound, channels that use either headphones or carefully placed speakers to create completely 3D sound experiences. In an IVE, virtual objects with attached sounds can move in 3D space, causing sounds to change based on a user's tracked movements. For example, sound volumes of an object can increase when approached by a user. Past research has shown that fully utilizing immersive sound channels can increase the immersive qualities of an IVE (Loomis, Herbert, & Cicinelli, 1990; Zahorik, 2002). *Haptic feedback*, or virtual touch, can also contribute to an IVE by providing forces or resistance with a physical device that resists a physical hand or finger. These devices can range in intricacy and price, but are usually meant to interact with the hands in some way (Basdogan, Ho, Srinivasan, & Slater, 2000; Lanier, 1992, 1997; Tan & Pentland, 1997). Vibration motors in smartphones and tablets can also be seen as a rudimentary form of haptic feedback. Olfactory senses are sometimes used in IVEs (Richard, Tijou, Richard, & Ferrier, 2006) but play less of a role in research due to their intricate and sometimes impractical device designs. One Japanese company plans to release a device that spatially moves smells around a television display with a fan system (Matsukura, Yoneda, & Ishida, 2013).

IVE Research Methods

Since its inception as a new technological tool that early pioneers projected would create an entirely new type of social space (Lanier, 1992; Rheingold, 1991), IVEs have since permeated many different research areas. This chapter focuses on reviewing the literature that uses IVEs in a social science context as a tool to not only study traditional social phenomena and theories (Loomis, Blascovich, & Beall, 1999) but to also examine the new types of experiences that IVEs create that would never have existed otherwise.

As a research tool, IVEs offer many advantages to social scientists over traditional lab-based experimental research that psychology labs have used for decades. Early work by Blascovich outlines three advantages that IVEs offer over their real-world counterparts: the mundane realism tradeoff, representative sampling, and experimental replication (Blascovich et al., 2002). The first advantage, the mundane realism tradeoff, deals with providing authentic stimuli in an IVE, as opposed to asking users to visualize these events or experiences (Blascovich et al., 2002). For example, in an IVE a virtual shark can approach a user in a threatening way, an experience that might be difficult for participants that have never physically seen a real shark. Another advantage that IVEs offer is solving nonrepresentative sampling issues by networking experiments across geographic bounds, including participants from multiple universities, states, or even countries (Blascovich et al., 2002). Recently, the online retailer Amazon has released a platform for crowdsourcing tasks called Mechanical Turk, which allows users to select specific demographic traits of online workers, such as age, native language, or educational background. Although this platform is more intended to collect data such as self-report surveys, some tasks involve navigating virtual worlds or playing an online game. One can imagine a future where it becomes even easier to recruit participants that own IVE technology at home using VR2GO or an Oculus Rift. The last advantage that Blascovich offers is ease of experimental replication, where IVEs can guarantee a consistent experience between participants, solving internal validity problems that social science experiments sometimes experience (Blascovich et al., 2002). This is especially important when dealing with *confederates*, or trained human actors, which can sometimes lack consistency based on the intricacy of the deception. Virtual actors, however, can be programmed specifically to behave the exact same way each time (Bailenson, Blascovich, Beall, & Loomis, 2001). Additionally, an IVE can transport a user to an environment that would be almost impossible to create in a traditional laboratory setting, such as an ocean, redwood forest, or even a different historical time period. These IVEs maintain consistency between uses. For example, a user that walks through a busy city sidewalk can encounter virtual agents in a completely replicated manner that would be extremely difficult, if not impossible, to achieve in a real-world setting without certain interference (Fox, Arena, & Bailenson, 2009).

In our current world, where users are constantly being tracked by browsers and smartphone applications, all actions online leave a *digital footprint* that can be used to build elaborate predictive algorithms about those users (Blascovich & Bailenson, 2011). Similarly, everything that a user does in an IVE can be recorded and output to a massive data file, allowing for more precise data analysis that can examine nonverbal behavioral variables such as spatial movement patterns, gaze, and gestures (Bailenson et al., 2001; Bailenson, Blascovich, Beall, & Loomis, 2003; Bente, Petersen, Krämer, & Ruier, 2001). For example, a study that analyzed participant proximity behaviors between *avatars*, virtual representations controlled and animated by physical humans, and *agents*, virtual representations programmed with a static set of animations or behaviors, found differences in tracking data where self-report yielded none (Bailenson, Aharoni et al., 2004). This format of large-scale data collection, a new digital trend called *big data*, is garnering attention from major companies like Amazon, Google, and Netflix that aim to track spending and browsing habits online to better predict what products users are likely to purchase next. One study looked at this in an experimental setting, where facial features of participants were tracked and processed through algorithm classifiers that predicted buyer intent above chance (Ahn, Jabon, & Bailenson, 2008). Other virtual environments on bigger scales, such *massively multiplayer online* games (MMOs), allow for custom programming scripts that can track users' gaming habits and behaviors (Friedman, Steed, & Slater, 2007; Yee & Bailenson, 2008; Yee, Bailenson, Urbanek, Chang, & Merget, 2007). One study collected massive amounts of data over a period of 10 weeks in the MMO Second Life to show the decline of user engagement over time (Harris, Bailenson, Nielsen, & Yee, 2009). The movement in a digital space for big data has much promise for educational models that seek to better understand and predict success in a more comprehensive way than traditional standardized testing, such as the SAT, but must collect data in a way that feels natural to the user. IVEs offer a solution to this in that users are often unaware that they are being tracked at all due to the engaging nature of the technology.

The big data and digital footprint models utilized in IVEs present possible solutions toward assessment problems that MOOC's face. Instead of merely assessing students based on problem sets or quizzes, machine learning can be utilized as a better predictor on how students arrive at answers based on their work (Piech, Sahami, Koller, Cooper, & Blikstein, 2012). In this technique, massive amounts of data are collected about students and then algorithms are constructed to better predict which students will struggle with the material. Because IVE constructs allow for collection of many more rich variables, such as nonverbal behavior, additional inputs can train an even better assessment method. Students in a MOOVE would capitalize on this data-driven strategy, receiving continual feedback about their performance based on not just their answers, but their problem solving methods. This approach also provides a solution to plagiarism in that digital footprint strategies are able to collect much continuous data about a user and verify that the students logging into a MOOVE are who they claim. One example of such a detection device is the Microsoft Kinect sensor, which can already detect gender based on the nonverbal data it collects about a user (Won, Yu, Janssen, & Bailenson, 2012). Tracking sensors and machine learning algorithms implemented in a MOOVE could potentially detect subtle nuances in physical movements or word syntax that would flag administrators of plagiarism. Obviously this data-driven approach is not perfect, and there is a potential for error in that students could be falsely accused of plagiarism for unexplained deviances in the collected measures. Still, big data and digital footprint measures offer potential solutions for current assessment problems in MOOCs.

Transformed Social Interaction

Because IVEs deal with a feedback loop between tracking and rendering, it is possible to intercept that process and manipulate elements in a way that a user is unaware of the changes. *Transformed social interaction* (TSI) is an area of research that examines the new social constructs this process creates, offering new situations that would not be possible in the real world (Bailenson, 2006; Bailenson, Beall, Blascovich, Loomis, & Turk, 2005; Bailenson, Beall, Loomis, Blascovich, & Turk, 2004; Bailenson, Yee, Blascovich, & Guadagno, 2008). This unique disruption of a computer-mediated process like an IVE allows for three advantages over traditional real-world interactions: enhancing normal perceptions, manipulating time and space, and identity capture of a speaker.

The first advantage TSI leverages is the ability of an IVE to augment the information of other users in the environment, such as displaying names of students or a quantified score of their current engagement level (Bailenson & Beall, 2006). A consumer example of this is a new project called Google Glass, which uses *augmented reality* (AR) to add a virtual heads-up display layer over glass frames that users wear. By connecting to the internet, wearable computing such as this provides the ability to combine facial recognition technology with available social media information to provide a real-time information system that enhances users' perceptions to recall others' names, interests, or even relationship statuses. Additionally, an IVE can allow for *multilateral perspective taking*, where a user can move to different spots in the environment (Bailenson & Beall, 2006), which can be especially useful in a lecture hall or a classroom where a student might prefer a certain location (McCall, Bailenson, Blascovich, & Beall, 2009). A user in an IVE teaching role can benefit from this enhanced sensory process, such as using an algorithm that changes the visual opaqueness of students in an IVE based on how much eye gaze is delivered to other students. As a user fails to make eye contact with other students, their opacities would decrease, reminding that user to spread gaze more evenly (Bailenson, Yee, Blascovich, Beall, et al., 2008).

The second advantage of TSI in an IVE is manipulation of time and space. For example, student users in an IVE can pause, slow down, or fast forward the speed of the content to process more difficult material at an individualized pace (Bailenson & Beall, 2006). If a lecture is entirely prerecorded, then users in hypothetical MOOVE could completely fast forward through the material, similar to current video lecture formats. If the lecture in the MOOVE is live, it can be paused and then sped up to the current broadcast point, similar to how current cable television controls work. This feature is especially important for courses with complex material and limited prerequisites, where students come from a myriad of educational backgrounds. For example, a huge computer-science MOOVE course could contain both highly technical students with intentions to thoroughly learn all aspects of the content and beginning students that aim to only learn the fundamentals. Space manipulations allow for users in an IVE to control both the environment of a classroom or the appearance of other rendered co-learners. Each individual user might choose to change the rendered environment to appear either like a large open lecture hall with hundreds of other students, or a small intimate classroom with only a dozen other students, all based on personal preference or learning style. The manipulation of virtual agents between attentive and distracted animations has been shown to affect the amount of material learned in an IVE lecture, where participants absorbed more material in a setting with other attentive co-learners (McCall et al., 2009). TSI components allow for users of all skill levels to process and control the material and environment in an efficient and individualized way to maximize learning.

In the introduction of this chapter, a learning scenario was proposed where a dull teacher was transformed into an animated and entertaining one. The third advantage of TSI, identity capture, is that this is entirely possible, allowing for digital manipulations from a teacher that other student users can then customize to fit their learning style (Bailenson & Beall, 2006). This means that attributes from a teacher such as voice pitch, volume, or even language translation are all possible in an IVE. Even further, the user's physical appearance can be subtly captured using available photographs and morphed into another user's face in an IVE without conscious recognition of the change, increasing the likability of the morphed individual (Bailenson, Garland, Iyengar, & Yee, 2006; Bailenson, Iyengar, Yee, & Collins, 2008). Although this action brings up ethical implications of persuasion, it offers additional benefits of simple changes that can change learning outcomes of interactions in a digital space.

Thus far, TSI has been shown to provide many unique opportunities to the IVE tracking and rendering process, but there are other similar technologies in place that resemble these interactions. Google has a service called Hangouts, which uses multiple webcams from users and aligns them side-by-side to allow for a live multi-user visual conference call. Other research has looked at replicating an experience like this in an IVE with avatars (Benford, Greenhalgh, Rodden, & Pycock, 2001; Joslin, Di Giacomo, & Magnenat-Thalmann, 2004; Reeves, Malone, & O'Driscoll, 2008). Already on the consumer market is a free application developed for the Microsoft Xbox 360 called Avatar Kinect that utilizes a Kinect sensor to track users' gestures, and places them in a virtual conference room where they can chat with friends. Environments can be changed, avatars can be customized, and multimedia elements such as sounds and objects can be added to enhance the communication experience. Although Avatar Kinect does not allow for complete control over the rendering process that is required to fully utilize TSI elements, it presents a novel example of a less immersive technology system already available for console owners. Discussion sections, an important element of most physical university classes where students meet in person to expand on lecture material, could use similar IVE technology to host virtual sections where users control avatars and go over course material in an engaging way. A version of a MOOVE could even expand to consoles, meaning devices like the Microsoft Xbox 360 and Kinect could be heavily utilized to track users and construct a more interactive and customizable learning platform.

When it comes to TSI, certain affordances of an IVE also offer solutions for accountability and personalization issues in MOOCs. In a current MOOC platform, students interact passively with lectures and do so fairly anonymously. However, in a hybrid MOOVE platform students could partake in a lecture by using IVE equipment. The rendering and tracking decoupling that TSI creates allows each student in the virtual lecture receive a personal callout from the lecturer, such as addressing the student by his or her name and asking an essential question. Lecturers are not able to respond to all students simultaneously, but a MOOVE could play back a collection of standardized responses based on the student's response. This unique interaction allows for students to engage with the material in a way that would be impossible in the current MOOC model, possibly solving high attrition rates and making the learning environment feel more personal.

Social Learning Theory

As humans, we often look to others to model what behaviors are acceptable for ourselves. *Social cognitive theory* conjectures that when a person refers to a model, similarity between the two can determine how likely behavior change is to occur (Bandura, 2001; Bandura & Huston, 1961). Characteristics such as gender (Bussey & Perry, 1982), age (Kazdin, 1976), skill aptitude (Meichenbaum, 1971), background similarity (Hilmert, Kulick, & Christenfeld, 2006), and prior behaviors (Andsager, Bemker, Choi, & Torwel, 2006) all can contribute toward the chance that a model will be effective in creating behavior change. For example, an individual watching a video of oneself can be a useful tool in using the self as a model but is limited to actions the individual has already completed, making it difficult to provide models for previously uncompleted or impossible actions (Dowrick, 1999). More specifically, *social learning theory* expands on this notion by detailing three types of effective models that can lead to behavior change: a live model, such as a physical person performing an action; a verbal instruction model, such as explaining a set of actions to someone; or a symbolic model, such as a digital character performing an action through a technology device (Bandura, 1977). For example, if a person is watching a public service announcement commercial on television, the extent to which that person feels similar to the character giving the announcement will predict if a behavior change occurs (Bandura, 1977). In an IVE, the social learning theory has been studied through three manipulation techniques called the Proteus effect, the chameleon effect, and doppelgangers, all of which offer opportunities to leverage affordances of the digital platform to enhance learning and provide more effective models.

The Proteus effect

Self-perception theory has shown that attitudes and beliefs are formed by examining behaviors of the self, instead of forming behaviors from those initial attitudes (Bem, 1967). *The Proteus effect*, inspired by self-perception theory and named after the Greek god Proteus that could change his form, encompasses an area of research that examines the behavioral effects of changing a user's embodied avatar in an IVE (Yee & Bailenson, 2007; Yee, Bailenson, & Ducheneaut, 2009). For example, when a user wears an avatar that is more attractive or taller, that user is more willing to reveal personal details and decrease interpersonal distance when interacting with a confederate (Yee & Bailenson, 2007). Other studies show that embodying an older avatar decreases stereotypes about elderly groups (Yee & Bailenson, 2006) and increases retirement savings allocations (Hershfield et al., 2011). Additionally, female users that embody or view sexualized avatars, meaning more scantily clad in dress, reported more body-related thoughts (Fox & Bailenson, 2009b; Fox, Bailenson, & Tricase, 2013).

Applications for the Proteus effect in MOOVE's where users are allowed to customize their digital selves are numerous. For example, an older student could embody a younger avatar to feel more comfortable in a virtual classroom dominated by other younger students. Women users in science, technology, engineering, and math

(STEM) courses, commonly dominated by male students, could similarly embody avatars of the opposite gender to decrease the barriers of stereotype threat. There are some dangers in this customizable option, however, since some studies have also shown that embodying an avatar of another race can actually increase implicit racial bias (Groom, Bailenson, & Nass, 2009). Still, the benefits that the Proteus effect offer for MOOVEs creates an exciting possibility of a virtual classroom where no student has to feel disadvantaged due to physical appearances.

Digital chameleons

Nonverbal cues in social interactions between humans show to be important drivers in communication interactions. When confederates in a virtual social interaction mimicked the nonverbal behaviors of an individual, that confederate was reported to be more likable (Chartrand & Bargh, 1999). In an IVE setting, *the chameleon effect* continued this finding to show that a virtual agent mimicking a user's tracked head movements at a 4 s delay was more persuasive and received more positive trait ratings than control conditions (Bailenson & Yee, 2005). Morphing a user's face into other virtual agents to increase facial similarity, similar to the TSI work explained earlier, increased problem-solving task performance (Bailenson & Yee, 2006). This effect also extends to other actions, such a virtual handshake where users reported agents that mimicked their handshakes as more likeable (Bailenson & Yee, 2007).

Similar to TSI applications outlined earlier, utilizing this digital chameleon effect in an IVE is easily done since a user's actions are very accurately tracked and recorded. Leveraging this in a MOOVE would allow for enhanced teacher ability, where a student's tracked movements could be mimicked at an undetectable delay, creating a more likable and persuasive lecturer. Facial similarity morphing could also be leveraged in a MOOVE so that a student user's face could morph not only into the teacher, but other attending students, as well. Again, these bring up ethical implications of swaying a user to more easily accept material without their consent of the manipulation. Still, the potential for creating a more successful learning environment for students that might be more introverted could benefit from such a manipulation in an IVE, allowing for better collaborations on peer assignments or group work.

Doppelgangers

The *doppelganger* is a virtual agent modeled with a user's captured visual identity, which can be animated asynchronously from the user's actual physical movements. This allows for expansion of the social learning theory, constructing virtual behaviors and actions that a user might not be able to actually physically perform. Users that interacted in front of their doppelganger in an IVE were more willing to perform embarrassing acts in front of that character as opposed to a dissimilar agent (Bailenson, Beall, Blascovich, Weisbuch, & Raimmundo, 2001; Bailenson, Yee, Blascovich, & Guadagno, 2008) and rated those agents with higher attractiveness and liking (Bailenson et al., 2008). Similar effects also translate to pro-social behavior change, where users in an IVE shown their doppelgangers exercising actually performed more physical exercise the following day (Fox & Bailenson, 2009). However, when a user's doppelganger endorsed a product in an IVE, the user later rated that product with higher association and more positive ratings (Ahn & Bailenson, 2011), leading to a potentially influencing opportunity for advertising giants to exploit the digital chameleon effect for profit. Children are also particularly vulnerable to this effect. For example, a study found that elementary children constructed false memories from an IVE where they saw their doppelganger swim with orca, thinking that they later thought they had actually encountered real orca whales (Segovia & Bailenson, 2009).

In a potential MOOVE, doppelgangers have the opportunity to create both positive and negative outcomes in an educational context. On the one hand, a doppelganger can be used virtually to encourage pro-social habits such as the effects of eating healthy in a lesson on nutrition. Users in a MOOVE listening to a lecture on proper nutrition could view how certain behaviors such as eating junk food or smoking cigarettes affect the body, by vividly and personally showing those changes on a virtual self. To boost confidence, users giving a speech in a MOOVE could also have the option to change every character in the audience to be a doppelganger, decreasing the anxiety of performing an embarrassing act in front of a crowd. Users could also choose to change the teacher in a lesson to be their doppelgangers, essentially receiving a lecture from their digital selves. All of these IVE affordances make a MOOVE feel more personal to the student's own learning style. This is the ultimate extension of the social learning theory, constructing a virtual model of the self that enables the user to be confident that they too can become an expert at the learning material.

Embodied Cognition Theory

Recent technologies such as the Microsoft Kinect and the Nintendo Wii signal a trend of commercial video games that encourages users to interactively move their bodies. Past research has looked at the role bodily states and actions play in perceptual processes (Barsalou, 2008). *Embodied cognition* looks at the effects that active body movements have on higher-order cognition, where perceptual information creates certain mental schemas. For example, past research has shown that nodding while listening causes people to more likely agree with statements (Wells & Petty, 1980) and that sitting upright as opposed to slumping causes more pride in achievement when people receive performance feedback (Stepper & Strack, 1993).

In an IVE, the multisensory and highly active nature of the technology allows for complex tracking and haptic systems to test embodied experiences. When receiving stimuli, users in an IVE can actively move around interact with those stimuli. For example, users that controlled a haptic device in an IVE to saw down a virtual tree changed their conservation behaviors by reducing measured paper usage as opposed to just reading about deforestation (Ahn, 2011). In another study, users in an IVE that took the perspective of a colorblind individual and performed a difficult task with a haptic device were more willing to volunteer their time later to help a confederate pretending to be colorblind (Ahn, Le, & Bailenson, 2013). This altruistic finding also extended to a study where users that embodied a superhero by actively moving their arms were more willing to help the researcher clean up from an accident (Rosenberg, Baughman, & Bailenson, 2013).

Embodied cognition findings also extend to education. When participants actively swung their arms in a manner that resembled the solution to a physics problem, those participants were more likely to come up with the correct answer than those that just stretched randomly (Thomas & Lleras, 2009). In a proposed MOOVE, limb tracking and active movement can be digitally integrated in a way that is beneficial for the type of educational content. For example, if students are learning about ocean currents in an environmental science lecture, they can use their hands to change current direction and speeds of an interactive model, all while viewing effects on the underwater ecosystems. Similar to the superhero experiment, students can also choose to navigate through the virtual environment by swimming with their hands, increasing the realism and making the MOOVE lesson more engaging. Since users are being tracked, feedback systems can be built to dynamically encourage user behaviors, such as physically sitting up straight or nodding along with the teacher, to maximize learning effectiveness. In an earlier TSI example, one where a student was actually called on personally by a lecturer, would have the chance to respond with voice recognition software. This act of using the voice to talk is one that is sparse in current MOOC models and has been shown to actually increase feelings of presence in an IVE (Aymerich-Frank, Karutz & Bailenson, 2012). If a student was able to respond to a question in a MOOVE, it could keep that student more engaged and accountable for fully completing a course.

Conclusion

The balance between positive and negative outcomes of digitally mediated technologies is what Blascovich and Bailenson (2011) call the "virtual yin and yang," referring to a confluence of forces that can both help and harm. IVEs are the same way, when used appropriately they offer physically impossible advantages, such as traveling to a virtual scene that a user may never experience otherwise. Yet when used to exploit, IVEs can also offer a platform for subtle mind control, which propaganda media have utilized historically to sway the public on certain issues. Imagine the danger if certain extremist political groups were able to leverage IVE findings to persuade users of radical agendas. One solution to the negative possibilities IVEs create is awareness. Blascovich and Bailenson advocate this (2011) by encouraging the public to watch their digital footprint and reconsider things like posting high-resolution photos online in social media or other platforms where user behavior is closely tracked. Like all things, healthy media consumption is a balance between analog and digital technologies, especially when it comes to more realistic media such as IVEs. As this immersive technology becomes more ubiquitous with devices like wireless smartphone HMDs and infrared tracking sensors, it is important to be mindful of the possibility of addiction. If an IVE is easily accessible by a device such as a smartphone or a tablet, what is to stop users from choosing to spend more and more time in an extremely engaging and vivid virtual world? New immersive video game devices such as the Nintendo 3DS handheld encourage users to take breaks every 30 min to avoid becoming sick (Nunneley, 2011). But addiction is a powerful force, and merely encouraging users to take breaks might not be enough.

Digital balance also applies to virtual selves. Since eternal digital life is entirely feasible through an avatar and user identity capture, what are the implications for a virtual self that never ages or dies? New technologies such as Kinect Fusion, which uses the Kinect sensor as a 3D scanner, can cheaply and quickly construct a lifelike 3D model of a person. Used positively, this could allow for children to interact virtually with grandparents that have long since passed away, or to preserve especially talented professors digitally. Although those professors might waver in their ability to give an effective lecture in reality, their digital doppelgangers can live on to lecture undergraduates forever. But postmortem, users need to consider the consequences regarding who has the rights to their digital doppelgangers. At Coachella in 2011, a popular music festival, the deceased rapper Tupac's doppelganger was digitally constructed and used to give a live performance, an experience that both excited and chilled viewers (Bailenson, 2012). Similarly, Orville Redenbacher's doppelganger has been featured in many of the company's commercials to endorse popcorn (Blascovich & Bailenson, 2011). Even now, Google is catching on to this eternal digital data, offering a new service called Inactive Account Manager that handles who owns the rights to users' data after they die (Rosen, 2013).

Studying the psychological effects behind these emergent technologies is incredibly difficult since social scientists have to work at following the blistering pace at which they develop. Future research directions still need to examine the longitudinal effects of IVEs, as well as how healthy new technologies such as VR2GO or the Oculus Rift are for users. Effects found for much of the research reviewed in this chapter look at behavior change from IVEs in a much more immediate context. But when IVE technology becomes more pervasive, how will lasting effects change behaviors many years later, and for better or for worse?

In the context of education, the proposed hybrid MOOVE model offers many new applications of IVE theory, few of which have been studied experimentally in an educational setting. Even the MOOC model, which is still at its nascent state, lacks comprehensive assessment of its effectiveness due to its still sprawling and evolving form. Assessment methods are still formulating, and what follows completion of a MOOC is still in question, such as receiving certifications and badges instead of official university credit. Software is still a challenge with the MOOC model since there is no current universal platform but a handful of startups building their own versions. For example, NovoEd is a startup evolved out of Stanford University that aims to solve currently high attrition rates in MOOCs, which hovers around 5% (Corcoran, 2013), by dividing students up into small groups and keeping them accountable for each other (Empson, 2013). With IVE technology, student collaboration in this manor could be improved even further. Tackling universal hardware across students is also another challenge. With so many devices to choose from, how can students in a MOOVE ensure that they have a similar experience? Especially when many areas of the world still lack high speed internet, IVE's require a system of low latency sensors and fast rendering rates in order to avoid making users sick.

Regardless of the technological challenges, the goal to radically change traditional educational environments is established and growing. As MOOCs find success and failure in universities, students will be the driving force in choosing just how they want their education delivered. Customizability and individualized education in an online platform provides an enticing package that public schools will have a challenging time replicating with continued budget woes. As IVE technology becomes cheaper and more widespread, users will become more and more comfortable interacting in multisensory virtual worlds. It makes sense that the marriage of the two technologies, MOOCs and IVEs, has the potential to provide an extremely high-quality and affordable educational platform. One where your professor, be it Mark Ruffalo's avatar or your digital doppelganger, might end up teaching future generations in the classrooms of tomorrow.

References

Ahn, S. J. (2011). *Embodied experiences in immersive virtual environments: Effects on pro-environmental attitude and behavior* (Doctoral dissertation). Stanford University, Stanford, CA.

Ahn, S. J., & Bailenson, J. N. (2011). Self-endorsing versus other-endorsing in virtual environments: The effect on brand attitude and purchase intention. *Journal of Advertising*, *40*(2), 93–106.

Ahn, S. J., Jabon, M. E., & Bailenson, J. N. (2008). Facial expressions as predictors of online buying intention. *Proceedings of the 58th Annual International Communication Association Conference*, May 22–26. Montreal, Canada.

Ahn, S. J., Le, A. M. T., & Bailenson, J. N. (2013). The effect of embodied experiences on self-other merging, attitude, and helping behavior. *Media Psychology, 16,* 17–38.

Alzieu, V. (2013). *Leap motion controller coming to HP computers. Digital Versus.* Retrieved fromhttp://www.digitalversus.com/motion-sensors/leap-motion-the-leap-p15766/leap-motion-controller-coming-hp-computers-n29173.html

Andsager, J. L., Bemker, V., Choi, H. L., & Torwel, V. (2006). Perceived similarity of exemplar traits and behavior effects on message evaluation. *Communication Research*, *33*(1), 3–18.

Aymerich-Franch, L., Karutz, C., & Bailenson, J. N. (2012). Effects of facial and voice similarity on presence in a public speaking virtual environment. *Proceedings of the International Society for Presence Research Annual Conference*, October 24–26. Philadelphia, PA.

Bailenson, J. N. (2006). Transformed social interaction in collaborative virtual environments. In P. Messaris, & L. Humphreys (Eds.), *Digital media: Transformations in human communication* (pp. 255–264). New York, NY: Peter Lang.

Bailenson, J. N. (2012). *Holograms are here, from Tupac to Marilyn Monroe. Will you be next? The Washington Post.* Retrieved from http://www.washingtonpost.com/opinions/holograms-are-here-from-tupac-to-marilyn-monroe-will-you-be-next/2012/11/30/92ff5d02-38d3-11e2-8a97-363b0f9a0ab3_story.html

Bailenson, J. N., Aharoni, E., Beall, A. C., Guadagno, R. E., Dimov, A., & Blascovich, J. (2004). Comparing behavioral and self-report measures of embodied agents' social presence in immersive virtual environments. *Proceedings of the 7th Annual International Workshop on PRESENCE*, Valencia, Spain.

Bailenson, J. N., & Beall, A. C. (2006). Transformed social interaction: Exploring the digital plasticity of avatars. In R. Schroeder, & A.-S. Axelsson, (Eds.), *Avatars at work and play: Collaboration and interaction in shared virtual environments* (pp. 1–16). Dordrecht, The Netherlands: Springer-Verlag.

Bailenson, J. N., Beall, A. C., Blascovich, J., Loomis, J., & Turk, M. (2005). Transformed social interaction, augmented gaze, and social influence in immersive virtual environments. *Human Communication Research, 31*, 511–537.

Bailenson, J. N., Beall, A. C., Blascovich, J., Weisbuch, M., & Raimmundo, R. (2001). Intelligent agents who wear your face: Users' reactions to the virtual self. *Lecture Notes in Artificial Intelligence, 2190*, 86–99.

Bailenson, J. N., Beall, A. C., Loomis, J., Blascovich, J., & Turk, M. (2004). Transformed social interaction: Decoupling representation from behavior and form in collaborative virtual environments. *Presence: Teleoperators and Virtual Environments, 13*(4), 428–441.

Bailenson, J. N., Blascovich, J., Beall, A. C., & Loomis, J. M. (2001). Equilibrium theory revisited: Mutual gaze and personal space in virtual environments. *Presence: Teleoperators and Virtual Environments, 10*, 583–598.

Bailenson, J. N., Blascovich, J., Beall, A. C., & Loomis, J. M. (2003). Interpersonal distance in immersive virtual environments. *Personality & Social Psychology Bulletin, 29*, 819–833.

Bailenson, J. N., Garland, P., Iyengar, S., & Yee, N. (2006). Transformed facial similarity as a political cue: A preliminary investigation. *Political Psychology, 27*, 373–386.

Bailenson, J. N., Iyengar, S., Yee, N., & Collins, N. (2008). Facial similarity between voters and candidates causes influence. *Public Opinion Quarterly, 72*(5), 935–961.

Bailenson, J. N., & Yee, N. (2005). Digital chameleons: Automatic assimilation of nonverbal gestures in immersive virtual environments. *Psychological Science, 16*, 814–819.

Bailenson, J. N., & Yee, N. (2006). A longitudinal study of task performance, head movements, subjective report, simulator sickness, and transformed social interaction in collaborative virtual environments. *Presence: Teleoperators and Virtual Environments, 15*(6), 699–716.

Bailenson, J. N., & Yee, N. (2007). Virtual interpersonal touch and digital chameleons. *Journal of Nonverbal Behavior, 31*, 225–242.

Bailenson, J. N., Yee, N., Blascovich, J., Beall, A. C., Lundblad, N., & Jin, M. (2008). The use of immersive virtual reality in the learning sciences: Digital transformations of teachers, students, and social context. *The Journal of the Learning Sciences, 17*, 102–141.

Bailenson, J. N., Yee, N., Blascovich, J., & Guadagno, R. E. (2008). Transformed social interaction in mediated interpersonal communication. In E. Konijn, M. Tanis, S. Utz, & A. Linden (Eds.), *Mediated interpersonal communication* (pp. 77–99). New York (NY): Lawrence Erlbaum Associates.

Bandura, A. (1977). *Social learning theory*. Englewood Cliffs, NJ: Prentice-Hall.

Bandura, A. (2001). Social cognitive theory of mass communication. *Media Psychology, 3*, 265–299.

Bandura, A., & Huston, A. C. (1961). Identification as a process of incidental learning. *Journal of Abnormal and Social Psychology, 63*(2), 311–318.

Barsalou, L. W. (2008). Grounded cognition. *Annual Review of Psychology, 59*, 617–645.

Basdogan, C., Ho, C.-H., Srinivasan, M. A., & Slater, M. (2000). An experimental study on the role of touch in virtual environments. *ACM Transactions on Computer-Human Interaction, 7*, 443–460.

Bem, D. J. (1967). Self-perception: An alternative interpretation of cognitive dissonance phenomena. *Psychological Review, 74*(3), 183–200.

Benford, S., Greenhalgh, C., Rodden, T., & Pycock, J. (2001). Collaborative virtual environments. *Communications of the ACM*, *44*(7), 79–85.

Bente, G., Petersen, A., Krämer, N. C., & de Ruiter, J. P. (2001). Transcript-based computer animation of movement: Evaluating a new tool for nonverbal behavior research. *Behavior Research Methods, Instruments, & Computers*, *33*, 303–310.

Biocca, F. (1997, August). The cyborg's dilemma: Embodiment in virtual environments. In J. P. Marsh (Ed.), *Proceedings of the Second International Conference on Cognitive Technology, 1997, Humanizing the Information Age*, August 25–28, 1997, Aizu-Wakamatsu City, Japan (pp. 12–26). Los Alamitos, CA: IEEE.

Blascovich, J., & Bailenson, J. N. (2011). *Infinite reality—Avatars, eternal life, new worlds, and the dawn of the virtual revolution*. New York, NY: William Morrow.

Blascovich, J., Loomis, J., Beall, A. C., Swinth, K. R., Hoyt, C. L., & Bailenson, J. N. (2002). Immersive virtual environment technology as a methodological tool for social psychology. *Psychological Inquiry*, *13*, 103–124.8–9.

Bussey, K., & Perry, D. G. (1982). Same-sex imitation: The avoidance of cross-sex models or the acceptance of same-sex models? *Sex Roles*, *8*(7), 773–784.

Chartrand, T. L., & Bargh, J. A. (1999). The chameleon effect: The perception-behavior link and social interaction. *Journal of Personality & Social Psychology*, *76*, 893–910.

Chung, J. C., Harris, M. R., Brooks, F. P., Fuchs, H., Kelley, M. T., Hughes, J., … Pique, M. (1989). Exploring virtual worlds with headmounted displays. *Proceedings of SPIE*, *1083*, 42–52.

Cooper, S., & Sahami, M. (2013). Reflections on Stanford's MOOCs. *Communications of the ACM*, *56*(2), 28–30.

Corcoran, B. (2013). *Adding collaboration to massive online courses. edSurge*. Retrieved from https://www.edsurge.com/n/2013-04-15-adding-collaboration-to-massive-online-courses

Cruz-Neira, C., Sandin, D. J., DeFanti, T. A., Kenyon, R. V., & Hart, J. C. (1992). The CAVE: Audio visual experience automatic virtual environment. *Communications of the ACM*, *35*(6), 64–72.

Dowrick, P. W. (1999). A review of self modelling and related interventions. *Applied & Prevention Psychology*, *8*, 23–39.

Empson, R. (2013). *Stanford's NovoEd brings collaboration and group learning to MOOCs to help fight attrition. TechCrunch*. Retrieved from http://techcrunch.com/2013/04/15/stanfords-novoed-brings-collaboration-and-group-learning-to-moocs-to-help-fight-attrition/

Fini, A. (2009). The technological dimension of a massive open online course: The case of the CCK08 course tools. *The International Review of Research in Open and Distance Learning*, *10*(5), 1–26.

Fowler, G. (2013). *Online education lifts pass rates at university. The Wall Street Journal*. Retrieved from http://online.wsj.com/article_email/SB10001424127887323741004578414861572832182-lMyQjAxMTAzMDEwMDExNDAyWj.html

Fox, J., Arena, D., & Bailenson, J. N. (2009). Virtual reality: A survival guide for the social scientist. *Journal of Media Psychology*, *21*(3), 95–113.

Fox, J., & Bailenson, J. N. (2009a). Virtual self-modeling: The effects of vicarious reinforcement and identification on exercise behaviors. *Media Psychology*, *12*, 1–25.

Fox, J., & Bailenson, J. N. (2009b). Virtual virgins and vamps: The effects of exposure to female characters' sexualized appearance and gaze in an immersive virtual environment. *Sex Roles*, *61*(3–4), 147–157.

Fox, J., Bailenson, J. N., & Tricase, L. (2013). The embodiment of sexualized virtual selves: The Proteus effect and experiences of self-objectification via avatars. *Computers in Human Behavior*, *29*, 930–938.

Friedman, D., Steed, A., & Slater, M. (2007). Spatial social behavior in second life. In C. Pelachaud (Ed.), *Intelligent virtual agents* (pp. 252–263). New York, NY: Springer.

Friedman, T. (2013). *The professors' big stage. The New York Times.* Retrieved from http://www.nytimes.com/2013/03/06/opinion/friedman-the-professors-big-stage.html?_r=0

Furness, T. (1987). Designing in virtual space. In W. B. Rouse & K. R. Boff (Eds.), *System design: Behavioral perspectives on designers, tools, & organization* (pp. 127–143). New York, NY: North-Holland.

Gaudiosi, J. (2012). *Riot games' league of legends officially becomes most played PC game in the World. Forbes.* Retrieved from http://www.forbes.com/sites/johngaudiosi/2012/07/11/riot-games-league-of-legends-officially-becomes-most-played-pc-game-in-the-world/

Groom, V., Bailenson, J. N., & Nass, C. (2009). The influence of racial embodiment on racial bias in immersive virtual environments. *Social Influence*, *4*(1), 1–18.

Harris, H., Bailenson, J. N., Nielsen, A., & Yee, N. (2009). The evolution of social behavior over time in second life. *Presence: Teleoperators & Virtual Environments*, *18*(6), 294–303.

Hershfield, H. E., Goldstein, D. G., Sharpe, W. F., Fox, J., Yeykelis, L., Carstensen, L. L., & Bailenson, J. N. (2011). Increasing saving behavior through age-progressed renderings of the future self. *Journal of Marketing Research*, *48*, S23–S37.

Hilmert, C. J., Kulik, J. A., & Christenfeld, N. J. (2006). Positive and negative opinion modeling: The influence of another's similarity and dissimilarity. *Journal of Personality and Social Psychology*, *90*(3), 440–452.

Hoberman, P., Krum, D. M., Suma, E. A., & Bolas, M. (2012, March). Immersive training games for smartphone-based head mounted displays. In *Virtual Reality Short Papers and Posters (VRW), 2012 IEEE* (pp. 151–152). IEEE.

Joslin, C., Di Giacomo, T., & Magnenat-Thalmann, N. (2004). Collaborative virtual environments: From birth to standardization. *IEEE Communications*, *42*(4), 28–33.

Kaiser Family Foundation. (2010). *Generation M2: Media in the lives of 8–18-year olds.* Menlo Park, CA: Rideout, V., Foehr, U., & Roberts, D.

Kazdin, A. E. (1976). Effects of covert modeling, multiple models, and model reinforcement on assertive behavior. *Behavior Therapy*, *7*(2), 211–222.

Kizilcec, R. F., Piech, C., & Schneider, E. (2013). Deconstructing disengagement: Analyzing learner subpopulations in massive open online courses. In D. Suthers, K. Verbert, E. Duval, & X. Ochoa (Eds.), *Proceedings of the Third International Conference on Learning Analytics and Knowledge* (pp. 170–179). New York, NY: ACM.

Lanier, J. (1992). Virtual reality: The promise of the future. *Interactive Learning International*, *8*, 275–279.

Lanier, J. (1997). Beyond goggles and gloves. *Byte*, *22*(9), 32.

Loomis, J. M. (1992) Distal attribution and presence. *Presence: Teleoperators and Virtual Environments*, *1*, 113–119.

Loomis, J. M., Blascovich, J. J., & Beall, A. C. (1999). Immersive virtual environment technology as a basic research tool in psychology. *Behavior Research Methods, Instruments, and Computers*, *31*, 557–564.

Loomis, J. M., Hebert, C., & Cicinelli, J. G. (1990). Active localization of virtual sounds. *Journal of the Acoustical Society of America*, *88*, 1757–1764.

Martin, F. G. (2012). Will massive open online courses change how we teach? *Communications of the ACM*, *55*(8), 26–28.

Matsukura, H., Yoneda, T., & Ishida, H. (2013). Smelling screen: Development and evaluation of an olfactory display system for presenting a virtual odor source. *Visualization and Computer Graphics, IEEE Transactions on*, *19*(4), 606–615.

McAuley, A., Stewart, B., Siemens, G., & Cormier, D. (2010). *The MOOC model for digital practice.* Retrieved from, http://www.elearnspace.org/Articles/MOOC_Final.pdf

McCall, C., Bailenson, J. N., Blascovich, J., & Beall, A. C. (2009). Leveraging collaborative virtual environment technology for inter-population research on persuasion in a classroom setting. *Presence: Teleoperators & Virtual Environments, 18*(5), 361–369.

Meichenbaum, D. H. (1971). Examination of model characteristics in reducing avoidance behavior. *Journal of Personality and Social Psychology, 17*(3), 298.

Meyer, K., Applewhite, H. L., & Biocca, F. A. (1992). A survey of position trackers. *Presence: Teleoperators and Virtual Environments, 1*(2), 173–200. (ISSN 1054-7460).

Nunneley, S. (2011). *3DS: Nintendo recommends a break every 30-minutes when playing games on the handheld. VG24/7.* Retrieved from http://www.vg247.com/2011/02/23/3ds-nintendo-recommends-a-break-every-30-minutes-when-playing-games-on-the-handheld/?

Olson, J. L., Krum, D. M., Suma, E. A., & Bolas, M. (2011, March). A design for a smartphone-based head mounted display. In *Virtual Reality Conference (VR)*, 2011 IEEE (pp. 233–234). New York, NY: IEEE.

Piech, C., Sahami, M., Koller, D., Cooper, S., & Blikstein, P. (2012, February). Modeling how students learn to program. In L. A. Smith King, D. R. Musicant, T. Camp, & P. T. Tymann (Eds.), *Proceedings of the 43rd ACM Technical Symposium on Computer Science Education* (pp. 153–160). New York, NY: ACM.

Reeves, B., Malone, T. W., & O'Driscoll, T. (2008). Leadership's online labs. *Harvard Business Review, 86*(5), 58–66.

Rheingold, H. (1991). *Virtual reality.* New York, NY: Simon & Schuster.

Richard, E., Tijou, A., Richard, P., & Ferrier, J. L. (2006). Multi-modal virtual environments for education with haptic and olfactory feedback. *Virtual Reality, 10*(3), 207–225.

Rodriguez, C. O. (2012). *MOOCs and the AI-Stanford like Courses: Two Successful and Distinct Course Formats for Massive Open Online Courses. European Journal of Open and Distance Learning, 15.* Retrieved March 10, 2013 from http://www.eurodl.org/index.php?p=current&article=516

Rosen, R. (2013). *Google death: A tool to take care of your Gmail when you're gone. The Atlantic.* Retrieved from http://www.theatlantic.com/technology/archive/2013/04/google-death-a-tool-to-take-care-of-your-gmail-when-youre-gone/274934/

Rosenberg, R. S., Baughman, S. L., & Bailenson, J. N. (2013). Virtual superheroes: Using superpowers in virtual reality to encourage prosocial behavior. *PLOS One, 8*(1), 1–9.

Segovia, K. Y. & Bailenson, J. N. (2009). Virtually true: Children's acquisition of false memories in virtual reality. *Media Psychology, 12*, 371–393.

Stepper, S., & Strack, F. (1993). Proprioceptive determinants of emotional and nonemotional feelings. *Journal of Personality and Social Psychology, 64*, 211–211.

Suma, E. A., Lange, B., Rizzo, A., Krum, D. M., & Bolas, M. (2011, March). Faast: The flexible action and articulated skeleton toolkit. In *Virtual Reality Conference (VR)*, 2011 IEEE (pp. 247–248). New York, NY: IEEE.

Sutcliffe, A., Gault, B., Fernando, T., & Tan, K. (2006). Investigating interaction in CAVE virtual environments. *ACM Transactions on Computer-Human Interaction, 13*, 235–267.

Tan, H. Z., & Pentland, A. (1997). Tactual displays for wearable computing. *Personal Technologies, 1*, 225–230.

Thomas, L. E., & Lleras, A. (2009). Swinging into thought: Directed movement guides insight in problem solving. *Psychonomic Bulletin & Review, 16*(4), 719–723.

Wang, X., Stolka, P., Boctor, E., Hager G., & Choti M. (2012). The Kinect as an interventional tracking system. Medical Imaging 2012: Image-guided procedures. *Robotic Interventions, and Modeling, 8316*, 1–6.

Welch, G. F. (2009). History: The use of the Kalman filter for human motion tracking in virtual reality. *Presence: Teleoperators and Virtual Environments, 18*(1), 72–91.

Wells, G. L., & Petty, R. E. (1980). The effects of over head movements on persuasion: Compatibility and incompatibility of responses. *Basic and Applied Social Psychology, 1*(3), 219–230.

Witmer, B. G., & Singer, M. J. (1998). Measuring presence in virtual environments: A presence questionnaire. *Presence: Teleoperators and Virtual Environments, 7*(3), 225–240.

Woods, A. (2009). 3-D displays in the home. *Information Display, 362*, 8–12.

Won, A. S., Yu, L., Janssen, J. H., & Bailenson, J. N. (2012). Tracking gestures to detect gender. *Proceedings of the International Society for Presence Research Annual Conference,* October 24–26. Philadelphia, PA.

Yee, N., & Bailenson, J. N. (2006). Walk a mile in digital shoes: The impact of embodied perspective-taking on the reduction of negative stereotyping in immersive virtual environments. *Proceedings of PRESENCE 2006: The 9th Annual International Workshop on Presence,* August 24 & 26, Cleveland, OH.

Yee, N., & Bailenson, J. N. (2007). The Proteus effect: Self transformations in virtual reality. *Human Communication Research, 33*, 271–290.

Yee, N., & Bailenson, J. N. (2008). A method for longitudinal behavioral data collection in Second Life. *Presence: Teleoperators and Virtual Environments, 17*(6), 594–596.

Yee, N., Bailenson, J. N., & Ducheneaut, N. (2009). The Proteus effect: Implications of transformed digital self-representation on online and offline behavior. *Communication Research, 36*(2), 285–312.

Yee, N., Bailenson, J. N., Urbanek, M., Chang, F., & Merget, D. (2007). The unbearable likeness of being digital: The persistence of nonverbal social norms in online virtual environments. *Cyberpsychology and Behavior, 10*, 115–121.

Zahorik, P. (2002). Assessing auditory distance perception using virtual acoustics. *Journal of the Acoustical Society of America, 111*, 1832–1846.

Part IV

Adoption, Use and Abuse of Communication Technologies

14

The Psychology of the Diffusion and Acceptance of Technology

Arun Vishwanath

Department of Communication, University at Buffalo, SUNY at Buffalo, NY, USA

Exactly a quarter of a century since the publication of the first edition of *Diffusion of Innovations* (Rogers, 1962), the theory remains the dominant paradigm applied to the understanding of how a new technology spreads through a social system. The diffusion of innovations is a middle range theory (Merton, 1968), one that provides empirically derived and testable propositions about the generic diffusion process, its constituents, antecedents, and consequences, all of which could be easily adapted to advance the study of the diffusion of any innovation (Vishwanath & Barnett, 2011).

What makes the theory particularly attractive is the breadth of its scope and explanations. Beginning with a broad definition of what constitutes an innovation—any idea, practice, product, technique, or technology that diffuses through a social system—the theory includes as predictors proximate factors such as change agent efforts and the perceived attributes of the innovation and ultimate factors such as cultural norms and types of communication channels that exist in the social system. The key to understanding the psychology of adoption is in the examination of perceived attributes of a new technology, how these beliefs emerge, and how they influence the individual's decision to adopt (or reject) the technology.

There are, however, close to 50,000 articles that cite Everett Rogers on Google Scholar, and thousands more that focus on these processes, making it impossible to synthesize and present the entire literature. Thus, a convenient way to distill the overall research is to delineate them based on how they approach the measurement and modeling of the adopters' beliefs and expectations from a new technology. Using this premise, three broad approaches can be identified. One is a more traditional approach commonly seen in the field of communication; a second is a more contemporary approach common to the field of management information systems (MIS); and a third is a newer approach that takes a cognitive orientation. Each approach fundamentally differs in the focal psychological constructs of interest, the relationship

The Handbook of the Psychology of Communication Technology, First Edition. Edited by S. Shyam Sundar.

between the constructs and how they are measured and modeled, and in the breadth of technologies whose likely adoption they explain.

This chapter presents an overview of each perspective along with the methods, measures, and models that are indicative of each perspective, followed by the inherent advantages of each approach. It concludes by examining the emerging lines of inquiry in diffusion research and avenues for future exploration.

The Traditional Approach to the Study of Technology Adoption

The traditional approach to the study of technology adoption is grounded in the principles established by Rogers (1962, 2003) in the diffusion of innovations theory. Reflecting Everett Rogers's training as a sociologist, the theory focuses on the forces that influence individuals' decisions to adopt a new technology. These include macro societal forces such as culture, system norms, mass media influence, and change agent efforts, to micro level influences such as the decision context, the degree of autonomy in the decision, and the perceived beliefs about the technology.

Beginning with how a technology is defined to how the process of adoption is construed, diffusion theory emphasizes the role of communication, uncertainty reduction, and information exchange in the technology adoption process. Technology is defined in terms of the uncertainty it produces owing to its newness and the inability of adopters to predict the consequences of its use. To reduce this uncertainty, potential adopters are driven to seek information from the mass media, from opinion leaders, and through interpersonal exchanges with other individuals. The information received from these sources shapes adopters' perceptions or expectations from the new technology. It is these perceptions that ultimately determine the individuals' decision to adopt or reject the technology.

Rogers (2003) identified five perceived attributes of technologies that influenced individual adoption decisions across a range of innovations. These were the relative advantage of the technology over preceding technologies; its compatibility with other existing technologies; the complexity in learning to use the technology; the perceived visibility or ability to vicariously observe its consumption; and the ability to test the technology on a limited basis. Together, these perceptions explain close to fifty percent of the variance in the rate of adoption of any technology, making them the most significant psychological determinants of individual decisions regarding new technologies.

Measurement and modeling

Rogers (2003) presented a broad model to explain the generic diffusion process that forms the basis for research on technology adoption within the field of communication. An important aspect of the Rogers's (2003) diffusion model is that it neither prescribes the relationships among the major independent variables nor specifies the extent to which the factors within each major independent variable might influence each other. Thus, it is conceivable that societal norms might influence how the mass media portrays a new technology in a culture that in turn influences the perceptions about the compatibility of a new technology. Likewise, it is also conceivable that individual-level perceptions such as the perceived observability of a new technology could influence perceptions of its complexity and its relative advantage, that is, perhaps seeing others use a technology makes people believe it might be easy to learn and valuable. Many such relationships are theoretically plausible and the model leaves these decisions to the individual researcher. Most researchers, however, tend to ignore the relationships among the psychology predictors, and instead, researchers model all the independent variables as linear predictors and estimate their relative influence using exploratory techniques such as OLS regression (e.g., Leung & Wei, 1999).

Further, the model provides limited guidance about the specific predictors that might be pertinent within a given adoption context. Hence, researchers often hand-pick the factors they believe are likely to influence adoption based on how they define the boundaries of the system in which the technology is being introduced. For instance, a scholar examining the adoption of a technology within an organization might consider advertising about the technology in the mass media as an independent variable in the model, while another scholar might argue that it is outside the scope of the model and exclude it.

Furthermore, drawing from the social-psychological approach to behavioral measurement espoused by models such as the Fishbein and Ajzen's (1975) Theory of Reasoned Action (TRA), diffusion theory does not provide a set of measures for any of the predictors of adoption in the model. Hence, most scholars derive the measures on a study specific basis by using projective techniques, interviews, and focus groups, and then empirically validate the items.

Overall, there are at least three inherent advantages to the diffusion theoretic approach to the study of technology adoption. First, the general nature of the model makes it possible to easily adapt it to study the diffusion of virtually any technology in any system. Extant communication research applying diffusion theory to examine the adoption of hardware based technologies ranging from cell phones (Wei, 2001) to high-definition TV (Dupagne, 1999) and software-based technologies ranging from electronic messaging systems (Rice, Grant, Schmitz, & Torobin, 1990) to patient health record systems (Vishwanath, 2009a) are evidence of the breadth and scope of this approach.

Second, the descriptive nature of the relationships between the predictors in the model makes it possible to study various components of the diffusion process in isolation of each other. For instance, some scholars have focused on just the influence of interpersonal interactions on adopters using social network analysis (Rice & Aydin, 1991), others have focused on the influence of system norms (Vishwanath, 2011) and group dynamics (Vishwanath, 2006), while some other have focused on impact of change agent efforts (Singhal, 2011) on the adoption process.

Third, the diffusion theoretic measures tend to have high content validity, particularly with regard to the measurement of key psychological attributes, because the measures are drawn directly from the study population. This also has the added advantage of uncovering new dimensions and discovering new, unexpected factors that might influence the adoption of a technology. For instance, while examining the adoption of electronic medical records among physicians, Vishwanath and Scamurra (2007) discovered additional factors such as concern over system customization, return on investment, system-integration concerns, and workflow issues that along with the known attributes of technological innovations also had a significant impact on the physicians' likelihood of adoption of the technology. Thus, overall, the broad, descriptive, and exploratory nature of the model allows for a richer understanding of the diffusion process.

The Contemporary Approach to the Study of Technology Acceptance

The contemporary approach to the study of technology adoption can be traced to three studies from the MIS field. The first was a study by Moore and Benbasat's (1991) that focused on improving the measurement of the perceived attributes of innovations. Dissatisfied with the study-specific approach to measurement advocated by diffusion theory, and with an eye toward the successful implementation of software based information technology (IT) within organizations, the authors conducted a series of scale development studies. In addition to measures for the five perceived the attributes of innovations, their research uncovered three additional dimensions that were important for potential end users of IT within organizational settings: the degree to which the technology was believed to enhance the status of the user or its image; the degree to which the use of the technology was perceived as being voluntary; and the tangibility or demonstrability of the outcomes of it use. The final outcome of the study was a 38-item instrument for the eight dimensions that spurred a newer approach to the measurement of the perceived attributes of innovations.

The second was a study by Davis, Bagozzi, and Warshaw (1989) that resulted in the technology acceptance model (TAM). Their research focused on further refining the measurement and modeling of the adoption process. The outcome was a parsimonious model that explained the intention and use of IT within organizations. TAM is grounded in social-psychological theories such as TRA, based on which it draws a distinction between the end users' perceived beliefs, attitudes, and their intention to adopt and subsequently use IT. Unlike TRA where the beliefs are derived on a study specific basis, in TAM the perceived attributes are measured using a priori scales, as advocated by Moore and Benbasat (1991). TAM shifted the focus of adoption research toward the use of confirmatory models. TAM also marks the beginning of the use of the term "technology acceptance" to distinguish between the diffusion theoretic approach to the study of the adoption of any technological innovation and the MIS focus on the successful implementation and use of software based IT within organizations.

Due to its focus on the acceptance of just software-based IT within organizations, TAM identified a fewer number of psychological factors as being pertinent to technology users. According to TAM, two instrumental attributes of IT best predict the end users' intent to accept and use the technology: the perceived ease of use and perceived usefulness. Perceived ease of use refers to the degree of effort required to learn to use the technology and is similar to the diffusion theoretic concept of complexity. Perceived usefulness refers to the user's subjective judgment of the utility of the prospective technology in increasing his or her job performance within the organization; it closely parallels the relative advantage factor from diffusion theory. Because complex technologies with steep learning curves reduce the perceived utility of the technology, in TAM, perceived ease of use influences usefulness. Together, these two instrumental beliefs mediate the influence of external factors on the user's intent to adopt the technology that in turn predicts actual adoption.

There were many subsequent modifications and additions to TAM from within the MIS field that culminated in the Unified Theory of Acceptance and Use of Technology (UTAUT) (Venkatesh, Morris, Davis, & Davis, 2003). According to UTAUT, four beliefs about IT are core determinants of an individual's intent to adopt: performance expectancy, effort expectancy, social influence, and facilitating conditions. Performance expectancy and effort expectancy are similar to the innovation-diffusion constructs of relative advantage and perceived complexity, respectively; social influence is similar to the construct of image, while facilitating conditions taps into the innovation-diffusion construct of compatibility. Of these, facilitating conditions is considered the least important because its influence is not significant in the presence of performance expectancy and effort expectancy. Factors such as gender, age, experience, and voluntariness of use are thought to moderate the influence of these four core determinants on behavioral intent.

Measurement and modeling

This contemporary approach to measurement relies on a priori scales that are modified by a researcher to suit the context of the software being examined. It is a confirmatory approach that uses structural equation modeling to examine the fit between the psychological measures of technology acceptance that are empirically derived from potential end users within an implementation context against a theoretical model such as TAM or UTAUT.

In many ways, this model-fit based approach is restrictive because it emphasizes parsimony over richness of explanation. Hence, beyond stating that a handful of factors influence the likelihood of end users' intention use of IT, there is limited guidance provided by the results of studies applying this approach. Consider the case of a consultant who utilizes TAM to explain to an organization that the technology they implement should be easy to use and useful. These instrumental features are but the most obvious, minimum requirements that users would expect from anything that they adopt, be that new IT or a consumer technology such as a vacuum cleaner.

Further, the contemporary approach limits the focus of researchers to only the model specific determinants of user acceptance of IT. All other factors that are excluded in the original models are either treated as exogenous predictors, or considered outside the scope of the model and ignored. For instance, scholars applying TAM consider factors such as age and gender as external factors that indirectly influence behavioral intent by influencing the end users' perceptions of the ease of use and usefulness of the technology while ignoring social influence. In contrast, UTAUT includes social influence as a mediator, individual differences in age and gender as moderators, but the influence of the personality of end users is ignored.

Furthermore, the contemporary approach results in contextually bound models that explain a small subset of the larger technology acceptance process. Understanding the adoption of a different technology or examining a facet of the adoption process that the original model was not intended to explain requires the development of new measures and alternative models. For instance, Ahuja and Thatcher (2005) required new constructs such as autonomy and overload along with new measures for these constructs to model the postimplementation usage of IT within organizations. Similarly, Venkatesh and Brown (2001) developed a new model of adoption of technology in households (MATH) to explain the adoption of personal computers in households that needed newer constructs such as utilitarian outcomes, hedonic outcomes, social outcomes, social influences, and adoption barriers, along with new measures for each construct.

Overall, this contemporary approach has resulted in theoretical silos where researchers are constantly reinventing the original diffusion constructs and in some cases, ignoring diffusion theory altogether. For instance, Ahuja and Thatcher's (2005) model deals with the reinvention of IT by end users but there is no mention of Roger's (1962, 2003) perspectives on this issue. Likewise, many of the constructs in the MATH model (Venkatesh & Brown, 2001) are identical to the diffusion theoretic constructs such as observability, relative advantage, and compatibility, while factors such as the influence of the mass media, technology ownership or technology clusters, and concepts such as change agent efforts that have been shown to influence the adoption of innovations in the communication literature are altogether absent.

This contemporary approach to diffusion modeling, however, has a number of advantages. First, the models are robust and parsimonious. They account for a large percentage of intention (close to 70%) while using just a handful of measures. This makes the models particularly popular and appealing from an applied standpoint. Second, the measures developed are readily available and easily adapted to varying study contexts. This makes it not only easy to apply these measures but also justify their use—the latter being an important expectation of reviewers and journal editors. Not surprisingly, TAM and UTAUT are among the most widely used models of technology adoption that currently dominate the diffusion literature.

The Cognitive Perspective

The cognitive perspective extends to the study of technology adoption a fundamental view of human cognition—that the human mind organizes all stimuli based on their conceptual similarity with other objects, such that objects that are similar are more closely associated than those that are dissimilar—a view that has existed since the time of Aristotle (Anderson, 1983) and frequently used as a basis for social and cognitive science research (Barnett, 1988; Carley, 1986; Shah, Kwak, Schmierbach, & Zubric, 1991; Vishwanath & Chen, 2006; Woelfel, 1975).

The cognitive perspective begins with the idea that any technology is essentially a concept that exists among a constellation of concepts in a person's mind. The concepts could range from ideas about the technology's intrinsic and extrinsic features, its real or perceived attributes, or its relationships with other technologies. The concepts could also include other technologies and their features, attributes, and relationships. The concepts are oriented and ordered in a certain way based on information exchanges in the social system and through the individual's experiences. Technologies that are similar share concepts, are closely linked, and have similar conceptual structures, while technologies that are dissimilar share fewer concepts, are farther apart in the cognitive space, and result in different conceptual arrangements. Thus, a map of the cognitive associations between the concepts would represent the individual's attitudes, values, beliefs, and likely behaviors with regard to various technologies. When aggregated across individuals, such a map would represent the shared, cognitive representations of the technologies among those individuals within that social system.

In this view, the process of adoption of any technology is viewed as a cognitive process, where a new technology provides an opportunity for new linkages or rearrangements of elements that have not been previously associated in the minds of individuals and in the collective consciousness. As the technology spreads though a social system, the conceptual configuration shared by the members change thereby altering their associative structures. The degree of reorganization of the associative elements is a function of the amount of information members of that system share and receive about the innovation (Woelfel & Barnett, 1982). That is, the rearrangement of concepts about a technology would be quicker when the new technology is less disruptive or requires a less severe change in concept relationships; when the culture does not eschew new innovations and ideas; and when there is less friction in the flow of information within the culture. The diffusion of the innovations is much faster in all such cultures and for such technologies.

For instance, prior to the advent of touchscreen phones, most cellphones had static screens that presented information and required inputs using keypads on the screen. Knowledge on how to use these cellphones was diffused through the mass media and learnt through vicarious observation. This created an arrangement of concepts such as screen, static, input, and keypad in users' cognition that resulted in them interacting with screens and using their cellphones in certain ways. Users extended the same schema while operating other technologies such as navigation systems and laptops, which at that point in time had similar static screens. Later on when touch screen devices were developed and information about their use was diffused through the system, it resulted in a new arrangement of the concepts in the individual's minds, with concepts such as keypad being replaced by other concepts such as swipe, gestures, and such, which then became the dominant schema utilized to understand these technologies. In the cognitive view of the diffusion of innovations, mapping the cognitive structure and tracking the changes in the conceptual configurations about a technology among its present and potential users within a system is thus considered key to understanding the process of adoption of any technology.

Measurement and modeling

Cognitive mapping of the technology adoption process requires a system of measurement that meets the following criteria (Barnett & Siegel, 1988; Vishwanath & Chen, 2006): (a) the links among the set of concepts or stimuli must be measured in terms of the similarity (or dissimilarity) among the elements. The measurement system needs to be capable of relating technologies to each other; (b) the measurement system needs to be holistic and capable of simultaneously measuring all integrated dimensions and producing a complete description of the relationships between technologies within and across different concepts; (c) it must involve consensual measures that allow for predictions at both the individual and the aggregated or system level; and (d) in order to make possible calculations of the rate of change or the degree of differences between associations of adopters and nonadopters, the measurement scheme must ideally be at the ratio level.

The Galileo system developed by Woelfel and Fink (1980) satisfies all these demands. The Galileo system is an integrated methodology for metric multidimensional scaling (MDS) using paired distance judgment data. Galileo measurement begins by measuring dissimilarities among a set of concepts describing a domain, defined by those concepts, using paired comparison (every possible pair of the concepts' dissimilarities are measured). So given N concepts, the dissimilarity among $N(N-1)/2$ pairs of concepts is estimated by respondents. In Galileo measurement, the "self" or "me" is also included as a concept since it provides a reference frame to understand the position of the ego relative to the other concepts. Respondents are provided with a known distance estimate and instructed to judge the distances between pairs of concepts in proportion to this standard distance. The instruction is generally of the form, "If the distance between *x* and *y* are *u* units, how far apart are concepts *a* and *b*?"

The empirical distance data collected yields a matrix of concept dissimilarities measured at the ratio level. This matrix is transformed into a scalar products matrix that is double centered to establish the origin at the centroid of the distribution. The matrix is then decomposed to achieve a coordinate matrix whose columns are orthogonal axes (dimensions) and rows are the projections of the concepts on the dimensions (Woelfel & Fink, 1980). Since the coordinates for the items represents the least-square balance points for the items relative to each other, the distances between objects (their coordinates) can be taken to indicate the conceptual similarity between the items. Thus, the greater the reported distance the greater the conceptual differentiation, and vice versa.

A vast body of communication, linguistic, and cross-cultural research has demonstrated the reliability and theoretical validity of the Galileo System (Barnett & Woelfel, 1988; Kincaid, Yum, Woelfel, & Barnett, 1983). Rice and Rogers (1984) recommended its use to describe the changes in conceptual structures of adopters of information technologies. To date, it has been used to map various aspects of the diffusion process ranging from how individuals form cognitive connections with communication technologies (Vishwanath & Chen, 2008), how early and later technology adopters differ in their cognitive structures (Vishwanath & Chen, 2006), to how the

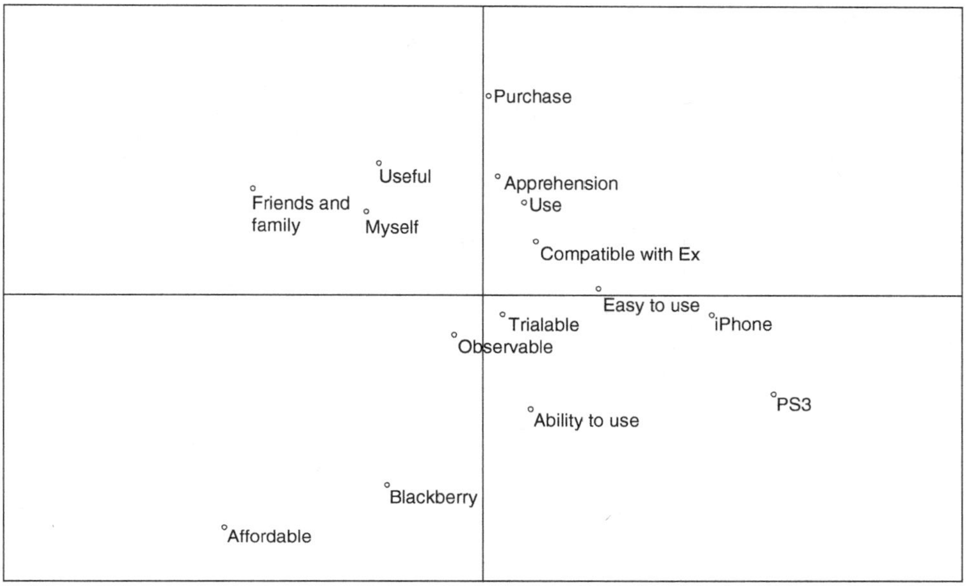

Figure 14.1 A Cognitive map of early adopters' technology preferences based on the relational model of adopter choice (Vishwanath & Chen, 2011).

cognitive associations among adopters change overtime as a technology diffuses through a social system (Barnett & Siegel, 1988).

Recently, Vishwanath and Chen (2011) developed a relational model of adopter choice by applying the Galileo system to map the differences in cognitive structure between early and later technology adopters and predict their relative choice among technologies. The model included the perceived attributes of innovations such as compatibility with existing technology, trialability, and observability from the diffusion literature; ease of use and usefulness from the MIS literature; social-psychological factors shown to influence adoption such as uncertainty (concept: apprehension) (Vishwanath, 2005); technological efficacy (concept: ability to use) (Agarwal, Sambamurthy, & Stair, 2000); and concepts measuring affordability, behavioral intent (purchase) and actual use along with the self-concept (myself). Reflecting the blurring of boundaries between technology hardware, software, and brands, the focus was on three technology concepts, Blackberry, iPhone, and PS3.

Figure 14.1 presents the cognitive maps for early technology adopters from their relational model. The mapping process showed that early and later adopters differed significantly in their cognitive structures and in the evaluative schema they used during the innovation-decision process. For instance, affordability and ease of use anchored the first dimension, and observability and affordability anchored the third dimension for both early and later adopters, indicating that affordability, ease of use, and observability or the image of the product were primary considerations for technology adopters. The second dimension for early adopters was anchored by purchase and affordability with the self leaning toward purchase indicating intent to adopt based on the affordability of the technology. In contrast, for later adopters the second dimension was anchored by affordability and friends and family, with the self leaning toward friends and family, indicating a higher need for social approval for their decision. The relational model predicted that early adopters had a 72% likelihood of choosing an iPhone while later adopters had a 65% likelihood of choosing an iPhone. Reflecting the student composition of the sample and the period of the data collection (November 2007), both segments showed an equal likelihood (around 70%) of choosing a PS3, while later adopters had a slightly higher (62%) likelihood of choosing a Blackberry over early adopters (56%).

The cognitive approach overcomes a number of fundamental limitations imposed by the prior approaches to studying the psychology of adopters. An important limitation of the traditional and contemporary approaches is they assume that the same sets of finite, unidimensional valuations of a new innovation applies regardless of when, where, and by whom the technology is potentially adopted. In other words, only perceived ease of use, usefulness, or one of the diffusion theoretic attributes of technology matter to a technology adopter. Empirical evidence, however, suggests otherwise. Research in communication has shown that early adopters and later adopters cognitively conceptualize technologies they adopt differently (Vishwanath & Chen, 2006). Likewise, within the MIS realm, studies that have tracked user preferences have found that users use different valuations prior to and after implementation (Karahanna, Straub, & Chervany, 1999). Hence, not only do different types of adopters use different dimensions, they also use different valuation schemes throughout the innovation decision process. The multidimensional nature of the cognitive mapping approach to the study of technology adoption overcomes this limitation because it makes it possible to empirically derive and track the changing dimensions of the evaluative schemes used by different adopters, in different systems, and over time.

Another limitation of the traditional and contemporary models of adoption is that they focus on the potential adopters' decision to adopt a singular innovation that is often preselected for them by the researcher prior to model testing. That is, all the models are absolute models of intent rather than models of relative choice. In almost all human decision-making scenarios, however, individuals choose between multiple alternatives and utilize a personal valuation scheme to cognitively process and reduce the set of alternatives to one final choice (Simmons, 1974). The primary reason for ignoring the choice processes in the manner in which choice is measured and modeled. While behavioral intent and beliefs are easily measured using metric measures, choice is always a categorical variable. Hence, choice between more than a few innovations while easy to measure poses a challenge to modelers who use standard regression–based approaches that require metric data. The cognitive approach overcomes this limitation because of the use of ratio-level measures that allows the estimation of absolute and relative distances between technologies. For instance, in their relational model of adopter choice, Vishwanath and Chen (2011) used the coordinates for the relative distance of the self from each technology across all the dimensions to predict the adopters' likely choice of one technology over another.[1]

Finally, the use of the traditional or contemporary approaches to the measurement and modeling of adoption involves a trade-off. In the traditional approach, scholars have to choose between methodological rigor and the ability to measure nuanced variations in adopters' intentions. In the contemporary approach, the trade-off is between a parsimonious, good fitting research model, and an explanation that is anything beyond the obvious. The cognitive approach, however, allows for a synthesis of both the perspectives. Concepts could be derived using projective techniques or generated from the existing literature. The approach can be applied to ascertain the likely acceptance and use of a technology by end users in an organization, or the diffusion of a technology within a social system. It could be used to predict the intent to adopt a singular technology or the relative choice of a technology from among a basket of technologies as well as understand and confirm the stability of the evaluative dimensions used by adopters while making such decisions. Thus, overall, the cognitive perspective allows for a more comprehensive evaluation of the psychology behind the diffusion and acceptance of technology.

The Future of Diffusion Research

The traditional, contemporary, and cognitive perspectives approach the study of the adoption of technology in distinct ways. Each perspective has evolved differently and has been influenced by other perspectives and by each other. Everett Rogers's work and his background in sociology heavily influences the communication research tradition; the MIS tradition draws from social-psychological theories; and the cognitive approach draws from older associational theories and synthesizes the diffusion and MIS perspectives. Individually, each explain anywhere from 50% to 70% of likelihood of adoption of a technology. With such high variances explained, scholars have wondered whether there is anything left to study about the adoption of technology? Has the study of the diffusion of technology reached a saturation point?

We clearly have an advanced understanding of the psychology of adoption, with numerous publications and citations as proof of its academic maturity. This entire body of knowledge could not, however, predict many of the recent advances in technology. None of these perspectives could predict the rise in popularity of Facebook, the fall of MySpace, or the stagnation of Google Plus. Few could foretell why, in less than 5 years, a product such as Blackberry with close to 50% of the mobile market share among business users could become unpopular and be considered as obsolete as a telegraph machine (Surowiecki, 2012). Fewer still can predict, with any amount of certainty, whether Apple or Facebook would remain dominant in the next 5 years. The answers to these and other intriguing questions such as why highly skilled physicians in the United States would resist the adoption of a technology such as electronic medical records (EMR) when it meets every criterion that the diffusion perspectives suggest (Vishwanath & Scamurra, 2007), remains elusive.

Answering such questions requires a multidisciplinary approach that combines the best of the various perspectives to the study of adoption. Three emerging lines of inquiry are indicative of such synthesis in the field of communication.

The first stream of research amalgamates the diffusion theoretic approach with the MIS perspective. Vishwanath and Goldhaber (2003) present a theoretical model that

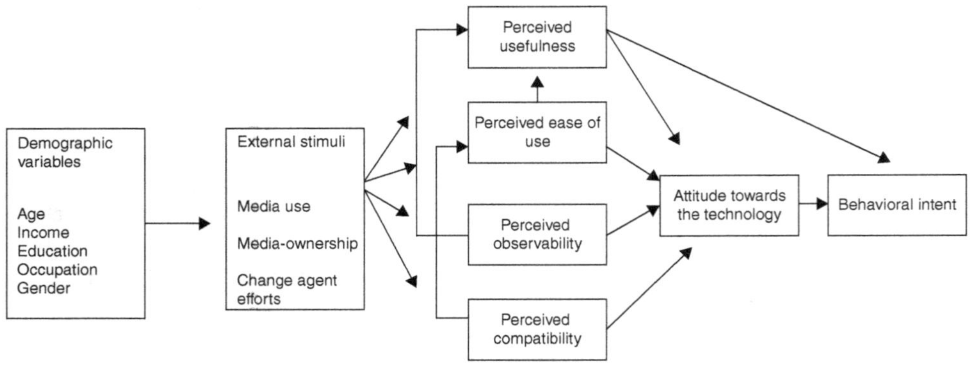

Figure 14.2 A research model combining diffusion and MIS perspectives.
Source: Adapted with permission from Vishwanath and Goldhaber (2003). © Sage.

integrates the psychological factors from diffusion research such as perceived compatibility and observability, along with MIS constructs such as perceived ease of use, perceived usefulness, and attitudes toward technology. Figure 14.2 presents their research model. The model is tested using a priori measures for the MIS constructs and context-specific measures for the diffusion constructs. Based on the communication literature, constructs such as media use, media ownership, and change agent contacts are included, while the relationships between and among the constructs are grounded in the MIS literature.

This research is noteworthy because it incorporates the communication perspective on diffusion, particularly the influence of information exchange through change agent contacts and mass media exposure on shaping micro-level, individual perceptions about the technology, into the existing MIS models. It also unifies the two theoretical perspectives by adding the diffusion constructs of perceived compatibility and observability to the MIS constructs of perceived ease of use and usefulness. Drawing from models such as TAM and TRA, this research takes a confirmatory, model testing approach and delineates the direct and indirect effects of internal and external influence factors on individual psychological orientations toward the innovation. Hence, the individuals' perceived beliefs and attitudes are seen as being influenced by external influence factors such as change agent contacts, and mass media ownership and use; these influences mediate the effect of user demographics on their perceived beliefs about innovations and the individuals' affective orientations toward the technology that in turn predict their intent to adopt the innovation. Thus, the overall outcome is a comprehensive model that incorporates the communication perspective of information exchange; the sociological perspective of external and internal influence on individual behavior; the social-psychological perspective of distinguishing between attitudes, beliefs, and behaviors; and the MIS perspective of using a priori scales and confirmatory models to the study of the diffusion of technology. Other noteworthy research that take a similar approach include Peters (2010), who combines two widely accepted communication models, the expectancy-value model premised on the uses and gratifications perspective along with UTAUT, to examine the stages in the acceptance of a technology, and Cho (2011), who extends the UTAUT and TAM models by incorporating the diffusion construct of perceived critical mass into the predictive model.

The second stream of emerging inquiry combines the cognitive perspective with the diffusion approach. Vishwanath and Scamurra (2007) utilize mixed-methods approaches to map the barriers to the adoption of EMRs within the embedded context of health care users. This research examined how physicians conceptualized the barriers to the adoption of a new EMR system and began by using focus groups and one-on-one interviews of physicians to identify all the potential factors that potentially hinder their utilization of a new technology. This was followed by a quantitative measurement exercise, where physicians independently ranked and rated the relative influence of each of these factors on their adoption decision. The ranking data were then analyzed using MDS and the rating data were used for a cluster analyses that was overlaid on the MDS maps to ultimately reveal how physicians conceptualized the proposed EMR technology. Figure 14.3 presents the two-dimensional conceptual map of the barriers to the adoption of EMRs among physicians. Using a similar approach, Vishwanath, Singh, and Winkelstein (2010) generated cognitive maps and longitudinally tracked the preimplementation expectations and postimplementation utilization of a new EMR and its overall impacts on workflows within a hospital.

This line of work is noteworthy because it focuses on a specific part of the technology sense-making processing, particularly how individuals conceptualize and understand a new technology and how this conceptualization changes as individuals utilize the technology. Hence, while most research on diffusion and technology adoption focuses on behavioral models, this line of work focuses on structuring and mapping the cognitive schema used by individuals during the process of adoption. Other research that similarly provides newer perspectives include the multitheory, multilevel model of adoption (Wijngaert & Bouwman, 2012) that combines qualitative and quantitative methods to study the process of adoption over time, and the positive deviance approach (Singhal, 2011) that reconceptualizes the process of diffusing innovations.

A final stream of research combines the cognitive approach with the MIS approach. Vishwanath (2009a, b) uses experimental techniques to evaluate how information framing influences end user perceptions about a new technology's ease of use, usefulness, and social influence. This line of research focuses on how the features in a technology work as triggers, interact with information, and shape user expectations; how these expectations become the lens through which the technology is experienced; and how overtime these experience dominate sense making. This stream of research directly examines how cognitive schemas are shaped by the interaction between communicated information and technology features and is hence one of the most promising avenues of emerging research on the psychology of technology adoption.

This stream of work extends theories from political communication on issue framing to the understanding of how individual schemas about a technology are shaped by information inputs from the mass media and the individuals' social networks. Research here reveals that individuals utilize different cognitive orientations during different stages of the adoption-decision process. When individuals first encounter a technology,

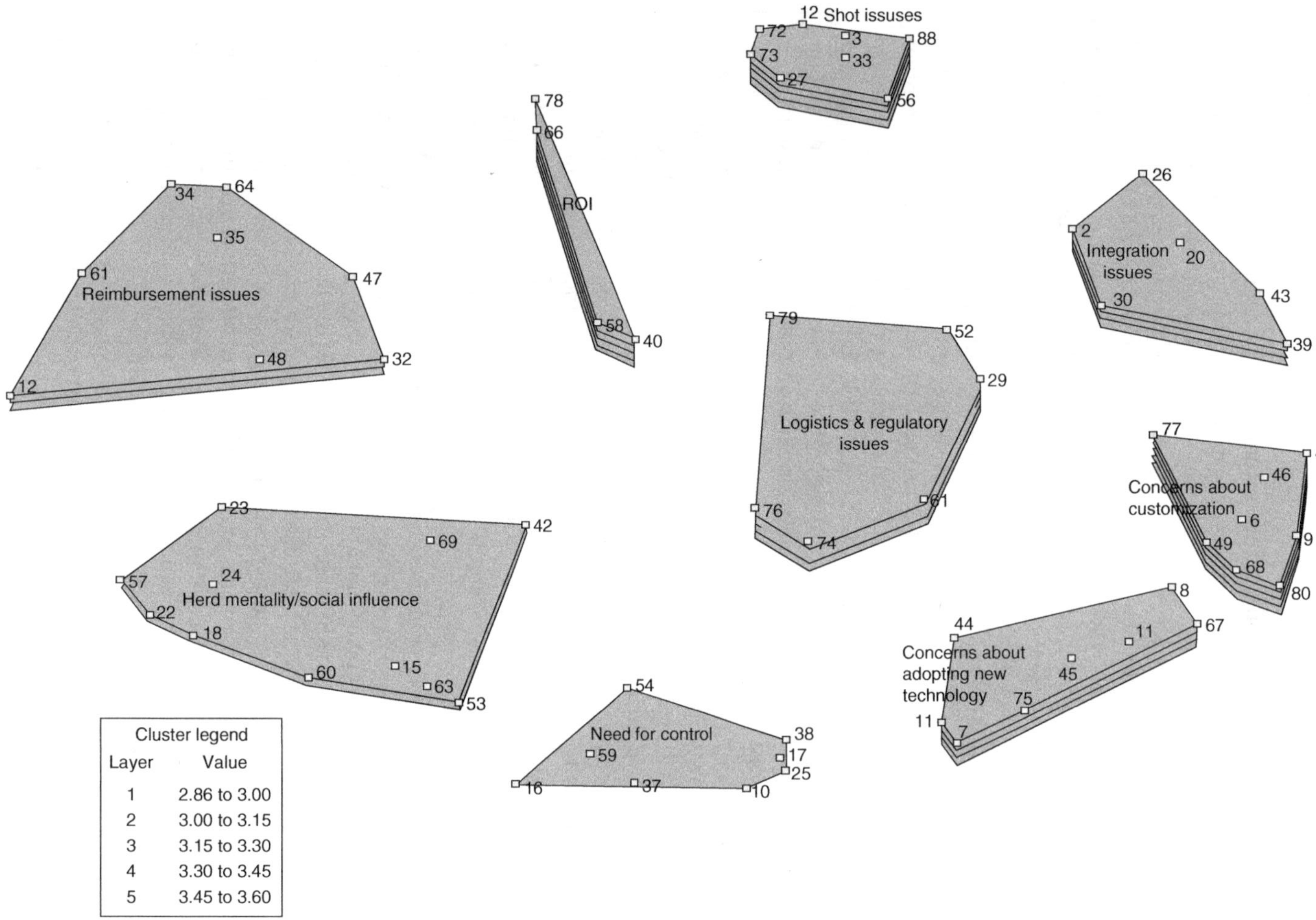

Figure 14.3 A two-dimensional cognitive map of the barriers to the adoption of EMRs among physicians.

Source: Adapted with permission from Vishwanath and Scamurra (2005). © Sage.

they tend to focus on the functional aspects of its use. Technology features tend to trigger and dominate sensemaking (Griffith, 1999) during this stage. Consequently, information about how to use a technology has a limited framing effect at this stage because individuals attempt to understand the technology by applying existing schemas that they posses from prior experience with related technologies that are triggered by the salient features in the new innovation. For instance, when presented with a new GPS system with a large screen, individuals might attempt to touch and interact with it with using the "touch-screen" schema that they acquired while using smartphones. Touching the screen and figuring out how to program the device using available schemas would thus dominate cognition during these early stages of encounter with the new technology.

When individuals have achieved some understanding of the functions of the technology, they tend to shift their attention to extrinsic factors such as the perceived usefulness and social cachet of the technology. Because these extrinsic factors require information inputs from outside the technology for verification, in this stage, information frames seems to have a significant influence on individual perceptions. For instance, the social value or image of a technology needs to be verified through social comparison and by examining how others view the technology. Hence, frames that present the social value of utilizing the technology is usually attended to by the individual and tends to shape how "cool" the technology appears to the individual; frames that present the usefulness of the technology in a positive or negative light, likewise, influence how the individual values the utility of the technology. These frames seem to have an enduring effect and become the lens through which the innovation is not only interpreted but also subsequently experienced.

This stream of research thus disentangles the relationships between communicated information, beliefs, and the mediating role of cognitive schemas used to explore and experience a technology. Research applying these perspectives has also evaluated how individuals form opinions about technology (Druckman & Blosen, 2011), and how the individual's personality based innovativeness influence the types of cognitive biases they demonstrate when dealing with the failure of a technology (Vishwanath & LaVail, in press). While research on framing focuses on the sense making process, other emerging research in this realm looks into the active role that adopters play in shaping innovations (Dearing & Myers, 2012), and the activities of the adopter throughout the diffusion process that leads to the reinvention of technology (Karnowski, Pape, & Wirth, 2010). These lines of inquiry also open a number of new areas for future exploration that examines how individuals cognitively process information when encountering a new technology, how technology affordances (Sundar, 2008) and the feature-based sensemaking triggers interact and trigger evaluations about the technology's intrinsic and extrinsic features (Griffith, 1999), the extent to which specific feature-based cues dominate cognitive processing, and how individual cognitive biases in the attribution of causation, ego enhancement motives, and personality-based factors influence the postadoption evaluation of a technology and other competing technologies.

Overall, these emerging lines of Communication research amalgamate existing approaches, incorporate newer theories, and present a fresh way of thinking about the psychology of technology adoption. In doing so they provide fertile ground for further theoretical and methodological development, and lay the groundwork for the next generation of scholarship on the diffusion and adoption of technology.

Note

1 In the relational model of adopter choice (Vishwanath & Chen, 2011), the adopter's likelihood of choosing a technology from among technologies is computed using the following formula:

$$P(x) = \frac{1 - Sx}{Sx + Sy + Sz} \tag{14.1}$$

where
"P(x)" represents the probability of choosing technology x
Sx, Sy, Sz are the aggregate distances between the self and technology x, y, z, respectively, across all dimensions.

References

Agarwal, R., Sambamurthy, V., & Stair, R. M. (2000). Research report: The evolving relationship between general and specific computer self-efficacy: An empirical Assessment. *Information Systems Research*, *11*(4), 418–430. doi:10.1287/isre.11.4.418.11876

Ahuja, M. K., & Thatcher, J. B. (2005). Moving beyond intention and toward the theory of trying: Effects of work environment and gender on post-adoption information technology use. *MIS Quarterly*, *29*(3), 427–459.

Anderson, J. R. (1983). *The architecture of cognition*. Cambridge, MA: Harvard University Press.

Barnett, G. A. (1988). An associational model for the diffusion of complex innovations. In G. A. Barnett & J. Woelfel (Eds.), *Readings in the Galileo system theory, methods and applications* (pp. 55–74). Dubuque, IA: Kendall/Hunt.

Barnett, G. A., & Siegel, G. (1988). The diffusion of computer assisted legal research systems. *Journal of the American Society for Information Science*, *39*(4), 224–234. doi:10.1002/(SICI)1097-4571(198807)39:4<224::AID-ASI2>3.0.CO;2-O

Barnett, G. A., & Woelfel, J. (1988). *Readings in the Galileo system*. Dubuque, IA: Kendall/Hunt.

Carley, K. A. (1986). An approach for relating social structure to cognitive structure. *Journal of Mathematical Sociology*, *12*(2), 137–189. doi:10.1080/0022250X.1986.9990010

Cho, H. (2011). Theoretical intersections among social influences, internal beliefs, and intentions in the context of 3G mobile phone services in Singapore: Decomposing perceived critical mass and subjective norms. *Journal of Communication*, *61*(2), 283–306. doi:10.1111/j.1460-2466.2010.01532.x

Davis, F. D., Bagozzi, R. P., & Warshaw, P. R. (1989). User acceptance of computer technology: A comparison of two theoretical models. *Management Science*, *35*(8), 982–1003. doi:10.1287/mnsc.35.8.982

Dearing, J. W., & Myers, G. (2012). The active adopter in the diffusion of innovations. In A. Vishwanath & G. A. Barnett (Eds.), *Advances in the study of the diffusion of innovations: A communication science perspective* (pp. 207–230). New York, NY: Peter Lang.

Druckman, J. N., & Bolsen, T. (2011). Framing, motivation reasoning, and opinions about emergent technologies. *Journal of Communication*, *61*(4), 659–688. doi:10.1111/j.1460-2466.2011.01562.x

Dupagne, M. (1999). Exploring the characteristics of the potential high-definition televisions adopters. *Journal of Media Economics*, *12*(1), 35–50.

Fishbein, M., & Ajzen, I. (1975). *Belief, attitude, intention, and behavior: An introduction to theory and research.* Reading, MA: Addison-Wesley.

Griffith, T. L. (1999). Technology features as triggers for sensemaking. *Academy of Management Review*, *24*(3), 472–488.

Karahanna, E., Straub, D. W., & Chervany, N. L. (1999). Information technology adoption across time: A cross-sectional comparison of pre-adoption and post-adoption beliefs. *MIS Quarterly*, *23*(2), 183–213.

Karnowski, V., Pape, V. T., & Wirth, W. (2012). Overcoming the binary logic of adoption. In A. Vishwanath & G. A. Barnett (Eds.), *Advances in the study of the diffusion of innovations: A communication science perspective* (pp. 57–76). New York, NY: Peter Lang.

Kincaid, D. L., Yum, J. O., Woelfel, J., & Barnett, G. A. (1983). The cultural convergence of Korean immigrants in Hawaii: An empirical test of a mathematical theory. *Quality and Quantity*, *18*, 59–78.

Leung, L., & Wei, R. (1999). Who are the mobile phone have-nots? Influences and consequences. *New Media & Society*, *1*(2), 209–226. doi:10.1177/1461444899001002003

Merton, R. K. (1968). *Social theory and social structure.* New York, NY: Free Press.

Moore, G. C., & Benbasat, I. (1991). Development of an instrument to measure the perceptions of adopting an information technology innovation. *Information Systems Research*, *2*(3), 192–220. doi:10.1287/isre.2.3.192

Peters, O. (2010). Three theoretical perspectives on communication technology adoption. In A. Vishwanath & G. A. Barnett (Eds.), *Advances in the study of the diffusion of innovations: A communication science perspective* (pp. 33–56). New York, NY: Peter Lang.

Rice, R. E., & Aydin. C. (1991). Attitudes towards new organizational technology: Network proximity as a mechanism for social information processing. *Administrative Science Quarterly*, *36*(2), 219–244.

Rice, R. E., Grant, A. E., Schmitz, J., & Torobin. J. (1990). Individual and network influences on the adoption and perceived outcomes of electronic messaging. *Social Networks*, *12*(1), 27–55. doi:10.1016/0378-8733(90)90021-Z

Rice, R. E., & Rogers, E. M. (1984). New methods and data for the study of new media. In R. E. Rice (Ed.), *The new media: Communication, research, and technology* (pp. 81–99). Beverly Hills, CA: Sage.

Rogers, E. M. (1962). *Diffusion of innovations* (1st ed.). New York, NY: Free Press.

Rogers, E. M. (2003). *Diffusion of innovations* (5th ed.). New York, NY: Free Press.

Shah, D. V., Kwak, N., Schmierbach, M., & Zubric, J. (2004). The interplay of news frames on cognitive complexity. *Human Communication Research*, *30*(1), 102–120. doi:10.1111/j.1468-2958.2004.tb00726.x

Simmons, P. J. (1974). *Choice and demand.* New York, NY: Wiley.

Singhal, A. (2011). Turning diffusion of innovations paradigm on its head. In A. Vishwanath & G. A. Barnett (Eds.), *The diffusion of innovations: A communication science perspective* (pp. 193–206). New York, NY: Peter Lang.

Sundar, S. S. (2008). The MAIN model: A heuristic approach to understanding technology effects on credibility. In M. J. Metzger & A. J. Flanagin (Eds.), *Digital media, youth, and credibility*. The John D. and Catherine T. MacArthur Foundation Series on Digital Media and Learning (pp. 73–100). Cambridge, MA: The MIT Press. doi:10.1162/dmal.9780262562324.073

Surowiecki, J. (2012, February 13). *Blackberry season. The New Yorker*. Retrieved from http://www.newyorker.com/talk/financial/2012/02/13/120213ta_talk_surowiecki

Venkatesh, V., & Brown, S. A. (2001). A longitudinal investigation of personal computers in homes: Adoption determinants and emerging challenges. *MIS Quarterly*, *25*(1), 71–102. doi:10.2307/3250959

Venkatesh, V., Morris, M. G., Davis, G. B., & Davis, F. D. (2006). User acceptance of information technology: Toward a unified view. *MIS Quarterly*, *27*(3), 425–478. doi:10.2307/30036540

Vishwanath, A. (2005). Impact of personality on technology adoption: An empirical model. *The Journal of the American Society for Information Science and Technology*, *56*(8), 803–811. doi:10.1002/asi.v56:8

Vishwanath, A. (2006). Effect of the number of opinion seekers and leaders on technology adoption. *Human Communication Research*, *32*(3), 322–345. doi:10.1111/j.1468-2958.2006.00278.x

Vishwanath, A. (2009a). Using frames to influence consumer willingness to pay for the Patient Health Record (PHR): A randomized experiment. *Health Communication*, *24*(5), 473–482. doi:10.1080/10410230903025407

Vishwanath, A. (2009b). From beliefs to intentions: The impact of framing on the adoption of innovations. *Communication Monographs*, *76*(2), 177–206. doi:10.1080/0363775 0902828438

Vishwanath, A. (2011, November 20). Global differences in internet diffusion: An empirical investigation into the effects of culture on the diffusion of the internet. *Communication and Science Journal*. Retrieved from http://www.galileoco.com/comSciJliterature/vishwanath11.pdf

Vishwanath, A., & Barnett, G. A. (2011). *The diffusion of innovations: A communication science perspective*. New York, NY: Peter Lang.

Vishwanath, A., & Chen, H. (2006). Technology clusters: Using multidimensional scaling to evaluate and structure technology clusters. *Journal of the American Society for Information Science and Technology*, *57*(11), 1451–1460. doi:10.1002/asi.v57:11

Vishwanath, A., & Chen, H. (2008). Personal communication technologies as an extension of the self: A cross-cultural comparison of people's associations with technology and their symbolic proximity with others. *The Journal of the American Society for Information Science and Technology*, *59*(11), 1761–1775. doi:10.1002/asi.20892

Vishwanath, A., & Chen, H. (2011). Towards a comprehensive understanding of the innovation-decision process: A relational model of adopter choice. In A. Vishwanath & G. A. Barnett (Eds.), *Advances in the study of the diffusion of innovations: A communication science perspective* (pp. 9–32). New York, NY: Peter Lang.

Vishwanath, A., & Goldhaber, G. M. (2003). An examination of the factors contributing to adoption decisions among late diffused technology products. *New Media & Society*, *5*(4), 547–572. doi:10.1177/146144480354005

Vishwanath, A., & LaVail, K. H. (2013). The role of attributional judgments when adopted computing technology fails: a comparison of Microsoft Windows PC user perceptions of Windows and Macs. *Behaviour & Information Technology*, *32*(11), 1155–1167.

Vishwanath, A., & Scamurra, S. (2007). Barriers to the adoption of electronic health records: Using concept mapping to develop a comprehensive empirical model. *Health Informatics Journal, 13*(2), 119–134. doi:10.1177/1460458207076468

Vishwanath, A., Singh, R. S., & Winkelstein, P. (2010). The impact of electronic medical records on outpatient workflows: A longitudinal evaluation of its workflow effects. *International Journal of Medical Informatics, 79*(11), 778–791. doi:10.1016/j.ijmedinf.2010.09.006

Wei, R. (2001). From luxury to utility: A longitudinal analysis of cell phone laggards. *Journalism and Mass Communication Quarterly, 78*(4), 702–719. doi:10.1177/107769900107800406

Wijngaert, L. V. D., & Bouwman, H. (2012). Theoretical and methodological starting points for a situational approach towards the understanding of the adoption and use of new technologies and services. In A. Vishwanath & G. A. Barnett (Eds.), *Advances in the study of the diffusion of innovations: A communication science perspective* (pp. 77–102). New York, NY: Peter Lang.

Woelfel, J. (1975). Theoretical issues and alternatives: A theory of occupational choice. In J. S. Picou & R. E. Campbell (Eds.), *Career behavior of special groups* (pp. 41–61). Columbus, OH: Merrill.

Woelfel, J., & Barnett, G. A. (1982). Multidimensional scaling in Riemann space. *Quality and Quantity: A European Journal of Methodology, 16*(6), 469–491.

Woelfel, J., & Fink, E. L. (1980). *The measurement of communication processes: Galileo theory and method.* New York, NY: Academic Press.

15

Adolescent Development and Psychological Mechanisms in Interactive Media Use

Elly A. Konijn, Jolanda Veldhuis, Xanthe S. Plaisier, Marloes Spekman, and Anouk den Hamer

Department of Communication Science, VU University Amsterdam, Amsterdam, Netherlands

"A Day not Wired is a Day not Lived"

Introduction

There is an urgent need for a better understanding of how new communication technology and adolescent development intersect and how this may impact contemporary youth—that is what this chapter is about. While adolescents today live in a media-saturated world and the media play a major role in adolescent development (Brown & Bobkowski, 2011; Mitchell & Ybarra, 2009; Subrahmanyam, Greenfield, & Tynes, 2004), there is an important lack of knowledge regarding the psychological mechanisms in so called new, digital, or interactive media use among adolescents. In particular, understanding the cross-overs between their abundant media use and developmental needs is important to inform the public health debate regarding children and adolescents (Moreno, Parks, Zimmerman, Brito, & Christakis, 2009; Strasburger, Jordan, & Donnerstein, 2010). Research indicates that adolescents seem most susceptible when it comes to media's impact because of their developmental stage in which parental mediation strategies are less effective than with younger children and peers becoming more important (Cantor & Wilson, 2003; Nathanson, 2004). Thus far, quite some work has been done with media use among younger children and (young) adults (e.g., mostly college students). However, much less is known about how adolescents use and process media content, especially in view of the newer, interactive communication technology. We assume that adolescents' media use and processing intersect with their developmental stage and may impact psychological well-being in this crucial phase of life in which they become increasingly independent. Clearly, new communication technology is important in many youngsters' daily lives, while not all are affected in

The Handbook of the Psychology of Communication Technology, First Edition. Edited by S. Shyam Sundar.

similar ways. With this chapter, we provide a glimpse of our current research program at VU University Amsterdam that aims to fill part of this gap.[1]

Thus far, research within the domains of (a) new media and communication technology and (b) adolescent development have largely evolved as separate strands of scholarship. In our program, we integrate these strands by studying media use from an adolescent developmental perspective. In the following, we outline part of this program in focusing on how today's new communication technology intersects with adolescent developmental issues. This chapter consists of three parts. A first part outlines how newer communication technology appeals to adolescents' developmental needs, such as the role of user-generated content in the so-called imaginary audience, identity development, and peers' influence. From here, in part 2, several underlying mechanisms are defined that are specifically relevant in how new communication technology may sort effects in adolescents, underscored with empirical evidence. Obviously, this chapter can only sketch several aspects of new communication technology in how they tap into adolescent developmental issues and sketch only a few underlying mechanisms at work in adolescents' new media use such as wishful identification, perceived realism, and emotion regulation. A third section describes how creative use of newer communication technology provides opportunities to improve and enhance the methodology of media research. For example, through newly developed measurements that apply social media features (e.g., YouTube) to construct instruments that are more appealing to the target group than traditional instruments. Moreover, such an approach allows to include user-generated content, for example, to (experimentally) investigate contextual factors that operate while using newer communication technology. Finally, the conclusions are summarized and discussed highlighting the importance of a multidisciplinary approach to media scholarship to better understand adolescent media use within the context of their psychosocial development. In tying together the various research lines as presented in this chapter, we then outline a schematic representation of a theoretical model representing a Cyclic Process Model on Adolescent Media and Peer Communication (CPM-A-MPC).

Features of New Media Technology that Intersect with Needs Fulfillment in Adolescence

Today's adolescents use their media devices so actively that it seems as if their motto sounds like "A day not wired is a day not lived." Research shows that Facebook is the most widely used social networking site (Special & Li-Barber, 2012), especially popular among adolescents who spent about an hour per day on average, with a mean number of 220 "friends" (Christofides, Muise, & Desmarais, 2012a). Over half of teenagers check their social media sites every day, while almost a quarter (22%) check their Facebook 10 times a day (Festl, Scharkow, & Quandt, 2013). In adolescents' media-saturated world, one may wonder how the newer communication technology and adolescent development intersect. Although it might be hard to draw a clear line between the old media (e.g., books, radio, movies, magazines, and television) and the new media, the latter is usually understood as including digital and interactive communication technology (e.g., games, smartphones, social media, blogs, and the internet) (e.g., Brown & Bobkowski, 2011). However, a lot of the newer media formats may also be used in similar ways as older media because practically all media fare is available through the internet these days. For example, when just watching a digitally recorded "old movie" on YouTube, the difference with the old media format is fuzzy. One may argue that YouTube has user-generated content and affordances added (e.g., likes or feedback); however, this may just be ignored by a particular user and perhaps not even noticed when only interested in an "old media" experience.

Important differences between the older and newer media in how they are generally referred to not only concern a more active and interactive media use (e.g., customization, blogging, and agency; Sundar, 2008; Sundar & Marathe, 2010) but also the technologically advanced mobility of new media devices. Hence, on the one hand, new media technology evokes higher levels of involvement in the act of communication, while on the other hand, allows for a continuous connectedness, anywhere, any time. This may have important implications for how today's youngsters will develop. In fact, there is an important lack of knowledge regarding the psychological mechanisms in so-called new, digital, and interactive media use among adolescents (Brown & Bobkowski, 2011; Mitchell & Ybarra, 2009; Subrahmanyam et al., 2004). In particular, understanding the cross-overs between developmental aspects of adolescence and their abundant media use is important to inform the public health debate regarding childrens' and adolescents' well-being (Moreno et al., 2009; Strasburger et al., 2010). In the following, we will discuss how features of newer communication technology may tap into developmental needs of adolescents. We assume that the fulfillment of many needs in adolescents are afforded by features of new technology, or keep promises of fulfillment the least.

The developmental stage of adolescence is characterized by a relatively strong egocentric perspective due to cognitive limitations because the prefrontal cortex in the brain has not yet fully developed (Casey, Getz, & Galvan, 2008a; Choudhury, Blakemore, & Charman, 2006). Adolescent egocentrism comes with phenomena like an "imaginary audience" and "personal fable" (Cingel & Krcmar, 2012; Elkind, 1967, 1985; Frankenberger, 2000; Martin & Sokol, 2011; Ryan, Hicks, & Midgley, 1997). Interestingly, the interactive and user-generated tools of new media seem to closely connect to these phenomena. Likewise, the formation of one's own independent identity is a primary developmental task during adolescence for which virtual worlds and game environments provide a good match. Social development is another key feature of adolescence wherein the role of peers becomes increasingly important. Perhaps even more so with the newer communication technology through which peers find their own and never-lasting ways to exert their influence.

New media and adolescents' belief in an imaginary audience and personal fable

The concept of "imaginary audience" refers to the adolescent's belief that he or she is continuously watched and evaluated by others. Indeed, as teenagers continuously monitor themselves, they are highly occupied with their own looks and behaviors, and highly critical upon themselves while no actual audience is out there; this seems primarily an imagined audience (Cingel & Krcmar, 2012; Elkind, 1967, 1985; Frankenberger, 2000; Martin & Sokol, 2011; Ryan et al., 1997). However, new media technology allows substantiating such an audience on social media platforms by creating their personal profile on social networking sites like Facebook, enhancing this status by posting updates, sharing information, and seeking followers. Subsequently, the adolescent may bathe in his self-created celebrity status while receiving "likes" on Facebook or by adding new "fans."

Theoretically, the concept of imaginary audience is an interesting one—how imaginary is a virtual audience? The virtual audience may be perceived as real by the adolescent in exposing himself/herself on social network sites. This poses interesting questions in regard to the virtual status of such networked friends—how real is a virtual audience in the perception of the adolescent high in "imaginary audience?" While the communication on social media does occur by real, existing humans, mainly peers, the roles in which they appear and how they present themselves may be highly artificial, fake, or pretended. So, how "real" has the imaginary audience become for the target and with what consequences?

A similar line of reasoning holds for the concept of "personal fable." Related to the imaginary audience, the adolescent believes that he or she is highly special and unique (Alberts, Elkind, & Ginsberg, 2007; Elkind, 1985; Martin & Sokol, 2011). The more followers she/he may have been able to create or attract on social media, the more strongly his/her personal fable seems to be fed. The many options new media technology offers to give expression to such a personal fable may further underscore his/her uniqueness and make the fable more tangible. The possibilities offered by contemporary social media to act out one's personal fable and further support the adolescent beliefs in such a fable may also lend further support to his/her imaginary audience. In fact, these imaginary concepts are made even more concrete and alive by filling in one's personal fable and creating one's virtual audience through user-generated content such as postings, personal blogs, pictures, video clips, and the like on personal pages like Facebook or MySpace. Likewise, one can easily create (virtual) friends on Facebook and even believe that one has a lot of friends by adding friends to one's personal pages. Certainly, the "materialized" personal fables on social media will be strongly guided by impression management (cf. Kalyanaraman & Sundar, 2008; Krämer & Winter, 2008). In fact, it appears that most individuals present themselves more positively on personalized websites than they in fact are, especially on dating sites (Ellison, Heino, & Gibbs, 2006)—idealized selves are portrayed through virtually materialized personal fables.

The newer communication technology may thus tap into these phenomena of adolescent development in that social networking sites provide rich audiences and opportunities to fulfill their needs. Perhaps, lonely or depressed adolescents may comfort themselves with such substantiated imaginary audiences and acting our one's personal fable. Thus far, studies have shown mixed results. On the one hand, studies found that adolescents who feel lonely, depressed, or socially anxious are more likely to use new communication technologies to fill their social life with people they do not meet in real life (Gross, Juvonen, & Gable, 2002; Ybarra, Alexander, & Mitchell, 2005).

On the other hand, youngsters benefit highly from sharing their daily affairs, worries, and concerns on social networking sites and they appear to facilitate friendship formation (e.g., Jung Oh, Lauckner, Boehmer, Fewins-Bliss, & Li, 2013; Mitchell & Ybarra, 2009; Peter, Valkenburg, & Schouten, 2005). In sum, social media may provide a rich platform to fulfill the need for a personal fable and to give expression to one's felt uniqueness and having "all eyes on me."

Furthermore, the generally increased levels of sensation seeking during adolescence as well as a strong need to explore the "forbidden" in their search for independence (Arnett, 1992, 1999; Boyer, 2006; Kirsh, 2003) coincides with the personal fable in facing risky challenges in the belief that one's unique strengths will survive. The internet provides many relatively safe challenges through which these youngsters may perform risky behaviors and act out norm-crossing exposure, for example, by posting hazardous pictures on Partypeeps or extravagant clips on YouTube. They may not realize, however, that once exposed on the internet, it is exceptionally hard to get this information off (Christofides, Muise, & Desmarais, 2012b). And *if* they are informed or aware of possible risks, they may wave them away due to their conviction of being invulnerable in accordance to the adolescent personal fable of being extremely unique and special. The concepts of imaginary audience and personal fable thus explain the willingness and abundant exposure of private information on social media. In the longer run, this may alter one's understanding of social reality and privacy. In all, it is unclear how this may influence the self-concept and behavior later in life.

Identity formation and new media technology

Another strong need in adolescent development that may intersect with the affordances of newer communication technology concerns the process in search of gender and *identity formation*, which drives adolescents to look for attractive role models with whom they can identify. They will find them abundantly in the media, which includes social networking sites and media like Facebook, MySpace, and LinkedIn, but also in video games and virtual worlds. In the relatively safe environments of the "virtuality" of the internet, the adolescents may feel safe to experiment with and try out different roles or explore a variety of "possible selves" (Kapidzic & Herring, 2011; Konijn & Nije Bijvank, 2009; Subrahmanyam et al., 2004). In traditional media offerings, one could search through the many models presented on stage or screen, and once in a while a "click" with an attractive role model would occur. Posters and images of these models would then appear on the walls of the teenager's bedroom and on personal belongings to express some possible or ideal self (Steele & Brown, 1995). However, only in using one's fantasy and in imagination one could project oneself onto such media figures. New media technology increased the potential for exploring role models enormously, not only in numbers but, more importantly, also in allowing to create one's own possible self in much more integrated and dynamic ways. Identity construction is supported not only through the exposure of one's "personal fable" and showing favorite role models on social networking sites, but also by the possibility of actually "acquiring an identity" through creating an avatar (Bers, 2001; Grasmuck, Martin, & Zhao, 2008; You & Sundar, 2013). Creating personalized avatars in games and virtual worlds in particular provide a lot of opportunities to explore possible selves and enact behaviors as if one "is" the role he/she created in virtual space. In contemporary digital games and virtual environments, the player not only "passively" watches heroes, actions, and scenery, but the interactive nature allows more active involvement in vicariously enacting in-game behavior. Indeed, sophisticated technology in modern games include more immersion in virtual environments and evoke more involvement with characters and actions than the older versions and so-called "passive media" (e.g., Carnagey & Anderson, 2004; Carnagey, Anderson, & Bartholow, 2007; Kim, Park, Sundar, & del Pobil, 2012; Konijn, Nije Bijvank, & Bushman, 2007; Weibel, & Wissmath, 2011).

In acting as the avatar, the creator or player behind the avatar, experiences how others respond to that type of character and how well this fits with one's own felt personality or aspirations. On the one hand, one may wish to experience a variety of creatures in building one's identity, while on the other hand, an avatar may be used to train one's aspired identity. For example, Breivik (the shooter from Norway, Breivik, 2011) is said to have trained himself through excessive video game play in preparation for his deed (Melle, 2013). However, role playing games may certainly also be used for positive and beneficial purposes, such as in health care, both for the benefit of patients and training specialists (Kato, 2010). Often encountered identity explorations in virtual space concern gender identity. For some, a confusing time commences during adolescence in accepting and sometimes figuring out one's gender identity or sexual orientation (Kafai, Heeter, Denner, & Sun, 2008). Research to date has revealed that virtual environments and new technology have provided new means to explore one's gender identity and experiment in different gender roles. For example, in Second Life and World of Warcraft, many males presented themselves as females (admitted, not only for reasons of gender identity exploration; Hussain & Griffiths, 2008; Paika & Shia, 2013).

Peer influence in new media environments

When studying adolescence, the increasingly important role of peers cannot be overlooked, who then often become more important than the adolescent's parents (Christofides et al., 2012b; Eccles, Wigfield, Harold, & Blumenfeld, 1993; Martin & Sokol, 2011; Plaisier & Konijn, 2013; Steinberg & Morris, 2001). In this respect, the newer communication technologies provide ample opportunities and in fact new ways through which peers communicate and influence each other, more so than in traditional media. Not only through the enormous amounts of self-exposure on social networking sites, but also through the options of direct feedback on what is posted online. In particular on social networking sites and blogs, peers directly comment on each other's postings. The abundant exposure to a large variety of relevant others will evoke continuous comparisons with peer group members (e.g., Brown & Larson, 2009). In particular with the introduction of social media and user-generated audio-visual content, adolescents today have continuous access to how relevant others look through the widely available pictures and clips, have access to how many friends and "likes" others have, what leisure activities, how they look at parties, and so on and so forth. Thus, adolescents are not only exposed to the more traditional media models like celebrities and fashion models as in magazines and movies, but also to a large number of daily life "age-mates," more so than one could ever have met in real life. Because such role models are much closer (i.e., psychologically less distant; Konijn & Hoorn, 2005) than the more traditional media models, these online accessible role models will likely have more impact. The effect this may have on adolescent development is still unclear. However, given the increased importance of peers during the developmental stage of adolescence, this will most likely enhance processes of peer group comparisons.

Adolescents strongly depend on their peers for emotional security and normative behavior while secure relationships with peers are associated with social and emotional competence (Laible, 2007; Laible, Carlo, & Raffaelli, 2000; Nickerson, 2005). Clearly, the newer communication technology may support adolescents in building and sustaining secure relationships—often by complementing existing social bonds but sometimes also by establishing new friendships and finding a match (e.g., Antheunis, Valkenburg, & Peter, 2012; Peter et al., 2005). While a lot of youngsters benefit from the opportunities provided to chat, to being able to talk all the time and to find their own ways of keeping in touch with their friends, a number have also experienced the backside of these newer communication technologies in having been negatively approached, insulted, out-casted, or cyberbullied. Though most of these undesired behaviors are incidental (e.g., stalking, sexual intimidation; Calvete, Orue, Estévez, Villardón, & Padilla, 2010), cyberbullying behaviors and social rejection via new media technology occur more frequently (Calvete et al., 2010; Tokunaga, 2010).

Cyberbullying behavior (in which bullying behavior takes place through electronic media such as internet and mobile phones) has attracted considerable research attention over the last few years because it is prevalent and harmful (Feldman Hertz, Donato, & Wright, 2013). A meta-analysis showed that cyberbullying behavior is especially prevalent among adolescents, and between 20% and 40% of them have been a victim of cyberbullying behavior (Tokunaga, 2010), while a more recent study in the Netherlands showed that over 60% reported having been involved in cyberbullying incidentally and approximately 30% "several times" (den Hamer, Konijn, & Keijer, 2014; Table 5). The majority of the victims do not tell their parents about the victimization (Tokunaga, 2010) and victims of cyberbullying behavior have higher suicidal ideation rates than non-victims (Hinduja & Patchin, 2010; Kim & Leventhal, 2008; Li, 2007). These numbers underpin the emergency to determine the factors that stimulate cyberbullying behavior. It seems that the ease with which insulting messages are created and sent through the internet, as well as the virtuality and anonymity, further instigates online bullying behaviors (Wright, 2013), though most victims report they know the bully or the sender of these harmful messages (Feldman Hertz et al., 2013; Sumter & Valkenburg, 2011). These features then may coincide with impulse behavior stemming from anger and frustration upon victimization—one does not need to think twice to send a message through social networking sites, apparently.

A closely related issue where peers exert a strong influence on adolescents is the feeling to not belong and be rejected by peers. Research states that peer-rejected youngsters form a vulnerable group for negative developmental outcomes, including aggression, antisocial behavior, adjustment problems, and are more likely to drop out of school (Coie, Dodge, & Kupersmidt, 1990; Juvonen, Graham, & Schuster, 2003). During adolescence, being rejected by peers is extremely painful and leads to intensely felt negative emotions (e.g., anger, frustration, and anxiety; Davey, Yücel, & Allen, 2008; Prinstein & Aikins, 2004). In our own research, we have showed the importance of including social rejection in studying how youngsters may become attracted to media content portraying antisocial and risk behaviors. In mimicking online rejection in an experimental setting, we demonstrated adolescents who felt socially rejected in a ball tossing game were drawn towards YouTube clips with antisocial content (Plaisier & Konijn, 2013). On social media like Facebook, it is very easy and often occurs that "friends" are rejected, excluded, and ostracized (Christofides et al., 2012b; Feldman Hertz et al., 2013). Clearly, the occurrence and consequences of such virtual peer rejection can be experienced as quite real, yet the "perceived realism" of virtual encounters needs further study.

In a similar vein, the pressure peers pursue on each other to conform to in-group standards, norms, and values-highly common among adolescents-may be further increased through the explicit expressions of individual opinions and postings of exemplars that are available to many at once and created upon wish over and over again through new media technology. The massive portrayals of specific types of young characters in the online portals may perhaps even push toward standardization of specifically held opinions. For example, the overwhelming portrayal of partying, good-looking youngsters may cause peer pressure on others. This may invoke increased awareness and a felt need to expose oneself in some, while it may invoke enhanced objectified consciousness in others and perhaps also decreased self-esteem in those who feel inferior compared to all those optimally exposed identities (Veldhuis, Konijn, & Seidell, 2014a; 2012b). However, to what extent would the young visitor exposing himself/herself to all those "glorifying personal fables" on the internet realize how much of "distorted reality" is reflected in these virtual identities? Hence, it is highly relevant to include the role of peer influence, in particular social rejection and peer pressure, in studying how new media technology affects adolescent development.

In brief, adolescence comes with an apparent need to express one's likes and dislikes, show one's favorite models, assemble music, images, items, and friends that somehow reflect and support the direction into which one's identity is developing—in the desire to establish their very own selves. Thus, identity construction may be highly supported through the expression and creation of one's "personal fable" on social networking sites and soliciting for a dialogue to further enhance the narrative of the fable. However, when overstretched, successful exposure on social media may feed narcissism or entitlement which is known for its detrimental effects on interpersonal behavior (e.g., Campbell, Bonacci, Shelton, Exline, & Bushman, 2004; Leung, 2013; McKinney, Kelly, & Duran, 2012).

In conclusion, the appeal of new communication technology to youngsters may be partly explained from a developmental perspective in the needs for egocentrism and identity formation that can be fulfilled in relatively easy and safe ways through new media technology as well as in a much broader variety than ever before in real life or through the traditional media. Thus, adolescents may freely experiment with their egocentrism and act out their personal fables and idealized selves through self-enactment and virtual impression management in their own-created media environments, and hence mediated self-enactment. Their peers are overwhelmingly present in virtual space to support or hinder this process, which may place additional pressures on the maturing adolescent to find his/her own way.

Underlying psychological mechanisms adding to communication technology's effectiveness

Following the intersection of new media technology's offerings and the broader developmental context of adolescence, we will now look a step further into several underlying mechanisms that are specifically relevant in view of how newer communication technology may sort effects in adolescents from a psychological view on information processing. Within the developmental context as described earlier, the newer communication technology gives rise to another look at the role of identification. Furthermore, contemporary highly graphical communication technology asks for a reconsideration of fiction–reality perceptions. Finally, we will discuss how newer communication technology may serve the developmentally relevant tasks of emotional regulation and morality in adolescent development.

Wishful identification with ideal selves and others

As argued earlier, the interactive nature of video games, and virtual worlds in particular, provides an even more attractive environment for vicariously trying out different roles than the more traditional media. Especially because early adolescence is a time of increased risk taking and novelty seeking (Kirsh, 2003), virtual environments lend themselves to explore a lot of risk and deviant behavior, trying out unexplored territories and behaviors, meeting virtual strangers, and peek into others' lives to encounter novel experiences. This holds true for digital games and virtual worlds, but occurs in part also through social media like Facebook. The extended nature and enhanced opportunities to "identify" with another character compared to the more traditional media puts the well-known concept of identification in a different perspective.

In studying the more traditional media, identification has most often been interpreted as similarity-based—in terms of similarities observed in age, gender, and contextual factors such as social indicators of wealth or living circumstances. For example, in TV shows, the observer identifies more strongly with characters whom she/he believes is from a similar developmental stage and social background (e.g., Hoffner, 1996; Hoffner & Buchanan, 2005; Tian & Hoffner, 2010). Given a number of similarities between the character and the observer (i.e., similarity-based identification), the observer may more easily take over the perspective of the character and feel as if she/he is the character herself/himself; looking at the world through the eyes of the character. In newer media like role-playing games or first-person shooters, however, the player often "inhabits" the character, thus posing another view on the interpretation of "identification." Identification of the player ("observer") with a virtual character usually takes place through actual "embodiment" in the form of an avatar (Yee, 2006; Yee, Bailenson, & Ducheneaut, 2009). Most avatars can be personalized to a considerable extent. Hence, adolescents will model their avatars according to attractive qualities they already have and, quite likely, in accordance with their idealized selves or ideal models that possess qualities they do not have but wish they had or dream of possessing in the near future—an ideal to look up to and which may serve as a model to build on one's identity. Therefore, interactive communication technology makes it particularly salient to distinguish between *similarity identification* and *wishful identification* (cf. Hoffner & Buchanan, 2005; Konijn & Hoorn, 2005; Konijn et al., 2007; Von Feilitzen & Linné, 1975).

In similarity identification, the observer and the "subject of identification" (e.g., a role model, game hero) share salient characteristics, and similarity in feeling and thinking follows from an imaginary merging of observer and character. However, in wishful identification, the desire to emulate the character is key; the wish to be like or become like an admired other—someone who possesses qualities one wishes to have oneself, while the self-other distinction is clear. This can occur in general terms (e.g., a role model for future action) or in specific terms (e.g., imitating a particular behavior; Hoffner & Buchanan, 2005; Hoffner & Cantor, 1991; Von Feilitzen & Linné, 1975). Through wishful identification, playing roles in games or acting out in virtual worlds, "in imagination" inhabiting these roles, provide a glimpse of "what if." This may act as a powerful vehicle for adolescent boys and girls who wish to enact their idealized selves in a safe environment (Konijn & Nije Bijvank, 2009). Hence, establishing identification of the "observer" with the character (i.e., avatar) can be self-created in terms of "outer appearances" like gender, age, and sociodemographic factors. This supports the similarity-based identification. However, the driving force in role-playing games (and related digital games, virtual worlds, and social media) is not so much similarity identification, but rather wishful identification, or "self-ideal identification" for that matter.

Adolescents would be particularly interested in experimenting with heroes who possess attractive qualities they do not (yet) have themselves but dream of possessing in the near future, more so than similar roles. For example, adolescent boys and girls are attracted to much older heroes in games rated for 18+ or M+ (Nije Bijvank, Konijn, Bushman, & Roelofsma, 2009). Most male models that adolescent boys encounter in the games that are popular among them are tough male warriors who succeed in solving problems in ways typical for a masculine identity (Carnagey & Anderson, 2004; Dietz, 1998; Jansz, 2005; National Television Violence Study, 1998). Therefore, adolescent boys would be especially attracted to play such games in trying out and stepping into the "virtual skin" of their brave heroes who are stronger, more stable, and more courageous than they consider themselves. Accordingly, our research showed that wishful identification acted as a significant underlying mechanism in exerting effects in the behavior of adolescent boys after playing (violent vs. non-violent) interactive games (Konijn et al., 2007; Nije Bijvank, Konijn, & Bushman, 2012).

Similar reasoning applies to adolescent girls, being highly interested in wishfully identifying with models that portray attractive features they dream to once possess themselves. For example, many adolescent girls may be prone to an idealization effect (Veldhuis, Konijn, & Seidell, 2014b) when exposed to the thin-body ideal in triggering the wish to become as thin as those "beautiful and successful" models portrayed on many media sites (Harrison, Taylor, & Marske, 2006). Given the high number of eating disorders and girls subjecting themselves to stringent dieting and plastic surgery, they clearly wish to become as thin as the models portrayed on screen. Moreover, their peers may increase this wish due to the option of providing feedback on models presented on the internet through the option of adding user-generated content. In our research (Veldhuis, Konijn, & Seidell, 2014b), we showed how user-generated content from peers on YouTube images portraying the thin-body ideal increased awareness of one's body when peers indicated that the portrayed ideal body is a goal within reach of adolescent girls.

In sum, the abundant portrayals and intense search of adolescents for idealized selves, including "glorifying personal fables" (see earlier texts), feed the wish to identify with the ideals as displayed via new media technology. However, this poses a challenge to the perceptions of realism of such ideals and brings them into balance with the realities of one's offline life.

Perceived realism of interactive and virtual environments

New communication technology poses challenges to the perception of realism in what is fictional, virtual, or inflicted reality. As we briefly touched upon the question of how real the "materialized imaginary audiences" are, as in the earlier text, the current section will take a closer look into the concept of perceived realism in view of the graphically rich newer communication technology. In addition to wishfully identifying with virtual heroes, we consider the degree to which adolescents perceive the characters, behaviors, and events as portrayed in virtual worlds (including games and social media) as a realistic representation of the real physical world, an important psychological mechanism underlying media use and effects (cf., Busselle, 2001; Konijn, Walma van der Molen, & Van Nes, 2009; Potter & Tomasello, 2003; Shapiro & Chock, 2003). Current trends in digital game production are highly sophisticated in making games and virtual environments look and feel realistic in their graphical portrayals (Hoorn, 2012; Hoorn, Konijn, & van der Veer, 2003). Increasing the level of realism is highly popular among game designers (e.g., Van Vugt, Konijn, Hoorn, Eliëns, & Keur, 2007) and, accordingly, players rate realistic video games more favorably than unrealistic ones (Wood, Griffiths, Chappell, & Davies, 2004).

The extent to which games were *perceived* as realistic by adolescent boys significantly affected them in the effects of playing violent games (Konijn et al., 2007; Nije Bijvank et al., 2012), while we did not find a significant effect of realism as an experimental factor through comparing more realistic games versus games high in fantasy. That is, the perception of realism from the player's perspective is of relevance here, not per se reflecting the producer's claims or an outsider's evaluation of whether a game is fantasy or reality-based. Thus, even a fantasy game may be perceived as somehow representative of real-world events from the perspective of a particular player. This is more likely in younger players than in older ones (Konijn et al., 2009; Mares, 1996; Potter, 1982) and more likely in those with lower cognitive skills than in those higher in cognitive abilities (Nije Bijvank et al., 2012).

We found that the extent to which the level of realism was ***perceived*** significantly contributed to the wish to be like the hero in the game (Konijn et al., 2007). Characters perceived as more realistic would more likely be chosen as models to identify with than characters perceived as plain fantasy with whom it might be difficult to (wishfully) identify with. Furthermore, we found that those who perceived more realism in fictional content were also inclined to attribute higher levels of information value to such media portrayals (Konijn et al., 2009). Hence, ***perceived realism*** of virtual and mediated content may act as an important underlying mechanism in affecting the user of new media. Specifically, perceived realism would increase the likelihood that popular models and optimally portrayed "ideal selves" and information on networking sites increase wishful identification and are taken as guidelines for future behaviors.

The distinction between the real, physical worlds in which the adolescents live may get increasingly blurred with the virtual worlds in which they communicate, extending the real world in cyberspace and the other way around. This requires increasingly sophisticated judgments to determine what is fictional, fabricated, and factual. Various dimensions and operationalizations of perceived realism have been studied extensively, varying from perceptual, psychological, graphical, behavioral realism to many other variations with mixed results (Busselle & Greenberg, 2000; Konijn & Bushman, 2007; Konijn et al., 2009; Shapiro & Chock, 2003; van Vugt, Bailenson, Hoorn, & Konijn, 2010; van Vugt et al., 2007). In our research (Konijn et al., 2009), perceived realism was defined as a subjective interpretation of the extent to which media portrayals (including situations, characters, behavior, stories, etc.) contain a reflection of daily life realities, irrespective of the program makers' intentions. In brief, how representative of real-life affairs a particular (segment of) media content is evaluated. Despite the highly complex definition issue and the highly subjective and varying interpretations of the "epistemics of the virtual" (Hoorn, 2012), the concept of perceived realism is highly relevant in view of adolescents' perceptions of virtual worlds.

In specifically addressing adolescents, cognitive skills are still developing and those who are further developed are probably better able to reflect on the fabricated nature of digital games and virtual worlds (yet, note that development of the higher cognitive functions in the brain is not necessarily equal to educational ability level; Shaw et al., 2006). Thus, they would perhaps be less inclined to perceive the represented worlds as representative of real life. Consequently, adolescents with further advanced cognitive skills will be less inclined to identify with (violent) game characters as role models. In fact, research showed that boys of lower educational ability are more inclined to perceive the represented acts in games as realistic and to wishfully identify with violent game heroes, more so than boys of higher educational ability levels (Nije Bijvank et al., 2012). Our research further shows that the more emotional people are when processing media, the more likely they will believe it. Individuals experiencing higher levels of negative emotions (i.e., annoyance in Study 1 and empathetic sadness in Study 2 in Konijn et al., 2009) while presented with audiovisual materials taken from a television documentary, attributed higher levels of perceived realism and information value to the contents than those who were less emotionally moved. Adolescents in particular were more strongly inclined to have their perceptions of realism guided by their emotional state than adults.

In line with emotion psychology, processing media messages in an emotional state may give the viewer the impression that something real, psychologically real, or of real importance is going on (Konijn, 2012; Konijn & Ten Holt, 2010; Konijn et al., 2009). The lower parts of the brain in fact respond to mediated images on screens in similar ways as to real-life events (e.g., through so-called "mirror neurons"; Rizzolatti & Craighero, 2004). Likewise, media images that only roughly match a real-world event may trigger the amygdala to respond as if real, due to the fast lower pathway in our brains which allows us to react before we have fully cognitively processed the information (Damasio, 1994; LeDoux, 2000). Therefore, it may be hard for the lower pathway to distinguish between imagined and perceived images when using graphically rich media, having trouble distinguishing between the fictional and factual (Konijn, 2012; Konijn et al., 2009; Kosslyn, Thompson, & Alpert, 1997; Murray, 2008; Zillmann, 2010). The lower pathway runs to the amygdala, which is part of the limbic system and is involved in the processing of emotions. Thus, emotions may blur perceptions of the factual and fictional, in particular in the highly interactive and graphically rich envrionments of new media technology.

Following the aforementioned reasoning, the emotional state of adolescents may play a significant role in how they use new media technology. Therefore, the following text will discuss the intersection of the emotional developmental needs and new media technology.

Emotion regulation needs and morality in adolescents' use of new media

The interactive nature of newer communication technology in particular seems to provide a wealth of opportunities for exploring, experimenting, and coping with emotions from the viewpoint of adolescents' developmental needs. Especially, in dealing with emotions that are felt as unpleasant, disturbing, or too intense, adolescents seem to seek relief in using media. Such emotional needs may act as an underlying mechanism in being affected by that media, because emotion regulation needs may drive the adolescent to use a specific type of media. For example, qualitative research has indicated that adolescent boys preferred to play violent video games in coping with experienced anger and frustration (Arnett, 1991; Olson & Cheryl, 2010; Olson, Kutner, & Warner, 2008). These ideas have been empirically supported by our own experiments and cross-sectional research (e.g., den Hamer et al., 2014; Plaisier & Konijn, 2013).

Results of our research have shown that being rejected by peers induced intense feelings of anger, frustration, and negative affect, both as induced in experiments and assessed in the classroom. These negative emotions made the rejected adolescents morally more tolerant toward the antisocial media content in terms of a lowered moral judgment. That is, the rejection-based negative emotions led adolescents to judge antisocial media content as more tolerable and acceptable than those who were not rejected. Subsequently, a lower moral judgment was directly related to a preference for this type of antisocial media content. In contrast, these findings were not established for a comparing group of young adults, while the rejection–manipulation was successful in both the adolescent and young adult samples in terms of perceived feelings of belonging (Plaisier & Konijn, 2013). The subsequent steps in the process of being rejected by peers and a preference to watch YouTube clips portraying antisocial and risk behavior content were thus evidenced for the adolescents only, not the young adults.

We assume that differences in emotion regulation skills between adolescents and young adults may explain the aforementioned findings. Young adults' emotion regulation skills are more elaborate and effective, while adolescents are still developing these skills (Eisenberg, 2000; Garnefski & Kraaij, 2006; Spear, 2000). Adolescents are not yet well-equipped to deal effectively with intensely felt emotions because specific brain areas related to acquiring cognitive skills that are needed for emotion regulation are still developing during adolescence (Casey, Jones, & Hare, 2008b). Related, moral reasoning still develops during adolescence due to the structural and functional maturation of the prefrontal cortex and the limbic subcortical regions (Casey et al. 2008b; Choudhury et al., 2006; Somerville, Hare, & Casey, 2011; Steinberg, 2005). Moral reasoning requires cognitive reappraisal, perspective taking, and reflective cognitive skills. Pre- and early adolescents lack those skills and tend to focus on themselves, in particular in angry moods, thereby inhibiting prosocial moral reasoning (Eisenberg, 2000; Eisenberg et al., 1994) and instigating a more lenient moral judgment and moral permissibility (Ugazio, Lamm, & Singer, 2012). Therefore, this relation between anger and moral judgment, subsequently affecting their media use, may be more outspoken for adolescents than for other age groups. Hence, both adolescents and young adults are influenced by social rejection, even in a virtual or online environment; however, adolescents lack the skills to effectively regulate the associated emotions.

We conclude that it is the specific combination of the developmental stage of adolescence, an intensely felt unpleasant emotional state, and affordances of new communication technology which draws peer-rejected adolescents to consume antisocial media content, further guided by an increased tolerance toward morally adverse media content. In our mediated society in which many parents, educators, and the public at large are concerned about the omnipresence of new media technology including its potential risks abundant in the lives of their offsprings, these considerations bring highly relevant and up-to-date research questions to the fore. In particular, given the increased independency of adolescents' media use occurring outside of parents' control. Future research should examine how emotions associated with peer influences may further affect adolescents' use of new media and affect their moral development.

In line with this research, we hypothesized and found that being a victim of face-to-face bullying would instigate feelings of anger and frustration, which would then draw the victimized adolescent to higher levels of exposure to media with antisocial content than without those victimization-related feelings. In turn, such media use would facilitate or amplify cyberbullying behavior (den Hamer et al., 2014). Thus, victims of bullying were drawn to media with antisocial content via victimization-related anger and frustration. This research confirmed and extended previous research showing that adolescents who are face-to-face bullies tend to be cyberbullies as well (Estévez, Villardón, Calvete, Padilla, & Orue, Estévez, Villardon, Calvete, Padilla and

Orue, 2010; Vandebosch & Van Cleemput, 2009) and victims of cyberbullying appear to become a cyberbully themselves (Li, 2007; Walrave & Heirman, 2011). Our research revealed the underlying mechanisms in between being victimized and becoming a cyberbully oneself, highlighting media's role. For the victimized, contemporary communication technology offers plenty of easily accessible opportunities to (imaginary) "fight back" and "feel in control" while staying on the safe side of the (assumed) anonymity of online communication. Future research may more specifically test this assumption.

In brief, these rejected or victimized adolescents end up in a cyclic process that may explain why victims often become cyberbullies themselves (den Hamer et al., 2014). The idea of such a cyclic process is further supported by previous research showing that individuals who exposed themselves to antisocial media content in order to cope with anger were even more aggressive after antisocial media usage (Bushman, 2002; Bushman, Baumeister, & Phillips, 2001). Longitudinal research is needed to further test the assumed *Cyclic Process Model.* In order to further study the multifaceted interactions between affordances of newer communication technology and adolescent emotional and moral development, including the dominant influence of peers, up-to-date research methodology is also needed. Features of newer social media offer a number of promising opportunities to improve methodologies and measurement instruments to further enhance the field of mediated communication research.

Methodological opportunities of newer communication media

In pursuing the research reported in this chapter, we implemented several methodological innovations while taking advantage of new communication technology. To study how adolescents use new media technology and how this may affect them, research tools that make use of and that can be integrated into such highly popular media would be very helpful. Not only might this be more attractive to the target group of adolescents than traditional ways of measurement, but it also widens the research possibilities in a closer fit with new media environments (Konijn, Veldhuis, & Plaisier, 2013). This section will therefore describe several promising features and illustrate how we have developed new measurement instruments to assess content-based media exposure (i.e., the C-ME scale) and to assess moral judgment and media preferences for (antisocial) media content (i.e., the Media, Morality and Youth Questionnaire: MMaYQue). Furthermore, we describe how the feature of user-generated content in social media was applied to "authentically" manipulate peer feedback on media content in a controlled study design (i.e., the Peer Influence on Media System: PIM-S).

The content-based media exposure scale: C-ME scale

In examining the role of media use in adolescents' cyberbullying behaviors, we needed a measurement device to specifically assess the type of content to which one has been exposed. Thus far, the extant measurement devices rely heavily on a specific medium or format and most commonly ask for hours of exposure to a particular device (e.g., hours of exposure to TV or the frequency of video game play) in a given time period (e.g., per day, week, or weekend). The underlying assumption is that a high level of media exposure in general would imply a high level of exposure to violence. Others have been asking for participants' most favorite video games and have them coded afterward for violent or non-violent content (e.g., Anderson, Gentile, & Buckley, 2007; Nije Bijvank et al., 2012), which is quite labor-intense. However, in the contemporary media landscape where all types of media content are easily accessible through any modern device, it is key to ask for the specific type of *content* one has been exposed to, not to a particular device (Konijn et al., 2014). Therefore, we developed C-ME scale (den Hamer et al., 2014; den Hamer, Konijn, Plaisier, Krabbendam, & Bushman, under review).

To specifically assess content-based media exposure in a way suitable for sophisticated statistical analyses, a Likert-type scale was preferred with items that all together would provide an index for the extent to which one is exposed to specific media content throughout various devices. Because we presumed that exposure to media with an antisocial nature would particularly relate to cyberbullying behavior, we focused on constructing a scale that would primarily assess exposure to media with antisocial content. That is, media portraying antisocial behavior in general as well as including risk and norm-crossing behaviors (e.g., binge drinking, unsafe sex, and drug abuse). In order to construct the items, we extensively scrutinized YouTube clips as well as TV series, movies, and games popular among youth on the specific behaviors portrayed. In so doing, we generated a list of behaviors and items commonly portrayed in contemporary media fare covering four types of antisocial and risk behaviors as described in Gardner and Steinberg (2005). The selected clips and episodes were described as briefly as possible in one sentence as is commonly encountered on YouTube sites.

After extensive examination and pretesting of items, the resulting C-ME-scale consists of 17 items and measures how often someone watches various portrayals of antisocial and risk behavior portrayals in media content. In addition, several filler items were included as well (den Hamer et al., 2014). Currently, the filler items are further developed into a prosocial dimension of the C-ME scale to increase its usability (den Hamer et al., under review). Each item is followed by a 5-point rating scale (varying from never to very often) to indicate the extent to which one encounters the specific behaviors described. The instruction states that questions pertain to media content one encounters "on television, in games, or on the internet". Sample items are "How often do you watch people fighting?" and "How often do you watch people destroy someone else's property?" Psychometric analyses revealed a reliable scale to measure exposure to media with antisocial and risk behavior content (Cronbach's alpha = 0.84). A subsequent validation study with various samples shows that the C-ME-scale also possesses good criterion, discriminant, and convergent validity (den Hamer et al., under review). As described in the above, the C-ME has been successfully applied in demonstrating the amplifying role of selective media exposure in the process of cyberbullying behaviors. Future research will further test the C-ME in a range of media-related studies, both surveys, experiments, and longitudinal research and has a wide applicability in many media-related research questions.

The Media Morality and Youth questionnaire: MMaYQue

Media content portraying antisocial and risk behavior (e.g., reckless driving, binge drinking, drug abuse, and bullying) attracts large adolescent audiences (Brown & Witherspoon, 2002; Funk & Buchman, 1996; Strasburger et al., 2010). Earlier, we have argued that emotional experiences resulting from peer influence may, in part, drive adolescents to use this type of media content. In order to study the underlying mechanisms that might elicit adolescent preferences for antisocial, violent, and risky media content, a new measurement device has been constructed to assess the adolescents' moral judgment of antisocial media content and their momentary preferences for specific content. As common types of assessment of media exposure ask for frequency of exposure in general and not for preferences at a specific moment in time, these could not be used for our research purposes. Furthermore, common measurements of morality usually focus on identifying the developmental stage of moral reasoning in using scenario-based measurement instruments (e.g., the Prosocial Reasoning Objective Measure, PROM; Carlo, Eisenberg & Knight, 1992). The meanwhile popular moral foundations questionnaire (Graham, Haidt, & Nosek, 2009) is meant to analyze the various intuitive domains of morality (e.g., fairness, and authority), and not moral judgment per se. In addition, the extant measurement devices could not be used in an experimental design, while we aimed at investigating changes in moral judgment and media preferences to establish causality of emotional states as a result of (manipulated) peer influence (Konijn et al., 2013; Plaisier & Konijn, 2013). To overcome these shortcomings, we developed a new measurement device named MMaYQue for which we used new media technology.

The MMaYQue contains 15 descriptions of popular YouTube clips which were selected based on their popularity among adolescents and four domains of antisocial and risk-taking behaviors: violence/aggression, sexual harassment/sexual self-exposure, substance abuse/binge drinking, and reckless driving (Gardner & Steinberg, 2005). The items were formatted such that they looked like original YouTube clips; very brief content descriptions presented on a typical YouTube black opening screen with the "play" button in the middle (see Konijn et al., 2013). A sample item is: "Youngsters scolding at a police officer and pushing him off his motorbike." In addition, we included six filler items that described neutral/prosocial media content (e.g., "Cat playing a song on the piano"). To assess *media preference*, participants were asked to indicate how much they would like to watch each of the clips on 5-point rating scales. A clear advantage of such Likert-type items is the possibility to calculate a clear index from the participants' scores for further statistical analyses.

To measure *moral judgment* of media content, the same materials were used. That is, participants indicated how acceptable or "okay" he/she judged the behavior as portrayed in the clip on 5-point rating-scales. Several validation studies showed good psychometric qualities of the MMaYQue as well as good criterion, discriminant, and convergent validities. For example, the Cronbach's alpha for antisocial media preference ranges from $\alpha = 0.85$ to 0.94 and the Cronbach's alpha for moral judgment of antisocial media content ranges from $\alpha = 0.74$ to 0.93. In addition, adolescents high on trait aggressiveness, among other variables, were also high on preferences for antisocial media content and showed a more lenient moral judgment of antisocial media content, in contrast to adolescents high in empathy (Konijn et al., 2013; Plaisier & Konijn, 2013, under review).

Future research may apply the MMaYQue to further study how differential media content preferences are generated to provide important insights in how specific media effects are established. Given the wide availability of all types of content through the internet, it is relevant to investigate when and why specific media content is preferred by whom under which circumstances as well as how specific content is morally evaluated. For example, research shows that negative developmental outcomes might result from media exposure to morally adverse content as well as from peer pressure and rejection (Juvonen et al., 2003; Stense, Scholte, Salmivalli, & Voeten, 2007; Strasburger, 2009). However, how media use among adolescents and peer influence mutually affect each other is largely unclear and understudied. Peer influence on adolescents' media use may be studied in combination with the MMaYQue by including peer feedback on (antisocial) media content in the interactive environments using the user-generated content options. This will be illustrated in the following.

The peer influence on media system: PIM-S

New media environments open possibilities to examine the causal and interaction effects of media portrayals and peer feedback in an integrated and systematic way. Thus far, such research efforts could hardly be accomplished with traditional media formats but are now within reach with the newer communication technology. We exploited the function of user-generated content by peers in studying the *combined* influence of peers' feedback and media content on adolescents, and hence termed *Peer Influence on Media System: PIM-System*. In particular, the feature of social media to directly post comments and feedback to online materials opens a wide array of unprecedented research opportunities (Konijn et al., 2013). Our research on negotiating body ideals and body perceptions in adolescent girls may just serve as an example. Clearly, the thin-body ideal prevails in today's appearance-focused society, and models and celebrities with ideally-toned bodies dominate the media landscape (e.g., Grabe, Ward, & Hyde, 2008; Groesz, Levine, & Murnen, 2002). Not only traditional media but also the newer and social media strongly favor the portrayal of the thin-body ideal, and the images adolescent girls post on their personal websites are obviously in line with such an ideal. Connecting to the aforementioned need of adolescents to materialize their "personal fable" and "ideal selves," media formats like Facebook and girl-specific websites (e.g., Girlscene) seem to appeal to this need. Clearly, peers contribute to how media-generated ideals are perceived and modeled after (e.g., Dohnt & Tiggemann, 2006). Hence, we studied the combined effects of media and peers.

We applied the feature of user-generated content as a research tool to examine how peer feedback on the thin-body ideal would affect the body perceptions of adolescent girls. The *PIM-S* allowed for a systematic integration of peer feedback on media portrayals to study their combined influence in a systematic way. Many studies have shown that the thin-body ideal prompts negative psychosocial responses in girls about their body image, such as increased body dissatisfaction, social comparison, and objectified body consciousness (see meta-analysis of Grabe et al., 2008 and an extensive review of Lopèz-Guimerà, Levine, Sánchez-carracedo, & Fauquet, López-Guimerà, Levine, Sánchez-carracedo and Fauquet, 2010). Subsequently, this might put vulnerable individuals at risk of developing eating disorders from (extreme) dieting behavior (e.g., Harrison, 2003; Lee, Damhorst, & Ogle, 2009) and at risk for being overweight and obese in a later stage in life (Neumark-Sztainer et al., 2006). Therefore, we looked for a means to negotiate the effects of the thin-body ideal, and to restore more healthy perceptions of an adolescent girls' body size, in applying features of new media technology. In manipulating and mimicking peer feedback on such thin-ideal portrayals, we developed the PIM-S.

Searching for a realistic real-life setting to experimentally combine thin-ideal images *and* textual comments on these images with peers presented as sender of this information, we proposed YouTube (Konijn et al., 2013; Veldhuis et al., 2014b). Through the PIM-S, we implemented the principle of labeling media models' body shape into user-generated content on YouTube websites and presented this as "authentic" feedback from peers. Each image of a media model was accompanied with 10 peer comments, varying in style and jargon, and relating to one of the peer comment conditions combined with neutral filler comments. All materials as they commonly appear on a YouTube site, including a media model, the side clips, ads, typically adolescent nicknames (e.g., Rosie2U) and peer feedback, were adjusted to meet our study requirements and the results were integrated in graphical software to look like an original full-color YouTube-screenshot (see Konijn et al., 2013; Veldhuis et al., 2014b).

The content of the peer feedback was based on previous research showing that simple textual labels are effective in *counteracting* negative psychosocial responses in adolescent girls (Veldhuis, Konijn, & Seidell, 2012, 2014a). That is, we found that information labels (compared to warning labels) are most effective in reducing body dissatisfaction when these labels accurately informed about the underweight status of extremely thin media models (Veldhuis et al., 2014a). Furthermore, a *normalization* effect could be established as a result of informing participants that a model is of "normal weight" (Veldhuis et al., 2012). Apparently, this induced a reference value in terms of what body shape is considered "normal" with concurrent effects. The specific direction of a normalization effect is dependent upon the specific combination of a model's body shape and normalization information. For example, if the added information stated that an extremely thin model was considered of a normal weight, this raised significantly more body dissatisfaction than normal-weight information on moderately shaped media models (Veldhuis et al., 2012). Hence, we manipulated such information as "authentic feedback" on media models stemming from peers.

We transformed the information labels into peer comments reflecting on the models in a YouTube clip. For example, under the image of an extremely thin media model in a YouTube screenshot, we included "as-if" user-generated content from peers who commented on this model. Their comments varied according to the experimental condition, claiming the model's weight status being "too thin," "extremely underweight," or, in contrast, of "normal weight" and "good looking." Hence, the peer feedback reflected the model's (under)weight status from an adolescent's perspective.

An additional finding from the study including the feedback from peers was that the comments of peers, more so than just anonymous information labels, could instigate an *idealization effect*. This further extends the normalization effect in not only providing a reference value but also indicating what is ideal and within reach (Veldhuis et al., 2014b). That is, peer feedback telling the adolescent girl that an ideally-shaped thin body of a model presented in the media is within reach because it is only slightly underweight induced negative feelings in terms of objectified body consciousness. This is in accordance with the finding that many girls wish to lose some kilos (Field et al., 1999; Neumark-Sztainer & Hannan, 2000). In conclusion, this study illustrates that the sender of the information plays an important role, which naturally varies on social media, but can easily be mimicked and manipulated as "authentic" user-generated content via our PIM-S to fit research purposes.

A large difference with traditional media is that much of the newer communication technology allows to provide direct feedback to what is portrayed on screen. As such, media effects resulting from new media environments should be studied within the context of how peers respond to it, likely sorting combined effects. Our research illustrates how such an endeavor can be reached: new media content can be effectively manipulated in applying the PIM-S and used in ecologically valid, experimentally controlled research designs.

Conclusion and discussion

Summary and conclusions This chapter reviewed how interactive communication technology and adolescent development intersect in appealing contemporary youth. Adolescent egocentric needs are exemplified through phenomena like an imaginary audience and personal fables, indicating the idea that one is continuously monitored and highly special and unique. Social networking sites in particular offer many possibilities to materialize these imaginary audiences and personal fables. Additionally, virtual worlds and digital games provide safe environments for identity formation, including gender identity and norm-crossing explorations. Given the increased importance of peers during adolescence, we also discussed how peers may sort an unprecedented norm-setting and pressing influence through user-generated content in new media environments. Our research demonstrated how victimization and being rejected by peers, even if "only virtual," may instigate preferences for media with antisocial content and facilitate or amplify to become a cyberbully oneself. Despite possible negative effects (on others), new interactive media offer many advantages for adolescents to express and explore attractive role models, try out ideal or possible selves, and act out peer influence and undesired emotional states in a more active and involved way than ever possible before in the more traditional, lean backward media. Clearly, developmental needs during adolescence are fulfilled through interactive communication technology and contribute to our understanding why today's adolescents' are so strongly attracted to new, interactive media like Facebook, MySpace and YouTube, as well as to virtual worlds and digital games.

In a second part, we reviewed several underlying psychological mechanisms that add to communication technology's effectiveness, following from the intersection of developmental needs and new technology features. Specifically, we discussed how impression management attempts in materializing personal fables and idealized selves on social media may further instigate the wish to be like them. That is, we argued that wishful identification more appropriately denotes one of the mechanisms underlying the effectiveness of new media communication than the more traditional similarity-based concept of identification. We also discussed how the increasingly blurring borders between virtuality and reality poses challenges to the understanding of realism of virtual friends, idealized others, and social reality in adolescents who seem to use newer communication technology as an extension of their daily-life realities. Given the inherent subjectivity of what realism entails anyway, the role of what is perceived as real may need a renewed focus. From here, we presented research showing that adolescents are particularly prone to bias their perception of what is real in accordance to what they feel. In addition, we showed that unpleasant or negative emotions may draw adolescents to use media with antisocial content in expecting a so-called Cathartic effect of media use. We argued that underdeveloped emotion regulation skills may guide adolescents in such media use due to an important lack of maturation of cognitive functions in the adolescent brain. Furthermore, underdeveloped emotion regulation skills are closely related to a lack of social skills to take the perspective of suffering others and constrain ethical moral reasoning. Hence, a lenient moral judgment of morally abject media content facilitates the appeal of such media.

In order to investigate media use and processing of adolescents in more detail, we need more up-to-date research methodology that we then illustrated in a third section of this chapter. In applying methodological opportunities of newer communication media, we presented two new measurement intruments. Given that any type of media content is available anywhere and anytime through new media technology (e.g., smartphones and YouTube), traditional media exposure measurements are not adequate anymore as they rely heavily on a particular device (e.g., frequency of playing video games). In constructing and validating the C-ME, we aimed at a widely applicable measurement instrument to asses media exposure from a content-based perspective. Likewise, we developed the MMaYQue to assess how adolescents themselves morally judge portrayals of antisocial, unhealthy, and risk behavior content in contemporary media fare as well as their own, momentary preferences to watch specific media content. Both measurement instruments can be used in experiments, cross-sectional, and longitudinal research designs. Finally, we discussed how we exploited the user-generated content options on social networking sites to systematically study PIM-S. Illustrated by our research into the interplay of media's thin-body ideal and peer feedback on such idealized images, we emphasize the importance of studying adolescent media use within the developmental context of the relevance of peer feedback during adolescence.

Based on our research thus far, we present a tentative model highlighting a cyclic process of adolescent media use and peer influence in the CPM-A-MPC. In all, this model brings together the various perspectives as presented in this chapter as well as the empirical support for various steps in this model. We believe that the theoretical and empirical underpinnings of the CPM-A-MPC have promising outlooks for future investigation in increasing our understanding of what drives adolescents in their abundant use of new media technology and how specific media use and effects may be explained.

Cyclic process model on adolescent media and peer communication: CPM-A-MPC The results of our studies show that peer influence and adolescent media use act together in affecting adolescents. We not only established how an important part of adolescents' lives, that is, "peer status," impacts their preferences to watch specific media content, but we also provided evidence for important underlying mechanisms. Given that social rejection among peers occurs quite often, for example, in relation to (cyber) bullying, victimization, and ostracism (cf., Wang, Iannotti, & Nansel, 2009), our findings are significant for understanding adolescents' media use. Even more so when taking into account that previous research showed that peer rejection significantly affects negative developmental trajectories among adolescents (Buhs & Ladd, 2001; Juvonen et al., 2003) as well as that individuals exposing themselves to antisocial media content in order to cope with anger were even more aggressive after antisocial media usage (Bushman, 2002; Bushman et al., 2001). Likewise, our research showed that peer rejection in adolescents induces moral tolerance toward antisocial media content and also that media with antisocial content plays a facilitating role in cyberbullying behavior. Therefore, the wide availability and selection of antisocial media content as a means of regulating one's feelings of anger induced by peer rejection can lead to detrimental effects. Media content influences beliefs, attitudes, and behavior, which in turn predicts media selection in the future.

Following this process of reciprocal influence, individuals selecting antisocial media content will adjust their beliefs, attitudes, and behavior in a negative, antisocial way (cf. Slater, 2007; Slater, Henry, Swaim, & Anderson, 2003; Slater, Henry, Swaim, & Cardador, 2004) and extend their perceptions of a mean world to social reality conceptualizations (e.g., Hawkins & Pingree, 1990; Signorielli, 1990), including a hostile attribution bias (Gunther, Miller, & Liebhart, 2009). Furthermore, aggression is related to peer-rejection (Stense et al., 2007). Consequently, antisocial media use in response to being rejected by one's peers may further increase peer rejection, resulting in a cyclic process of combined negative media and peer influences. In a longitudinal study, this cyclic process will be further examined.

Likely, such a cyclic process of media and peer influence will not hold for all adolescents, probably not even a majority. Nevertheless, if an adolescent gets caught in a cyclic process of exclusion, negative emotional state, and antisocial media use, the consequences can be pretty serious even if it holds for only a limited number of them. At the same time, the reverse of such a process may keep promises for positive uses of media content, even if it concerns antisocial, risk behavior, or morally adverse content—some may benefit in reflecting on the issues portrayed in media or not wanting to become like "those despisable creatures."

Figure 15.1 visualizes a schematic representation of our tentative theoretical model representing the CPM-A-MPC. Various paths of the represented cyclic process have been empirically supported in adolescents (mainly the causal effects of peer and media influence on moral judgment, media use, cyberbullying, and the thin-body ideal), while others are currently under investigation (mainly the longitudinal and reciprocal

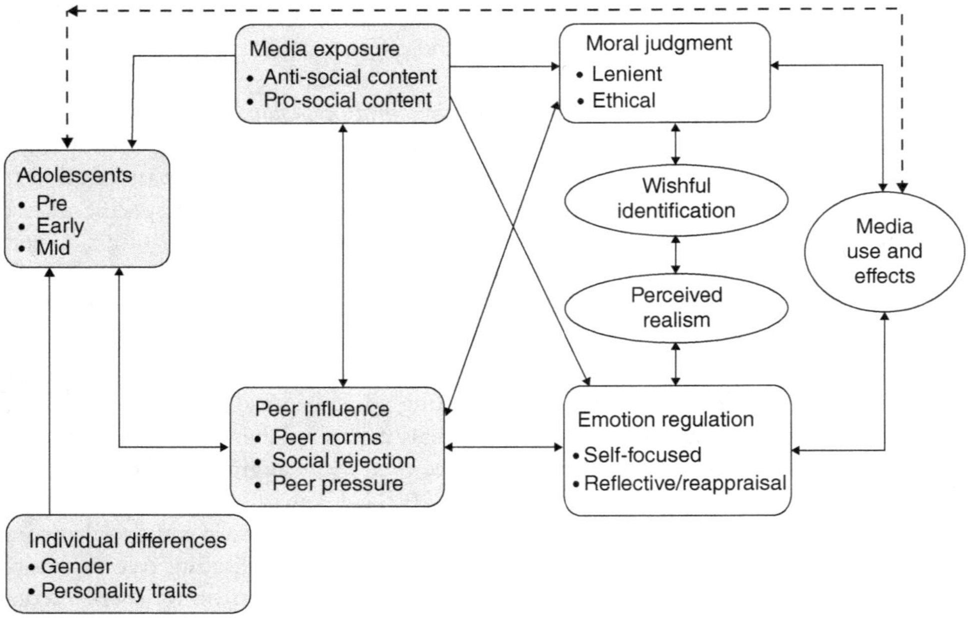

Figure 15.1 Schematic representation of a theoretical model representing the CPM-A-MPC.

interplay of adolescents' media use and peer influence over time). The CPM-A-MPC advances multidisciplinary theorizing about the interplay of adolescents' media use and peer influence from an adolescent developmental perspective in understanding their abundant use of new communication technology.

Adolescents strongly depend on their peers for emotional security and normative behavior and are guided by egocentric beliefs in an imaginary audience and personal fable. New communication technology provides unprecedented opportunities to materialize imaginary audiences, personal fables, and an explicit role of peer influence through user-generated content. The ease of freely posting anything, anywhere and anytime, has become a common habit among youngsters. Many benefit from these opportunities in fulfilling their adolescent needs and extend their identity seeking experiences. However, the backside is that bullying and rejecting others has also become more commonplace than before.

Peer-rejected youngsters form a vulnerable group for negative developmental outcomes, including aggression, antisocial behavior, adjustment difficulties, and dropping out of school. Social rejection is especially painful during adolescence and is associated with intense feelings of anger and frustration and non-constructive coping. Due to the structural and functional maturation of the prefrontal cortex and the limbic subcortical regions, pre- and early adolescents lack cognitive control required for cognitive reappraisal and reflective cognitive skills needed for moral reasoning. Negative emotions constrain reflective reappraisal strategies for emotion regulation and narrow the focus to the self, to one's own needs instead of the needs of others, thereby limiting prosocial moral judgments. Therefore, it seems hard for early adolescents to regulate anger and frustration from peer rejection in an adequate way and it affects their moral judgment in a self-focused way. Hence, they feel attracted to antisocial and emotion-arousing media in search of a means to regulate emotions and explore moral boundaries. Adolescents' emotional states and a lenient moral judgment may feed the wish to be like the courageous heroes they encounter in the media and increase the perceived realism of represented acts and events.

Media and peers act together in establishing emotion regulation beliefs (e.g., media use to cope with anger) and moral judgment (e.g., how tolerable they judge gossiping in cyberspace). Likewise, media and peers act together in negotiating who the attractive role models are to be taken as ideals for future behavior as well as negotiating the levels of realism of their virtual encounters. The various lines and steps in the model are put to empirical tests as to validate and further refine the theoretical model in answering the question *why adolescents* in particular are attracted to violent, antisocial, and risky media contents and *how* such content is processed in view of their developmental stage.

Note

1 This chapter describes part of Konijn's scientific research program in Media Psychology at the VU University Amsterdam. This program moves along three main lines: (a) media use among adolescents; (b) emotions and reality-fiction perceptions in entertainment media; and (c) virtual humans and virtual environments. Application areas are human well-being and health, media education, and creative industry. This chapter merely covers research line (a). Papers on the second and third line can be found elsewhere. About research line 2, for example, see Konijn and Hoorn (2005); Konijn et al. (2009); Konijn (2012); Plaisier & Konijn (2013). Examples of research line 3 can be found in: Konijn and Van Vugt (2008); Van Vugt, Konijn, Hoorn, & Veldhuis (2009); Van Vugt et al. (2010), Konijn (2012); Pontier, Hoorn, Konijn (under review), and the SELEMCA-program: http://crispplatform.nl/projects/selemca for our e-health research including Caredroids.

References

Alberts, A., Elkind, D., & Ginsberg, S. (2007). The personal fable and risk-taking in early adolescence. *Journal of Youth and Adolescence, 36*, 71–76.

Anderson, C. A., Gentile, D. A., & Buckley, K. E. (2007). *Violent video game effects on children and adolescents: Theory, research, and public policy.* Oxford, UK: Oxford University Press.

Antheunis, M. L., Valkenburg, P. M., & Peter, J. (2012). The quality of online, offline, and mixed-mode friendships among users of a social networking site. *Cyberpsychology: Journal of Psychosocial Research on Cyberspace, 6*(3), article 6. doi:10.5817/CP2012-3-6

Arnett, J. (1991). Adolescents and heavy metal music: From the mouths of metalheads. *Youth and Society, 23*(1), 76–98. doi:10.1177/0044118X91023001004

Arnett, J. (1992). Reckless behavior in adolescence: A developmental perspective. *Developmental Review, 12*, 339–373.

Arnett, J. (1999). Adolescent storm and stress, reconsidered. *American Pcychologist, 54*, 317–326.

Bers, M. U. (2001). Identity construction environments: Developing personal and moral values through the design of a virtual city. *Journal of the Learning Sciences*, *10*(4), 365–415. doi:10.1207/S15327809JLS1004new

Boyer, T. W. (2006). The development of risk-taking: A multi-perspective review. *Developmental Review*, *26*, 291–345. doi:10.1016/j.dr.2006.05.002

Breivik, A. B. (2011). *2083: a European declaration of independence*. Available from ULR http://www.kevinislaughter.com/wp-content/uploads/2083+-+A+European+Declaration+of+Independence.pdf [Downloaded 2011-07-23]

Brown, J. D., & Bobkowski, P. S. (2011). Older and newer media: Patterns of use and effects on adolescents' health and well-being. *Journal of Research on Adolescence*, *21*(1), 95–113. doi:10.1111/j.1532-7795.2010.00717.x

Brown, B. B., & Larson, J. (2009). Peer relationships in adolescence. In R. M. Lerner & L. Steinberg (Eds.), *Handbook of adolescent psychology* (3rd ed., pp. 74–103). New York, NY: Wiley. doi:10.1002/9780470479193

Brown, J. D., & Witherspoon, E. M. (2002). The mass media and American adolescents' health. *Journal of Adolescent Health*, *31*(6)(Suppl), 153–170.

Buhs, E. S., & Ladd, G. W. (2001). Peer rejection as an antecedent of young children's school adjustment: An examination of mediating processes. *Developmental Psychology*, *37*(4), 550–560.

Bushman, B. J. (2002). Does venting anger feed or extinguish the flame? Catharsis, rumination, distraction, anger, and aggresive responding. *Personality and Social Psychology Bulletin*, *28*(6), 724–731. doi:10.1177/0146167202289002

Bushman, B. J., Baumeister, R. F., & Phillips C. M. (2001). Do people aggress to improve their mood? Catharsis beliefs, affect regulation opportunity, and aggressive responding. *Journal of Personality and Social Psychology*, *8*(1), 17–32. doi:10.1037/0022-3514.81.1.17

Busselle, R. W. (2001). Television exposure, perceived realism, and exemplar accessibility in the social judgment process. *Media Psychology*, *3*(1), 43–67. doi:10.1207/S1532785XMEP0301_03

Busselle, R. W., & Greenberg, B. S. (2000). The nature of television realism judgments: A re- evaluation of their conceptualization and measurement. *Mass Communication and Society*, *3*(2/3), 249–268. doi:10.1207/S15327825MCS0323_05

Calvete, E., Orue, I., Estévez, A., Villardón, L., & Padilla, P. (2010). Cyberbullying in adolescents: Modalities and aggressors' profile. *Computers in Human Behavior*, *26*, 1128–1135. doi:10.1016/j.chb.2010.03.017

Campbell, W. K., Bonacci, A. M., Shelton, J., Exline, J. J., & Bushman, B. J. (2004). Psychological entitlement: Interpersonal consequences and validation of a self-report measure. *Journal of Personality Assessment*, *83*(1), 29–45. doi:10.1207/s15327752jpa8301_04

Cantor, J., & Wilson, B. J. (2003). Media and violence: Intervention strategies for reducing aggression. *Media Psychology*, *5*, 363–403. doi:10.1207/S1532785XMEP0504_03

Carlo, G., Eisenberg, N., & Knight, G. P. (1992). An objective measure of prosocial moral reasoning. *Journal of Research on Adolescence*, *2*, 331–349.

Carnagey, N. L., & Anderson, C. A. (2004). Violent video game exposure and aggression: A literature review. *Minerva Psichiatrica*, *45*(1), 1–18.

Carnagey, N. L., Anderson, C. A., & Bartholow, B. D. (2007). Media violence and social neuroscience: New questions, new opportunities. *Current Directions in Psychological Science*, *16*, 178–182.

Casey, B. J., Getz, S., & Galvan, A. (2008a). The adolescent brain. *Developmental Review*, *28*, 62–77.

Casey, B. J., Jones, R. M., & Hare, T. A. (2008b). The adolescent brain. *Annals of the New York Academy of Sciences*, *1124*, 111–126. doi:10.1196/annals.1440.010

Choudhury, S., Blakemore, S. J., & Charman, T. (2006). Social cognitive development during adolescence. *Social Cognitive and Affective Neuroscience*, *1*(3), 165–174. doi:10.1093/scan/nsl024

Cingel, D. P., & Krcmar, M. (2012). Understanding the experience of imaginary audience in a social media environment: Implications for adolescent development. *Journal of Research on Adolescence*.

Christofides, E., Muise, A., & Desmarais, S. (2012a). Hey Mom, What's on Your Facebook? Comparing Facebook disclosure and privacy in adolescents and adults. *Social Psychological and Personality Science*, *3*(1), 48–54. doi:10.1177/1948550611408619

Christofides, E., Muise, A., & Desmarais, S. (2012b). Risky disclosures on Facebook: The effect of having a bad experience on online behavior. *Journal of Adolescent Research*, *27*(6), 714–731. doi:10.1177/0743558411432635

Coie, J. D., Dodge, K. A., & Kupersmidt, J. B. (1990). Peer group behavior and social status. In S. R. Asher & J. D. Coie (Eds.), *Peer rejection in childhood* (pp. 17–59). New York, NY: Cambridge University Press.

Damasio, A. R. (1994). Descartes' error: Emotion, rationality and the human brain. New York: Putnam, 352.

Davey, C. G., Yücel, M., & Allen, N. B. (2008). The emergence of depression in adolescence: Development of the prefrontal cortex and the representation of reward. *Neuroscience and Biobehavioral Reviews*, *32*(1), 1–19. doi:10.1016/j.neubiorev.2007.04.016

den Hamer, A., Konijn, E. A., & Keijer, M. G. (2014). Cyberbullying Behavior and Adolescents' Use of Media with Antisocial Content: A Cyclic Process Model. *Cyberpsychology, Behavior, and Social Networking*, *17*(2), 74–81.

den Hamer, A., Konijn, E. A., Plaisier, X. S. Krabbendam, L. C. & Bushman, B. J. (under review). *Validation study on the Content based Media Exposure-scale (CobME)*.

Dietz, T. L. (1998). An examination of violence and gender role portrayals in video games: Implications for gender socialization and aggressive behavior. *Sex Roles*, *38*(5/6), 425–442. doi:10.1023/A:1018709905920

Dohnt, H., & Tiggemann, M. (2006). The contribution of peer and media influences to the development of body satisfaction and self-esteem in young girls: A prospective study. *Developmental Psychology*, *42*(5), 929–936. doi:10.1037/0012-1649.42.5.929

Eccles, J. S., Wigfield, A., Harold, R., & Blumenfeld, P. B. (1993). Age and gender differences in children's self- and task perceptions during elementary school. *Child Development*, *64*(3), 830–847. doi:10.1111/j.1467-8624.1993.tb02946.x

Eisenberg, N. (2000). Emotion, regulation, and moral development. *Annual Review of Psychology*, *51*, 665–697. doi:10.1146/annurev.psych.51.1.665

Eisenberg, N., Fabes, R. A., Murphy, B., Karbon, M., Maszk, P., Smith, M., … Suh, K. (1994). The relations of emotionality and regulation to dispositional and situational empathy-related responding. *Journal of Personality and Social Psychology*, *66*(4), 776–797. doi:10.1037/0022-3514.66.4.776

Elkind, D. (1967). Egocentrism in adolescence. *Child Development*, *38*(4), 1025–1034.

Elkind, D. (1985). Egocentrism redux. *Developmental Review*, *5*(3), 218–226.

Ellison, N., Heino, R., & Gibbs, J. (2006). Managing impressions online: Self-presentation processes in the online dating environment. *Journal of Computer-Mediated Communication*, *11*(2), article 2.

Estévez, A., Villardon, L., Calvete, E., Padilla, P., & Orue, I. (2010). Adolescent victims of cyberbullying: Prevalence and characteristics. *Psicología Conductal*, *18*(1), 73–89. doi:10.1016/j.chb.2010.03.017

Feldman Hertz, M., Donato, I., & Wright, J. (2013). Bullying and suicide: A public health approach (editorial). *Journal of Adolescent Health*, *53*, S1–S3.

Festl, R., Scharkow, M., & Quandt, T. (2013. Peer influence, internet use and cyberbullying: A comparison of different context effects among German adolescents. *Journal of Children and Media*, *7*(4), 446–462.

Field, A. E., Cheung, L., Wolf, A. M., Herzog, D. B., Gortmaker, S. L., & Colditz, G. A. (1999). Exposure to the mass media and weight concerns among girls. *Pediatrics*, *103*(3), 36–40. doi:10.1542/peds.103.3.e36

Frankenberger, K. D. (2000). Adolescent egocentrism: A comparison among adolescents and adults. *Journal of Adolescence*, *23*(3), 343–354. doi:10.1006/jado.2000.0319

Funk, J. B., & Buchman, D. D. (1996). Playing violent video and computer games and adolescent self-concept. *Journal of Communication*, *46*(2), 19–32.

Gardner, M., & Steinberg, L. (2005). Peer influence on risk taking, risk preference, and risky decision making in adolescence and adulthood: An experimental study. *Developmental Psychology*, *41*(4), 625–635. doi:10.1037/0012-1649.41.4.625

Garnefski, N., & Kraaij, V. (2006). Relationships between cognitive emotion regulation strategies and depressive symptoms: A comparative study of five specific samples. *Personality and Individual Differences*, *40*(8), 1659–1669. doi:10.1016/j.paid.2005.12.009

Grabe, S., Ward, L. M., & Hyde, J. S. (2008). The role of the media in body image concerns among women: A meta-analysis of experimental and correlational studies. *Psychological Bulletin*, *134*(3), 460–476. doi:10.1037/0033-2909.134.3.460

Graham, J., Haidt, J., & Nosek, B. (2009). Liberals and conservatives rely on different sets of moral foundations. *Journal of Personality and Social Psychology*, *96*(5), 1029–1046.

Grasmuck, S., Martin, J., & Zhao, S. (2008). Identity construction on Facebook: Digital empowerment in anchored relationships. *Computers in Human Behavior*, *24*(5), 1816–1836. http://dx.doi.org/10.1016/j.chb.2008.02.012

Groesz, L. M., Levine, M. P., & Murnen, S. K. (2002). The effect of experimental presentation of thin media images on body satisfaction: A meta-analytic review. *International Journal of Eating Disorders*, *31*(1), 1–16. doi:10.1002/eat.10005

Gross, E. F., Juvonen, J., & Gable, S. L. (2002). Internet use and well-being in adolescence. *Journal of Social Issues*, *58*(1), 75–90. doi:10.1111/1540-4560.00249

Gunther, A. C., Miller, N., & Liebhart, J. L. (2009). Assimilation and contrast in a test of the hostile media effect. *Communication Research*, *36*(6), 747–764. doi:10.1177/0093650209346804

Harrison, K. (2003). Television viewers' ideal body proportions: The case of the curvaceously thin woman. *Sex Roles*, *48*(5/6), 255–264. doi:10.1023/A:1022825421647

Harrison, K., Taylor, L. D., & Marske, A. L. (2006). Women's and men's eating behavior following exposure to ideal-body images and text. *Communication Research*, *33*, 507–529.

Hawkins, R. P., & Pingree, S. (1990). Divergent psychological processes in constructing social reality from mass mectia content. In N. SignorieIli & M. Morgan (Eds.), *Cultivation analysis: New directions in media effects research* (pp. 35–50). Newbury Park, CA: Sage.

Hinduja, S., & Patchin, J. W. (2010). Bullying, cyberbullying, and suicide. *Archives of Suicide Research*, *14*(3), 206–221. doi:10.1080/13811118.2010.494133

Hoffner, C., & Cantor, J. (1991). Perceiving and responding to mass media characters. In J. Bryant & D. Zillmann (Eds.), *Responding to the screen: Reception and reaction processes* (pp. 63–101). Hillsdale, NJ: Erlbaum.

Hoffner, C. (1996). Children's wishful identification and parasocial interaction with favorite television characters. *Journal of Broadcasting and Electronic Media*, *40*, 389–402.

Hoffner, C., & Buchanan, M. (2005). Young adults' wishful identification with television characters: The role of perceived similarity and character attributes. *Media Psychology*, *7*(4), 325–351. doi:10.1207/S1532785XMEP0704_2

Hoorn, J. F. (2012). *Epistemics of the virtual.* Philadelphia, PA: John Benjamins.

Hoorn, J. F., Konijn, E. A., & Van der Veer, G. (2003). Virtual reality: Do not augment realism, augment relevance. *Upgrade: Human-Computer Interaction*, *IV*(1), 18–26.

Hussain, Z., & Griffiths, M. D. (2008). Gender swapping and socializing in cyberspace: An exploratory study. *CyberPsychology and Behavior*, *11*(1), 47–53. doi:10.1089/cpb.2007.0020

Jansz, J. (2005). The emotional appeal of violent video games for adolescent males. *Communication Theory*, *15*(3), 219–241. doi:10.1111/j.1468-2885.2005.tb00334.x

Jung Oh, H., Lauckner, C., Boehmer, J., Fewins-Bliss, R., & Li, K. (2013). Facebooking for health: An examination into the solicitation and effects of health-related social support on social networking sites. *Computers in Human Behavior*, *29*(5), 2072–2080.

Juvonen, J., Graham, S., & Schuster, M. A. (2003). Bullying among young adolescents: The strong, the weak, and the troubled. *Pediatrics*, *112*(6), 1231–1237.

Kafai, Y. B., Heeter, C., Denner, J., & Sun, J. Y. (Eds.). (2008). *Beyond Barbie & Mortal Kombat: New perspectives on gender and gaming.* Cambridge, MA: MIT Press.

Kalyanaraman, S., & Sundar, S. S. (2008). Impression formation effects in online mediated communication. In E. Konijn, S. Utz, M. Tanis, & S. Barnes (Eds.), *Mediated interpersonal communication* (pp. 217–233). New York, NY: Routledge.

Kapidzic, S., & Herring, S. C. (2011). Gender, communication, and self-presentation in teen chatrooms revisited: Have patterns changed? *Journal of Computer-Mediated Communication*, *17*(1), 39–59. doi:10.1111/j.1083-6101.2011.01561.x

Kato, P. M. (2010). Video games in health care: Closing the gap. *Review of General Psychology*, *14*(2), 113–121.

Kim, Y. S., & Leventhal, B. (2008). Suicide and bullying. A review. *International Journal of Adolescent Medicine and Health*, *20*, 133e54.

Kim, K. J., Park, E., Sundar, S. S., & del Pobil, A. P. (2012, March 5–8). The effects of immersive tendency and need to belong on human-robot interaction. *Proceedings of the 7th ACM/IEEE International Conference on Human-Robot Interaction (HRI'12)* (pp. 207–208). Boston, MA.

Kirsh, S. J. (2003). The effects of violent video games on adolescents: The overlooked influence of development. *Aggression and Violent Behavior*, *8*(4), 377–389.

Konijn, E. A. (2012). The role of emotion in media use and effects. In K. Dill (Ed.), *The Oxford handbook of media psychology* (pp. 186–211). New York, NY: Oxford University Press.

Konijn, E. A., & Bushman, B. J. (2007). World leaders as movie characters? Perceptions of G.W. Bush, T. Blair, O. Bin Laden, and S. Hussein at the eve of Gulf War II. *Media Psychology*, *9*(1), 157–177. doi:10.1080/15213260709336807

Konijn, E. A., & Hoorn, J. F. (2005). Some like it bad. Testing a model for perceiving and experiencing fictional characters. *Media Psychology*, *7*(2), 107–144. doi:10.1207/S1532785XMEP0702_1

Konijn, E. A., & Nije Bijvank, M. (2009). Doors to another me. Identity construction through video game play. In U. Ritterfeld, M. Cody, & P. Vorderer (Eds.), *Serious games: Mechanisms and effects* (pp. 177–201). New York, NY: Routledge.

Konijn, E. A., Nije Bijvank, M., & Bushman, B. J. (2007). I wish I were a warrior: The role of wishful identification in the effects of violent video games on aggression in adolescent boys. *Developmental Psychology*, *43*, 1038–1044.

Konijn, E. A., & Ten Holt, J. (2010). From noise to nucleus: Emotion as key construct in processing media messages. In K. Döveling, C. Von Scheve, & E. A. Konijn (Eds.), *The Routledge handbook of emotions and mass media* (pp. 37–60). New York, NY: Routledge.

Konijn, E. A., Van Vugt, H. C. (2008). Emotions in mediated interpersonal communication. *Mediated interpersonal communication*, 100.

Konijn, E. A., Veldhuis, J., & Plaisier, X. S. (2013). YouTube as research tool—Three approaches. *Cyberpsychology, Behavior, and Social Networking*, *16*(9), 695–701.

Konijn, E. A., Walma van der Molen, J. H., & Van Nes, S. (2009). Emotions bias perceptions of realism in audiovisual media. Why we may take fiction for real. *Discourse Processes*, *46*(4), 309–340. doi:10.1080/01638530902728546

Kosslyn, S. M., Thompson, W. L., & Alpert, N. M. (1997). Neural systems shared by visual imagery and visual perception: A positron emission tomography study. *NeuroImage*, *6*(4), 320–334.

Krämer, N., & Winter, S. (2008). Impression management 2.0: The relationship of self-esteem, extraversion, self-efficacy, and self-presentation within social networking sites. *Journal of Media Psychology: Theories, Methods, and Applications*, *20*(3), 1864–1105.

Laible, D. (2007). Attachment with parents and peers in late adolescence: Links with emotional competence and social behavior. *Personality and Individual Differences*, *43*(5), 1185–1197. doi:10.1016/j.paid.2007.03.010

Laible, D. J., Carlo, G., & Raffaelli, M. (2000). The differential relations of parent and peer attachment to adolescent adjustment. *Journal of Youth and Adolescence*, *29*(1), 45–59.

LeDoux, J. E. (2001). Emotion circuits in the brain. *The Science of Mental Health: Fear and anxiety*, 259.

Lee, H., Damhorst, M. L., & Ogle, J. P. (2009). Body satisfaction and attitude theory: Linkages with normative compliance and behaviors undertaken to change the body. *Family and Consumer Sciences Research Journal*, *37*(4), 466–488. doi:10.1177/1077727X09333165

Leung, L. (2013). Generational differences in content generation in social media: The roles of the gratifications sought and narcissism. *Computers in Human Behavior*, *29*, 997–1006.

Li, Q. (2007). New bottle but old wine: A research of cyberbullying in schools. *Computers in Human Behavior*, *23*(4), 1777–1791. doi:10.1016/j.chb.2005.10.005

López-Guimerà, G., Levine, M. P., Sánchez-carracedo, D., & Fauquet, J. (2010). Influence of mass media on body image and eating disordered attitudes and behaviors in females: A review of effects and processes. *Media Psychology*, *13*(4), 387–416. doi:10.1080/15213269.2010.525737

Mares, M. L. (1996). The role of source confusions in television's cultivation of social reality judgments. *Human Communication Research*, *23*(2), 278–297. doi:10.1111/j.1468-2958.1996.tb00395.x

Martin, J., & Sokol, B. (2011). Generalized others and imaginary audiences: A neo-Meadian approach to adolescent egocentrism. *New Ideas in Psychology*, *29*(3), 364–375. http://dx.doi.org/ 10.1016/j.newideapsych.2010.03.006

McKinney, B. C., Kelly, L., & Duran, R. L. (2012). Narcissism or openness? College students use of Facebook and Twitter. *Communication Research Reports*, *29*, 108–118.

Melle, I. (2013). The Breivik case and what psychiatrists can learn from it. *World Psychiatry*, *12*(1), 16–21. doi:10.1002/wps.20002

Mitchell, K. J., & Ybarra, M. (2009). Social networking sites. Finding a balance between their risks and benefits. *Archives of Pediatrics and Adolescent Medicine*, *163*(1), 87–93. doi:10.1001/archpediatrics.2008.534

Moreno, M. A., Parks, M. R., Zimmerman, F. J., Brito, T. E., & Christakis, D. A. (2009). Display of health risk behaviors on MySpace by adolescents. *Archives of Pediatrics and Adolescent Medicine*, *163*(1), 27–34. doi:10.1001/archpediatrics.2008.528

Murray, J. P. (2008). Media violence: The effects are both real and strong. *American Behavioral Scientist, 51*, 1212–1230.

Nathanson, A. I. (2004). Factual and evaluative approaches to modifying children's responses to violent television. *Journal of Communication, 54*(2), 321–336. doi:10.1111/j.1460-2466.2004.tb02631.x

National Television Violence Study. (1998). *National television violence study* (Vol. 3). Santa Barbara, CA: Center for Communication and Social Policy, University of California.

Neumark-Sztainer, D., & Hannan, P. J. (2000). Weight-related behaviors among adolescent girls and boys. *Archives of Pediatrics and Adolescent Medicine, 154*(6), 569–577. doi:10.1001/archpedi.154.6.569

Neumark-Sztainer, D., Wall, M., Guo, J., Story, M., Haines, J., & Eisenberg, M. (2006). Obesity, disordered eating, and eating disorders in a longitudinal study of adolescents: How do dieters fare 5 years later? *Journal of the American Dietetic Association, 106*(4), 559–568. doi:10.1016/j.jada.2006.01.003

Nickerson, A. B. (2005). Parent and peer attachment in late childhood and early adolescence. *The Journal of Early Adolescence, 25*(2), 223–249. doi:10.1177/0272431604274174

Nije Bijvank, M., Konijn, E. A., & Bushman, B. J. (2012). "We don't need no education": Video game preferences, video game motivations, and aggressiveness among adolescent boys of different educational ability levels. *Journal of Adolescence, 35*(1), 153–162. doi:10.1016/j.adolescence.2011.04.001

Nije Bijvank, M., Konijn, E. A., Bushman, B. J., & Roelofsma, P. H. M. P. (2009). Age and violent-content labels make video games forbidden fruits for youth. *Pediatrics, 123*(3), 870–876.

Olson, C. K., & Cheryl, K. (2010). Children's motivations for video game play in the context of normal development. *Review of General Psychology, 14*(2), 180–187. doi:10.1037/a0018984

Olson, C. K., Kutner, L. A, & Warner, D. E. (2008). The role of violent video game content in adolescent development: Boys' perspectives. *Journal of Adolescent Research, 23*(1), 55–75. doi:10.1177/0743558407310713

Paika, P. C. & Shia, C.-K. (2013). Playful gender swapping: User attitudes toward gender in MMORPG avatar customisation. *Digital Creativity, 24*(4), 310–326, doi:10.1080/14626268.2013.767275

Peter, J., Valkenburg, P. M., & Schouten, A. P. (2005). Developing a model of adolescent friendship formation on the Internet. *CyberPsychology and Behavior, 8*(5), 423–430. doi:10.1089/cpb.2005.8.423

Plaisier, X. S., & Konijn, E. A. (2013). Rejected by peers—Attracted to antisocial media content: Rejection-based anger impairs moral judgment among adolescents. *Developmental Psychology, 49*(6), 1165–1173. doi:10.1037/a0029399

Plaisier, X. S., & Konijn, E. A. (under review). Constructing and validating the Media, Morality, and Youth Questionnaire (MMaYQue). Presented at the International Communication Conference (ICA), May 2013, Seattle.

Potter, W. J. (1992). How do adolescents' perceptions of television reality change over time? *Journalism Quarterly, 69*(2), 392–405. doi:10.1177/107769909206900213

Potter, W. J. & Tomasello, T. K. (2003). Building upon the experimental design in media violence research: The importance of including receiver interpretations. *Journal of Communication, 53*(2), 315–329. doi:10.1111/j.1460-2466.2003.tb02593.x

Prinstein, M. J., & Aikins, J. W. (2004). Cognitive moderators of the longitudinal association between peer rejection and adolescent depressive symptoms. *Journal of Abnormal Child Psychology, 32*(2), 147–158. doi:10.1023/B:JACP.0000019767.55592.63

Rizzolatti, G., & Craighero, L. (2004). The mirror-neuron system. *Annual Review of Neuroscience, 27*, 169–192.

Ryan, A. M., Hicks, L., & Midgley, C. (1997). Social goals, academic goals, and avoiding seeking help in the classroom. *The Journal of Early Adolescence, 17*(2), 152–171. doi:10.1177/0272431697017002003

Sentse, M., Scholte, R., Salmivalli, C., & Voeten, M. (2007). Person–group dissimilarity in involvement in bullying and its relation with social status. *Journal of abnormal child psychology, 35*(6), 1009–1019.

Shapiro, M. A., & Chock, T. M. (2003). Psychological processes in perceiving reality. *Media Psychology, 5*(2), 163–198. doi:10.1207/S1532785XMEP0502_3

Shaw, P., Greenstein, D., Lerch, J., Clasen, L., Lenroot, R., Gogtay, N., … Giedd, J. (2006). Intellectual ability and cortical development in children and adolescents. *Nature, 440*, 676–679. doi:10.1038/nature04513

Signorielli, N. (1990). Television's mean and dangerous world: A continuation of the cultural indicators perspective. In N. Signorielli & M. Morgan (Eds.), *Cultivation analysis: New directions in media effects research* (pp. 85–106). Newbury Park, CA: Sage.

Slater, M. D. (2007). Reinforcing spirals: The mutual influence of media selectivity and media effects and their impact on individual behavior and social identity. *Communication Theory, 17*(3), 281–303.

Slater, M. D. Henry, K. L., Swaim, R. C., & Anderson, L. L. (2003). Violent media content and aggressiveness in adolescents a downward spiral model. *Communication Research, 30*(6), 713–736.

Slater, M. D. Henry, K. L., Swaim, R. C., & Cardador, J. M. (2004). Vulnerable Teens, Vulnerable Times How Sensation Seeking, Alienation, and Victimization Moderate the Violent Media Content–Aggressiveness Relation. *Communication Research, 31*(6), 642–668.

Somerville, L. H., Hare, T., & Casey, B. J. (2011). Frontostriatal maturation predicts cognitive control failure to appetitive cues in adolescents. *Journal of Cognitive Neuroscience, 23*, 2123–2134.

Spear, L. P. (2000). The adolescent brain and age-related behavioral manifestations. *Neuroscience and Biobehavioral Reviews, 24*(4), 417–463. doi:10.1016/S0149-7634(00)00014-2

Special, W. P., & Li-Barber, K. T. (2012). Self-disclosure and student satisfaction with Facebook. *Computers in Human Behavior, 28*(2), 624–630.

Steele, J. R., & Brown, J. D. (1995). Adolescent room culture: Studying media in the context of everyday life. *Journal of Youth and Adolescence, 24*(5), 551–576.

Steinberg, L. (2005). Cognitive and affective development in adolescence. *Trends in Cognitive Sciences, 9*(2), 69–74. doi:10.1016/j.tics.2004.12.005

Steinberg, L., & Morris, A. S. (2001). Adolescent development. *Annual Review of Psychology, 52*, 83–110. doi:10.1146/annurev.psych.52.1.83

Strasburger, V. C. (2009). Why do adolescent health researchers ignore the impact of the media? *Journal of Adolescent Health, 44*, 203–205.

Strasburger, V. C., Jordan, A. B., & Donnerstein, E. (2010). Health effects of media on children and adolescents. *Pediatrics, 125*, 756–767. doi:10.1542/peds.2009-2563

Subrahmanyam, K., Greenfield, P. M., & Tynes, B. (2004). Constructing sexuality and identity in an online teen chat room. *Journal of Applied Developmental Psychology, 25*, 651–666. doi:10.1016/j.appdev.2004.09.007

Sundar, S. S. (2008). Self as source: Agency and customization in interactive media. In E. Konijn, S. Utz, M. Tanis, & S. Barnes (Eds.), *Mediated interpersonal communication* (pp. 58–74). New York, NY: Routledge.

Sundar, S. S., & Marathe, S. S. (2010). Personalization vs. customization: The importance of agency, privacy and power usage. *Human Communication Research*, *36*, 298–322.

Tian, Q., & Hoffner, C. A. (2010). Parasocial interaction with liked, neutral, and disliked character on a popular TV series. *Mass Communication and Society*, *13*(3), 250–269. doi:10.1080/15205430903296051

Tokunaga, R. S. (2010). Following you home from school: A critical review and synthesis of research on cyberbullying victimization. *Computers in Human Behavior*, *26*, 277–287. doi:10.1016/j.chb.2009.11.014

Ugazio, G., Lamm, C., & Singer, T. (2012). The role of emotions for moral judgments depends on the type of emotion and moral scenario. *Emotion*, *12*(3), 579–590. doi:10.1037/a0024611

Vandebosch, H., & Van Cleemput, K. (2009). Cyberbullying among youngsters: Profiles of bullies and victims. *New Media Society*, *11*(8), 1349–1371. doi:10.1177/1461444809341263

Van Vugt, H. C., Bailenson, J. N., Hoorn, J. F., & Konijn, E. A. (2010). Effects of facial similarity on user responses to embodied agents. *ACM Transactions on Computer-Human Interaction*, *17*, 2 (Article 7, May 2010), 27. doi:http://doi.acm.org/10.1145/1746259.1746261

van Vugt, H. C., Konijn, E. A., Hoorn, J. F., Eliëns, A., & Keur, I. (2007). Realism is not all! User engagement with task-related interface characters. *Interacting with Computers*, *19*(2), 267–280. doi:10.1016/j.intcom.2006.08.005

van Vugt, H. C., Konijn, E. A., Hoorn, J. F., Eliëns, A., & Veldhuis, J. (2009). When too heavy is just fine: Creating trustworthy e-health advisors. *International Journal of Human-Computer Studies*, *67*(7), 571–583.

Veldhuis J., Konijn E. A., & Seidell J. C. (2012). Weight information labels on media models reduce body dissatisfaction in adolescent girls. *Journal of Adolescent Health*, *50*, 600–606. doi:10.1016/j.jadohealth.2011.10.249

Veldhuis, J., Konijn, E. A., & Seidell, J. C. (2014a). Counteracting media's thin-body ideal for adolescent girls: Informing is more effective than warning, *Media Psychology*, *1*, 1–31.

Veldhuis, J., Konijn, E. A., & Seidell, J. C. (2014b). Peer feedback modifies effects of media's thin-body ideal on adolescent girls. *Appetite*, *73*, 172–82.

Von Feilitzen, C., & Linné, O. (1975). Identifying with television characters. *Journal of Communication*, *25*(4), 51–55. doi:10.1111/j.1460-2466.1975.tb00638.x

Walrave, M., & Heirman, W. (2011). Cyberbullying: Predicting victimisation and perpetration. *Children and Society*, *25*(1), 59–72. doi:10.1111/j.1099-0860.2009.00260.x

Wang, J., Iannotti, R. J., & Nansel, T. R. (2009). School bullying among adolescents in the United States: Physical, verbal, relational, and cyber. *Journal of Adolescent health*, *45*(4), 368–375.

Weibel, D., & Wissmath, B. (2011). Immersion in computer games: The role of spatial presence and flow. *International Journal of Computer Games Technology*, Article ID 282345, *14*. http://dx.doi.org/10.1155/2011/282345

Wood, R. T. A., Griffiths, M. D., Chappell, D., & Davies, M. N. O. (2004). The structural characteristics of video games: A psycho-structural analysis. *CyberPsychology and Behavior*, *7*(1), 1–10. doi:10.1089/109493104322820057

Wright, M. F. (2013). The relationship between young adults' beliefs about anonymity and subsequent cyber aggression. *Cyberpsychology, Behavior, and Social Networking*, *16*(12), 853–857. doi:10.1089/cyber.2013.0009.

Ybarra, M. L., Alexander, C., & Mitchell, K. J. (2005). Depressive symptomatology, youth Internet use, and online interactions: A national survey. *Journal of Adolescent Health*, *38*(1), 9–18. doi:10.1016/j.jadohealth.2003.10.012

Yee, N. (2006). The demographics, motivations and derived experiences of users of massively-multiuser online graphical environments. *Presence: Teleoperators and Virtual Environments, 15*, 309–329.

Yee, N., Bailenson, J. N., & Ducheneaut, N. (2009). The Proteus effect: Implications of transformed digital self-representation on online and offline behavior. *Communication Research, 36*, 285–312.

You, S., & Sundar, S. S. (2013, April). I feel for my avatar: embodied perception in VEs. In *Proceedings of the SIGCHI Conference on Human Factors in Computing Systems* (pp. 3135–3138). ACM.

Zillmann, D. (2010). Mechanisms of emotional reactivity to media entertainments. In K. Döveling, C. Von Scheve, & E. A. Konijn, (Eds.), *Handbook of emotions and mass media* (pp. 101–115). New York, NY: Routledge.

16

The Psychology of Interactive Media Habits

Robert LaRose
Department of Media and Information, Michigan State University, MI, USA

Media habits, defined as a form of automaticity in media consumption that develops as people repeat media behavior in stable circumstances (LaRose, 2010a), have recently revived as a topic of interest in communication studies. Here, the origins of the media habit construct, its current status, and its implications for ongoing research traditions in media research are examined in the context of interactive media.

Discovering Interactive Media Habits

The influence of interactive media habits is discovered in the everyday experiences of media consumers. When starting their days, do they actively consider whether they have social, information, or entertainment needs and then thoughtfully select "online newspaper" from among the many alternatives available (Palmgreen, Wenner, & Rosengren, 1985)? Do they carefully weigh the positive and negative expected outcomes of their actions before clicking on their favorite news site (Ajzen, 1991; Bandura, 1991)? Do somber moods (Zillmann & Bryant, 1985) compel them to check the comic strips at the same point in their morning routines, every day, for years? If so, the endless media choices would soon exhaust scarce cognitive resources. Habits produce the cognitive efficiency necessary to navigate the myriad of choices and carry out the many actions involved in daily living with a minimum expenditure of cognitive resources.

Habits have periodically appeared, faded from view, and reappeared in the media literature (LaRose, 2004, 2010a). The most recent revival of the construct grew from a long-recognized limitation of the dominant Uses and Gratifications paradigm, which typically accounted for only about 10% of the variance in media behavior

The Handbook of the Psychology of Communication Technology, First Edition. Edited by S. Shyam Sundar.

(Palmgreen et al., 1985), suggesting that additional explanatory variables were yet to be uncovered.

Interactive media crazes at the turn of the century that seemed to defy rational explanation through conscious decision-making models, including Internet browsing (LaRose, Lin, & Eastin, 2003) and electronic commerce (LaRose & Eastin, 2002), called for renewed attention to media habits. Internet addiction (Young, 1999) was by then an emerging explanation for interactive media usage but the disease metaphor was arguably an exaggeration. The self-reported, self-administered inventories of symptoms did not require the necessary determination (Shaffer, Hall, & Vander Bilt, 2000) that serious life consequences such as dropping out of school or losing one's job or one's family were involved.

Drawing from the same diagnostic criteria for pathological gambling and substance abuse that others had associated with Internet addiction, LaRose (2004) interpreted these as indicators of deficient self-regulation. The self-regulation mechanism of socio-cognitive theory (Bandura, 1991) was inverted to consider states of mind in which self-regulation is weak and self-control over behavior is lax. Social cognitive theory pointed away from personality traits that were then the focus of media addiction research (e.g., McIlwraith, 1998). Rather, social cognitive theory explains behavior in terms of the expected outcomes of future behavior based on the history of reinforcement that each individual either experiences for themselves or observes through others. Expected outcomes work together with self-regulatory processes to guide—but in some cases fail to guide—behavior. Consistency in behavior across situations or behavioral domains is the result of consistent patterns of reinforcement or consistent applications of self-regulatory influence. Thus, instead of positing a compulsive buying trait to explain online shopping, LaRose and Eastin (2002) proposed a measure of habit strength.

Drawing from Marlatt, Baer, Donovan, and Kivlahan's (1988) account of behavioral addictions, self-reactive outcomes, in which individuals come to rely on media use to relieve dysphoric moods, were conceptualized as the source of habit formation (LaRose et al., 2003). Pathological use could result if the Internet became the primary means of relieving dysphoric states. However, pathology that met the criteria of causing serious life consequences was extremely rare in the normal populations that were the subjects of that research.

LaRose (2010a) integrated research focusing on interactive media habits with ongoing discussions about habits in neuroscience, information systems, and social psychology. As LaRose and Eastin (2004) did for Internet use working from socio-cognitive theory, Ouellette and Wood (1998) demonstrated that habits were significant predictors of television viewing after accounting for conscious determinants of behavior from the theory of planned behavior. Among those with a consistent context for viewing, habits but not conscious intentions predicted subsequent viewing behavior. Meta analyses (Conner & Armitage, 1998; Limayem, Hirt, & Cheung, 2007;Ouellette & Wood, 1998) provided assurance that habits were important determinants of behavior across domains. Wood, Quinn, and Kashy (2002) found that habits accounted for over half of all media behaviors.

Similarly, information systems researchers working within the technology acceptance model tradition (Kim & Malhotra, 2005) discovered that habit strength was such a strong predictor of Internet use that behavioral intentions were insignificant, when the two variables were considered together. Limayem et al. (2007) found that strong habits suppressed the effect of behavioral intentions on continuing Web use.

Verplanken and Orbell (2003) introduced the Self-Report Habit Index (SRHI), which closely paralleled measures of deficient self-regulation developed at the same time in the media psychology literature (LaRose et al., 2003). This approach overcame the problem with self-reports of past behavior as measures of habit strength (Ajzen, 2002) by incorporating dimensions of automaticity (awareness, intention, efficiency). The problem was that the frequency of past behavior did not distinguish those that were habitual from those that were consciously enacted upon each repetition.

Also, functional magnetic resonance imaging indicated a region of the brain (the basal ganglia) through which control over behavior is transferred from the prefrontal cortex (Yin & Knowlton, 2006). This supports a dual process model in which behaviors that are initially controlled by conscious intentions gradually fall under the aegis of nonconscious processes as they are repeated and reinforced. And, habitual behaviors did not stimulate accelerated cortical activity as if they were enacted by more rapid processing of conscious thoughts as a result of repetition (Saling & Phillips, 2007).

In their review, Saling and Phillips (2007) concluded that the four dimensions of automaticity (lack of awareness, efficiency, intention, and control, Bargh, 1994) were independent and might occur in varying combinations. This led to drawing a distinction between habit and deficient self-regulation, recognizing that they could be two dimensions of automatic, habitual behavior: one is a diminished ability to accurately observe and judge one's own behavior that is present in both benign and harmful habits, henceforth called deficient self-observation. The second is an inability to control undesirable habits, or deficient self-reaction (LaRose, 2010a).

Recent Research

Since LaRose (2010a) was written in early 2007, interactive media habits have continued as a topic of interest. Here we review research relevant to consumer-oriented interactive media technologies and applications.

Interactive media habits in communication research

Habits have been observed to determine usage of a growing array of interactive media, including specific types of Internet use (online shopping, downloading of entertainment files, social networking, and online games, LaRose, Kim, & Peng, 2008), video game play (Hartmann, Jung, & Vorderer, 2012; Lee & LaRose, 2007), and cell phone use (Bayer & Campbell, 2012; Chuang, 2011; Peters, 2009). In a meta-analysis, Tokunaga and Rains (2010) found support for a causal relationship between deficient self-regulation and the amount of time spent using the Internet.

Interactive media habits have been extended to understanding the effects of interactive media. Deficient self-reaction, although not deficient self-observation, was directly related to negative outcomes of Internet use such as missing school, work, or social engagements (LaRose, Kim, & Peng, 2010). Deficient self-reaction predicted poor academic adjustment among first year college students (LaRose, Wohn, Ellison, & Steinfield, 2011). On the positive side, habitual use of dial-up Internet was positively related to intentions to adopt broadband (LaRose et al., 2012).

Interactive media habits in information systems research

Expanding upon a theory of information system continuance (i.e., continuing use after initial adoption, Bhattacherjee, 2001), Limayem et al. (2007) introduced comprehensiveness of use (i.e., use of a number of applications found in an information system) as an antecedent to habits, along with user satisfaction and the frequency of past behavior, also finding that habit strength moderated the relationship between intentions and continuing web usage. Similarly, frequency of past behavior, satisfaction, and enjoyment predicted the strength of instant messenger and search engine habits (Song & Zhang, 2011). Limayem and Cheung (2008) found that habit and past use predicted continuing use of an interactive learning system but that conscious intentions had no direct effect after accounting for a significant interaction between habit and intention. Habit was a more powerful predictor of mobile data services than conscious intentions (Kim, 2011) and perceived monetary value, satisfaction, and the variety of uses were in turn predictors of habit strength. Habit predicted intentions to use Web portals better than of perceived usefulness and perceived playfulness and habit was also an antecedent to perceived usefulness in Liao, To, Liu, Kuo, and Chung (2011). Satisfaction, prior experience, and the importance of an interactive academic application predicted habit strength, which in turn predicted continuing use (Lankton, Wilson, & Mao, 2010). However, habit was a significant but relatively weak predictor of continuance intentions for Twitter and was itself predicted by perceived critical mass and frequency of past use but not the comprehensiveness of use (e.g., for communication, coordination, and/or cooperation) in Barnes and Bahringer (2011).

In an empirical test of UTAUT2 (Unified Theory of the Acceptance and Utilization of Technology, version 2; Venkatesh, Thong, & Xu, 2012), an adaptation of the technology acceptance model, habits predicted both intentions to use mobile Internet and actual use. After accounting for a host of interactions among demographic variables and prior experience, they found that habit, but neither the conscious influence of behavioral intentions nor awareness of facilitating conditions, was a significant predictor of usage. There were significant interactions with prior experience: it moderated the effect of behavioral intentions, habit, and facilitating conditions on continuing usage, in combination with various demographic variables.

The psychology of habits: recent developments

In social psychology recent attention has focused on altering habits, much of it relating to food consumption. Given the many positive and negative effects interactive media have been associated with, these efforts might enrich communication effects research. For example, Robinson and Borzekowski (2006) demonstrated that disruptions of TV and video game habits resulted in declines in child obesity.

Verplanken and Wood (2006) argued that "upstream" interventions that change the context in which habits are activated are the most likely to succeed. To change the interactive media habits of their children parents should disable the players (as did Robinson & Borzekowski, 2006). Interventions coinciding with naturally occurring life changes (e.g., moving to a new home) or public policy interventions that change the environment (e.g., prohibiting smoking in public areas) are also effective. "Downstream" interventions that require the willpower of the user are generally much less effective.

However, interactive media habits appear to be fleeting ones and might be amenable to "downstream" intervention. Most teens initially addicted to online video games were no longer so a year later (Van Rooij, Shoenmakers, Vermulst, Van Den Eijnden, & Van De Mheen, 2011). In a Taiwanese sample, the remission rate was nearly 50% (Ko, Yen, Yen, Lin, & Yang, 2007).

One strategy for changing habits is self-monitoring to restore effective self-observation. Lally, Chipperfield, and Wardle (2008) offered weight-loss tips along with a self-monitoring checklist, resulting in formation of healthy eating habits. Quinn, Pascoe, Wood, and Neal (2010) found that vigilant monitoring with self-administered admonitions (e.g., "don't do it!") when an unwanted habitual behavior was observed suppressed habits. A mindfulness intervention (Chatzisarantis & Hagger, 2007) shielded the expression of bad habits and made good habits more mentally accessible.

Deficient self-reaction can be overcome through implementation intentions if individual self-regulatory styles (e.g., either promoting exercise or preventing video game play) are matched with the implementation strategy (i.e., promoting alternative behaviors or preventing undesirable ones, Tam, Bagozzi, & Spanjol, 2010). Combining mental contrasting (e.g., imagining a happy family dinner without texting and contrasting it with frequent dinner interruptions) with implementation intentions can also be effective (Adriaanse, Gollwitzer, De Ridder, de Wit, & Kroese, 2011). Shielding against the intrusion of habitual responses by forming intentions to perform nonhabitual behaviors that achieve the same goal (e.g., playing a board game instead of a video game to achieve enjoyment) also shows promise (Danner, Aarts, Papies, & de Vries, 2011).

Wood and Neal (2007) argued that an error detection mechanism (in the anterior cingulate cortex) is activated when there are mismatches between actual responses and intended ones, re-engaging conscious control. Complex tasks, the association of triggering stimuli with multiple responses, and purposeful attention to one's actions may inhibit the transformation from goal directed to habitual behavior. Neal, Wood, Wu, and Kurlander (2011) inhibited habits by changing the context in which they were performed and interrupting the flow of performance (by using a nondominant hand to execute it).

Strengthening desirable habits, such as exercise, has also been a focus of recent research. A meta-analysis integrating self-determination theory with the theory of planned behavior produced a path model in which habits, defined as the frequency of past behavior, had a direct effect on future behavior when introduced as a control variable. However, habit also had direct effects on perceived autonomy support, self-determination motivation, and the theory of planned behavior variables of subjective norms and perceived behavioral control (Hagger & Chtazisarantis, 2009).

Self-determined regulation stemming from inherent interest in an activity or its desirable expected outcomes had a stronger positive relationship to habit strength (Gardner & Lally, 2013) in comparison with external motivation, an effect that was independent of the frequency of past behavior. These studies suggest that the original source of motivation for performing a behavior, whether intrinsic or extrinsic, affects the ultimate strength of habits.

Thus, the facilitating conditions for habit formation and performance recognized by social psychologist are distraction, time pressure, intrinsic motivation, and depleted self-regulatory resources. Satisfaction with on-going habits is recognized as an inhibitor of habit change. However, rather than an active thought process, satisfaction is believed to be the result of distorted inferences about the desirability of behavior (Wood & Neal, 2009), a type of deficient self-observation in present terms. The psychology of self-regulation (Baumeister, Schmeichel, & Vohs, 2007) conceptualizes self-regulation as a resource similar to muscle strength. The resource can be temporarily depleted by over exertion or conditions (such as stress) that sap it but strengthened over the long run by repeated exercise, strength that is then available to other domains of behavior. So, strategies that yield incremental strengthening, such as resistance to smartphone check-ins, might prove effective.

Recent neuroscience scholarship (Ashby, Turner, & Horvitz, 2010;Graybiel, 2008) provides a fuller picture of the neurological mechanisms of habits. Both cognitions and behaviors are recognized to be subject to the same learning processes involving the basal ganglia that result in the "chunked" representations of habitual sequences of actions. The development of habits can be described as a transition from an exploratory to an exploitive mode of activity, or in different terms, from action-outcome behaviors that pursue a current goal to nonconscious stimulus-response behaviors that were learned in response to a previous goal. There are several different cortico-basal ganglia learning circuits in the brain that can produce habits and they might account for different types of habits once fully understood. For now, it appears that there is a similar pattern for both normal habits and drug addictions (see also Wood & Neal, 2009). Procedural learning (e.g., learning how to kill zombies in a first person shooter game) and habit learning (e.g., learning to play a favorite game every day after work) also follow similar patterns.

Toward a Theory of Interactive Media Habits

Are interactive media habits different from others? It is unlikely that interactive media habits have different underlying mechanisms but they might differ with respect to the conditions that trigger them and the speed and strength with which they develop.

Anytime, anywhere, anything, anyway, anonymously

Accessibility, availability, and anonymity were proposed to explain online sex addictions (Cooper, 1998) but apply to all forms of interactive media. Accessibility and availability provide abundant opportunities to encounter trigger stimuli. The sight of one's own interactive media devices or the vicarious experience of seeing others use theirs may trigger habits anywhere at any time. Anonymity facilitates solitary consumption undetectable by significant others, weakening the effect of social norms that might moderate habits. Information systems researchers discovered comprehensiveness as a determinant of habits (e.g., Limayem et al., 2007) which we designate with a fourth "A-word," anything, to indicate the wide range of applications (and thus habit-triggering cues and digital routines) accessible through our interactive devices. We might add a fifth, anyhow, to account for cross-platform applications.

Collectively, the Five As contribute to frequent opportunities to perform certain interactive media behaviors. Since priming the contexts in which habits are performed stimulates the performance of habits (Neal, Wood, Labrecque, & Lally, 2012), then the ubiquitous presence of interactive media outlets can be expected to stimulate repeated performances of their consumption. A recent experience sampling study of desires in everyday life (Hofmann, Vohs, & Baumeister, 2012) found that media-related desires lag behind only sleeping and eating with respect to frequency of occurrence throughout the day and self-control failure rates (i.e., yielding to desires that individuals attempted to resist) were by far the highest for media desires. High-opportunity behaviors may produce stronger habits than low-opportunity ones (Ouellette & Wood, 1998), although low-frequency behaviors may also be controlled by habits (Lankton et al., 2010). Social media may be highly habit forming in that the brain circuits that govern habits are especially sensitive to social cues (Graybiel, 2008).

The Five As raise questions about the stability of contexts required to activate habits. Early on, stable contexts were equated with sequences of immediately preceding behaviors (Ouellette & Wood, 1998) and consistent locations (Wood et al., 2002). Later (Ji & Wood, 2007), contextual conditions included time, place, social conditions, and mood states. But also, mental states such as goals to travel to work or, perhaps, to obtain an evening's relaxation can trigger the expression of habits (Verplanken, Aarts, & van Knippenberg, 1997; Wood & Neal, 2007). Since habits include cognitive routines as well as sequences of observable behaviors (Graybiel, 2008; Verplanken, 2006) it stands to reason that the stimuli that trigger them can be related thoughts that are primed by daily activities, although many performances of habitual behavior are preceded by unrelated thoughts as well (Wood et al., 2002). Since there are few daily activities that are not in some way connected to our interactive media devices and to popular applications such as Facebook, the interactive media are themselves perhaps supply a "stable context" for evoking habitual behaviors.

Thus, hyperlinked connections may at any time activate mental triggers for interactive media habits, many of which can be immediately enacted with a click. LaRose (2010a) proposed slightly different mechanisms for media habit formation, thought to require a stable context, and media habit expression, which could be triggered by a wide range of related stimuli connected in some way to the context in which the habit was originally formed. This model remains untested.

However, the Five As might also account for volatility in certain interactive media habits while strengthening others. The association of trigger stimuli (e.g., the home screen of one's smartphone) with multiple responses (e.g., opening Facebook or Angry Birds apps) may weaken habitual performance (Wood & Neal, 2007) even as the habit of smartphone checking is strengthened through the comprehensiveness of an information system (i.e., the smartphone) with so many applications available (Limayem et al., 2007).

Newell (2003) produced the only known research directly comparing interactive (Web, email, and instant messenger) with conventional (television) media habits in a doctoral dissertation in which a measure of stable context called environmental stability was included. Habit strength measured by the SRHI was about equally well correlated with usage intentions for the night following the survey for all four media (r = 0.49–0.54), although the correlation with actual consumption behavior on the target night, assessed through a follow-up survey, was somewhat higher for the Web (r = 0.45) and for instant messaging (r = 0.48) than for email (r = 0.33) or television (r = 0.37). There was no evidence that environmental stability was necessary. On the contrary, there were higher correlations between habit strength and typical patterns of media consumption among those with low levels of environmental stability than among those reporting high levels of stability for both instant messenger use (r = 0.52 for high stability environments compared to 0.63 for low stability) and Web use (r = 0.46, 0.61). The correlations were about equal for television and email.

If one credits instant messenger and Web surfing as "more interactive" than email in that the former entail tight loops of stimulus and response while the latter allows asynchronous deliberation, then Newell's results might be interpreted to indicate that the "more interactive" media are more habit prone than others. Then, the activation of interactive media habits leads individuals to overcome contextual conditions that inhibit their expression. Interactive media habits may be examples of instances in which goals (i.e., to connect with a friend in IM or update what is happening on the Web) activate habits, one of the possibilities proposed in the examination of the goal–habit interface by Wood and Neal (2007). Similarly, Pahnila and Sipponen (2010) found that implementation intentions caused habitual use of online newspapers.

Behavioral conditioning

These results also lend credence to the suggestion (LaRose, 2004) that interactive media are electronic versions of the so-called "Skinner box" in which behaviorist B. F. Skinner (1938) taught pigeons to peck at buttons to receive food pellets, a position that is apparently consistent with current neuropsychology (Ashby et al., 2010; Graybiel, 2008). Mouse clicks, video game controller manipulations, and touch screen taps are instrumental behaviors that are cued by interactive media stimuli and followed by immediate administrations of reinforcement that may rapidly shape interactive media habits. Habits that form in this way follow a distinctive pattern (Yin & Knowlton, 2006) in which the rate of performance initially increases linearly in response to reinforcement (the learning phase) but then reaches a plateau and continues at a constant rate independent of continuing reinforcement (the overlearning phase). Mathematically, the rate of behavior reaches an asymptote. Lally, van Jaarsveld, Potts, and Wardle (2010) observed this pattern in ratings of the SRHI applied to daily behaviors, with individuals reaching an asymptotic plateau anywhere from 18 to 254 days.

These processes are to some extent amenable to the survey methods used in most interactive media habit studies and are reflected in the models that those investigations have produced. The executive, prefrontal cortex control present in the learning phase may be tapped by outcome expectations (e.g., LaRose & Eastin, 2004) and behavioral intentions (e.g., Verplanken & Orbell, 2003). Individuals are also aware of their habitual behaviors through reflection, even though awareness and attention are not required at the instant they are enacted, and those reflections impact self-images (Wood & Neal, 2008).

Predictions about interactive media habits

To answer the question posed at the outset of this section, interactive media habits differ from conventional media habits with respect to the frequency of opportunities to perform them (the Five As) that provide an abundance of trigger stimuli as well as the tightness of the stimulus-response loops they foster. These distinctions lead to the prediction that compared to conventional media habits, and feedback. Interactive media habits should be less reliant on consistent interactive media habits will develop more quickly (i.e., reach their asymptote more rapidly) due to more easily accessible opportunities for repeated behavior and tight loops of performance contexts defined in terms of places or times of day due to the ubiquity and portability of interactive platforms that themselves provide the contexts for performance. Interactive habits related to multimedia platforms (e.g., overall smartphone or Internet use) or multi-function applications (e.g., Facebook) should be more persistent and take longer to extinguish due to the wide availability of cues that trigger them. However, interactive media habits focused on a single application (e.g., Angry Birds) will be relatively volatile and short-lived. Interactive media habits will more readily displace other important life activities than conventional media habits due to the intense performance-feedback loops that engage consumers in activities over which they sense a lack of control. Finally, since the brain circuits governing habits are especially sensitive to social rewards (Graybiel, 2008), interactive *social* media should form stronger habits, more swiftly, with more persistence, and greater resistance to change than either conventional media or interactive media that do not afford social rewards.

Implications for Theories of Media Attendance

Next, critical examinations of current theories of media attendance are undertaken. Ways in which recent developments in the psychology of media habits might further research within the existing paradigms by adding new variables or by reconceptualizing their key mechanisms and underlying explanations of media behavior are proposed.

Mood management theory

Mood management theory (Zillmann & Bryant, 1985) and the theory of media attendance (LaRose & Eastin, 2004) both propose that dsyphoric moods play a role in media consumption. The former describes the accumulation of "traces" left behind by mood adjustments that foster future reliance on genres of media content. LaRose et al. (2003) uncovered a mechanism that further describes the process: depression was directly related to expectations that Internet use would relieve dysphoria, which were in turn directly related to both deficient self-observation and deficient self-reaction, both of which were powerful determinants of Internet usage.

Thus, habit activation emerges as a rival explanation. "Management" perhaps does not characterize the process since no conscious managing takes place in most real world media selections, which are habit-driven (Wood et al., 2002). Consciously processed needs lose their influence after habits are formed since the individual no longer has to inventory herself to decide if she "needs" a comedy to cheer herself up. Pre-existing habits might explain why respondents often fail to select the "right" choice that would logically regulate their moods. Future research might benefit from controlling pre-existing habits or manipulating them (e.g., Quinn et al., 2010; Verplanken, 2006).

Uses and gratifications

Nonconscious habits are at odds with the basic uses and gratifications premise of active media selection. Early models (Palmgreen et al., 1985) incorporated habits, although not as direct influences on media consumption, and enduring lists of gratifications refer to them (Rubin, 1984), although in insufficient numbers to form a reliable measure. Inclusion of measures of habit strength in future research should provide a more robust account of media behavior (as in LaRose & Eastin, 2004).

However, uses and gratifications researchers have arguably been studying a determinant of media habits for decades. Once behaviors become habitual they are determined by long run average outcome expectations. Typical constructions of uses and gratifications items (e.g., "I use television because…" from Rubin, 1984) reflect the long-term accumulation of experience rather than immediate outcome expectations. However, if gratifications are long run average outcome expectations, they may be caused by inferences drawn from reflections on repeated behavior, rather than causes of it (Newell, 2003; Wood & Neal, 2009), thus reversing the direction of the "causal arrow" between media selection and uses and gratifications variables. Uses and gratifications should also be powerful predictors when habits are unlikely to overwhelm active selection processes, such as adoption of novel technologies (Lin, 1999).

Action theories

Here we consider general models of human behavior that have been applied to media consumption including social cognitive theory (LaRose & Eastin, 2004), the theory of planned behavior (Ouellette & Wood, 1998), and models of interactive media use derived from the latter: TAM and UTAUT2. While each now incorporates habit strength as an explanatory variable, there are differences in the way the influence of habit is conceptualized. Some have modeled the relationship of habit strength to behavior as a direct effect in parallel to conscious influences such as intentions and expected outcomes (e.g., LaRose & Eastin, 2004; Ouellette & Wood, 1998; Verplanken

& Orbell, 2003) while others have specified a moderator role for habits (e.g., Ji & Wood, 2007; Limayem et al., 2007; Venkatesh et al. 2012). Direct, parallel effects of conscious and nonconscious influences perhaps better reflect the theoretical notion of a dual process model (see section "Modeling Habits"). The possibility of varieties of habits, including the distinction between deficient observation and deficient self-reaction advocated earlier, might also be further investigated, including the nature of the relationship between the two, which remains an unanswered question (LaRose & Eastin, 2004).

Addiction theories

While all addictions are habits (Graybiel, 2008; Marlatt et al., 1988), not all habits are pathological addictions. Pathological cases are seldom found among the normal populations typically surveyed in interactive media addiction studies. For example, in Caplan (2002) we can estimate (with a mean = 1.24, S.D. = 0.77) from the areas under a normal distribution, that it is probable that none of the 386 respondents truly had "a problem" by strongly agreeing on a five-point scale that they had experienced the moderately negative outcomes (e.g., losing a grade in a class) in the survey. Still, measures of interactive media addictions are positively related to interactive media use. But what are they, really, if not mental illnesses or even "problems"?

Interactive media addictions are habits with low to moderate (but seldom high) levels of deficient self-observation and/or deficient self-reaction. Caplan (2002), exploring the dimensions of problematic Internet use, operationally defined "compulsivity" with items that reflect the controllability dimension of automatic behavior (e.g., "unable to control use") and thus deficient self-reaction. The "excessive time" dimension included items indicating awareness and intentionality (e.g., lose track of time online, I use the Internet for longer than I expect to) while others paralleled the SRHI (miss being online if I can't go on it) fell on a withdrawal dimension and bespeak deficient self-observation. Similarly, the Internet Addiction Test yielded "lack of control" and "excessive use" factors (Widyanto & McMurran, 2004) corresponding to deficient self-reaction and deficient self-observation, respectively. In the Game Addiction Scale (Lemmens, Valkenburg, & Peter, 2009), deficient self-reaction is represented by the "relapse" factor and deficient self-observation by "tolerance" and "withdrawal," whereas in the Compulsive Internet Use Scale (Meerkerk, van den Ejinden, Vermulst, and Garretsen, 2009) one finds "loss of control" and "withdrawal."

The various scales contain further dimensions relevant to an understanding of interactive media habits and their effects including subscales for mood control, craving/preoccupation, and negative consequences. Byun et al. (2009) found 39 quantitative studies of Internet addiction through 2006 and by 2011 the present author estimated that number had swollen to 200, by then including studies of video games and cell phones as well as subtypes of Internet use. There lies a potential treasure trove of habit research examining demographics and lifestyles, online and real world behaviors, media use, uses and gratifications, psychological states such as depression and self-esteem, and a plethora of personality variables.

Unfortunately, with rare exception, interactive media addiction researchers have not broken out results for the specific dimensions of addiction. Also, most studies report mean differences between groups classified as either addicted or nonaddicted instead of correlations among variables. It is hoped that future interactive media addiction researchers will rethink both the terminology and the disease paradigm that they use when studying normal populations and provide correlational results that break out subdimensions for meta-analyses. Media scholars might provide interactive media addiction researchers with a missing link in their thus far unsuccessful arguments for a new disease classification; namely, an understanding of the etiology of the syndrome (LaRose, 2010b).

Future Directions for Interactive Media Habit Research

Here conceptual and operational challenges confronting interactive media habit research are considered. Research on interactive media habits might clarify broader issues about the influence of habits on human behavior.

What are habits?

Psychologists disagree as to whether habits should be strictly defined as repeated behaviors performed in stable contexts (Wood & Neal, 2009) or viewed more expansively as mental constructs (Graybiel, 2008; Verplanken, 2006). The former view is problematic when examining interactive media habits since interactive devices and applications themselves provide the context. It might be possible to integrate interactive media into the stable context perspective by examining (as in Ji & Wood, 2007) interactive devices as contexts. Alternatively, sequences of preceding actions (e.g., checking one's smartphone screen for alerts) could be considered as "contexts" for later ones (e.g., selling a stock that dropped below the level that triggered the alert). Specific applications (e.g., Facebook) should also be considered since they provide stable contexts for repeated routines (e.g., profile updating, "liking") as well.

Are habits repeated behaviors or mental constructs? Information systems researchers have distinguished the frequency of past behavior from habits on an empirical level, consistent with Verplanken's (2006) finding that the two could be distinguished conceptually and experimentally. Weighing the frequency of past behavior by the consistency of the context (Ji & Wood, 2007) does not overcome the basic conceptual issue with the frequency of past behavior in that the former cannot distinguish repeated behaviors that are made through conscious decisions.

If habits are mental constructs or, if one prefers, if sequences of preceding actions can provide cues for the performance of habitual behavior, it is useful to link habits to scripts. Scripting (Abelson, 1981; Fujii & Garling, 2008) describes how routines can be embedded in elaborate sequences of behavior that are partially under conscious control but that incorporate habitual components within them. It appears that, like scripts, habits undergo continual reorganization for the sake of obtaining cognitive efficiency (Saling & Phillips, 2007), implying that the sequences and cues that trigger them may not be rigidly context-specific.

Consequently, the extent of interactive media habits should be further explored. Early research in the stable context stream (Wood et al., 2002) may have underestimated the incidence of media habits, having been conducted before interactive media so fully permeated daily life. Also, behaviors were identified as habits only if participants were *not* thinking of them immediately beforehand. That indicates a form of automaticity in which awareness and attention are not required to initiate the behavior. However, this approach may ignore strong habits in which rewards are anticipated with the presentation of the triggering cue, producing a craving that directly precedes the behavior. It is now thought likely (Saling & Phillips, 2007) that the four facets of automaticity are independent, so that one might have in mind that it is "time" to check in with Facebook or email but then the behavior is executed without further formation of a conscious intention to do so or without exercising self-control over the behavior. Even assuming that the initial decision to "check in" is effortful, carefully considered and explicitly intended, and executed with full awareness and control, the actions that follow may be incorporated in a script that has many automatic components mixed in with conscious ones.

Varieties of habits should also be further explored, beginning with distinguishing those rooted in deficient self-observation from those indicating deficient self-reaction and extending the distinction to all four dimensions of automaticity. Many studies employing survey measures of habits (e.g., Limayem et al., 2007; Verplanken & Orbell, 2003) may have focused on relatively weak habits (see also below) that do not engender deficient self-reaction. As new brain circuits governing habits are uncovered (Graybiel, 2008), yet further varieties of habits may be uncovered.

Antecedents and consequences of habits

Information systems research emphasizes one set of antecedents of habits to explain the continued use of interactive media (satisfaction, comprehensiveness, frequency of past use) and social psychology (type of initial motivation, distraction, time pressure, and depleted self-regulatory resources) and communication research (expected outcomes, expected self-reactive outcomes, depression) have contributed their own. Each tradition recognizes the amount of consumption as the consequence of habits, but communication research has begun to include indicators of psychological well-being, including disruptions to other life activities and academic adjustment. The interactive media addiction literature potentially offers an immense selection of both antecedents and consequences.

The scripting concept can also spawn further exploration of the effects of habits in communication research. There, the role of priming has been incorporated in theories of media effects. It appears that scripts prime behaviors and so may offer a further mechanism for understanding the effects of media that extend beyond considering the amount of exposure. For example, are habitual media behaviors more likely than nonhabitual ones to evade executive control over antisocial behavior?

Alternatively, could mastery of interactive media habits strengthen self-regulatory resources in other behavioral domains affected by media consumption, such as food consumption?

The obvious task that lies ahead is to compare, contrast, and evaluate the competing antecedents and consequences. More fundamentally, a comprehensive model of antecedent and consequent conditions and their causal relationships to habits, perhaps building on LaRose (2010a), has yet to be developed.

Modeling habits

The structural models that have guided information systems and communication research thus far leave fundamental questions about the direction of causation (e.g., does habit cause behavior or does behavior cause habit, or both?) and the nature of the relationships (e.g., do intentions and habits have direct effects on behavior or does one moderate the other and if so which one?). Information system researchers model habit strength as a moderator of conscious influences rather than as a direct cause of behavior that acts in parallel with conscious intentions. This position is supported by comparing competing structural equation models of direct and moderated effects (Limayem et al., 2007) but is seemingly at odds with the dual process model in which cortical control contends with nonconscious processes (Graybiel, 2008). Moreover, experimental studies of habit change induced through conscious processes of self-monitoring and implementation intentions argue that intentions also moderate habits.

Experimental manipulations of the frequency of past behavior (Verplanken, 2006), of habit strength (Quinn et al., 2010), and conscious influences on habits (e.g., Quinn et al., 2010; Tam et al., 2010) have been developed that can be applied to interactive media habits and move the discussion forward. Alternatively, time series studies can be employed to study causal relationships in real-world interactive media environments.

Measuring habits

Measurement continues to plague habit research. Despite dissatisfaction with the frequency of past behavior (Ajzen, 2002), two current research streams (Hagger & Chatzisarantis, 2009; Wood & Neal, 2008) continue it. The SRHI has achieved wide use but is criticized for confounding self-identity, the frequency of behavior, and the duration of the habit and for making no reference to context (Sniehotta & Presseau, 2012). LaRose (2010a) noted that while attention, awareness, and intentionality are addressed in the SRHI there are no items that tap the controllability dimension of automaticity, leaving deficient self-reaction unexamined.

Following Verplanken (2006), the two SRHI items addressing the frequency and duration of the habit should be deleted when the frequency of performance (e.g., interactive media use) is the dependent variable. The self-identity item should also be dropped, although self-identity is perhaps worthy of study in its own right. The SRHI should be supplemented with items measuring the stability of the context (e.g., Ji & Wood, 2007; Newell, 2003). Deficient self-reaction can be measured with scales developed by the author (e.g., LaRose et al., 2003, 2010) or with subscales drawn from the interactive media addiction studies. Information systems employs two slightly different measures, one of which (Lankton et. al., 2010;Limayem & Hirt, 2003; Venkatesh et al., 2012) confounds deficient self-reaction by asking respondents if they are "addicted" to the technology in question. Others (Limayem et al., 2007) employed a version that dropped the confounding item. It would be desirable to "factor out" the various measures in a single study in hopes of arriving at a standardized measure.

However, all introspective self-reports of habit strength can be criticized on the grounds that introspections about of automatic, nonconscious behaviors may be inherently subject to error, especially a bias toward assuming that behavior is goal-directed rather than automatic (Neal et al., 2012). Naturalistic methods such as diary records of daily behaviors, the contexts in which they are performed, and the thought processes that precede them (e.g., Wood et al., 2002) offer an alternative. Laboratory methods that measure habit strength based on reaction times in word association tasks (e.g., associating a context cue and a related behavior) have also been developed (Neal et al., 2012) and represent a promising future direction for research on media habits that, for example, might be able to distinguish between the contexts represented by interactive and noninteractive media habits.

Conclusion

Habits are well established as powerful determinants of interactive media consumption. Interactive media habits may differ from others in that the devices and applications that comprise them act as their own contexts for cuing the performance of repeated behavior. Continuing communication, information systems, and psychological research can develop a better understanding of the antecedents and consequences of habits and of ways to manipulate them to encourage positive habits and disrupt harmful ones.

References

Abelson, R. P. (1981). The psychological status of the script concept. *American Psychologist, 36*, 715–729.

Adriaanse, M. A., Gollwitzer, P. M., De Ridder, D. T. D., de Wit, J. B. F., & Kroese, F. M. (2011). Breaking habits with implementation intentions: A test of underlying processes. *Personality and Social Psychology Bulletin, 37*(4), 502–513.

Ajzen, I. (1991). The theory of planned behavior. *Organizational Behavior and Human Decision Processes, 50*, 179–211.

Ajzen, I. (2002). Residual effects of past on later behavior: Habituation and reasoned action perspectives. *Personality and Social Psychology Review, 6*, 107–122.

Ashby, F. G., Turner, B. O., & Horvitz, J. C. (2010). Cortical and basal ganglia contributions to habit learning and automaticity. *Trends in Cognitive Sciences, 14*(5), 208–215.

Bandura, A. (1991). Social cognitive theory of self-regulation. *Organizational Behavior and Human Decision Processes, 50*, 248–287.

Bargh, J. A. (1994). The four horsemen of automaticity: Awareness, intention, efficiency, and control in social cognition. In R. S. Wyer & T. K. Srull (Eds.), *Handbook of social cognition* (Vol. 1, pp. 1–40). Hillsdale, NJ: Lawrence Erlbaum.

Barnes, S. J., & Bohringer, M. (2011). Modeling use continuance behavior in microblogging services: The case of Twitter. *Journal of Computer Information Systems, 51*(4), 1–10.

Baumeister, R. F., Schmeichel, B. J., & Vohs, K. D.(2007). Self-regulation and the executive function: The self as controlling agent. In A. W. Kruglanski & E. T. Higgins (Eds.), *Social psychology: Handbook of basic principles* (2nd ed., pp. 516–540), New York, NY: Guilford Press.

Bayer, J. B., & Campbell, S. W. (2012). Texting while driving on automatic: Considering the frequency-independent side of habit. *Computers in Human Behavior, 28,* 2083–2090.

Bhattacherjee, A. (2001). Understanding information systems continuance: An expectation–confirmation model. *MIS Quarterly, 25*(3), 351–370.

Byun, S., Ruffini, C., Mills, J.E., Douglas, A.C., Niang, M., Stepchenkova, S.,..., Blanton, M. (2009). Internet addiction: Metasynthesis of 1996–2006 quantitative research. *CyberPsychoogy & Behavior, 12*(2), 203–207.

Caplan, S.E. (2002). Problematic internet use and psychosocial well-being: Development of a theory-based cognitive-behavioral measurement instrument. *Computers in Human Behavior, 18,* 553–575.

Chatzisarantis, N. L. D., & Hagger, M. S. (2007). Mindfulness and the intention-behavior relationship within the theory of planned behavior. *Personality and Social Psychology Bulletin, 33*(5), 663–676.

Chuang, Y.-F. (2011). Pull-and-suck effects in Taiwan mobile phone subscribers switching intentions. *Telecommunications Policy, 35*(2), 128–140.

Conner, M., & Armitage, C. (1998). The theory of planned behavior: A review and avenues for future research. *Journal of Applied Social Psychology, 28,* 1429–1464.

Cooper, A. (1998). Sexuality and the Internet: Surfing into the new millennium. *Cyberpsychology & Behavior, 1,* 181–187.

Danner, U. N., Aarts, H., Papies, E. K., & de Vries, N. K. (2011). Paving the path for habit change: Cognitive shielding of intentions against habit intrusion. *British Journal of Health Psychology, 16,* 189–200.

Fujii, S., & Garling, T. (2008). Habit formation as generalization of script-based choice. *International Journal of Psychology, 43*(3–4), 26–27.

Gardner, B.,& Lally, P. (2013). Does intrinsic motivation strengthen physical activity habit? Modeling relationships between self-determination, past behaviour, and habit strength. *Journal of Behavioral Medicine, 36,* 488–497.

Graybiel, A. M. (2008). Habits, rituals, and the evaluative brain. *Annual Review of Neuroscience, 31,* 359–387.

Hagger, M. S., & Chatzisarantis, N. L. D. (2009). Integrating the theory of planned behaviour and self-determination theory in health behaviour: A meta-analysis. *British Journal of Health Psychology, 14*(2), 275–30.

Hartmann, T., Jung, Y., & Vorderer, P. (2012). What determines video game use? The impact of users' habits, addictive tendencies, and intentions to play. *Journal of Media Psychology, 24*(1), 19–30.

Hofmann, W., Vohs, K. D., & Baumeister, R. F.(2012). What people desire, feel conflicted about, and try to resist in everyday life. *Psychological Science, 23*(6), 582–588.

Ji, A. W., & Wood, W. (2007). Purchase and consumption habits: Not necessarily what you intend. *Journal of Consumer Psychology, 17,* 261–276.

Kim, B. (2011). The diffusion of mobile data services and applications: Exploring the role of habit and its antecedents. *Telecommunications Policy, 36*(1), 69–81.

Kim, S. S., & Malhotra, N. K. (2005). A longitudinal model of continued IS use: An integrative view of four mechanisms underlying postadoption phenomena. *Management Science, 51*, 741–755.

Ko, C. H., Yen, J. Y., Yen, C. F., Lin, H. C., & Yang, M. J. (2007). Factors predictive for incidence and remission of Internet addiction in young adolescents: A prospective study. *Cyberpsychology & Behavior, 10*(4), 545–551.

Lally, P., Chipperfield, A., & Wardle, J. (2008). Healthy habits: Efficacy of simple advice on weight control based on a habit formation model. *International Journal of Obesity, 32*, 700–707.

Lally, P., van Jaarsveld, C. H. M., Potts, H. W. W., & Wardle, J.(2010). How are habits formed: Modelling habit formation in the real world. *European Journal of Social Psychology, 40*(6), 998–1009.

Lankton, N. K., Wilson, E. V., & Mao, E. (2010). Antecedents and determinants of information technology habit. *Information & Management, 47*(5–6), 300–307.

LaRose, R. (2004). Cybercompulsions: Media habits, media addictions and the Internet. In P. Lee, L. Leung, & C. So (Eds.), *Impact and issues in new media: Toward intelligent societies* (pp. 103–130). Cresskill, NJ: Hampton Press.

LaRose, R. (2010a). The problem of media habits. *Communication Theory, 20*, 194–222.

LaRose, R. (2010b). The Uses and Gratifications of Internet addiction. In K. Young & C. N. de Abreu (Eds.) *Internet addiction handbook* (pp. 55–72). New York, NY: Wiley.

LaRose, R., DeMaagd, K., Chew, H. E., Tsai, H.-Y. S., Steinfield, C., Wildman, S. S., & Bauer, J. M. (2012). Measuring sustainable broadband adoption: An innovative approach to understanding broadband adoption and use. *International Journal of Communication, 6*, 2576–2600.

LaRose, R., & Eastin, M. S. (2002). Is on-line buying out of control? Electronic commerce and consumer self-regulation. *Journal of Broadcasting and Electronic Media, 45*(4), 549–564.

LaRose, R., & Eastin, M. S. (2004). A social cognitive theory of internet uses and gratifications: Toward a new model of media attendance. *Journal of Broadcasting and Electronic Media, 48*, 358–377.

LaRose, R., Kim, J., & Peng, W. (2008, November). *The problem of problematic internet use: Disease, dangerous habit, or just one of our favorites?* Paper presented at the Annual Conference of the National Communication Association, San Diego, CA.

LaRose, R., Kim, J. H., & Peng, W. (2010). Social networking: Addictive, compulsive, problematic, or just another media habit? In Z. Pappacharissi (Ed.), *Anetworked self: Identity, community, and culture on social network sites* (pp. 59–81). New York, NY: Routledge.

LaRose, R., Lin, C. A., & Eastin, M. S. (2003). Unregulated internet usage: Addiction, habit, or deficient self-regulation? *Media Psychology, 5*, 225–253.

LaRose, R., Wohn, D. Y., Ellison, N., & Steinfield, C. (2011, August). Facebook fiends: Compulsive social networking and adjustment to college. *Proceedings of the International Association for the Development of the Information Society*, Rome, Italy.

Lee, D. W., & LaRose, R. (2007). A socio-cognitive model of video game usage. *Journal of Broadcasting and Electronic Media, 51*(4), 632–650.

Lemmens, J. S., Valkenburg, P. M., & Peter, J. (2009). Development and validation of a game addiction scale for adolescents. *Media Psychology, 12*(1), 77–95.

Liao, C., To, P. L., Liu, C. C., Kuo, P. Y., & Chuang, S. H. (2011). Factors influencing the intended use of web portals. *Online Information Review, 35*(2), 237–254.

Limayem, M., & Cheung, C. M. K. (2008). Understanding information systems continuance: The case of Internet-based learning technologies. *Information & Management, 45*(4), 227–232.

Limayem, M., & Hirt, S. G. (2003). Force of habit and information systems usage: Theory and initial validation. *Journal of the AIS, 4*(1), 65–97.

Limayem, M., Hirt, S. G., & Cheung, C. M. K. (2007). How habit limits the predictive power of intention: The case of information systems continuance. *MIS Quarterly, 31*, 705–738.

Lin, C. A. (1999). Online-service adoption likelihood. *Journal of Advertising Research, 39*, 79–89.

Marlatt, G. A., Baer, J. S., Donovan, D. M., & Kivlahan, D. R. (1988). Addictive behaviors: Etiology and treatment. *Annual Review of Psychology, 39*, 223–252.

McIlwraith, R. D. (1998)."I'm addicted to television": The personality, imagination and TV watching pattern of self-identified TV addicts. *Journal of Broadcasting and Electronic Media, 42*, 371–386.

Meerkerk, G. J., Van Den Ejinden, R. J. J. M., Vermulst, A. A., & Garretsen, H. F. L. (2009). The Compulsive Internet Use Scale (CIUS): Some psychometric properties. *CyberPsychology & Behavior, 12*(1), 1–6.

Neal, D. T., Wood, W., Labrecque, J. S., & Lally, P. (2012). How do habits guide behavior? Perceived and actual triggers of habits in daily life. *Journal of Experimental Social Psychology, 48*(2), 492–498.

Neal, D. T., Wood, W., Wu, M., & Kurlander, D. (2011). The pull of the past: When do habits persist despite conflict with motives? *Personality and Social Psychology Bulletin, 37*, 1428–1437,

Newell, J. (2003). *The role of habit in the selection of electronic media (Doctoral dissertation)*. Michigan State University, East Lansing.

Ouellette, J. A., & Wood, W. (1998). Habit and intention in everyday life: The multiple processes by which past behavior predicts future behavior. *Psychological Bulletin, 124*, 54–74.

Pahnila, S., & Siponen, M. (2010). *Implementation intentions explain how a behavior becomes habitual: The use of online newspapers.* Paper presented at the 43rd Hawaii International Conference on Social Systems, Koloa, Hawaii.

Palmgreen, P., Wenner, L., & Rosengren, K. (1985). Uses and gratifications research: The past ten years. In K. Rosengren, L. Wenner, & P. Palmgreen (Eds.), *Media gratifications research* (pp. 11–37). Beverly Hills, CA: Sage.

Peters, O. (2009). A social cognitive perspective on mobile communication technology use and adoption. *Social Science Computer Review, 27*(1), 76–95.

Quinn, J. M., Pascoe, A., Wood, W., & Neal, D. T. (2010). Can't control yourself? Monitor those bad habits. *Personality and Social Psychology Bulletin, 36*(4), 499–511.

Robinson, T. N., & Borzekowski, D. L. G. (2006). Effects of the SMART classroom curriculum to reduce child and family screen time. *Journal of Communication, 56*(1), 1–26.

Rubin, A. M. (1984). Ritualized and instrumental television viewing. *Journal of Communication, 34*, 67–77.

Saling, L. L., & Phillips, J. G. (2007). Automatic behaviour: Efficient not mindless. *Brain Research Bulletin, 73*, 1–20.

Shaffer, H. J., Hall, M. N., & Vander Bilt, J. (2000)."Computer addiction": A critical consideration. *American Journal of Orthopsychiatry, 70*, 162–168.

Skinner, B. F. (1938). *The behavior of organisms: An experimental analysis.* New York, NY: Appleton-Century.

Sniehotta, F., & Presseau, J. (2012). The habitual use of the Self-report Habit Index. *Annals of Behavioral Medicine, 43*(1), 139–140.

Song, P., & Zhang, C. (2011). Understanding the role of intrinsic motivations in information technology usage habit: Evidence from instant messenger and search engine. *International Journal of Organisational Design and Engineering, 1*(3), 163–184.

Tam, L., Bagozzi, R. P., & Spanjol, J. (2010). When planning is not enough: The self-regulatory effect of implementation intentions on changing snacking habits. *Health Psychology, 29*(3), 284–292.

Tokunaga, R. S. & Rains, S. A. (2010). An evaluation of two characterizations of the relationships between problematic Internet use, time spent using the Internet, and psychosocial problems. *Human Communication Research, 36*, 512–545.

Van Rooij, A. J., Schoenmakers, T. M., Vermulst, A. A., Van Den Eijnden, R. J. J. M., & Van De Mheen, D. (2011). Online video game addiction: Identification of addicted adolescent gamers. *Addiction, 106*(1), 205–212.

Venkatesh, V., Thong, J. Y. L., & Xu, X. (2012). Consumer acceptance and use of information technology: Extending the unified theory of acceptance and use of technology. *MIS Quarterly, 36*(1), 157–178.

Verplanken, B. (2006). Beyond frequency: Habit as mental construct. *British Journal of Social Psychology, 45*, 639–656.

Verplanken, B., Aarts, H., & van Knippenberg, A. (1997). Habit, information acquisition, and the process of making travel mode choices. *European Journal of Social Psychology, 27*, 539–560.

Verplanken, B., & Orbell, S. (2003). Reflections on past behavior: A self-report index of habit strength. *Journal of Applied Social Psychology, 33*, 1313–1330.

Verplanken, B., & Wood, W. (2006). Interventions to break and create consumer habits. *Journal of Public Policy & Marketing, 25*, 90–103.

Widyanto, L., & McMurran, M.(2004). The psychometric properties of the Internet addiction test. *CyberPsychology & Behavior, 7*, 443–50.

Wood, W., & Neal, D. T. (2007). A new look at habits and the habit-goal interface. *Psychological Review, 114*, 843–863.

Wood, W., & Neal, D. T. (2009). The habitual consumer. *Journal of Consumer Psychology, 19*(4), 579–592.

Wood, W., Quinn, J. M., & Kashy, D. (2002). Habits in everyday life: Thought, emotion and action. *Journal of Personality and Social Psychology, 83*, 1281–1297.

Yin, H. H., & Knowlton, B. J. (2006). The role of the basal ganglia in habit formation. *Nature Reviews Neuroscience, 7*, 464–476.

Young, K. S. (1999). Evaluation and treatment of Internet addiction. In L. Vande Creek & T. Jackson (Eds.), *Innovations in clinical practice: A source book* (Vol. 17, pp. 19–31). Sarasota, FL: Professional Resource Press.

Zillmann, D., & Bryant, J. (1985). Affect, mood and emotion as determinants of selective media exposure. In D. Zillmann & J. Bryant (Eds.), *Selective exposure to communication* (pp. 157–190). Hillsale, NJ: Lawrence Erlbaum.

17

Online Addictions
Gambling, Video Gaming, and Social Networking

Mark D. Griffiths and Daria J. Kuss

International Gaming Research Unit, Psychology Division, Nottingham Trent University, Nottingham, UK

Introduction

The internet has become a fundamental part of many people's day-to-day lives. As with the introduction of other mass communication technologies, issues surrounding use, abuse, and addiction have surfaced (Griffiths, 2009). It has been claimed that excessive internet use can be pathological and addictive (Widyanto & Griffiths, 2006) and that such behavior comes under the more generic label of "technological addiction" (Griffiths, 1995, 1998). It has been argued that behavioral addictions are no different from chemical addictions (e.g., alcoholism, heroin addiction, etc.) in terms of the core components of addiction such as salience, tolerance, withdrawal, mood modification, conflict, and relapse. Reliance on internet use has most commonly been conceptualized as a behavioral addiction, which operates on a modified principle of classic addiction models, but the validity and clinical usefulness of such claims have also been questioned (Holden, 2001). Other studies have also supported the concept that problematic internet use might be associated with features of DSM-IV impulse control disorder (King, Haagsma, Delfabbro, Gradisar, & Griffiths, 2013; Shapira, Goldsmith, Keck, Khosla, & McElroy, 2000). Young (1999) claims that internet addiction is a broad term that covers a wide variety of behaviors and impulse control problems. She claims there are five specific subtypes:

- *Cybersexual addiction:* Compulsive use of adult web sites for cybersex and cyberporn.
- *Cyber-relationship addiction:* Over-involvement in online relationships.
- *Net compulsions:* Obsessive online gambling, shopping or day trading.
- *Information overload:* Compulsive web surfing or database searches.
- *Computer addiction:* Obsessive computer game playing.

The Handbook of the Psychology of Communication Technology, First Edition. Edited by S. Shyam Sundar.

However, Griffiths (2000a) argued that many of these excessive users are not "internet addicts" but just use the internet excessively as a medium to fuel other addictions. He argued that most of the individuals who use the internet excessively are not addicted to the internet itself, but use it as a medium to fuel other addictions. Griffiths says that there is a need to distinguish between addictions *to* the internet and addictions *on* the internet. He gives the example of a gambling addict who chooses to engage in online gambling, as well as a computer game addict who plays online, stressing that the internet is just the place where they conduct their chosen (addictive) behavior. These people display addictions on the internet. However, there is also the observation that some behaviors engaged on the internet (e.g., cybersex, cyberstalking, etc.) may be behaviors that the person would only carry out on the internet because the medium is anonymous, non-face-to-face, and disinhibiting (Griffiths, 2000b, 2001a, 2012a).

Based on previous research, Kuss and Griffiths (2012a) have argued that a combination of individual, situational, and structural characteristics determine whether and to what extent individuals engage in various online activities. For instance, it is believed that access, affordability, and anonymity are critical factors that make the internet viable for the acquisition, development, and maintenance of online addictions (Kuss & Griffiths, 2012a). Because the current scientific knowledge of online addiction is copious in scope, this chapter examines the two most researched online addictions (i.e., online gambling addiction and online gaming addiction) and the online addiction that has come to most prominence in recent years (i.e., online social networking addiction). There are of course other online addictions that have been reviewed in depth elsewhere including online sex addiction, online shopping, and online auction use, although the number of empirical studies is relatively few (e.g., see Griffiths, 2012; Peters & Bodkin, 2007; Wang & Yang, 2006).

Online Gambling Addiction

It has been claimed that remote types of gambling have provided the biggest cultural shift in gambling in the past decade (Griffiths, Parke, Wood, & Parke, 2006) and that the introduction of internet gambling has the potential to lead to increased levels of problematic gambling behavior (Griffiths, 2003; Griffiths & Parke, 2002; Kuss & Griffiths, 2012b). To date, knowledge and understanding of how the medium of the internet affects gambling behavior is at an early stage. Globally speaking, proliferation of internet access is still an emerging trend and it will take some time before the effects on gambling behavior surface. However, there is a strong foundation to speculate on the potential hazards of internet gambling. The impact of gambling technology has been widespread, and there are many observed trends around the world that appear to have resulted from technological innovation (e.g., gambling coming out of gambling environments, gambling becoming a more asocial activity, widespread deregulation, and increased opportunities to gamble) (Griffiths, 2011).

There is now an increasingly growing number of studies on internet gambling and problem internet gambling that range widely in focus and methodological scope.

These have included national studies on adult internet gambling (e.g., Gambling Commission, 2008; Griffiths, 2001b; Griffiths, Wardle, Orford, Sproston, & Erens, 2009; Wardle, Moody, Griffiths, Orford, & Volberg, 2011), national studies on adolescent internet gambling (e.g., Griffiths & Wood, 2007), regional studies of internet gamblers (e.g., Ialomiteanu & Adlaf, 2001; Wood & Williams, 2007), survey studies on self-selected samples of internet gamblers (e.g., Griffiths & Barnes, 2008; International Gaming Research Unit, 2007; Matthews, Farnsworth, & Griffiths, 2009; McCormack, Shorter, & Griffiths, 2013; Wood, Griffiths, & Parke, 2007), studies examining behavioral tracking data of internet gamblers from online gaming sites (e.g., Broda et al., 2008; Dragicevic, Tsogas, & Kudic, 2011; LaBrie, Kaplan, LaPlante, Nelson, and Shaffer, 2008; LaBrie, LaPlante, Nelson, Schumann, and Shaffer, 2007;), qualitative studies of internet gamblers (e.g., McCormack & Griffiths, 2012a; Parke & Griffiths, 2011, 2012), internet gambling case studies (Griffiths & Parke, 2007), studies examining very specific forms of gambling such as online poker (Griffiths, Parke, Wood, & Rigbye, 2010; Hopley & Nicki, 2011; McCormack & Griffiths, 2012b; Wood & Griffiths, 2008; Wood et al., 2007), and studies examining internet gambling and social responsibility features (Auer & Griffiths, 2013; Griffiths, Wood, & Parke, 2009; Smeaton & Griffiths, 2004).

Griffiths, Wardle, Orford, Sproston, and Erens (2009, 2011) provided the first ever analysis of a representative national sample of internet gamblers. Using participant data from the 2007 British Gambling Prevalence Survey (N=9,003 adults aged 16 years and over), all participants who had gambled online, bet online, and/or who had used a betting exchange in the last 12 months (N=476) were compared with all other gamblers who had not gambled via the internet. Overall, results showed a number of significant sociodemographic differences between internet gamblers and noninternet gamblers. When compared to noninternet gamblers, internet gamblers were more likely to be male, relatively young adults, single, well educated, and in professional/managerial employment.

Further analysis of the problem gambling Diagnostic and Statistical Manual of Mental Disorders (DSM-IV; American Psychiatric Association, 2000) scores showed that the problem gambling prevalence rate was significantly higher among internet gamblers (5%) than noninternet gamblers (0.5%). It was also found that some items on the DSM-IV were more heavily endorsed by internet gamblers including gambling preoccupation and gambling to escape. Griffiths et al.'s (2009, 2011) results suggest that the medium of the internet may be more likely to contribute to problem gambling than gambling in offline environments. More specifically, this finding suggests that for vulnerable individuals (e.g., problem gamblers), the medium of the internet may be more problematic than offline gambling environments. There are a number of studies that have reported exactly the same finding (i.e., that problem gamblers are more likely to be internet gamblers than those who do not gamble via the internet) although all these studies have either used self-selected samples, examined only one form of internet gambling such as online poker, and/or examined particular subsamples of the population such as students (e.g.,Griffiths & Barnes, 2008; Ladd & Petry, 2002; Wood, Griffiths, & Parke, 2007; Wood & Williams, 2007).

More importantly, Wardle and Griffiths (2011) have asked what exactly is an "online gambler"? Very few people only gamble online and most online gamblers also gamble offline (Griffiths et al., 2009). In the 2007 British Gambling Prevalence Survey (Wardle et al., 2007), there were 476 people (out of 9003 people who participated in the survey) who reported gambling online in the past year. Of these, only nine people did not report also participating in some kind of offline gambling activity. In other words, over 98% of online gamblers also gambled offline. These data suggest that "pure" online gamblers (i.e., gamblers who gamble online and online only) are relatively rare.

According to secondary analysis from the latest BGPS study (Wardle, Moody, Spence, et al., 2011), the number of "online only gamblers" had slightly increased to 2% but the data suggest there were a number of distinct ways to categorize gamblers based on the medium in which they gamble and what activities they gamble on in those mediums. The 2011 BGPS report surveyed 7,756 adult gamblers. Approximately one in seven respondents (14%) had gambled online in the past year (i.e., had gambled on at least one gambling activity such as gambling at online casinos and/or playing the lottery online). However, for the first time ever, the authors created four new groups of gamblers for comparison. These were those that

- Gambled offline only (i.e., had gambled on at least one activity such as buying a lottery ticket in a shop or playing roulette at an offline casino but had not gambled online in the past year)
- Gambled online only (i.e., had gambled on at least one activity such as gambling on a betting exchange or gambling at an online casino but had not gambled offline in the past year)
- Gambled both online and offline but on different activities (i.e., had gambled on at least one activity online and one activity offline, but these were different activities such as gambling on a slot machine in an amusement arcade and playing blackjack in an online casino)
- Gambled both online and offline but on the same activities (i.e., had gambled on at least one activity both online and offline, such as gambling at both an online and offline casino)

Perhaps unsurprisingly, of all gamblers, the largest group were those who gambled offline only (80.5%) and the smallest group were those who gambled online only (2.1%). Of far more interest were the rates of problem gambling among these four groups. The highest prevalence rates of problem gambling were among mixed mode gamblers who gambled on different activities (4.3%), followed by mixed mode gamblers who gambled on the same activities (2.4%), those who only gambled offline (0.9%), and those who only gambled online (0%).

The most interesting statistic is arguably the fact that *there was not a single case of problem or pathological gambling among those gamblers who only gambled online.* Extreme caution must be given as the player base for "online only" gamblers was very small when compared to the other groups. However, this certainly opens up an area for future research as to whether those who only gamble online are more resilient to developing gambling problems than those who engage in mixed modes of gambling. Sociodemographic information from the BGPS studies suggest that those who gamble online are more educated and in better occupations than those who have never gambled online. Maybe, these demographic factors are also protective factors when it comes to the development of gambling problems? The limited data (to date) suggest that it is not the medium of gambling that is more problematic *per se*, but that to vulnerable people (e.g., problem gamblers), the internet may be providing easily accessible "convenience" gambling that perhaps explains why problem gambling prevalence rates among online gamblers appear to be much higher than nononline gamblers.

Online Gaming Addiction

The internet offers a wide variety of games to play, which are distributed across a variety of game genres. These include (but are not restricted to) casual Browser Games such as *DarkOrbit*, First-Person or Ego-Shooters such as *Counterstrike*, Massively Multiplayer Online Role-Playing Games (MMORPGs) such as *World of Warcraft* (WoW), and Simulation Games such as *Second Life*, to name but the most important. Most research into online gaming addiction has been carried out on those who play MMORPGs. This type of game is played by hundreds of thousands of users throughout the world simultaneously. Research has consistently shown that a small minority of people who play online games do so excessively (Griffiths, Kuss, & King, 2012; Kuss & Griffiths, 2012a). As with online gambling, research suggests that this minority may experience symptoms traditionally associated with substance-related addictions, such as mood modification, tolerance, and behavioral salience (Hsu, Wen, & Wu, 2009; King et al., 2013; Ko et al., 2009; Mehroof & Griffiths, 2010; Wölfling, Thalemann, & Grüsser-Sinopoli, 2008; Young, 2009).

To date, approximately 70 studies have assessed pathological characteristics of online gaming and a number of recent literature reviews have examined the key findings (e.g., Griffiths et al., 2012; King et al., 2013; Kuss & Griffiths, 2012a, 2012c). Estimated prevalence rates of problematic online gaming range from 1.7% to over 10%. These studies also indicate that, in general, males are significantly more likely than females to report problems relating to their online gaming. According to King, Delfabbro, Griffiths, and Gradisar (2011; King et al., 2013), the differences in methods of assessing game-based problems may partly account for differences in prevalence rates. Furthermore, many studies fail to assess prior problems (i.e., lifetime prevalence). King, Delfabbro, Griffiths, et al. (2011) also noted that some studies do not consider subclinical cases (i.e., meeting some but not all criteria for problematic use), and the presence of comorbid psychopathology is not routinely assessed.

There are some generalizations that can be made with regard to the demographic characteristics of problematic online gamers. The literature suggests that adolescent males and young male adults appear to be at greater risk of experiencing problematic online gaming. However, the course and severity of these problems is not well known (King, Delfabbro, Griffiths, et al., 2011) and the finding that this group is more at risk may be a consequence of sampling bias and the fact that this group plays online video games more frequently than other sociodemographic groups. It has also been suggested that university students may be vulnerable to developing problematic online gaming. Reasons for this include their flexible tuition and study hours, ready access to high-speed broadband on a 24/7 basis, and multiple stressors associated with adjusting to new social obligations and/or living out-of-home for the first time (King, Delfabbro, & Zajac, 2011; Young, 1998).

Irrespective of whether problematic online gaming can be classed as an addiction, there is now a relatively large number of studies all indicating that excessive video game play can lead to a wide variety of negative psychosocial consequences for a minority of affected individuals. These were recently summarized by Griffiths et al. (2012) and include (a) sacrificing work, education, hobbies, socializing, time with partner/family, and sleep, (b) an absence of real life relationships, (c) lower psychosocial well-being and loneliness, (d) poorer social skills, (e) decreased academic achievement, (f) increased inattention, (g) aggressive/oppositional behavior and hostility, (h) maladaptive coping, (i) decreases in verbal memory performance, (j) maladaptive cognitions, and (k) suicidal ideation. In addition to the reported negative psychosocial consequences, Griffiths et al. (2012) also reported that there were many potential health and medical consequences that may result of excessive gaming. These include (a) epileptic seizures, (b) auditory hallucinations, (c) obesity, (d) wrist, neck, and/or elbow pain (including repetitive strain injuries), (e) blisters, calluses, sore tendons, and numbness of fingers, (f) hand-arm vibration syndrome, (g) sleep abnormalities, and (h) psychosomatic challenges. Taken together, this relatively long list of potential psychosocial and medical negative consequences clearly indicates that excessive online gaming is a problem irrespective of whether it is an addiction. It also suggests that more extensive recognition is needed of the wide range of potential negative and life-limiting consequences of excessive video play.

A number of studies have examined the role of different personality factors, comorbidity factors, and biological factors, and their association with online gaming addiction. In relation to personality traits, online gaming addiction has been shown to be associated with psychopathological traits using the Minnesota Multiphasic Personality Inventory-2 (Spekman, Konijn, Roelofsma, & Griffiths, 2013), neuroticism (Mehroof & Griffiths, 2010; Peters & Malesky, 2008), aggression and hostility (Caplan, Williams, & Yee, 2009; Chiu, Lee, & Huang, 2004; Kim, Namkoong, Ku, & Kim, 2008; Mehroof & Griffiths, 2010), avoidant and schizoid interpersonal tendencies (Allison, von Wahlde, Shockley, & Gabbard, 2006), loneliness and introversion (Caplan et al., 2009), social inhibition (Porter, Starcevic, Berle, & Fenech, 2010), boredom inclination (Chiu et al., 2004), sensation-seeking (Chiu et al., 2004; Mehroof & Griffiths, 2010), diminished agreeableness (Peters & Malesky, 2008), diminished self-control and narcissistic personality traits (Kim et al., 2008), low self-esteem (Ko, Yen, Chen, Chen, & Yen, 2005), state and trait anxiety (Mehroof & Griffiths, 2010), and low emotional intelligence (Parker, Taylor, Eastabrook, Schell, & Wood, 2008). It is hard to assess the etiological significance of these associations with online gaming addiction as they may not be unique to the disorder. Further research is therefore needed.

Research has also shown gaming addiction to be associated with a variety of comorbid disorders. This includes attention deficit hyperactivity disorder (Allison et al.,

2006; Batthyány, Müller, & Wölfling, 2009; Chan & Rabinowitz, 2006; Han et al., 2009), symptoms of generalized anxiety disorder, panic disorder, depression, social phobia (Allison et al., 2006), school phobia (Batthyány et al., 2009), and various psychosomatic symptoms (Batthyány et al., 2009). Through use of fMRI, biological research has shown that online gaming addicts show similar neural processes and increased activity in brain areas associated with substance-related addictions and other behavioral addictions, such as pathological gambling (significant activation in the left occipital lobe, parahippocampal gyrus, dorsolateral prefrontal cortex, nucleus accumbens, right orbitofrontal cortex, bilateral anterior cingulate, medial frontal cortex, and the caudate nucleus) (Han, Hwang, & Renshaw, 2010; Hoeft, Watson, Kesler, Bettinger, & Reiss, 2008; Ko et al., 2009). It has also been reported that gaming addicts (like substance addicts) have a higher prevalence of two specific polymorphisms of the dopaminergic system (i.e., Taq1A1 allele of the dopamine D2 receptor and the Val158Met in the Catecholamine-*O*-Methyltransferase) (Han et al., 2007).

One of the major problems in the online gaming addiction field is that different terminologies are often applied for similar phenomena, ranging from compulsive Internet use (van Rooij, Schoenmakers, van de Eijnden, & van de Mheen, 2010), problem video game playing (Salguero & Moran, 2002), and problematic online game use (Kim & Kim, 2010) to video game addiction (Skoric, Teo, & Neo, 2009) and online gaming addiction (Charlton & Danforth, 2007; Griffiths, 2010). Similarly, a variety of measurement instruments have been used in order to assess the specified problematic/addictive behaviors, namely the Compulsive Internet Use Scale (Meerkerk, Van Den Eijnden, Vermulst, & Garretsen, 2009), the Problematic Video Game Playing Test (adapted from Young, 1998), the Problem Video Game Playing Scale (Salguero & Moran, 2002), the Problematic Online Game Use Scale (Kim & Kim, 2010), an assessment of addiction tendencies (based on American Psychiatric Association, 2000), the Addiction-Engagement Questionnaire (modified from Charlton, 2002), the Game Addiction Scale (based on Lemmens, Valkenburg, & Peter, 2009), and the Problematic Online Gaming Questionnaire (Demetrovics et al., 2012).

The results of the studies indicate that internet gaming addiction appears to be a viable construct worthy of individual and independent investigation (van Rooij et al., 2010). Furthermore, it was emphasized that addiction cannot be equated with problematic use. Some of the studies suggest that problematic game playing lies on a continuum toward addiction, as it can result in addiction symptoms, namely salience, mood modification, tolerance, withdrawal, conflict, and relapse (Griffiths, 2010; King et al., 2009; Salguero & Moran, 2002). Others adopt more detailed approaches to classification, claiming that problematic use is characterized by investing much time and energy in the game, euphoria, tolerance, denial, and a preference for online relationships (Kim & Kim, 2010). This finding is in line with the result that addiction core criteria (conflict, withdrawal symptoms, relapse, reinstatement, and behavioral salience) must be distinguished from peripheral criteria (cognitive salience, tolerance, and euphoria), for only the former were found to load on an addiction factor (Charlton & Danforth, 2007). In a similar vein, it was found that addiction does not equal excessive engagement: only when significant negative consequences of excessive gaming occur can one speak of an addiction (Griffiths, 2010; Skoric, 2009). Arguably the most significant development in the field is that online gaming addiction (in the form of *internet gaming disorder*) has been included in the appendix of the new DSM-5 (American Psychiatric Association, 2013) in order to encourage research to determine whether this particular condition should be added to the manual as a disorder in the future.

Online Social Networking Addiction

More recently, it has been speculated that online social networking may also be particularly addictive to young people (Luscombe, 2009; Griffiths, Kuss, & Demetrovics, 2014). It has been claimed that those people addicted to using social network sites (SNSs) such as *Facebook* experience symptoms similar to those experienced by those who suffer from addictions to substances or other behaviors (Luscombe, 2009). In addition to this, scholars have hypothesized that young vulnerable people with narcissistic tendencies are particularly prone to engaging with SNSs in an addictive way (Campbell, Bosson, Goheen, Lakey, & Kernis, 2007). A recent review by Griffiths et al. (2014) categorized the literature on SNS addiction as falling into one of four types: (a) self-perception studies of social networking addiction, (b) studies of social networking addiction utilizing a social networking addiction scale, (c) studies examining the relationship between social networking and other online addictions, and (d) studies examining social networking addiction and interpersonal relationships.

Self-perception studies (e.g., Cabral, 2011; Cha, 2010; Machold et al., 2012; Olowu & Seriu, 2012) have all tended to show that either a significant minority or slight majority of young adults claim they are addicted to social networking or admit to having addiction-like behaviors (e.g., constantly checking social networks, engaging in the activity longer than intended, feeling stressed if unable to access social networks, etc.). However, Griffiths et al. (2014) asserted that all of the self-perception studies identified in their review had severe methodological weaknesses particularly as their reliance on self-report data, small sample sizes, and convenience sampling. Furthermore, none of studies actually assessed social networking addiction as no assessment scale was actually used.

A growing number of studies have assessed SNS addiction and/or its relationship to other psychological factors using adapted and/or self-devised social networking addiction scales. These include studies using an adapted version of the Internet Addiction Test (Young, 1998) modified to specifically assess the addiction to SNSs in 1,257 university students (Cam & Isbulan, 2012) and 335 college students (Wan, 2009), an adapted version of the Addictive Tendencies Scale (Walsh, White, & Young, 2007) to examine the relationship between personality and SNS addiction in 201 university students (Wilson, Fornasier, & White, 2010), the self-devised Facebook Addiction Symptoms Scale to assess the prevalence of SNS addiction among 1,000 university students (Alabi, 2012), the self-devised Bergen Facebook Addiction Scale to examine the relationship between personality and Facebook addiction in 423 university students (Andraessen, Tosheim, BrunBerg, & Pallesen, 2012), an adapted version of the Internet-Related Problem Scale (Armstrong, Phillips, & Saling, 2000) to examine the relationship between social networking dependency and mood modification in 343 university students (Cheak, Goh, & Chin, 2012), the Online Cognitions Scale (Davis et al., 2002) to examine the relationship between adolescent social networking motives, parenting styles and cognitions in 1,971 high school students (Floros & Siomos, 2013), an adapted version of the Online Gaming Addiction Scale (Charlton & Danforth, 2007) to examine the relationship between online enjoyment and social networking addiction in 194 university students (Turel & Serenko, 2012), a self-devised SNS addiction scale to examine Facebook addiction among 380 female university students (aged 19–28 years (Sofiah, Zobidah, Bolong, & Osman, 2011), an adapted version of the Internet Addiction Questionnaire (Echeburua, 1999) to examine the association between Facebook dependence and poor sleep quality in a sample of 418 undergraduate university students (Wolniczak et al., 2013), and the self-devised Facebook Addiction Scale to explore the relationship with Facebook addiction and associated behavioral, demographic, and psychological health predictors in 447 college students (Koc & Gulyagci, 2013).

Results from these studies have shown that (a) there was wide variation (1.6% to 34%) in the prevalence of SNS addiction (e.g., Alabi, 2012; Wan, 2009), (b) personality traits such as high extraversion, high neuroticism, and low conscientiousness significantly predicted both addictive tendencies and the time spent using an SNS (e.g., Andraessen et al., 2012; Wilson et al., 2010), and (c) that excessive SNS can lead to a number of addiction-like symptoms including salience, tolerance, conflict (with relationships, education), negative consequences (e.g., withdrawal symptoms, poor sleep/insomnia, anxiety), escape, and modification (e.g., Andraessen et al., 2012; Cam & Isbulan, 2012; Cheak et al., 2012; Floros & Siomos, 2013; Koc & Gulyagci, 2013; Turel & Serenko, 2012; Wolniczak et al., 2013). However, these studies suffer from a variety of methodological limitations. They attempted to assess SNS addiction, but mere assessment of addiction tendencies does not suffice to demarcate real pathology (Griffiths et al., 2014). Furthermore, most of the samples were generally small, specific, self-selected, convenient, and skewed with regard to young adults and female gender. Empirical studies need to ensure that they are assessing addiction rather than excessive use and/or preoccupation.

A number of studies have examined the relationship between social networking and other online addictions. Kittinger, Correia, and Irons (2012) examined how the use of Facebook relates to problematic internet use (as measured by Young's (1996, 1998) Internet Addiction Test [IAT]) among 281 American undergraduate students. Results showed that IAT scores were significantly correlated with both daily use of Facebook, and self-reports by participants that they felt addicted. Using regression analyses, including age, gender, amount of time spent online per day (including use of Facebook), the number of times Facebook accounts were accessed per day, and the minutes spent on Facebook per day accounted for 34% of the variance in IAT scores. The authors concluded that the use of Facebook contributes to IAT scores, even after accounting for demographic variables and general patterns of internet use.

Zhou (2010) assessed SNS game addiction via the IAT in 342 Chinese college students. In this study, SNS game addiction referred specifically to being addicted to the

SNS game *Happy Farm*. Students were defined as addicted to using this SNS game when they endorsed a minimum of five (out of eight) items of the IAT. Using this cut-off, 24% of the participants were classified as addicted. Zhou also investigated rewards of SNS game use, loneliness, leisure boredom, and self-esteem. The findings indicated that there was a weak positive correlation between loneliness and SNS game addiction and a moderate positive correlation between leisure boredom and SNS game addiction. The rewards of "inclusion" (in a social group) and "achievement" (in game), leisure boredom, and being male significantly predicted SNS game addiction.

Karaiskos, Tzavellas, Balta, and Paparrigopoulos (2010) reported the case of a 24-year-old female who used SNS to such an extent that her behavior significantly interfered with her life. As a consequence, she was referred to a psychiatric clinic. She used *Facebook* excessively for at least 5 hr a day and was dismissed from her job because she continuously checked her SNS instead of working. Indicatively, during the clinical interview, she used her mobile phone to access *Facebook*. In addition to excessive use that led to significant impairment in a variety of areas in the woman's life, she developed anxiety symptoms as well as insomnia. Such extreme cases have led some researchers to conceptualize SNS addiction as internet spectrum addiction disorder (Wilson et al., 2010). This indicates that first, SNS addiction can be classified within the larger framework of online addictions, and second, that it is a specific type of internet addiction, alongside other addictive internet applications such as online gaming addiction, online gambling addiction, and online sex addiction.

In light of this qualitative case study, it appears that from a clinical perspective, SNS addiction is a mental health problem that may require professional treatment. Unlike the quantitative studies, the case study emphasizes the significant individual impairment that is experienced by individuals that spans a variety of life domains, including their professional life as well as their psychosomatic condition. Future researchers are therefore advised to not only investigate SNS addiction in a quantitative way but to further our understanding of this new mental health problem by analyzing cases of individuals who suffer from excessive SNS usage.

Finally, there have been a couple of studies examining social networking addiction and interpersonal relationships. Elphinston and Noller (2011) carried out a study examining the relationship between Facebook intrusion, jealousy in romantic relationships, and relationship outcomes among 342 undergraduate students involved in a romantic relationship. The authors' self-constructed measure—the 8-item Facebook Intrusion Questionnaire (FIQ)—was to all intents and purposes a measure of "Facebook addiction" as the eight statements were based on the core components of addiction (Griffiths, 2005). The authors claimed that a single-factor structure for the FIQ was supported (i.e., it was a unidimensional scale) and that the internal consistency of the measure was high. Results showed that Facebook intrusion was associated with relationship dissatisfaction (via jealous cognitions and surveillance behaviors), although time spent on Facebook was not. Facebook intrusion was only moderately associated with time spent on Facebook and supports Griffiths' (2010) assertions that online excessive use does not necessarily equate with addiction.

Porter, Mitchell, Grace, Shinosky, and Gordon (2012) also examined the relationship between social media use, interpersonal relationship satisfaction, and addiction in

219 young adult social media users aged of 18–25 years. They used Mitchell and Beard's (2010) Internet Dependency Scale (IDS) to measure time spent using social media and social media withdrawal. Results showed that participants would feel some withdrawal from social media after a certain period of time. The results did not support their hypothesis that there would be a negative correlation between social media use and relationship satisfaction. As with other survey studies highlighted in this chapter, the samples sizes for these two studies were small, specific, used self-selected convenience samples.

Discussion

This chapter has demonstrated that research into online gambling addiction, online gaming addiction, and (even more recently) online social networking addiction has proliferated over the last few years. These three online addictions were found to have negative consequences, that allow for the behavior to be classified as pathological as based on established clinical standards (American Psychiatric Association, 2000, 2013). In line with this, all of these online addictions may require professional treatment (see King, Delfabbro, & Griffiths, 2012 for a recent overview).

On a neuronal and biochemical level, these online addictions appear to be similar to other substance-related addictions, thus supporting the assumption that it is an addiction, although a behavioral one (Batthyány & Pritz, 2009; Grüsser & Thalemann, 2006). The efficacy of psychopharmacological interventions that may alleviate the addiction symptoms in these behaviors therefore supports its biochemical, cognitive, and behavioral basis. The genetic polymorphisms found in both online gambling and gaming addicts (e.g., Comings et al., 1996) are similar to those associated with reward dependence in alcohol (Blum et al., 1990), and cocaine addictions (Noble et al., 1993). Online gambling, gaming, and social networking addictions are behavioral addictions that appear to be similar to substance-related addictions and thus support the idea of a syndrome model of addiction. Put simply, Shaffer et al. (2004) suggest that each addiction—whether it be to gambling, drugs, sex, or the internet—might be a distinctive expression of the same underlying syndrome (i.e., addiction is a syndrome with multiple opportunistic expressions). These findings emphasize the pathological status of various online addictions and demarcate the latter as a mental health concern that is increasingly gaining recognition.

Another important finding from the studies is the distinction that has been made between excessive engagement and addiction. This is in line with adopting either monothetic or polythetic formats for addiction diagnosis, as set forth by Lemmens et al. (2009). In the former, all addiction criteria must be met in order to diagnose someone with an addiction, whereas in the latter, only half of them need to be endorsed. What follows as a consequence of the utilization of these dissimilar frameworks is the apparent discrepancy between prevalence estimates. From the studies reviewed, it furthermore appears that online addicts (whether addicted to gambling, gaming, or social networking) particularly experience symptoms associated with addiction, such as tolerance, mood modification, and negative psychosocial consequences.

With regard to psychopathological status, it seems fruitful to distinguish between excessive engagement and addiction as suggested by past research (Charlton & Danforth, 2007). This is a requirement particularly when taking into consideration the context of the American Psychiatric Association's definition of what constitutes a mental disorder worthy of professional treatment. According to the APA, a mental disorder is a "clinically significant behavioral or psychological syndrome or pattern that occurs in an individual and that is associated with present distress (…) or disability (i.e., impairment in one or more important areas of functioning) or with a significantly increased risk of suffering death, pain, disability, or an important loss of freedom" (American Psychiatric Association, 2000, p. xxxi). Accordingly, only when the condition is experienced as significantly impairing can one speak of an addiction, which is clearly not the case for excessive gamblers, gamers, and social networkers who enjoy themselves while playing their games and for whom their gaming does not result in significant negative consequences. In line with this, researchers must be cautious in deploying the label "addiction" for it does not merely denote the extreme utilization of substances or engagement in certain behaviors, but it demarcates a genuine mental health problem.

For that reason, future researchers are advised to properly investigate what they claim to be an addiction in order to make sure that their identification of pathology is commensurate with clinical parlance even within the confinements of research using surveys for diagnosis only. Related to this is the utilization of a wide variety of assessment instruments to diagnose addiction, most of which have not been validated (particularly in the area of gaming and social networking addictions). Likewise, many studies (particularly in gaming and social networking studies) used nonrepresentative self-selected samples and small sample sizes. This obstructs the comparability of results, but it also puts into question the validity of diagnosis. Clearly, future researchers are advised not to develop additional measurement instruments, but to assess the validity and reliability of those already constructed against the official criteria of substance dependence as established by the American Psychiatric Association (2000, 2013).

It should also be acknowledged that there are some case studies that seem to report an addiction to the internet itself (Griffiths, 2000c; Young, 1996). Most of these individuals use functions of the internet that are not available in any other medium, such as chat rooms. These are people addicted to the internet. However, despite these differences, there seem to be some common findings, most notably reports of the negative consequences of excessive internet use (neglect of work and social life, relationship breakdowns, loss of control, etc.), which are comparable to those experienced with other, more established addictions (e.g., Kuss, Griffiths, & Binder, 2013; Kuss, van Rooij, Shorter, Griffiths, & van de Mheen, 2013).

Research into addiction to the internet (rather than addictions on the internet) suggests that it does indeed exist but that it affects only a very small minority of users (Widyanto & Griffiths, 2006, 2009). These are usually people who use internet chat rooms—activities that they would not engage in except on the internet itself. To some extent, these internet users are engaged in text-based virtual realities and take on other social personas and social identities as a way of making them feel good about themselves. In such cases, the medium of the internet may provide an alternative reality to the user and allow them feelings of immersion and anonymity, feelings that may lead to an altered state of consciousness for the user. This in itself may be highly psychologically and/or physiologically rewarding. There appear to be many people who use the internet excessively but are not addicted as measured by addiction criteria. Most people researching in the field have failed to use stringent criteria for measuring addiction (Widyanto & Griffiths, 2006). The main problems with much of the research to date is that the measures used have (a) no measure of severity; (b) no temporal dimension; (c) a tendency to overestimate the prevalence of problems; and (d) no account of the context of internet use.

It should also be noted that the medium of the internet may also enhance and/or facilitate excessive behavior (Griffiths, 2003). Research has shown that online environments have the potential to provide short-term comfort, excitement, and/or distraction—all of which can be highly reinforcing to online users and may enhance the psychological rewards and feelings provided by specific activities such as gambling, gaming, and social networking. Research has consistently shown that there are many generic factors that facilitate online use including accessibility, anonymity, affordability, convenience, escape, immersion, interactivity, disinhibition, and simulation (Griffiths, 2003). These all appear to have influence on the specific online behaviors outlined in this chapter.

Taking into account the diversity of the functions offered by the internet, it seems highly unlikely that "internet addiction" exists to any great extent. "Internet addiction" implies that an individual is addicted to the experience of being online, despite whatever function they are using. What seems to be more probable is for an individual's addiction to be associated to a function that happens to be online. Making the distinction between viewing the internet as one general application and breaking it down into the specific functions available on it might have important implications for future research. The internet consists of so many diverse functions, with different uses and therefore, different types of users who have different reasons for using a particular function. Rather than focusing on a unified concept of internet addiction, it may be more helpful to conceptualize and study disturbed patterns separately according to specific internet activities (Morahan-Martin, 2005). Using the current empirical knowledge, it is argued that addictions on the internet (e.g., gambling, gaming, and social networking) follow a continuum, with antecedents in etiology and risk factors, through to the development of a "full-blown" addiction, followed by ramifications in terms of negative consequences (e.g., conflict with education, job and/or relationships) and potential treatment (e.g., pharmacotherapy, cognitive-behavioral-therapy, psychological counseling).

References

Alabi, O. F. (2012). A survey of Facebook addiction level among selected Nigerian University undergraduates. *New Media and Mass Communication*, *10*, 70–80.

Allison, S. E., von Wahlde, L., Shockley, T., & Gabbard, G. O. (2006). The development of the self in the era of the Internet and role-playing fantasy games. *American Journal of Psychiatry*, *163*(3), 381–385.

American Psychiatric Association. (2000). *Diagnostic and statistical manual of mental disorders: Text revision*. Washington, DC: Author.

American Psychiatric Association. (2013). *Diagnostic and statistical manual of mental disorders* (5 ed.). Washington, DC: Author.

Andraessen, C. S., Tosheim, T., BrunBerg, G. S., & Pallesen, S. (2012). Development of a Facebook addiction scale. *Psychological Reports, 110*, 501–517.

Armstrong, L., Phillips, J. G., & Saling, L. L. (2000). Potential determinants of heavier Internet usage. *International Journal of Human-Computer Studies, 53*, 537–550.

Auer, M., & Griffiths, M. D. (2013). Limit setting and player choice in most intense online gamblers: An empirical study of online gambling behaviour. *Journal of Gambling Studies, 29*, 647–660. doi:10.1007/s10899-012-9332-y

Batthyány, D., Müller, K. W., Benker, F., & Wölfling, K. (2009). Computer game playing: Clinical characteristics of dependence and abuse among adolescents. *Wiener Klinsche Wochenschrift, 121*(15–16), 502–509.

Batthyány, D., & Pritz, A. (Eds.). (2009). *Rausch ohne Drogen*. Wien, Austria: Springer.

Blum, K., Noble, E. P., Sheridan, P. J., Montgomery, A., Ritchie, T., Jagadeeswaran, P., ..., Cohn, J. B. (1990). Allelic association of human dopamine D2 receptor gene in alcoholism. *Journal of the American Medical Association, 263*, 2055–2060.

Broda, A., LaPlante, D. A., Nelson, S. E., LaBrie, R. A., Bosworth, L. B., & Shaffer, H. J. (2008). Virtual harm reduction efforts for Internet gambling: Effects of deposit limits on actual Internet sports gambling behaviour. *Harm Reduction Journal, 5*, 27.

Cabral, J. (2011). Is Generation Y addicted to social media? *Elon Journal of Undergraduate Research in Communications, 2*(1), 5–13.

Cam, E., & Isbulan, O. (2012). A new addiction for teacher candidates: Social networks. *Turkish Online Journal of Educational Technology, 11*(3), 14–19.

Campbell, W. K., Bosson, J. K., Goheen, T. W., Lakey, C. E., & Kernis, M. H. (2007). Do narcissists dislike themselves "deep down inside?" *Psychological Science, 18*, 227–229.

Caplan, S. E., Williams, D., & Yee, N. (2009). Problematic Internet use and psychosocial well-being among MMO players. *Computers in Human Behavior, 25*, 1312–1319.

Cha, J. (2010). Factors affecting the frequency and amount of social networking site use: Motivations, perceptions, and privacy concerns. *First Monday, 15*(12). Retrieved from http://firstmonday.org/ojs/index.php/fm/article/viewArticle/2889/2685

Chan, P. A., & Rabinowitz, T. (2006). A cross-sectional analysis of video games and attention deficit hyperactivity disorder symptoms in adolescents. *Annals of General Psychiatry, 5*(1), 16–26.

Charlton, J. P. (2002). A factor-analytic investigation of computer 'addiction' and engagement. *British Journal of Psychology, 93*, 329–344.

Charlton, J. P., & Danforth, I. D. W. (2007). Distinguishing addiction and high engagement in the context of online game playing. *Computers in Human Behavior, 23*, 1531–1548.

Cheak, A. P. C., Goh, G. G. G., & Chin, T. S. (2012). Online social networking addiction: Exploring its relationship with social networking dependency and mood modification among undergraduates in Malaysia. *Proceedings of the International Conference on Management, Economics and Finance* (pp. 247–262). Sarawak, Malaysia: Global Research.

Chiu, S. I., Lee, J. Z., & Huang, D. H. (2004). Video game addiction in children and teenagers in Taiwan. *CyberPsychology & Behavior, 7*, 571–581.

Comings, D. E., Rosenthal, R. J., Lesieur, H. R., Rugle, L. J., Muhleman, D., Chiu, C., ..., Gade, R. (1996). A study of the dopamine D2 receptor gene in pathological gambling. *Pharmacogenetics, 6*, 223–234.

Davis, R. A., Flett, G. L., & Besser, A. (2002). Validation of a new scale for measuring problematic Internet use: Implications for pre-employment screening. *CyberPsychology & Behavior, 5*, 331–345.

Demetrovics, Z., Urbán, R., Nagygyörgy, K., Farkas, J., Griffiths, M. D., Pápay, O., & Oláh, A. (2012). The development of the problematic online gaming questionnaire (POGQ). *PLoS ONE, 7*(5), e36417. doi:10.1371/journal.pone.0036417

Dragicevic, S., Tsogas, G., & Kudic, A. (2011). Analysis of casino online gambling data in relation to behavioural risk markers for high-risk gambling and player protection. *International Gambling Studies, 11*, 377–391.

Echeburua, E. (1999). *Adicciones...Sin drogas? Las nuevas adicciones: Juego, sexo, comida, compras, trabajo, Internet.* Bilbao, Espana: Desclee-de-Brouwer.

Elphinston, R. A., & Noller, P. (2011). Time to face it! Facebook intrusion and the implications for romantic jealousy and relationship satisfaction. *Cyberpsychology, Behavior, & Social Networking, 14*, 631–635.

Floros, G., & Siomos, K. (2013). The relationship between optimal parenting, Internet addiction and motives for social networking in adolescence. *Psychiatry Research, 209*, 529–534. Retrieved from http://dx.doi.org/10.1016/j.psychres.2013.01.010

Gambling Commission (2008). *Survey data on remote gambling participation.* Birmingham: Gambling Commission.

Griffiths, M. D. (1995). Technological addictions. *Clinical Psychology Forum, 76*, 14–19.

Griffiths, M. D. (1998). Internet addiction: Does it really exist? In J. Gackenbach (Ed.), *Psychology and the Internet: Intrapersonal, interpersonal and transpersonal applications* (pp. 61–75). New York, NY: Academic Press.

Griffiths, M. D. (2000a). Internet addiction: Time to be taken seriously. *Addiction Research, 8*, 413–418.

Griffiths, M. D. (2000b). Excessive internet use: Implications for sexual behavior. *Cyberpsychology & Behavior, 3*, 537–552.

Griffiths, M. D. (2000c). Does internet and computer 'addiction' exist? Some case study evidence. *Cyberpsychology & Behavior, 3*, 211–218.

Griffiths, M. D. (2001a). Sex on the internet: Observations and implications for sex addiction. *Journal of Sex Research, 38*, 333–342.

Griffiths, M. D. (2001b). Internet gambling: Preliminary results of the first UK prevalence study. *Journal of Gambling Issues, 5*. Retrieved from http://www.camh.net/egambling/issue5/research/griffiths_article.html

Griffiths, M. D. (2003). Internet gambling: Issues, concerns and recommendations. *CyberPsychology & Behavior, 6*, 557–568.

Griffiths, M. D. (2005). A 'components' model of addiction within a biopsychosocial framework. *Journal of Substance Use, 10*, 191–197.

Griffiths, M. D. (2009). Internet gambling in the workplace. *Journal of Workplace Learning, 21*, 658–670.

Griffiths, M. D. (2010). The role of context in online gaming excess and addiction: Some case study evidence. *International Journal of Mental Health and Addiction, 8*(1), 119–125.

Griffiths, M. D. (2011). Technological trends and the psychosocial impact on gambling. *Casino and Gaming International, 7*(1), 77–80.

Griffiths, M. D. (2012a). Internet sex addiction: A review of empirical research. *Addiction Research and Theory, 20*, 111–124.

Griffiths, M. D. (2012b). Facebook addiction: Concerns, criticisms and recommendations. *Psychological Reports, 110*(2), 518–520.

Griffiths, M. D., & Barnes, A. (2008). Internet gambling: An online empirical study among student gamblers. *International Journal of Mental Health and Addiction, 6*, 194–204.

Griffiths, M. D., Kuss, D. J., & Demetrovics, Z. (2014). Social networking addiction: An overview of preliminary findings. In K. Rosenberg & L. Feder (Eds.), *Behavioral addictions: Criteria, evidence and treatment* (pp. 119–141). New York, NY: Elsevier.

Griffiths, M. D., Kuss, D. J., & King, D. L. (2012). Video game addiction: Past, present and future. *Current Psychiatry Reviews, 8*, 308–318.

Griffiths, M. D., & Parke, J. (2002). The social impact of internet gambling. *Social Science Computer Review, 20*, 312–320.

Griffiths, M. D., & Parke, J. (2007). Betting on the couch: A thematic analysis of Internet gambling using case studies. *Social Psychological Review, 9*(1), 29–36.

Griffiths, M. D., Parke, A., Wood, R. T. A., & Parke, J. (2006). Internet gambling: An overview of psychosocial impacts. *Gaming Research and Review Journal, 27*(1), 27–39.

Griffiths, M. D., Parke, J., Wood, R. T. A., & Rigbye, J. (2010). Online poker gambling in university students: Further findings from an online survey. *International Journal of Mental Health and Addiction, 8*, 82–89.

Griffiths, M. D., Wardle, J., Orford, J., Sproston, K., & Erens, B. (2009). Socio-demographic correlates of internet gambling: Findings from the 2007 British Gambling Prevalence Survey. *CyberPsychology & Behavior, 12*, 199–202.

Griffiths, M. D., Wardle, J., Orford, J., Sproston, K., & Erens, B. (2011). Internet gambling, health. Smoking and alcohol use: Findings from the 2007 British Gambling Prevalence Survey. *International Journal of Mental Health and Addiction, 9*, 1–11.

Griffiths, M. D., & Wood, R. T. A. (2007). Adolescent Internet gambling: Preliminary results of a national survey. *Education and Health, 25*, 23–27.

Griffiths, M. D., Wood, R. T. A., & Parke, J. (2009). Social responsibility tools in online gambling: A survey of attitudes and behaviour among Internet gamblers. *CyberPsychology & Behavior, 12*, 413–421.

Grüsser, S. M., & Thalemann, C. N. (Eds.). (2006). *Verhaltenssucht - Diagnostik, Therapie, Forschung*. Bern, Switzerland: Hans Huber.

Han, D. H., Hwang, J. W., & Renshaw, P. F. (2010). Bupropion sustained release treatment decreases craving for video games and cue-induced brain activity in patients with Internet video game addiction. *Experimental and Clinical Psychopharmacology, 18*, 297–304.

Han, D. H., Lee, Y. S., Na, C., Ahn, J. Y., Chung, U. S., Daniels, M. A., ..., Renshaw, P. F. (2009). The effect of methylphenidate on Internet video game play in children with attention-deficit/hyperactivity disorder. *Comprehensive Psychiatry, 50*(3), 251–256.

Han, D. H., Lee, Y. S., Yang, K. C., Kim, E. Y., Lyoo, I. K., & Renshaw, P. F. (2007). Dopamine genes and reward dependence in adolescents with excessive internet video game play. *Journal of Addiction Medicine, 1*(3), 133–138.

Hoeft, F., Watson, C. L., Kesler, S. R., Bettinger, K. E., & Reiss, A. L. (2008). Gender differences in the mesocorticolimbic system during computer game-play. *Journal of Psychiatric Research, 42*, 253–258.

Holden, C. (2001). Behavioral' addictions: Do they exist? *Science, 294*(5544), 980–982.

Hopley, A. A. B., & Nicki, R. M. (2010). Predictive factors of excessive online poker playing. *Cyberpsychology, Behavior & Social Networking, 13*, 379–385.

Hsu, S. H., Wen, M. H., & Wu, M. C. (2009). Exploring user experiences as predictors of MMORPG addiction. *Computers & Education, 53*(2), 990–999.

International Gaming Research Unit (2007). The global online gambling report: An exploratory investigation into the attitudes and behaviours of internet casino and poker players. Report for *eCOGRA (e-Commerce and Online Gaming Regulation and Assurance)*.

Ialomiteanu, A., & Adlaf, E. (2001). Internet gambling among Ontario adults. *Journal of Gambling Issues, 5*, doi: 10.4309/jgi.2001.5.9.

Karaiskos, D., Tzavellas, E., Balta, G., & Paparrigopoulos, T. (2010). Social network addiction: A new clinical disorder? *European Psychiatry, 25*, 855.

Kim, M. G., & Kim, J. (2010). Cross-validation of reliability, convergent and discriminant validity for the problematic online game use scale. *Computers in Human Behavior, 26*, 389–398.

Kim, E. J., Namkoong, K., Ku, T., & Kim, S. J. (2008). The relationship between online game addiction and aggression, self-control and narcissistic personality traits. *European Psychiatry, 23*, 212–218.

King, D. L., Delfabbro, P. H., & Griffiths, M. D. (2012). Clinical interventions for technology-based problems: Excessive Internet and video game use. *Journal of Cognitive Psychotherapy: An International Quarterly, 26*, 43–56.

King, D. L., Delfabbro, P. H., Griffiths, M. D., & Gradisar, M. (2011). Assessing clinical trials of Internet addiction treatment: A systematic review and CONSORT evaluation. *Clinical Psychology Review, 31*, 1110–1116.

King, D. L., Delfabbro, P. H., & Zajac, I. T. (2011). Preliminary validation of a new clinical tool for identifying problem video game playing. *International Journal of Mental Health and Addiction, 9*, 72–87.

King, D. L., Haagsma, M. C., Delfabbro, P. H., Gradisar, M. S., & Griffiths, M. D. (2013). Toward a consensus definition of pathological video-gaming: A systematic review of psychometric assessment tools. *Clinical Psychology Review, 33*, 331–342.

Kittinger, R., Correia, C. J., & Irons, J. G. (2012). Relationship between Facebook use and problematic internet use among college students *Cyberpsychology, Behavior, & Social Networking, 15*, 324–327.

Ko, C. H., Liu, G. C., Hsiao, S. M., Yen, J. Y., Yang, M. J., Lin, W. C., …, Chen, C. S. (2009). Brain activities associated with gaming urge of online gaming addiction. *Journal of Psychiatric Research, 43*, 739–747.

Ko, C. H., Yen, J. Y., Chen, C. C., Chen, S. H., & Yen, C. F. (2005). Gender differences and related factors affecting online gaming addiction among Taiwanese adolescents. *Journal of Nervous and Mental Disease, 193*, 273–277.

Koc, M., & Gulyagci, S. (2013). Facebook addiction among Turkish college students: The role of psychological health, demographic, and usage characteristics. *Cyberpsychology, Behavior, & Social Networking, 16*, 279–284. doi:10.1089/cyber.2012.0249

Kuss, D. J., & Griffiths, M. D. (2012a). Online gaming addiction: A systematic review. *International Journal of Mental Health and Addiction, 10*, 278–296.

Kuss, D. J., & Griffiths, M. D. (2012b). Internet gambling behavior. In Z. Yan (Ed.), *Encyclopedia of cyber behavior* (pp. 735–753). Hershey, PA: IGI Global.

Kuss, D. J., & Griffiths, M. D. (2012c). Online gaming addiction in adolescence: A literature review of empirical research. *Journal of Behavioral Addiction, 1*, 3–22.

Kuss, D. J., Griffiths, M. D., & Binder, J. (2013). Internet addiction in students: Prevalence and risk factors. *Computers in Human Behavior, 29*, 959–966.

Kuss, D. J., van Rooij, A. J., Shorter, G. W., Griffiths, M. D., & van de Mheen, D. (2013). Internet addiction in adolescents: Prevalence and risk factors. *Computers in Human Behavior, 29*, 1987–1996.

LaBrie, R. A., Kaplan, S., LaPlante, D. A., Nelson, S. E., & Shaffer, H. J. (2008). Inside the virtual casino: A prospective longitudinal study of Internet casino gambling. *European Journal of Public Health, 18*(4), 410–416.

LaBrie, R. A., LaPlante, D.A., Nelson, S. E., Schumann, A., & Shaffer, H. J. (2007). Assessing the playing field: A prospective longitudinal study of internet sports gambling behavior. *Journal of Gambling Studies, 23*, 347–363.

Ladd, G. T., & Petry, N. M. (2002). Disordered gambling among university-based medical and dental patients: A focus on Internet gambling. *Psychology of Addictive Behaviors, 16,* 76–79.

Lemmens, J. S., Valkenburg, P. M., & Peter, J. (2009). Development and validation of a game addiction scale for adolescents. *Media Psychology, 12*(1), 77–95.

Luscombe, B. (2009). Social norms. Facebook and divorce. *Time, 173,* 93–94.

Machold, C., Judge, G., Mavrinac, A., Elliott, J., Murphy, A. M., & Roche, E. (2012). Social networking patterns/hazards among teenagers. *Irish Medical Journal, 105,* 151–152.

Matthews, N., Farnsworth, W. F., & Griffiths, M. D. (2009). A pilot study of problem gambling among student online gamblers: Mood states as predictors of problematic behaviour. *CyberPsychology and Behavior, 12,* 741–746.

McCormack. A., & Griffiths, M. D. (2012a). Motivating and inhibiting factors in online gambling behaviour: A grounded theory study. *International Journal of Mental Health and Addiction, 10,* 39–53.

McCormack. A., & Griffiths, M. D. (2012b). What differentiates professional poker players from recreational poker players? A qualitative interview study. *International Journal of Mental Health and Addiction, 10,* 243–257.

McCormack, A., Shorter, G., & Griffiths, M. D. (2013). An examination of participation in online gambling activities and the relationship with problem gambling. *Journal of Behavioral Addictions, 2*(1), 31–41.

Meerkerk, G. J., Van Den Eijnden, R., Vermulst, A. A., & Garretsen, H. F. L. (2009). The compulsive Internet use scale (CIUS): Some psychometric properties. *CyberPsychology & Behavior, 12,* 1–6.

Mehroof, M., & Griffiths, M. D. (2010). Online gaming addiction: The role of sensation seeking, self-control, neuroticism, aggression, state anxiety, and trait anxiety. *CyberPsychology & Behavior, 13,* 313–316.

Mitchell, K., & Beard, F. (2010). Measuring Internet dependence among college students: A replication and confirmatory analysis. *Southwestern Mass Communication Journal, 25*(2), 15–28.

Morahan-Martin, J. (2005). Internet abuse: addiction? Disorder? Symptom? Alternative explanations? *Social Science Computer Review, 23,* 39–48.

Noble, E. P., Blum, K., Khalsa, M. E., Ritchie, T., Montgomery, A., Wood, R. C., …, Sparkes, R. S. (1993). Allelic association of the D2 dopamine receptor gene with cocaine dependence. *Drug and Alcohol Dependence, 33,* 271–285.

Olowu, A. O., & Seri, F. O. (2012). A study of social network addiction among youths in Nigeria. *Journal of Social Science and Policy Review, 4,* 62–71.

Parke, A., & Griffiths, M. D. (2011). Effects on gambling behaviour of developments in information technology: A grounded theoretical framework. *International Journal of Cyber Behaviour, Psychology and Learning, 1*(4), 36–48.

Parke, A., & Griffiths, M. D. (2012). Beyond illusion of control: An interpretative phenomenological analysis of gambling in the context of information technology. *Addiction Research and Theory, 20,* 250–260.

Parker, J. D. A., Taylor, R. N., Eastabrook, J. M., Schell, S. L., & Wood, L. M. (2008). Problem gambling in adolescence: Relationships with internet misuse, gaming abuse and emotional intelligence. *Personality and Individual Differences, 45,* 174–180.

Peters, C., & Bodkin, C. D. (2007). An exploratory investigation of problematic online auction behaviors: Experiences of eBay users. *Journal of Retailing and Consumer Services, 14*(1), 1–16.

Peters, C. S., & Malesky, L. A. (2008). Problematic usage among highly-engaged players of massively multiplayer online role playing games. *CyberPsychology & Behavior*, *11*, 480–483.

Porter, K., Mitchell, J., Grace, M., Shinosky, S., & Gordon, V. (2012). A study of the effects of social media use and addiction on relationship satisfaction. *Meta-Communicate*, *2*(1), 50–54. Retrieved from http://journals.chapman.edu/ojs/index.php/mc/article/view/340

Porter, G., Starcevic, V., Berle, D., & Fenech, P. (2010). Recognizing problem video game use. *Australian and New Zealand Journal of Psychiatry*, *44*, 120–128.

Salguero, R. A. T., & Moran, R. M. B. (2002). Measuring problem video game playing in adolescents. *Addiction*, *97*, 1601–1606.

Shaffer, H. J., LaPlante, D. A., LaBrie, R. A., Kidman, R. C., Donato, A. N., & Stanton, M. V. (2004). Toward a syndrome model of addiction: Multiple expressions, common etiology. *Harvard Review of Psychiatry*, *12*, 367–374.

Shapira, N., Goldsmith, T., Keck, P., Jr., Khosla, D., & McElroy, S. (2000). Psychiatric features of individuals with problematic internet use. *Journal of Affective Disorders*, *57*, 267–272.

Skoric, M. M., Teo, L. L. C., & Neo, R. L. (2009). Children and video games: Addiction, engagement, and scholastic achievement. *CyberPsychology & Behavior*, *12*, 567–572.

Smeaton, M., & Griffiths, M. D. (2004). Internet gambling and social responsibility: An exploratory study. *CyberPsychology & Behavior*, *7*, 49–57.

Sofiah S., Zobidah O. S., Bolong J. N., & Osman, M. (2011). Facebook addiction among female university students. *Public Administration and Social Policy Review*, *2*, 95–109.

Spekman, M. L. C., Konijn, E. A., Roelofsma, P. H. M. P., & Griffiths, M. D. (2013). Gaming addiction, definition, and measurement: A large-scale empirical study. *Computers in Human Behavior*, *29*, 2150–2155.

Turel, O., & Serenko, A. (2012). The benefits and dangers of enjoyment with social networking websites. *European Journal of Information Systems*, *21*, 512–528.

van Rooij, A. J., Schoenmakers, T. M., van de Eijnden, R., & van de Mheen, D. (2010). Compulsive Internet use: The role of online gaming and other Internet applications. *Journal of Adolescent Health*, *47*, 51–57.

Walsh, S. P., White, K. M., & Young, R. M. (2007). Young and connected: Psychological influences of mobile phone use amongst Australian youth, Mobile Media 2007. In G. Goggin & L. Hjorth (Eds.), *Proceedings of an International Conference on Social and Cultural Aspects of Mobile Phones, Media, and Wireless Technologies, Sydney, 2007.* (pp. 125–134). Sydney, Australia: University of Sydney.

Wan, C. (2009). *Gratifications & loneliness as predictors of campus-SNS websites addiction & usage pattern among Chinese college students* (Master's thesis). Chinese University of Hong Kong, Hong Kong, China.

Wang, C. C., & Yang, H. W. (2006). Passion and dependency in online shopping activities. *CyberPsychology & Behavior*, *10*, 296–298.

Wardle, H., & Griffiths, M. D. (2011). Defining the 'online gambler': The British perspective. *World Online Gambling Law Report*, *10*(2), 12–13.

Wardle, H., Moody, A., Griffiths, M. D., Orford, J., & Volberg, R. (2011). Defining the online gambler and patterns of behaviour integration: Evidence from the British Gambling Prevalence Survey 2010. *International Gambling Studies*, *11*, 339–356.

Wardle, H., Sproston, K., Orford, J., Erens, B., Griffiths, M. D., Constantine, R., & Pigott, S. (2007). *The British Gambling Prevalence Survey 2007.* London, UK: The Stationery Office.

Widyanto, L., & Griffiths, M. D. (2006). Internet addiction: A critical review. *International Journal of Mental Health and Addiction*, *4*, 31–51.

Widyanto, L., & Griffiths, M. D. (2009). Unravelling the web: Adolescents and internet addiction. In R. Zheng, J. Burrow-Sanchez, & C. Drew (Eds.), *Adolescent online social communication and behavior: Relationship formation on the Internet* (pp. 29–49). Hershey, PA: Idea Publishing.

Wilson, K., Fornasier, S., & White, K. M. (2010). Psychological predictors of young adults' use of social networking sites. *Cyberpsychology, Behavior & Social Networking, 13*, 173–177.

Wölfling, K., Thalemann, R., & Grusser-Sinopoli, S. M. (2008). Computer game addiction: A psychopathological symptom complex in adolescence. Psychiatrische Praxis, 35, 226–232.

Wölfling, K., Thalemann, R., & Grusser-Sinopoli, S. M. (2008). Computer game addiction: A psychopathological symptom complex in adolescence. *Psychiatrische Praxis, 35*, 226–232.

Wolniczak, I., Caceres-DelAguila, J. A., Palma-Ardiles, G., Arroyo, K. J., Solís-Visscher, R., Paredes-Yauri, S., …, Bernabe-Ortiz, A. (2013). Association between Facebook dependence and poor sleep quality: A study in a sample of undergraduate students in Peru. *PLoS ONE, 8*(3), e59087. doi:10.1371/journal.pone.0059087

Wood, R. T. A., & Griffiths, M. D. (2008). Why Swedish people play online poker and factors that can increase or decrease trust in poker websites: A qualitative investigation. *Journal of Gambling Issues, 21*, 80–97.

Wood, R. T. A., Griffiths, M. D., & Parke, J. (2007). The acquisition, development, and maintenance of online poker playing in a student sample. *CyberPsychology & Behavior, 10*, 354–361.

Wood, R. T., & Williams, R. J. (2007). Problem gambling on the Internet: Implications for internet gambling policy in North America. *New Media and Society, 9*, 520–542.

Young, K. (1996). Psychology of computer use: XL. Addictive use of the internet: A case that breaks the stereotype. *Psychological Reports, 79*, 899–902.

Young, K. (1998). *Caught in the net.* New York, NY: Wiley.

Young, K. (1999). Internet addiction: Evaluation and treatment. *Student British Medical Journal, 7*, 351–352.

Young, K. (2009). Understanding online gaming addiction and treatment issues for adolescents. *American Journal of Family Therapy, 37*, 355–372.

Zhou, S. X. (2010). *Gratifications, loneliness, leisure boredom and self-esteem as predictors of SNS-game addiction and usage pattern among Chinese college students* (Master's thesis). Chinese University of Hong Kon, Hong Kong, China.

Part V

Exposure, Experience and Evaluations of Digital Media

18

Selective Exposure in the Communication Technology Context

Silvia Knobloch-Westerwick, Axel Westerwick, and Benjamin K. Johnson

The Ohio State University, Columbus, OH, USA

Classic studies and theories in communication have emphasized the pivotal role of selective exposure for any media impact (Lazarsfeld, Berelson, & Gaudet, 1944; McGuire, 1989). The phenomenon affects almost any type of mediated communication, as "captive audience" (Hovland, 1959) situations hardly ever occur in real-life circumstances. Developments in communication technology in particular have created high-choice media landscapes. For instance, American households can commonly access hundreds of TV channels and a virtually inexhaustible number of Internet outlets at any given moment. Communication technologies allow users to customize content and access it whenever they wish. For example, online news outlets encourage their readers to implement settings that filter certain news sections or topics; digital recording technologies have revolutionized how audiovisual materials are selected and viewed as well as shared with other media users.

This chapter argues that the new communication technology context (a) has brought a virtually inexhaustible extent of message choices upon us, which underlines the key role of selective exposure even more so than in traditional media use contexts, (b) involves the emergence of new media message cues, with some signaling classic characteristics (e.g., source credibility) in new fashion and others yielding inherently new cue types (e.g., download numbers reflecting message popularity), and (c) provides a context that greatly facilitates empirical research into the pivotal phenomenon of selective exposure. The chapter will first explicate selective exposure and describe an important paradigm in this domain of research. Linked to the selective exposure paradigm (Knobloch-Westerwick, 2015), implications of communication technology developments will be considered before we review theoretical frameworks and empirical findings—pertaining to confirmation bias, credibility, informational utility, mood management, social comparison, and technological cues. Finally, we will offer conclusions and ideas on where this line of research needs to go next.

The Handbook of the Psychology of Communication Technology, First Edition. Edited by S. Shyam Sundar.

Selective Exposure as a Concept

In light of developments in communication technologies, the concept of selective exposure deserves more scholarly attention than ever before. Indeed, the past decade has seen an upsurge of research activity pertaining to politically motivated selective exposure—an area that has been dormant for an extended period. In this specific domain, a certain use has defined the term *selective exposure*. Specifically, the term became synonymous with an exposure pattern reflecting a preference for attitude-aligned messages and an avoidance of attitude-discrepant content. This specific understanding reflects the strong reliance of historical selective exposure research on Festinger's theory of cognitive dissonance (1957), which suggests that individuals prefer messages that converge with their preexisting attitudes to avoid unpleasant dissonance arising from cognitive inconsistencies. Nevertheless, some of the earliest writings on selective exposure pointed out that the phenomenon has much broader implications. In the classic, groundbreaking book *The People's Choice*, the matter was described as follows: "The fact that people select their exposure along the lines of their political predispositions is only a special case of a more general law which pervades the whole field of communication research. Exposure is always selective" (Lazarsfeld et al., 1944, p. 164).

Indeed, communication researchers have extended selective exposure beyond an understanding grounded in cognitive dissonance theory (Festinger, 1957), according to which individuals prefer attitude-consistent messages and avoid attitude-discrepant messages. For instance, Sears and Freedman (1967, p. 195) talked about "any systematic bias in audience composition" as a phenomenon that can be labeled selective exposure. More recent contributions also argued for this conceptualization of selective exposure "as denoting *any* bias of exposure to available communication content" (Knobloch-Westerwick, 2015). To illustrate the notion, consider a website with 10 news reports available for in-depth reading—users will be *unlikely* to allocate 10% of their site visit time to each of the 10 reports and instead will exhibit preferences or avoidance. The selective exposure patterns may result from the reports' topics, tone, or featured individuals, and will be moderated by users' trait and situational characteristics such as biological sex, habitual interests, and moods. This understanding of selective exposure (or selective avoidance as the flipside) can be found in a number of recent publications in communication research (e.g., Aubrey, 2006; Segrin & Nabi, 2002) and will also guide the present discussion of selective exposure in the context of communication technologies.

Selective Exposure Research Paradigm

With this very generic conceptualization of selective exposure as any bias of exposure to available communication content, it appears that much research pertaining to media use would inform our knowledge on this phenomenon. However, the argument can be made that important limitations exist on theoretical and methodological grounds. Regarding theoretical frameworks, selective exposure research can be seen as distinct from research guided by the uses-and-gratifications approach because this approach rests on the assumption that media users have full awareness of their use motives (Katz, Blumler, & Gurevitch, 1974). This assumption also has important methodological implications because, if media users have full introspection, survey measures can capture exposure to media messages. Yet it has frequently been argued that media users are not aware of processes governing their media selections and that usually very little cognitive elaboration takes place (Knobloch, 2003; Zillmann, 1985, 1988). In fact, learning along the lines of operant conditioning has been said to play an important role in selection processes (Zillmann & Bryant, 1985), more so than deliberate action. Especially in the online media context and in an age of abundant television channels, where many media use selections occur with great speed and in an automatic manner (e.g., Lohr, 2012), a compelling argument can be made that media users are often not fully aware of their own selection motives and thus are not in a position to report them through survey measures. Moreover, impaired recall, imprecise response scales, and social desirability likely undermine the validity of self-report data further (Knobloch-Westerwick, 2015). Accordingly, self-reports are not the first choice of measurement of selective exposure phenomena. Based on these considerations, the case for a dominant selective exposure research paradigm (Knobloch-Westerwick, 2015) building on unobtrusive measures can be made. Pertinent investigations are reviewed later.

The Context of Communication Technologies

A surge of diversification in communication technologies over recent decades has fueled selective exposure research. Although the problem of "coping with plenty" (Becker & Schoenbach, 1989) media messages already presented itself to the audience when the number of TV channels sky-rocketed through cable technology, digital media and the Internet along with broadband transmission led to an immense abundance of messages from which media users may choose. The fundamental importance of selectivity in media use has become even more obvious in light of seemingly infinite numbers of URLs, web pages, and public communication players, including non-media institutions and individuals, who can now reach out directly without involving traditional media channels. Thus, new questions arose regarding source perceptions and what message characteristics attract selective exposure. While generic discussions of an "active audience" (Webster, 1998) existed before, they usually pertained more to audience motives and interpretation of media messages. More recent research has further aimed to predict specifically what communication elements are selected and what the consequences of the resulting selective exposure are.

Fortunately, the new communication technologies provide a context that fosters empirical selective exposure research by offering new methodological pathways. They facilitate unobtrusive exposure measures, which have been applied in much of the recent rigorous research. With regard to methods to capture selective exposure as an observed behavior, researchers may tape reading behavior (e.g., Zillmann, Knobloch, & Yu, 2001), but switching TV channels (Bryant & Zillmann, 1984;

Rossmann, 2006) or online message selections (Hastall & Knobloch-Westerwick, 2013a) can be better recorded via technology. Indeed, computerized approaches have been widely used in this area of study (e.g., Garrett, 2009a; Knobloch, 2003; Knobloch-Westerwick, Sharma, Hansen, & Alter, 2005; Zillmann, Chen, Knobloch, & Callison, 2004). While some designs utilize recorded page visits to derive fruitful information (Tewksbury, 2003), this approach presents limitations in that the causes of selective exposure are not identifiable without further knowledge on user characteristics and without experimental message manipulations (Hastall & Knobloch-Westerwick, 2013a).

Regarding new conceptual challenges, to begin with, the levels of choice need to be reconsidered. Media users can choose from media channels (e.g., TV, radio, newspapers, magazines, websites), then on a subsequent level choose from specific TV channels or specific newspaper titles, next go further by picking specific TV programs or newspaper sections, and on an even more refined level select TV episodes or scenes or news articles and paragraphs, and so on. Defining media content units available for selection becomes a new challenge in the new media context because of their interactive nature (December, 1996). A website will often have numerous subsections, articles, hyperlinks, and interactive features that make structuring the content available for media users' selection more complex. Further, the actual selective exposure could be defined per selection exerted per mouse click, per page visit time, or even more specifically by glancing at specific components displayed on the screen, which would require eyeball tracking (e.g., Garcia & Stark, 1991).

In light of these complexities, researchers who are interested in selective exposure need to make a number of decisions—such as length of examined time period, number of available choices, differentiation of available choices—when they tackle the phenomenon in actual empirical research. These decisions should be based on theoretical frameworks that pertain to selective exposure research. A number of theoretical approaches that have been applied to study selective exposure in the new communication technology context will be reviewed in the following, along with selected empirical findings.

Theoretical Tenets and Empirical Evidence

Confirmation bias

The earliest usage of *selective exposure* (Lazarsfeld et al., 1944) characterized partisans' disproportionate attendance to ideologically sympathetic messages. A strong preference for attitude-consistent information would allow individuals not only to confirm their preexisting beliefs and reinforce their opinions, but would also act as a barrier to persuasion or other communication effects (Klapper, 1960). The development of cognitive dissonance theory (Festinger, 1957) fostered even greater interest in selective exposure research, using attitude-consistent and attitude-discrepant messages on a wide range of topics. Yet, by the late 1960s, inconsistent support in experimental studies cast doubt on the extent to which selectivity operated as a dependable bias in message exposure (Sears & Freedman, 1967). Subsequent reviews of selective exposure research have taken more care to account for differences in study methodology, and have come to the conclusion that attitude-consistent selective exposure is a clear and reliable effect (Cotton, 1985; D'Alessio & Allen, 2002; Hart et al., 2009).

Changes in communication technology have brought confirmation bias to the forefront of debates over the future of both communication research and the media landscape itself. The proliferation of information brought on by the Internet, along with the customization and control Internet technologies offer over that information, create enhanced opportunities for confirmation bias in selective exposure (Sunstein, 2001). Sunstein's concerns about polarization and extremism driven by partisan echo chambers have been influential, reinvigorating research and debate into the extent and consequences of confirmation bias.

Building on this issue, Bennett and Iyengar (2008) argued that information technology has ensured greater selectivity among media users, ushering in a "new era" of limited effects (cf. Klapper, 1960), in which persuasion, knowledge gains, or other outcomes of receiving mediated messages are likely to be minimal or nonexistent because audiences are able to develop media routines that only tell them what they already believe and know. However, while attitude-consistent selective exposure is pervasive, some evidence suggests that selective avoidance is not nearly as strong and that most media users do regularly encounter unfamiliar and disagreeable information (Frey, 1982; Garrett, 2009b; Knobloch-Westerwick & Kleinman, 2012). Furthermore, selectivity does not rule out the possibility of media effects, as attitude-consistent exposure bears effects of its own (Holbert, Garrett, & Gleason, 2010; Knobloch-Westerwick & Meng, 2011).

In the recent surge of selective exposure research in the context of political communication (see overview by Knobloch-Westerwick, 2014, 2015; for examples in the online context, see Johnson, Zhang, & Bichard, 2011; Jun, 2012), much of the evidence at hands stems from survey-based research. It may be peculiar to study the phenomenon of a bias through self-report measures that will very likely also be subject to related biases, in addition to other impairments from lay theories, social desirability, imperfect recall, and drive to provide internally consistent responses. In light of these problems, the present review of related research focuses on studies that utilized unobtrusive behavioral measures of selective exposure.

A number of studies have shown that online users of political information exhibit a confirmation bias (Graf & Aday, 2008; Knobloch-Westerwick, 2007; Knobloch-Westerwick & Meng, 2009). However, utility of the information (see further details on this concept given later) may override the confirmation bias in online users' exposure to political information (Knobloch-Westerwick & Kleinman, 2012; Valentino, Banks, Hutchings, & Davis, 2009). Further, issue importance affects the confirmation bias, such that online users who attach greater importance to an issue show a stronger bias toward attitude-consistent information (Knobloch-Westerwick, 2007; Westerwick, Kleinman, & Knobloch-Westerwick, 2013). Moreover, situational goals regarding accurate versus defensive processing may have implications for selective exposure (Kim, 2007), although some work yielded that, even when anticipating to discuss and to defend their viewpoint regarding counter-arguments, participants preferred attitude-consistent information (Westerwick et al., 2013). Most recently, a number of studies have demonstrated impacts on political attitudes, partisanship, and political participation from attitude-consistent selective exposure to political information online (Knobloch-Westerwick, 2014; Knobloch-Westerwick & Johnson, 2014; Knobloch-Westerwick & Meng, 2011; Westerwick et al., 2013).

Most important for the context of this chapter, the impact of Internet use on the extent of a confirmation bias in political communication is still subject to further testing. Some evidence suggests that greater habitual online news use reduces the confirmation bias in online exposure to political information (Knobloch-Westerwick & Johnson, 2014; Knobloch-Westerwick & Kleinman, 2012). However, further research is needed to elucidate how the use of new communication technologies affects a confirmation bias and what the implications are. At the front of technology design, some scholars have found that formal emphasis of attitude-consistent and attitude-discrepant messages affects user responses to an online site (Munson & Resnick, 2010) (see also section "Technology Cues") and work on algorithms to build online aggregators of news and op-ed items that present users with greater diversity without turning them off, in the interest of societal benefits of increased diversity of information and opinion (Munson, Zhou, & Resnick, 2009).

Credibility

Today, an increasing number of people use new communication technologies not only for entertainment but to decide on important matters for work or their private lives. Most of all, the Internet is providing abundant information to be used in different contexts with utmost relevance—political discussions, working on career-related assignments, or developing personal health plans. On the other hand, the number of information sources and authors of online information increased dramatically over the past years, which not only provides an almost endless pool of information about the same topics but also raises questions about information credibility. If new communication technology is used for retrieval of vital information, one should assume that users are concerned about information credibility to avoid being misled by incorrect information.

Classic persuasion research (e.g., Hovland & Weiss, 1951; Wilson & Sherrell, 1993) has long suggested that source expertise has a significant influence on the persuasion of an audience, which is closely related to source credibility. Westerwick et al. (2013) investigated if source credibility and attitude toward political messages influence users' selection of online content. The results suggest that people who are more involved in a political topic select online content that converges with their own attitudes significantly more often than counter-attitudinal messages, regardless of the credibility of the source. In contrast, less involved people select messages from high credibility sources significantly more often. Thus, source credibility emerges as one factor that influences selective exposure, but it may be less influential if other factors such as attitudes with great personal importance come into play.

Other studies investigated how users select Google search results to find online content such as health information (Mager, 2009; Pan et al., 2007; Westerwick,

2013). The results suggest that users do not only use search engines for convenience reasons but they also trust that especially the top-ranked search results provide the best information about a search topic. Thus, trusting search engines such as Google affects which websites users select and which websites they probably do not consider in their search, namely the lower-ranked search results. While Westerwick (2013) shows that online information users still consider source credibility of a website when evaluating the credibility of the posted information, search result rankings also have significant influence on website selection and credibility perceptions. Westerwick's (2013) experiment further demonstrated that technological cues—such as tabs, navigation bar, and search function—foster the credibility of online information, which in turn should affect selective exposure.

Indeed, a few studies have examined how credibility influences selection of online information. Winter and Krämer (2012) found that blogs about science information were more often selected if the associated blogger was said to have professional expertise in the topic area. Westerwick et al. (2013) examined the political attitudes and attitude importance in combination with source credibility for online search results and found that those users who attached low importance to the topics spent more time with information from high-credibility sources, whereas those who attached high importance to the topics favored attitude-aligned messages even from low-credibility sources. In the health context, however, a general preference for high-credibility sources emerged in selective exposure to online information search results (Knobloch-Westerwick, Johnson, & Westerwick, 2013).

Furthermore, technological cues likely affect user perceptions and, subsequently, selective exposure—for example, personalization of a webpage may influence credibility perceptions and subsequent channel browsing activity (Kalyanaraman & Sundar, 2006). Another possible area to investigate the influence of information credibility on selective exposure is consumer product ratings. While the authors of the online ratings are mostly anonymous, message credibility—for example, based on message design, writing style, or recency—and attitude toward the product or the manufacturer may be the determining factors that influence which ratings are considered for further reading and decision making. However, there is currently very limited insight on those factors in regard to selective exposure. A study by Winter, Krämer, Appel, and Schielke (2011) did not yield much influence from community ratings of authorship expertise in the context of an online science blog.

Informational utility

Informational utility is a factor in selective exposure that had been previously conceptualized (Atkin, 1973; Freedman & Sears, 1965) but was only sporadically examined until the Internet era. The informational utility of a message refers to the practical use that information conveyed in the message offers the individual in reducing uncertainty and guiding behavior (Atkin, 1973). The surplus of news and information made available by the Internet has facilitated attention to, and methods for studying, the effect of these message characteristics on selective exposure (Knobloch-Westerwick, 2008).

The message characteristics that constitute informational utility have been conceptualized as magnitude, likelihood, immediacy, and efficacy (Knobloch, Dillman Carpentier, & Zillmann, 2003; Knobloch-Westerwick, Hastall, Grimmer, & Brück, 2005). The presence of any of these four dimensions contributes additively to perceived message utility. Respectively, magnitude refers to the size or importance of an issue's consequences, likelihood refers to the probability that the individual will be affected, immediacy refers to the speed with which the issue will produce consequences, and efficacy refers to the ease and effectiveness of possible individual responses to the issue at hand.

Empirical investigations into the effect of informational utility on selective exposure have largely focused on online news. Readers are more likely to select and spend more time reading online news with greater informational utility (Arpan & Tüzünkan, 2011; Kim, 2008; Knobloch, Dillman Carpentier, et al., 2003; Knobloch-Westerwick, Hastall, et al., 2005), regardless of issue valence (Knobloch-Westerwick, Dillman Carpentier, Blumhoff, & Nickel, 2005). The impact on selective exposure has also been demonstrated across cultures (Knobloch-Westerwick, Dillman Carpentier, et al., 2005). Recent research has continued to explore the nature of informational utility, finding that increased anxiety heightens selective exposure to online news with greater utility (Valentino et al., 2009) and that magnitude and efficacy can interact such that discrepancies between the intensity of the two dimensions may foster greater selective exposure in the health information context (Hastall & Knobloch-Westerwick, 2013b). Further, in the political context, informational utility has been suggested to override the confirmation bias in selective exposure to political online messages (Knobloch-Westerwick & Kleinman, 2012).

Mood management

Despite much talk of the information age due to the advent of the Internet (e.g., Vorderer, 2001), media use for entertainment purposes is still a grand attraction for the audience. Although new media technologies often initially raise hopes for better education through media (e.g., consider VCRs), they typically have their breakthrough as entertainment equipment. Similarly, digital and online media serve entertainment purposes when their users, for instance, surf the web to distract themselves or to seek out what they find surprising, bizarre, amusing, or horrifying. More obviously, electronic games are made to entertain and to play on their users' emotions. But before engaging in this interactive entertainment, gamers need to select a game and theoretically have 20,000 options according to games market researchers (EEDAR, n.d.). Hence, theorizing that has guided selective exposure research in the entertainment realm of traditional media may also yield insights into selection patterns in the new media context.

The most influential theory in this context is Zillmann's (1988) mood management theory. At its core, the theory suggests that media users select content to improve their moods. Mood states may be described along the lines of arousal and valence: when bored, exciting content will be preferred to increase personal arousal states to more agreeable levels. On the other hand, individuals under stress are said to prefer calming messages. Generally speaking, media users will seek out materials they anticipate will help to enhance the hedonic tone of their mood and to distract them from sources of a negative mood. While a number of studies have supported these notions for traditional media (Knobloch-Westerwick, 2006), a few extended it to the context of new communication technologies as described in the following.

Two studies applied mood management theory to general web use. Surfing the World Wide Web has been found to be influenced by mood valence, as Knobloch (2002) demonstrated that experimentally induced mood predicted subsequent web browsing behavior such that individuals in a negative mood spent more time with positively valenced pages than individuals in a "neutral" mood. Further, in the same study, negative mood led to longer exposure time with web pages perceived as "entertainment" than positive mood. Mastro, Eastin, and Tamborini (2002) also induced different excitation levels experimentally and then observed subsequent web use; their findings showed that bored participants selected to view more pages than stressed participants.

One study built off of empirical work that had examined how mood management applied to selective news reading but additionally included a social interaction component. Knobloch-Westerwick and Alter (2006) had a session supervisor provoke participants and either led them to anticipate they would get to retaliate against the provoker or not. After the provocation, participants browsed an online news platform with negative and positive news. Among the participants who anticipated a retaliation opportunity, the selective exposure pattern differed by participant sex because men spent more time with reading negative news whereas women avoided reading negative online news. This pattern suggested that men aimed to sustain their anger from the provocation if they anticipated a chance to retaliate later, whereas women aimed to dissipate their anger if they anticipated a chance to retaliate after being provoked. Hence, selective exposure did not merely serve to optimize and enhance mood (Zillmann, 1988) but to adjust mood (Knobloch, 2003) for an anticipated situation.

Additional studies utilized tenets of mood management theory to study video games use (Bowman & Tamborini, 2012; Reinecke & Trepte, 2008) but did not examine actual selective exposure; Bushman and Whitaker (2010) applied anger regulation to study selection of violent video games with an interest survey measure. However, work by Reinecke et al. (2012) specifically examined choice of video game difficulty level: participants first had a chance to try a flight simulator game in three different difficulty levels and then went through a mood induction procedure (the same as used by Knobloch & Zillmann, 2002, and slightly varied by Knobloch-Westerwick & Alter, 2006). In the negative affect condition, participants chose a more demanding setting when they were asked to play the flight simulator game again compared to the positive affect condition. This finding aligns with the mood management tenet that media users will seek out messages that distract them from sources of a negative mood (the authors, however, combined mood management theory and self-determination theory to interpret these results).

Most recent work has begun to extend mood management to the social networking site users while connecting it to social comparison theory, further addressed in the following section.

Social comparisons and affiliations

The use of social networking sites has been sky-rocketing (Madden & Zickuhr, 2011), and researchers turn to social-psychological theories to account for the phenomenon (boyd & Ellison, 2007; Wilson, Gosling, & Graham, 2012). Given that the content presented in most social media is largely created by and concerned with one's interpersonal connections, theories of social comparison and affiliation can begin to account for selective exposure to content on social networking sites and other web content that focuses on depictions of individuals.

A fundamental aspect of social life is that people frequently evaluate their own performance and opinions by comparing themselves to similar others in their social environment (Festinger, 1954). In addition to the need for self-evaluation, social comparison also takes place for purposes of self-enhancement, in which the individual makes self-serving comparisons to others. These can be either downward social comparisons that make the individual feel superior by contrast (Wills, 1981) or comparisons in which the individual identifies with or aspires toward more upward social comparison targets (Lockwood & Kunda, 1997). The wealth of a diverse range of depictions of other people online, from online news to social networking sites, allows Internet users much latitude to exercise selectivity that would facilitate self-enhancing social comparisons.

Selective exposure studies of depictions of others online have shown that Internet users are certainly discerning with regard to choosing content that can be used for social comparison purposes. Men and women reading online news about the positive or negative achievements of others tend to make selective comparisons on topics that are sex-specific or fit their gender role schemata (Knobloch-Westerwick & Alter, 2007; Knobloch-Westerwick, Brück, & Hastall, 2006). With regard to aspirational upward social comparisons, in a study employing a digital magazine, readers exhibited less avoidance of advertisements with "ideal body" depictions when the surrounding magazine content stressed the attainability of body ideals (Knobloch-Westerwick & Romero, 2011). Additionally, upward and downward comparison strategies when reading online news depictions can also be influenced by the individual's sex and self-esteem (Knobloch-Westerwick & Hastall, 2006).

Group membership influences selective exposure to depictions of ingroup and outgroup members in online news, especially for minority group members (Knobloch-Westerwick, Appiah, & Alter, 2008). With regard to social comparison, a study of numeric majority and minority groups found that for those with low to moderate self-esteem, majority members avoided positive ingroup depictions in Internet news while minority members selected positive over negative ingroup depictions (Knobloch-Westerwick & Westerwick, 2011). Although comparisons to same-sex persons of the same age are generally preferred (Knobloch-Westerwick & Hastall, 2006), selective exposure to negative outgroup depictions may occur. For example, Knobloch-Westerwick and Hastall (2010) found that older adults reading online news articles exhibited a strong preference for negative news about young individuals, which in turn led to increased self-esteem among the older readers.

In addition to an effect on self-esteem, downward social comparison—in this case, forced exposure to social networking site profiles—was observed to increase positive affect (Haferkamp & Krämer, 2011), which suggests that selective exposure to social networking site content can improve affect and could be used to facilitate mood management. Similarly, a study of mood management using popular music selected from a digital media player established that romantically dissatisfied young adults avoided songs that entailed upward comparisons (to love-celebrating songs by a performer of the same sex), preferring love-lamenting music by a same-sex performer (Knobloch & Zillmann, 2003).

Technology cues

There is a variety of research with a focus on the influence of technology cues on the behavior and perceptions of Internet users. Website design, interactivity, and information recency are some of the cues that a technology provides (Sundar, 2008), and these cues should influence selective exposure to Internet content. Most relevant studies investigated if users engage differently in political discussions, perceive the provided information as more credible, or select a website or the posted information more often if certain technology cues are salient or not. The mixed findings suggest that especially design cues and recency information about a posted message influence users' perceptions and selection of online information (Fogg et al., 2001; Shon, Marshall, & Musen, 2000; Sundar, Knobloch-Westerwick, & Hastall, 2007; Wathen & Burkell, 2002). Metzger, Flanagin, and Medders (2010) used expectancy violations theory to explain their findings about how website design and features influence users' information credibility perceptions. They suggest that websites with a design that meets or exceeds users' expectations are likely to receive higher credibility ratings than those websites with poor design and features.

However, it is also suggested that source credibility can be the determining factor which information users select if the reason to search for information is a more serious one and not entertainment, which may decrease the importance of technology cues such as website design significantly (Westerwick, 2013). All of these findings show that different factors may influence Internet content selection but that work is still required to disentangle which roles different technology cues in different use contexts really play. To guide these endeavors, it will be of great relevance to identify a strong theoretical framework. As of now, it is unclear how the selective exposure impacts of technology cues would differ from the impacts from cues to be found in traditional media, for example, imagery (Knobloch, Hastall, Zillmann, & Callison, 2003; Zillmann et al., 2001).

Consequences of Selective Exposure in the Context of New Communication Technologies

While the review of theories and findings given earlier focused on research on actual selective exposure patterns as they occur in use of communication technologies, a recent set of studies has taken this approach a step further by considering implications of selective media use. It could be argued that much media effects research has tackled the same overarching question, but the bulk of research has drawn on self-report measures to capture selective media use, which involves some methodological issues (see earlier text). In contrast, observing selective exposure in controlled settings together with capturing subsequent impacts offers valuable insights in the psychological processes at work in everyday media use situations.

Most research along these lines has been conducted in relation to the confirmation bias, with recent inclusion of the credibility concept. An analysis by Knobloch-Westerwick and Meng (2011) showed that selective exposure to political content strengthened students' political self-concept. Similar results emerged from data collected right before the 2008 presidential election from a nonstudent sample. Westerwick et al. (2013) were able to demonstrate the selective exposure to online search results affected attitudes depending on the credibility of the sources associated with search result message.

Additional research on selective exposure impacts pertains to mood management and social comparisons in the context of online and gaming use: For instance, Knobloch-Westerwick and Hastall (2010) found that older media users' self-esteem was boosted by selective reading of negative online news about younger individuals because it allowed for favorable comparisons with a relevant outgroup. Reinecke et al. (2012) found that mood repair indeed occurred through video gaming for which users could select the difficulty level.

In brief, a number of studies have been able to demonstrate the important implications of selective exposure for media effects. Some of the findings may come at no surprise and confirm classic ideas that were proposed fifty years ago along the lines of a "limited effects" paradigm in media research (Klapper, 1960). Other findings are truly innovative and break new ground—for example, Westerwick et al.'s (2013) study showed that if exposure to attitude-discrepant messages from high-credibility sources occurs, related attitudes are weakened.

The Future of Selective Exposure Research in the Communication Technology Context

Existing research demonstrated that selective exposure in the context of new communication technologies is affected by a confirmation bias, informational utility considerations, mood management motivations, and social comparison processes. Selective exposure research that examines cues that are specific for the new communication technologies context is still scarce. Future work could marry Sundar's (2008) MAIN model with the selective exposure paradigm (Knobloch-Westerwick, 2015) to address this gap.

Although important steps have been accomplished to advance our understanding of selective exposure and its implications in the current new media era, much remains to be done. The list of technological cues that could be examined regarding their influences on exposure is sheer endless. Future work will need to elaborate and extend theoretical frameworks in order to truly accelerate insight into new media use processes. Oftentimes, this theorizing will involve how offline social-psychological processes project upon communication technology use situations. For example, along the lines of Knobloch-Westerwick, Sharma, et al.'s (2005) study, technological cues may serve as indications of what other people (communicators or other audience members) think of media messages. Accordingly, perceptions regarding who contributed to those cues with what motivations should channel selective exposure during online media use. Further, technological cues may render certain social identities salient and/or create the impression of minority versus majority groups. Such social-psychological phenomena likely have immense impact on what content is attended to by users and subsequently influences them.

References

Arpan, L. M., & Tüzünkan, F. (2011). Photographic depiction of normative deviance and informational utility as predictors of protest news exposure, related perceptions, and story comprehension. *Mass Communication and Society, 14*(2), 178–195. doi:10.1080/15205431003650486

Atkin, C. K. (1973). Instrumental utilities and information seeking. In P. Clarke (Ed.), *New models for mass communication research* (pp. 205–242). Beverly Hills, CA: Sage.

Aubrey, J. S. (2006). Effects of sexually objectifying media on self-objectification and body surveillance in undergraduates: Results of a 2-year panel study. *Journal of Communication, 56*(2), 366–386. doi:10.1111/j.1460-2466.2006.00024.x

Becker, L. B., & Schoenbach, K. (Eds.). (1989). *Audience responses to media diversification: Coping with plenty.* Hillsdale, NJ: Erlbaum.

Bennett, W. L., & Iyengar, S. (2008). A new era of minimal effects? The changing foundations of political communication. *Journal of Communication, 58*(4), 707–731. doi:10.1111/j.1460-2466.2008.00410.x

Bowman, N. D., & Tamborini, R. (2012). Task demand and mood repair: The intervention potential of computer games. *New Media & Society, 14*, 1339–1357. doi:10.1177/1461444812450426

boyd, d. m., & Ellison, N. B. (2007). Social networking sites: Definition, history, and scholarship. *Journal of Computer-Mediated Communication, 13*(1), 210–230. doi:10.1111/j.1083-6101.2007.00393.x

Bryant, J., & Zillmann, D. (1984). Using television to alleviate boredom and stress: Selective exposure as a function of induced excitational states. *Journal of Broadcasting, 28*(1), 1–20. doi:10.1080/08838158409386511

Bushman, B. J., & Whitaker, J. L. (2010). Like a magnet: Catharsis beliefs attract angry people to violent video games. *Psychological Science, 21*(6), 790–792. doi:10.1177/0956797610369494

Cotton, J. L. (1985). Cognitive dissonance in selective exposure. In D. Zillmann & J. Bryant (Eds.), *Selective exposure to communication.* Mahwah, NJ: Erlbaum.

D'Alessio, D., & Allen, M. (2002). Selective exposure and dissonance after decisions. *Psychological Reports, 91*(2), 527–532. doi:10.2466/pr0.2002.91.2.527

December, J. (1996). Units of analysis for Internet communication. *Journal of Communication, 46*(1), 14–38. doi:10.1111/j.1460-2466.1996.tb01459.x

EEDAR. (n.d.) About EEDAR. Retrieved from http://www.eedar.com/About/Default.aspx

Festinger, L. (1954). A theory of social comparison processes. *Human Relations, 7*(2), 117–140. doi:10.1177/001872675400700202

Festinger, L. (1957). *A theory of cognitive dissonance.* Stanford, CA: Stanford University Press.

Fogg, B. J., Marshall, J., Laraki, O., Osipovich, A., Varma, C., Fang, N., ... Treinen, M. (2001). What makes Web sites credible? A report on a large quantitative study. In M. Beaudouin-Lafon & R. J. K. Jacob (Eds.), *Proceedings of ACM CHI 2001 Conference on Human Factors in Computing Systems, Seattle, Washington, USA* (pp. 61–68). New York, NY: ACM. doi:10.1145/365024.365037

Freedman, J. L., & Sears, D. O. (1965). Selective exposure. In L. Berkowitz (Ed.), *Advances in experimental social psychology* (Vol. 2, pp. 57–97). New York, NY: Academic Press. doi:10.1016/S0065-2601(08)60103-3

Frey, D. (1982). Different levels of cognitive dissonance, information seeking, and information avoidance. *Journal of Personality and Social Psychology, 43*(6), 1175–1183. doi:10.1037/0022-3514.43.6.1175

Garcia, M. R., & Stark, P. (1991). *Eyes on the news.* St. Petersburg, FL: Poynter Institute for Media Studies.

Garrett, R. K. (2009a). Echo chambers online? Politically motivated selective exposure among Internet news users. *Journal of Computer-Mediated Communication, 14*(2), 265–285. doi:10.1111/j.1083-6101.2009.01440.x

Garrett, R. K. (2009b). Politically motivated reinforcement seeking: Reframing the selective exposure debate. *Journal of Communication, 59*(4), 676–699. doi:10.1111/j.1460-2466.2009.01452.x

Graf, J., & Aday, S. (2008). Selective attention to online political information. *Journal of Broadcasting & Electronic Media, 52*(1), 86–100. doi:10.1080/08838150701820874

Haferkamp, N., & Krämer, N. C. (2011). Social comparison 2.0: Examine the effects of online profiles on social-networking sites. *Cyberpsychology, Behavior, and Social Networking, 14*(5), 309–314. doi:10.1089/cyber.2010.0120

Hart, W., Albarracín, D., Eagly, A. H., Brechan, I., Lindberg, M. J., & Merrill, L. (2009). Feeling validated versus being correct: A meta-analysis of selective exposure to information. *Psychological Bulletin, 135*(4), 555–588. doi:10.1037/a0015701

Hastall, M. R., & Knobloch-Westerwick, S. (2013a). "Caught in the act": Measuring selective exposure to experimental online stimuli. *Communication Methods and Measures, 7*(2), 94–105. doi:10.1080/19312458.2012.761190

Hastall, M. R., & Knobloch-Westerwick, S. (2013b). Severity, efficacy, and evidence type as determinants of health message exposure. *Health Communication, 28*(4), 378–388. doi:10.1080/10410236.2012.690175

Holbert, R. L., Garrett, R. K., & Gleason, L. S. (2010). A new era of minimal effects? A response to Bennett and Iyengar. *Journal of Communication, 60*(1), 15–34. doi:10.1111/j.1460-2466.2009.01470.x

Hovland, C. I. (1959). Reconciling conflicting results derived from experimental and survey studies of attitude change. *The American Psychologist, 14*(1), 8–17. doi:10.1037/h0042210

Hovland, C. I., & Weiss, W. (1951). The influence of source credibility on communication effectiveness. *Public Opinion Quarterly, 15*(4), 635–650. doi:10.1086/266350

Johnson, T. J., Zhang, W., & Bichard, S. L. (2011). Voices of convergence or conflict? A path analysis investigation of selective exposure to political websites. *Social Science Computer Review, 29*(4), 449–469. doi:10.1177/0894439310379962

Jun, N. (2012). Contribution of Internet news use to reducing the influence of selective online exposure on political diversity. *Computers in Human Behavior, 28*(4), 1450–1457. doi:10.1016/j.chb.2012.03.007

Kalyanaraman, S., & Sunder, S. S. (2006). The psychological appeal of personalized content in web portals: Does customization affect attitudes and behavior? *Journal of Communication, 56*(1), 110–132. doi:10.1111/j.1460-2466.2006.00006.x

Katz, E., Blumler, J. G., & Gurevitch, M. (1974). Utilization of mass communication by the individual. In J. G. Blumler & E. Katz (Eds.), *The uses of mass communications: Current perspectives on gratifications research* (pp. 19–32). Beverly Hills, CA: Sage.

Kim, Y. M. (2007). How intrinsic and extrinsic motivations interact in selectivity. *Communication Research, 34*(2), 185–211. doi:10.1177/0093650206298069

Kim, Y. M. (2008). Where is my issue? The influence of news coverage and personal issue importance on subsequent information selection on the web. *Journal of Broadcasting & Electronic Media, 52*(4), 600–621. doi:10.1080/08838150802437438

Klapper, J. T. (1960). *The effects of mass communication.* New York, NY: The Free Press.

Knobloch, S. (2002). 'Unterhaltungsslalom' bei der WWW-Nutzung: Ein Feldexperiment ['Entertainment slalom' while surfing the WWW: A field experiment]. *Publizistik, 47*(3), 309–318. doi:10.1007/s11616-002-0068-z

Knobloch, S. (2003). Mood adjustment via mass communication. *Journal of Communication, 53*(2), 233–250. doi:10.1111/j.1460-2466.2003.tb02588.x

Knobloch, S., Dillman Carpentier, F. R., & Zillmann, D. (2003). Effects of salience dimensions of informational utility on selective exposure to online news. *Journalism & Mass Communication Quarterly, 80*(1), 91–108. doi:10.1177/107769900308000107

Knobloch, S., Hastall, M., Zillmann, D., & Callison, C. (2003). Imagery effects on the selective reading of Internet newsmagazines. *Communication Research, 30*(1), 3–29. doi:10.1177/0093650202239023

Knobloch, S., & Zillmann, D. (2002). Mood management via the digital jukebox. *Journal of Communication, 52*(2), 351–366. doi:10.1111/j.1460-2466.2002.tb02549.x

Knobloch, S., & Zillmann, D. (2003). Appeal of love themes in popular music. *Psychological Reports, 93*(3), 653–658. doi:10.2466/pr0.2003.93.3.653

Knobloch-Westerwick, S. (2006). Mood management theory, evidence, and advancements. In J. Bryant & P. Vorderer (Eds.), *Psychology of entertainment* (pp. 239–254). Mahwah, NJ: Erlbaum.

Knobloch-Westerwick, S. (2007). Kognitive Dissonanz 'Revisited': Selektive Zuwendung zu einstellungskonsistenten und -inkonsistenten politischen Informationen [Cognitive dissonance revisited: Selective exposure to attitude-consistent and counter-attitudinal political information]. *Publizistik, 52*(1), 51–62. doi:10.1007/s11616-007-0005-2

Knobloch-Westerwick, S. (2008). Informational utility. In W. Donsbach (Ed.), *The international encyclopedia of communication* (pp. 2273–2276). Malden, MA: Blackwell Publishing.

Knobloch-Westerwick, S. (2014). Selection, perception, and processing of political messages. In C. Reinemann (Ed.), *Political communication* (Vol. 18 of Handbook of Communication Sciences) (pp. 507–526). Mouton de Gruyter, Berlin.

Knobloch-Westerwick, S. (2015). *Choice and preference in media use: Advances in selective exposure theory and research.* New York, NY: Routledge.

Knobloch-Westerwick, S., & Alter, S. (2006). Mood adjustment to social situations through mass media use: How men ruminate and women dissipate angry moods. *Human Communication Research, 32*(1), 58–73. doi:10.1111/j.1468-2958.2006.00003.x

Knobloch-Westerwick, S., & Alter, S. (2007). The gender news use divide: Americans' sex-typed selective exposure to online news topics. *Journal of Communication, 57*(4), 739–758. doi:10.1111/j.1460-2466.2007.00366.x

Knobloch-Westerwick, S., Appiah, O., & Alter, S. (2008). News selection patterns as a function of race: The discerning minority and the indiscriminating majority. *Media Psychology, 11*(3), 400–417. doi:10.1080/15213260802178542

Knobloch-Westerwick, S., Brück, J., & Hastall, M. R. (2006). The gender news use divide: Impacts of sex, gender, self-esteem, achievement, and affiliation motive on German newsreaders' exposure to news topics. *Communications, 31*(3), 329–345. doi:10.1515/COMMUN.2006.021

Knobloch-Westerwick, S., Dillman Carpentier, F. R., Blumhoff, A., & Nickel, N. (2005). Informational utility effects on selective exposure to good and bad news: A cross-cultural investigation. *Journalism & Mass Communication Quarterly, 82*(1), 181–195. doi:10.1177/107769900508200112

Knobloch-Westerwick, S., & Hastall, M. R. (2006). Social comparisons with news personae: Selective exposure to news portrayals of same-sex and same-age characters. *Communication Research, 33*(4), 262–284. doi:10.1177/0093650206289152

Knobloch-Westerwick, S., & Hastall, M. R. (2010). Please your self: Social identity effects on selective exposure to news about in- and out-groups. *Journal of Communication, 60*(3), 515–535. doi:10.1111/j.1460-2466.2010.01495.x

Knobloch-Westerwick, S., Hastall, M. R., Grimmer, D., & Brück, J. (2005). "Informational Utility": Der Einfluss der Selbstwirksamkeit auf die selektive Zuwendung zu Nachrichten. *Publizistik, 50*(4), 462–474. doi:10.1007/s11616-005-0144-2

Knobloch-Westerwick, S., & Johnson, B. K. (2014). Selective exposure for better or worse: Its mediating role for online news' impact on political participation. *Journal of Computer-Mediated Communication, 19*, 184–196.

Knobloch-Westerwick, S., Johnson, B. K., & Westerwick, A. (2013). To your health: Self-regulation of health behavior through selective exposure to online health messages. *Journal of Communication, 63*, 807–829.

Knobloch-Westerwick, S., & Kleinman, S. B. (2012). Preelection selective exposure: Confirmation bias versus informational utility. *Communication Research, 39*(2), 170–193. doi:10.1177/0093650211400597

Knobloch-Westerwick, S., & Meng, J. (2009). Looking the other way: Selective exposure to attitude-consistent and counterattitudinal political information. *Communication Research, 36*(3), 426–448. doi:10.1177/0093650209333030

Knobloch-Westerwick, S., & Meng, J. (2011). Reinforcement of the political self through selective exposure to political messages. *Journal of Communication, 61*(2), 349–368. doi:10.1111/j.1460-2466.2011.01543.x

Knobloch-Westerwick, S., & Romero, J. P. (2011). Body ideals in the media: Perceived attainability and social comparison choices. *Media Psychology, 14*(1), 27–48. doi:10.1080/15213269.2010.547833

Knobloch-Westerwick, S., Sharma, N., Hansen, D. L., & Alter, S. (2005). Impact of popularity indications on readers' selective exposure to online news. *Journal of Broadcasting & Electronic Media, 49*(3), 296–313. doi:10.1207/s15506878jobem4903_3

Knobloch-Westerwick, S., & Westerwick, A. (2011). Social comparisons at your fingertips: The importance of ingroup/outgroup status. *Proceedings of the 3rd International ACM/ICA Web Science Conference, Koblenz, Germany.* New York, NY: ACM. doi: 10.1145/2527031.2527036

Lazarsfeld, P. F., Berelson, B., & Gaudet, H. (1944). *The people's choice: How the voter makes up his mind in a presidential campaign.* New York, NY: Columbia University Press.

Lockwood, P., & Kunda, Z. (1997). Superstars and me: Predicting the impact of role models on the self. *Journal of Personality and Social Psychology, 73*(1), 91–103. doi:10.1037/0022-3514.73.1.91

Lohr, S. (2012, February 29). For impatient web users, an eye blink is just too long to wait. *New York Times*, p. A1.

Madden, M., & Zickuhr, K. (2011, August 26). *65% of online adults use social networking sites.* Pew Internet and American Life Project. Retrieved from http://www.pewinternet.org/Reports/2011/Social-Networking-Sites.aspx

Mager, A. (2009). Mediated health: Sociotechnical practices of providing and using online health information. *New Media & Society, 11*(7), 1123–1142. doi:10.1177/1461444809341700

Mastro, D. E., Eastin, M. S., & Tamborini, R. (2002). Internet search behaviors and mood alterations: A selective exposure approach. *Media Psychology, 4*(2), 157–172. doi:10.1207/S153278XMEP0402_03

McGuire, W. J. (1989). Theoretical foundations of campaigns. In R. E. Rice & C. K. Atkin (Eds.), *Public communication campaigns* (2nd ed., pp. 43–65). Newbury Park, CA: Sage.

Metzger, M. J., Flanagin, A. J., & Medders R. B. (2010). Social and heuristic approaches to credibility evaluation online. *Journal of Communication, 60*(3), 413–439. doi:10.1111/j.1460-2466.2010.01488.x

Munson, S. A., & Resnick, P. (2010). Presenting diverse political opinions: How and how much. In E. D. Mynatt et al. (Eds.), *Proceedings of the 28th International Conference on Human Factors in Computing Systems, Atlanta, Georgia, USA*. New York, NY: Association for Computing Machinery. doi:10.1145/1753326.1753543

Munson, S. A., Zhou, D. X., & Resnick, P. (2009). Designing interfaces for presentation of opinion diversity. In D. R. Olsen et al. (Eds.), *Proceedings of the 27th International Conference Extended Abstracts on Human Factors in Computing Systems, Boston, MA, USA*. New York, NY: Association for Computing Machinery. doi:10.1145/1520340.1520552

Pan, B., Hembrooke, H., Joachims, T., Lorigo, L., Gay, G., & Granka, L. (2007). In Google we trust: Users' decisions on rank, position, and relevance. *Journal of Computer-Mediated Communication, 12*(3), 801–823. doi:10.1111/j.1083-6101.2007.00351.x

Reinecke, L., Tamborini, R., Grizzard, M., Lewis, R., Eden, A., & Bowman, N. D. (2012). Characterizing mood management as need satisfaction: The effects of intrinsic needs on selective exposure and mood repair. *Journal of Communication, 62*(3), 437–453. doi:10.1111/j.1460-2466.2012.01649.x

Reinecke, L., & Trepte, S. (2008). In a working mood? The effects of mood management processes on subsequent cognitive performance. *Journal of Media Psychology: Theories, Methods, and Applications, 20*(1), 3–14. doi:10.1027/1864-1105.20.1.3

Rossmann, R. (2006). *Selective-exposure-designs in der Zappingforschung [Selective exposure designs in zapping research]*. Munich, Germany: Dr. Hut.

Sears, D. O., & Freedman, J. L. (1967). Selective exposure to information: A critical review. *Public Opinion Quarterly, 31*(2), 194–213. doi:10.1086/267513

Segrin, C., & Nabi, R. L. (2002). Does television viewing cultivate unrealistic expectations about marriage? *Journal of Communication, 52*(2), 247–263. doi:10.1111/j.1460-2466.2002.tb02543.x

Shon, J., Marshall, J., & Musen, M. A. (2000). The impact of displayed awards on the credibility retention of Web site information. *Proceedings of AMIA Symposium 2000* (pp. 794–798). American Medical Informatics Association, Bethesda, MD.

Sundar, S. S. (2008). The MAIN model: A heuristic approach to understanding technology effects on credibility. In M. Metzger & A. Flanagin (Eds.), *Digital media, youth, and credibility* (pp. 73–100). The John D. and Catherine T. McArthur foundation series on digital media and learning. Cambridge, MA: The MIT Press.

Sundar, S. S., Knobloch-Westerwick, S., & Hastall, M. R. (2007). News cues: Information scent and cognitive heuristics. *Journal of the American Society for Information Science and Technology, 58*(3), 366–378. doi:10.1002/asi.20511

Sunstein, C. R. (2001). *Republic.com*. Princeton, NJ: Princeton University Press.

Tewksbury, D. (2003). What do Americans really want to know? Tracking the behavior of news readers on the Internet. *Journal of Communication, 53*(4), 694–710. doi:10.1111/j.1460-2466.2003.tb02918.x

Valentino, N. A., Banks, A. J., Hutchings, V. L., & Davis, A. K. (2009). Selective exposure in the Internet age: The interaction between anxiety and informational utility. *Political Psychology, 30*(4), 591–613. doi:10.1111/j.1467-9221.2009.00716.x

Vorderer, P. (2001). It's all entertainment—sure. But what exactly is entertainment? Communication research, media psychology, and the explanation of entertainment experience. *Poetics, 29*(4–5), 247–261. doi:10.1016/S0304-422X(01)00037-7

Wathen, C. N., & Burkell, J. (2002). Believe it or not: Factors influencing credibility on the Web. *Journal of the American Society of Information Science and Technology, 53*(2), 134–144. doi:10.1002/asi.10016

Webster, J. G. (1998). The audience. *Journal of Broadcasting & Electronic Media, 42*(2), 190–207. doi:10.1080/08838159809364443

Westerwick, A. (2013). Effects of sponsorship, Web site design, and Google ranking on the credibility of online information. *Journal of Computer-Mediated Communication, 18*(2), 80–97. doi:10.1111/jcc4.12006

Westerwick, A., Kleinman, S., & Knobloch-Westerwick, S. (2013). Turn a blind eye if you care: Impacts of attitude consistency, importance, and credibility on seeking of political information and implications for attitude. *Journal of Communication, 63*, 432–453. doi:10.1111/jcom.12028

Wills, T. A. (1981). Downward comparison principles in social psychology. *Psychological Bulletin, 90*(2), 245–271. doi:10.1037/0033-2909.90.2.245

Wilson, R. E., Gosling, S. D., & Graham, L. T. (2012). A review of Facebook research in the social sciences. *Perspectives on Psychological Science, 7*(3), 203–220. doi:10.1177/1745691612442904

Wilson, E. J., & Sherrell, D. L. (1993). Source effects in communication and persuasion research: A meta-analysis of effect size. *Journal of the Academy of Marketing Science, 21*(2), 101–112. doi:10.1007/BF02894421

Winter, S., & Krämer, N. C. (2012). Selecting science information in Web 2.0: How source cues, message sidedness, and need for cognition influence users' exposure to blog posts. *Journal of Computer-Mediated Communication, 18*(1), 80–96. doi:10.1111/j.1083-6101.2012.01596.x

Winter, S., Krämer, N. C., Appel, J., & Schielke, K. (2011, May). *A question of expertise: How heuristic source cues influence selective exposure to online science information.* Paper presented at the 61st Annual Conference of the International Communication Association, Boston, MA.

Zillmann, D. (1985). The experimental exploration of gratifications from media entertainment. In K. E. Rosengren, L. A. Wenner, & P. Palmgreen (Eds.), *Media gratifications research: Current perspectives* (pp. 225–305). Beverly Hills, CA: Sage.

Zillmann, D. (1988). Mood management through communication choices. *American Behavioral Scientist, 31*(3), 327–340. doi:10.1177/000276488031003005

Zillmann, D., & Bryant, J. (Eds.). (1985). *Selective exposure to communication.* Mahwah, NJ: Erlbaum.

Zillmann, D., Chen, L., Knobloch, S., & Callison, C. (2004). Effects of lead framing on selective exposure to Internet news reports. *Communication Research, 31*(1), 58–81. doi:10.1177/0093650203260201

Zillmann, D., Knobloch, S., & Yu, H. (2001). Effects of photographs on the selective reading of news reports. *Media Psychology, 3*(4), 301–324. doi:10.1207/S1532785XMEP0304_01

19

Affording Control
How Customization, Interactivity, and Navigability Affect Psychological Responses to Technology

Sriram "Sri" Kalyanaraman[1] and Bartosz W. Wojdynski[2]

[1] *University of Florida, Gainesville, FL, USA*
[2] *University of Georgia, Athens, GA, USA*

In the two decades since Reeves and Nass' (1996) program of research on Social Responses to Communication Technology (SRCT) laid the edifice for systematic investigation on the psychology of new communication technologies, digital information has become a ubiquitous and device-independent currency, placing a premium on being able to efficaciously locate, disseminate, and consume information. Research across disciplines has examined several elements endemic to new technologies including their influence on users, the properties of emerging (and newer) media platforms, the nuances of specific venues, and the processes governing the interplay between—and among—the technology, content, and the user. While scholarship in "new media" is abundant, such prolificity has not been accompanied by concomitant theoretical progress. We propose that three elements—customization, interactivity, and navigability—are especially salient in charting conceptual growth in psychological research on emerging communication technologies for several reasons. First, these elements have elicited enduring interdisciplinary interest. Second, these elements are laden with theoretical richness that can subsume diverse technological platforms. Third, they have the potential to lend themselves to isomorphic operationalizations. Fourth, users are able to psychologically discriminate between and among different levels of each of these elements. Finally, these elements are best equipped to inspire ecumenical theoretical advancement by their inclusion in integrated models, as has been proposed by a small albeit influential group of researchers (see Sundar, Jia, Waddell, & Huang, 2015).

In this chapter, we attempt to make several contributions. First, while acknowledging the existing zeitgeist, we propose the essential conceptual underpinnings of each of the three elements. In particular, we recommend the consideration of those conceptualizations that can be applied across an array of media platforms and environments. Next, we offer clear illustrations of how the conceptual value of these

The Handbook of the Psychology of Communication Technology, First Edition. Edited by S. Shyam Sundar.

attributes can be harnessed operationally, both in isolation and in concert. Finally, we attempt to maintain the vision of this edited literary enterprise by offering our perspectives on future, programmatic research centered on these variables. We showcase the utility of these three elements for examination in concert, while proposing the importance of key parameters to inform the discernment of cognitive, affective, and behavioral outcomes to encourage theoretical advancement.

Customization

In the last decade or so, the concept of customization has occupied an enviable position in the technological firmament. Perhaps more so than any other defining feature of new media technologies, customization has attracted scholarly attention as well as implementation across almost every Web-based venue or application. As even a cursory inspection of the Web suggests, the range of online venues offering customized options is dizzying: users can opt to receive news catered to their interests; they can receive customized product recommendations; online learners can choose to pursue customized modules to enhance their educational experience; users searching for information can receive search results that are customized; videogame players can choose to customize their avatar appearance. Indeed, it would not be inaccurate to proclaim that customization—to varying degrees—is a ubiquitous feature of modern media environments.

Not surprisingly, the literature on customization is vast and varied, with scholars from several disciplines devoting attention to obtain a better understanding of the concept. These include researchers in communication, information science, marketing, psychology, and public health, among others. Although the nomenclatures that scholars have adopted reflect disciplinary allegiances to some extent (and, in some cases, disciplinary discord as well), the fundamental assumption is strikingly similar: that the labels of customization, personalization, tailoring, or matching essentially reflect the practice of designing messages based on a consideration of individual desires and wants (see Kalyanaraman & Sundar, 2006). In such a scenario, traditional notions of mass audiences are obliterated as the message is intended solely for a so-called "audience of one."

The methodology of studies exploring the psychology of customization has generally centered on creating content according to inimitable tastes and preferences based on participants' responses to a questionnaire. In examining the outcomes of customized messages, scholars have focused on typical decision-making variables such as attitudes, cognitions, and intentions (and, in rarer circumstances, behavioral consequences). Among these outcomes, findings pertaining to attitudes have been remarkably consistent: across studies and venues, empirical evidence has corroborated conceptual expectation by demonstrating that higher levels of customization engender positive attitudes. Under such circumstances, to sustain conceptual richness and avoid the dangers of saturation, what should programmatic scholarship on customization include? We believe the answer to this question lies in a fundamental consideration of customization as reflecting the self and a customized message as embodying the essence of matching the message to some aspect of the self (see Petty, Barden, & Wheeler, 2002).

The consideration of the self as a critical edifice of customization opens up intriguing vistas for exploration. Related to the agency of the self, for example, an obvious question that crops up is whether individuals exhibit different responses when exposed to customized options created based on their stated preferences vis-à-vis an experimental scenario wherein they are able to create such options themselves (based on their preferences)? The latter experience may be more compelling as it falls under the rubric of the "I-designed-it-myself" effect (Franke, Schreier, & Kaiser, 2010; Sundar & Marathe, 2010) while also kindling questions about cognitive demands imposed on the user in the process of active customization. For instance, preliminary data on agency effects in customization (i.e., examining differences between users who customize their own websites versus those users who have the sites customized for them) offer telling insights (Kalyanaraman, 2013). While self-customization procedures are perceived as tedious with resulting lukewarm perceptions of the (customized) interface, users have a healthy appreciation of interface-customized content. However, over time, this pattern undergoes substantial alteration: at subsequent time periods, the former group appears to demonstrate an increasing liking for the interface (that they customize), whereas the latter group appears to suffer from such habituation (while showing either indistinguishable—or decreased—liking over time). Also, users in the self-customization condition perceive increased control over their experience and even ascribe higher monetary worth to their (customized) content, relative to their counterparts in the interface-customization condition.

As an example of another question that warrants scrutiny consider a long-held assumption of customization: that messages tailored according to individual preferences will *always* be seen as more desirable (compared to less-customized content). This assumption, while having received consistent empirical support, prompts closer investigation based on the specific consideration of the self. With most of the evidence on the psychology of customization based on data obtained with Western participants who espouse the individualist self (i.e., celebrating the idea of independence), it is not surprising that highly customized messages would be appreciated by such participants. On the other hand, participants who signify a collectivist orientation (e.g., people from Eastern societies) are typically uncomfortable with situations that celebrate individual uniqueness and instead prefer the notion of the interdependent self. For these participants, then, messages that target them as part of a representative community, rather than as unique individuals, will likely exert greater persuasive effects. Indeed, this supposition has been examined experimentally and has received clear support (Li & Kalyanaraman, 2013).

The role of the self is also an important dimension in exploring how consumers respond to messages that are customized according to inherent personality traits. For instance, a recent study found that participants responded more favorably to advertisements customized according to their scores on the Big Five personality index. Specifically, participants reported a preference for ads that accentuated a motivational element associated with each personality trait when that element corresponded with their own personality type (Hirsh, Kang, & Bodenhausen, 2012). Personality variables can also offer further insights into how users process customized messages. Kalyanaraman, Oliver, and Magee (2010) explored the role of individuals' predisposition to engage in varying levels of cognitive activity, as measured by need for cognition (NFC; Cacioppo, Petty, & Kao, 1984), and found that high-NFC individuals appeared to treat customization as an argument, whereas low-NFC individuals tend to treat it as a cue. This finding offers some preliminary support to the possibility that customization might fulfill the tenets of the multiple-role hypothesis, which states that the same variable can function both as an argument and a cue under different circumstances (Petty et al., 2002).

Although the literature on the self enjoys sacred status in terms of the sheer volume of scholarship surrounding it, we believe that two considerations of the self are particularly relevant to scholars interested in pursuing a research agenda on the psychosocial effects of customization. Baumeister's (1998) landmark thesis on the self serves as the benchmark for extracting the most important tenets of the self and delineating them to the customization realm. In the interest of parsimony, we propose that the notion of reflexive consciousness and the allied characteristics of self-referencing, self-awareness, self-knowledge are worthy of appraisal. First, reflexive consciousness presumes a requisite human proclivity to be cognizant of experiences and events that are part of our environment, and to then have the ability to deliberate on those experiences to construct a thorough understanding of oneself (Baumeister, 1998). While reflexive consciousness is believed to be an amalgam of both "self" and "other" awareness, we suggest that, for clarity of purpose in communication research, the external media environment constitute the "other." Hence, an awareness of one's customized self would also presumably be accompanied by an awareness of the external environment. In addition to self-awareness, self-referencing and self-knowledge are also pertinent. The former points to increased cognitive activity in the face of exposure to customized messages with its supposition that content relevant to the self would be scrutinized and processed in greater depth (Baumeister, 1998; see the original postulation by Rogers, Kuiper, & Kirker, 1977). Self-knowledge posits that humans construct a "self-digest" (Higgins, 1996) based on an assimilation of knowledge about their selves and that is then used to situate their relationship with the larger environment surrounding them. The ways in which such perceptions or relationships are constructed depend on whether people adopt an appraisal motive, a self-enhancement motive, or a consistency motive (Baumeister, 1998; see also Chen & Chaiken, 1999). Extrapolating our fundamental understanding of reflexive consciousness, self-awareness, self-referencing, and self-knowledge, users exposed to customized messages would exhibit cognitive and affective responses. Furthermore, such psychological manifestations would not just be restricted to the external (media) environment but also include the evaluation of those parameters that inform the make-up of the individual as well as the role of the individual in that mediated environment. Programmatically, these considerations imply that research on individuals' exposure to customized content could also be examined under the aegis of what we label as the "selfless" self and the "selfish" self. The types of questions that would fit under the former relate to the possibility that customization can compel an individual to seek greater meaning ("beyond myself"), enable efforts for establishing connectedness, enhance compassion, and inculcate sharing behaviors. The selfish self, in contrast, would relate to the relationship between customization and perceptions of narcissism and materialism while also providing a framework to validate warnings of polarized communities (Sunstein, 2009) and filer bubbles (Pariser, 2011).

Another consideration of the interaction between individuals and customized media environments relies on broadening the framework of the extended self. In his seminal article, Belk (1988) proposed the thesis that humans regard their possessions as parts of themselves—the idea that such possessions are an extension of the self. In forwarding his arguments, Belk drew upon William James' (1890) conception of the self: "a man's Self is the sum total of ALL that he can call his..." (pp. 291–292). Belk proposed that the extended self includes categories such as body parts, internal processes (e.g., the conscience), ideas, and experiences, among others. In the context of immersive virtual environments (IVEs), for example, the ability to sensorially experience a "real" simulation (see Blascovich & Bailenson, 2011) should lead to the entire process being perceived as part of one's extended self. Belk also drew upon ways in which possessions can be assimilated into the extended self. Here, he turns to Sartre (1943), who proposed that individuals can incorporate objects into their self in one of three ways: by "appropriating or controlling an object for personal use; by creating an object; or by knowing an object" (see Belk, 1988, pp. 150–151). These methods would be applicable in the context of media technologies such as personal digital assistants (e.g., smartphones) and IVEs as they allow users to not only create and disseminate information, but also offer such features that enhance control and allow users to comprehend the technology. As Beaglehole (1932) suggested, a personal mastery of objects (e.g., media) not only makes them "ours" but also becomes part of the self. Thus, a smartphone or IVE will be seen as an extension of an individual's self and lead to enhanced knowledge (about oneself), which could presumably result in increased self-referencing, and ultimately result in more informed decision making.

We suggest that the ability to perceive one's extended self is important in discerning outcomes. Indeed, such ability is likely to enhance the sense of control that users have over their digital environment and result in greater knowledge of oneself, especially in relation to specific issues. Virtual doppelgangers, for instance, can lead to more acute perceptions and awareness of users' selves (Kalyanaraman, Penn, Ivory, & Judge, 2010). The projection of individuals' virtual selves can be valuable to explore if it can lead to improved decision making, especially in those situations where traditional perspective-taking scenarios may not be as effective. For instance, a smoker's extended virtual self may include bodily changes that are graphically observable, compared to an imagined scenario of what the harms of smoking could be. Such discernment has promise for intended consequences (e.g., cessation).

In summary, we believe that a sense of control and increased knowledge can be magnified by customizing virtual doppelgangers. Theoretically, an enhanced sense of control is likely to be realized when the customized or personalized self is based on dispositional, enduring preferences, rather than transient, situational preferences. Such dispositional matching—what we label as the "salient self"—is likely to result in greater control over the surrounding mediated environment. Of greater import, of course, is the type of outcome that can be examined as a result of increased control. That is, enhanced user control by itself would not be perceived as a crucial ingredient unless it could explain changes in attitudes, cognitions, and behaviors. As some emerging research has shown, enhanced user control can influence behavioral manifestations (Kalyanaraman, 2013). Future research would benefit by not only testing the various dispositions that would lend themselves to effective message tailoring, but also examine how they influence attitudes and behaviors in a variety of situations, and over time.

Interactivity

Among several technological elements that have been examined in experimental studies of communication technology, arguably, no other variable has received as much attention as has interactivity. Historically, while there have been several attempts to establish a typology of interactivity (or classification of interactive systems), many of the existing conceptualizations suffer from reification due to empirical lacunae, which has impeded discernment of ***psychological*** meaning. Indeed, with the exception of a few labs (see Sundar, 2007), the shortcomings in existing interactivity research may be attributed to an excess of attention being showered on identification of features that result in a system (or interface) being classified as "interactive" while only allowing for perfunctory investigation of its psychological import.

We propose that for the potential of interactivity to be harnessed, communication scholars need to adopt a programmatic mindset of inquiry, especially from an experimental perspective. Such adoption will not only facilitate unraveling of deeper, underlying constructs, but also provide unambiguous evidence for the futility (or usefulness) of merely classifying a system as "interactive." For instance, we may find through experimentation that greater (lesser) use of system (user) activity does not enhance perceptions of interactivity. Then, it forces us to go back to the drawing board and assume one of three roles: (a) either reject the conceptually accepted, but empirically disproven supposition that "interactivity is the ratio of system activity to user activity" (see Paisley, 1983, p. 155); (b) modify the existing definition based on empirical substantiation and test the new meaning empirically; or (c) test the existing supposition in diverse venues and situations before final pronouncement. Irrespective of the specific choice that one may accept, such a methodical approach would rid the current terrain of numerous, ill-defined, conceptualizations. Over time, it may lead to organization of terminology that has withstood the rigors of empirical inquiry.

At a fundamental level, we agree with other scholars who have considered interactivity as a process-oriented variable (Ariely, 2000; Sundar, Xu, & Bellur, 2010). To stimulate scholarship in terms of processes and effects, the psychological import of interactivity can be best gleaned by contemplating it in terms of adaptive engagement. Adaptive, because it delineates that the environment must provide the capacity of modification and control to interactants. Engagement, because it stipulates the importance of a meaningful dialog that must essentially exist between either two (or more) users or between user and interface. Adapativity and engagement can be closely intertwined in a process and have the capacity to exert mutual influence. That is, if users believe that the interface is not being adaptive enough (i.e., not allowing the user to perform the task that he/she set out to accomplish), it may lead to the realization that they are unable to engage in a meaningful dialogue with the interface. The final result, then, could be one where the entire process of accomplishing or evaluating a task is rated as dissatisfactory. These facets holds great promise to capture the quintessence of interactivity in terms of information control.

The interactivity-as-information control perspective is particularly useful as it coalesces facets of pre-eminent viewpoints of interactivity—those that are conceptually grounded and have been examined in experimental investigations. Two species of interactivity that have received considerable attention are those that have considered interactivity as a feature of the medium (functional view) or as a message feature (contingency view). Essentially, the functional view regards interactivity as a feature of the medium, with perceptions of interactivity being realized by the number of "functional" features (graphics, audio/video features, downloadable options, etc.) that are present in the interface. The contingency view, on the other hand, considers interactivity to be a message feature in that perceptions of interactivity are realized depending on how well future messages relate to—or are contingent upon—present and past messages (see Sundar, Kalyanaraman, & Brown, 2003, for a detailed review of the distinction between the two views).

The most compelling conceptualization of interactivity as an affordance of a medium treats interactivity as a measure of the degree to which the user can exercise control over the content offered by that medium (Kalyanaraman & Sundar, 2008; Teo, Oh, Liu, & Wei, 2003). Steuer's (1992, p. 14) definition of interactivity, "the extent to which users can participate in modifying the form and content of a medium in real time," adds the component of near-instant system response, or "active" control, which has been shown to influence subsequent decision making (Ariely, 2000). The idea that interface elements with which the user can engage through a medium's input devices (i.e., clicking, hapticity, etc.) are a form of information control has been at the bedrock of a number of studies that have created varying operational levels of interactivity (Teo et al., 2003). Consistent with this view, Sundar et al. use the term "modality interactivity" to refer to "interface tools that afford users greater activity, resulting in greater depth and breadth of mentally representing and experiencing mediated content" (2010, p. 2248). Some studies have also claimed to manipulate an interface's interactivity by concurrently adding features that promote more effective communication (Liu & Shrum, 2009; Tao & Bucy, 2007). We believe that varying the presence of features is a valuable component of operationalizing interactivity to the extent that the features allow the user to manipulate some aspect of information presentation.

The interactivity-as-information-control approach is also valuable in that it subsumes central elements of message contingency (Rafaeli, 1988; Sundar et al., 2003). The contingency view has conceptual appeal in its intuitive mirroring of nonmediated interpersonal interaction. It has been effectively used to create two or three levels of stimuli that lead to variance in both perception of interactivity and some expected outcomes of interactivity (Sundar et al., 2003; Sundar & Kim, 2005). However, this approach scales up somewhat ineffectively to interfaces that offer a variety of opportunities for users to send input to a system and receive input back, or that may send input beyond three message exchanges. With the Web's increasing reliance on dynamically served content, the potential conversation loops between users and systems may be perpetual. Message contingency is as much a function of the content being communicated as of the technology that affords such communication.

The degree to which digital messages can control media is the product of two components: first, the means of information input available to the user, and second, the aspects of message content that vary as a result of this input. Input modality can vary along a number of dimensions, including hardware (e.g., mouse, touchpad) and natural mapping of devices (e.g., radio button, slider). The aspects of message content that can be can controlled also vary along several lines, among them granularity from binary elements (captions on vs. off) to continuous elements (volume, playhead position), and relevance to content. Conceptually defining interactivity as control over information leaves sizable room for discussion on how to operationalize ordinal levels of control. One approach to arranging aspects of information control on a continuum ranging from less interactive to more interactive would be to adopt a compelling framework proposed by Kristof and Satran (1995). Rather than simply increasing the number of features present, Kristof and Satran's hierarchy of interactivity in Web design proposes that interactivity increases as users gain control over additional aspects of the information, ranging from control only over pace at the low end to control over a simulation at the high end. Such an approach positions the level of control afforded by a medium as the central defining feature of interactivity. A growing number of experimental studies that have operationalized interactivity according to these principles have shown that varying levels of information control affects attitudes and cognitions (Kalyanaraman, Ito, Malik, & Ferris, 2009; Teo et al., 2003; Wojdynski & Kalyanaraman, 2012). These studies have generally manipulated the ***number of dimensions*** of the content over which control is afforded. For instance, both Teo et al. (2003) and Wojdynski and Kalyanaraman (2012) created three orthogonal dimensions of interactivity by adding features to each successive level that allow the user to vary additional aspects of the content. At the lowest levels, users had control over one or two dimensions of content presentation, pace, and/or sequence. At the highest, users had control over (between) five and seven dimensions of content, including control over media format, and control over variables that govern what content is displayed, and how. In both studies, the authors added specific technological features (e.g., a search engine, chat room) to manipulate some of the control levels. We believe that future applications of this approach would do well to move beyond solely manipulating the number of dimensions, and to incorporate both variance in the granularity of control afforded and the number of unique opportunities to control each dimension.

The merits of an empirical approach that situates interactivity within affordances of control make this a worthy approach for continued examination of interactivity's effects. First, this approach treats interactivity as an aspect of the technology, rather than relying on effect-based variable definitions (O'Keefe, 2003), and allows the examination of specific message manipulations on perceptions, the latter of which may or may not mediate additional effects on outcomes. Second, rigorous operationalization can identify the salient dimensions or domains of content that can inform knowledge of *how* interactivity can be most effectively deployed by developers in the pursuit of outcomes, such as maximizing efficiency, learning, or persuasion.

Besides methodological precision, the consideration of interactivity in terms of information control also provides cogent theoretical structure for understanding the effects of interactivity. The notion that humans have an innate motivation to control the environment around them is well established (e.g., Heider, 1958; Rothbaum, Weisz, & Snyder, 1982; Seligman, 1975), and desire to control content viewed has been shown as a strong factor underlying the use of the Internet over other media (Korgaonkar & Wolin, 1999). Users of more interactive media may benefit both from a *feeling* of control, and from the consequences of that control in minimizing the effort needed to achieve a task (Dholakia, Miao, Dholakia, & Fortin, 2000). We believe that for psychological research on interactivity to flourish, those interface features that offer control need to be examined thoroughly to determine optimum combinations. Such careful experimentation may also reveal threshold effects and allow us to adopt operationalizations that reflect theoretical distinction, rather than manipulations based on trial-and-error iterations. A second avenue that we recommend for future research on interactivity is the examination of varying levels of content. Presumably, users would desire greater control when faced with a wide array of choices as the cognitive constraints that would have to be overcome would be formidable (relative to fewer choice sets). The interplay between information control and magnitude of content could be especially important, given the entropic nature of human-information seeking. The detection of time periods when users seek greater control over the information at their disposal and the specific technological elements that they deploy to exert such control would shed further light on interactivity.

Navigability

Both explicitly and indirectly, navigability has since the early days of hypertext been a central concept in theories of communication technology. Navigability is a term usually used to describe the quality of the site structure and presentation with respect to their facilitation of information access. As Balakrishnan and Sundar (2011) pointed out, the study of how individuals locate information in digital media is shaped by differences between individuals and differences between interfaces, and "navigability" is used to refer to the variable characteristics of the interface component. Interfaces that permit information access can be seen as more navigable, while those that obfuscate the content available, convey little evaluable information about the content, or impede access are less navigable. Although there is no widely agreed-upon definition of navigability, the definitions that are present in trade and scholarly definitions converge around the ease of access to information (see Fang, Chau, Hu, Yang, & Sheng, 2006; Palmer, 2002; Zhang, Zhu, & Greenwood, 2004). These definitions generally converge in their emphasis on the layout, underlying structure, and qualities of hyperlinked elements, all of which are components of interface design that vary widely, and are applicable beyond websites to a broad range of digital media, content, and devices. Thus, we propose a definition of navigability as the degree to which the design of interface elements facilitates ease and efficiency of locating of information.

An inspection of the existing scholarship on navigability reveals that the most popular approach has assessed navigability as a function of quantifying hyperlinks (Fang et al., 2006). While this approach has provided a foundation, it does not quite account for the actual use of the site, and as such is of limited value in understanding effects. Balakrishnan and Sundar (2011) offered an explication of navigability rooted in the role of technology characteristics that contribute to spatial presence in virtual environments. They focused on two components: a "travel" component based on locomotion in virtual spaces, and a "wayfinding" component that refers to cognitive processes associated with goal-directed behaviors, specifically information seeking, wayfinding, and attaining and preserving special knowledge (see Jul, 2004).

In a recent paper (Wojdynski & Kalyanaraman, 2014), we outlined an explication of navigability focused primarily on elements related to wayfinding, which breaks the construct down into three functions that effective navigational elements fulfill. By focusing on informational and not simulation environments, this conceptualization pinpoints elements that help users orient themselves with regard to informational organization rather than the literal representation of space. The three dimensions are discussed in greater detail later, along with applications to a broader scope of media.

Logic of structure refers to the degree to which certain information may be accessed from others. Information itself may have pre-existing geospatial, taxonomic, or hierarchical attributes. Users of information systems bring their understanding of those relationships to bear, and their expectations based on these organizational structure influence the way they process a novel structural scheme. Sites that convey a high logic of structure are designed so that with each step down a navigational path, users can narrow the scope of the visible information to a subset that is more relevant to their current task.

Some common means of organizing the structure of a site rely on alphabet, locations, time, continuums, numbers, categories, or randomness (Shedroff, 1999); the appropriateness of one over the other depends on the information being organized. It should be less taxing for users to select their state from a list—or map—of all 50 states than to choose their physician from a list of 50 doctors. User familiarity with the content domain and pre-existing schemata make the former a more logical structure than the latter. Conversely, a hierarchical design in which the doctors are organized by specialty may be more logical—and thus navigable—than one in which the states are organized by their primary industry. Hierarchical navigation schema have been found to lead to shorter task completion times and reduced uncertainty while browsing (Mohageg, 1992). With respect to breadth and depth of menus, the effectiveness of one or the other may also vary on the basis of the tasks required, with more complex tasks being better served by broader menu navigation (see Van Oostendorp, Madrid, & Melguizo, 2009).

A second dimension of navigability is *clarity of structure*. This dimension addresses the degree to which an interface conveys to the user what portion of the available information is being viewed, and the scope of the other available information. For hierarchical digital information, such as static, multinode websites, clarity of structure is often conveyed through the presence of hierarchical menus that highlight the user's present position. An alternate design convention is that of breadcrumbs (Bernstein, 1988; Instone, 2002) or a series of links that allow the user to directly access higher levels of navigation contain the current page. Various interface elements have been developed and tested that allow users to visualize sequential levels of structure of a site, either scoping out more broadly (Hochheiser & Shneiderman, 2000) or narrowing further (Ahmed & Blustein, 2006; Pardue, Landry, Kyper, & Lievano, 2009) from the user's present location.

The third dimension of navigability is *clarity of target.* Rather than referring to aspects of the overall structure of an information space, this dimension deals with an aggregate quality of the specific navigational elements. From both the programmer's perspective and the user's perspective, a hyperlinked object consists of a representation—some text or visual element that provides the shape of the object—and the target (some information that accounts for what will happen when the user chooses to activate that link). In the case of websites, the target is most often a URL that loads a new content page, although it may also be an internal anchor linking to elsewhere in the page, or a cue to run some programming script that changes only some of the content on the screen.

Clarity of target deals with the degree to which the representation conveys to the user what the target of the link is, an aspect of navigation design frequently mentioned as important to outcomes (e.g., Farkas & Farkas, 2000). Most typically, the information the user obtains is the link anchor text (visible in the Web browser) and the URL of the target, which is visible in the status bar of most browsers upon hover (Weinreich & Lamersdorf, 2000). In terms of effects, while information spaces contain a number of discrete opportunities for hyperlink navigation, the impact of clarity of target on user outcomes need not be limited to characteristics of links actually selected by the user. Any navigational element that is evaluated as a possible selection demands some resources, and the degree to which the nature of the element facilitates a quick assessment of potential value should decrease cognitive demands.

Several frameworks of information-seeking provide insights into the mechanisms related to effects of navigability. Information foraging theory (Pirolli & Card, 1999) suggests that users rely on environmental cues throughout the information-seeking process. This process is one based on optimization so that temporal and processing costs are minimized, while the volume and rate of information consumption are sought to be maximized. Similarly, Bates's berrypicking model of information retrieval (1989) suggests that users modify their search strategies and goals based on information encountered in the process of navigating information spaces, with users adapting their search need by filtering and evaluating available information. These two theories highlight the dual nature of navigation vis-à-vis content in information environments. Navigational elements may compete for cognitive resources with content, and the process of interpreting and acting on navigation may tax cognitive resources in a way that detracts from users' ability to process the end content. On the other hand, at its best, navigation may not only guide users to the desired result efficiently, but also provide meaningful contextual information that may aid in the speed or quality of judgments (Kalyanaraman & Ivory, 2009).

While a swift perusal of an interface's navigability may provide some instant cues for site evaluation, we believe that navigability primarily affects psychological outcomes due to its impact on users' cognition. Effective navigation minimizes search costs by requiring users to process less information, resulting in greater cognitive resources to evaluate the desired content. Studies on disorientation in Web browsing or the "lost in hyperspace" phenomenon (e.g., Ahuja & Webster, 2001) suggest that under high-navigability conditions, fewer resources can be devoted to orientation within the site, which frees those resources up for processing content. Secondly, effective navigation can provide contextual clues about the ultimate information that may aid users in classifying and encoding it. By focusing on the quality and clarity of how such information is structured, interconnected, and conveyed to the user, navigability plays a crucial role in understanding the effects of digital information spaces. While customization and interactivity refer to opportunities provided within a digital message to control information, navigability describes the degree to which the message makes these opportunities for control visible, comprehensible, and predictable to the user. Each of the three dimensions of navigability detailed earlier—logic of structure, clarity of structure, and clarity of target—may make unique contributions to message use and effects, and the extent of their contributions may vary situationally based on characteristics of the message format and content domain. As a whole, we believe that they are a crucial link between the way digital information is presented and the way its users experience the process of message consumption.

Directions for Future Research

In summary, the consideration of customization, interactivity, and navigability (CIN) refers to a variable property of digital mediated messages that is independent of the semantic content of the message. For instance, a mediated message may be customized to the user based on any number of dimensions, and pertaining to any topic. Any type of content can be made more or less interactive by manipulating the opportunities the users has to select, choose, identify, and modify the content, and more or less navigable by varying how the structure and interconnectedness of elements is presented. Also, while some theoretical elements loom large with specific variables (e.g., the self and customization), the notion of information/user control as a defining aspect permeates each of the three variables. Future scholarship would benefit by explicating the idea of control in greater detail, and to test its impact in a variety of situations employing customization, interactivity, and navigability as experimental manipulations.

Finally, in this concluding section, we offer some recommendations for scholars pursuing a program of research centered on these concepts and issues. Of course, while such recommendations could be manifold, we forward those that we believe would be particularly useful to impel theoretical progress and spur a catholic understanding of these variables.

Testing in concert

At the conceptual level, the case for treating customization, interactivity, and navigability as separate and distinct characteristics of mediated messages is formidable, and there is compelling reason to believe that each variable uniquely affects how that message is processed. However, the empirical evidence is largely predicated on experiments that manipulated just one of the variables. In order to examine the effects of the CIN variables in concert, it is important to consider the impact of conceptually driven manipulations on the other variables, treating them as independent, but potentially integrated, dependent on the specific operationalizations employed.

Operationalizing interactivity necessitates either varying the interface features that allow the user to exercise control over content, or holding such interface features constant but varying the degree to which a feature allows control. However, manipulating the presence of features that allow user control such as hyperlinks, sliders, selection menus, and visual scrolling screens makes it difficult to hold other aspects of the content constant. When dealing with nonlinear information (e.g., a group of thumbnail images that represent different video clips on a site), removing the links to the content that allow the user to access each clip individually changes the nature of the information. To provide the same content without the individual links, the designer either has to make only one of the original video clips accessible or has to create a linear video that combines the original clips into one. The alternate option consists of controlling the presence of features across conditions, but varying the degree to which the features allow control over content. The choice of adopting a particular option (or a combination of the two) would depend on the specificity of the research question and stimulus interface chosen for study.

The potential operational overlap between interactivity and navigability may occur as a result of variations in interface elements that allow the user to understand the structure of the underlying site. Most often, a site's structure is made accessible through visual representations—images or text that serves as a hyperlink—that vary across conditions. Varying the presence of those elements necessarily changes the opportunities a user has to interact with or control the site content. The interactivity operationalization employed by Sundar et al. (2003) serves as a useful example here. To create three ordinal levels of interactivity, the researchers varied the structure of the content of a political candidate's website. In the low-interactivity condition, the site content was available all on one page. In the medium-interactivity condition, the main page offered links to four subsections of content, grouped by political issue. In the high-interactivity condition, each of the four subsection pages contained further links to separate pages for each policy position. Thus, from low to medium to high interactivity, the content was spread across 1, 4, and 16 pages of the site. Such an operationalization offered the user increasing levels of control in the form of removing unwanted content from the screen as the user clicked, but also required the user to scan the hyperlink text on the initial page content, decide which of the four links present allowed the optimal path, and make a choice by clicking that link (once in the "medium" condition, and twice in the "high" condition).

Interactivity and customization may likewise be intertwined in their implementation (see Ito, Kalyanaraman, Ford, Brown, & Miller, 2008). Although some interfaces may offer customized content entirely without requiring user input, many others do so based on user activity and selection (of preferences). In order to receive content that matches their individual dispositions more closely, users are required to interact with the system at least initially, and their input is stored for future usage sessions. Indeed, several implementations of interactivity can be seen as providing a semblance of customization; the nature of effective user interactions with content is motivated by seeking an experience that is tailored to the user's stated preferences. In an effectively designed interactive system, a user can express an iterative series of decisions by inputting information, each of which yields an increasingly customized set of decision options.

The dual influence of the CIN variables

In specifying the phase when the CIN variables shape communication outcomes, we suggest that they can do so at two junctures. First, they can serve as heuristics in initial evaluation of content, and second, during the course of the mediated experience, they can impact user engagement.

The existing body of scholarship points to some situational heuristic effects of technological variables. Affordances of newer technologies such as interactivity and navigability cue cognitive heuristics that trigger evaluations of the quality or credibility of the message (Sundar, 2008). It is also worth noting that these heuristic effects may be delimited by differences between individual users as well as types of content. Such heuristics may be closely linked to positive evaluations, as in the case of a highly interactive interface being preferred to a less-interactive one by cuing notions of control that engender perceptions of credibility. Other heuristics may foster negative evaluations, while still others may wield positive or negative effects based on contextual factors (see Sundar, 2008).

In accordance with dual-process models of persuasion, heuristics can affect evaluations through several pathways. According to Chen and Chaiken (1999), heuristics can evoke instantaneous judgment, or they can bias systematic processing. In either case, the heuristic is a perception or mental shortcut that serves as a lens affecting how the information is processed. Of course, the extent to which specific perceptions of communication technologies serve as heuristics remains ambiguous. Sundar (2008) proposes a number of possible heuristics invoked by technological features—bells and whistles, coolness, bandwagon, choice, and interaction heuristics, among others. Whether all of these are distinct heuristics, whether they are triggered independently of each other or interdependently, and the magnitude of their influence in evaluative tasks is an avenue for further research.

We would like to emphasize that conceptualizing the roles of technological variables in information processing solely in terms of heuristics misses a large chunk of the puzzle. There is also evidence that, beyond serving as a heuristic, the CIN variables can shape users' engagement with content in the course of the mediated experience, and this can manifest itself in the form of different patterns of behavior and variations in allocating cognitive resources. Changes in these three variables can fundamentally alter the user experience, and not merely the way users perceive the content they are exposed to. So, the CIN variables can directly facilitate greater engagement throughout the user experience by decreasing the extent to which users must allocate resources to content that is irrelevant to their situational (or perhaps even dispositional) preferences and goals. For example, if the content is customized, irrelevant content is filtered out from display; if the content is interactive, the user has nonlinear access to information; if it is navigable, the user can locate and process the content with relative efficiency. The careful examination of the dual influence propositions will further advance our understanding of the CIN elements.

Longitudinal studies

Although our interactions with technology are enduring and involve iterations, scholars studying the effects of such interactions rarely examine the consequences over a period of time. Indeed, the bulk of the evidence from experimental research resides in data from cross-sectional studies. Of course, systematic evidence from such studies has catapulted the study of the psychology of technology to prominence, but we believe future progress would benefit by examining the role of these technological variables over time. Typically, participants are exposed to a technology (usually a website) that features a manipulation of (say) interactivity. Once they go through the site, their affective and cognitive perceptions are assessed. While such investigations have helped lay the foundation for experimental work and have also advanced conceptual progress, examining these effects over time is suffused with numerous benefits. For one, they would offer a more stable picture of user responses to technology and account for the limitations imposed by transient factors. Indubitably, interacting with a website over a period of time is different from situations calling for evaluations based on short-term exposure. The former accounts for improved familiarity with the site, an enhanced understanding of the structural elements characterizing the site, and the user's propensity to include or discard the specific technological element in decision making. By encompassing a range of information-processing demands—from simple information to scenarios where the complexity of information to be negotiated is high (see Ariely, 2000), data from such studies can produce more discerning patterns of user responses to technology. The results can be provocative.

As pointed out earlier, longitudinal studies about agency effects of customization have revealed that initial positive perceptions of matched messages taper over time. These findings challenge our fundamental belief that customization is a ubiquitous panacea and instead suggest that other factors (in this case, agency) warrant close consideration. Furthermore, longitudinal designs are also useful in clarifying theoretical issues (such as the means via which a variable like customization can function in persuasion processes). Such perspectives would remain latent if we restricted our examination of technological effects to cross-sectional designs.

The role of content

Scholars pursuing experimental research in human–technology interaction are typically concerned with manipulation of some technological element, while normally ignoring the effects of content. Typically, researchers suggest that a particular technological element must be examined in different content situations, but these suggestions are generally included in discussion sections to point out the limited external validity of such studies. We submit that content has an important role to play in information-processing situations, especially when examined in conjunction with the CIN variables.

For example, relevance of content is inextricably tied to the underlying premise of customization. As we also pointed out earlier, the specificity of content would go a long way toward creating more meaningful customized interfaces as it would address the make-up of the "pertinent" self. The role of content is also of vital importance in elucidating the impact of interactivity and navigability. As some scholars have proposed, interactivity bestows a thorough inspection of content (Sundar, 2007). Also, consistent with the information control perspective, users are likely to seek greater control over information that they are invested in (see Kalyanaraman & Sundar, 2008). Moreover, changes in content (qualitatively or quantitatively) will, in all likelihood, lead to users processing content according to the degree of information control at their disposal (Ariely, 2000). Users' navigational strategies are also often dictated by their familiarity with a particular site. Such familiarity is presumably contingent on the content of the site. In a similar vein, the key criteria for optimum navigability would be impacted by content—navigational elements would interact with relevance or absorption potential of content.

Meaningful measures

The final recommendation we offer to advance programmatic work on CIN is related to measurement of outcomes and processes. The preponderance of psychological studies has generally relied on how manipulations of CIN variables influence affective (e.g., attitudes) and cognitive (e.g., memory) dimensions. These of course are necessary, but we propose that the scope of outcome measures be expanded to provide more meaning to the psychological distinction. For one, inclusion of behavioral measures would be useful. These would not just include browsing behaviors (such as number of links clicked on) but other indicators of how users are influenced to act such as sharing or forwarding information, posting comments, and so on. Also, if attitudes and memory are indeed believed to be induced by experimental manipulations, their corresponding behavioral manifestations also need to be scrutinized. For example, what does increased positive attitudes as a function of interactivity really translate to? Such questions would clearly depend on the context (e.g., persuasion, health) but they would be eminently worthy of inclusion. For instance, if interactive websites are effective in increasing knowledge about sexual health, it would also be meaningful to examine if they translate into safer sex practices as well. Similarly, exposure to a customized health message touting the benefits of regular physical activity should not just demonstrate a willingness to exercise regularly (or more), but may ideally result in actual (increased) exercise behaviors. In circumstances where it may not be possible to gauge behavioral responses, we recommend the inclusion of behavioroid (i.e., the tendency to assess behavioral commitment) measures (see Aronson, Wilson, & Brewer, 1998). The challenges of including such outcomes are many, but we hope that the inclusion of more meaningful measures in CIN research comes to be seen as the rule rather than the exception.

In addition to analyzing how affect and cognition may impact behavioral responses, we also urge the adoption of more process-centered measures in experimental studies. Some explanatory variables would perhaps resonate across many scenarios (e.g., the engenderment of increased involvement as a function of customization), whereas other processes would be best discerned in specific circumstances (e.g., customization fostering a greater sense of belonging in community portals), but sustained and systematic experimental research should juxtapose theory-centered work with data-driven investigations to specify the various processes and mechanisms that drive the CIN variables. Existing scholarship also points to the intertwining between and among the CIN elements—hence, our earlier call for experimental work to examine the CIN elements in concert would assume greater purport as they would begin to clarify the unique and combined impact of different mechanisms.

We conclude this essay by pointing out that our confidence in the viability of the CIN elements as the bedrock of the variable-centered approach to communication technology validates and complements similar pronouncements (see Sundar, 2009). As a broad recommendation, we call for diligent experimental research to further our awareness of these variables. Additionally, we hope that our suggestions for programmatic research will inspire coherent inquiry. By investigating the CIN variables in concert, in various content domains, with the deployment of process-centered mechanisms and meaningful outcomes, and appraising these over time, scholars would be better positioned to postulate—and build on—existing models of theoretical integration (see Sundar et al., 2015). Such investigations would pave the way for theoretical perspicuity and clear identification of boundary conditions, while also specifying methodological criteria. Ultimately, we hope these guidelines stimulate programmatic research on the psychology of customization, interactivity, and navigability, and enrich the wider landscape of communication technology.

References

Ahmed, I., & Blustein, J. (2006). Influence of spatial ability in navigation: Using breadcrumbs on the Web. *International Journal of Web Based Communities, 2*(2), 183–197.

Ahuja, J. S., & Webster, J. (2001). Perceived disorientation: An examination of a new measure to assess Web design effectiveness. *Interacting With Computers, 14*, 15–29.

Ariely, D. (2000). Controlling the information flow: Effects on consumers' decision making and preferences. *Journal of Consumer Research, 27*(2), 233–248.

Aronson, E., Wilson, T. D., & Brewer, M. B. (1998). Experimentation in social psychology. In D. T. Gilbert, S. T. Fiske, & G. Lindzey (Eds.), *The handbook of social psychology* (4th ed., pp. 99–142). New York, NY/Oxford, UK: Guilford/Oxford University Press.

Balakrishnan, B., & Sundar, S. S. (2011). Where am I? How can I get there? Impact of navigability and narrative transportation on spatial presence. *Human-Computer Interaction, 26*(3), 161–204.

Bates, M. J. (1989). The design of browsing and berrypicking techniques for the online search interface. *Online Information Review, 13*(5), 407–432.

Baumeister, R. F. (1998). The self. In D. T. Gilbert, S. T. Fiske, & G. Lindzey (Eds.), *The handbook of social psychology* (4th ed., pp. 680–740). New York, NY: McGraw-Hill.

Beaglehole, E. (1932). *Property: A study in social psychology.* New York, NY: Macmillan.

Belk, R. W. (1988). Possessions and the extended self. *Journal of Consumer Research, 15*, 139–168.

Bernstein, M. (1988). The bookmark and the compass: Orientation tools for hypertext users. *ACM SIGOIS Bulletin, 9*(4), 34–45.

Blascovich, J., & Bailenson, J. N. (2011). *Infinite reality: Avatars, eternal life, new worlds, and the dawn of the virtual revolution.* New York, NY: William Morrow.

Cacioppo, J. T., Petty, R. E., & Kao, C. F. (1984). The efficient assessment of need for cognition. *Journal of Personality Assessment, 48*, 306–307.

Chen, S., & Chaiken, S. (1999). The heuristic-systematic model in its broader context. In. S. Chaiken & Y. Trope (Eds.), *Dual process theories in social psychology* (pp. 73–96). New York, NY: Guilford.

Dholakia, R., Miao, Z., Dholakia, N., & Fortin, D. (2000). *Interactivity and revisits to websites: A theoretical framework. RITIM Working Paper.* Available from http://ritim.cba.uri.edu/wp/

Fang, X., Chau, M., Hu, P. J., Yang, Z., & Liu Sheng, O. R. (2006). Web mining-based objective metrics for measuring Web site navigability. *Proceedings of the Twenty-Seventh International Conference on Information Systems*, Milwaukee, Wisconsin, USA, 195–208.

Farkas, D. K., & Farkas, J. B. (2000). Guidelines for designing Web navigation. *Technical Communication, 3*, 341–358.

Franke, N., Schreier, M., & Kaiser, U. (2010). The "I designed it myself" effect in mass customization. *Management Science, 56*(1), 125–140.

Heider, F. (1958). *The psychology of interpersonal relations.* New York, NY: Wiley.

Higgins, E. T. (1996). The "self-digest": Self-knowledge serving self-regulatory functions. *Journal of Personality and Social Psychology, 71*, 1062–1083.

Hirsh, J. B., Kang, S. K., & Bodenhausen, G. V. (2012). Personalized persuasion: Tailoring persuasive appeals to recipients' personality traits. *Psychological Science, 23*(6), 578–581.

Hochheiser, H., & Shneiderman, B. (2000). Performance benefits of simultaneous over sequential menus as task complexity increases. *International Journal of Human-Computer Interaction, 12*(2), 173–192.

Instone, K. (2002). *Location, path and attribute breadcrumbs.* Poster presented at Proceedings of the 3rd Annual Information Architecture Summit, Baltimore, March 16–17. Retrieved from http://www.instone.org/files/KEI-Breadcrumbs-IAS.pdf

Ito, K. E., Kalyanaraman, S., Ford, C. A., Brown, J. B., & Miller, W. C. (2008). Let's talk about sex: Pilot study of an interactive CD-ROM to prevent HIV/STIs in female adolescents. *AIDS Education & Prevention, 28*(1), 78–89.

James, W. (1890). *The principles of psychology.* New York, NY: Henry Holt.

Jul, S. (2004). *From brains to branch points: Cognitive constraints in navigational design* (Unpublished doctoral dissertation). University of Michigan, Ann Arbor, MI.

Kalyanaraman, S. (2013). *iCreate: Longitudinal effects of agency in customized portals.* Unpublished manuscript.

Kalyanaraman, S., Ito, K., Malik, C., & Ferris, E. (2009, May). *Control ergo cogito: An experimental investigation of the interactivity-as-information control perspective.* Paper presented to the Information Systems Division at the 59th Annual Convention of the International Communication Association, Chicago, IL.

Kalyanaraman, S., & Ivory, J. (2009). Enhanced information scent, selective discounting, or consummate breakdown: The psychological effects of Web-based search results. *Media Psychology, 12*, 295–319.

Kalyanaraman, S., Oliver, M. B., & Magee, R. (2010). *The interplay between customization and cognition in Web portals: The multiple role hypothesis explanation.* Paper presented to the Information Systems Division at the 60th Annual Convention of the International Communication Association, Singapore.

Kalyanaraman, S., Penn, D., Ivory, J., & Judge, A. (2010). The virtual doppelganger: Effects of a virtual reality simulator on perceptions of schizophrenia. *Journal of Nervous and Mental Disease, 198*(6), 437–443.

Kalyanaraman, S., & Sundar, S. S. (2006). The psychological appeal of personalized content in web portals: Does customization affect attitudes and behavior? *Journal of Communication, 56*(1), 110–132.

Kalyanaraman, S., & Sundar, S. S. (2008). Impression formation effects in online mediated communication. In E. A. Konijn, S. Utz, M. Tanis, & S. B. Barnes (Eds.), *Mediated interpersonal communication* (pp. 217–233). New York, NY: Routledge.

Korgaonkar, P. K., & Wolin, L. D. (1999). A multivariate analysis of web usage. *Journal of Advertising Research, 39*, 53–68.

Kristof, R., & Satran, A. (1995). *Interactivity by design: Creating and communicating with new media.* Mountain View, CA: Adobe Press.

Li, C., & Kalyanaraman, S. (2013). "I, me. Mine" or "us, we, ours?" The influence of cultural psychology on Web-based customization. *Media Psychology, 16*(3), 272–294.

Liu, Y., & Shrum, L. J. (2009). A dual-process model of interactivity effects. *Journal of Advertising, 38*(2), 53–68.

Mohageg, M. F. (1992). The influence of hypertext linking structures on the efficiency of information retrieval. *Human Factors: The Journal of the Human Factors and Ergonomics Society, 34*(3), 351–367.

O'Keefe, D. J. (2003). Message properties, mediating states, and manipulation checks: Claims, evidence, and data analysis in experimental persuasive message effects research. *Communication Theory, 13*(3), 251–274.

Paisley, W. (1983). Computerizing information: Lessons of a videotext trial. *Journal of Communication, 33*, 153–161.

Palmer, J. W. (2002). Web site usability, design, and performance metrics. *Information Systems Research, 13*(2), 151–167.

Pardue, J. H., Landry, J. P., Kyper, E., & Lievano, R. (2009). Look-ahead and look-behind shortcuts in large item category hierarchies: The impact on search performance. *Interacting with Computers, 21*(4), 235–242.

Pariser, E. (2011). *The filter bubble: What the internet is hiding from you.* New York, NY: The Penguin Press.

Petty, R. E., Barden, J., & Wheeler, S. C. (2002). The elaboration likelihood model of persuasion. In R. J. DiClemente, R. A. Crosby, & M. Kegler (Eds.), *Emerging theories in health promotion practice and research* (pp. 71–99). San Francisco, CA: Jossey-Bass.

Pirolli, P., & Card, S. K. (1999). Information foraging. *Psychological Review, 106*(4), 643–675.

Rafaeli, S. (1988). Interactivity: From new media to communication. In R. Hawksins, J. Wiemann, & S. Pingree (Eds.) *Advancing communication science: Merging mass and interpersonal processes* (pp. 110–134). Newbury Park, CA: Sage.

Reeves, B., & Nass, C. (1996). *The media equation: How people treat computers, television, and new media like real people and places.* Stanford, CA: CSLI Publications.

Rogers, T. B., Kuiper, N. A., & Kirker, W. S. (1977). Self-reference and the encoding of personal information. *Journal of Personality and Social Psychology, 35*, 677–688.

Rothbaum, F., Weisz, J. R., & Snyder, S. S. (1982). Changing the world and changing the self: A two-process model of perceived control. *Journal of personality and social psychology, 42*(1), 5–37.

Sartre, J, P. (1943). *Being and nothingness: A phenomenological essay on ontology.* New York, NY: Philosophical Library.

Seligman, M. E. (1975). *Learned helplessness: On depression, development and death.* San Francisco, CA: Freeman.

Shedroff, N. (1999). Information interaction design: A unified field theory of design. In R. Jacobson (Ed.), *Information design* (pp. 267–292). Cambridge, MA: MIT Press.

Steuer, J. (1992). Defining virtual reality: Dimensions determining telepresence. *Journal of Communication*, *42*(4), 73–93.

Sundar, S. S. (2007). Social psychology of interactivity in human-website interaction. In A. N. Joinson, K. Y. A. McKenna, T. Postmes, & U.-D. Reips (Eds.), *The Oxford Handbook of Internet Psychology* (pp. 89–104). Oxford, UK: Oxford University Press.

Sundar, S. S. (2008). The MAIN model: A heuristic approach to understanding technology effects on credibility. In M. J. Metzger & A. J. Flanagan (Eds.), *The John D. and Catherine T. MacArthur foundation series on digital media and learning* (pp. 73–100). Cambridge, MA: MIT Press.

Sundar, S. S. (2009). Media effects 2.0: Social and psychological effects of communication technologies. In R. L. Nabi & M. B. Oliver (Eds.), *The SAGE handbook of media processes and effects* (pp. 545–560). Thousand Oaks, CA: Sage.

Sundar, S. S., Kalyanaraman, S., & Brown, J. (2003). Explicating Web site interactivity: Impression formation effects in political campaign sites. *Communication Research*, *30*(1), 30–59.

Sundar, S. S. Jia, H., Waddell, T. F., & Huang, Y. (2015). Toward a theory of interactive media effects (TIME): Four models for explaining how interface features affect user psychology. In S. S. Sundar (Eds.), *The handbook of the psychology of communication technology*. Malden, MA: John Wiley & Sons, Inc.

Sundar, S. S., & Kim, J. (2005). Interactivity and persuasion: Influencing attitudes with information and involvement. *Journal of Interactive Advertising*, *5*(2), 6–29.

Sundar, S. S., & Marathe, S. S. (2010). Personalization versus customization: The importance of agency, privacy, and power usage. *Human Communication Research*, *36*(3), 298–322.

Sundar, S. S., Xu, Q., & Bellur, S. (2010). Designing interactivity in media interfaces: A communications perspective. In *Proceedings of the 28th International Conference on Human Factors in Computing Systems (CHI'10)* (pp. 2247–2256). New York, NY: ACM.

Sunstein, C. R. (2009). *Republic.com 2.0*. Princeton, NJ: Princeton University Press.

Tao, C., & Bucy, E. P. (2007). Conceptualizing media stimuli in experimental research. *Human Communication Research*, *33*, 397–426.

Teo, H. H., Oh, L. B., Liu, C., & Wei, K. K. (2003). An empirical study of the effects of interactivity on Web user attitude. *International Journal of Human Computer Studies*, *58*, 281–305.

Van Oostendorp, H., Madrid, R., & Melguizo, M. C. P. (2009). The effect of menu type and task complexity on information retrieval performance. *Ergonomics Open Journal*, *2*, 64–71.

Weinreich, H., & Lamersdorf, W. (2000). Concepts for improved visualization of Web link attributes. *Computer Networks*, *33*(1), 403–416.

Wojdynski, B. W., & Kalyanaraman, S. (2012, May). *Parsing the effects of Web site interactivity and navigability*. Paper presented at the 62nd Annual Conference of the International Communication Association, Phoenix, AZ, USA.

Wojdynski, B. W., & Kalyanaraman, S. (2004). The three dimensions of Web site navigability: Explication and effects. *Journal of the American Society for Information Science and Technology*. Advance online publication. doi:10.1002/asi.23395

Zhang, Y., Zhu, H., & Greenwood, S. (2004, September). Web site complexity metrics for measuring navigability. *Proceedings of the Fourth International Conference on Quality Software*, 172–179. doi:10.1109/QSIC.2004.1357927

20

Psychological Approaches to Credibility Assessment Online

Miriam J. Metzger and Andrew J. Flanagin
Department of Communication, University of California, Santa Barbara, Santa Barbara, CA, USA

As networked information and communication technologies have enabled nearly instant access, from virtually anywhere, to almost inconceivably large information repositories, people have come to rely upon web-based information resources in a wide variety of ways. For example, even as early as 2006, nearly half of users in the United States said that the Internet played a crucial or important role in at least one major decision in their lives, such as attaining additional career training, helping themselves or someone else with a major illness or medical condition, or making a major investment or financial decision (Horrigan & Rainie, 2006). Yet, the *credibility* of information obtained online can be problematic, which raises concerns about how people obtain, interpret, and evaluate this information.

Defining Credibility

The study of credibility is highly interdisciplinary and implicates a number of allied concepts. For example, in information science, the emphasis has been on information evaluation, where perceived quality affects information seekers' judgments about the relevance of information (Rieh & Danielson, 2007). In this context, quality has been conceptualized as a user criterion (made up of accuracy, reliability, currency, comprehensiveness, and reliability) invoked when an individual evaluates a piece of information. Authority can also indicate credibility, even without being sanctioned by official credentials, because people may be recognized as "cognitive authorities in the sphere of their own experience, on matters they have been in a position to observe or undergo" (Wilson, 1983, p. 15).

Trust is also closely related to credibility and is a core concern of work in management information systems (MIS), where the focus has been on interpersonal

The Handbook of the Psychology of Communication Technology, First Edition. Edited by S. Shyam Sundar.

or interorganizational trust in technical contexts, including the domains of e-commerce (Gefen, Karahanna, & Straub, 2003; McKnight, Choudhury, & Kacmar, 2002) and technology acceptance (Wu & Chen, 2005; Wu, Zhao, Zhu, Tan, & Zheng, 2011). In organizational studies, trust has also been a critical concern, for instance, in the study of interorganizational relations (Lane & Bachmann, 1998) or interactions and behaviors in virtual teams (Ridings, Gefen, & Arinze, 2002; Sarker, Valacich, & Sarker, 2003). In these contexts, trust refers to one exchange partner's confidence, belief, and expectation that another will act in their best interest.

In the fields of communication and psychology, credibility is traditionally defined as the believability of information, and it rests largely on the trustworthiness and expertise of the information source or message, as interpreted by the information receiver (Hovland, Janis, & Kelley, 1953; Rieh & Danielson, 2007; Tseng & Fogg, 1999). This differs from trust in that although trustworthiness is one dimension of believability, credibility typically also involves a judgment of source expertise or information accuracy as well. Similarly, a key difference from definitions of credibility that focus on information accuracy is in the subjective nature of credibility—trustworthiness is a receiver judgment, based primarily on subjective factors. Expertise can also be subjectively perceived but includes relatively objective characteristics of the source or message as well (see Metzger, Flanagin, Eyal, Lemus, & McCann, 2003; Tseng & Fogg, 1999 for extended discussions of these points). Thus, the fields of communication and psychology treat credibility as a perceptual variable: credibility is not an objective property of a source or a piece of information. Instead, it is a subjective perception on the part of the information receiver (Fogg & Tseng, 1999). As such, the credibility of the same source or piece of information may be judged differently by different people.

Digital media are complicating notions of credibility today (Metzger et al., 2003). Socio-technical changes in the last few decades have spawned new concerns, and amplified existing ones, for people seeking credible information. For example, a number of challenges have accompanied the web's rise to prominence as an information repository, which have become even more pronounced as the Internet has evolved into a vibrant, interactive information sharing environment. The specific nature of these challenges suggests the importance of understanding the psychological dimensions of people's information evaluation experiences.

Challenges to Credibility in the Contemporary Media Environment

Perhaps the greatest change in the information environment in the last 15 years or so is that digital media have provided access to an unprecedented amount of information for public consumption. As digital network technologies have lowered the cost and complexity of producing and disseminating information, the nature of information providers has shifted. Rather than a small number of sources, each with a substantial investment in the information production and delivery processes, information is increasingly provided by a wide range of sources, many of whom can readily create and deliver information to large audiences worldwide. One consequence of this evolution in information production is an almost incomprehensibly vast information repository in the form of the Web and other online resources.

Accompanying this proliferation is the fact that much online information is not subject to filtering through professional gatekeepers and, as a result, it may be more prone to being poorly organized, out of date, incomplete, or inaccurate (Danielson, 2005; Flanagin & Metzger, 2000, 2007; Metzger et al., 2003; Rieh & Danielson, 2007). Through what he calls "disintermediation," Eysenbach (2008) notes that networked digital media remove many traditional information intermediaries such as opinion leaders, experts, and information arbiters, thereby forcing individuals to evaluate the vast amount of information online on their own. Moreover, Callister (2000) argues that traditional strategies for information evaluation, such as trusting a source believed to provide reliable information (e.g., the government) or to possess appropriate credentials (e.g., an expert), works only when there is a limited number of sources and when there are high barriers for public dissemination of information. Thus, while under conditions of information scarcity it is possible for gatekeepers to filter much of the information available (and gatekeepers have incentive to uphold credibility standards), the Internet presents an environment of information abundance that makes traditional models of gatekeeper oversight and quality control untenable due to the sheer volume of information that would have to be vetted.

Others have noted that digital media tools sometimes reduce the reliability of standard authority indicators such as author identity or reputation (Danielson, 2005; Fritch & Cromwell, 2002), and yet source information is crucial to credibility because it is the primary basis upon which credibility judgments rest (Sundar, 2008). In some cases, source information is unavailable, masked, or missing online. In other cases, source information is provided, but hard to interpret, such as when information is co-produced, re-purposed from one site, channel, or application to another, or when information or news aggregators display information from multiple sources in a centralized location that may itself be perceived as the source. These issues have prompted concerns about the credibility of online information because they create uncertainty regarding who is responsible for information and, thus, whether it should be believed (Rieh & Danielson, 2007).

Relatedly, understanding credibility online is especially problematic since there are many potential "targets" of credibility evaluation that often are at work simultaneously. Taking the quintessential example of Wikipedia, credibility judgments can be made at the website level (is Wikipedia a credible source of information?), at the content level (is any specific entry within Wikipedia credible?), or regarding specific information author(s) (are specific contributors to Wikipedia credible?). Of course, these various targets can work in concert or at odds with one another, suggesting that in combination the complexity of their credibility judgment is even more pronounced. Research has confirmed this complexity, as past studies have shown that the source and content of information interact in intricate ways, across various user audiences (e.g., Flanagin & Metzger, 2007, 2011).

Finally, the dramatic rise in "user-generated content" in recent years, where individuals are increasingly responsible not just for consuming, but also for producing, many of the information resources available online, has prompted a number of concerns with regard to information and source credibility. Although user-provided information offers exciting promise, it can also be problematic. The "unfiltered social web" can result in the dissemination of information that is not reliable, and it can be difficult to undo damage from erroneous or biased information distributed via such technologies (see Hesse et al., 2011). In addition, it may be unclear whether the information source is who or what is claimed, and whether the information is original or has been repurposed or altered at some point. Moreover, user-generated mechanisms work best when a diversity of opinions are represented, when individual opinions are independent of others' influence, when community members are able to draw upon specialized local knowledge, and when the convergence of judgments within a community is facilitated (Surowiecki, 2005). Although these conditions are present in some social media venues, they are by no means universal. Group processes are quite often subject to biases through processes of bandwagon effects, groupthink, and other problematic group dynamics (Forsyth, 2010).

In light of the foregoing concerns, we argue that it is critical to understand the psychological dimensions of people's information evaluation experiences. We propose several classes of explanatory mechanisms that help to explain how people navigate the online information environment today, including information processing views, personality-based characteristics, and information assessment perspectives that explicitly take into account the dynamics of social interaction.

Information Processing and Credibility Evaluation

Studies of credibility evaluation across disciplines such as psychology, communication, education, information science, and others have found that people rely on a wide variety of factors to decide whether to believe the information they obtain online. Early work by Fogg and colleagues (2003), for example, identified over 18 classes of cues that people used to determine the credibility of websites across ten content categories (e.g., news, health, commercial, financial, etc.). These included such cues as author reputation, site design and navigability, writing tone, and users' past experience with the site, for example. Subsequent research added several more cues to this list (Flanagin & Metzger, 2010; Hilligoss & Rieh, 2008; Metzger, 2007), and categorized these in terms of whether each is a feature of the source, message, medium, or receiver (Del Giudice, 2010). For example, author credentials constitute an author cue that information consumers might rely on to assess the relative credibility of information emanating from a source. Information currency, by contrast, is a message cue that has been shown to influence credibility perceptions. The presence or absence of website advertising is an example of a site or source cue affecting perceived credibility, and the degree of an individual's experience with a medium or a source is a receiver characteristic that has been shown to influence people's credibility evaluations. Figure 20.1 presents a list of factors that have been shown to be relevant to the evaluation of the credibility of online information.

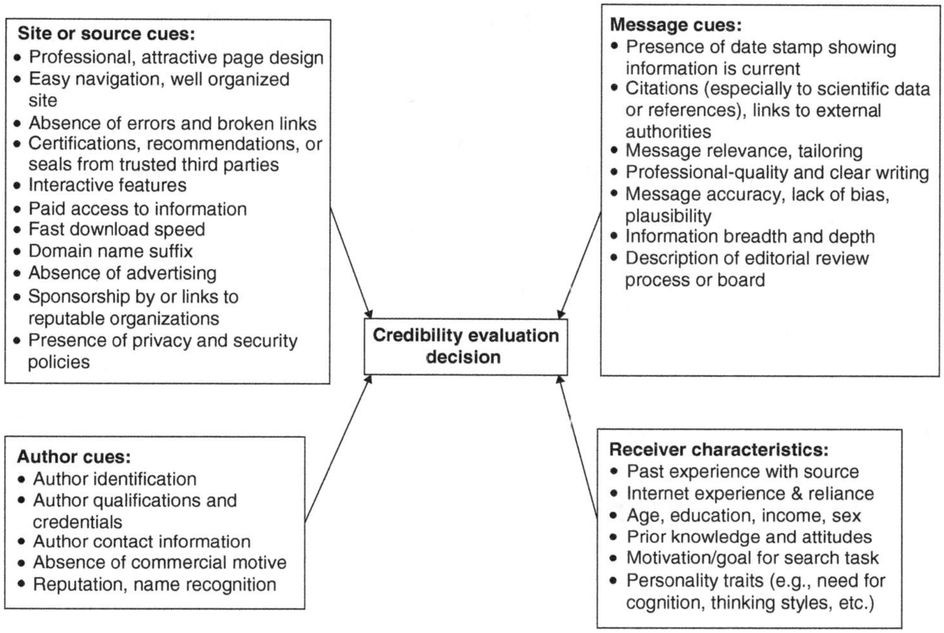

Figure 20.1 Schematic representation of potential factors relevant to credibility evaluation of online information.

With a dizzying array of credibility cues to choose from, an important question is how information seekers wade through all of the available cues to arrive at a credibility judgment? Information processing theories have been applied to help to understand this process. For example, the Limited Capacity Model of message processing (Lang, 2000) finds that because people do not have unlimited cognitive capacity, they cannot process all aspects of all messages they receive, and so they instead select only some salient features to encode, store, and retrieve.

Fogg's (2003) Prominence-Interpretation Theory of web credibility similarly suggests that not all elements of a website can be attended to by users, and so not all elements will enter into users' credibility evaluations. This theory states that credibility evaluations are a function of users (a) noticing a cue (prominence) and (b) making a judgment about that cue (interpretation) in an ongoing and iterative fashion until the user reaches an overall credibility decision or encounters a constraint, such as lack of time that bars examination of more cues. Prominence is affected by user characteristics (e.g., involvement, experience, ability to process), contextual factors (e.g., time pressures or task type, such as seeking information vs. amusement), and the artifact being evaluated (e.g., the type or content of a website or its navigability). Interpretation refers to the value or meaning that users assign to a credibility cue, typically whether a cue is a positive or negative credibility indicator. Several factors affect interpretation, including users' assumptions, cultural background, prior experience or knowledge, skill, goals, and information seeking context, including norms and expectations. Any particular credibility cue will only impact an information seeker's credibility assessment if it is perceptually prominent and is both interpretable and interpreted (Fogg, 2003).

Dual processing models of information processing and evaluation, such as the Elaboration Likelihood Model (Petty & Cacioppo, 1981) and the Heuristic-Systematic Model (Chaiken, 1980) (see Chaiken & Trope, 1999 for an overview of dual processing models), have also been invoked to answer the question of how and when people use available credibility cues to evaluate information online (e.g., Metzger, 2007; Sundar, 2008; Wathen & Burkell, 2002). Dual processing models emphasize the role of motivation and cognitive ability in guiding information assessment and decision making, and theorize that people will process messages in more or less depth depending upon the message receiver's motivation and ability to do so.

Metzger's (2007) dual processing model of credibility assessment posits that information seekers' motivation and ability are key to whether and to what degree users will evaluate Web information for its credibility. According to this model, the degree to which online messages will be scrutinized for their credibility depends on individual users' *ability* to evaluate the message, which may be a function of their knowledge or training in how to evaluate information, critical thinking skills, time constraints, or other factors, and their *motivation* or purpose for seeking the information, which involves their awareness and salience of the consequentiality of receiving low-quality or inaccurate information. The dual processing model of credibility assessment posits two general strategies that reflect greater and lesser degrees of cognitive rigor: The "analytic" strategy involves a more systematic attempt to discern credibility by considering more deeply a wider range of author, message, or medium cues. By contrast, the "heuristic" strategy relies on a faster and more cursory examination of credibility cues and often focuses primarily or exclusively on surface characteristics of the information (e.g., the visual design elements of a website) or on a user's gut feelings about the credibility of a piece of information.

Chen and Chaiken (1999) found that individuals' "accuracy goals" vary depending on the situation, such that people are more or less motivated to reach accurate judgments across different contexts. Internet users' accuracy goals likely vary from search to search, and their information seeking may be more or less casual or purposive depending on their goals. While some Web browsing is certainly highly motivated by a person's need to find accurate information (e.g., information sought for an important medical decision), other information seeking situations may not require such a high level of credibility checking (e.g., looking up information on a music group). The dual processing model of credibility evaluation argues that users look to different aspects of Web-based information to assess its credibility such that under conditions of high accuracy motivation, online information seekers will likely pay more attention to a broader array of credibility cues and perform more rigorous information evaluation compared with when motivation is lower (for similar logic, see Fogg et al., 2003; Sundar, 2008; Wathen & Burkell, 2002). Consistent with this model, credibility assessment has been shown to vary by information type, such that people exert significantly more effort to verify the credibility of factual or reference information than they do for entertainment information (Flanagin & Metzger, 2000).

Others have confirmed that user ability also plays a role in predicting online information seekers' credibility assessment strategies. For example, studies have found that Internet experience and Internet self-efficacy are positively related to the amount of effort users exert to verify the credibility of information they find online (Flanagin & Metzger, 2000; Hong, 2006), and other research has shown that topic experts employ more cues to evaluate the quality of a website's information and an information source's credentials, and are less likely to rely exclusively on simple visual appeal of the site to assess its credibility, compared to less knowledgeable users (Flanagin & Metzger, 2007; Fogg et al., 2003). Parallel results were found in these studies when the information obtained online was higher in personal salience and consequentiality (see also Byerly & Brodie, 2005; Sillence, Briggs, Harris, & Fishwick, 2007), which lends further support to the contention that that people with different levels of ability and motivation pay attention to different criteria when judging the credibility of websites.

Although dual processing theories tend to imply that only one strategy may be used in any information evaluation situation, recent studies have shown that both analytic and heuristic strategies for credibility evaluation can operate simultaneously. In a survey of 3,991 adults in the United States, respondents reported using both analytic and heuristic strategies when evaluating credibility across a variety of online information seeking contexts (Metzger et al., 2011). Some of the most common analytic means of evaluation included double-checking facts, checking other websites for supporting information, and considering all views on a topic. Frequent heuristic strategies included relying on gut feelings and making quick credibility decisions. Similar results were found in a nationally representative survey of US children between the ages of 11 and 18 (Flanagin & Metzger, 2010).

Although analytic means of credibility evaluation are traditionally thought to be superior and result in better decisions, heuristic strategies appear to be quite common (Metzger, 2007; Metzger, Flanagin, & Medders, 2010). Fogg and colleagues (2003), for example, found that online information consumers' main consideration in credibility assessment was the visual design elements of websites, rather than any content or source information. They argue that because web users typically spend little time at any given website, they likely develop quick strategies for assessing credibility. They say, "one could argue that people typically process web information in superficial ways, that using peripheral cues is the rule of web use, not the exception" (p. 15). Other research similarly shows that people rarely engage in effortful information evaluation tasks, opting instead to base decisions on factors like website design and navigability, while Internet users report verifying the information they found online for its credibility only "rarely" to "occasionally" and tend to verify using strategies that require the least time and effort (see Metzger, 2007, for a review). While in some views cognitive heuristics are thought to lead to biased or faulty information processing (Tversky & Kahneman, 1974), growing evidence suggests that heuristics can be equally effective as more cognitively demanding information processing strategies in inference and decision-making contexts (Gigerenzer & Todd, 1999; Gladwell, 2005).

The use of heuristic strategies may be explained by the fact that online information seekers often perform credibility evaluations within the limits of bounded rationality (Del Giudice, 2010, Metzger et al., 2010). Bounded rationality operates on the principle of least effort and acknowledges that decision makers "must arrive at their inferences using realistic amounts of time, information, and computational resources" (Gigerenzer & Todd, 1999, p. 24). Consequently, people are not always able to act perfectly rationally due to limitations imposed by the human mind (e.g., noninfinite cognitive resources) and external conditions (e.g., noninfinite time) (Simon, 1955). One form of bounded rationality in information processing is *satisficing*, whereby people use not all but rather just enough of their cognitive resources to obtain optimal outcomes for the situation or context. Gigerenzer and Todd (1999) argue that information processing employing highly rational, true optimizing strategies is actually quite rare, and research on information foraging theory finds evidence that satisficing is a frequent strategy of Internet information seekers (Pirolli, 2005).

These perspectives suggest that Internet information seekers are likely to cope with the perceived costs of information overload and credibility evaluation by using strategies that minimize their cognitive effort and time. A number of recent studies have identified several cognitive heuristics, or mental short cuts or rules-of-thumb used as a judgment rule that help people make quick evaluations (Sundar, Oledorf-Hirsh, & Xu, 2008), that online information seekers invoke to assist in the process of credibility evaluation (e.g., Hilligoss & Rieh, 2008; Metzger at al., 2010; Sundar, 2008). For example, Metzger et al. (2010) found six cognitive heuristics used to evaluate credibility: *reputation*, *endorsement*, *consistency*, *self-confirmation*, *expectancy violation*, and *persuasive intent*.

The *reputation* heuristic says that when choosing between sources, people should favor a source whose name they recognize as more credible compared to an unfamiliar source. The rationale underlying the reputation heuristic is that reputation or name recognition is earned via an accumulated record of transactions over time, the results of which are spread through people's social networks. The reputation heuristic may also be a subset of the authority heuristic, or whether a source is recognized as an official authority or not, which has been found to impact Internet users' credibility assessments (Sundar, 2008).

The *endorsement* heuristic suggests that people are inclined to believe information and sources if others do so also. People tend to trust sites and sources that are trusted and recommended by *known* others, or that come from *unknown* persons in aggregated form (Metzger et al., 2010). Trust derived from known others may be underpinned by another form of heuristic reasoning known as the "liking/agreement heuristic" (Chaiken, 1987), which says that people often agree with those they like. Trust derived from aggregated information from unknown others (e.g., testimonials or ratings) likely stems from bandwagon heuristics, whereby people assume that something is correct if many others think it is correct (see Sundar, 2008). The *consistency* heuristic is a close relative of the endorsement heuristic because it also operates on the principle that people tend to believe things if others believe them. The consistency heuristic comprises a credibility evaluation stopping rule that says a piece information should be judged as credible when it is found to agree with information from another independent source. The implication of a heuristic "stopping rule" illustrates the potential ambiguity between heuristic strategies relying on mental shortcuts and more cognitively laborious processing efforts.

The *self-confirmation* heuristic reflects the human tendency to notice and place greater value on information that supports one's beliefs, while discounting information that refutes those beliefs. In terms of credibility, it is the tendency for people to view information as credible if it confirms their preexisting beliefs and not credible if it counters their existing beliefs, regardless of how well argued, exhaustively researched, comprehensive, or appropriately sourced it is. The self-confirmation heuristic likely stems from the false consensus effect: research in cognitive psychology finds that people tend not only to believe that their own opinions are right but that they are widely shared by others (Ross, Greene, & House, 1977). Such cognitive biases serve as ego defense mechanisms, resulting in a tendency for people to evaluate ambiguous information in a way that is beneficial to their own needs or interests.

The *expectancy violation* heuristic in credibility evaluation says that if a website fails to meet users' expectations in some way, for example, if a site asks for more personal information than is necessary or contains typos or grammatical errors, they will immediately judge it as *not* credible without further inspection of its content. The reverse could also be true: positive surprises that contradict negative expectations might bolster perceived credibility. The expectancy-violation heuristic may be underpinned in part by the "effort heuristic," which is the human tendency to value objects based on how much effort went into producing them (Kruger, Wirtz, van Boven, & Altermatt, 2004). Finally, the *persuasive intent* heuristic is the tendency to feel that information that may be biased—typically for commercial purposes—is not credible. For example, the presence of advertising on websites where it is unexpected seems to elicit an immediate defense mechanism that leads people to mistrust information without further scrutiny (Metzger et al., 2010; for similar results, see also Fogg et al., 2003; Walther, Wang, & Loh, 2004).

Individual Differences in Credibility Assessment

Because credibility judgments are a subjective perception on the part of the information receiver, individual differences naturally impact users' credibility evaluations (Del Giudice, 2010; Hong, 2006; Flanagin & Metzger, 2010, 2013; Metzger et al., 2011), in part by making certain cues more salient to certain users (Fogg, 2003). Many types of individual-level differences have been discussed in the literature, including, for example, demographic differences, patterns of Internet use and past experience, information literacy skills, personality traits, familiarity with the topic, cultural norms and values, and levels of cognitive development. Beyond mentioning these as possible factors, however, little empirical data exist showing how each of these individual factors affects credibility decision making.

To fill this void, Flanagin and Metzger conducted two large-scale surveys using representative samples of Internet users in the United States in which several individual-level differences were measured and analyzed for their impact on Internet users' credibility evaluations and verification behaviors (Flanagin & Metzger, 2010; Metzger et al., 2011). They specifically studied how user demographics, types of Internet use, past experience, information literacy skills, and personality traits affect information seekers' concern about credibility and the degree to which they believe information obtained online.

Although digital divide perspectives and past research suggest that information evaluation strategies and opportunities may vary across demographic characteristics (see Hargittai, 2002; van Dijk, 2006), the studies by Metzger and Flanagin found little evidence that sex, income, education, or race greatly affected users' levels of concern about credibility or the amount and types of online information that people believe to be credible, although some small differences did emerge. Among adult Internet users, older Internet users were slightly more concerned about credibility, as were females and users with greater education. At the same time, older adults and females also felt that slightly more of the information on the Internet was believable (Metzger et al., 2011). Among children, race played a very minor role, such that nonwhite youth expressed somewhat higher concern about credibility, and children from families of higher income said they believed more information on the Internet. Both younger kids and girls were more likely to believe the information they find online compared to older kids and boys, respectively, although all of these differences explained only a small portion of the variance in credibility perceptions (Flanagin & Metzger, 2010).

Individuals' patterns of Internet usage, access, and past negative experiences were found to influence credibility judgments, presumably by leading different types of users to different types of information and by influencing their level of skepticism. For adults, greater use of the Internet for certain activities, such as social networking and electronic commerce, resulted in less concern about credibility and more trust of online information, whereas people who used the Internet to post content reported less trust in online information credibility (Metzger et al., 2011). Adults who reported spending more time online and possessing greater Internet skill expressed less concern about credibility and trusted more of the information available online, although children with higher skill and who spent more time online reported greater concern (Flanagin & Metzger, 2010). Users' online experiences and training in digital literacy also influenced their credibility beliefs: having had a bad experience with misinformation personally or even vicariously in the past, as well as having had formal instruction in credibility evaluation, contributed to greater concern about and less trust in the credibility of information on the Internet. With one exception as noted earlier, findings for children were similar to those for adults.

Several *personality traits* were also explored in these studies for their contribution to people's credibility beliefs and evaluation practices, including cognitive dispositions or "thinking styles" that have been shown to influence how people approach information. *Need for cognition*, for example, reflects the degree to which people engage in and enjoy thinking deeply about problems or information and, thus, may be willing to exert effort to critically evaluate information. *Flexible thinking* refers to people's willingness to consider opinions different from their own, which might impact how users process contradictory or contrasting information when judging credibility online. *Faith in intuition* reflects a tendency to trust based on first impressions, instincts, and feelings. And, *social trust*, or the propensity to trust strangers, is likely to affect the degree to which people judge information provided by those they do not know online to be trustworthy.

The studies by Flanagin and Metzger found that trait-based individual differences do affect credibility perceptions. Adults who were higher in need for cognition, and lower in social trust of others online expressed greater concern about credibility online (Metzger et al., 2011). Among children, flexible thinking style and social trust were related to children's level of credibility concern: Children who reported being more flexible in attending to information that runs counter to their own beliefs and who were less trusting of others online expressed greater concern about credibility (Flanagin & Metzger, 2010). Users of all ages with higher faith in intuition and lower need for cognition believed that more of the information available on the Internet is credible. In addition, higher need for cognition predicted greater analytic and lower heuristic credibility evaluation strategies in both younger and older Internet users.

Differences in findings between the adult and child samples noted earlier suggest that cognitive development may be another important individual-level variable that affects credibility beliefs and evaluations. Although children may be talented and comfortable users of technology, they may lack tools and cognitive abilities critical to effectively evaluate information. Youth also have less life experience than adults to help them compare information they find online or to discern relative reputational cues across sources. In addition, they may not have the same level of experience with or knowledge about media institutions, which can make it difficult for them to understand differences in editorial standards across various media channels and outlets compared to adults (Metzger & Flanagin, 2008).

Research in this area finds that as children mature, they show increased sophistication in interaction with information (Gross, 1999; Kafai & Bates, 1997; Solomon, 1993). That said, children often accept information presented online at face value and fail to apply critical criteria in their evaluation of the source or quality of online materials (Wallace & Kupperman, 1997). Many children report trusting the accuracy of information on the Internet simply because it exists online (Kafai & Bates, 1997), and many attribute altruistic intent to authors based on surface qualifications without considering possible bias or motives (Brem, Russell, & Weems, 2001). Eastin, Yang, and Nathanson (2006) found that children judged websites to be similarly credible regardless of whether the sites contained useful source information, were dominated by advertising, or simply offered dynamic features. In fact, children rated sites with advertising and no source as highest in credibility—even though they often believed the author of the page was the advertiser—indicating a potential misunderstanding of the relationship between content provider and advertiser. These findings reflect a level of confusion on the part of younger information consumers about the production processes, and possible ignorance regarding the ease with which any Internet user can publish online.

Additional individual traits are likely to be important in credibility assessment as well. For example, users with high topical knowledge have been found to be more critical of the information they find, as well as to assess qualifying characteristics of the source or the message more quickly than people with less knowledge on the topic about which they seek information (Fink-Shamit & Bar-Ilan, 2008; Fogg & Tseng, 1999; Gugerty, Billman, Pirolli, & Elliott, 2007). Similarly, early studies showed that users with greater Internet experience were better able to determine which features of websites indicated higher information quality and were more likely to lead to an accurate assessment of credibility (Flanagin & Metzger, 2007; Freeman & Spyridakis, 2004; Johnson & Kaye, 2000, 2002). Del Giudice (2010) argues that traits such as field dependence, locus of control, evaluation apprehension, and task orientation affect the likelihood of a person relying on available cues as guides for attitudes, perceptions, and behavior, and thus may influence credibility evaluation strategies and decisions. She further points out that cultural values such as power distance, subjective norms, individualism or collectivism, and uncertainty avoidance may affect how users determine the credibility of information they find online because these factors influence adoption of information technologies and affect the information contained within those technologies.

Social Interaction and Information Assessment

A transformative development in recent years has been the shift toward the increasingly *social* nature of information available online. The Internet environment is notable for its capacity to promote, maintain, and sustain collective endeavors among disaggregated individuals. A large class of tools and applications, including blogs, social bookmarking, wikis, social networking sites, and a range of ratings, recommendation, reputation, and credentialing systems, enables diverse opinions, experiences, and knowledge to be combined across individuals online. These tools have resulted in historically unparalleled information resources for people, who are now able to locate others with shared interests across a tremendous range of domains, and to efficiently share relevant information with one another in a timely manner.

In such an environment, people routinely defer to external sources of knowledge on a very large scale, resulting in a "radical externalization of the processes involved in trust assessment" (Taraborelli, 2007, p. 1). One form that this externalization takes is heavy reliance on more social means of online information processing and assessment through electronic networks, which are used to harness collective intelligence for evaluating information and sources online (Flanagin & Metzger, 2008, 2011, 2013; Metzger et al., 2010). In this manner, social media tools and applications have the potential to "replace the authoritative heft of traditional institutions with the surging wisdom of crowds" (Madden & Fox, 2006, p. 2).

The core advantage underlying user-generated information online is the capacity to share the experiential aspects of people's information, making aggregated individual experiences available to many. In spite of their relative lack of official authority, individuals may possess relevant expertise due to their firsthand knowledge or experience with a topic or situation, and thus may be accurately perceived by others as having a great deal of *experiential credibility* (Flanagin & Metzger, 2008, 2013; Pure et al., 2013). Indeed, shifts in information dissemination challenge longstanding models of information provision by suggesting circumstances under which sources that are not understood as "experts" in the traditional sense are in fact in the best position to provide the most credible information. Under conditions where knowledge is esoteric, diffused among many individuals, and depends on specific, situational understanding, it is often the case that the most reliable information is gleaned not from a traditional source imbued with authority by virtue of position or status, but rather from a diversity of individuals lacking official credentials or a widely recognized reputation.

A number of theoretical perspectives suggest why aggregated user-generated information rooted in experiential credibility, such as that contained in consumer ratings or reviews online, might be viewed as a credible source of information. The warranting principle, for example, suggests that people's judgments online, where key personal and relational information is often missing, are more aptly based on information that cannot be easily manipulated by an information source (Walther & Parks, 2002). Relatedly, signaling theory (Donath, 2007) argues that certain signals available about information sources—particularly those signals that are difficult to fake, are supported by the rule of law or social convention, or are costly to obtain or to mimic—are most reliable for assessing credibility. Signaling theory suggests that, when aggregated, user-provided information, can signal information credibility because it is difficult to for any one person to manipulate an average when there are many inputs. The warranting principle also implies that user-generated information may be perceived to be credible because the aggregation process makes it unlikely that any particular contributor has control over the collective opinion ultimately represented. Aggregated instances of users' experiential credibility thus serve as a warrant that information is valid and reliable.

Tests of the warranting principle and signaling theory show that cues that are difficult to fake online are indeed seen as more credible. For example, friends' comments on Facebook profiles were more influential in assessments of physical attractiveness than self-comments were (Walther, Van Der Heide, Hamel, & Shulman, 2009) and information generated by others had a higher impact than self-generated information on communal orientation (Utz, 2010). Also, consumer reviews have been found to be a stronger predictor of the trustworthiness of an online store than either store reputation or assurance seals, both of which are more readily controlled by the store itself (Utz, Kerkhof, & van den Bos, 2012). A variety of signals (i.e., seller reputation, product condition, and argument quality) have also been shown to predict important outcomes in online auctions (e.g., number of bids, auction success, and willingness to pay; Shen, Chiou, & Kuo, 2011), and people also put greater emphasis on signals that are costly to fake when locating experts online (Shami, Ehrlich, Gay, & Hancock, 2009).

Explicitly *social* influence processes also affect people's credibility assessments online. Social influence can take several forms. ***Informational social influence***, for instance, is the tendency to "accept information obtained from another as *evidence* about reality" (Deutsch & Gerard, 1955, p. 629; emphasis in original). Particularly in the absence of firsthand experience, people tend to believe that others' interpretations are more correct than their own, and rely on others to help them choose the appropriate course of action. In this manner, informational social influence is a means of gaining information under circumstances when people are uncertain about their own perceptions and may be the underlying theoretical driver of bandwagon effects. Information provided by others becomes influential in this view as a means of removing ambiguity and establishing subjective validity.

A related but complementary form of social influence is *referent informational influence* (Turner, Wetherell, & Hogg, 1989). Stemming from theories of social identity and self-categorization, referent informational influence explains opinion and behavior conformity by processes of group identification and the subsequent adherence to standards of group behavior. The most reliable and cognitively available sources of information about group norms are the attitudes or behaviors of fellow group members themselves, which serve to reduce the uncertainty of a social situation. Referent informational influence, therefore, operates by priming a social identity or category within the available social context. Individuals then rely on a cognitive representation of the group norm as a guide for their attitudes and behaviors.

Many studies have demonstrated the effects of social influence online. For example, people viewing movie ratings online tend to rate movies consistent with the ratings they have been shown and people's choices online are swayed by others' views in recommender systems (Cosley, Lam, Albert, Konstan, & Riedl, 2003; Zhu, Huberman, & Luon, 2011). In addition, people tend to find information contributed by similar others to be more credible (Flanagin, Hocevar, & Samahito, 2014) and online adoption of user-created content has been shown to be particularly prone to the social influences of friends (Bakshy, Karrer, & Adamic, 2009), suggesting that referent informational influence is a particularly important influence on Internet users' attitudes and behaviors. Both informational social influence and referent informational influence emphasize the role that socially available others play in prompting opinion and behavioral conformity, either through the information resources they provide or via the perceived norms of the identifiable group, both of which can serve to disambiguate complex information environments.

These perspectives suggest that others' opinions may be an important source of information when users evaluate the credibility of user-generated information, especially if information evaluators identify with the people who post their opinions. Supporting this view, research by Sundar and colleagues (2008) found that online consumers' opinions of product quality were highly influenced by others' opinions and found stronger attitude conformity as social information (i.e., the number of customer reviews) increased. Flanagin and Metzger (2013) similarly found that users' personal evaluations of movies tended to agree with online user-generated movie reviews as both the quantity of reviews and the degree to which participants generally identified with others' taste in movies increased.

Through these mechanisms, social media tools and applications can assist users in locating credible information online, and yet they pose difficulties for credibility assessment as well. Anonymous and multiple authors make the concept of "source" difficult to understand or authenticate and, as such, users know little about the expertise, qualifications, and potential biases that may be infused into the information they obtain from these resources. The proliferation of sources also amplifies concerns about the lack of gatekeepers, disintermediation, and source ambiguity. Also, while there is evidence that information aggregated across users can produce enhanced collective intelligence via the "wisdom of crowds," it is also true that crowds are not always wise, particularly when bandwagons can develop easily such that popularity can trump quality, when talent is assumed to be equally distributed across all contributors, and when the specific training and expertise of contributors are undervalued (Keen, 2007).

Research on the dynamics of user-provided online information is just beginning to address these realities by examining the tensions information consumers are now faced with when evaluating these newer forms of online information and sources. With Wikipedia, for instance, where information consumers must assess the credibility of information that originates primarily from largely unknown, fellow users, there is evidence that many people are skeptical of Wikipedia content (Flanagin & Metzger, 2011; Menchen-Trevino & Hargittai, 2011), and that people tend to find traditional venues (e.g., Encyclopaedia Britannica) to be more credible (Flanagin & Metzger, 2011). Interestingly, however, adults also find information (i.e., the content of the encyclopedia entries themselves, independent from the source) originating from Encyclopaedia Britannica and Wikipedia to be equally credible, and moreso than parallel encyclopedia entries on the hybrid user- and expert-produced content on Citizendium, demonstrating that they attend to credibility cues in the entries themselves. Moreover, children (ages 11–18) find encyclopedia entries from Wikipedia to be significantly more believable than those from other encyclopedias, but only when they think they are actually from more reputable sources (e.g., Encyclopaedia Britannica), suggesting an intriguing sort of social desirability effect, where internalized knowledge about what is and is not an acceptable information source exists in tension with credibility cues gleaned from the content of the message (Flanagin & Metzger, 2011).

Additionally, both children and adults report that other people should believe information on Wikipedia less than they themselves report believing it (Flanagin & Metzger, 2011), indicating a form of the "optimistic bias" effect (Weinstein, 1980) based on social comparison, whereby people tend to see themselves as less likely than others to experience negative life events. Research on optimistic bias has demonstrated its stability across a wide range of demographic variables, including age, sex, and education (Weinstein, 1987), but little research has focused on the occurrence of the optimistic bias in a digital media environment. The surveys described earlier by Metzger and Flanagin, however, found that children and adults believed that their ability to figure out which information is good and bad online was superior to that of "typical" Internet users, and reported being more likely to question information they find on the Internet, and slightly less likely to believe false information online compared to others (Flanagin & Metzger 2010; Metzger et al., 2011). These data provide support that for an optimistic bias in users' perceptions of their information literacy and credibility evaluation skills.

Compounding the problem of assessing the credibility of user-generated content online is that it may be difficult for people to interpret users' "experiential-based information" correctly, absent knowledge about basic statistical principles like sampling (Eysenbach, 2008). For example, research on user-generated commercial ratings shows that Internet users attend to the average "star" ratings for a product without also attending to the number of such ratings provided, potentially at the expense of critical information about the influence of a small number of opinions on the aggregate rating (Flanagin, Metzger, Pure, Markov, & Hartsell, 2014). Under such circumstances, people's ability to take appropriate advantage of user-generated information to make accurate credibility decisions is clearly suboptimal.

Research Directions and Conclusions

Features of the contemporary information and communication environment highlight the complex reality that information consumers confront when evaluating online information today. The opportunities and challenges presented by this environment suggest the importance of the psychological dimensions of people's information evaluation experiences, including their information processing activities, the personality-based characteristics that influence information evaluation, and the dynamics of information assessment that unfold in the context of social interaction online. A number of research directions naturally arise from these perspectives.

For example, the use of credibility cues and cognitive heuristics in decision making about credibility raises a number of questions that can serve to guide future research. As mentioned earlier, some evidence suggests that heuristic means of credibility assessment may be ***equally*** effective as analytic strategies in terms of the quality of credibility judgments that are rendered. This is because cognitive heuristics are based on rational principles and are honed through both evolutionary and individual learning processes (Gigerenzer & Todd, 1999). An avenue for research, therefore, is to address the questions of whether and under what circumstances the use of heuristic evaluation strategies leads to good or bad credibility decisions. Taraborelli (2007) provides a useful model for conducting this type of research in an experimental setting that may be helpful for researchers pursuing work in this area.

It is also likely that many instances of online information involve multiple cues that bear on a user's credibility evaluation, so another important question for future research concerns how people cope with conflicting credibility cues. An example of this is when bandwagon effects appear to validate information that violates users' expectations for professionalism because, for example, it is presented in a sloppy manner (see also Sundar, 2008; Sundar, Xu, & Oeldorf-Hirsch, 2009 for other examples). Wikipedia may also evoke heuristics based on authority cues that compete with bandwagon heuristics and notions about the wisdom of crowds. In these circumstances, which cue is privileged and how do people reconcile the dissonance they might experience when interpreting conflicting credibility cues or heuristics? Credibility cues may be ordered hierarchically in cognition, may interact with specific individual-level traits or attitudes to be more or less perceptually salient, or may be based on the relative prominence of cues in the particular interface, but future research is needed to determine the validity of these as well as other potential explanations.

Social influence processes may also help to explain how users deal with the tension between believing those with credentialed authority, such as acknowledged experts or traditionally recognized gatekeepers, and those with high experiential credibility, such as laypeople with esoteric or firsthand knowledge, when information provided by each is in conflict. These views suggest that this cognitively demanding task is guided by a host of factors, including individual information processing differences and experiences, social cues, and relevant group identities.

Informational social influence processes, for example, suggest that both types of information might be relied upon to guide information consumers' opinions, particularly if they lack their own firsthand knowledge. Referent social influence further suggests that these influences might be particularly pronounced under conditions where group identification is high. To unravel the specific influences on people's credibility assessments, research will need to locate the conditions under which social influence processes are most likely online (e.g., those where individuals lack their own concrete knowledge, such as evaluative situations where subjective opinions are prevalent), relevant individual-level differences (e.g., social trust), and the group identifications that might be germane to each situation (e.g., specific group-based alliances such as political party identification or identification with particular groups' information sharing practices). Research in this domain might illuminate the social influences and cognitive processes individuals undergo when faced with complex, and occasionally competing, information influences online. Related to this, signaling theory and the warranting principle suggest that lay users may be viewed as equally or even more credible than experts, at least in situations where sufficient evidence shows that user-generated information is unlikely to have been manipulated by a small number of biased sources (e.g., see Flanagin & Metzger, 2013).

Because the tension between expert- and user-generated information is situation-specific, future research should also account for particular information contexts, and their psychological impacts. For instance, during times of crisis where accurate and timely information is critical (e.g., a rapidly unfolding natural disaster, such as a hurricane or wildfire), users may face added pressure to locate specific and credible information, which might not be available in sufficient degree from traditional sources, such as the mainstream news media. Whereas mainstream media tend to rely on a small number of credentialed sources (e.g., the fire or police chief), information that originates from individuals reporting on their own firsthand observations has the potential to be superior since, using social media tools, people can provide specific experiential information in real time to large forums. Research to date, though, has not addressed the basic issues implied in these scenarios, such as the tradeoff between urgency and accuracy (perhaps as understood through the theoretical lens of motivation), differences in comfort levels with or acceptance of fellow citizens as external information sources, whether some users (e.g., those with particular facility in using social media tools) are more comfortable in trusting this information than others, and generational or developmental differences along these lines.

Finally, a good deal of work is also needed to identify and inventory the various cues, heuristics, and strategies that people use during credibility assessment within and across various online environments and information-seeking contexts because it is an important first step in developing theoretical explanations of information evaluation behavior online. Moreover, a thorough understanding of the processes used in online information assessment can help educators and others to design intervention strategies to improve Internet users' information evaluation skills. In the end, the application of research findings in this manner can help people avoid deception, manipulation, and persuasion by misinformation in the Internet environment, which is critical as people increasingly turn to online sources for information that matters in their lives.

References

Bakshy, E., Karrer, B., & Adamic, L. A. (2009). Social influence and the diffusion of user-created content. *Proceedings of the Tenth ACM Conference on Electronic Commerce (EC09)* (pp. 325–334). Stanford, CA: ACM.

Brem, S. K., Russell, J., & Weems, L. (2001). Science on the web: Student evaluations of scientific arguments. *Discourse Processes, 32*(2–3), 191–213.

Byerly, G., & Brodie, C. S. (2005, April). *Internet (and/or institutional) credibility and the user.* Paper presented at the Internet Credibility and the User Symposium at the University of Washington, Seattle, WA.

Callister, T. A., Jr. (2000). Media literacy: On-ramp to the literacy of the 21st century or cul-de-sac on the information superhighway. *Advances in Reading/Language Research, 7*, 403–420.

Chaiken, S. (1980). Heuristic versus systematic processing and the use of source versus message cues in persuasion. *Journal of Personality and Social Psychology, 39*, 752–766.

Chaiken, S. (1987). The heuristic model of persuasion. In M. P. Zanna, J. M. Olsen, & C. P. Herman (Eds.), *Social influence: The Ontario symposium* (Vol. 5, pp. 3–39). Hillsdale, NJ: Erlbaum.

Chaiken, S., & Trope, Y. (1999). *Dual-process theories in social psychology.* New York, NY: Guilford Press.

Chen, S., & Chaiken, S. (1999). The heuristic-systematic model in its broader context. In S. Chaiken & Y. Trope (Eds.), *Dual-process theories in social psychology* (pp. 73–96). New York, NY: Guilford Press.

Cosley, D., Lam, S. K, Albert, I., Konstan, J. A., & Riedl, J. (2003). Is seeing believing? How recommender system interfaces affect users' opinions. *Proceedings of the SIGCHI conference on Human Factors in Computing Systems (CHI '03)*, Vancouver, Canada.

Danielson, D. R. (2005). Web credibility. In C. Ghaoui (Ed.), *Encyclopedia of human-computer interaction* (pp. 713–721). Hersey, PA: Idea Group.

Del Giudice, K. V. (2010). *Trust on the web: The impact of social consensus on information credibility* (Unpublished doctoral dissertation). University of Central Florida, Orlando, FL. Retrieved from http://purl.fcla.edu/fcla/etd/CFE0003240

Deutsch, M., & Gerard, H. B. (1955). A study of normative and informational social influences upon individual judgment. *Journal of Abnormal and Social Psychology, 51*, 629–636.

Donath, J. (2007). Signals in social supernets. *Journal of Computer-Mediated Communication, 13*(1), Article 12. Retrieved from http://jcmc.indiana.edu/vol13/issue1/donath.html

Eastin, M. S., Yang, M. S., & Nathanson, A. I. (2006). Children of the net: An empirical exploration into the evaluation of Internet content. *Journal of Broadcasting & Electronic Media, 50*(2), 211–230.

Eysenbach, G. (2008). Credibility of health information and digital media: New perspective and implications for youth. In M. J. Metzger & A. J. Flanagin (Eds.), *Digital media, youth, and credibility* (pp. 123–154). Cambridge, MA: MIT Press.

Fink-Shamit, N., & Bar-Ilan, J. (2008). Information quality assessment on the web—An expression of behaviour. *Information Research, 13*(4), 357. Retrieved from http://informationr.net/ir/13-4/paper357.html

Flanagin, A. J., Hocevar, K., & Samahito, S. (2014). Connecting with the user-generated web: How shared social identity impacts online information sharing and evaluation. *Information, Communication, and Society, 17*, 6, 683–694.

Flanagin, A. J., & Metzger, M. J. (2000). Perceptions of internet information credibility. *Journalism and Mass Communication Quarterly, 77*, 515–540.

Flanagin, A. J., & Metzger, M. J. (2007). The role of site features, user attributes, and information verification behaviors on the perceived credibility of Web-based information. *New Media & Society, 9*, 319–342.

Flanagin, A. J., & Metzger, M. J. (2008). Digital media and youth: Unparalleled opportunity and unprecedented responsibility (pp. 5–27). In M. J. Metzger & A. J. Flanagin (Eds.), *Digital media, youth, and credibility*. Cambridge, MA: MIT Press.

Flanagin, A. J., & Metzger, M. J. (2010). *Kids and credibility: An empirical examination of youth, digital media use, and information credibility*. Cambridge, MA: MIT Press.

Flanagin, A. J., & Metzger, M. J. (2011). From Encyclopedia Britannica to Wikipedia: Generational differences in the perceived credibility of online encyclopedia information. *Information, Communication, and Society, 14*, 355–374.

Flanagin, A. J., & Metzger, M. J. (2013). Trusting expert- versus user-generated ratings online: The role of information volume, valence, and consumer characteristics. *Computers in Human Behavior, 29*, 1626–1634.

Flanagin, A. J., Metzger, M. J., Pure, R., Markov, A., & Hartsell, E. (2014). Mitigating risk in ecommerce transactions: Perceptions of information credibility and the role of user-generated ratings in product quality and purchase intention. *Electronic Commerce Research, 14*, 1–23.

Fogg, B. J. (2003). Prominence-interpretation theory: Explaining how people assess credibility online. *Proceedings of CHI'03, Extended Abstracts on Human Factors in Computing Systems* (pp. 722–723). New York, NY: ACM.

Fogg, B. J., Soohoo, C., Danielson, D. R., Marable, L., Stanford, J., & Trauber, E. R. (2003). How do users evaluate the credibility of Web sites? A study with over 2,500 participants. *Proceedings of the 2003 Conference on Designing for User Experiences*, San Francisco, CA. Retrieved from http://portal.acm.org/citation.cfm?doid=997078.997097

Fogg, B. J., & Tseng, H. (1999). The elements of computer credibility. *Proceedings of the CHI '99, Human Factors in Computing Systems*, Pittsburgh, PA (pp. 80–87). New York, NY: ACM.

Forsyth, D. R. (2010). *Group dynamics* (5th ed.). Belmont, CA: Wadsworth.

Freeman, K. S., & Spyridakis, J. H. (2004). An examination of factors that affect the credibility of online health information. *Technical Communication, 51*, 239–263.

Fritch, J. W., & Cromwell, R. L. (2002). Delving deeper into evaluation: Exploring cognitive authority on the Internet. *Reference Services Review, 30*, 242–254.

Gefen, D., Karahanna, E., & Straub, D. (2003). Trust and TAM in online shopping: An integrated model. *MIS Quarterly, 27*, 51–90.

Gigerenzer, G., & Todd, P. M. (1999). *Simple heuristics that make us smart*. New York, NY: Oxford University Press.

Gladwell, M. (2005). *Blink: The power of thinking without thinking*. New York, NY: Little, Brown & Company.

Gross, M. (1999). Imposed queries in the school library media center: A descriptive study. *Library & Information Science Research, 21*, 501–521.

Gugerty, L., Billman, D., Pirolli, P., & Elliott, A. (2007). An exploratory study of the effect of domain knowledge on internet search behavior: The case of diabetes. *Proceedings of the Human Factors and Ergonomics Society, 51*, 775–779.

Hargittai, E. (2002). Second-level digital divide: Differences in people's online skills. *First Monday, 7*(4), np.

Hesse, B. W., O'Connell, M., Augustson, E. M., Chou, W., Shaikh, A. R., & Finney Rutton, L. J. (2011). Realizing the promise of web 2.0: Engaging community intelligence. *Journal of Health Communication, 16*, 10–31.

Hilligoss, B., & Rieh, S. Y. (2008). Developing a unifying framework for credibility assessment: Construct, heuristics, and interactions in context. *Information Processing & Management, 44*, 1467–1484.

Hong, T. (2006). The Internet and tobacco cessation: The roles of Internet self-efficacy and search task on the information-seeking process. *Journal of Computer-Mediated Communication, 11*(2), Article No. 8. Retrieved from http://jcmc.indiana.edu/vol11/issue2/hong.html

Horrigan, J., & Rainie, L. (2006). *When facing a tough decision, 60 million Americans now seek the internet's help: The internet's growing role in life's major moments.* Retrieved from http://www.pewinternet.org/2006/04/19/the-internets-growing-role-in-lifes-major-moments/

Hovland, C. I., Janis, I. L., & Kelley, H. H. (1953). *Communication and persuasion.* New Haven, CT: Yale University Press.

Johnson, T. J., & Kaye, B. K. (2000). Using is believing: The influence of reliance on the credibility of online political information among politically interested internet users. *Journalism and Mass Communication Quarterly, 77*, 865–879.

Johnson, T. J., & Kaye, B. K. (2002). Webelievability: A path model examining how convenience and reliance predict online credibility. *Journalism and Mass Communication Quarterly, 79*, 619–642.

Kafai, Y., & Bates, M. (1997). Internet web-searching instruction in the elementary classroom: Building a foundation for information literacy. *School Library Media Quarterly, 25*, 103–111.

Keen, A. (2007). *The cult of the amateur: How today's Internet is killing our culture and assaulting our economy.* London, UK: Nicholas Brealey Publishing.

Kruger, J., Wirtz, D., Van Boven, L., & Altermatt, T. W. (2004). The effort heuristic. *Journal of Experimental Social Psychology, 40*, 91–98.

Lane, C., & Bachmann, R. (Eds.). (1998). *Trust within and between organizations: Conceptual issues and empirical applications.* Oxford, UK: Oxford University Press.

Lang, A. (2000). The limited capacity model of mediated message processing. *Journal of Communication, 50*, 46–70.

Madden, M., & Fox, S. (2006). Riding the waves of Web 2.0. Retrieved from http://pewresearch.org/pubs/71/riding-the-waves-of-web-20

McKnight, D. H., Choudhury, V., & Kacmar, C. (2002). Developing and validating trust measures for e-commerce: An integrative typology. *Information Systems Research, 13*, 334–359.

Menchen-Trevino, E., & Hargittai, E. (2011). Young adults' credibility assessment of Wikipedia. *Information, Communication and Society, 14*, 24–51.

Metzger, M. J. (2007). Making sense of credibility on the Web: Models for evaluating online information and recommendations for future research. *Journal of the American Society of Information Science and Technology, 58*, 2078–2091.

Metzger, M. J., & Flanagin, A. J. (Eds.). (2008). *Digital media, youth, and credibility.* Cambridge, MA: MIT Press.

Metzger, M. J., Flanagin, A. J., Eyal, K., Lemus, D. R., & McCann, R. (2003). Credibility in the 21st century: Integrating perspectives on source, message, and media credibility in the contemporary media environment. In P. Kalbfeisch (Ed.), *Communication yearbook 27* (pp. 293–335). Mahwah, NJ: Lawrence Erlbaum.

Metzger, M. J., Flanagin, A. J., & Medders, R. B. (2010). Social and heuristic approaches to credibility evaluation online. *Journal of Communication, 60*, 413–439.

Metzger, M. J., Flanagin, A. J., Pure, R., Medders, R., Markov, A., Hartsell, E., & Choi, E. (2011). *Adults and credibility: An empirical examination of digital media use and information credibility.* Research report prepared for the John D. and Catherine T. MacArthur Foundation. University of California, Santa Barbara, CA.

Petty, R. E., & Cacioppo, J. T. (1981). *Attitudes and persuasion: Classic and contemporary approaches.* Dubuque, IA: Brown.

Pirolli, P. (2005). Rational analyses of information foraging on the Web. *Cognitive Science, 29,* 343–373.

Pure, R. A., Markov, A. R., Mangus, J. M., Metzger, M. J., Flanagin, A. J., & Hartsell, E. H. (2013). Understanding and evaluating source expertise in an evolving media environment. In T. Takševa (Ed.), *Social software and the evolution of user expertise: Future trends in knowledge creation and dissemination* (pp. 37–51). Hershey, PA: Information Science Reference. doi:10.4018/978-1-4666-2178-7.ch003

Ridings, C., Gefen, D., & Arinze, B. (2002). Some antecedents and effects of trust in virtual communities. *The Journal of Strategic Information Systems, 11*(3–4), 271–295.

Rieh, S. Y., & Danielson, D. R. (2007). Credibility: A multidisciplinary framework. In B. Cronin (Ed.), *Annual review of information science and technology* (Vol. 41, pp. 307–364). Medford, NJ: Information Today.

Ross, L., Greene, D., & House, P. (1977). The false consensus effect: An egocentric bias in social perception and attribution processes. *Journal of Experimental Social Psychology, 13,* 279–301.

Sarker, S., Valacich, S. J., & Sarker, S. (2003). Virtual team trust: Instrument development and validation in an IS educational environment. *Information Resources Management Journal, 16*(2), 35–55.

Shami, N. S., Ehrlich, K., Gay, G., & Hancock, J. T. (2009). Making sense of strangers' expertise from signals in digital artifacts. *Proceedings of the 27th International Conference on Human Factors in Computing Systems.* New York, NY: ACM.

Shen, C., Chiou, J., & Kuo, B. (2011). Remedies for information asymmetry in online transaction: An investigation into the impact of web page signals on auction outcome. *Internet Research, 21,* 154–170.

Sillence, E., Briggs, P., Harris, P., & Fishwick, L. (2007). How do patients evaluate and make use of online health information? *Social Science & Medicine, 644,* 1853–1862.

Simon, H. A. (1955). A behavioral model of rational choice. *Quarterly Journal of Economics, 69,* 99–118.

Solomon, P. (1993), Children's information retrieval behavior: A case analysis of an OPAC. *Journal of the American Society for Information Science, 44,* 245–264.

Sundar, S. S. (2008). The MAIN model: A heuristic approach to understanding technology effects on credibility. In M. Metzger & A. Flanagin (Eds.), *Digital media, youth, and credibility* (pp. 73–100). Cambridge, MA: MIT Press.

Sundar, S. S., Oeldorf-Hirsch, A., & Xu, Q. (2008). The bandwagon effect of collaborative filtering technology. *Proceedings of CHI'08 Extended Abstracts on Human Factors in Computing Systems, 26,* 3453–3458.

Sundar, S. S., Xu, Q., & Oeldorf-Hirsch, A. (2009). Authority vs. peer: How interface cues influence users. *Proceedings of the 27th International Conference Extended Abstracts on Human Factors in Computing Systems (CHI'09), 27,* 4231–4236.

Surowiecki, J. (2005). *The wisdom of crowds.* New York, NY: Anchor Books.

Taraborelli, D. (2007, June). *Soft deference: How the Web is changing the way we trust.* Paper presented at the fifth European Computing and Philosophy Conference (ECAP 2007), Twente, The Netherlands.

Tseng, S., & Fogg, B. J. (1999). Credibility and computing technology. *Communications of the ACM, 42*(5), 39–44.

Turner, J. C., Wetherell, M. S., & Hogg, M. A. (1989). Referent informational influence and group polarization. *British Journal of Social Psychology, 28,* 135–147.

Tversky, A., & Kahneman, D. (1974). Judgment under uncertainty: Heuristics and biases. *Science, 185*, 1124–1131.

Utz, S. (2010). Show me your friends and I will tell you what type of person you are: How one's profile, number of friends, and type of friends influence impression formation on social network sites. *Journal of Computer-Mediated Communication, 15*, 314–335.

Utz, S., Kerkhof, P., & van den Bos, J. (2012). Consumers rule: How consumer review influence perceived trustworthiness of online stores. *Electronic Commerce Research and Applications, 11*, 49–58.

van Dijk, J. (2006). Digital divide research, achievements and shortcomings. *Poetics, 34*, 221–235.

Wallace, R., & Kupperman, J. (1997, March). *On-line search in the science classroom: Benefits and possibilities.* Paper presented at the Annual Meeting of the American Educational Research Association, Chicago, IL.

Walther, J. B., & Parks, M. R. (2002). Cues filtered out, cues filtered in: Computer-mediated communication and relationships. In M. L. Knapp & J. A. Daly (Eds.), *Handbook of interpersonal communication* (3rd ed., pp. 529–563). Thousand Oaks, CA: Sage.

Walther, J. B., Van Der Heide, B., Hamel, L., & Shulman, H. (2009). Self-generated versus other-generated statements and impressions in computer-mediated communication: A test of warranting theory using Facebook. *Communication Research, 36*, 229–253.

Walther, J. B., Wang, Z., & Loh, T. (2004). The effect of top-level domains and advertisements on health web-site credibility. *Journal of Medical Internet Research, 6*(3), np.

Wathen, C. N., & Burkell, J. (2002). Believe it or not: Factors influencing credibility on the Web. *Journal of the American Society for Information Science and Technology, 53*, 133–144.

Weinstein, N. D. (1980). Unrealistic optimism about future life events. *Journal of Personality and Social Psychology, 39*, 806–820.

Weinstein, N. D. (1987). Unrealistic optimism about susceptibility to health problems: Conclusions from a community-wide sample. *Journal of Behavioral Medicine, 10*, 481–500.

Wilson, P. (1983). *Second-hand knowledge: An inquiry into cognitive authority.* Westport, CT: Greenwood Press.

Wu, I., & Chen, J. (2005). An extension of trust and TAM model with TPB in the initial adoption of on-line tax: An empirical study. *International Journal of Human-Computer Studies, 62*,784–808.

Wu, K., Zhao, Y., Zhu, Q., Tan, X., & Zheng, H (2011). A meta-analysis of the impact of trust on technology acceptance model: Investigation of moderating influence of subject and context type. *International Journal of Information Management, 31*, 572–581.

Zhu, H., Huberman, B. A., & Luon, Y. (2011). To switch or not to switch: Understanding social influence in recommender systems. Retrieved from http://arxiv.org/abs/1108.5147

Part VI

Good Technology for Better Health

21

Trust and Engagement in Online Health
A Timeline Approach

Elizabeth Sillence and Pam Briggs
Faculty of Health and Life Sciences, PaCT Lab, Department of Psychology, Northumbria University, Newcastle upon Tyne, UK

Introduction

Online resources are now established as a major route for patients' health information and support (Ziebland & Wyke, 2012). The volume of available material, however, and the unregulated nature of health information on the Internet create potential issues for users with reviews of e-health material across a range of contexts typically concluding that quality is a problem (e.g., Eysenbach, Powell, Kuss, & Sa, 2002). How do people navigate through the health resources online, how do they decide where to go for information and advice and which sites to trust and engage with further? This chapter aims to explore the role of the Internet in health behavior placing particular emphasis on the issue of trust and trusting behaviors as this is seen as key to determining the impact of the Internet on health outcomes and behaviors. The rest of this chapter is organized as follows. In the next section, we discuss the role of the Internet in the context of health information and advice, exploring issues of user motivation, types of online health resources available and potential outcomes in terms of perceived health benefits as well as interactions with healthcare professionals. In the section "Trust in e-Health," we examine the context of trust in relation to online health advice and information and present a staged model of trust that helps reconcile differences in the literature. The next section presents a timeline approach to the study of trust and engagement in online health. This section details the way in which health consumers are initially drawn to online resources through to their longer-term engagement and their influence upon health-based decision making. The section "Discussion and Practical Implications" draws together the literature and the results of the work to present some conclusions and considerations for future research.

The Handbook of the Psychology of Communication Technology, First Edition. Edited by S. Shyam Sundar.

The Role of the Internet in Health Information and Advice

Dealing with any health condition requires patients to engage with information and advice. The source of that advice varies depending on the nature of the condition, the support networks surrounding the patient and the preferences of the patient themselves. One original source of information is medical expertise. Visits to the clinician, nurse or other health care professional are an important source of information and advice. Typically the kind of information provided by medical experts takes the form of facts and figures, statistical accounts of likely treatment outcomes. Large-scale evidence-based medicine must of course play a role in the development of appropriate information and advice to patients be that via face-to-face meetings with clinicians or as is increasingly the case via the Internet. Eighty percent of Internet users look online for health information (Fox, 2011). Using the Internet to supply the medical expertise allows users to access information at their own convenience as well as extending the transfer of information that is possible within the space of the short consultation session. While this medical facts and figures approach has formed the traditional basis for online health information, an increasingly popular alternative source of information is that provided by the layperson. Accounts of disease experience and treatment decisions authored by patients themselves represent another kind of expertise. These accounts exist within discussion forums, chat rooms, blogs and commentaries, as well as patients' home pages. The patient as expert provides information and advice not directly available from any other source. As Preece (1999) put it: "Physicians can provide the facts, but other patients can tell you what it really feels like and what to expect next, in a way that only someone with personal experience can" (p. 63). This kind of testimonial account provides people with more than scientific facts. It allows them to know more about the experiences of other people with the same condition and be able to share their own experience with others and can be viewed as a more trusted source of information than doctors themselves (Schaffer, Kuczynski, & Skinner, 2008). Knowing other people's experience has the potential to affect decision making, reduce ones sense of isolation, and improve adjustment to the disease. It can also contribute to a shifting understanding of a person's chronic condition (Preece & Maloney-Krichmar, 2005). Ziebland and Wyke (2012) identified seven possible effects of exposure to patient experiences including, use of health services, learning to relate the story and visualization.

Online health resources are developed and hosted online by a range of providers from charities and support groups through to government departments, pharmaceutical companies, and commercial organizations (Mo, Sillence, Briggs, & Harris, 2011; Ziebland & Wyke, 2012). As more providers enter the space, it is interesting to note the changes that have taken place over the last decade. A series of large-scale questionnaire studies conducted by the authors looking at online health advice shows a trend toward the rise in less regulated sites, that is, those owned by an individual or support group (Sillence, Mo, Briggs, & Harris, 2011). These kinds of sites, along with others, typically contain patient experiences. As previously mentioned, patient experiences are becoming an increasingly important feature of health resources online. Table 21.1 indicates the range of different sites available and provides an

Table 21.1 Health websites and range of features including patient experiences

	Website		
	Netdoctor	Healthtalkonline	NHS choices
	www.netdoctor.co.uk	www.healthtalkonline.org	www.nhs.uk
Health condition	General health	For example, high blood pressure	General health
Owner	Web provider	Charity	Government
Type of advice	Patient fact sheets	Collections of video and audio clips on health experiences and illnesses	Health encyclopedia
			Self-help guide
	Ask the doctor service		Clinical evidence–based research reports
	News	Medical information	Short films in which people talk about their health conditions
	Online forums	Frequently asked questions	
Commercial aspects	Health-related advertising	Donations requested	None
Patient experience features	Online forums—readers can share their experiences	Online collections of videos	Films and written stories
		Online forums	

overview of three different UK-based sites highlighting their main features and the range of services available on each site including how they incorporate patient experiences material.

The three sites described in Table 21.1 highlight some of the different ways in which patient experiences are being incorporated into health websites. Others include blogs that chronicle patient journeys, social networking sites, which allow patients to share ideas and knowledge directly without the need for intermediaries and home-made videos and testimonials that allow people to visualize health, illness and treatment. This "social software" (Giustini, 2006) also allows people to share and obtain recommendations about health services, hospitals, and providers at websites such as RateMDs (https://www.ratemds.com) or collaborate on data collection using sites like Patients Like Me (www.patientslikeme.com).

Our own work examining the kinds of patient experiences people report using online indicates that typically the sites they visit contain more than one example of patient experience material. Sites, for example, often contain blogs, forums, and videos and also vary in the extent to which they wholly or partly comprise of patient experiences. Some larger general health information sites contain approximately 10% patient experiences while others such as a support group for women facing hysterectomy are made up of about 90% patient experiences.

Trust in e-Health

Trust—"an expectancy held by an individual or group that the word, promise, verbal or written statement of another individual or group can be relied upon" (Rotter, 1967, p. 651)—is widely considered to be pivotal to the process of engaging with online health resources. A key point to note is that trust is intimately associated with risk—indeed, it is possible to argue that in the absence of risk, trust is meaningless (Brien, 1998). Even within seemingly simple acts of trust complex sets of judgments are invoked and a risk assessment is involved and in fact some of the trust models that have been developed in recent years have explicitly included risk (Corritore, Kracher, & Wiedenbeck, 2003; Corritore, Wiedenbeck, Kracher, & Marble, 2007). In online advice, there is no doubt that people are more willing to trust a site if the perceived risk is low. In a study by Briggs, Burford, De Angeli, and Lynch (2002), those that sought advice in relatively high-risk domains (e.g., finance) were less likely to trust and subsequently act on the advice than those who sought advice in low-risk domains (e.g., entertainment).

Surveys of Internet users suggest that trust is an important issue in the health domain (Williams, Huntington, & Nicholas, 2003) and online users appear to trust the health advice offered to them (Sillence, Briggs, Harris, & Fishwick, 2007b) yet attempts to unpack the concept of trust with respect to online information and advice reveal a multifaceted, complex concept. Acceptance of online health advice will depend upon people's trust in online transactions, in the technology used for the transactions, and in organizations (or individuals) as the other parties in the transactions (Beldad, de Jong, & Steehouder, 2010). The e-commerce literature provides a starting point for examining the ways in which people evaluate the trustworthiness of health information and advice online (see Grabner-Krauter & Kaluscha, 2003 for a review). This literature points to a number of key factors for trust development. Firstly, the look and feel of the site is important with people trusting, for example, those sites rated high in visual appeal and mistrusting those sites with poor visual design, unprofessional errors, or poor color appeal (Cyr, Head, & Larios, 2010; Fogg et al., 2002; Lindgaard, Dudek, Sen, Sumegi, & Noonan, 2011; Stanford, Tauber, Fogg, & Marable, 2002). Secondly, they may be influenced by the branding of the site or by the presence of familiar images, photographs or trusted logos, seals of approval, privacy policies, or third-party certificates (Riegelsberger, Sasse, & McCarthy, 2003; Wang & Emurain, 2005). Thirdly, they may be influenced by the quality of information available on the site, trusting those sites with greater perceived expertise, integrity, benevolence or information credibility (Bhattacherjee, 2002; Corritore et al., 2003, 2007; McKnight & Chervany, 2001); and fourthly, a number of authors have highlighted the importance of personalization or social feedback in the formation of trust judgments (Briggs, de Angeli, & Simpson, 2004; Giudice, 2010) or the notion of good relationship management (Egger, 2000).

While these different factors appear to be important, researchers disagree over their relative importance in fostering trust with some models emphasizing the design factors (Fogg et al., 2001) and others credibility of information, reputation, and risk (Corritore et al., 2007; Cugelman, Thelwall, & Dawes, 2009). One feature that some

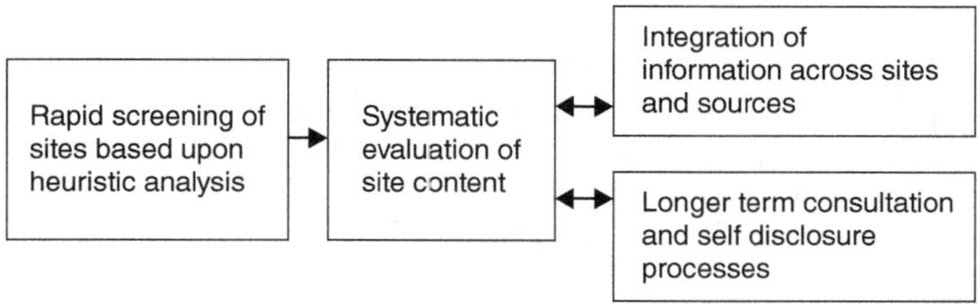

Figure 21.1 Staged model of trust. From Briggs et al. (2002). © Sage.

of the models allude to although few if any detail explicitly is the notion of the way trust develops over time. Riegelsberger, Sasse, and McCarthy (2005) in their model of trust describe different stages of trust that develop over time: early, medium, and mature forms of trust. These concepts however are not included in their model diagram; however, it is worth distinguishing between the kinds of trust that support transient interactions and those that support longer-term relationships (Meyerson, Weick, & Kramer, 1996). If one considers trust in this developmental context, then some of the findings in the literature make more sense. In particular, consideration of a developmental context helps to reconcile the tension between those models of trust that suggest that it is a concept grounded in careful judgment of institution and process factors such as vendor expertise and experience, predictability, degree of personalization and communication integrity, and those models that suggest trust decisions rely more heavily on the attractiveness and professional feel of a site.

A staged or developmental model of trust sees users engage in a fast preliminary assessment of a site before moving on to a more in-depth evaluation of the information available in a selected few sites, and finally developing a long-term trusting relationship with one or more particular sites. Such a model (Figure 21.1) was proposed by Briggs et al. (2002).

According to this model, most users engage initially in a rapid screening of the large numbers of sites accessed using general search engines and do so using relatively superficial aspects of a website. Subsequently, they are presumed to spend longer exploring a handful of sites in greater depth using more considered judgments of content. The first stage is thus a heuristic or affect-based stage in which an initial trust impression is formed based upon more salient or easily processed information; the second stage is an analytic stage in which a decision is made to engage more systematically with the material. Such a strategy is consistent with dual process models, such as those developed in the persuasion literature (Chaiken, 1980; Petty & Cacioppo, 1986). Chaiken (1980) identified two processing strategies by which an evaluation of trustworthiness may be made: firstly a heuristic strategy that follows a "cognitive miser" principle—where people base decisions on only the most obvious or apparent information; and secondly, a systematic strategy that involves the detailed processing of message content. A number of other studies in the persuasion literature support the two-process model, namely that people use cognitively intense analytical processing when the task is an important or particularly engaging one, whereas they use affect or other simple heuristics to guide their decisions when they lack the motivation or capacity to think properly about the issues involved. Such different processing strategies also reflect the distinction between the preliminary stage of (a) intention to trust and the later stage of (b) trusting activity (McKnight & Chervany, 2001). Briggs et al. (2002) conducted two different studies in order to provide information at both of these processing stages. In the first, qualitative, study, they investigated house hunters attempting to locate useful information online and explored those factors that influence first impressions of a website and thus intention to trust. In the second study, they carried out a survey of over two and a half thousand people who had actively sought advice over the Internet, thus exploring those factors that are more predictive of the uptake of advice, that is, the factors that influence trusting behavior.

The trusting process, however, does not stop after these two stages and a more realistic assessment of the development of trust should include a third stage in which a trusting relationship develops between the consumer and the website. Egger (2000, 2001) describes trust acquired over time (related to experienced trustworthiness). This, he proposes, is based on repeated interactions and a constant monitoring of the other party's trustworthiness. In e-commerce terms, this can be seen, for example, in the handling of inquires or orders over time; whether communication happens before or after ordering and the responsiveness and the quality of the help provided. Although the nature of the specific enquiries may differ within a health context, good communication and relationship management will play a role in the maintenance of trust between a consumer and the website over time.

The concept of a trusting relationship is once again center stage with the development of web 2.0 and online social networks. Here the technology, the website itself, and the other network participants can all be considered as objects of trust (Grabner-Kräuter, 2010). People involved in these kinds of trusting relationships are typically engaging in identification-based trust emerging from a history of personal interactions with the other (Van der Zee, Vos, & Luijters, 2009). Trust is influenced by joint values, tasks and goals, a sense of collective identity, and emotional closeness (Grabner-Kräuter, 2010). Within e-health, the peer-to-peer support group movement has been pivotal in its use of the Internet to develop long-term supportive relationships, providing empathy and alleviating patients' sense of isolation (Rozmovits & Ziebland, 2004; Ziebland & Wyke, 2012). The online community literature also suggests that long-term relationships online are based upon a sense of trust that develops through the exchange of information that is often highly personal. The ability to personalize content and to build personalized content is important in developing empathy within health-based communities (Preece & Ghozahti, 2001). Reading other peoples' reviews can affect credibility judgments of online health information with differing effects for older and younger adults (Liao, 2011).

A Timeline-Based Approach to Understanding Health Consumers' Use of the Internet

The staged model of trust (Figure 21.1) provides a framework for understanding how genuine health consumers search for and appraise online health information and advice. Based on this framework, a series of novel techniques were devised to examine people's decision-making and engagement processes, in particular with respect to

Table 21.2 Example of methodology from HRT study in phase 1

Fifteen women at various stages of the menopause participated in the study (mean age 49 years). All the women were interested in finding out more about the menopause and all used the Internet at least once a week. Each participant attended a total of four 2-hr sessions held in an Internet café in Newcastle-upon-Tyne, UK. During all four sessions, participants used the Internet to search for information and advice on the menopause for approximately 1 hr followed by a break and then a 45 min group discussion with a facilitator. Participants were told to freely surf the web during sessions one and four for information about the menopause and were directed to a number of specific websites during sessions two and three. These five to six specific sites were chosen for their trust design elements and contained a range of provider, content, and design features varied for trust. The group discussion guide covered the following main areas: (a) current information sources, (b) search strategies, (c) trusted and mistrusted websites, (d) first impressions, and (e) revisiting websites. Following the completion of the 4-week period, the participants were given a 6-month diary in which they were asked to record their ongoing information and advice searches both online and offline. At the end of the 6-month period, participants were invited to take part in a final, follow-up interview. This structured telephone interview examined the ways in which the Internet sources affected the patients' experiences of the menopause both in terms of decision making and communications with physicians

trust, through the first few seconds of interaction with a website to 6 months of engagement time. In addition, a number of ethnographic studies focusing on health-based online communities were carried out in order to examine longer-term disclosure and consultation processes with respect to decision making.

Next we present some findings from this series of studies drawing on examples from our research looking at a broad range of health websites as well as those containing patient experiences in particular. During phase 1, we focused on people faced with risky health decisions. Three longitudinal studies examined groups of people with decisions to make: (a) menopause and hormone replacement therapy (HRT), (b) hypertension, and (c) healthy living. In addition, a small-scale study examining the measles, mumps, rubella vaccination (MMR) was also carried out. The methodology for each health group was similar and is outlined in Table 21.2. During phase 2, we concentrated on online patient experiences. We sought to (a) undertake a systematic comparison of the impact of online experiences on patients' attention and (b) explore patient preferences for different types of experience. Participants selected for these separate but related studies were drawn from different health conditions in order to triangulate the findings in relation to different stages of engagement with patient experiences. The sample consisted of participants from three specific health groups—smokers wanting to quit, asthma sufferers, and MS carers—with variation across demographic and condition variables (e.g., severity of asthma). These three patient groups were chosen to reflect different types of health issue (i.e., patients with a chronic health condition, individuals looking for a health behavior change, and individuals who are responsible for someone with a serious health condition). Once again the studies within phase 2 followed a similar agenda and an overview example of the methodology is provided in Table 21.3. In phase 3, we focused on online communities of people

Table 21.3 Example of methodology from asthma study in phase 2

Fifteen people with a diagnosis of asthma (7 males and 8 females) participated in the first part of the study (mean age = 34; range 22–60). Having established the factors important in the initial selection and rejection of patient experience websites in an earlier study, this study undertook a systematic comparison of the impact of patient experiences on patient attention. To those ends we wanted to understand to what extent people chose to engage with websites containing patient experiences over sites that did not. Participants attended a 2-hr focus group held in an internet café in Newcastle-Upon-Tyne, UK. The session was divided into two parts. Firstly, participants were asked to look at seven specified health websites relating to asthma, some containing patient experience material and some without. Participants then had to choose one of the websites to look at in more detail before taking part in a group discussion. Participants were then asked to look at two further specified sites both of which contained patient experience material. Participants noted their thoughts in logbooks during the tasks and then took part in a group discussion. Follow-up telephone interviews where then carried out 2 weeks later.

Having established preferences for websites containing patient experiences, we then carried out a second study with a different group of asthma patients in order to explore patient preferences for different types of experience. This second group *N* = 14 (5 male, 9 female, mean age = 28; range 19–64) were invited to attend a 2½ hr individual testing session in the lab. During the session, participants viewed eight specific health relevant websites containing different types of experiences (e.g., blogs, patient testimonials, and message boards) and were directed to choose one site to focus on in more detail before taking part in a group discussion. As previously, a telephone interview was carried out 2 weeks later

with prostate cancer as they utilized message forums in order to ask for and receive advice about treatment decisions. This phase examined how people manage the process of advice giving in a cancer support group. A discourse analysis of data collected from an online message forum explored how participants manage to ask for and offer advice within a peer setting.

Data analysis

Logging data: This automatically generated data was subjected to simple descriptive analysis to establish the length of time each participant spent on the different websites.

Verbal protocols and log books: A content analysis of the protocols and log books sought to identify and quantify comments relating to the initial impressions of the websites.

Group discussions, interviews, and diaries: These were subjected to a thematic analysis in which participants comments were coded against a number of anticipated trust markers in addition to emergent themes.

Message analysis: The message posts were taken from three of the online cancer support website's forums and contained entire thread sequences, that is, the full set of messages and replies around a particular posting. Fifteen threads containing over 350 messages were retrieved and examined. A discursive analysis of two messages threads was carried out.

Findings from these studies are presented next and are divided into three areas: (a) rapid screening and initial selection/rejection, drawing on the logging data, verbal protocols and log books; (b) systematic evaluations including patient experiences, drawing on the group discussions and interviews; and (c) longer-term trust and engagement including online health communities, drawing on the diaries, and message analysis.

Rapid screening and early mistrust of sites

First, the automatically logged data for each of the three study groups in phase 1 was used to explore the processes of accepting or rejecting sites during the first few minutes. Such analysis typically revealed that health consumers disliked general information portals and were sensitive to sales sites, rejecting them quickly. Many sites were rejected within a ten second window mirroring findings that suggest that people form an opinion about the visual appeal of a website within the first 50 ms (Lindgaard, Fernandes, Dudek, & Brown, 2006). The speed with which consumers reject information provides a salutary lesson regarding the importance of providing the right cues to site content in a highly visible manner. Participants missed out on some potentially good sites with useful information because of poor design.

Second, those factors that led to early mistrust of sites were identified from the verbal protocols and log books of individual participants. Categorization of such factors was made in terms of design and content with the majority of comments relating to the design look and feel of the websites. In a study on menopause and HRT, for example, 94% of the total number of comments about rejection concerned design factors including inter alia an inappropriate name for the website, complex busy layout, lack of navigation aids, and a corporate look and feel. For more details, see Sillence, Briggs, Fishwick, and Harris (2004).

Developing trust through the systematic evaluation of content

The qualitative data from group discussions was analyzed in relation to the trust markers of the selected sites. In stage 1, the participants mentioned a number of factors in terms of the sites they had selected to explore in more depth. The participants liked sites that contained a great deal of information that was presented in such a manner that an individual could quickly pinpoint their own specific areas of interest. Participants trusted the selected sites because they demonstrated an in-depth knowledge of a wide variety of relevant topics and put forward unbiased clear information. Participants were more likely to trust the information if they could verify it and cross check it with other websites. Most individuals preferred sites that were run by reputable organizations or had a medical or expert feel about them. They trusted the information on such websites especially when the credentials of the site and its authors were made explicit. Most participants showed some distrust of the advice and information on websites sponsored by pharmaceutical companies or those explicitly selling products. Participants were looking for sites that were written by people similar to themselves, who shared similar interests.

This point was explored further in the phase 2 studies focusing on patient experience material online. In those studies (described in Table 21.3), engagement with that specific form of e-health website was predicated on the participants being able to find a trustworthy voice within the patient experiences (Sillence, Hardy, Briggs, & Harris, 2013) The “voice” of the experience consisted of a source and a message and both elements were considered by participants when deciding whether or not to trust the patient experience and engage with it further. Trustworthy sources were typically individuals who resonated with the participant in terms of age, gender and location as illustrated by the smoker below:

> I like reading that I mean it’s the same age, my age, my background, she’s smoked for 30 years I’ve smoked for 30 years! Its probably the same and I’m pretty active go to the gym so balancing, when you looked at the age oh it could be me! And that’s really good. (SmokingFG3P1)

Trustworthy and engaging messages again matched participants in terms of health issue (length since diagnosis, severity, treatment regime) or outlook and attitude (participants favored positive upbeat messages over those which were simply “venting”). Participants then compared themselves to the experiences they encountered and either rejected them or chose to trust them and consider engaging further as the extract next from one MS carer demonstrates:

> He took me in with his writing, yeah I thought it was quite poignant, it got to me a bit… I think he’s wrote it as he’s felt it, yeah I’d write something similar… I can totally empathise with him because I know his job inside out… if I saw somebody like that bloke there, if I was to read that again maybe later maybe I might write to him with a view to perhaps to put in my tupence worth and cheer him up I don’t know. (MSIntP4)

Longer-term trust and engagement

The 2-week and 6-month follow-up studies indicated a number of ways in which consumers engaged with the material they read online. In particular, diary studies and interviews identified the ways in which online health resources had assisted with decision making. To those ends participants indicated they felt the Internet had influenced their decision making in a number of distinct ways. First, it had improved their communications with health professionals and interestingly allowed them to act much like “scientists,” developing and testing their own hypotheses about the way the menopause or hypertension medication affected them. Participants used the Internet to investigate specific concerns, to compare sites and to assess risk. As well as testing out ideas online, the participants were also keen to test out ideas on themselves. This included finding answers to questions such as “could I manage with a lower dose” and “will I feel better if I stop taking HRT?” As the following extract from one of the follow-up telephone calls indicates:

> Well I was on HRT when I started out when we came to the end of our session I sort of looked at the options and decided for various reasons that I was going to try to come off HRT my main reason to be perfectly honest was just to find out what my body was doing because I had been on for about 5 years and there were all sorts of scare stories umm

> which I tend to take with a pinch of salt I must admit but there were there anyway and I was looking at it in such detail that I thought I'd try without so I was actually off for six months and at the end of the six months I couldn't stand it any longer cos I have really severe sweats and so I am back on and I am absolutely fine. (HRTP2)

The following example illustrates how one participant used the information online in conjunction with health professionals to bring about changes in his medication regime.

> TT took part in the hypertension study. He was taking medication for his high blood pressure and was unhappy with what he saw as the debilitating side effects. Discussions prior to the study had left him frustrated since his doctor had advised him that side effects were very unusual for this medication. TT was convinced his symptoms were not normal. During the sessions he came across one site in particular which detailed other people's experiences of side effects whilst taking this particular medication. During the group discussions he told everyone of his findings and his intention to show these results to his doctor. The diaries indicated that he returned to the website several times before visited his doctor. In consultation with his doctor he had agreed a change in his medication.

The follow-ups while useful in assessing how information was integrated and to some extent how decisions were made they did not provide any detailed information on the self disclosure processes involved and longer-term consultation. For this, we used an alternative methodology and observed postings to health message forums. In the case of ongoing, complex health issues, it became possible to examine interactions with health advice and information online over the longer term to see the development of trust between posters. Conveying trust to other online posters is not straightforward in a reduced cues environment. We have found that posters in prostate cancer forums use a variety of techniques in order to convey their trustworthiness to other readers (Sillence, 2010). First, the community itself draws on the medical discourse to add to its credibility. They carry out "a thorough medical history" of the poster as a doctor would do with his or her patient demanding certain pieces of information from a poster before any advice is provided. Second, members who respond to requests for advice use a number of warranting strategies (Richardson, 2003) to indicate their credibility and suitability for the task. These include presenting their own history and experiences through the profiles and the message posts, presenting their own key statistics and making reference to other websites and sources of information.

Men suffering from prostate cancer already possess some information about the condition and possible treatments before approaching the message forums in order to assist with their health decision making. They have already received at the very least, information and advice from their doctor. Once online they are exposed to a range of decision-making processes some based on a systematic evaluation of treatment options, success rates, and side effects others based on a more heuristic or rule of thumb approach such as simply having faith in the doctor's advice or following a lead from a friend or family member (Sillence & Mo, 2014). While narrative information in particular first-person narratives in experimental settings appear to influence health decision making, the mechanisms are not well understood (Winterbottom, Bekker, Conner, & Mooney, 2008). Our observational studies show that people usually come to such message forums with an initial decision already in mind. In the case of prostate cancer posters, the initial post to the message board was important. In this first "positioning" message, the newcomer detailed their situation, provided some clues as to their current thinking or viewpoint, and set the parameters by which advice givers could operate. During their interactions, online advice seekers were exposed to information and advice that did not fit with their current decision making and thus can be seen as uncongenial. In one case, the advice seeker simply ignored messages they did not want to hear and only responded to messages conveying congenial advice. They also guarded against uncongenial information through humor and through a purposeful adaptation of medical terminology. As in the HRT study (Sillence, Briggs, Harris, & Fishwick, 2007a), women were often motivated to seek information and advice that supported their own viewpoint and used this to build confidence in their decision making. For example, one participant reported "I wasn't too keen on HRT before and I'm certainly not now" after reading information on a number of websites. Such identity preferences are consistent with self-verification theory (Swann & Read, 1981), which explicitly predicts that people will seek the advice of others whose experiences and beliefs overlap with their own.

Discussion and Practical Implications

Our studies provide support for a staged model of trust in which trust develops over time and is influenced by different factors at each stage of the process. Taking a timeline approach reconciles differences in the literature and allows a closer inspection of time-dependent trust issues. Initially, the design look and feel of a site is incredibly important to the user. Within the first few seconds they make rapid judgments about which sites to trust and select and which they mistrust and so reject. The way in which health websites are designed, their use of color, graphics, and their underlying navigation structures all provide clues to the user about whether or not this is a suitable site for them. In consequence, some poorly designed yet credible information sites are rejected without further exploration. Design issues have been criticized in the past by medical informatics reviews as being a shallow evaluation factor. While genuine health consumers are concerned with the content of e-health resources, the sheer volume of sites returned makes it almost inevitable that they will fall back on some simple heuristics, involving low cognitive effort, based on visual appeal to sift out and reject poor sites (Peterson, Aslani, & William, 2003; Sillence et al., 2004). Rather than viewing design issues as shallow, they should be perhaps regarded as an important feature with real implications. If people cannot move beyond the poor design, then the quality of the content becomes irrelevant. Furthermore, those design-based markers of trust have been shown to have an impact upon the effectiveness of online health messages. In some recent experimental work, we showed that a breast cancer and alcohol health message on a website containing positive design trust markers (i.e., trust seals) exerted more influence over a group of female participants with high-to-moderate drinking levels than the nontrustworthy equivalent website, that is, the one containing adverts—to the extent that they reported drinking less (Harris, Sillence, & Briggs,

2009). Some health portals are recognizing the importance of design issues in developing trust. They have chosen to build the most user-friendly website as a way of improving the trustworthiness of the portal as opposed to focusing purely on the best medical content (Luo & Najdawi, 2004).

Where particular types of e-health resources are on offer, for example, patient experiences, it is the design of the site in which those experiences are embedded that is first considered by the user. If the "host" or "parent" site is professionally designed, has a good reputation, and is seen as being impartial, then there is a transfer of trust and the user moves on to consider the voice of the site. A comparison process begins in which the user reflects on the extent to which the experiences online resonate with their own story, thoughts, and feelings. As the trust process develops, so users begin to share as well as read the experiences. Results from our work on online communities provide insights into the way trust develops and is maintained within such settings with lay experts providing warranting strategies to prove their credibility.

Acknowledging the interest in patient authored material an increasing number of e-health providers have begun to incorporate this resource into their online provision. The range of providers includes charity groups, government departments, as well as more commercial ventures such as pharmaceutical companies (Mo et al., 2011) and e-commerce sites. Inevitably, the depth, breadth, and quality of patient experiences online becomes highly variable with participants sometimes questioning whether or not the experiences on commercial sites are indeed genuine. This variable provision may also account for our finding that while participants like sites containing patient experiences they do not necessarily trust them (Sillence et al., 2011). Our qualitative work on patient experiences and online health communities indicates that it takes users time to "probe" and ask the necessary questions of the content in relation to themselves in order to build up a sense of trust. Similarly assessing any form of online health information often involves a process of corroboration in which other websites and other sources are checked for consistency of information before the user decides whether to trust and act upon the site (Harris, Sillence, & Briggs, 2011).

To summarize, our research suggests that various elements of online health websites make it more or less likely that users will be able to engage further with the content, trust the information and advice they encounter, and perhaps over time develop a trusting relationship both with the website and with the other users or content providers on the site. Providing a sufficient depth and breadth to the website's content is a key factor. Sites containing multiple format information, downloadable factsheets, information straplines, referenced material, and markers of credibility including organizational logos and indicators of security and privacy policies all encourage greater systematic processing on the part of the health consumer. In terms of patient-authored material, more specifically features such as blogs, biographies, FAQs, multiple format communication, moderation and variation in expertise levels and member backgrounds all encourage greater engagement with the online material. Links to other pages and websites, cross-referencing, and "about us" information allow users to integrate information across sites and sources while the speed of interactivity, the choice of message types and communication channels, as well as photographs and feedback via other systems, for example, email all promote longer-term engagement and disclosure with health websites.

Despite our confidence in health consumers as increasingly careful evaluators of content, is there any evidence to suggest that reading online health information in whatever form leads to better health outcomes? Some studies have indicated that information gained from an online search helps the majority of users to understand more about an illness or injury and in some of these cases the information is sufficient for patients to act upon to improve their health (Nicholas, Huntington, Gunter, Russell, & Withey, 2003). Reviews of quantitative web-based intervention studies have generally indicated beneficial effects of the Internet on health outcomes, although issues concerning the methodological quality of such studies remain (Bessell et al., 2002). A meta-analysis by Wantland, Portillo, Holzemer, Slaughter, and McGhee (2004) concluded that in the majority of studies, knowledge and/or behavioral outcomes improved for participants using web-based health interventions. However, other studies have indicated that while the Internet can increase patients' knowledge about their health conditions, they can be left feeling too overwhelmed by the information available online to be able to make an informed decision about their own health care (Hart, Henwood, & Wyatt, 2004; Schwartz, 2004). We know that hearing or reading about other peoples' experiences can help patients to recognize that there are decisions to be made and can assist in clarifying options (Entwistle et al., 2011) and that personal narratives are often more powerful in affecting decision making than straightforward health facts and figures (Winterbottom et al., 2008). It is important to remember, however, that people may respond differently to sites containing information and advice they find unwelcome or threatening than they respond to sites containing information they find congenial and comforting (Good & Abraham, 2007; Harris & Epton, 2009).

Directions for Future Work

Guidelines for the design of trustworthy websites can be found in Sillence, Briggs, Harris, and Fishwick (2007c); we are currently in the process of updating these based on the findings from our work examining online patient experiences. Going forward we are keen to further our understanding of the potential moderators of online trust in health information. Researchers in health psychology have developed a number of models to account for the ways in which people respond to health-risk information, including information the individual may find uncomfortable or threatening (e.g., protection motivation theory (PMT) (Rogers, 1983; Rogers & Prentice-Dunn, 1997); the extended parallel process model (EPPM) (Witte, 1992)). Our own work has assessed whether adding measures of threat, control, and coping affected perceptions of trust (Harris et al., 2011). In the health domain, people often have strong initial expectations and show preferences for particular sorts of information and these may shape their search strategies (Joinson & Banyard, 2002) and influence which sites they trust. We are interested in examining these and other factors in relation to more specific forms of online health information, for example, patient experiences.

Experimentally we are investigating social moderators of patient experiences in order to understand the importance of different moderating factors on the acceptance

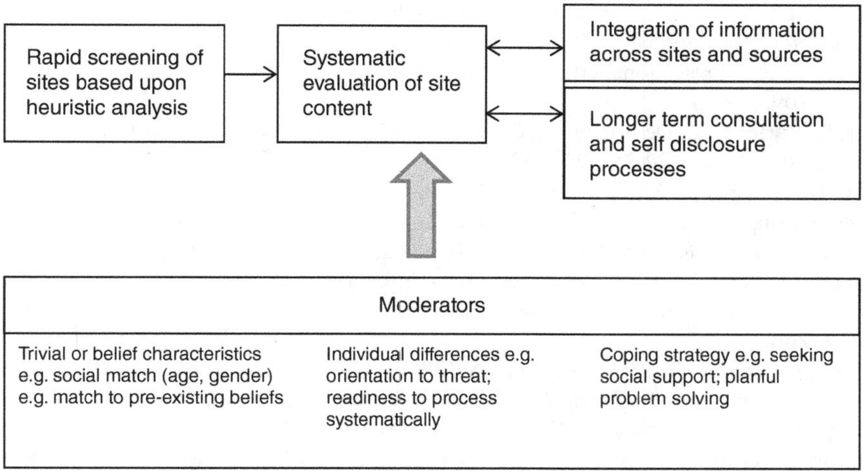

Figure 21.2 Overview of current research programme including staged model of trust development and the key moderators of interest.

or rejection of health messages. We are investigating whether the format of health messages is important in the reduction of defensive processing. Furthermore, we are keen to explore whether these effects are moderated by matching to trivial or belief characteristics such as age and gender and by individual differences, for example, in terms of orientation to threat. An overview of this research program is shown in Figure 21.2. In the future we are keen to explore the mechanisms of decision making in more detail and unpack the factors that make personal experiences such a powerful resource. This will encompass blogs and forum messages as well as reputation mechanisms and other forms of social filtering.

Acknowledgments

This publication presents independent research commissioned by the National Institute for Health Research (NIHR) under its Programme Grants for Applied Research program (RP-PG-0608-10147). The views expressed in this publication are those of the author(s) and not necessarily those of the NHS, the NIHR, or the Department of Health.

References

Beldad, A., de Jong, M., & Steehouder, M. (2010). How shall I trust the faceless and the intangible? A literature review on the antecedents of online trust. *Computers in Human Behavior*, *26*(5), 857–869. doi:10.1016/j.chb.2010.03.013

Bessell, T., McDonald, S., Silagy, S., Anderson, J., Hiller, J., & Sansom, L. (2002). Do internet interventions for consumers cause more harm than good? A systematic review. *Health Expectations*, *5*(1), 28–37.

Bhattacherjee, A. (2002). Individual trust in online firms: Scale development and initial test. *Journal of Management Information Systems, 19*(1), 211–241.

Brien, A. (1998). Professional ethics and the culture of trust. *Journal of Business Ethics, 17*(4), 391–409.

Briggs, P., Burford, B., De Angeli, A., & Lynch, P. (2002). Trust in online advice. *Social Science Computer Review, 20*(3), 321–332.

Briggs, P., de Angeli, A., & Simpson, B. (2004). Personalisation and trust: A reciprocal relationship? In K. Karat, J. O. Blom, & J. Karat (Eds.), *Designing personalised user experiences in e-commerce* (pp. 39–55). Norwell, MA: Kluwer Academic.

Chaiken, S. (1980). Heuristic versus systematic information processing and the use of source versus message cues in persuasion. *Journal of Personality and Social Psychology, 39*, 752–766.

Corritore, C., Kracher, B., & Wiedenbeck, S. (2003). On-line trust: Concepts, evolving themes, a model. *International Journal of Human-Computer Studies, 58*, 737–758.

Corritore, C. L., Wiedenbeck, S., Kracher, B., & Marble, R. P. (2007). Online trust and health information websites. In W. Hong et al. (Eds.), *Proceedings of the Sixth Annual Workshop on HCI Research in MIS* (pp. 25–29). Montreal, QC: AIS.

Cugelman, B., Thelwall, M., & Dawes, P. (2009). Communications of the association for information systems the dimensions of web site credibility and their relation to active trust and behavioural impact their relation to active trust and behavioural. *Communications of the Association for Information Systems, 24*(26), 455–472.

Cyr, D., Head, M., & Larios, H. (2010). Colour appeal in website design within and across cultures: A multi-method evaluation. *International Journal of Human-Computer Studies, 68*(1–2), 1–21. doi:10.1016/j.ijhcs.2009.08.005

Egger, F. N. (2000). "Trust me, I'm an online vendor": Towards a model of trust for e-commerce system design. In G. Szwillus & T. Turner (Eds.), *Proceedings of the SIGCHI 2000 Extended Abstracts: Conference on Human Factors in Computing Systems,* The Hague, The Netherlands (pp. 101–102). New York, NY: ACM.

Egger, F. N. (2001). Affective design of e-commerce user interfaces: How to maximise perceived trustworthiness. In M. Helander, H. M. Khalid, & M. P. Tham (Eds.), *Proceedings of the Conference on Affective Human Factors Design, Singapore* (pp. 317–324). New York, NY: ACM.

Entwistle, V. A., France, E. F., Wyke, S., Jepson, R., Hunt, K., Ziebland, S., & Thompson, A. (2011). How information about other people's personal experiences can help with healthcare decision-making: A qualitative study. *Patient Education and Counseling, 85*(3), 291–298.

Eysenbach, G., Powell, J., Kuss, O., & Sa, E.-R. (2002). Empirical studies assessing the quality of health information for consumers on the World Wide Web, a systematic review. *The Journal of the American Medical Association, 287*(20), 2691–2700.

Fogg, B. J., Kameda, T., Boyd, J., Marchall, J., Sethi, R., Sockol, M., & Trowbridge, T. (2002). *Stanford-Makovsky web credibility study: Investigating what makes web sites credible today.* Retrieved from Stanford University, Stanford, CA, Stanford Persuasive Technology Lab & Makovsky & Company website http://www.webstrat.fr/sites/www.webstrat.fr/files/Stanford-MakovskyWebCredStudy2002-prelim.pdf

Fogg, B. J., Marshall, J., Laraki, O., Osipovish, A., Varma, C., Fang, N., … Treinen, M. (2001). What makes web sites credible? A report on a large quantitative study. *Proceedings of SIGCHI.* Conference on Human Factors in Computing Systems 2001, 61–68, ACM, NY, USA.

Fox, S. (2011). *The social life of health information.* Retrieved from Pew Research Center, Internet & American Life Project website http://pewinternet.org/~/media/Files/Reports/2011/PIP_Social_Life_of_Health_Info.pdf

Giudice, K. D. (2010, October). *Crowdsourcing credibility: The impact of audience feedback on Web page credibility.* Paper presented at ASIST 2010, Pittsburgh, PA.

Giustini, D. (2006). How Web 2.0 is changing medicine. *British Medical Journal, 333*, 1283.

Good, A., & Abraham, C. (2007). Measuring defensive responses to threatening messages: A meta-analysis of measures. *Health Psychology Review, 1*, 208–229. doi:10.1080/17437190802280889

Grabner-Kräuter, S. (2010). Web 2.0 social networks: The role of trust. *Journal of Business Ethics, 90*(S4), 505–522. doi:10.1007/s10551-010-0603-1

Grabner-Krauter, S., & Kaluscha, E. (2003). Empirical research in online-trust: A review and critical assessment. *International Journal of Human-Computer Studies, 58*, 783–812.

Harris, P. R., & Epton, T. (2009). The impact of self-affirmation on health-related cognition and health behaviour: Issues and prospects. *Social and Personality Psychology Compass, 4*, 439–454. doi:10.1111/j.1751-9004.2010.00270.x

Harris, P. R., Sillence, E., & Briggs, P. (2009). The effect of credibility-related design cues on responses to a web-based message about the breast cancer risks from alcohol: randomized controlled trial. *Journal of Medical Internet Research, 11*(3), e37. doi:10.2196/jmir.1097

Harris, P. R., Sillence, E., & Briggs, P. (2011). Perceived threat and corroboration: Key factors that improve a predictive model of trust in internet-based health information and advice. *Journal of Medical Internet Research, 13*(3), e51. doi:10.2196/jmir.1821

Hart, A., Henwood, F., & Wyatt, S. (2004). The role of the Internet in patient practitioner relationships: Findings from a qualitative research study. *Journal of Medical Internet Research, 6*(3), e36.

Joinson, A. N., & Banyard, P. (2002). Psychological aspects of information seeking on the internet. *Aslib Proceedings, 54*, 95–102.

Liao, V. (2011). How user reviews influence older and younger adults' credibility judgments of online health information. In *Proceedings of CHI EA '11* (pp. 893–898). New York, NY: ACM.

Lindgaard, G., Dudek, C., Sen, D., Sumegi, L., & Noonan, P. (2011). An exploration of relations between visual appeal, trustworthiness and perceived usability of homepages. *ACM Transactions on Computer-Human Interaction, 18*(1), 1–30. doi:10.1145/1959022.1959023

Lindgaard, G., Fernandes, G., Dudek, C., & Brown, J. (2006). Attention web designers: You have 50 milliseconds to make a good first impression! *Behaviour Information Technology, 25*(2), 115–126. doi:10.1080/01449290500330448

Luo, W., & Najdawi, M. (2004). Trust-building measures: A review of consumer health portals. *Communications of the ACM, 47*(1), 108–113.

McKnight, D., & Chervany, N. (2001). Trust and distrust definitions: One bite at a time. In R. Falcone, M. Singh, & Y.-H. Tan (Eds.), *Trust in cyber-s* (pp. 27–54). Berlin, Germany: Springer-Verlag.

Meyerson, D., Weick, K. E., & Kramer, R. M. (1996). Swift trust and temporary groups. In R. M. Kramer & T. R. Tyler (Eds.), *Trust in organizations: Frontiers of theory and research* (pp. 166–195). Thousand Oaks, CA: Sage.

Mo, P., Sillence, E., Briggs, P., & Harris, P. (2011, July). *Patient experience information online: A content analysis of smoking related websites.* Paper presented at the workshop on Online Patient Experience and its Role in E-health at the 25th Human Computer Interaction conference, Northumbria University, Newcastle-upon-Tyne, UK.

Nicholas, D., Huntington, P., Gunter, B., Russell, C., & Withey, R. (2003). The British and their use of the web for health information and advice: A survey. *Aslib Proceedings*, *55*(5/6), 261–276.

Oostveen, A. (2011). The Internet as an Empowering Technology for Stigmatized Groups: A Case Study of Weight Loss Bloggers. Paper presented at the 25th BCS Conference on "Health, Wealth & Happiness." *Human Computer Interaction (HCI2011)*, Northumbria University, Newcastle-upon-Tyne, July 4–8.

Peterson, G., Aslani, P., & Williams, K. A. (2003). How do consumers search for and appraise information on medicines on the internet? A qualitative study using focus groups. *Journal of Medical Internet Research*, *5*(4), e33.

Petty, R. E., & Cacioppo, J. T. (1986). *Communication and persuasion: Central and peripheral routes to attitude change*. New York, NY: Springer-Verlag.

Preece, J. (1999). Empathic communities: Balancing emotional and factual communication. *Interacting with Computers*, *12*(1), 63–77.

Preece, J., & Ghozati, K. (2001). Observations and explorations of empathy online. In R. R. Rice & J. E Katz (Eds.), *The internet and health communications: Experience and expectations* (pp. 237–260). Thousand Oaks, CA: Sage.

Preece, J., & Maloney-Krichmar, D. (2005). Online communities: Design, theory, and practice. *Journal of Computer-Mediated Communication*, *10*(4), 1.

Richardson, K. P. (2003). Health risks on the internet: Establishing credibility on line. *Health, Risk and Society*, *5*(2), 171–184.

Riegelsberger, J., Sasse, M., & McCarthy, J. (2003). Shiny happy people building trust? Photos on e-commerce websites and consumer trust. *Proceedings of the CHI 2003*. New York, NY: ACM.

Riegelsberger, J., Sasse, M. A., & McCarthy, J. D. (2005). The mechanics of trust: A framework for research and design. *International Journal of Human-Computer Studies*, *62*(3), 381–422.

Rogers, R. W. (1983). Cognitive and physiological processes in fear appeals and attitude change: A revised theory of protection motivation. In B. L. Cacioppo & L. L. Petty (Eds.), *Social psychophysiology: A sourcebook* (pp. 153–176). London, UK: Guildford.

Rogers, R. W., & Prentice-Dunn, S. (1997). Protection motivation theory. In D. S. Gochman (Ed.), *Handbook of health and behavior research* (pp. 113–132). New York, NY: Plenum Press.

Rotter, J. B. (1967). A new scale for the measurement of interpersonal trust. *Journal of Personality*, *35*, 651–665.

Rozmovits, L., & Ziebland, S. (2004). What do patients with prostate or breast cancer want from an internet site? A qualitative study of information needs. *Patient Education and Counselling*, *53*, 57–64.

Schaffer, R., Kuczynski, K., & Skinner, D. (2008). Producing genetic knowledge and citizenship through the internet: mothers, pediatric genetics, and cybermedicine. *Sociology of Health & Illness*, *30*(1), 145–159. doi:10.1111/j.1467-9566.2007.01042.x

Schwartz, B. (2004). *The paradox of choice: Why more is less*. New York, NY: Ecco.

Sillence, E. (2010). Seeking out very like-minded others: Exploring trust and advice issues in an online health support group. *International Journal of Web Based Communities*, *6*(4), 376–394. doi:10.1504/IJWBC.2010.035840

Sillence, E., Briggs, P., Fishwick, L., & Harris, P. (2004). Trust and mistrust of online health sites. *Proceedings of the SIGCHI*, *6*(1), 663–670.

Sillence, E., Briggs, P., Harris, P. R., & Fishwick, L. (2007a). How do patients evaluate and make use of online health information? *Social Science & Medicine*, *64*(9), 1853–1862. doi:10.1016/j.socscimed.2007.01.012

Sillence, E., Briggs, P., Harris, P. R., & Fishwick, L. (2007b). Going online for health advice: Changes in usage and trust practices over the last five years. *Interacting with Computers, 19*(3), 397–406. doi:10.1016/j.intcom.2006.10.002

Sillence, E., Briggs, P., Harris, P. R., & Fishwick, L. (2007c). Developing trust practices for e-health. In R. Song, L. Korba, & G. Lee (Eds.), *Trust in e-services: Technologies, practices and challenges* (pp. 235–258). Hershey, PA: Idea Group Publishing.

Sillence, E., Hardy, C., Briggs, P., & Harris, P. R. (2013). How do people with asthma use Internet sites containing patient experiences? *Patient Education and Counseling, 93*(3), 439–443.

Sillence, E., & Mo, P. (2014). Communicating health decisions: An analysis of messages posted to online prostate cancer forums. *Health Expectations, 17*(2), 244–253. doi:10.1111/j.1369-7625.2011.00745.x

Sillence, E., Mo, P., Briggs, P., & Harris, P. (2011). The changing face of trust in health websites. *Oxford Symposium ICS-OII "A decade in Internet time."* Retrieved from http://ssrn.com/abstract=1920317 or http://dx.doi.org/10.2139/ssrn.1920317

Stanford, J., Tauber, E., Fogg, B.J. & Marable, L. (2002). How Do People Evaluate a Web Site's Credibility? Retrieved from Consumers Union (September 2014). http://consumersunion.org/research/how-do-people-evaluate-a-web-sites-credibility/

Swann, W. B., Jr., & Read, S. J. (1981). Self-verification processes: How we sustain our self-conceptions. *Journal of Experimental Social Psychology, 17*, 351–372.

Van der Zee, K., Vos, M., & Luijters, K. (2009). Social identity patterns and trust in demographically diverse work teams. *Social Science Information, 48*(2), 175–198.

Wang, Y. D., & Emurain, H. (2005). An overview of online trust: Concepts, elements and implications. *Computers in Human Behavior, 21*, 105–125.

Wantland, D., Portillo, J., Holzemer, W., Slaughter, R., & McGhee, E. (2004). The effectiveness of web-based vs. non-web based interventions: A meta-analysis of behavioral change outcomes. *Journal of Medical Internet Research, 6*(4), e40.

Williams, P., Huntington, P., & Nicholas, D. (2003). Health information on the internet: A qualitative study of NHS Direct Online users. *Aslib Proceedings, 55*(5/6), 304–312.

Winterbottom, A., Bekker, H. L., Conner, M., & Mooney, A. (2008). Does narrative information bias individual's decision making? A systematic review. *Social Science & Medicine, 67*(12), 2079–2088. doi:10.1016/j.socscimed.2008.09.037

Witte, K. (1992). Putting the fear back into fear appeals: The extended parallel process model. *Communication Monographs, 59*, 329–349. doi:10.1080/03637759209376276

Ziebland, S., & Wyke, S. (2012). Health and illness in a connected world: How might sharing experiences on the internet affect people's health? *Millbank Quarterly, 90*(2), 219–249.

22

Computer-Mediated Support for Health Outcomes: Psychological Influences on Support Processes

Kevin B. Wright

Department of Communication, George Mason University, Fairfax, VA, USA

The growth and popularity of computer-mediated support groups and other, less formal, types of online social support (i.e., social support that occurs within Facebook and other social media) has gained the attention of social support researchers in recent years who are interested in the underlying psychological processes of social support within this context (e.g., Barrera, Glasgow, McKay, Boles, & Feil, 2002; Gustafson et al., 2005; Lieberman & Goldstein, 2005; Tanis, 2008). Scholars have been intrigued by the potential of online support networks to supplement or replace traditional face-to-face social support networks (Rains & Young, 2009; Wright & Miller, 2010).

Communication scholars and other social scientists have begun to examine the implications of these innovations in terms of their impact on the process of how and why individuals obtain online support, advantages and disadvantages of computer-mediated support, and the relationship between online support and health-related outcomes (i.e., reductions in physical symptoms, stress, depression, etc.). More recently, scholars have begun to focus on computer-mediated support interventions (see Gustafson et al., 2005), particularly among individuals who may lack access to adequate support for their health concerns in their traditional face-to-face support networks. Despite great strides in research during the past 15 years, many questions regarding the psychological predispositions of computer-mediated support group participants, social support processes within this context, and the relationships among social support, behavioral change, and key health outcomes remain to be explored.

The purpose of this chapter is to provide an overview of the state of research in the area of computer-mediated support and health outcomes with a focus on key psychological variables and processes. Toward that end, the chapter examines the growth and prevalence of computer-mediated support in the past two decades, research on the perceived advantages and disadvantages of computer-mediated support, key variables, and theoretical frameworks that are useful for understanding

The Handbook of the Psychology of Communication Technology, First Edition. Edited by S. Shyam Sundar.

psychological processes related to computer-mediated support, common methods for studying computer-mediated support, limitations of this body of research, and suggestions for future theoretical and applied work in this area.

Prevalence and Growth of Computer-Mediated Support

Over the past two decades, paralleling the growth of the Internet, mobile technologies, and new communication applications, we have witnessed a tremendous growth in online social support activity. In recent years, various types of social media have changed patterns of communication by increasing individuals' connectivity, interactivity, and direct participation in computer-mediated support (see Chou, Hunt, Beckjord, Moser, & Hesse, 2009; Wright & Bell, 2003).

Fifty-nine percent of US adults, constituting 8 in 10 Internet users, seek health information on the Internet (Fox, 2011). Of those online health information seekers, 60% have performed at least 1 of 11 social media activities assessed related to health and health care, such as reading someone else's commentary or experience about health issues (Fox, 2011). Moreover, in addition to formal, disease-specific, computer-mediated support groups, social support appears to be a common activity that occurs within social networking sites, such as Facebook and Twitter (Steinfield, Ellison, & Lampe, 2008; Wright, 2012).

Motivations for Participating in Computer-Mediated Support Groups

Researchers have found that people have a variety of different motivations for participating in computer-mediated support groups and other types of online support (McKenna, 2008; McKenna & Green, 2002; Wright, 2002). Uses-and-gratifications theory (Katz, Blumler, & Gurevitz, 1974) is a helpful framework in terms of examining motives for using various types of media, including online communities (Courtois, Merchant, De Marez, & Verleye, 2009; Ebersole, 2006; Wright, 2002). These researchers have found motives to be important antecedent variables in terms of understanding cognitive processes behind various communication-related behaviors, such as the amount of time individuals spent online, relational development and maintenance activities, and they ways in which individuals communicate with others in virtual communities.

Researchers have identified a number of motives for using computer-mediated support groups as well as differences in interaction based upon a persons' motives. One such motivation is the amount/quality of health information available within online support communities. According to Purcell, Rainie, Mitchell, Rosenstiel, and Olmstead (2010), 39% of Americans say there is not enough coverage of health and medical news. This may in part account for the increasing number of people going online for health information. Perceived risk and efficacy (as they relate to health issues) have been shown to motivate information-seeking behaviors

(Rimal & Real, 2003), including online health information seeking (e.g., Cline & Haynes, 2001).

A second motive that has been identified in the online support literature is the desire for emotional support and relationship formation (Buchanan & Coulson, 2007; Coulson & Knibb, 2007; Leimeister, Schweizer, Leimeister, & Krcmar, 2008). Leimeister et al. (2008) found that both informational motives and interpersonal motives for joining a virtual cancer support community were linked to the formation of virtual relationships, and subsequently receiving support. Butler, Sproull, Kiesler, and Kraut (2007) found that individuals who value the social benefits of an online community are more likely to engage in community building activities (e.g., posting, providing encouragement) compared to those who turn to the community for informational benefits. By contrast, Nonnecke, Andrews, and Preece (2006) found that informational motives for joining an online support group in the absence of socio-emotional support motives were linked to "lurking" (i.e., reading but not posting messages) in the online community. Wright (2002) found that interpersonal and social integrative motives predicted increased interpersonal communication, relational development and maintenance, and social involvement in online support communities. These findings suggest that members who are motivated solely by a strong desire for information will be less active in providing support to others.

Perceived Advantages and Disadvantages of Computer-Mediated Support

Early studies of computer-mediated support largely focused on participant perceptions of advantages and disadvantages of using these groups. While largely descriptive, these studies helped to identify key variables and processes that were subsequently investigated in later work. The following section provides a brief overview of the findings related to perceived advantages and disadvantages of obtaining social support in computer-mediated contexts.

Perceived advantages

According to studies of individuals who seek support in computer-mediated environments, the Internet (and related new technologies) provides several advantages over traditional face-to-face support, including broad reach, 24-hr availability, increased interactivity with others, international participation, multimedia capabilities, and greater anonymity (Cline & Haynes, 2001; Neuhauser & Kreps, 2008; Wright, 2002). Researchers have also identified many other advantages of computer-mediated support, including (a) reduced feeling of being stigmatized due to one's health condition (which I will discuss in greater depth later in the chapter), (b) it allows participants to be evaluated on their contributions to the group rather than their physical appearance or disabilities (Wright, 2000), (c) convenience, (d) increased opportunities for social comparisons, and (e) access to in-depth and diverse information about a health condition (Wright, 2002). Moreover, the greater anonymity in online social interaction allows people who are reluctant to seek support in a face-to-face setting to communicate with others in a straightforward and honest way about their conditions and concerns (Caplan & Turner, 2007; Tidwell & Walther, 2002), and participation within online support groups enhances the possibility that individuals get incidental exposure to health information in addition to their active information seeking online (Kontos, Emmons, Puleo, & Viswanath, 2010).

Perceived disadvantages

However, computer-mediated support groups have also been found to suffer from a variety of problems, such as off-topic discussion, flaming, sporadic membership, and other negative aspects, such as deception and cyberstalking (Caplan & Turner, 2007; Wright, 2002; Wright & Bell, 2003).

Essentially, there are risks involved with obtaining social support online that may potentially undermine perceptions regarding the credibility of support providers and ultimately lead to dissatisfaction with the support that is offered within computer-mediated support groups. For example, the anonymity of participants within these groups may lead to flaming and other forms of unsupportive behavior (Barak, Boniel-Nissim, & Suler, 2008) and can make it more difficult than in face-to-face settings for individuals to detect deception, insincerity, and alternative motives for using the group (Hancock, 2007; Wright, 2002). This includes people within the groups who are voyeurs, profiteers who hope to sell health-related products to participants, or people who may have other motives for using the group (besides giving and receiving support). These factors, as well as others, can potentially undermine the credibility of online support providers and the usefulness of computer-mediated support groups for obtaining adequate support.

Key Variables That Influence Participation in Computer-Mediated Support

Health-related stigma

One variable that appears to be an important predictor of participating in some form of computer-mediated support is the degree to which individuals perceive they are stigmatized because of the issues they face (particularly health-related issues). Health-related stigma is a significant problem that many individuals facing health concerns have to deal with on a daily basis (Herek & Glunt, 1988). It has been linked reductions in the size of individuals' support networks, problems discussing concerns with others, dissatisfaction with one's support network, reduced compliance with treatment recommendations, and increased health problems (Rosman, 2004; Vanable, Carey, Blair, & Littlewood, 2006). Individuals with stigmatized health issues report increased stress and depression (Riggs, Vosvick, & Stallings, 2007; Wolitski, Pals, Kidder, Courtenay-Quirk, & Holtgrave, 2008), both of which have been linked to a multitude of health problems, including alcohol and substance abuse, increased usage of tobacco products, anxiety and related mental health problems, reduced immune system functioning, and increased physical health problems (Duncan, Hart, Scoular, & Bigrigg, 2001).

Computer-mediated support groups can provide people with stigmatized health problems greater access to online weak ties and weak-tie support (Wright & Bell, 2003; Wright, Rains, & Banas, 2010). These weak ties offer several advantages relative to strong ties including being less judgmental and more objective, offering unique information, and a reduced potential for role conflict (Wright & Miller, 2010). Applied communication researchers (and applied scholars from other disciplines) have become increasingly interested in computer-mediated support groups as a way to supplement (or replace) traditional face-to-face support networks in the development of interventions for individuals facing a variety of health concerns (Houston, Cooper, & Ford, 2002; Shaw, Hawkins, McTavish, Pingree, & Gustafson, 2006). McKenna (2008) found that becoming actively involved in a virtual community that focuses on a stigmatized or secret aspect of one's identity that was previously kept hidden leads individuals to become more self-accepting of this aspect of themselves and to feel more comfortable sharing that identity with others, thus becoming less socially isolated and more socially connected with their face-to-face community.

Homophily

A second variable that appears to influence participation in computer-mediated support is homophily, or the degrees to which members of a social network are similar to one another in certain attributes or characteristics. Homophily predisposes people to more attraction, trust, and understanding than one would find in dissimilar individuals. Close, personal networks tend to be homophilous, although weaker ties online can exhibit situational homophily (Walther & Boyd, 2002; Wright, 2000) in terms of stressful situations that online communicators have in common. Similarity between a sender and receiver may increase the persuasiveness of messages, and may be especially important when new attitudes and beliefs are formed. For example, Wang, Walther, Pingree, and Hawkins (2008) showed that perceived similarity of support group members influenced perceptions of their credibility and, in turn, the evaluation of health information they provided. Moreover, Wright (2000) and Campbell and Wright (2002) found that homophily was a key perception that influences social support satisfaction and perceptions of credibility of support providers within health-related online support groups.

Perceptions of support provider credibility

A third important set of variables is perceptions of online support provider credibility. For example, Robinson, Patrick, Eng, and Gustafson (1998) argue that several characteristics of the Internet influence perceptions of source/information credibility, including (a) improved opportunity to tailor messages; (b) the increased possibility for users to remain anonymous, which may increase their willingness to engage in more honest discussions; and (c) increased access to information and support on demand. The increased interactivity of computer-mediated communication (e.g., the ability to post comments and receive feedback within on-line communities, etc.) is also important to consider since individual users are able to ask direct questions of information sources, which is generally impossible with traditional media. In addition, the ability of members within the larger online community to read and respond to online posts provides an opportunity for increased quality control of information (Esquivel, Meric-Bernstam, & Bernstam, 2006). For example, information that does not resonate with the experiences of members of the larger community may be subject to greater scrutiny (Rieh & Belkin, 2000), a process similar to peer review.

Research conducted to date has shown that perceptions of online support group credibility are associated with several noteworthy outcomes. Wright (2000) found that two dimensions of source credibility, perceived competence and character, were associated with the perceived similarity of other online support group members, online support group network size and satisfaction with the support they received. Similarly, Campbell and Wright (2002) found that these dimensions of source credibility were associated with perceptions of situational similarity among online support group members. Moreover, both dimensions of source credibility were associated with perceptions of social support providers' receptivity and status equality within these groups.

Coping strategy predispositions

Finally, a fourth variable that is likely to influence social support processes within computer-mediated contexts is the coping styles/predispositions of individuals who seek support within this context. Stemming from Lazarus and Folkman's (1984) theory of psychological stress and coping, coping styles have received a substantial amount of scholarly attention, particularly as they relate to social support (Billings & Moos, 1981; Kohn, 1996). Coping can be defined as a persons' ongoing cognitive and behavioral efforts to manage external or internal demands that are appraised as taking or exceeding a person's resources (Lazarus & Folkman, 1984). Studies suggest that problem-focused coping, which is "directed at remedying a threatening or harmful external situation" (Kohn, 1996, p. 189), is often linked with positive adaptation to stressful situations (Heady & Wearing, 1990; Lazarus & Folkman, 1984), while emotion-focused coping, which is defined as "ventilating, managing, or palliating an emotional response to a situation" (Kohn, 1996, p. 189), is typically associated with negative adaptation (Billings & Moos, 1981; Kohn, 1996). However, problem-focused coping strategies are more likely to be used when a situation is appraised as changeable whereas emotion-focused strategies tend to be used when a situation is assessed as unchangeable (Folkman, Lazarus, Dunkel-Schetter, DeLongis, & Gruen, 1987). As a result, sometimes emotion-focused coping can lead to positive outcomes, although to a lesser extent than we typically see with problem-focused coping. In the context of research on computer-mediated support groups, Wright (1999, 2000) found that support group participants who were more satisfied with the support provided by other group members were more likely to use problem-focused coping strategies than individuals who had lower support network satisfaction scores.

Prominent Theoretical Frameworks in Computer-Mediated Support Research

Theory plays an important role in terms of understanding social support processes and outcomes within computer-mediated contexts. This section presents several theories that have been applied to the study of computer-mediated social support. While many early studies of computer-mediated support tended to lack strong theoretical frameworks, more recent work has become more theoretically rich, with scholars testing various hypotheses derived from theory in many studies of computer-mediated support. These theories appear to have utility in terms of providing explanations of online support processes as well as helping scholars to predict relationships among variables.

Social support and the optimal matching model

One theoretical framework that has received a good deal of attention in the computer-mediated support literature is the optimal matching model (Cutrona & Russell, 1990). For many years, social scientists have known that people often differ in terms of the types of support they find useful due to factors such as the context of the stressful situation they are facing, their perceived coping skills, and their relationship with the support provider (Cutrona & Russell, 1990). The optimal matching model (Cutrona & Russell, 1990) suggests that matching the specific type of support offered with the dimensions of a stressor (e.g., desirability, controllability, life domain, and duration of consequences) produces the most positive outcomes. More generally, the model is grounded in the notion that a match between the needs of support seekers and the resources/abilities of support providers is important in terms of coping with the many relational challenges associated with communicating social support. For example, if an individual is seeking emotional support for a health concern and he or she perceives that members of his or her support network have competently listened, expressed empathy, and acknowledged the severity of the issue, then this would be considered an example of an optimal match between the support seeker and support providers.

People tend to make decisions regarding approaching potential supporters based on the perception that members of their network will be able to meet their specific need(s) for support as well as their relational needs. Goldsmith (2004) contends that optimal matches in supportive episodes may lead to more positive perceptions of relational partners and the type of support that is being offered, and this, in turn, may ultimately influence positive health outcomes. Yet it is also possible that a recipient of support may perceive some types of support negatively (which is often the case among strong ties who react negatively to stigmatized health conditions), and this may negate the positive effects of the supportive attempt or it may actually have a negative impact on health or quality of life (Dakof & Taylor, 1990). There is evidence to suggest that weak ties accessed via computer-mediated communication might be particularly valuable when strong ties are unable or unwilling to provide support.

This perspective may help to provide important insights into the supportive needs of individuals who seek computer-mediated support. For example, drawing upon an optimal matching theory framework, Eichhorn (2008) found that informational support through shared experiences was the most common type of support for members of an online support groups for eating disorders (followed by emotional support). Moreover, Sullivan (2003) found that men were more likely to seek informational support within online support groups, whereas women were more likely to seek emotional support and validation.

Social comparison theory

Social comparison theory (Festinger, 1954) has been useful in terms of examining perceptions of people within our support networks and for understanding why support from these individuals may not always lead to positive outcomes. According to social comparison theory, individuals make assessments about their own health and coping mechanisms by comparing them to others in their social network (Helgeson & Gottlieb, 2000).

Helgeson and Gottlieb (2000) mention that *lateral comparisons*, comparisons to similar others, may normalize people's experiences and reduce uncertainty and stress for those dealing with health concerns. However, when individuals compare themselves to others, their self-assessment could be either positive or negative. For example, if a person with cancer feels that he is coping with problems less effectively than others in his network (such as a friend or relative who has or had cancer or a similar life-threatening illness), this may create *upward comparisons*, which could produce feelings of frustration or serve as a source of inspiration to the person to cope more effectively by emulating the successful behaviors of those other members. Conversely, *downward comparisons* to others in the social network, such as when an individual feels that he or she is coping better than other members, can lead to positive self-assessments and/or to negative feelings about people if interaction with the other members is perceived as being unhelpful.

Participants often glean information about the status of their health issues through social comparisons that take place within supportive interactions (Davison, Pennebaker, & Dickerson, 2000). According to Sarasohn-Kahn (2008), the second most popular reason people seek health information through online social networks is to research other participants' experiences with a health condition. Such social comparison processes do not even require actual participation in the online group; rather, individuals may engage in these practices passively by reading the posted group discussions.

Social information processing theory

One computer-mediated communication theory that has been useful in terms of understanding the effects of computer-mediated channels on the perceptions of individuals who engage in online social support is social information processing theory (Walther, 1992; Walther, 2007). Walther (1992) asserts that within the context of computer-mediated communication, message senders portray themselves in a socially favorable manner to draw the attention of message receivers and foster anticipation of future interaction. Message receivers, in turn, tend to idealize the image of the sender due to overvaluing minimal, text-based cues. Idealized perceptions and optimal self-presentation in the computer-mediated communication process tend to intensify in the feedback loop, and this can lead to what Walther (1996) labeled as "hyperpersonal interaction," or a more intimate and socially desirable exchange than in face-to-face interactions.

Hyperpersonal interaction is enhanced when no face-to-face relationship exists, so that users construct impressions and present themselves "without the interference of environmental reality" (Walther, 1996, p. 33), and it appears to skew perceptions of relational partners in positive ways, and in some cases, computer-mediated relationships may exceed face-to-face interactions in terms of intensity (King & Moreggi, 1998; Walther, 1996; Wright & Bell, 2003). Moreover, according to Walther (1996), the reduced number of available nonverbal cues in CMC increases message-editing capabilities, and the temporal features of CMC allow communicators to be more selective and strategic in their in self-presentation, form idealized impressions of their partners, and, consequently, engage in more intimate exchanges than people in face-to-face situations. For instance, Walther, Slovacek, and Tidwell (2001) found that individuals rated online interaction partners as more socially attractive and affectionate when a photo was not present compared to those who did view a photo of the interaction partner. In addition, dyads in computer-mediated settings also appear to self-disclose more than face-to-face dyads (Tidwell & Walther, 2002).

Strength of weak ties

Several of the preceding theories suggest that some individuals may benefit from characteristics of online networks in terms of obtaining optimal support. One theory that links social support processes to social network characteristics is Granovetter's Strength of Weak Ties theory (1973). This theory posits that the spread of social support is dependent upon the structure of communication networks in communities. Specifically, personal networks are made up of strong ties (such as close friends and family), weak ties (such as coworkers, acquaintances, and people with whom one has infrequent contact), and absent ties. Small clusters made up of an individual and his or her strong ties may be linked to other strong tie clusters by weak ties. Without weak ties, communication can only flow among small clusters, and groups become information saturated. Weak ties reach larger numbers of people, and longer distances than strong ties. The strength of weak ties theory is essential in adding an element of structure to understanding online supportive interactions and for distinguishing the roles and relationships inherent in the different positions people hold within online networks.

Individuals often seek support through weak tie networks instead of within their strong tie network because weak tie networks can provide access to more diverse points of view and information that may not be available within more intimate relationships (Adelman, Parks, & Albrecht, 1987; Haythornthwaite, 2002). Typically, individuals form close relationships with others who are similar to them in terms of demographics, attitudes, and backgrounds. This homogeneous preference can limit the diversity of information and viewpoints obtained about topics, including health concerns. Access to more diverse viewpoints about health problems can provide individuals with more varied informational support about health issues, and interacting with varied types of people increases the number of social comparisons a person can make about his or her health condition vis-à-vis others (Adelman et al., 1987).

The advantage of weak tie sources are that there are more of them, they are more likely to be different from the receiver and from one another, and there's a greater likelihood of being able to find an expert in a particular area in weak tie rather than strong tie sources. Members of weak tie networks may be more willing to talk about illness since these individuals tend to be less emotionally attached to a person (Adelman et al., 1987). Weak tie network members are often able to provide more objective feedback about a problem since they are less emotionally attached to a person with health problems than family and friends. People with whom we have close relationships, in an effort to spare our feelings when talking about a health problem, may be more likely to "sugar coat" the advice they give us or not be completely honest when discussing how they feel about our illness. According to Goldsmith and Albrecht (2011), weaker ties tend to be perceived as helpful when a person is coping with an issue that requires new information or skills (that may be limited within a close-knit family or friendship social network).

In addition, studies have found that role obligations in close ties and reciprocity issues can lead to problems with the provision of social support. Support for a loved one who is ill in close ties can lead to increased conflict or negative feelings toward the person due to role obligations (Albrecht & Goldsmith, 2003; LaGaipa, 1990), despite the fact that people may care deeply for the person. According to LaGaipa (1990), "social obligations may override the positive effect of companionship and social support. Such constraints can have a negative effect on mental well-being that may not make up for the beneficial aspects of personal relationships" (p. 126). All of these features of weak ties can be beneficial to people who are coping with stressful situations that may be difficult to ameliorate in strong ties due to lack of information and relational problems in close relationships, and access to weak ties is facilitated by the features of computer-mediated communication (Wright & Miller, 2010).

Health Outcomes Related to Computer-Mediated Social Support

Decades of social support research have linked improved social support to a variety of positive physical and mental health outcomes, such as increased health literacy, reduced stress, increased coping abilities, lower depression levels, greater adherence to treatment, and reduced morbidity and mortality rates (Barrera et al., 2002; Rains & Young, 2009). Communication researchers have contributed to this body of literature by providing a broader understanding of social support processes by examining characteristics of supportive messages (Goldsmith, 2004), and the relationship between social support processes and health outcomes among individuals living with a variety of health conditions (Braithwaite, Waldron, & Finn, 1999; Query & Wright, 2003), and by demonstrating how several key aspects of supportive relationships play critical roles in both psychological and physical health.

Researchers have known for decades that perceived life stress is inversely related to many physical and psychological health outcomes, including increased blood pressure, reduced immune system response, cardiovascular disease, morbidity and mortality rates, depression, and loneliness (Aneshensel & Stone, 1982; Berg & McQuinn, 1989; Berkman & Syme, 1979; Houston et al., 2002). Perceived stress is also associated with perceived quality of life and general physical/psychological well-being.

In recent years, a growing number of studies have measured psychological and physical health outcomes related to participation in computer-mediated support groups (Barrera et al., 2002; Rains & Young, 2009; Wright, 2000). There is empirical evidence that these groups provide a wide variety of health benefits for users (such as individuals with cancer, diabetes, and substance abuse problems), including reduced stress, increased positive coping, increased quality of life, increased self-efficacy in terms of managing one's health problems, reduced depression, and increased physical health benefits (Beaudoin & Tao, 2007; Gustafson et al., 2005; Houston et al., 2002; Jones et al., 2008; Owen, Klapow, Roth, Shuster, & Bellis, 2005; Rains & Young, 2009; Shaw et al., 2006; Wright, 1999, 2000).

Wright (2000) provided evidence that participation in a computer-mediated support community for older adults predicted reduced stress and increased coping skills. Owen et al. (2005) discovered that participation within an online support group for cancer patients predicted higher quality of life and reduced stress. In addition, in a recent meta-analysis of 28 computer-mediated support group studies, Rains and Young (2009) found that participation in computer-mediated social support groups led to increased social support, decreased depression, increased quality of life, and increased self-efficacy in terms of managing health conditions. Two decades of research on the Comprehensive Health Enhancement Support System (CHESS) has demonstrated positive health outcomes for women with breast cancer and people with HIV/AIDS, such as reduced stress, fewer healthcare visits and hospitalizations, and fewer incidents of disease symptoms (Gustafson et al., 1999, 2001). Wright et al. (2010) found that characteristics of weak-tie support network preference among members of health-related, computer-mediated support groups were predictive of reduced stress surrounding health problems. Other researchers have found similar health benefits for individuals who obtain social support within social networking sites (see Ellison, Steinfield, & Lampe, 2007; Kim & Lee, 2011; Manago, Taylor, & Greenfield, 2012; Wright et al., 2013).

Finally, the act of helping others appears to be associated with reduced stress and an increase sense of well-being. For example, Caplan and Turner (2007) contend that effective comforting communication within computer-mediated support groups fosters adaptive cognitive reappraisals of upsetting experiences, which in turn ameliorate emotional distress. Caplan and Turner (2007) assert that comforting communication helps on-line support group members to reduce stress by "discursively constructing" adaptive, or less troublesome, cognitive reappraisals of a distressing experience or situation during the process of comforting other members.

Prominent Methods Used in Computer-Mediated Support Research

Researchers have used a variety of methods to study computer-mediated social support, including survey research, content analysis, experimental designs, and naturalistic research. My own research and work with colleagues (Campbell & Wright, 2002; Johnson et al., 2008; Wright, 1999, 2000, 2002; Wright et al., 2010) in this area has been largely limited to survey research (with some content analysis and ethnographic methods), as well as work by other online support group researchers (see Shaw & Gant, 2002; Walther & Boyd, 2002). While survey research has a number of limitations and it lends itself to more descriptive types of research, it will likely continue to be a prominent method of studying online support due to the convenience of online survey tools (such as Qualtrics and SurveyMonkey) and the ability to use online surveys to reach large numbers of individuals who share common characteristics despite being geographically dispersed (Wright, 2005). However, many researchers have engaged in experimental or quasi-experimental methods to examine several facets of computer-mediated social support for people facing health concerns (see Barrera et al., 2002; Shaw & Gant, 2002; Gustafson et al., 1999, 2001, 2005). These designs will continue to be important to our understanding of causal relationships among exogenous variables, key online support variables, moderating variables, and key health outcomes, which is particularly helpful in terms of designing and evaluating support interventions.

Finally, researchers have also relied on content analysis and naturalistic methods in the study of computer-mediated social support. For example, Preece (1999) content-analyzed postings in online support groups for people coping with serious health problems, such as diabetes and cancer, and she found that the highest percentage of messages were some form of empathic discourse followed by personal narratives. Rains and Keating (2011) used content analysis to assess supportive messages and self-reported health outcomes within health-related blogs. Rains and Young (2009) conducted a meta-analysis of over 20 computer-mediated support interventions, and they found moderate effect sizes when assessing the relationship between online support and a variety of health outcomes. Ley (2007) used ethnographic methods to study social support in an online pregnancy support group, and found that the groups' technical features influenced patterns of social support and, in some cases, had contradictory effects on the perceived helpfulness of support. Future computer-mediated support research will likely benefit from other types of naturalistic research, particularly studies that allow researchers to understand how meanings of supportive messages are co-constructed through interactions within the groups.

Limitations of Computer-Mediated Support Scholarship and Future Theoretical/Applied Directions for Research

Despite the promise of computer-mediated support, there are still many facets of individual support predispositions, support processes, and links between computer-mediated support and health outcomes that need to be investigated. This section briefly discusses several of these issues.

One issue concerns accounting for the influence of overlapping sources of social support on key outcome variables, such as stress and depression. According to Haythornthwaite (2002), both online and offline supportive exchanges influence health outcomes. In short, it becomes difficult to separate online supportive influences from offline influences. For example, we typically mix face-to-face contact with e-mail, searching the Internet with asking friends and reading books, and sending regular mail and e-mail with use of the telephone. Caplan and Turner (2007) contend that future computer-mediated support researchers should pursue whether online supportive communication employs messages that are more, less, or equally empathic to those conveyed in similar face-to-face encounters.

Future research in this area would benefit from the development of theories and methods that take into account a more ecological perspective of the influence of social support on health outcomes, including the main effects and interaction effects of online and offline sources of social support, psychological predispositions, mediated variables (i.e., the influence of CMC channels, contexts), and key demographic and environmental variables (such as whether a person lives in a city, rural location, etc.) on health outcomes.

Although computer-mediated channels are able to connect individuals with larger, more diverse, networks of individuals who may be able to offer types of social support that transcend the quantity and quality of support within traditional face-to-face networks (i.e., weak tie support), less is known about potential negative aspects of weak tie support, such as the potential for increased deception, manipulation, cybersurveillance, and other problems that can occur when communicating with relative strangers (Barak et al., 2008; Finn & Banach, 2000). Given the potential benefits of computer-mediated support groups and the risks associated with seeking support online, it is critical for computer-mediated support researchers to better understand how group members evaluate the credibility of supportive providers and supportive messages within these groups.

As we have seen in this chapter, the bulk of research on computer-mediated support has focused on more formal health-related online support groups, although everyday types of online social support, such as support within social networking sites and virtual organizations, have been studied more in recent years (Steinfield et al., 2008; Wright, 2012). Future research would benefit from studies that increase our understanding of the many types of daily, mundane social support people give and seek online, and the cumulative effects of this support on health outcomes.

Finally, relatively little is known about how minority groups and other populations facing health disparities use computer-mediated support groups. However, it appears that members of minority groups engage in a variety of online social support activities, and individuals within these groups may benefit from computer-mediated support interventions. For example, Fogel et al. (2003) found that while African-Americans, Hispanics, and Asian Americans tend to use the Internet less than whites, their Internet use was associated with greater ability to talk with someone about problems and to obtain other types of social support.

Researchers have also found that access to traditional face-to-face support networks (such as close family ties) often deteriorates among members of certain low-income and minority groups due to factors such as unemployment, transience, and substance abuse (Roschelle, 1997), underscoring the need to develop interventions targeting these groups. Weinert and Hill (2005) found that rural women (including a high percentage of minorities and individuals from lower socioeconomic backgrounds) using a computer-mediated support intervention had lower levels of depression and higher self-reported management of day-to-day chronic illness symptoms than a control group of similar rural women living with chronic illness. Irrizary, Downing, and West (2002) and Wright (2000) found that computer-mediated support communities were helpful in terms of helping isolated older adults facing health concerns to become better connected with other individuals with similar circumstances. In short, there is great potential for researchers to develop and test computer-mediated support interventions with underserved populations.

Conclusion

Paralleling the growth of the Internet and social media, computer-mediated support has become a common online activity in recent years, and it has gained the attention of many researchers interested in understanding the nature of social support processes and outcomes within this context. Building upon initial descriptive studies, scholars have identified a number of theoretical frameworks that help to explain computer-mediated support processes and their relationship to positive health outcomes for individuals who participate in computer-mediated social support. While early studies relied on survey research to understand the nature of computer-mediated social support, more recent studies have moved toward experimental designs and intervention approaches. The findings from these studies provide support for the idea that computer-mediated support interventions may benefit certain populations (e.g., people coping with a stigmatized health condition, individuals who lack support resources in the face-to-face world, etc.). Future research is certainly needed to gain a better understanding of intervening variables that may influence the relationship between computer-mediated social support and various outcomes, and more studies are needed that examine computer-mediated support among more diverse populations.

References

Adelman, M. B., Parks, M. R., & Albrecht, T. L. (1987). Beyond close relationships: Support in weak ties. In T. L. Albrecht & M. B. Adelman (Eds.), *Communicating social support* (pp. 126–147). Newbury Park, CA: Sage.

Albrecht, T. L., & Goldsmith, D. J. (2003). Social support, social networks, and health. In T. L. Thompson, A. M. Dorsey, K. I. Miller, & R. Parrott (Eds.), *Handbook of health communication*. Mahwah, NJ: Erlbaum.

Aneshensel, C. S., & Stone, J. D. (1982). Stress and depression: A test of the buffering model of social support. *Archives of General Psychiatry*, *39*, 1392–1396.

Barak, A., Boniel-Nissim, M., & Suler, J. (2008). Fostering empowerment in online support groups. *Computers in Human Behavior*, *24*, 1867–1883.

Barrera, M., Glasgow, R. E., McKay, H. G., Boles, S. M., & Feil, E. G. (2002). Do internet-based support interventions change perceptions of social support? An experimental trial of approaches for supporting diabetes self-management. *American Journal of Community Psychology*, *30*, 637–654.

Beaudoin, C. E., & Tao, C. C. (2007). Benefitting from social capital in online support groups: An empirical study of cancer patients. *CyberPsychology & Behavior*, *10*, 587–590.

Berg, J. H., & McQuinn, R. D. (1989). Loneliness and aspects of social support networks. *Journal of Social & Personal Relationships*, *6*, 359–372.

Berkman, L. F., & Syme, L. S. (1979). Social networks, host resistance, and mortality: A nine-year follow-up study of Alameda County residents. *Journal of Epidemiology*, *109*, 186–204.

Billings, A. G., & Moos, R. H. (1984). The role of coping responses and social resources in attenuating the impact of stressful life events. *Journal of Behavioral Medicine*, *4*, 139–157.

Braithwaite, D. O., Waldron, V. R., & Finn, J. (1999). Communication of social support in computer-mediated groups for people with disabilities. *Health Communication*, *11*, 123–151.

Buchanan, H., & Coulson, N. (2007). Accessing dental anxiety online support groups: An exploratory qualitative study of motives and experiences. *Patient Education and Counseling*, *66*, 263–269.

Butler, B., Sproull, L., Kiesler, S., & Kraut, R. (2007). Community effort in online groups: Who does the work and why? In S. Weisband & L. Atwater (Eds.), *Leadership at a distance: Research in technologically supported work* (pp. 171–194). Mahwah, NJ: Lawrence Erlbaum Associates.

Campbell, K., & Wright, K. B. (2002). On-line support groups: An investigation of relationships among source credibility, dimensions of relational communication, and perceptions of emotional support. *Communication Research Reports*, *19*, 183–193.

Caplan, S. E., & Turner, J. S. (2007). Bringing theory to research on computer-mediated comforting communication. *Computers in Human Behavior*, *23*, 985–998.

Chou, W. Y., Hunt, Y. M., Beckjord, E. B., Moser, R. P., & Hesse, B. W. (2009). Social media use in the United States: Implications for health communication. *Journal of Medical Internet Research*, *11*, e48. doi:10.2196/jmir.1249.

Cline, R. J. W., & Haynes, K. M. (2001). Consumer health information seeking on the Internet: The state of the art. *Health Education Research*, *16*, 671–692.

Coulson, N., & Knibb, R. (2007). Coping with food allergy: Exploring the role of the online support group. *CyberPsychology and Behavior*, *10*, 145–148.

Courtois, C., Merchant, P., De Marez, L., & Verleye, G. (2009). Gratifications and seeding behavior of on-line adolescents. *Journal of Computer-Mediated Communication*, *15*, 109–137.

Cutrona, C. E., & Russell, D. W. (1990). Type of social support and specific stress: Toward a theory of optimal matching. In B. R. Sarason, I. G. Sarason, & G. R. Pierce (Eds.), *Social support: An interactional view* (pp. 319–366). Oxford, England: Wiley.

Dakof, G. A., & Taylor, S. E. (1990). Victim's perceptions of social support: What is helpful from whom? *Journal of Personality and Social Psychology*, *58*, 80–89.

Davison, K. P., Pennebaker, J. W., & Dickerson, S. S. (2000). Who talks? The social psychology of illness support groups. *American Psychologist*, *55*, 205–217.

Duncan, B., Hart, G., Scoular, A., & Bigrigg, A. (2001). Qualitative analysis of psychosocial impact of diagnosis of Chlamydia trachomatis: Implications for screening. *BMJ*, *322*, 195–199.

Ebersole, S. (2006). Uses and gratifications of the web among students. *Journal of Computer-Mediated Communication*, *6*. Retrieved from http://www3.interscience.wiley.com/cgi-bin/fulltext/120837804/HTMLSTART

Eichhorn, K. C. (2008). Soliciting and providing support over the Internet: An investigation of online eating disorder groups. *Journal of Computer-Mediated Communication*, *14*, 67–78.

Ellison, N. B., Steinfield, C., & Lampe, C. (2007). The benefits of Facebook "friends": Social capital and college students' use of online social network sites. *Journal of Computer-Mediated Communication*, *12*, 1143–1168.

Esquivel, A., Meric-Bernstam, F., & Bernstam, E. V. (2006). Accuracy and self- correction of information received from an Internet breast cancer list: Content analysis. *British Medical Journal, 332*, 939–943.

Festinger, L. (1954). A theory of social comparison processes. *Human Relations, 7*, 117–140.

Fogel, J., Albert, S., Schnabel, F., Ditkoff, B., & Neugut, A. (2003). Racial/ethnic differences and potential psychological benefits in use of the internet by women with breast cancer. *Psycho-Oncology, 12*, 107–117.

Finn, J., & Banach, M. (2000). Victimization online: The downside of seeking human services for women on the Internet. *Cyberpsychology & Behavior, 3*, 785–796.

Folkman, S., Lazarus, R. S., Dunkel-Schetter, C., DeLongis, A., & Gruen, R. J. (1987). Dynamics of a stressful encounter: Cognitive appraisal, coping, and encounter outcomes. *Journal of Personality and Social Psychology, 50*, 992–1003.

Fox, S. (2011, March 11). *Pew Internet & American life project: Health topics.* Retrieved from http://www.pewinternet.org/Reports/2011/HealthTopics.aspx

Goldsmith, D. J. (2004). *Communicating social support.* New York, NY: Cambridge University Press.

Goldsmith, D. J., & Albrecht, T. L. (2011). Social support, social networks, and health. In T. L. Thompson, R. Parrott, & J. F. Nussbaum (Eds.), *The Routledge handbook of health communication* (pp. 335–348). New York, NY: Routledge.

Granovetter, M. S. (1973). The strength of weak ties. *American Journal of Sociology, 78*, 1360–1380.

Gustafson, D. H., Hawkins, R., Boberg, E., Pingree, S., Serlin, R. E., Graziano, F., & Chan, C. L. (1999). Impact of a patient-centered, computer-based health information/support system. *American Journal of Preventative Medicine, 16*, 1–9.

Gustafson, D. H., Hawkins, R., Pingree, S., McTavish, F., Arora, N. K., Mendenhall, J., … Salner, A. (2001). Effect of computer support on younger women with breast cancer. *Journal of General Internal Medicine, 16*, 435–445.

Gustafson, D. H., McTavish, F. M., Stengle, W., Ballard, D., Hawkins, R., Shaw, B. R., … Landucci, G. (2005). Use and impact of ehealth system by low-income women with breast cancer. *Journal of Health Communication, 10*, 195–218.

Hancock, J. (2007). Digital deception: When, where and how people lie on-line. In A. N. Joinson, K. Y. A. McKenna, T. Postems, & U. Reips (Eds.), *Oxford handbook of internet psychology* (pp. 287–301). Oxford, UK: Oxford University Press.

Haythornthwaite, C. (2002). Strong, weak, and latent ties and the impact of new media. *The Information Society, 18*, 385–401.

Heady, B. W., & Wearing, A. J. (1990). Subjective well-being and coping with adversity. *Social Indicators Research, 22*, 327–349.

Helgeson, V. S., & Gottlieb, B. H. (2000). Support groups. In S. Cohen, L. G. Underwood, & B. H. Gottlieb (Eds.), *Social support measurement and intervention* (pp. 221–245). New York, NY: Oxford University Press.

Herek, G., & Glunt, E. K. (1988). An epidemic of stigma: Public reactions to AIDS. *American Psychologist, 43*, 886–891.

Houston, T. K., Cooper, L. A., & Ford, D. E. (2002). Internet support groups for depression: A 1-year prospective cohort study. *American Journal of Psychiatry, 159*, 2062–2068.

Irrizary, C., Downing, A., & West, D. (2002). Promoting modern technology and internet access for under-represented older populations. *Journal of Technology in Human Services, 19*, 13–29.

Johnson, A. J., Wright, K. B., Craig, E. A., Gilchrist, E. S., Lane, L. T., & Haigh, M. M. (2008). A model for predicting stress levels and marital satisfaction for stepmothers utilizing a stress and coping approach. *Journal of Social and Personal Relationships, 25*, 119–142.

Jones, M., Luce, K. H., Osborne, M. L., Taylor, K., Cunning D., Doyle, A. C., ... Taylor, B. (2008). Randomized, controlled trial of an Internet-facilitated intervention for reducing overeating and overweight in adolescents. *Pediatrics, 121*, 453–462.

Katz, E., Blumler, J. G., & Gurevitz, M. (1974). Utilization of mass communication by the individual. In J. G. Blumler & E. Katz (Eds.), *The uses of mass communication: Current perspectives on gratifications research* (pp. 19–32). Beverly Hills, CA: Sage.

Kim, J., & Lee, J. E. (2011). The Facebook paths to happiness: Effects of the number of Facebook friends and self-presentation on subjective well-being. *Cyberpsychology, Behavior, and Social Networking, 6*, 359–364.

King, S. A., & Moreggi, D. (1998). Internet therapy and self-help groups: The pros and cons. In J. Gakenbach (Ed.), *Psychology of the Internet: Intrapersonal, interpersonal, and transpersonal implications* (pp. 77–109). San Diego, CA: Academic Press.

Kohn, P. M. (1996). On coping adaptively with daily hassles. In M. Zeidner & N. S. Endler (Eds.), *Handbook of coping* (pp. 181–201). New York, NY: Wiley.

Kontos, E. Z., Emmons, K. M., Puleo, E., & Viswanath, K. (2010). Communication inequalities and public health implications of adult social networking site use in the United States. *Journal of Health Communication: International Perspectives, 15*(Suppl 3), 216–235.

LaGaipa, J. L. (1990). The negative effects of informal support systems. In S. Duck & R. C. Silver (Eds.), *Personal relationships and social support* (pp. 122–139). Newbury Park, CA: Sage.

Lazarus, R. S., & Folkman, S. (1984). *Stress, appraisal, and coping*. New York, NY: Springer.

Leimeister, J., Schweizer, K., Leimeister, S., & Krcmar, H. (2008). Do virtual communities matter for the social support of patients: Antecedents and effects of virtual relationships in online communities. *Information, Technology, & People, 21*, 350–374.

Ley, B. L. (2007). Vive les rose!: The architecture of commitment in an online pregnancy and mothering group. *Journal of Computer-Mediated Communication, 12*, 1388–1408.

Lieberman, M. A., & Goldstein, B. A. (2006). Not all negative emotions are equal: The role of emotional expression in online support groups for women with breast cancer. *Psycho-Oncology, 15*, 160–168.

Manago, A. M., Taylor, T., & Greenfield, P. M. (2012). Me and my 400 friends: The anatomy of college students' Facebook networks, their communication patterns, and well-being. *Developmental Psychology, 48*(2), 369–380.

McKenna, K. (2008). Influences on the nature and functioning of online groups. In A. Bark (Ed.), *Psychological aspects of cyberspace: Theory, research, applications* (pp. 228–242). New York, NY: Cambridge University Press.

McKenna, K., & Green, A. (2002). Virtual group dynamics. *Group Dynamics: Theory, Research and Practice, 6*, 116–127.

Neuhauser, L., & Kreps, G. L. (2003). Rethinking communication in the e-health era. *Journal of Health Psychology, 8*, 7–23.

Nonnecke, B., Andrews, D., & Preece, J. (2006). Non-public and public online community participation: Needs, attitudes, and behavior. *Electronic Commerce Research, 6*, 7–20.

Owen, J. E., Klapow, J. C., Roth, D. L., Shuster, J. L., & Bellis, J. (2005). Randomized pilot of a self-guided Internet coping group for women with early-stage breast cancer. *Annals of Behavioral Medicine, 30*, 54–64.

Preece, J. (1999). Empathic communities: Balancing emotional and factual information. *Interacting with Computers, 12*, 63–77.

Purcell, K., Rainie, L., Mitchell, A., Rosenstiel, T., & Olmstead, K. (2010, March). *Pew Internet & American life project: Understanding the participatory news consumer*. Retrieved from http://www.pewinternet.org/Reports/2010/Online-News.aspx

Query, J. L., Jr., & Wright, K. B. (2003). Assessing communication competence in an on-line study: Towards informing subsequent interventions among older adults with cancer, their lay caregivers, and peers. *Health Communication, 15*, 205–219.

Rains, S. A., & Keating, D. M. (2011). The social dimension of blogging about health: Health blogging, social support, and well-being. *Communication Monographs, 78*, 511–534.

Rains, S. A., & Young, V. (2009). A meta-analysis of research on formal computer-mediated support groups: Examining group characteristics and health outcomes. *Human Communication Research, 35*, 309–336.

Rieh, S. Y., & Belkin, N. J. (2000). Interaction on the web: Scholars' judgment of information quality and cognitive authority. In D. H. Kraft (Ed.), *Proceedings of the 63rd ASIS Annual Meeting* (pp. 25–33). Silver Spring, MD: American Society for Information Science.

Riggs, S. A., Vosvick, M., & Stallings, S. (2007). Attachment style, stigma and psychological distress among HIV+ adults. *Journal of Health Psychology, 12*, 922–936.

Rimal, R. N., & Real, K. (2003), Perceived risk and efficacy beliefs as motivators of change. *Human Communication Research, 29*, 370–399.

Robinson, T. N., Patrick, K., Eng, T. R., & Gustafson, D. (1998). An evidence-based approach to interactive health communication: A challenge to medicine in the information age. Science panel on interactive communication and health. *The Journal of the American Medical Association, 280*, 1264–1269.

Roschelle, A. (1997). *No more kin: Race, class, and gender in family networks.* Thousand Oaks, CA: Sage.

Rosman, S. (2004). Cancer and stigma: Experience of patients with chemotherapy-induced alopecia. *Patient Education and Counseling, 52*, 333–339.

Sarasohn-Kahn, J. (2008). *The wisdom of patients: Health care meets online social media.* Oakland, CA: California HealthCare Foundation.

Shaw, L. H., & Gant, L. M. (2002). In defense of the Internet: The relationship between Internet communication and depression, loneliness, self-esteem, and perceived social support. *Cyberpsychology & Behavior, 5*, 157–171.

Shaw, B. R., Hawkins, R., McTavish, F., Pingree, S., & Gustafson, D. H. (2006). Effects of insightful disclosure within computer-mediated support groups on women with breast cancer. *Health Communication, 19*, 133–142.

Steinfield, C., Ellison, N. B., & Lampe, C. (2008). Social capital, self-esteem, and use of online social network sites: A longitudinal analysis. *Journal of Applied Developmental Psychology, 29*, 434–445.

Sullivan, C. F. (2003). Gendered cybersupport: A thematic analysis of two on-line cancer support groups. *Journal of Health Psychology, 8*, 83–103.

Tanis, M. (2008). Health-related online forums: What's the big attraction. *Journal of Health Communication, 13*, 698–714.

Tidwell, L. C., & Walther, J. B. (2002). Computer-mediated communication effects on disclosure, impression, and interpersonal evaluations: Getting to know one another a bit at a time. *Human Communication Research, 28*, 317–348.

Vanable, P. A., Carey, M. P., Blair, D. C., & Littlewood, R. A. (2006). Impact of HIV-related stigma on health behaviors and psychological adjustment among HIV-positive men and women. *AIDS and Behavior, 10*, 473–482.

Walther, J. B. (1992). Interpersonal effects in computer-mediated interaction: A relational perspective. *Communication Research, 19*, 52–90.

Walther, J. B. (1996). Computer-mediated communication: Impersonal, interpersonal, and hyperpersonal interaction. *Communication Research, 23*, 3–43.

Walther, J. B. (2007). Selective self-presentation in computer-mediated communication: Hyperpersonal dimensions of technology, language, and cognition. *Computers in Human Behavior*, *23*, 2538–2557.

Walther, J. B., & Boyd, S. (2002). Attraction to computer-mediated social support. In C. A. Lin & D. Atkin (Eds.), *Communication technology and society: Audience adoption and uses* (pp. 153–188). Cresskill, NJ: Hampton Press.

Walther, J. B., Slovacek, C., & Tidwell, L. C. (2001). Is a picture worth a thousand words? Photographic images in long term and short term virtual teams. *Communication Research*, *28*, 105–134.

Wang, Z., Walther, J. B., Pingree, S., & Hawkins, R. P. (2008). Health information, credibility, homophily, and influence via the Internet: Web sites versus discussion groups. *Health Communication*, *23*, 358–368.

Weinert, C., & Hill, W. G. (2005). Rural women with chronic illness: Computer use and skill acquisition. *Women's Health Issues*, *15*, 230–236.

Wolitski, R. J., Pals, S. L., Kidder, D. P., Courtenay-Quick, C., & Holtgrave, D. R. (2008). The effects of HIV stigma on health, disclosure of HIV status, and risk behavior of homeless and unstably housed persons living with HIV. *AIDS Behavior*, *13*, 1222–1232.

Wright, K. B. (1999). Computer-mediated support groups: An examination of relationships among social support, perceived stress, and coping strategies. *Communication Quarterly*, *47*, 402–414.

Wright, K. B. (2000). Perceptions of on-line support providers: An examination of perceived homophily, source credibility, communication and social support within on-line support groups. *Communication Quarterly*, *48*, 44–59.

Wright, K. B. (2002). Social support within an on-line cancer community: An assessment of emotional support, perceptions of advantages and disadvantages, and motives for using the community. *Journal of Applied Communication Research*, *30*, 195–209.

Wright, K. B. (2005). Researching Internet-based populations: Advantages and disadvantages of on-line survey research, on-line questionnaire authoring software packages, and web survey services. *Journal of Computer-Mediated Communication*, *10*(3), Article 11. Retrieved form http://jcmc.indiana.edu/vol10/issue3/wright.html

Wright, K. B. (2012). Emotional support and perceived stress among college students using Facebook.com: An exploration of the relationship between source perceptions and emotional support. *Communication Research Reports*, *29*, 1–10.

Wright, K. B., & Bell, S. B. (2003). Health-related support groups on the Internet: Linking empirical findings to social support and computer-mediated communication theory. *Journal of Health Psychology*, *8*, 37–52.

Wright, K. B., & Miller, C. H. (2010). A measure of weak tie/strong tie support network preference. *Communication Monographs*, *77*, 502–520.

Wright, K. B., Rains, S., & Banas, J. (2010) Weak tie support network preference and perceived life stress among participants in health-related, computer-mediated support groups. *Journal of Computer-Mediated Communication*, *15*, 606–624.

Wright, K. B., Rosenberg, J., Egbert, N., Ploeger, N., Bernard, D. R., & King, S. (2013). Communication competence, social support, and depression among college students. *A model of Facebook and face-to-face support network influence. Journal of Health Communication*, *18*, 41–47.

23

Using Digital Games to Promote Health Behavior Change

Debra A. Lieberman

Department of Communication, University of California, Santa Barbara, CA, USA

More people play digital games than ever before and the games they play are becoming more networked, mobile, interactive, and social (Entertainment Software Association, 2013; Lieberman, Biely, Fisk, et al., 2014). In addition to games made for entertainment and shared fun, some are works of art that are deeply moving and inspiring, some offer very difficult yet compelling challenges to overcome, some are vehicles for persuasion and propaganda, and some aim to influence players' knowledge, attitudes, and behaviors in areas such as learning, health, and social change.

Games with purposes beyond entertainment have been called, for example, "serious games," "games for impact," "games for change," and "meaningful play." This chapter focuses on the research and design of serious games made for, or repurposed for, health behavior change. It begins with a definition of the term "digital game," then discusses the variety of genres, content, formats, and technologies in today's games, as background for a discussion of theory and research findings in three major areas that contribute to health outcomes with digital games: (a) motivation, (b) learning, and (c) behavior change. In addition to presenting psychological processing of health games and associated effects on behavior, some history of the field is interspersed throughout the chapter, from the earliest computer-based games for learning in the early 1960s to the annual Games for Health Conference that began in 2004, to the field-building role of the Health Games Research national program, funded by the Robert Wood Johnson Foundation's Pioneer Portfolio from 2007 to 2013 to advance the research, design, and effectiveness of health games.

People have played entertainment games—and serious games—on computer-based technologies for more than 50 years. Digital games for learning were first used in classrooms in the early 1960s, about two decades before video game consoles and desktop computers became popular mass market consumer products. The first digital

The Handbook of the Psychology of Communication Technology, First Edition. Edited by S. Shyam Sundar.

games and simulations were delivered to US classrooms via mainframe computers and teletype machines, and research found that these computer-based interactive experiences led to significant gains in student learning (Coleman, 1971; Hansen, 1966; Suppes, 1967). In 1970, Clark Abt introduced the term "serious games" to describe computer-based learning games, in his book called *Serious Games*, and he defined them as games that "have an explicit and carefully thought-out educational purpose" (Abt, 1970, p. 9). A designer and creator of some of the first computer-based games for military and diplomacy training in the 1960s, Abt was optimistic that serious games could be effective for learning and training because "they are highly motivating," "they communicate very efficiently the concepts and facts of many subjects," and they offer us "a rich field for a risk-free, active exploration of serious intellectual and social problems" (p. 13). His book discussed computer-based games, and games played in person, for learning, problem solving, job training, economics, government, and industry.

Fast-forward to today, and serious games and game technologies have advanced tremendously. Digital games are available on consoles and computers in the home; on Internet-connected mobile technologies such as mobile phones and tablets that can be used on-the-go; in physical locations such as K-12 and college classrooms, workplace training environments, hospitals, clinics, community centers, and gyms; and in settings such as op-ed pages of online news media, virtual worlds such as Second Life, and augmented reality environments in which digital content and data, from networks and sensors, are seamlessly displayed and integrated into the player's social and physical world. Along with today's increased and more widely varied use of digital games for work and play, a growing body of research is identifying theory-driven, evidence-based principles of effective game design and demonstrating that digital games can be designed to influence players' thoughts and emotions in ways that enhance learning and health behavior change. (For research overviews, see Blumberg, 2014; Noar & Harrington, 2012; Ritterfeld, Cody, & Vorderer, 2009; Vorderer & Bryant, 2006.)

Research has also investigated potential harmful effects of video games on players' thoughts, emotions, behaviors, and health, regarding violent content (Anderson & Bushman, 2001; Anderson, et al., 2010), stereotyped portrayals of characters (Mou & Peng, 2009), advergames designed to advertise and sell unhealthy products (Cauberghe & De Pelsmacker, 2010; Harris, Speers, Schwartz, & Brownell, 2012), sedentary lifestyle due to excessive video game playing (Rey-Lopez, Vicente-Rodriguez, Biosca, & Moreno, 2007), and video game addiction (Griffiths, Kuss, & King, 2012). It is important that developers of serious games use these research findings as guides to help them avoid creating potentially harmful content. It is also important that players of serious games to develop media literacy skills (Lieberman, 1980) that can enable them to be critical and selective consumers of video games, balance their game playing with a healthy mix of other worthwhile activities, limit their game playing so that it does not become excessively sedentary or all-consuming, and, even within a game-play session, include some active games to get up off the couch and move.

Defining Digital Games

Digital games are media that potentially have any of the characteristics that other electronic or cinematic or digital media might have, such as video, film, audio, animation, characters and stories, interactions, transactions, simulations, communication channels, search, input from sensors, programmability, and the connectivity and functionality of the Internet. However, games have distinct characteristics that lead to new ways of interacting with media and new cognitive, emotional, social, and even physiological responses to media. In contrast to other media, digital games involve challenge to reach a goal. Fundamentally, a game is a rule-based activity that involves challenge to reach a goal and provides feedback on the player's progress toward the goal (see Gee, 2009; Lieberman, 2006, 2009). The rules, challenges, and feedback of games produce experiences that are not possible with nongame media formats.

Digital games are among the most interactive media available today because they take into account player inputs and past game events so thoroughly that they are, or seem to be, very responsive to the player at any given moment as the game state unfolds. The player's game state and new messages from the game to the player are contingent upon all previous messages exchanged between the player and the game. Kiousis (2002), Rafaeli (1988), Rafaeli and Ariel (2007), and Sundar, Kalyanaraman, and Brown (2003) offer insightful explications and definitions of interactivity between people and digital media. Like other highly interactive digital media formats such as certain online ads, retail web sites, and search engines, digital games can be programmed to respond comprehensively to all (a) previous inputs from the player, (b) data about the player collected in the past from various sources (e.g., demographic characteristics, school records, health history, purchasing history, memberships, online search queries, online posts, and much more), and (c) real-time data about the player's current circumstances, location, vital signs, and physiological and emotional state, collected, for example, from cameras, GPS, biosensor technologies, medical devices, and people in the player's social network. In real time it is possible to measure geo-location, heart rate, blood pressure, stress, steps taken, distance covered, brain waves, eye gaze, body movements, and even the slight movements of facial muscles that indicate emotional state.

A game can be programmed to save many or all of the player's inputs, saved data, and real-time sensor data, cumulatively and in context, and can use the saved data to represent the player's game state. This individualized game state based on the player's previous choices and unique characteristics can enhance the player's sense of autonomy and control in the game environment, and when a game meets the player's basic needs for autonomy and control, it can increase engagement and enjoyment (Deci & Ryan, 2002). Two-way interaction between player and game is another feature that can boost the player's engagement and enjoyment, and it can motivate the player to persevere to achieve the goals of the game (Lieberman, 1982b; Sundar & Kim, 2005).

Digital games can be played on a variety of technology platforms that enable a range of game play styles, genres, and player inputs, and these platforms include computers, video game consoles, mobile phones, tablets, handheld game devices, electronic toys, electronic learning systems, arcade systems, gym equipment (such as stationary bikes and "smart walls"), 3D virtual displays, augmented reality displays, and robots, among others. Game playing can take place entirely through interaction with digital media (e.g., a game played entirely on a computer or console and displayed on a screen) or through a combination of events in the physical world and the virtual digital world (e.g., a fitness game that uses sensors to track a runner's distance and time, stress levels, vital signs, and sleep patterns, and provides these measures as game feedback; or an alternate reality game that adds fictional elements to experiences in the physical world and has an accompanying website that provides challenges, clues, news about fictional characters, and record-keeping). In digital games played on any of these technology platforms, there are features, interfaces, and affordances that have the potential to support learning and health behavior change (see Bransford, Brown, & Cocking, 2000; Gee, 2009; Lieberman, 2006, 2009, 2013; Lieberman, Fisk, & Biely, 2009; Shute & Ke, 2012; Shute, Rieber, & Van Eck, 2011; Vogel et al., 2006).

Game genres and interfaces range from pure puzzle or skill games, to simulation games, to physically challenging active games (such as dance pad games or camera-based games that require physical exertion in order to play, or games that track the player's fitness workouts or provide real-time workout or physical therapy coaching), to expressive games in which the player creates art or gives an artistic performance, to story-based dramatic adventure games in which the player's decisions and actions help move the plot forward. Players apply their skills and learn from experience as they focus on achieving the game goal, and typically they must repeat and rehearse the same actions many times until they eventually succeed. Story-based games immerse players in a virtual world where they advance the story line by gathering information, making decisions, overcoming obstacles, solving puzzles or mysteries, and using new skills, often interacting with objects, systems, and characters in environments that support learning-by-doing. When networked, digital games enable players to access online sources of information or interact with other players or with the characters (avatars) that other players control. Algorithms can analyze all the game inputs (data gathered from the player's actions in the game, data drawn from databases, and real-time data about the player gathered by sensors) to make inferences about the player that can enhance interactivity and customize a game to meet the player's needs, interests, and abilities.

As a result, digital games are a perfect storm of individualized and interactive entertainment, content, and virtual experiences that have become a favorite pastime for many. They are engaging environments for learning, motivation, attitude change, and behavior change; are widely available; and have a "cool" factor that attracts diverse demographic groups. For health behavior change, digital games can reach people who are at-risk and not likely to seek information, counseling, or clinical care. An exciting game that also, by the way, happens to deal with health topics and contains behavior change features, would be an effective way to reach those populations, just as soap operas and medical TV drama series have used entertainment-education approaches to integrate health messages and role modeling into popular media with entertaining dramatic story lines (Lieberman, 1982a; Lieberman, Chaffee, & Roberts, 1988; Singhal, Cody, Rogers, Sabido, & Pitt, 2006).

Digital games designed for health-related learning and behavior change can integrate behavioral health theories and principles into game play, to help players try out and rehearse new skills, develop expertise and self-confidence, gain deeper understanding of personal risks and susceptibilities to health problems, and virtually experience the consequences of healthy and unhealthy decisions and behaviors. For example, digital games have led to health behavior change in areas including nutrition with the game *Squire's Quest* (Baranowski et al., 2003) and with the game *Mindless Eating Challenge* (Byrne et al., 2012); adherence to one's treatment plan for cancer with the game *Re-Mission* (Kato, Cole, Bradlyn, & Pollock, 2008); reduction of diabetes-related urgent care and emergency visits with the diabetes self-management game *Packy & Marlon* (Brown et al., 1997); increases in physical activity and cardiovascular health with active games such as *Dance Dance Revolution* (Murphy et al., 2009), with robots that deliver active games such as *Robot Motivator* (Fasola & Mataric, 2011 and with alternate reality games such as *The Skeleton Chase* that immerses players in a fictional story that takes place in their everyday surroundings and is enhanced and supported by digital and traditional media (Johnston, Massey, & Marker-Hoffman, 2012); and improvement of older adults' cognitive skills with the driving simulation game *NeuroRacer* (Anguera et al., 2013) and with virtual racing games (cybercycling) that involves using a stationary bike and looking at a screen that immerses the player in various racing experiences (Snyder et al., 2010), to name a few.

Digital games also support the delivery of health care including diagnosis, treatment, clinical training, and administrative planning. Clinicians have used health games to diagnose and treat patients' conditions, such as attention deficit disorder (Pope & Bogart, 1996); autism (Tanaka et al., 2010); age-related reduction in multitasking ability and cognitive control (Anguera et al., 2013); mental health problems (Griffiths et al., 2012; Wilkinson, Ang, & Goh, 2008); lazy eye (Hess et al., 2012); cystic fibrosis (Bingham, Bates, Thompson-Figueroa, & Lahiri, 2010); and deficiencies in the strength, coordination, and balance of people needing physical therapy and rehabilitation (Fritz, Rivers, Merlo, & Duncan, 2009).

Motivation to Engage in Digital Games

What motivates people to play digital games with rapt concentration and to persist even when the game is difficult? A game's compelling challenge—be it a puzzle to solve, skill to employ, system to explore, or dramatic story to experience—draws players in and it is typical for players to rehearse game skills repeatedly until they have mastered them (Gee, 2009; Lieberman, 2013; Ritterfeld, Cody, & Vorderer, 2009; Vogel et al., 2006). Incremental accomplishments give players a feeling of progress as the game becomes increasingly difficult. Players expect to experiment, fail, and try again until they succeed, and receive help or coaching when needed. Many games adapt to players' abilities and keep the level of difficulty in a range that is highly challenging but not impossible. Flow theory points to the strong motivation that occurs when this state is optimized, keeping the player in a zone of intense concentration and effort that is continuously neither too difficult and frustrating nor too easy and boring (Csikszentmihalyi, 1990).

Game-delivered feedback shows players their progress, and they are able to see how their choices have enhanced or hindered the desired outcome. In games that use rewards, players learn what is valued when they see which actions accrue points, status, access to resources, or access to new game levels. In games that focus on a story, players enjoy being a game character who makes powerful events happen and makes the story advance in interesting ways. Games are immersive and experiential. Players are motivated to exert effort and persist until they have developed the skills they need to complete the game (Bingham et al., 2010; Brown et al., 1997; Snyder et al., 2010; Sundar & Kim, 2005; Vorderer & Bryant, 2006).

Emotion can enhance motivation. Emotional experiences in games can capture players' attention, engage them in a task, and improve their memory and recall (Baranowski et al., 2008; Pugh & Bergin, 2006). There is an emotional component to playing digital games involving, for example, players' empathy for the characters they encounter, their fear and trepidation about what's about to happen, the suspense of not knowing whether they will succeed, and their triumph when they overcome a difficult obstacle.

Social aspects of game playing can be highly motivating. Some players want to be admired for their expertise, and so they will strive to excel and be recognized as a top player in a leader board. Also, social networks and friends can provide encouragement, support, and praise for players who, for example, are playing a fitness game to increase the frequency and intensity of their workouts or a nutrition game to improve their food choices or weight management. In the digital environment, members of the social network can see their friend's tracked progress and share helpful feedback and acknowledgement. Games can also leverage social influence through collaboration and teamwork. In collaborative games, there is motivation to contribute to the group effort and not let one's partner or teammates down (Staiano, Abraham, & Calvert, 2012). If the team's score is highly dependent on the player's performance, the player will exert more effort than in a game that simply aggregates the scores of all players' contributions (Feltz, Irwin, & Kerr, 2012). Competition between individuals or teams can be very motivating, and is an element used in many digital games (Snyder et al., 2010).

Self-determination theory offers insights into motivation by identifying two broad categories: intrinsic motivation and extrinsic motivation (Deci & Ryan, 2002; Ryan, Rigby, & Przybylski, 2006). Intrinsic motivation is driven by the individual's innate personal interests and goals. A digital game is intrinsically motivating if it engages the skills or topics that are inherently interesting and satisfying to the player or provides challenges that the player already wants to become good at accomplishing.

However, if the player is not intrinsically motivated and not interested in the skills, topics, or challenges addressed in a particular game, extrinsic motivators such as points and level-ups may engage the player in simple game tasks for a short time, but they will not motivate the depth of processing or lasting engagement that often occur when a game is aligned with the player's intrinsic motivations. Research has even found that the use of extrinsic motivators can in some circumstances undermine and decrease a person's existing intrinsic motivation (Deci, 1971). This was found in the workplace, when a nongame task (work) was turned into a game, in a process called "gamification," and extrinsic rewards were given for work productivity. When employees were required to play in the gamified workplace without prior consent, it backfired and productivity decreased (Mollick & Rothbard, 2013).

According to self-determination theory, people are motivated by three deep-seated intrinsic needs, including the need for autonomy (also called control or agency), competence (also called mastery), and relatedness (also called social connectedness and social purpose). Studies have found that games designed to support these three fundamental human needs tend to increase players' game enjoyment and satisfaction, and motivate them to engage more deeply in the game (Peng, Lin, Pfeiffer, & Winn, 2012).

Learning with Digital Games

How do people learn with digital games? It is well established that digital games can enhance learning (see Blumberg, 2014; Bransford et al., 2000; Chen, Lieberman, & Paisley, 1985; Gee, 2003, 2009; Krendl & Lieberman, 1988; Lieberman, 1985, 2006, 2009, 2014; Lieberman, Bates, & So, 2009; Lieberman & Linn, 1991; Moreno & Mayer, 2004; Pillay, Brownlee, & Wilss, 1999; Pugh & Bergin, 2006; Shaffer, Squire, Halverson, & Gee, 2005; Shute et al., 2011). Research on some health games has found significant gains in players' knowledge (e.g., Brown et al., 1997; Kato et al., 2008). To help explain how learning occurs in games, theories and research on experiential learning, individualization, feedback, skill rehearsal, transfer of learning, and constructivist learning are useful, as follows.

Digital games can be immersive experiences similar to real life and players often respond emotionally and even physiologically to media and game experiences as if they were real events (Reeves & Nass, 1995). As noted earlier, players become emotionally involved in games and game characters, and engaged in compelling challenges. Immersiveness, emotional involvement, interactivity, and feedback are only a few of the features that make digital games such effective experiential learning environments. Game technologies bring the unique characteristics of networked, interactive media and can add stories, information, data, social interaction, reality-based or fantasy-based worlds and characters, and computing power to the digital world of the game. For example, learners can experiment in simulations to discover cause-and-effect relationships; take the perspective of others to gain insights; receive coaching or interact with a helper or co-learner character; zoom into the future to see how the consequences of their actions in a game would progress over time; and interact with other players or teammates. These game experiences can require the player to use and rehearse skills as wide ranging as creativity in its many forms, logic, attention, memory, social skills, leadership skills, clinical skills, counseling skills, persuasive skills, visual or auditory acuity, eye-hand coordination, balance, physical strength, and/or physical endurance.

Game environments put players in interactive simulated worlds they could never visit themselves, such as worlds in the past or future, or that involve many risks and dangers, or that exist in places that cannot support human life, or that are at the scale of molecules or galaxies, or that must slow down or speed up time to demonstrate processes that humans cannot ordinarily detect, or that are simplified so that players can more clearly see interrelationships between a limited number of elements in a complex system.

Digital games can be tailored and individualized to each player and can adapt to the player's changing abilities, adjusting the difficulty level in response to player success or failure, to keep the game extremely challenging but still just within the highest boundary of the player's ability. Visuals, sounds, and kinesthetic experiences can convey abstract concepts in multiple and juxtaposed modalities such as images, graphs, animations, text, narration, music, sound effects, and haptics (tactile feedback).

Feedback on progress and performance is a key component of digital games and occurs continuously. Game players expect and welcome ongoing assessment so they can learn from errors (Collins & Halverson, 2009; Gee, 2009). Assessment and feedback are central to motivation (Garris, Ahlers, & Driskell, 2002), especially when the player has high commitment to the goal and confidence in ultimate success, which the best digital games are designed to inspire (Bransford & Schwartz, 2001; Shute & Ke, 2012; Shute et al., 2011). It is not surprising that some K-12 classrooms and university courses are beginning to use elements of digital games to structure their lessons by engaging students in "quests" to learn about an issue or topic and by providing several forms of assessment and feedback that are not used as end points to learning in a final grade but instead are presented as nonthreatening opportunities to improve skills so learners can continue to make progress (Sheldon, 2011).

Digital games enhance the cognitive and motor skills that players use repeatedly while they play, sometimes dozens or hundreds of times in a single game-play session (Gee, 2003, 2009; Ritterfeld, Cody, & Vorderer, 2009; Shute et al., 2011). For example, studies have found that fast-paced action games can improve spatial and temporal resolution of vision, visual short-term memory, spatial cognition, pattern recognition, reaction time and speed-accuracy, selective attention, divided attention, sustained attention, metacognition about one's own process of learning, and aspects of executive functioning (Bavelier, Green, Pouget, & Schrater, 2012). The field of neurotherapeutics has begun to use specially designed digital games because they are highly motivating and effective tools for delivering cognitive rehearsal and therapy (Anguera et al., 2013).

There is growing evidence that players may transfer the skills they learn in games to nongame contexts. For example, flight school cadets who played a flight simulator game outperformed a no-game control group on actual flight performance even though the cockpit instrumentation layouts were not the same in the game and in the plane (see review by Boot, Kramer, Simons, Fabiani, & Gratton, 2008). Transfer is important because learners will spend their lives confronting new problems and will need the ability and the readiness to draw on their existing knowledge and skills in new contexts. Digital games constantly challenge players to tackle difficult problems, and they can be designed to give learners the skills and guided practice that would make them more inclined and more able to transfer (Lieberman, Biely, Thai, & Peinado, 2014; Salomon & Perkins, 1989). For example, game-based coaching can encourage learners to use mindfulness, cognitive elaboration, and self-monitoring to help them reflect on their own knowledge and their process of learning while it is taking place, and then make conscious efforts to transfer their knowledge and learning skills to new settings (Mayer 1999, 2003; Mayer & Wittrock, 1996; Singley & Anderson, 1989).

Constructivist learning occurs in an experiential and applied problem-solving environment, where learners take an active role in exploration and personally construct their own knowledge in authentic situations that allow them to build on what they already know (Honebein, Duffy, & Fishman, 1993; Prawat & Flowden, 1994). This form of learning provides choice and autonomy, and makes learning relevant and engaging. Related pedagogical approaches include experiential learning, scenario-based learning, learning-by-doing, problem-based learning, and authentic problem-solving. Health games can deliver and support constructivist learning experiences, either as standalone activities in a digital environment or integrated and blended with hands-on learning activities. One variant of constructivist learning is epistemic learning, which puts players in the virtual environment of a professional so they can confront the problems the professional might encounter and use the tools and resources the professional would use on the job. Many health games use epistemic learning when they challenge the player to play the role of an emergency room doctor treating patients, or a counselor detecting suicidal tendencies or drug use in interviews with virtual patients. In a similar way, many health games focusing on the player's prevention and self-care will put players into a constructivist environment in which they must care for a another character or for their own game character who has a health problem such as nutritional deficits (Byrne et al., 2012), asthma (Lieberman, 1999), type 1 diabetes (Brown et al., 1997; Lieberman, 2012b), or cancer (Kato et al., 2008).

Behavior Change with Digital Games

What are some of the psychological mechanisms that lead from game play to behavior change? The experiential, interactive, immersive nature of digital games makes them potentially rich environments for influencing players' self-concepts, self-efficacy, aspirations and goals, attitudes, risk perceptions, and many other factors known to improve health behaviors. There is a wealth of behavioral health theory and evidence to support a wide variety of game design strategies that might motivate and support health behavior change (Lieberman, 1992, 1997, 2001a, 2001b; 2012a, 2013; Lieberman & Brown, 1995). Some examples include processes of homophily and identification with game avatars, social cognitive theory (especially the components of this theory that focus on role modeling, vicarious experience, and self-efficacy), nurturing of game characters in simulated environments, the extended parallel process model and social support.

Homophily is the tendency of individuals to associate and bond with similar others (McPherson, Smith-Lovin, & Cook, 2001). People's personal networks tend to be homogeneous regarding a variety of demographic, behavioral, and personal characteristics. It follows that game players may seek out and engage more deeply with game characters who are a lot like them. Social cognitive theory finds that people attend to and identify more strongly with role models, in real life and in media, who are similar to them and, as a result, they are more likely to imitate the behaviors of those models (Bandura, 1997). It follows that a game character who looks a lot like the player is likely to be especially appealing and influential. This phenomenon has been found in studies of digital media avatars who are virtual characters that the player controls and who represent the player in a game world or a virtual environment. The research finds that game players and visitors to virtual worlds indeed do pay more attention to avatars who look like them (Blascovich & Bailenson, 2011).

When people embody their avatar character in the virtual world, they take the avatar's perspective and they see, hear, and feel as if they were having those sensory experiences in the real world. The implications for health behavior change are enormous. A few avatar studies have found that engaging in healthy behaviors in the virtual world while controlling an avatar that looks like the individual user can improve the user's own health behaviors related to, for instance, physical activity and healthy eating (Fox, Bailenson, & Ricciardi, 2012). Avatars can influence users' self-concepts. A study found that when users control an attractive avatar or a tall avatar, their perceptions of their own attractiveness, power, or leadership qualities continue from the virtual world into their real life, in a process called the "Proteus effect," and they change their interpersonal interaction style, as if they really had the characteristics of their avatar (Yee, Bailenson, & Ducheneaut, 2009). One study investigated effects of controlling an avatar that had disabilities and found that users who had that virtual experience became more empathetic and caring toward disabled people in the real world and they offered more help to people with disabilities (Ahn, Le, & Bailenson, 2013).

Social cognitive theory (Bandura, 1986, 1997) is one of the foundational theories that guide the design of health games. It identifies a variety of interrelated learning and behavior change processes including three that will be discussed here: observational learning, vicarious experience, and self-efficacy development. Games provide opportunities for observational learning when players watch a game character that they do not control. The character serves as a role model. Players learn through observation when they see the role model character enact a behavior. If the character is rewarded or punished for the behavior, players learn whether carrying out the behavior is a good or bad idea. For example, players can learn how to do something unhealthy when they see a role model do it, but if that role model is portrayed as receiving social disapproval or becoming sick, players are also likely to learn that the activity is not wise, and if portrayed as receiving social approval, players may want to emulate the behavior. Smoking, for example, has been portrayed as disgusting and worthy of ridicule in some story lines and as attractive and enhancing social status in others. When a healthy behavior is modeled, players may be encouraged to engage in it if the character experiences social approval and becomes more healthy, vigorous, and joyful. Observational learning becomes vicarious experience when players experience "self-other merging" and "self-modeling" by watching their own avatar carry out behaviors that they have controlled themselves. Not only do players see the rewarding or punishing consequences of those behavioral choices, they experience them vicariously through their avatar.

Observational learning and vicarious experience can improve a game player's self-efficacy, which is their self-confidence that they can carry out a specific behavior successfully. Behavior change is more likely to occur when self-efficacy for carrying out the desired behavior is high (Bandura, 1997). People must feel confident that they are capable of carrying out a behavior or task before they are likely to give it a try and to persist at it even when there are obstacles and setbacks. In well-designed health games, both observational learning and vicarious virtual experience via an avatar have led to improved self-efficacy for carrying out health behaviors and this has enhanced health behavior change (Blascovich & Bailenson, 2011; Kato et al., 2008; Lieberman, 2012; Thomas, Cahill, & Santilli, 1997). Digital games often involve the repeated rehearsal of skills and behaviors, in a nonthreatening environment, and this rehearsal increases players' self-efficacy for using those skills and performing those behaviors in their own lives. Higher levels of self-efficacy help behavior change to occur.

Games can be simulations of health regimens that players must carry out successfully in order to win the game or in order to keep their avatar healthy enough to win the game. Simulations in health games provide experiences that help players see the effects of health behaviors, and they challenge players to improve the prevention and self-care skills they use in the game, especially when health goals must be achieved in order to win. Nurturing another character, to bring it back to health or to help it maintain health, is also a type of simulation and it is an effective approach, used in several health games (e.g., Brown et al., 1997, Byrne et al., 2012; Kato et al., 2008). For a discussion of game design strategies involving game mechanics (player actions and game responses) that have been used to improve health, see Baranowski et al. (2013).

A useful approach for designing health games comes from the extended parallel process model (Witte, Meyer, & Martell, 2001), which focuses on messages that convey health risks as a way to change health behaviors. The model shows that the likelihood of health behavior change increases when individuals realize that they are susceptible to a health problem and that the problem is potentially severe; and they will be more likely to take action to prevent the threat of this health problem if they also have a strong sense of efficacy for dealing with it effectively. Efficacy has two components: self-efficacy for carrying out the desirable health behavior successfully and response efficacy, which is the strength of one's belief that the recommended health behavior will be effective enough to reduce the threat. For example, the model would predict that inactive people will become more physically active (behavior change) if they believe that (a) their current level of inactivity would harm their own cardiovascular health (perceived susceptibility), (b) the harm to their health would be seriously debilitating or deadly (perceived severity), (c) they are capable of increasing their physical activity (self-efficacy), and (d) their cardiovascular health would indeed improve if they increased their physical activity (response efficacy). All four of these extended parallel process model factors are essential for behavior change. Even if only one of the factors is weak, the model predicts that health behavior change is not likely to occur. Health games can be designed to improve any of these four factors that are known to motivate behavior change, focusing on one factor or on more than one, in one or more games, presented as a standalone game intervention or integrated into a coordinated transmedia health promotion campaign that conveys its health messages on a variety of media channels. Research on the target population for a health campaign can identify the factors that this population needs to improve. For example, a game for teens who think they are “invincible” when it comes to the potential dangers of smoking may benefit from a game that increases their perceptions of their own susceptibility to lung cancer if they smoke and strengthens their awareness of the severity of this disease.

Social interaction can be highly motivating in health games, to establish new norms for health behavior, engage people collectively in solving game challenges, provide support, and generate new ideas. It is well established that social support is strongly associated with better prevention behaviors, treatment adherence, and health outcomes (Peterson & Stunkard, 1989). Social networks and virtual worlds have presented health games in innovative ways. For example, in the virtual world Whyville, with content provided by the Centers for Disease Control and Prevention (CDC), children have learned how epidemics can spread and how to respond to them by seeing their own or others’ avatar characters catch the “whypox,” a virtual disease that infected many of the children’s avatars. This simulation of disease helped them discuss and discover how to prevent catching the disease, how to avoid transmitting it, and where to receive proper care in the virtual and actual world (Kafai, Quintero, & Feldon, 2010). Also, crowdsourcing (community-based design, participatory design) has been used in social games to gather ideas from the general public to discover new ways to address health problems. Two examples are the game *FoldIt*, where scientists invite players to come up with original ways to fold proteins so that players’ discoveries can be used to fight diseases such as cancer and HIV/AIDS, and the game *Breakthroughs to Cures*, in which players imagine and share new ways to accelerate medical research, knowing that scientists are interested and could potentially implement the best ideas that emerge from the game.

Social interactions and social networks are now used to promote health in a growing number of healthy lifestyle support games on web-connected mobile phones (Byrne et al., 2012; Lieberman, 2012). These games help people track and share their own data and then use the data for self-reflection, course correction, and coaching of friends. The games make healthy behaviors and better health outcomes the focus and purpose of the game challenge. Social connections are an essential component; friends and family members are enlisted to provide encouragement and support, and they have access to the player’s tracked data and health outcomes so they can provide motivation and acknowledgment. There are games that enable players to monitor their physical activity and food intake, calculate calories consumed and expended, and track other lifestyle habits and physiological measures such as stress, sleep, blood pressure, and heart rate. Since the games are on a network, it is easy to share the data with friends and family and with one’s doctors and clinical case managers.

Social support is often a key ingredient in active games (sometimes called “exergames” when physical exertion is the input to the game) that use, as their interface, dance pads, balance boards, gym and sports equipment, and sensors such as accelerometers, GPS, physiological monitors, and cameras. A study of teens who played active games found that cooperative and collaborative game play increased their use and effort expenditure with the games and led to better health and weight loss outcomes (Staiano et al., 2012). Games that require players to move their bodies vigorously to play the game, such as *Dance Dance Revolution* and certain games for the Wii console,

Wii Fit balance board, and Kinect camera-based console, can provide a moderate to vigorous cardiovascular workout (Graf, Pratt, Hester, & Short, 2009; Leatherdale, Woodruff, & Manske, 2010; Unnithan, Houser, & Fernhall, 2005). A study found that *Dance Dance Revolution* improved players' arterial and coronary functioning and weight management (Murphy et al., 2009). Mental health benefits were found in a study of older adults who experienced reductions in their subsyndromal depression after playing active games regularly and they improved their depressive symptoms, mental health-related quality of life (QoL), and cognitive performance (Rosenberg et al., 2010). See Lieberman et al. (2011) for an overview of research on active games.

This section of the chapter has discussed various approaches to the design of health games to improve players' health behaviors and outcomes, and it has cited many studies that provide supporting theories, models, and research evidence. In recent years, the quality and quantity of research on health games has grown remarkably. More researchers in a variety of disciplines are focusing on health games, new academic degree programs and scientific journals have appeared that focus on the research and design of serious games and specifically on health games, and funding sources in the public and private sectors have emerged. Most notably, regarding funding, the field received a great deal of support for research and scientific leadership when the Pioneer Portfolio of the Robert Wood Johnson Foundation (RWJF) funded a national program, called Health Games Research www.healthgamesresearch.org, from 2007 to 2013. RWJF began to support the health games field when it sponsored the first annual Games for Health Conference in 2004, led by Ben Sawyer, and continued to sponsor the conference and other meetings for ten years. Many of the game developers, health care providers, researchers, funding agencies, and investors who attended the first few annual conferences commented to RWJF that the health games field needed more research to demonstrate that well-designed health games could improve players' health behaviors and health outcomes. A more substantial body of research findings, they said, would provide evidence that health games work, and this would give the digital games industry the confidence to produce health games and would give health care providers the confidence to fund, purchase, recommend, and implement health games for patient care. Researchers in the field noted that it would also be important to investigate underlying psychological processes that occur with various game design features and player populations in order to improve the design and effectiveness of future health games.

RWJF decided to establish Health Games Research in 2007 to award research grants for the empirical study of health games and also to develop research resources, provide advocacy and leadership, and disseminate information and research findings to various stakeholder groups. Health Games Research was headquartered at the University of California, Santa Barbara, the home institution of its director, Debra Lieberman, who is the author of this chapter.

Work that emerged from Health Games Research has focused on processes of health behavior change with digital games. The national program awarded research grants and provided technical support to 21 research projects in the United States that tested a wide range of behavior change strategies in health games with various target populations, game genres, technology platforms, interfaces, health topics, and intended health outcomes. Findings from several of the funded projects are cited throughout this chapter (e.g., Anguera et al., 2013; Bingham et al., 2010; Byrne et al., 2012; Fasola & Mataric, 2011; Feltz et al., 2012; Fritz et al., 2009; Johnston et al., 2010; Peng et al., 2012; Snyder et al., 2010; Staiano et al., 2012; Tanaka et al., 2010) as are findings from studies conducted by the Health Games Research staff since 2007 (e.g., Lieberman, 2009, 2012a, 2012b, 2013; Lieberman, Bates, & So, 2009; Lieberman, Biely, Fisk, et al., 2014; Lieberman, Biely, Thai, & Peinado, 2014; Lieberman, Fisk, & Biely, 2009; Mayer & Lieberman, 2011). The research funded by Health Games Research, in these cited publications and many more from the 21 grantees and the staff, has launched new career directions, programs of research, resources for the field, graduate degree programs, academic centers focusing on digital games research (such as the Center for Digital Games Research at the University of California, Santa Barbara www.cdgr.ucsb.edu), and health game production companies (such as Akili Interactive Labs www.akiliinteractive.com). These entities are bringing more researchers into the field and are attracting new sources of funding to support research and to produce theory-driven, evidence-based health games.

Conclusion

Well-designed digital games are highly interactive and engaging, and they pose for the player a compelling challenge to reach a goal, bounded by a set of rules, while also providing feedback about progress made toward the goal. This combination of rules, challenge, and feedback, in a highly interactive virtual environment, offers unique opportunities to engage people cognitively, emotionally, socially, and even physiologically. As a result, digital games are powerful environments for health-related learning and behavior change. When designed by teams that include experts in the subject matter, the target population, the art of game design, and the science of learning and behavioral health, digital games can be very motivating and efficacious. In addition to design by an expert team, a game can benefit from formative research involving iterative rounds of usability testing and revision until the game appeals to the target population and demonstrably improves learning and antecedents of behavior change (Lieberman, 1999).

This chapter describes features of digital games and provides examples of foundational theories, concepts, models, and research findings that can guide the design of games for learning and health. The field is extensive and it draws from many disciplines. It will continue to need more research to discover principles of game-based learning and behavior change that can guide the development of future games. While human psychology does not change fundamentally, new questions about processes and effects will continue to arise as new game technologies and design innovations emerge. Games are improving in their capacity to sense the environment, detect emotions, provide compelling stories and challenges, involve players in social interactions, aggregate and use data from many players and other sources, vividly display the interactive game world on screens and in tangible objects that are networked and interactive, and individualize and tailor the game experience to match the player's interests, abilities, and personal characteristics. Further research on psychological processing in these areas would help improve game design for learning and health, and identify their impacts.

Digital games are media that involve challenge to reach a goal. Game experiences have become integrated into mainstream media and are often not differentiated from the flow of media content as separate entities. Instead, they are infused into the media landscape. For example, more people now use a tablet or mobile phone as their second screen during television viewing, and games are increasingly offered to supplement the television experience; games are often incorporated as transmedia elements in health campaigns that use many media channels to convey the same story and message; and there are numerous ways to participate in contests and crowdsourced games via digital media. To make good use of these emerging game formats and the more widely used formats of console and computer games, and to understand how games can help people learn and improve their health behaviors, more research is needed. Not only could research help improve the design and effectiveness of digital games, it could also demonstrate to decision-makers and the general public that games, when well designed and used wisely, are a viable, exciting, cross-generational, challenging, motivating, and highly worthwhile way to learn and to improve health.

Acknowledgment

Support for this chapter was provided by the Robert Wood Johnson Foundation's Pioneer Portfolio through a grant to its national program, Health Games Research: Advancing Effectiveness of Interactive Games for Health.

References

Abt, C. C. (1970). *Serious games.* New York, NY: Viking.

Ahn, S.J., Le, A.M.T., & Bailenson, J. (2013). The effect of embodied experiences on self-other merging, attitude, and helping behavior. *Media Psychology, 16*, 7–38.

Anderson, C., & Bushman, B.J. (2001). Effects of violent video games on aggressive behavior, aggressive cognition, aggressive affect, physiological arousal, and prosocial behavior: A meta-analytic review of the scientific literature. *Psychological Science, 12*(5), 353–359.

Anderson, C. A., Shibuya, A., Ihori, N., Swing, E. L., Bushman, B. J., Sakamoto, A., ...Saleem, M. (2010). Violent video game effects on aggression, empathy, and prosocial behavior in Eastern and Western countries: A meta-analytic review. *Psychological Bulletin, 136*(2), 151–173. doi:10.1037/a0018251

Anguera, J.A., Boccanfuso, J., Rintoul, J.L., Al-Hashimi, O., Faraji, O.F., Janowich, J., ...Gazzaley, A. (2013). Video game training enhances cognitive control in older adults. *Nature, 501*, 97–101.

Bandura, A. (1986). *Social foundations for thought and action: A social cognitive theory.* Englewood Cliffs, NJ: Prentice Hall.

Bandura, A. (1997). *Self-efficacy: The exercise of control.* New York, NY: W.H. Freeman and Company.

Baranowski, T., Baranowski, J., Cullen, K.W., Marsh, T., Islam, N., Zakeri, I., …deMoor, C. (2003). Squire's quest! Dietary outcome evaluation of a multimedia game. *American Journal of Preventive Medicine*, *24*(1), 52–61.

Baranowski, T., Baranowski, J., Cullen, K.W., Marsh, T., Islam, N., Zakeri, I., Noness-Morreale, L., & deMoor, C. (2003). Squire's Quest! Dietary outcome evaluation of a multimedia game. American Journal of Preventive Medicine, 24(1), 52–61.

Baranowski, T., Buday, R., Thompson, D.I., & Baranowski, J. (2008). Playing for real: Video games and stories for health-related behavior change. American Journal of Preventive Medicine, 34(1), 74–82.

Baranowski, T., Lieberman, D., Buday, R., Peng, W., Zimmerli, L., Weiderhold, B., & Kato, P.M. (2013). Videogame mechanics in games for health. *Games for Health Journal: Research, Development, and Clinical Applications*, *2*(4), 194–204.

Bavelier, D., Green, S.C., Pouget, A., & Schrater, P. (2012). Brain plasticity through the life span: Learning to learn and action video games. *The Annual Review of Neuroscience*, *35*, 391–416.

Bingham, P.M., Bates, J.H.T., Thompson-Figueroa, J., & Lahiri, T. (2010). A breath biofeedback computer game for children with cystic fibrosis. *Clinical Pediatrics*, *49*(4), 337–342.

Blascovich, J., & Bailenson, J.N. (2011). *Infinite reality—Avatars, eternal life, new worlds, and the dawn of the virtual revolution.* New York, NY: William Morrow.

Blumberg, F. (Ed.). (2014). *Learning by playing: Video gaming in education.* Oxford, UK: Oxford University Press.

Boot, W.R., Kramer, A.F., Simons, D.J., Fabiani, M., & Gratton, G. (2008). The effects of video game playing on attention, memory, and executive control. *Acta Psychologica*, *129*, 387–398.

Bransford, J. D., Brown, A. L., & Cocking, R. R. (2000). *Learning and transfer. How people learn: Brain, mind, experience, and school: Expanded edition.* Washington, DC: The National Academies Press.

Bransford, J.D., & Schwartz, D.L. (2001). Rethinking transfer: A simple proposal with multiple implications. In A. Iran-Nejad & P.D. Pearson (Eds.), *Review of research in education* (pp. 61–100). Washington, DC: American Educational Research Association (AERA).

Brown, S.J., Lieberman, D.A., Gemeny, B.A., Fan, Y.C., Wilson, D.M., & Pasta, D.J. (1997). Educational video game for juvenile diabetes: Results of a controlled trial. *Medical Informatics*, *22*(1), 77–89.

Byrne, S., Gay, G., Pollack, J. P., Gonzales, A., Retelny, D., Lee, T., & Wansink, B. (2012). Caring for mobile phone-based virtual pets can influence youth eating behaviors. *Journal of Children and Media*, *6*(1), 83–99.

Cauberghe, V., & De Pelsmacker, P. (2010). Advergames. *Journal of Advertising*, *39*(1), 5–18.

Chen, M., Lieberman, D., & Paisley, W. (1985). Microworlds of research. In M. Chen & W. Paisley (Eds.), *Children and microcomputers: Research on the newest medium* (pp. 276–296). Beverly Hills, CA: Sage.

Coleman, J. (1971). Learning through games. In E. Avedon & B. Sutton-Smith (Eds.), *The study of games* (pp. 322–329). New York, NY: Wiley.

Collins, A., & Halverson, R. (2009). *Rethinking education in the age of technology.* New York, NY: Teachers College Press.

Csikszentmihalyi, M. (1990). *Flow: The psychology of optimal experience.* New York, NY: Harper & Row.

Deci, E. L. (1971). Effects of externally mediated rewards on intrinsic motivation. *Journal of Personality and Social Psychology*, *18*, 105–115.

Deci, E., & Ryan, R. (Eds.). (2002). *Handbook of self-determination research*. Rochester, NY: University of Rochester Press.

Entertainment Software Association. (2013). Essential facts about the computer and video game industry: 2013 sales, demographic and usage data. Report accessed online March 25, 2014. Retrieved from http://www.theesa.com/facts/gameplayer.asp

Fasola, J., & Matarić, M. (2011). A Socially Assistive Robot Exercise Coach for the Elderly. *Journal of Human-Robot Interaction, 1*(1), 1–16.

Feltz, D.L., Irwin, B., & Kerr, N. (2012). Two-player partnered exergame for obesity prevention: Using discrepancy in players' abilities as a strategy to motivate physical activity. *Journal of Diabetes Science and Technology, 6*(4), 820–827.

Fox, J., Bailenson, J.N., & Ricciardi, T. (2012). Physiological responses to virtual selves and virtual others. *Journal of Cyber Therapy & Rehabilitation, 5*(1), 69–72.

Fritz, S.L, Rivers, E.D., Merlo, A.R., & Duncan, B.M. (2009). Examining the effects of commercially available video game systems, the Nintendo Wii and Sony Playstation 2, on balance and mobility in individuals with chronic stroke. *Journal of Neurologic Physical Therapy Abstracts, 33*(4), 224–232.

Garris, R., Ahlers, R., & Driskell, J.E. (2002). Games, motivation, and learning: A research and practice model. *Simulation and Gaming, 33*(4), 441–467.

Gee, J. P. (2003). *What digital games have to teach us about learning and literacy*. New York, NY: Palgrave Macmillan.

Gee, J. P. (2009). Deep learning properties of good digital games: How far can they go? In U. Ritterfeld, M. Cody, & P. Vorderer (Eds.), *Serious games: Mechanisms and effects* (pp. 65–80). New York, NY: Routledge.

Graf, D. L., Pratt, L. V., Hester, C. N., & Short, K. R. (2009). Playing active video games increases energy expenditure in children. *Pediatrics, 124*, 534–540.

Griffiths, M.D., Kuss, D.J., & King, D.L. (2012). Video game addiction: Past, present and future. *Current Psychiatry Reviews, 8*, 308–318.

Hansen, D.N. (1966). Computer assistance with the educational process. *Review of Educational Research, 36*(5), 588–603.

Harris, J.L., Speers, S.E., Schwartz, M. B., & Brownell, K.D. (2012). US food company branded advergames on the Internet: Children's exposure and effects on snack consumption. *Journal of Children and Media, 6*(1), 51–68.

Hess, R.F., Thompson, B., Black, J.M., Maehara, G., Zhang, P., Bobier, W.R., ...Cooperstock, J. (2012). An iPod treatment for Amblyopia: An updated binocular approach. *Optometry Online: Journal of the American Optometric Association, 83*(2), 87–94.

Honebein, P., Duffy, T., & Fishman, B. (1993). Constructivism and the design of learning environments: Context and authentic activities for learning. In T.M. Duffy, J. Lowyck, & D.H. Jonassen (Eds.), *Designing environments for constructive learning* (pp. 87–108). New York, NY: Springer-Verlag.

Johnston, J.D., Massey, A.P., & Marker-Hoffman, R.L. (2012). Using an alternate reality game to increase physical activity and decrease obesity risk of college students. *Journal of Diabetes Science and Technology, 6*(4), 828–838.

Kafai, Y.B., Quintero, M., & Feldon, D. (2010). Investigating the "Why" in Whypox: Casual and systematic explorations of a virtual epidemic. *Games and Culture, 5*(1), 116–135.

Kato, P.M., Cole, S.W., Bradlyn, A.S., & Pollock, B.H. (2008). A video game improves behavioral outcomes in adolescents and young adults with cancer: A randomized trial. *Pediatrics, 122*, e305–e317.

Kiousis, S. (2002). Interactivity: A concept explication. *New Media & Society, 4*(3): 355–383.

Krendl, K.A., & Lieberman, D.A. (1988). Computers and learning: A review of recent research. *Journal of Educational Computing Research, 4*(4), 367–389.

Leatherdale, S. T., Woodruff, S. J., & Manske, S. R. (2010). Energy expenditure while playing active and inactive video games. *American Journal of Health Behavior, 34*, 31–35.

Lieberman, D. (1980). A new high school curriculum: Critical television viewing skills. *National Association of Secondary School Principals Bulletin, 64*(2), 11–25.

Lieberman, D.A. (1982a). Federally funded television programming in the commercial marketplace. *Television and Children, 5*(3), 28–35.

Lieberman, D. (1982b). Interactive videodisc: Talking back to television. *Educational Technology, 1*(3), 8–16.

Lieberman, D. (1985). Research on children and microcomputers: A review of utilization and effects studies. In M. Chen & W. Paisley (Eds.), *Children and microcomputers: Research on the newest medium* (pp. 59–83). Beverly Hills, CA: Sage.

Lieberman, D.A. (1992). The computer's potential role in health education. *Health Communication, 4*(3), 211–225.

Lieberman, D.A. (1997). Interactive video games for health promotion: Effects on knowledge, self-efficacy, social support, and health. In R.L. Street, W.R. Gold, & T. Manning (Eds.), *Health promotion and interactive technology: Theoretical applications and future directions* (pp. 103–120). Mahwah, NJ: Lawrence Erlbaum Associates.

Lieberman, D.A. (1999). The researcher's role in the design of children's media and technology. In A. Druin (Ed.), *The design of children's technology* (pp. 73–97). San Francisco, CA: Morgan Kaufmann Publishers.

Lieberman, D.A. (2001a). Management of chronic pediatric diseases with interactive health games: Theory and research findings. *Journal of Ambulatory Care Management, 24*(1), 26–38.

Lieberman, D.A. (2001b). Using interactive media in communication campaigns for children and adolescents. In R. Rice & C. Atkin (Eds.), *Public communication campaigns* (3rd ed., pp. 373–388). Newbury Park, CA: Sage.

Lieberman, D.A. (2006). What can we learn from playing interactive games? In P. Vorderer & J. Bryant (Eds.), *Playing video games: Motives, responses, and consequences* (pp. 379–397). Mahwah, NJ: Lawrence Erlbaum Associates.

Lieberman, D.A. (2009). Designing serious games for learning and health in informal and formal settings. In U. Ritterfeld, M. Cody, & P. Vorderer (Eds.), *Serious games: Mechanisms and effects* (pp. 117–130). New York, NY: Routledge.

Lieberman, D.A. (2012a). Digital games for health behavior change: Research, design, and future directions. In S.M. Noar & N.G. Harrington (Eds.), *eHealth applications: Promising strategies for behavior change* (pp. 110–127). New York, NY: Routledge.

Lieberman, D.A. (2012b). Video games for diabetes self-management: Examples and design strategies. *Journal of Diabetes Science and Technology, 6*(4), 802–806.

Lieberman, D.A. (2013). Designing digital games, social networks, and mobile technologies to motivate and support health behavior change. In R.E. Rice & C.K. Atkin (Eds.), *Public communication campaigns* (4th ed., pp. 273–287). Thousand Oaks, CA: Sage.

Lieberman, D.A., Bates, C.H., & So, J. (2009). Young children's learning with digital media. *Computers in the Schools, 26*(4), 271–283.

Lieberman, D.A., Biely, E., Fisk, M.C., DeMartino, C.H., So, J., Thai, C.L., … Bondad-Brown, B.A. (2014). *Playful nation: A survey of video game play in the U.S., ages 1 to 93.* Santa Barbara, CA: University of California.

Lieberman, D.A., Biely, E., Thai, C.L., & Peinado, S. (2014). Transfer of learning from video game play to the classroom. In F. Blumberg (Ed.), *Learning by playing: Video gaming in education.* New York, NY: Oxford University Press.

Lieberman, D.A. & Brown, S.J. (1995). Designing interactive video games for children's health education. In K. Morgan, R.M. Satava, H.B. Sieburg, R. Mattheus, & J.P. Christensen (Eds.), *Interactive technology and the new paradigm for healthcare* (pp. 201–210). Amsterdam, The Netherlands: IOS Press.

Lieberman, D.A., Chaffee, S.H., & Roberts, D.F. (1988). Computers, mass media, and schooling: Functional equivalence in uses of new media. *Social Science Computer Review, 6*(2), 224–241.

Lieberman, D.A., Chamberlin, B., Medina, E., Jr., Franklin, B.A., McHugh Sanner, B.M., & Vafiadis, D.K. (2011). The power of play: Innovations in getting active summit 2011: A science panel proceedings report from the American Heart Association. *Circulation, 123*(21), 2507–2516. Retrieved from http://circ.ahajournals.org/content/123/21/2507.full

Lieberman, D.A., Fisk, M. C., & Biely, E. (2009). Digital games for young children ages three to six: From research to design. *Computers in the Schools, 26*(4), 299–313.

Lieberman, D.A., & Linn, M. C. (1991). Learning to learn revisited: Computers and the development of self-directed learning skills. *Journal of Research on Computing in Education, 23*(3), 373–395.

Mayer, R. E. (1999). Multimedia aids to problem solving transfer. *International Journal of Educational Research, 31*, 611–623.

Mayer, R. E. (2003). The promise of multimedia learning: Using the same instructional design methods across different media. *Learning and Instruction, 13*, 125–139.

Mayer, R.E. & Lieberman, D. A. (2011). Conducting scientific research on learning and health behavior change with computer-based health games. *Educational Technology, 51*(5), 3–14.

Mayer, R. E., & Wittrock, M. C. (1996). Problem-solving transfer. In D. Berliner & R. Calfee (Eds.), *Handbook of educational psychology* (pp. 47–62). New York, NY: Simon & Schuster Macmillan.

McPherson, M., Smith-Lovin, L., & Cook, J.M. (2001). Birds of a feather: Homophily in social networks. *Annual Review of Sociology, 27*, 415–444.

Mollick, E.R., & Rothbard, N. (2013). *Mandatory fun: Gamification and the impact of games at work.* The Wharton School Research Paper Series. Retrieved from http://ssrn.com/abstract=2277103 or http://dx.doi.org/10.2139/ssrn.2277103

Moreno, R., & Mayer, R. E. (2004). Personalized messages that promote science learning in virtual environments. *Journal of Educational Psychology, 96*(1), 165–173.

Mou, Y., & Peng, W. (2009). Gender and racial stereotypes in popular video games. In R. Ferdig (Ed.), *Handbook of research on effective electronic gaming in education* (pp. 922–937). Hershey, PA: Information Science Reference. doi:10.4018/978-1-59904-808-6.ch053

Murphy, E. C., Carson, L., Neal, W., Baylis, C., Donley, D., & Yeater, R. (2009). Effects of an exercise intervention using Dance Dance Revolution on endothelial function and other risk factors in overweight children. *International Journal of Pediatric Obesity, 4*, 205–214.

Noar, S.M., & Harrington, N.G. (Eds.). (2012). *eHealth applications: Promising strategies for behavior change.* New York, NY: Routledge.

Peng, W., Lin, J.-H., Pfeiffer, K. A., & Winn, B. (2012). Need satisfaction supportive game features as motivational determinants: An experimental study of a self-determination theory guided exergame. *Media Psychology, 15*(2), 175–196.

Peterson, C., & Stunkard, A. J. (1989). Personal control and health promotion. *Social Science and Medicine, 28*, 819–828.

Pillay, H., Brownlee, J., & Wilss, L. (1999). Cognition and recreational computer games: Implications for educational technology. *Journal of Research on Computing in Education, 32*(1), 203–216.

Pope, A.T., & Bogart, E.H. (1996). Extended attention span training system: Video game neurotherapy for attention deficit disorder. *Child Study Journal, 26*(1), 39–50.

Prawat, R.S., & Flowden, R. E. (1994). Philosophical perspectives on constructivist views of learning. *Educational Psychologist, 29*(1), 37–48.

Pugh, K.J., & Bergin, D.A. (2006). Motivational influences on transfer. *Educational Psychologist, 41*(3), 147–160.

Rafaeli, S. (1988). Interactivity: From new media to communication. In R. P. Hawkins, J. M. Wiemann, & S. Pingree (Eds.), *Sage annual review of communication research: Advancing communication science: Merging mass and interpersonal processes* (Vol. 16, pp. 110–134). Beverly Hills, CA: Sage.

Rafaeli, S., & Ariel, Y. (2007). Assessing interactivity in computer-mediated research. In A.N. Joinson, K. Y. A. McKenna, T. Postmes, & U. D. Rieps (Eds.), *The oxford handbook of internet psychology* (pp. 71–88). Oxford: Oxford University Press.

Reeves, B., & Nass, C. (1996). *The media equation: How people treat computers, television, and new media like real people and places.* New York, NY: Cambridge University Press.

Rey-Lopez, J.P., Vicente-Rodriguez, G., Biosca, M., & Moreno, L.A. (2007). Sedentary behaviour and obesity development in children and adolescents. *Nutrition, Metabolism, and Cardiovascular Diseases, 18*(3), 242–251.

Ritterfeld, U., Cody, M., & Vorderer, P. (Eds.), (2009). *Serious games: Mechanisms and effects.* New York, NY: Routledge.

Rosenberg, D., Depp, C.A., Bahia, I.V., Reichstadt, J., Palmer, B.W., Kerr, J., ... Jeste, D.V. (2010). Exergames for subsyndromal depression in older adults: A pilot study of a novel intervention. *American Journal of Geriatric Psychiatry, 18*, 221–226.

Ryan, R. M., Rigby, C. S., & Przybylski, A. (2006). The Motivational pull of video games: A self-determination theory approach. *Motivation and Emotion, 30*(4), 344–360.

Salomon, G., & Perkins, D. N. (1989). Rocky roads to transfer: Rethinking mechanisms of a neglected phenomenon. *Educational Psychologist, 24*(2), 113–142.

Shaffer, D. W., Squire, K. A., Halverson, R., & Gee, J. P. (2005). Digital games and the future of learning. *Phi Delta Kappan, 87*(2), 104–111.

Sheldon, L. (2011). *The multiplayer classroom: Designing coursework as a game.* Independence, KY: Cengage Learning.

Shute, V. J., & Ke, F. (2012). Games, learning, and assessment. In D. Ifenthaler, D. Eseryel, & X. Ge (Eds.), *Assessment in game-based learning: Foundations, innovations, and perspectives* (pp. 43–58). New York, NY: Springer.

Shute, V. J., Rieber, L., & Van Eck, R. (2011). Games and learning. In R. Reiser & J. Dempsey (Eds.), *Trends and issues in instructional design and technology* (3rd ed., pp. 321–332). Upper Saddle River, NJ: Pearson Education.

Singhal, A., Cody, M., Rogers, E., Sabido, M., & Pitt, C.E. (2006). Entertainment-education and socialchange: History, research, and practice. *Journal of Communication, 56*(1), 230–234.

Singley, M. K., & Anderson, J. R. (1989). *The transfer of cognitive skill.* Cambridge, MA: Harvard University Press.

Snyder, A., Gartenberg, A., Yerokhin, V., Griffiths, C., Rocha, T., Nimon, J.,...Anderson-Hanley, C. (2010). Social facilitation while cybercycling with avatars affects exercise behavior: Differential effects of competitiveness. *International Journal of Behavioral Medicine, 17*(Suppl 1), S161.

Staiano, A.E., Abraham, A.A., & Calvert, S.L. (2012). Adolescent exergame play for weight loss and psychosocial improvement: A controlled physical activity intervention. *Obesity, 21*(3), 598–601.

Sundar, S. S., Kalyanaraman, S., & Brown, J. (2003). Explicating website interactivity: Impression-formation effects in political campaign sites. *Communication Research, 30*(1), 30–59.

Sundar, S. S., & Kim, J. (2005). Interactivity and persuasion: Influencing attitudes with information and involvement. *Journal of Interactive Advertising, 5*(2), 5–18.

Suppes, P. (1967). On using computers to individualize instruction. In D.D. Bushnell & D.W. Allen (Eds.), *The computer in American education* (pp. 11–24). New York, NY: Wiley.

Tanaka, J.W., Wolf, J.M., Klaiman, C., Koenig, K., Cockburn, J., Herlihy, L., …Schultz, R.T. (2010). Using computerized games to teach face recognition skills to children with autism spectrum disorder: The let's face it! program. *Journal of Child Psychology and Psychiatry, 51*(8), 944–952.

Thomas, R., Cahill, J., & Santilli, L. (1997). Using an interactive computer game to increase skill and self-efficacy regarding safer sex negotiation: Field test results. *Health Education & Behavior, 24*, 71–86.

Unnithan, V. B., Houser, W., & Fernhall, B. (2005). Evaluation of the energy cost of playing a dance simulation video game in overweight and non-overweight children and adolescents. *International Journal of Sports Medicine, 26*, 1–11.

Vogel, J. F., Vogel, D. S., Cannon-Bowers, J., Bowers, C. A., Muse, K., & Wright, M. (2006). Computer gaming and interactive simulations for learning: A meta-analysis. *Journal of Educational Computing Research, 34*(3), 229–243.

Vorderer, P., & Bryant, J. (Eds.). (2006). *Playing video games: Motives, responses, and consequences.* Mahwah, NJ: Lawrence Erlbaum Associates.

Wilkinson, N., Ang, R.P., & Goh, D.H. (2008). Online video game therapy for mental health concerns: A review. *International Journal of Social Psychiatry, 54*(4), 370–382.

Witte, K., Meyer, G., & Martell, D. (2001). *Effective health risk messages.* Thousand Oaks, CA: Sage.

Yee, N., Bailenson, J.N., & Ducheneaut, N. (2009). The proteus effect: Implications of transformed digital self-representation on online and offline behavior. *Communication Research, 36*(2), 285–312.

24

Leveraging Psychology of Virtual Body for Health and Wellness

Giuseppe Riva[1,2], Antonios Dakanalis[3], and Fabrizia Mantovani[4]

[1] *Istituto Auxologico Italiano, Milan, Italy*
[2] *Università Cattolica del Sacro Cuore, Milan, Italy*
[3] *Università di Pavia, Pavia, Italy*
[4] *University of Milano Bicocca, Milan, Italy*

Introduction

Information and communication technologies are becoming more present in people's daily lives. But what is the final purpose of the mass proliferation of digital devices? Are they helpful in fostering personal growth and individual integration in the sociocultural environment, by promoting satisfaction, opportunities for action and self-expression?

Recently Riva and colleagues underlined that one of the fundamental objectives for our field—cyberpsychology—in the coming decade will be to create technologies that contribute to enhancement of happiness and psychological well-being (Botella et al., 2012; Riva, Banos, Botella, Wiederhold, & Gaggioli, 2012; Wiederhold & Riva, 2012). In particular, these authors suggested that it is possible to use technology to manipulate the quality of personal experience, with the goal of increasing wellness, and generating strengths and resilience in individuals, organizations, and society. But what is personal experience?

According to *Merriam Webster's Collegiate Dictionary* (http://www.merriam-webster.com/dictionary/experience), it is possible to define experience both as "a: direct observation of or participation in events as a basis of knowledge" (subjective experience) and "b: the fact or state of having been affected by or gained knowledge through direct observation or participation" (personal experience).

These definitions describe the two faces of our experience: On one side, we can intentionally control the contents of our experience (subjective experience); on the other side, its contents define our future emotions and intentions (personal experience).

The Handbook of the Psychology of Communication Technology, First Edition. Edited by S. Shyam Sundar.

However, there is a critical difference between them (Riva, 2012b). If subjective experience is the experience of being a subject (experience as subject), personal experience is the experience affecting a particular subject (experience as object). This difference suggests that, independently from the subjectivity of any individual, it is possible to alter the features of our personal experience from outside. In other words, personal experience becomes the dependent variable that may be manipulated by external researchers to improve health and wellness.

Specifically, this can be achieved in three separate but related ways (Riva, Banos et al., 2012):

- *By structuring the personal experience* using a goal, rules, and a feedback system (McGonigal, 2011): The goal provides subjects with a sense of purpose focusing attention and orienting his/her participation in the experience. The rules, by removing or limiting the obvious ways of getting to the goal, push subjects to see the experience in a different way. The feedback system tells players how close they are to achieving the goal and provides motivation to keep trying.
- *By augmenting the personal experience* to achieve multimodal and mixed experiences. Technology allows multisensory experiences in which content and its interaction is offered through more than one of the senses. It is even possible to use technology to overlay virtual objects onto real scenes (Rosenblum, 2000).
- *By replacing the personal experience* with a synthetic one. Using VR it is possible to simulate physical presence in a synthetic world that reacts to the action of the subject as if he/she was really there. More, as we have just, seen the replacement possibilities offered by technology even extend to the induction of an illusion of ownership over a virtual arm or a virtual body (Slater, Spanlang, Sanchez-Vives, & Blanke, 2010).

In this chapter we will try to demonstrate that virtual reality (VR), for its ability of addressing simultaneously all these three levels, is the key to open new opportunities for health and wellness. In particular, VR is able to influence our personal experience through its effects on our bodily experience. For example, different authors showed that it is possible to use VR both to induce an illusory perception of a fake limb (Slater, Perez-Marcos, Ehrsson, & Sanchez-Vives, 2009) or a fake hand (Perez-Marcos, Slater, & Sanchez-Vives, 2009) as part of our own body, and to produce an out-of-body experience (Lenggenhager, Tadi, Metzinger, & Blanke, 2007), by altering the normal association between touch and its visual correlate. It is even possible to generate a body transfer illusion (Slater et al., 2009): Slater and colleagues substituted the experience of male subjects' own bodies with a life-sized virtual human female body.

How are these experiences possible? What do they teach us about the potential of technology for health and wellness?

This chapter tries to answer these two questions by identifying a clear link between spatial perception and action (Riva, 2008). On one side, we conceive places in terms of the actions we could take toward them: we do not have a separate knowledge of the place's location relative to them, what they can do in it, and their purposes. On the other side, our spatial experience is the outcome of the interaction of different spatial representations whose integrated activity gives rise to spatial awareness. Then, we will introduce the concept of "presence" as the main outcome of these processes and we will discuss its effects on our personal experience. Finally, we suggest that it is possible to use technology to manipulate the features of our personal experience, with the goal of increasing well-being both in individuals and groups.

Our Body Is Virtual

We consider our body as the innermost part of our material self in each of us: we are our body. However, different studies and clinical experiences clearly show that our experience of the body is not direct as its perception may suggest. In particular, clinical disturbances such as anorexia nervosa—in which patients may have an intense fear of weight gain, even when they are underweight—or phantom limb—in which patients experience an amputated or missing limb as being still attached to the body and moving appropriately with other body parts—suggest that our physical body is matched by a virtual body created by our brain.

The importance of the link between the brain and the body is well reflected by a significant shift happening now in brain sciences. For a long time, brain sciences considered action, perception, and interpretation as separate activities. However, recently brain sciences started to describe cognitive processes as *embodied* (Prinz, 2006). As clearly outlined by Anderson (2003):

> For over fifty years… there has been a re-thinking of the nature of cognition. Instead of emphasizing formal operations on abstract symbols, this new approach focuses attention on the fact that most real-world thinking occurs in very particular (and often very complex) environments, is employed for very practical ends, and exploits the possibility of interaction with and manipulation of external props. It thereby foregrounds the fact that cognition is a highly embodied or situated activity—emphasis intentionally on all three—and suggests that thinking beings ought therefore be considered first and foremost as acting beings (p. 91).

The *Embodied Cognition* paradigm takes as its starting point the idea that cognition occurs in specific environments, and for specific ends (Clark, 1997, 2001; Haugeland, 1998). Moreover, the *Embodied Cognition* approach underlines the central role of body in shaping the mind (Clark, 2001, 2003; Gallagher, 2005; Gallese & Lakoff, 2005; Garbarini & Adenzato, 2004; Lakoff & Johnson, 1980; Ziemke, 2003). Specifically, the mind has to be understood in the context of its relationship to a physical body that interacts with the world. Hence human cognition, rather than being centralized, abstract, and sharply distinct from peripheral input and output modules, has instead deep roots in sensorimotor processing. In this picture, the experience of the body is not given, but it is the result of how the subject experiences his/her interaction with the environment. In simpler word, we may modify the experience of the body by modifying how the subject interacts with the environment.

Action and Space

The other main outcome of the Embodied Cognition research is the identification of a clear link between spatial perception, our bodily experience, and action.

On one side, we conceive places in terms of the actions we could take toward them: subjects have not a separate knowledge of the place's location relative to them, what they can do in it, and their purposes. An example can help in understanding this point. Retrieving an occluded object—for example, when we lift a box to retrieve a pencil from under it—is an action taken on the basis of a belief about where the object (pencil) is located relative to the self. It follows that to know that the pencil exists when it is occluded is a matter of knowing what can be done to make the pencil visible. More, if I want to grab the pencil, its spatial position will be represented in terms of the movements needed to reach for it. Further, its shape and size will be represented in terms of the type of handgrip it affords.

On the other side, our spatial experience is the outcome of the interaction of different spatial representations whose integrated activity gives rise to spatial awareness (Matelli & Luppino, 2001). In fact, evidence from clinical and experimental studies indicates that our spatial experience involves the integration of different sensory inputs within two different reference frames defined by our body model (see Figure 24.1) and related to its possibility of action (Galati et al., 2000; Longo, Azañón, & Haggard, 2010; Previc, 1998):

- *Subjective/near space* (body as reference of first-person experience). It is shaped around the *egocentric frame* and its primary sources are *somatoperceptions*: representations of the present state of the body and tactile stimuli from sensory inputs (Longo et al., 2010). In fact, it is based on the body of the observer; it is defined by its three axes: front-back, left-right, head-feet, and within this frame, the position of an object changes if the subject moves (Frith & de Vignemont, 2005). More, this frame defines the *peripersonal space*, the space immediately surrounding our bodies (Previc, 1998). Within this space the subject can directly grasp and manipulate objects and tools.
- *Objective/far space* (body as object in a world). It is shaped around the *extrapersonal space*, the space that occurs outside the reach of an individual (Previc, 1998). This space is defined by the spatial relations between the objects (*landmarks*) included in it. The extrapersonal space shapes the allocentric frame: within this frame, the position of an object does not change if the subject moves, and the object exists even if there is no relation with the self or another person (Frith & de Vignemont, 2005). In this frame, the body is an object like the other ones and its representation is the outcome of abstract knowledge, beliefs, and attitudes related to it as an object of third-person perception (looking at it from outside).

Being Present in a Space, Virtual, or Real

This picture suggests that we do not have a single global representation of space, even if we have a unitary spatial experience (Riva & Mantovani, 2012): We are *present* in a space even if we represent different spaces. But how are we present in a space? And what happens when the space is not real but virtual (telepresence)?

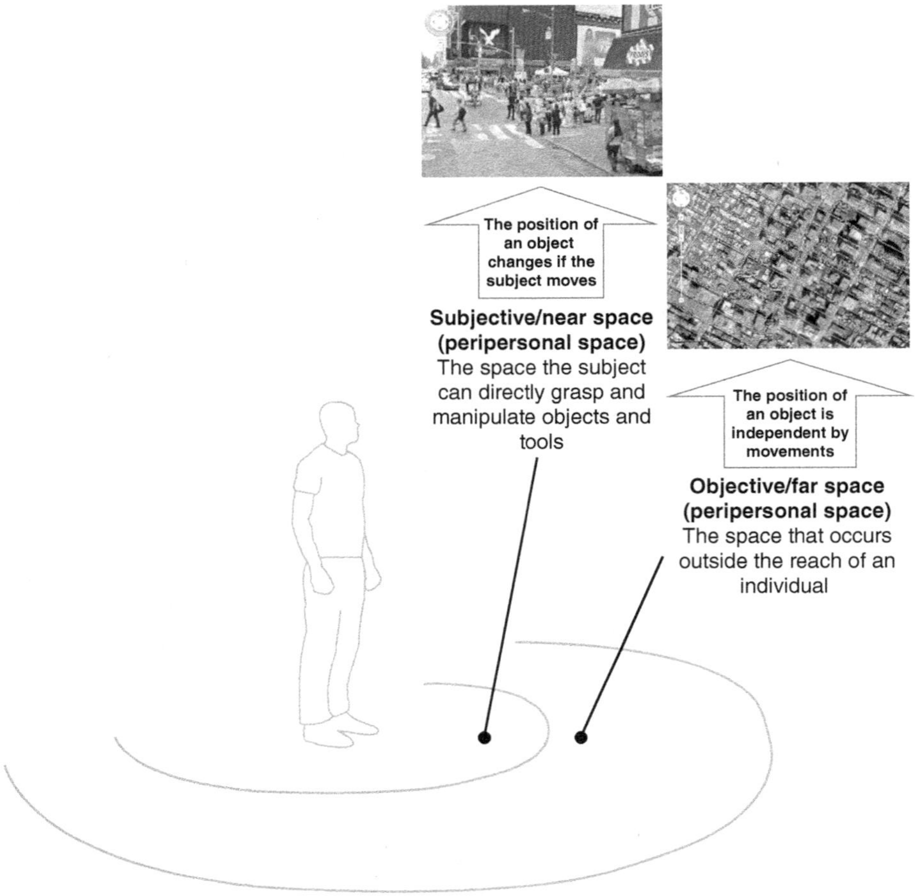

Figure 24.1 Reference frames and their cognitive representation.
Source: Photos: on the left, Times Square image taken from Google Maps Street View, 2014. © Google; on the right, Times Square image taken from Google Maps Satellite View, 2014. © Google BlueSky DigitalGlobe.

Unfortunately, the link between space, presence, and telepresence is underexplored in the recent scientific literature about this topic. And in most of the available research work, the term "presence" has referred to a widely reported sensation experienced during the use of VR (Riva, Waterworth, Waterworth, & Mantovani, 2011). However, as commented by Biocca (1997), and agreed by most researchers in the area, "while the design of virtual reality technology has brought the theoretical issue of presence to the fore, few theorists argue that the experience of presence suddenly emerged with the arrival of virtual reality."

To address this issue, a growing group of researchers is considering presence as a broad psychological phenomenon, not necessarily linked to the experience of a medium (Baños, Botella, & Perpiña, 1999; Baños et al., 2000; Lee, 2004a, 2004b; Mantovani & Riva, 1999; Marsh, Wright, & Smith, 2001; Moore, Wiederhold,

Wiederhold, & Riva, 2002; Riva & Davide, 2001; Riva, Davide, & IJsselsteijn, 2003; Riva et al., 2011; Schubert, Friedman, & Regenbrecht, 2001; Spagnolli & Gamberini, 2002; Spagnolli, Gamberini, & Gasparini, 2003; Waterworth & Waterworth, 2001, 2003; Waterworth, Waterworth, Mantovani, & Riva, 2010; Zahoric & Jenison, 1998).

In their recent paper, Riva and Mantovani (2012) detailed this vision describing presence as an implicit (not conscious) metacogntive process that tracks human volition. Koriat (2007) defines "metacognition" as "the processes by which people self-reflect on their own cognitive and memory processes (monitoring) and how they put their metaknowledge to use in regulating their information processing and behavior (control)" (p. 289). In their view, According to this vision, presence is characterized by the following features (Riva & Mantovani, 2012)

- It monitors pre-reflexively our activity processes;
- It is achieved using an embodied intuitive simulation of the intended action;
- Only when the subject acquires a motor skill, he/she is able to simulate its outcome intuitively (implicit learning);
- A break in presence is a violation of our intuitive simulation. As consequence, the subject is forced to shift to reasoning to understand and cope with the causes of the violation.

But what happens when an action is implemented using a tool? To answer this question, Riva and Mantovani distinguished between two different types of mediated action: *first order* or *second order*:

1. *First-order mediated action*: the body is used to control a proximal tool (an artifact present and manipulable in the peripersonal space) to exert an action upon an external object. An example is the videogame player using a joystick (proximal tool) to move an avatar (distal tool) to pick up a sword (external object).
2. *Second-order mediated action*: the body is used to control a proximal tool that controls a different distal one (a tool present and visible in the extrapersonal space, either real or virtual) to exert an action upon an external object. An example is the cranemen using a lever (proximal tool) to move a mechanical boom (distal tool) to lift materials (external objects).

In their view, which is in agreement with and supported by different scientific and research data (Balakrishnan & Sundar, 2011; Blanke, 2012; Clark, 2008; Herkenhoff Carijó, de Almeida, & Kastrup, 2013; Jacobs, Bussel, Combeaud, & Roby-Brami, 2009; Slater et al., 2010), these two mediated actions have different effects on our spatial and bodily experience (see Figure 24.1):

- *Incorporation*: the proximal tool extends the peripersonal space of the subject;
- *Incarnation*: the user experiences a second peripersonal space centered on the distal tool;
- *Embodiment*: the user experiences a new body in the distal tool.

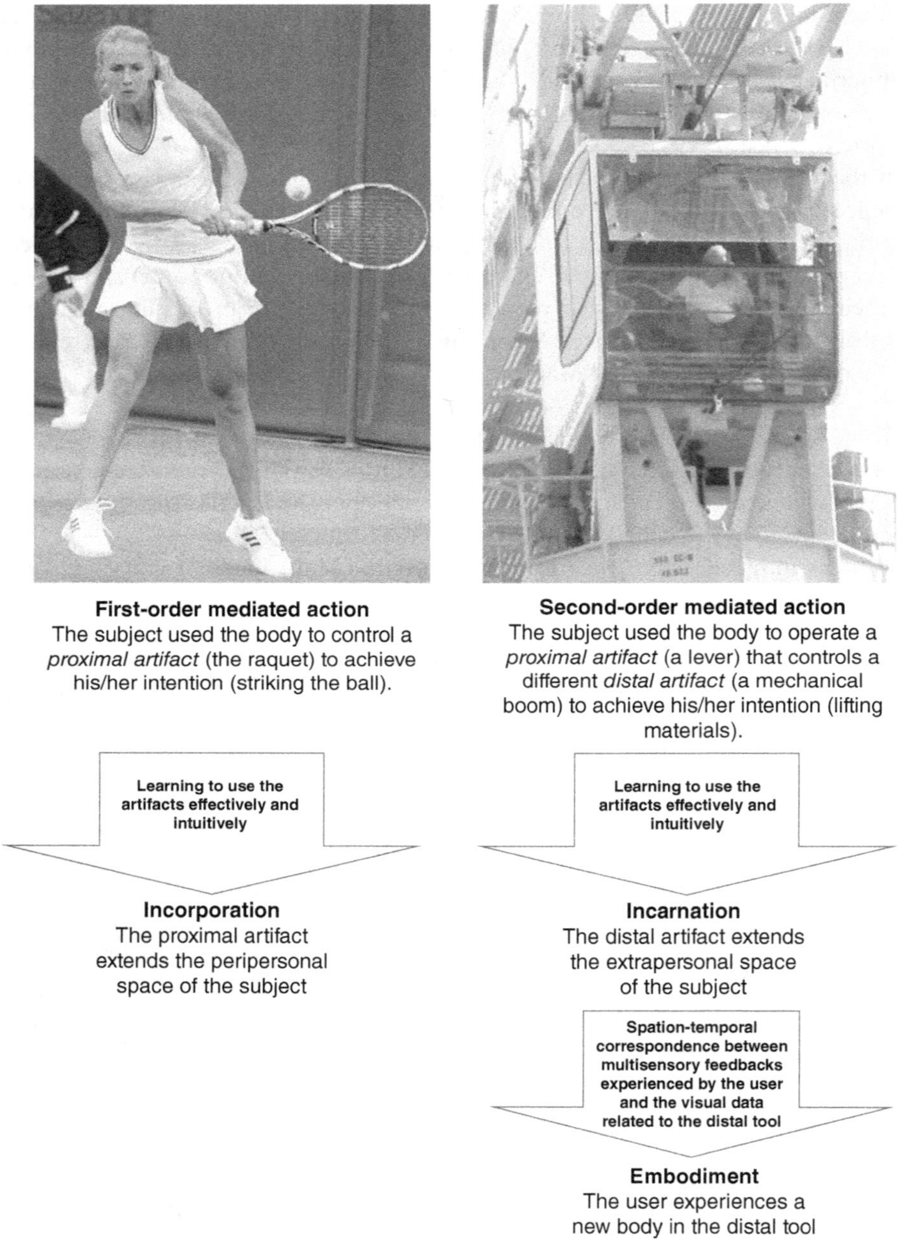

Figure 24.2 The effects of the intuitive use of an artifact on our spatial experience.
Source: Photo on left by si.robi: https://www.flickr.com/photos/sirobi/9498374388/.

A successfully learned first-order mediated action produces incorporation—the proximal tool extends the peripersonal space of the acting subject. In other words, the acquisition of a motor skill related to the use of a proximal tool extends the body model we use to define the near and far space. From a neuropsychological view point, the tool is incorporated in the near space, prolonging it till the end point of the tool. From a phenomenological view point, instead, we are now present in the tool and we can use it intuitively as we use our hands and our fingers.

A successfully learned second-order mediated action produces incarnation, too—a second body representation centered on the distal tool. In fact, second-order mediated actions are based on the simultaneous handling of two different body models—one centered on the real body (based on proprioceptive data) and a second centered on the distal tool (visual data)—that are weighted in a way that minimizes the uncertainty during the mediated action. In other words, this second peripersonal space centered on the distal tool competes with the one centered on the body to drive action and experience. Specifically, when the distal-centered peripersonal space becomes the prevalent one, it also shifts the extrapersonal space to the one surrounding the distal tool. From an experiential viewpoint, the outcome is simple: the subject experiences presence in the distal environment (telepresence).

Finally, a successfully learned second-order mediated action may lead to embodiment—the user experiences a new body in the distal tool—when there is a spatiotemporal correspondence between the multisensory signals and sensory feedback experienced by the user and the visual data related to the distal tool. For example, users are embodied in an avatar if the movements of the avatar are temporally synchronized with their own movements and there is a synchronous visuotactile stimulation of their's own and avatar's body.

This model has some contact points with the view of "Spatial Presence" introduced by Wirth and colleagues (Balakrishnan & Shyam Sundar, 2011; Wirth et al., 2007). According to these authors, Spatial Presence—the conviction of being located in a mediated environment (incorporation)—is the outcome of two processes:

1. The development of a media-based mental model of the distal space including spatial, temporal, and causal information.
2. The perceptual confirmation of this mental model is the primary ego-reference frame.

However, the presented model has three advantages over this vision of spatial presence. First, it suggests that this mental model is the outcome of an implicit simulation process related to the action affordances allowed by the proximal tool. In other words, through the use of the proximal tool, the subject implicitly learns how to interact with the distal tool and to predict the outcome of its actions. Second, according to it, the perceptual confirmation of the mental model is the outcome of a metacogntive process that tracks pre-reflexively our activity. This process compares an embodied intuitive simulation of the intended action (e.g., I have to move up the joystick to allow the avatar to jump the rock) with its actual outcome (e.g., the avatar jumps the rock). In this view, a break in presence is a violation of the intuitive simulation (e.g., I expected to move up the joystick for jumping but the avatar did not jump). Third, this model is an attempt to connect the concept of presence to established constructs in psychology and neuroscience. As a result, the model clearly relates the concept of presence to the spatial experience and clearly differentiates between the concepts of incarnation, incorporation, and embodiment that were formerly not so clearly connected.

Virtual Reality as Embodied and Positive Technology

VR is a popular technology that allows both incarnation and embodiment. But what is VR?

The basis for the VR idea is that a computer can synthesize a three-dimensional (3D) graphical environment from numerical data. Using visual, aural, or haptic devices, the human operator can experience the environment as if it were a part of the world. This computer-generated world may be a model of either a real-world object, such as a house; or an abstract world that does not exist in a real sense but is understood by humans, such as a chemical molecule or a representation of a set of data; or it might be in a completely imaginary science fiction world.

Usually VR is described as a particular collection of technological hardware. However, given the properties of distal tools described before, we can describe VR as an "embodied technology" for its ability of modifying the feeling of presence: the human operator can experience the synthetic environment as if it were "his/her surrounding world" (incorporation) or can experience the synthetic avatar as if it were "his/her own body" (incarnation).

In summary, VR provides a new human–computer interaction paradigm in which users are no longer simply external observers of images on a computer screen but are "present" within a computer-generated 3D virtual world (Riva, 1997). But what is the potential of these experiences for health and wellness?

In this process, a key role can be played by VR, the only technology that allows the simultaneous use of all these strategies. In fact, VR is starting to play an important role in clinical psychology and rehabilitation (Riva, 2005, 2009; Riva & Wiederhold, 2006). Clinical psychologists and rehabilitators use VR to provide a new human–computer interaction paradigm in which users are no longer simply external observers of images on a computer screen but are active participants within a computer-generated 3D virtual world. Within the virtual environment (VE), the patient has the possibility of learning to manage a problematic situation related to his/her disturbance. More, VEs are highly flexible and programmable. They enable the therapist to present a wide variety of controlled stimuli, such as a fearful situation, and to measure and monitor a wide variety of responses made by the user (Repetto & Riva, 2011).

For these advantages, the most common application of VR in behavioral health is the treatment of anxiety disorders (Emmelkamp, 2005). The rationale is simple: in VR the patient is intentionally confronted with the feared stimuli (VR exposure—VRE) while allowing the anxiety to attenuate. Avoiding a dreaded situation reinforces a phobia, and each successive exposure to it reduces the anxiety through the processes of habituation and extinction. VRE offers a number of advantages over in vivo or imaginal exposure. First, VRE can be administered in traditional therapeutic settings. This makes VRE to be more convenient, controlled, and cost-effective than in vivo exposure. Second, it can also isolate fear components more efficiently than in vivo exposure. For instance, in treating fear of flying, if landing is the most fearful part of the experience, landing can be repeated as often as necessary without having to wait for the airplane to take-off. Finally, the immersive nature of VRE provides a real-like experience that may be more emotionally engaging than imaginal exposure.

However, it seems likely that VR can be more than a tool to provide exposure and desensitization (Riva, 2008). In particular, we suggest that VR, for its ability of modifying the experience of the body, may be used to facilitate therapeutic change.

The Different Roles of Spatial Reference Frames in the Process of Change

How is it possible to change a patient? Even if this question has many different answers according to the specific psychotherapeutic approach, in general, change comes through an intense focus on a particular instance or experience (Wolfe, 2002): by exploring it as much as possible, the patient can relive all of the significant elements associated with it (i.e., conceptual, emotional, motivational, and behavioral) and make them available for a reorganization of his or her perspective.

Within this general model, we have the insight-based approach of psychoanalysis, the schema-reorganization goals of cognitive therapy or the enhancement of experience awareness in experiential therapies. What are the differences between them? According to Safran and Greenberg (1991), behind the specific therapeutic approach, we can find two different model of change: bottom-up and top-down. Bottom-up processing begins with a specific emotional experience and leads eventually to change at the verbal-representational and conceptual level, whereas top-down change usually involves exploring and challenging tacit rules and beliefs that guide the processing of emotional experience.

These two models of change are focused on two different cognitive systems, one for information transmission (top-down) and one for conscious experience (bottom-up), both of which may process sensory input (Brewin, 1989).

A possible way to conceptualize this difference is to use the egocentric/allocentric distinction we discussed before. As we have seen, egocentric processing allows people to locate objects relative to their body center and their bodies in 3D space. This body-centered system moves information from the primary visual processing areas at the back of the brain forward across the top-rear of the brain.

Allocentric processing, instead, enables people to handle relationships centered among objects and identify objects. The allocentric system moves information from the back of the head, forward along its sides, and closer to parts of the brain that verbally categorizes objects.

This dual encoding has been shown at the neural level in a distinction between two kinds of neurons in the parietal cortex (Sakata & Kusunoki, 1992). Some are dedicated to the position of the visual stimuli presented to an immobile observer in a frame of reference centered on the object. Others are dedicated to the stimuli in relation to the agent's actions in a frame of reference centered on the agent's body (relative to eye, head, and body position in space).

Given the strict link between action, perception, and concepts described by the embodied cognition approach, Frith and deVignemont (2005) suggest that this basic difference in the way we represent spatial information is a specific stance of our way of organizing representations. Specifically these authors suggest that we attribute mental states to our self and to the others by adopting either an egocentric or an allocentric stance:

> Egocentric representations of the self derive from a direct knowledge attached to the self, while allocentric representations of the self derive from a detached knowledge of the person that one happens to be. In the former case, one has privileged direct self-knowledge, which one cannot have about anybody else. In the latter case, one represents oneself as a person among others. Similarly, we would like to suggest that the other could also be represented according to an egocentric or an allocentric stance...Taking an egocentric stance means that others are represented only because they are related to the self in one way or another. Conversely, taking an allocentric stance means that the existence and/or mental states of others are completely independent from the self (p. 725).

Even if many therapeutic approaches are based on just one of the two change models, a therapist usually requires both (Wolfe, 2002). Some patients seem to operate primarily by means of top-down/allocentric information processing, which may then prime the way for corrective emotional experiences. For others, the appropriate access point is the intensification of their emotional experience and their awareness of it. Finally, different patients who initially engage the therapeutic work only through top-down/allocentric processing may be able later in the therapy to make use of bottom-up/egocentric emotional processing.

In sum, classical psychotherapy usually exploits the potential of a single frame of reference, even if the process of change may require both. A possible solution for this problem is the use of VR: in the next paragraph, we will discuss how the use of VR for addressing both egocentric and allocentric information processing may be used in the treatment of eating disorders.

Positive Technologies in Eating Disorders

A clinical area in which both change models are required is eating disturbances. According to the DSM IV-TR (APA, 2000) both anorexia and bulimia share an intense fear of gaining weight or becoming fat (bottom-up/egocentric emotional processing) and a distorted significance of body weight and shape are distorted in these individuals (top-down/allocentric information processing).

For this reason, the therapeutic approaches include cognitive–behavioral, interpersonal, and experiential therapies. However, the treatment of body image disturbances is the most difficult part of the therapy.

But what is body image? This term usually includes a complex set of intentional states and dispositions—perceptions, beliefs, and attitudes—in which the intentional object is one's own body (Gallagher, 2005). Specifically, within this term, we usually include:

1. *Body affect*: the emotional attitude toward the body;
2. *Body concept*: the conceptual description of the body in general;
3. *Body percept*: the perceptual experience of the body.

Some authors suggest that body image dissatisfaction is a form of cognitive bias (Thompson, 1996; Williamson, 1996). As noted by Williamson (1996):

> If information related to body is selectively processed and recalled more easily, it is apparent how the self-schema becomes so highly associated with body-related information. If the memories related to body are also associated with negative emotion, activation of negative emotion should sensitize the person to body-related stimuli causing even greater body size overestimation (pp. 49–50).

It is very difficult to counter a cognitive bias. In particular, Mou and McNamara's theory of spatial memory (Mou & McNamara, 2002; Mou, McNamara, Valiquette, & Rump, 2004) posits that people interpret spatial scenes by assigning a reference frame intrinsic to the layout and update their mental representation only if a later view provides a chance for superior encoding.

This suggests that there is a specific cognitive cost for changing perspective results from updating the stored view (i.e., the initial viewpoint)—which is activated when the spatial memory is brought into the focus of attention—to the view that is probed by the trial instruction.

For this reason, biased information processing occurs automatically, and the subjects are not aware about it. So, for them, the biased information is real. They are not able to distinguish between perceptions and biased cognitions. Moreover, attempts at persuasion are usually useless and may even elicit strong emotional defense.

According to the "Allocentric Lock Hypothesis—ALH" eating disorders and obesity may be the outcome of a primary disturbance in the way the body is experienced and remembered (Gaudio & Riva, 2013; Riva, 2012a; Riva & Gaudio, 2012): individuals with these disturbances may be locked to an allocentric (observer view) negative memory of the body that is no more updated by contrasting egocentric representations driven by perception. Using immersive VR, it is possible to induce a controlled sensory rearrangement that facilitates the update of the biased body image. This allows the differentiation and integration of new information, leading to a new sense of cohesiveness and consistency in how the self represents the body.

First, VR users become aware of their bodies during navigation: their head movements alter what they saw. The sensorimotor coordination of the moving head with visual displays produces a much higher level of sensorimotor feedback and first-person perspective (egocentric reference frame).

Second, in VR, the human operator's normal sensorimotor loops are altered by the presence of distortions, time delays, and noise (Riva, 1997). Such alterations, that are introduced unintentionally and usually degrade performance, affect body perceptions, too. The somesthetic system has a proprioceptive subsystem that senses the body's internal state, such the position of limbs and joints, and the tension of the muscles and tendons. Mismatches between the signals from the proprioceptive system and the external signals of a VE alter significantly body perceptions (Riva, 1998; Sadowsky & Massof, 1994).

As noted by Gallagher (2005) "[different] studies indicate that changes in various aspects of body schemas have an effect on the way subjects perceive their own body. More generally, changes in body schemas also affect spatial perception, the perception of objects, and intentional action" (p. 144).

Through the use of immersive VR, it is possible to induce a controlled sensory rearrangement that facilitates the update of the biased body image. When a particular event or stimulus violates the information present in our spatial memory (as occurs during a virtual experience), the information itself becomes accessible at a conscious level (Baars, 1998) and can be modified more easily. This facilitates the modification of the body image through the differentiation and integration of new information, leading to a new sense of cohesiveness and consistency in how the self represents the body. A possible strategy toward this goal is the adaptation to VR of the imagery rescripting method developed for the treatment of post-traumatic stress disorder (Riva, 2011; Smucker, Dancu, Foa, & Niederee, 1995). In particular, the user will experience twice the contents of the allocentric memory both in first person (experiencing his/her own body) and in third person (experiencing the body of his/her avatar). A tentative protocol, coming from the previous experience of coordinating partner is summarized in Table 24.1 (Riva, Cardenas-Lopez, Duran, Torres-Villalobos, & Gaggioli, 2012; Riva, Castelnuovo et al., 2012).

The results of this approach are very promising (Ferrer-Garcia & Gutierrez-Maldonado, 2012). Riva and his group (Riva, Castelnuovo et al., 2012) have recently tested it in a randomized controlled trial to date with 163 morbidly obese patients. This trial compared Experiential Cognitive Therapy (ECT)—the VR-based treatment for obesity—with nutritional (NT) and cognitive–behavioral (CBT) approaches along with waiting list controls. At the 12-month follow-up, ECT was significantly better in maintaining the results of the treatment than both CBT and NT. Riva and colleagues used ECT also in the treatment of anorexia, bulimia, and binge eating (Riva, Bacchetta, Baruffi, & Molinari, 2002; Riva, Bacchetta, Baruffi, Rinaldi, & Molinari, 1999) obtaining positive results. A similar approach was presented and tested successfully by Perpiña and colleagues (2003) in the treatment of eating disorders.

The embodied effects of VR may be used in other pathologies too. Viaud-Delmon and colleagues (2002) showed that subjects with high trait anxiety, for example, subjects with symptoms of panic and agoraphobia, have a strong dependence on a particular reference frame in which the sensory information is interpreted and in which the subject would remained anchored.

Table 24.1 The symbiotic rescripting protocol

Phase 1: Interview	During a clinical interview the patient is asked to relive the contents of the allocentric negative body image and the situation/s in which it was created and/or reinforced (e.g., being teased by my boyfriend at home) in as much detail as possible. The meaning of the experience for the patient was also elicited
Phase 2: Development of a VR scene	The clinician reproduces the setting of the identified situation (e.g., the corridor of the classroom where my boyfriend teased me) in virtual reality. A possibile tool for this goal is the free NeuroVR software (http://www.neurovr.org) developed by one of the authors of this chapter
Phase 3: First-person experience of the VR scene: I'm in the scene	The patient is asked to reexperience the event in VR from a first-person perspective (the patient does not see his/her body in the scene) expressing and discussing his/her feelings. The patient is then asked what is needed to change the feelings in a positive direction
	The main cognitive techniques used in this phase, if needed, are:
	Countering: Once a list of distorted perceptions and cognitions is developed, the process of countering these thoughts and beliefs begins *Label Shifting*: The patient first tries to identify the kind of negative words she uses to interpret situations in her life, such as bad, terrible, obese, inferior, and hateful. The situations in which these labels are used are then listed. The patient and therapist replace each emotional label with two or more descriptive words
Phase 4: Third-person experience of the VR scene: my avatar is in the scene	The patient is asked to reexperience the event in VR from a third-person perspective (the patient sees his/her body in the scene) intervening both to calm and reassure his/her virtual avatar and to counter any negative evaluation
	To improve the sense of presence included in the scene, we will also follow the touch/visual approach used by Slater and colleagues by stroking synchronously both the real subject and the avatar
	The therapist follows the Socratic approach, for example, "What would need to happen for you to feel better? How does it look through the eyes of a third person? Is there anything you as a third person like to do? How do the other people respond?"
	The main cognitive techniques used in this phase, if needed, are:
	Alternative Interpretation: The patient learns to stop and consider other interpretations of a situation before proceeding to the decision-making stage *Deactivating the Illness Belief*: The therapist first helps the client list her beliefs concerning weight and eating

Source: Adapted from Riva (2011). © Sage.

These research data suggest that a VR experience aimed at modifying the sensory reference frame may be useful in speeding up the process of change. Future studies are needed both to identify specific perceptive styles in different pathologies and to define the best protocols for changing them.

Conclusions

For a long time, cognitive science considered action, perception, and interpretation as separate activities. A recent trend in cognitive science is instead seeing cognition as *embodied:* our conceptual system produces dynamically contextualized representations (simulations) that support situated action in different contexts.

In this picture, what is the role of "VR"? The basis for the VR idea is that a computer can synthesize a 3D graphical environment from numerical data. Using visual, aural, or haptic devices, the human operator can experience the environment as if it were a part of the world.

For these features, VR is described as a "simulation technology" with, and within which, people can interact. In summary, VR provides a new human–computer interaction paradigm in which users are no longer simply external observers of images on a computer screen but are active participants within a computer-generated 3D virtual world.

Different studies demonstrated that it is possible to use VR to induce controlled changes to the experience of the body both to induce an illusory perception of a fake or a fake hand as part of our own body and to produce an out-of-body experience, by altering the normal association between touch and its visual correlate. It is even possible to generate a body transfer illusion.

How are these experiences possible? What do they teach us about the potential of technology for health and wellness?

This chapter offered an answer to these two questions by identifying a clear link between spatial perception and action. On one side, we conceive places in terms of the actions we could take toward them: we do not have a separate knowledge of the place's location relative to them, what they can do in it, and their purposes. On the other side, our spatial experience is the outcome of the interaction of different spatial representations whose integrated activity gives rise to spatial awareness. The link between these processes and VR is provided by the concept of "presence," the feeling of being in a real or virtual space. Using VR the subject is present in a VE that offers a realistic setting to manipulate problematic situations related to his/her problem. For this reason, the most common application of VR in this area is the treatment of anxiety disorders.

However, it seems likely that VR can be more than a tool to provide exposure and desensitization. As noted by Glantz and colleagues (1997), "VR technology may create enough capabilities to profoundly influence the shape of therapy." (p. 92). In particular, we suggested that VR, for its ability of modifying the experience of the body through the alteration of the sense of presence, may be used to facilitate therapeutic change.

The final part of the chapter discussed the possible use of this approach in the treatment of eating disturbances—obesity, anorexia, and bulimia. The results of the presented method are very promising as shown by different controlled trials.

DIfferent pathologies may benefit from a similar approach Lambrey and Berthoz (2003) showed that subjects use conflicting visual and non-visual information differently according to individual "perceptive styles" (bottom-up processes) and that these "perceptive styles" are made more observable by the subjects changing their perceptive strategy, that is, re-weighting (top-down processes).

For these reasons, a VR experience aimed at modifying the sensory reference frame may be useful in speeding up the process of change. Future studies are needed both to identify specific perceptive styles in different pathologies and to define the best protocols for changing them.

References

Anderson, M. L. (2003). Embodied cognition: A field guide. *Artificial Intelligence, 149*, 91–130.

APA. (2000). *Diagnostic and statistical manual of mental disorders* (4th ed.), *Text revision*. Washington, DC: Author.

Baars, B. J. (1998). *A cognitive theory of consciousness*. New York, NY: Cambridge University Press.

Balakrishnan, B., & Sundar, S. S. (2011). Where am I? How can I get there? Impact of navigability and narrative transportation on spatial presence. *Human–Computer Interaction, 26*(3), 161–204.

Baños, R. M., Botella, C., García-Palacios, A., Villa, H., Perpiñá, C., & Alcañiz, M. (2000). Presence and reality judgment in virtual environments: A unitary construct? *Cyberpsychology & Behavior, 3*(3), 327–355.

Baños, R. M., Botella, C., & Perpiña, C. (1999). Virtual reality and psychopathology. *CyberPsychology & Behavior, 2*(4), 283–292.

Biocca, F. (1997). The cyborg's dilemma: Progressive embodiment in virtual environments. *Journal of Computer Mediated-Communication, 3*(2). Retrieved from http://jcmc.indiana.edu/vol3/issue2/biocca2.html

Blanke, O. (2012). Multisensory brain mechanisms of bodily self-consciousness. [Research Support, Non-U.S. Gov't Review]. Nature reviews. *Neuroscience, 13*(8), 556–571. doi:10.1038/nrn3292

Botella, C., Riva, G., Gaggioli, A., Wiederhold, B. K., Alcaniz, M., & Banos, R. M. (2012). The present and future of positive technologies. *Cyberpsychology, Behavior and Social Networking, 15*(2), 78–84. doi:10.1089/cyber.2011.0140

Brewin, C. R. (1989). Cognitive change processes in psychotherapy. *Psychological Review, 96*(3), 379–394.

Clark, A. (1997). *Being there: Putting brain body and world together again*. Cambridge, MA: MIT Press.

Clark, A. (2001). Reasons, robots and the extended mind. *Mind & Language, 16*(2), 121–145.

Clark, A. (2003). *Natural born cyborgs: Minds, technologies, and the future of human intelligence*. Oxford, UK: Oxford University Press.

Clark, A. (2008). *Supersizing the mind: Embodiment, action and cognitive extension*. Oxford, UK: Oxford University Press.

Emmelkamp, P. M. (2005). Technological innovations in clinical assessment and psychotherapy. *Psychotherapy & Psychosomatics, 74*(6), 336–343.

Ferrer-Garcia, M., & Gutierrez-Maldonado, J. (2012). The use of virtual reality in the study, assessment, and treatment of body image in eating disorders and nonclinical samples: A review of the literature. *Body Image, 9*(1), 1–11. doi:S1740-1445(11)00140-9 [pii] 10.1016/j.bodyim.2011.10.001

Frith, U., & de Vignemont, F. (2005). Egocentrism, allocentrism, and Asperger syndrome. *Consciousness and Cognition*, *14*(4), 719–738.

Galati, G., Lobel, E., Vallar, G., Berthoz, A., Pizzamiglio, L., & Le Bihan, D. (2000). The neural basis of egocentric and allocentric coding of space in humans: A functional magnetic resonance study. *Experimental Brain Research*, *133*, 156–164.

Gallagher, S. (2005). *How the body shapes the mind*. Oxford, UK: Clarendon Press.

Gallese, V., & Lakoff, G. (2005). The brain's concept: The role of the sensory-motor system in reason and language. *Cognitive Neuropsychology*, *22*, 455–479.

Garbarini, F., & Adenzato, M. (2004). At the root of embodied cognition: Cognitive science meets neurophysiology. *Brain and Cognition*, *56*(1), 100–106.

Gaudio, S., & Riva, G. (2013). Body image disturbances in anorexia: The link between functional connectivity alterations and reference frames. *Biological Psychiatry*, *73*, e25–26.

Glantz, K., Durlach, N. I., Barnett, R. C., & Aviles, W. A. (1997). Virtual reality (VR) and psychotherapy: Opportunities and challenges. *Presence, Teleoperators, and Virtual Environments*, *6*(1), 87–105.

Haugeland, J. (1998). *Having thought: Essays in the metaphysics of mind*. Cambridge, MA: Harvard University Press.

Herkenhoff Carijó, F., de Almeida, M. C., & Kastrup, V. (2013). On haptic and motor incorporation of tools and other objects. *Phenomenology and the Cognitive Sciences*, *12*, 685–701. doi:10.1007/s11097-012-9269-8

Jacobs, S., Bussel, B., Combeaud, M., & Roby-Brami, A. (2009). The use of a tool requires its incorporation into the movement: Evidence from stick-pointing in apraxia. *Cortex*, *45*(4), 444–455.

Koriat, A. (2007). Metacognition and consciousness. In P. D. Zelaso, M. Moscovitch, & E. Thompson (Eds.), *Cambridge handbook of consciousness* (pp. 289–325). New York, NY: Cambridge University Press.

Lakoff, G., & Johnson, M. (1980). *Metaphors we live by*. Chicago, IL: University of Chicago Press.

Lambrey, S., & Berthoz, A. (2003). Combination of conflicting visual and non-visual information for estimating actively performed body turns in virtual reality. *International Journal of Psychophysiology*, *50*(1–2), 101–115.

Lee, K. M. (2004a). Presence, explicated. *Communication Theory*, *14*(1), 27–50.

Lee, K. M. (2004b). Why presence occurs: Evolutionary psychology, media equation, and presence. *Presence*, *13*(4), 494–505.

Lenggenhager, B., Tadi, T., Metzinger, T., & Blanke, O. (2007). Video ergo sum: Manipulating bodily self-consciousness. *Science*, *317*(5841), 1096–1099. doi:317/5841/1096 [pii] 10.1126/science.1143439

Longo, M. R., Azañón, L., & Haggard, P. (2010). More than skin deep: Body representation beyond primary somatosensory cortex. *Neuropsychologia*, *48*, 655–668.

Mantovani, G., & Riva, G. (1999). "Real" presence: How different ontologies generate different criteria for presence, telepresence, and virtual presence. *Presence, Teleoperators, and Virtual Environments*, *8*(5), 538–548.

Marsh, T., Wright, P., & Smith, S. (2001). Evaluation for the design of experience in virtual environments: Modeling breakdown of interaction and illusion. *Cyberpsychology & Behavior*, *4*(2), 225–238.

Matelli, M., & Luppino, G. (2001). Parietofrontal circuits for action and space perception in the macaque monkey. *Neuroimage*, *14*(1 Pt 2), S27–S32.

McGonigal, J. (2011). *Reality is broken. Why games make us better and how they can change the world*. New York, NY: The Penguin Press.

Moore, K., Wiederhold, B. K., Wiederhold, M. D., & Riva, G. (2002). Panic and agoraphobia in a virtual world. *Cyberpsychology & Behavior*, *5*(3), 197–202.

Mou, W., & McNamara, T. P. (2002). Intrinsic frames of reference in spatial memory. *Journal of Experimental Psychology: Learning, Memory, and Cognition*, *28*(1), 162–170.

Mou, W., McNamara, T. P., Valiquette, C. M., & Rump, B. (2004). Allocentric and egocentric updating of spatial memories. *Journal of Experimental Psychology: Learning, Memory, and Cognition*, *30*(1), 142–157.

Perez-Marcos, D., Slater, M., & Sanchez-Vives, M. V. (2009). Inducing a virtual hand ownership illusion through a brain-computer interface. *Neuroreport*, *20*(6), 589–594.

Perpiña, C., Botella, C., & Baños, R. M. (2003). Virtual reality in eating disorders. *European Eating Disorders Review*, *11*(3), 261–278.

Previc, F. H. (1998). The neuropsychology of 3-D space. *Psychological Bulletin*, *124*(2), 123–164.

Prinz, J. (2006). Putting the brakes on enactive perception. *PSYCHE*, *12*(1), 1–12. Retrieved from http://www.theassc.org/files/assc/2627.pdf

Repetto, C., & Riva, G. (2011). From virtual reality to interreality in the treatment of anxiety disorders. *Neuropsychiatry*, *1*(1), 31–43.

Riva, G. (1998). Modifications of body image induced by virtual reality. *Perceptual and Motor Skills*, *86*, 163–170.

Riva, G. (2005). Virtual reality in psychotherapy: Review. *CyberPsychology & Behavior*, *8*(3), 220–230; discussion 231–240.

Riva, G. (Ed.). (1997). *Virtual reality in neuro-psycho-physiology: Cognitive, clinical and methodological issues in assessment and rehabilitation*. Amsterdam, The Netherlands: IOS Press. Retrieved from http://www.cybertherapy.info/pages/book1.htm

Riva, G. (2008). From virtual to real body: Virtual reality as embodied technology. *Journal of Cybertherapy and Rehabiliation*, *1*(1), 7–22.

Riva, G. (2009). Virtual reality: An experiential tool for clinical psychology. *British Journal of Guidance & Counselling*, *37*(3), 337–345. doi:10.1080/03069880902957056

Riva, G. (2011). The key to unlocking the virtual body: Virtual reality in the treatment of obesity and eating disorders. *Journal of Diabetes Science and Technology*, *5*(2), 283–292.

Riva, G. (2012a). Neuroscience and eating disorders: The allocentric lock hypothesis. *Medical Hypotheses*, *78*, 254–257. doi:S0306-9877(11)00566-4 [pii] 10.1016/j.mehy.2011.10.039

Riva, G. (2012b). Personal experience in positive psychology may offer a new focus for a growing discipline. *American Psychologist*, *67*(7), 574–575.

Riva, G., Bacchetta, M., Baruffi, M., & Molinari, E. (2002). Virtual-reality-based multidimensional therapy for the treatment of body image disturbances in binge eating disorders: A preliminary controlled study. *IEEE Transactions on Information Technology in Biomedicine*, *6*(3), 224–234.

Riva, G., Bacchetta, M., Baruffi, M., Rinaldi, S., & Molinari, E. (1999). Virtual reality based experiential cognitive treatment of anorexia nervosa. *Journal of Behavioral Therapy and Experimental Psychiatry*, *30*(3), 221–230.

Riva, G., Banos, R. M., Botella, C., Wiederhold, B. K., & Gaggioli, A. (2012). Positive technology: Using interactive technologies to promote positive functioning. *Cyberpsychology, Behavior and Social Networking*, *15*(2), 69–77. doi:10.1089/cyber.2011.0139

Riva, G., Cardenas-Lopez, G., Duran, X., Torres-Villalobos, G. M., & Gaggioli, A. (2012). Virtual reality in the treatment of body image disturbances after bariatric surgery: A clinical case. *Studies in Health Technology and Informatics*, *181*, 278–282.

Riva, G., Castelnuovo, G., Cesa, G., Gaggioli, A., Mantovani, F., & Molinari, E. (2012). *Virtual reality for enhancing the cognitive behavioral treatment of obesity: A controlled study with one-year follow-up*. Paper presented at the Medicine 2.0'12, Boston, MA.

Riva, G., & Davide, F. (Eds.). (2001). *Communications through virtual technologies: Identity, community and technology in the communication age*. Amsterdam, The Netherlands: IOS Press. Retrieved from http://www.emergingcommunication.com/volume1.html

Riva, G., Davide, F., & IJsselsteijn, W. A. (Eds.). (2003). *Being there: Concepts, effects and measurements of user presence in synthetic environments*. Amsterdam, The Netherlands: IOS Press. Retrieved from http://www.emergingcommunication.com/volume5.html

Riva, G., & Gaudio, S. (2012). Allocentric lock in anorexia nervosa: New evidences from neuroimaging studies. *Medical Hypotheses*, *79*(1), 113–117.

Riva, G., & Mantovani, F. (2012). From the body to the tools and back: A general framework for presence in mediated interactions. *Interacting with Computers*, doi:10.1016/j.intcom.2012.1004.1007

Riva, G., Waterworth, J. A., Waterworth, E. L., & Mantovani, F. (2011). From intention to action: The role of presence. *New Ideas in Psychology*, *29*(1), 24–37.

Riva, G., & Wiederhold, B. K. (2006). Emerging trends in cybertherapy. Introduction to the special issue. *PsychNology Journal*, *4*(2), 121–128. Retrieved from http://www.psychnology.org/File/PNJ4%282%29/PSYCHNOLOGY_JOURNAL_4_2_RIVA.pdf

Rosenblum, L. (2000). Virtual and Augmented reality 2020. *IEEE Computer Graphics and Applications*, *20*(1), 38–39.

Sadowsky, J., & Massof, R. W. (1994). Sensory engineering: The science of synthetic environments. *John Hopkins APL Technical Digest*, *15*, 99–109.

Safran, J. D., & Greenberg, L. S. (1991). *Emotion, psychotherapy, and change*. New York, NY: The Guilford Press.

Sakata, H., & Kusunoki, M. (1992). Organization of space perception: Neural representation of three-dimensional space in the posterior parietal cortex. *Current Opinion in Neurobiology*, *2*(2), 170–174.

Schubert, T., Friedman, F., & Regenbrecht, H. (2001). The experience of presence: Factor analytic insights. *Presence: Teleoperators, and Virtual Environments*, *10*(3), 266–281.

Slater, M., Perez-Marcos, D., Ehrsson, H. H., & Sanchez-Vives, M. V. (2009). Inducing illusory ownership of a virtual body. *Front Neuroscience*, *3*(2), 214–220. doi:10.3389/neuro.01.029.2009

Slater, M., Spanlang, B., Sanchez-Vives, M. V., & Blanke, O. (2010). First person experience of body transfer in virtual reality. *PLoS One*, *5*(5), e10564. doi:10.1371/journal.pone.0010564

Smucker, M. R., Dancu, C., Foa, E. B., & Niederee, J. L. (1995). Imagery rescripting: A new treatment for survivors of childhood sexual abuse suffering from posttraumatic stress. *Journal of Cognitive Psychotherapy*, *9*, 3–17.

Spagnolli, A., & Gamberini, L. (2002, October). *Immersion/emersion: Presence in hybrid environments*. Paper presented at the Presence 2002: Fifth Annual International Workshop, Porto, Portugal.

Spagnolli, A., Gamberini, L., & Gasparini, D. (2003). Situated breakdown analysis for the evaluation of a virtual environment. *PsychNology Journal*, *1*(1). Retrieved from http://www.psychnology.org/File/PSYCHNOLOGY_JOURNAL_1_1_SPAGNOLLI.pdf

Thompson, J. K. (1996). *Body image, eating disorders and obesity*. Washington, DC:—American Psychological Association.

Viaud-Delmon, I., Berthoz, A., & Jouvent, R. (2002). Multisensory integration for spatial orientation in trait anxiety subjects: Absence of visual dependence. *European Psychiatry*, *17*(4), 194–199.

Waterworth, J. A., & Waterworth, E. L. (2001). Focus, locus, and sensus: The three dimensions of virtual experience. *Cyberpsychology and Behavior*, *4*(2), 203–213.

Waterworth, J. A., & Waterworth, E. L. (2003). The meaning of presence. *Presence-Connect*, *3*(2). Retrieved from http://www8.informatik.umu.se/~jwworth/PRESENCE-meaning.htm

Waterworth, J. A., Waterworth, E. L., Mantovani, F., & Riva, G. (2010). On feeling (the) present: An evolutionary account of the sense of presence in physical and electronically-mediated environments. *Journal of Consciousness Studies*, *17*(1–2), 167–178.

Wiederhold, B. K., & Riva, G. (2012). Positive technology supports shift to preventive, integrative health. *Cyberpsychology, Behavior and Social Networking*, *15*(2), 67-68. doi:10.1089/cyber.2011.1533

Williamson, D. A. (1996). Body image disturbance in eating disorders: A form of cognitive bias. *Eating Disorders*, *4*(1), 47–58.

Wirth, W., Hartmann, T., Bocking, S., Vorderer, P., Klimmt, C., Schramm, H., Saari, T., … Jancke, P. (2007). A process model of the formation of spatial presence experiences. *Media Psychology*, *9*(3), 493–525. doi:10.1080/15213260701283079

Wolfe, B. E. (2002). The role of lived experience in self- and relational observation: A commentary on Horowitz (2002). *Journal of Psychotherapy Integration*, *12*(2), 147–153.

Zahoric, P., & Jenison, R. L. (1998). Presence as being-in-the-world. *Presence, Teleoperators, and Virtual Environments*, *7*(1), 78–89.

Ziemke, T. (2003). *What's that thing called embodiment.* Paper presented at the Annual Meeting of the Cognitive Science Society, Boston, MA.

25

Treating Emotional Problems with Virtual and Augmented Reality

Rosa M. Baños Rivera[1,2], Cristina Botella Arbona[2,3], Azucena García-Palacios[2,3], Soledad Quero Castellano[2,3], and Juana Bretón López[2,3]

[1] *Universitat de Valencia, Valencia, Spain*

[2] *CIBER Fisiopatología Obesidad y Nutrición (CB06/03), Instituto Carlos III, Girona, Spain*

[3] *Universitat Jaume I. Castellón, Castellón, Spain*

Introduction

The term virtual reality (VR) was coined in 1989 by Jaron Lanier, although the literature usually cites Ivan Sutherland as the first author writing in the 1960s about this "computer-generated illusion." Some time and a lot of effort have been necessary for the computer science to develop useful virtual realities from a cost-benefit perspective. However, as it is true in computer science in general, as the technological advances develop and the costs become lower, VR systems are more and more available in the workplace, at home, and in most domains of our life. There is no doubt that VR is playing an important role in human–computer interaction and it is a valuable technological innovation. The advances achieved since the first works by Sutherland at the MIT and at the University of UTA are being fruitful now. Its impact in the field of health is remarkable.

VR allows generating three-dimensional (3D) computer-generated environments. The essential characteristic is being able to simulate reality; the user experiences the feeling of "being" in those virtual environments and interacts with the objects included in them. This feeling of "being immersed in the VR environment," of being there, of judging the experience as somewhat real, has been named as sense of "presence." From a psychological point of view, this feeling of being in a certain place and being able to have meaningful experiences there, becomes very important, given that it opens the possibility of using VR environments as powerful therapeutic tools helping the person to change. At the same time, this change is achieved in a safe place. In summary, it is possible to modify behaviors, thoughts, emotions, etc., by means of "special" virtual experiences, virtual experiences designed and adapted to the needs of the individual with the aim of promoting, facilitating, and enhancing the process of change (Botella, García-Palacios, Baños, & Quero, 2009). Because of

The Handbook of the Psychology of Communication Technology, First Edition. Edited by S. Shyam Sundar.

this, in the last years, the use of VR has been widely extended to the field of psychological treatments.

In the last decade of the twentieth century, a research line dedicated to the design and testing of VR applications for the treatment of several mental disorders started. The goal was to explore if the use of VR could enhance the efficacy of the psychological treatments or overcome some of their limitations. The first work describing the application of a VR program for the treatment of a mental disorder was published by Rothbaum et al. (1995). Since then, the interest of the scientific community for VR has grown all over the world, as well as the publication of studies testing the utility and efficacy of this tool in the treatment of several mental disorders. A remarkable number of scientific works focused in the therapeutic application of VR in Clinical Psychology has been carried out. There are evidences of the efficacy of VR procedures for several mental disorders (Meyerbröker & Emmelkamp, 2010; Powers & Emmelkamp, 2008). However, the field of anxiety disorders is the area where the application of VR has been more extended.

Anxiety disorders have an elevated prevalence among the mental disorders. With regard to the treatment of anxiety disorder, exposure therapy is one of the most effective techniques, given that a key feature of these disorders is avoidance. Exposure therapy involves confronting the feared situation or object with the help of the therapist in a repeated, systematic, and progressive way until the individual learns how to confront anxiety in the feared situation, with the result that anxiety diminishes. Exposure therapy could be applied imagining the feared situation (imaginal exposure) or confronting the real situation (in vivo exposure). Considering the features of exposure therapy, some authors in the 1990s saw the possibility of using VR as new way of delivering exposure therapy. VR would allow simulating reality and it could help the patient to confront the feared situation in an efficacious, safe, and controlled way. The pioneer work by Rothbaum et al. (1995) showed the utility of VR exposure in a case of acrophobia. Other works were conducted since then showing, most of them, positive results about the efficacy of this new tool for the treatment of phobias. Although only few studies show good results like Kamphuis, Emmelkamp, and Krijn (2002), review and meta-analysis studies show that VR exposure is effective in the treatment of anxiety disorders (Meyerbröker & Emmelkamp, 2010; Powers & Emmelkamp, 2008). The meta-analysis carried out by Powers and Emmelkamp (2008) demonstrated that in vivo exposure was not significantly more effective than VR exposure in the treatment of several anxiety disorders. In some cases, VR was even more effective than in vivo exposure.

Some studies have indicated not only the efficacy but the advantages of using VR as a therapeutic tool (e.g., Anderson, Jacobs, & Rothbaum, 2004). First, VR allows graduating the situation in a way that the patient can progress from easier tasks to harder ones. Besides, VR enhances the generalization of the therapeutic outcomes given that it is possible to work in different environments related with the problem. VR is flexible enough to design different scenarios where the patient could virtually confront the feared situation, and this confrontation can be done at different levels, even achieving overlearning, that is, going beyond the feared situation. Little by little, from the knowledge and master of the interactions with the virtual world, the individual is able to confront and master the real world. In this line, it is important to notice that VR allows structuring the treatment as a special and safe place, given that the individual is not confronting the "real" situation, so that it is easier to persuade him/her to take action and practice in the VR environment without feeling threatened. The virtual environment is a "secure base" that therapy makes available, because nothing bad can happen there given that it is virtual. This "as if" aspect of VR is very important given that it could be a key intermediate step between the therapist office (a totally safe environment) and the real world (a totally threatening environment for the patient). VR can be a fundamental step in order to overcome the problem, perhaps the step that prepares the patient to give more steps toward recovery. Besides, it is important to consider that VR allows even going beyond reality, making possible that the feared context changes at our convenience. VR does not have to recreate reality, but to create therapeutic contexts, that is, "create" aspects or features of the environment that are not available to the patient right now (probably because he/she has lost access due to his/her problem). In summary, VR becomes an important therapeutic tool that can be used for the patient to become familiar and interact with a situation that he/she considers threatening at his/her own pace.

Focusing in the specific use of VR as a means to deliver exposure therapy, it is important to notice that this tool can help to overcome some of the limitations of this therapeutic strategy.

- In vivo exposure has a high cost, given that many times it involves that the therapist goes with the patient to the feared situation. Related to this, the feared situation is not always easily available. VR technology allows generating different environments without traveling anywhere, even those that are not easily available.
- Some people are so afraid to confront the feared situation that refuse to get involved in in vivo exposure therapy or drop out from it. Even those people who accept the treatment can find it too aversive and feel not safe given that it is not possible to have control over the events in real life (e.g., stopping of an elevator or having technical problems in a plane, etc.). One of the advantages of VR is that it is safe. The individual and the therapist have control over the computer-generated environment without risks. Besides, VR allows graduating the feared situation with total accuracy, attending to individual differences. VR exposure can be "customized" for each patient or problem, for example, the particular hierarchy of each patient can be elaborated combining the different options offered by the virtual environment. The treatment can be applied in a progressive way attending to the specific feared stimuli by each patient. The treatment can progress in a controlled way, being focused on the particular target and the particular safety behaviors shown by each patient.
- When using imaginal exposure, there is an additional difficulty regarding individual differences in the ability to imagine. VR overcomes this limitation, given that it provides a sense of presence that is difficult to achieve with imagination.

- Although most patients improve significantly and overcome their fears, usually they are not totally free of symptoms after the treatment, presenting some "residual fears" in the feared situations. VR allows "mastering the feared environment" going beyond what a "real" situation would allow, generating high levels of self-efficacy. Being able to interact many times in a repeated way with a concrete key environment involves the possibility of re-experience and feeling many times the implications and consequences of that interaction. A good example is a well-known VR application, the flight simulator. It is possible to practice multiple situations, difficulties, unpredictable situations, mistakes, etc., although in the real world, nothing happens. The progress made in a feared virtual environment would help the patient to experience, feel, and live "reality" in a different way.

At this moment, there are many VR applications for the treatment of different mental disorders developed by several research teams all over the world. In this chapter, we will focus on the applications developed only by our research team for the treatment of anxiety disorders.

Open-Out: A Treatment for Claustrophobia

Claustrophobia is very common in the general population, it being one of the most prevalent specific phobias. The essential feature is a marked and persistent fear related to closed spaces. The exposure or anticipation of confrontation of these situations provokes an anxiety response, although the individual acknowledges that the fear is excessive or irrational. As a consequence of this marked fear, closed spaces are avoided or tolerated with a high level of discomfort, causing impairment in important life domains (family, work, social area, etc.). The treatment of choice for claustrophobia, as it is the case for other specific phobias, is exposure therapy. Our research team developed almost two decades ago a VR application named *Open-Out*, to facilitate the application of exposure therapy.

Open-Out includes two scenarios, the house and the elevator. The house has two rooms (the living room and the "magic room"). The living room has two windows and two doors, one of them takes to a balcony and the other one takes to the "magic room." In this environment, the individual can be exposed to different degrees of closure, open and closing windows, and doors. In the "magic room," there are no furniture and no windows. The ceiling and the floor have a texture designed to give a higher sense of closure. The door can be closed and locked. This room allows the overlearning feature that we mentioned in the former section because it allows going beyond reality. It is possible to make the space smaller and smaller, moving one of the walls (accompanied by a strong noise) toward the individual, which increases the sense of closure.

The elevator also offers different possibilities in order to simulate different degrees of threat, considering several parameters: size, position, possibility of sudden stop, and possibility of breakdown. Inside the elevator, there are two special buttons that the patient can use. One makes one of the walls move (with a strong noise) toward him/her making the space smaller; the other button makes the wall go back to the original position.

We designed a clinical protocol for the use of *Open-Out* that has been tested with success in case studies (e.g., Botella et al., 1998); and also in studies with a higher degree of experimental control like single case studies (Botella, Baños, Villa, Perpiñá, & García-Palacios, 2000). These works demonstrated that *Open-Out* is effective to reduce claustrophobic fear and avoidance, obtaining a maintenance of the improvement at three-month follow-up.

Virtual-Flight: A Treatment for Flight Phobia

Flight phobia is a specific phobia, characterized by a marked fear related to flight situations. Anticipatory anxiety can be present in several situations such as buying flight ticket, waiting in the boarding area, or even at home, and packing. The fear can be so intense that the individual avoids confronting the flight or any situation related with flying; sometimes the individual does not avoid the situations but tolerates them with a high level of discomfort.

Again, the best treatment for this specific phobia is in vivo exposure. However, in this case, the treatment could entail a high cost both economic and related to time. Rothbaum team was again one of the pioneers in considering this phobia one of the best candidates for the application of VR exposure (Rothbaum, Hodges, Watson, Kessler, & Opdyke, 1996). Our research team developed more than fifteen years ago the program *Virtual-Flight*, that includes three virtual scenarios to deliver exposure therapy for this problem. In all of them, the therapist and patient can have control over several events: time of the day (day or night); good or bad weather, and also to control certain changes in the environment in order to adjust the degree of threat. The first scenario (*getting ready for the flight*) focuses on addressing anticipatory anxiety before the flight. It is a room where the action of packing is simulated. The second scenario (*at the airport*) also addresses anticipatory anxiety but being already at the airport. In this environment, the individual cannot interact with any person or object. He/she is at the boarding area waiting for his/her flight. The environment provides information about updates regarding the next flights, see through a big window planes taking off and landing, and listen to conversations among other passengers waiting for the flight. Finally, the third scenario (*aboard*) simulates being inside an aircraft. In this scenario, the patients can interact with some objects, look around, but it is not possible to get up or walk around. This scenario is used to simulate key events like taking off, landing, or turbulences.

Our team developed a clinical protocol for the use of *Virtual-Flight*. This VR program has been tested in different studies showing always positive results regarding clinical efficacy (e.g., Baños et al., 2002; Botella, Osma, García-Palacios, Quero, & Baños, 2004). Recently, Tortella-Feliu et al. (2011) compared the efficacy of *Virtual-Flight* with two computerized treatment programs for flight phobia in a controlled study. This study showed that *Virtual-Flight* was effective for the reduction of flight fear and avoidance, achieving that the patients actually flew after the treatment. Besides, *Virtual-Flight* has demonstrated to be an efficient program, presenting advantages over traditional treatments related with saving time and money.

Going-Out: A Treatment for Panic Disorder with Agoraphobia

Panic disorder with agoraphobia (PDA) is one of the mental disorders with higher costs for health services. The main characteristic is the presence of panic attacks that consist of unexpected fear or intense discomfort attacks in which the person experiences intense bodily sensations (heart beatings, shakings, suffocation, dizziness, etc.). These attacks start abruptly and reach the maximum intensity in few minutes. They also go together with and immediate and imminent feeling of danger and a desire of escaping. Panic disorder, generally, goes together with agoraphobia, which consists of the experience of intense anxiety when the person comes across situations or places from where it is difficult to escape or obtain help in case of having a panic attack. This anxiety can make the person to avoid different situations: going out home alone, to be in crowded places such as football fields or concerts; traveling by train, bus, or car; and passing through bridges or tunnels.

After years of research, there is enough empirical evidence to affirm that cognitive–behavioral treatments are the most effective to treat this disorder. Many of these programs include the following therapeutic components: psychoeducation, cognitive restructuring in order to correct the catastrophic interpretations of the physical sensations, slow breathing training, relaxation, exposure to the feared physical sensations (interoceptive exposure), and exposure to the agoraphobic situations (in vivo exposure).

Following these guidelines, our group has designed an intervention protocol that includes those components applied along eight sessions. In order to apply both interoceptive and in vivo exposure using VR, the program *Going-out* is used. This program is made up of six virtual scenarios: a training scenario and five clinically significant scenarios for panic disordered patients (the house, the metro, the bus, the mall, and the tunnel). These scenarios include several modulators that allow to graduate and rank the difficulty of exposure for each patient: number of people, length of the trips, breakdown in the elevator, problems with the credit card payment, number of levels in the elevator, etc. In addition, all these scenarios, unlike other VR available programs currently, permit to simulate different bodily sensations (heart beatings, rapid breathing, tunnel vision, blurred vision, and double vision) through visual and auditory effects, allowing, therefore, to apply simultaneously both types of exposure. In sum, *Going-Out* is a flexible program that can be adapted to the patients' needs, offering the possibility of conducting the exposure technique in a gradual and controlled way.

The objective of "the house" is to diminish the anticipatory anxiety that patients with PDA have before facing an agoraphobic situation. The house includes a living room, an elevator, and the foyer. This scenario permits to work different typical situations present in this problem such as being alone, going out of home alone, and using the elevator. The main objective of "the metro" and "the bus" is to confront the patient to situations in which he/she experiences fear and uncontrollability. In both scenarios, the number of stops and people who enter or go out is modulated. The "mall" scenario allows the patient to be exposed to agoraphobic situations such as shopping in a mall, being in a crowded place, being in a queue, being in narrow spaces (corridors), and using the escalators. Finally, "the tunnel" permits to expose the patient to a very threatening agoraphobic situation like being in a place where to find the exit or escape is difficult in case of having a panic attack.

Regarding efficacy data of *Going-Out* program, several studies have demonstrated its efficacy and utility. A controlled study (Botella et al., 2007) compared the efficacy of VR exposure with in vivo exposure. Results showed that both treatment conditions were equally effective with no significant differences between them. Furthermore, patients' acceptability of the exposure component in both conditions was very high, considering the component very logic, satisfactory, advisable to others, useful for their problem and for other problems. Finally, in a recent study (Pérez-Ara et al., 2010), the utility of VR to conduct interoceptive exposure was analyzed. The efficacy of two treatment conditions that used virtual exposure to the agoraphobic situations was compared, but in one condition, participants were exposed simultaneously to the program sounds and visual effects, whereas in the other condition participants received interoceptive exposure in the traditional way, that is, separately from the VR exposure to the agoraphobic situations. Results showed that both conditions were capable to reduce the symptoms significatively, and this reduction was maintained at 3-month follow-up. Therefore, evoking physical sensations through VR was as powerful as evoking them by doing traditional in vivo exercises (such as hyperventilation, to go up and down stairs, to draw in a chair, etc.). In addition, participants in VR condition considered the interoceptive exposure more logical and useful to treat other problems than participants belonging to in vivo exposure condition.

In sum, the studies conducted with *Going-Out* program show that RV allows to apply the interoceptive exposure component and the exposure to the agoraphobic situations simultaneously, which might facilitate treatment, conducting the exposure in the clinician's consulting room, protecting patients' confidentiality and reducing costs related to the therapist's traveling expenses.

The EMMA's World: An Adaptive VR System for Stress-Related Disorders

Stress-related disorders like posttraumatic stress disorder (PTSD), adjustment disorders, and complicated grief have in common the forthcoming of clinically significant behavioral, emotional, and cognitive symptoms after being exposed or having experienced an event that has a negative impact in people's life. A common element in many therapeutic approaches for these disorders is the exposure and processing of stimuli related to the negative event. However, exposure technique for these disorders seems to be underused in clinical practice. The studies carried out by Becker, Zayfert, and

Anderson (2004) indicate that only few psychologists use this strategy. Several reasons explain this low rate of use. Regarding patients, for some of them, the imaginal exposure to the stressful event is a challenge too difficult and some patients are reluctant to remember the traumatic event, as well as the situations, objects, or people related to them. In other occasions, the patients are able to talk about the trauma, but they do so "detached" of the experience, they do not involve emotionally in the experience. This lack of emotional involvement may interfere in the anxiety reduction resulting in poor treatment outcomes. Negative treatment results may also be found due to the difficulty in imagination capacity for some people. But the reluctance comes from the clinicians as well. Some of them find the exposure technique procedure too aversive and painful and consider that this technique can "re-traumatize" patients or that the stress caused by its use could decompensate them or make them to abandon the treatment. VR can be a promising alternative to overcome these difficulties.

In the specific case of PTSD, several studies have shown that the use of VR exposure produce a significant reduction of symptoms. For example, Rothbaum's group has reported improvements in the clinical situation of war veterans (e.g., Rothbaum, Hodges, Ready, Graap, & Alarcon, 2001). Difede's group has treated several survivor of the September 11 terrorist attack in New York, using a simulation of the twin towers, finding positive results even in patients who had not improved with the use of imaginal exposure technique (Difede et al., 2007).

However, regardless of the continuous improvement and innovation of technologies, the VR systems used in the aforementioned studies only include one or several scenarios related to a specific problem, given that their main objective is addressed to represent the reality with the higher fidelity as possible. This issue has an important limitation since these scenarios can hardly be used to treat other problems different to those they were designed for or be applied in a population different to that they were thought of (for instance, the twin towers scenario developed by Diffede and Hoffman cannot be used in someone who suffer PTSD but who was not a victim of the September 11 attack). This shortcoming is especially important in the case of stress-related disorders due to the big amount and the wide range of traumatic or stressful events that may result in the development of these problems.

With the aim of overcoming these difficulties, our group developed the "EMMA's World" (Engaging Media for Mental Health Applications). The main objective was to build a VR application able to adapt itself in a flexible manner to the particular needs present in different problems (overcoming the limitation of the traditional VR systems that use the same scenario for everyone) providing, at the same time, significant environments capable of activating and enhancing the emotional processing of the experienced event. In addition, we also pretended to reduce the reluctances showed from both therapists and patients when applying exposure technique, trying to make the exposure less aversive and painful. The application, instead of building "realistic" scenarios, combines virtual elements that are capable of representing and evoking a symbolic way the meanings and emotional reactions related to the experienced event and which, at the same time, can be modified according to the changes that patients are experiencing along treatment. The objective is to offer the users a "physical" representation (virtual) of the event and their emotions and, through the

Figure 25.1 "EMMA's room" and its elements.

therapeutic work with such representation, to facilitate the emotional change, the improvement of the coping competences, and learning from the problems. In addition, the system maintains the advantages of the traditional VR systems, by permitting the control of the process at all time and by guarantying this occurs in a gradual and safe way. EMMA's World also allows creating and personalizing virtual environments that can be modified in real time and in which it is possible to introduce personalized multimedia elements (pictures, music, texts, videos or user's narratives). EMMA's World includes two virtual environments: "EMMA's room" and "EMMA's environments":

1. "EMMA's room" simulates an architectural structure where all therapeutic sessions start. It is an open space, with no walls, from where the surrounded landscape can be observed (see Figure 25.1). This room includes several

Figure 25.2 EMMA's environments.

therapeutic tools: (a) The database, which consists of a mobile screen with different elements that the user can manipulate (3D objects, pictures, music and sounds, colors and proverbs or sayings); the system also permits to introduce in the database elements from the particular patient, such as pictures or music. (b) The stands, a place where the elements chosen by the patient to work with are located, being possible different ways of mixing and combining them. (c) The inventory, a small viewer located in the right superior part of the screen with permits to carry the elements from the database to the stands. From there, can be stored in the book of life. (d) The book of life, which includes an index and a series of pages where the patient can write the chapters of his/her life related with the stressful event. In this book, the elements that are being processed and elaborated in therapy can also be stored; the book of life helps the patient to remember, analyze, and elaborate the events and allows him/her writing and rewriting (with words and with the database elements) the emotional experiences associated with the situation. (e) The Emotional Processor, which is a device that emerges from the floor when is activated. The elements related to the painful experience which have lost the power to emotionally disturb the patient can be located on this device so they can be transformed in a symbolic way.

2. "EMMA's environments" are five predefined landscapes that the patient can choose (see Figure 25.2): a meadow, a snow-covered town, an island, a threatening forest, and a desert. The landscapes can be selected and modified in real time with the aim of promoting the induction of emotions, reflecting the changes in the affect experienced by the patient during the session and helping him/her in his/her regulation and awareness work. The changes that can be done affect the environment conditions (e.g., to provoke wind, rain, snow, fog, a storm, an earthquake, or make the brambles grow) and the representation of time (e.g., to vary the time of the day or the speed with which time goes pass) and they can be made with different levels of intensity. All the elements and possibilities are designed with the purpose of reflecting the patient's mood state in the most accurate way as possible, aiming at helping the patient to experience, accept, and manage the suffered emotions and experiences that caused his/her discomfort.

Along with the development of the system, the design for the treatment protocols was developed. The treatment structure is similar for the three considered disorders, although the necessary modifications according to the specific characteristics of each disorder are included. In the case of PTSD, we made and adapted Foa and Rothbaum's (1998) treatment protocol and, for the adjustment disorders and complicated grief, we followed the guidelines stated by Neimeyer (2001). The common components in these protocols are (a) educational component; (b) exposure in EMMA's World, where the patient chooses the virtual environment and elements that represent the suffered event and his/her emotions that allows him/her to face them and to process and overcome progressively the experience; (c) the book of life, whose aim is to help the patient to recall, order, and save his/her memories, thoughts, feelings, and emotions related with the suffered experience; this book helps to the elaboration of the event, permits to extract learning from the event itself, and contributes to the process of incorporating the suffered experienced joint with the remaining life experiences; (d) in vivo exposure that states an exposure hierarchy or a graded list of situations, objects, or persons feared by the patient to which he/she has to confront progressively, starting with the easiest situation and moving forward to the most feared ones; (e) cognitive restructuring, focused on the identification, evaluation, and modification of negative thoughts and dysfunctional beliefs the patient has related to the stressful event; (f) acceptance strategies whose aim is that the patients become aware and "allow themselves" experiencing emotions, having thoughts, or producing behaviors, although these are aversive and they manage, finally, to accept the disturbing event; and (g) relapse prevention.

Patients are explained that, apart from the therapist, the system will accompany, support, and guide them along the whole process. They are told that EMMA's World is a place where they can express their problems, thoughts, and emotions freely and with no pressure, that the whole system is thought to help, support, and guide them and that the environment is a place where they can stop, rest, and think. Once the patient is immersed in the scenarios, the therapist ask him/her to narrate his/her memories about the disturbing experience in present time and with the highest level of detail as possible, reflecting at the same time the experience in the virtual environment. The selected elements that have been chosen and worked in session by the patient are saved in the book of life whose aim is to help the patient to order his/her life, to represent how he/she is solving the emotional and behavioral problems as he/she is confronting the disturbing elements of his/her experience, and to store those elements.

The utility and efficacy of this system has been tested in several works. Results obtained in case studies (Andreu-Mateu, Botella, Quero, Guillén, & Baños, 2012; Botella, Osma, García-Palacios, Guillén, & Baños, 2008) showed an improvement in the patients' clinical condition with different stress-related disorders. Other between-group studies were also conducted (e.g., Botella, Baños et al., 2006), finding that patients reported very high scores regarding treatment satisfaction, even higher than the expectations generated by the patients before treatment. It was also found that the condition that used EMMA's World evoked lower levels of aversiveness over the traditional condition. Finally, controlled studies have been conducted (e.g., Baños et al.,

2011) and the obtained results so far support the efficacy of the VR system developed by our group for the treatment of stress-related disorders, concretely for PTSD, adjustment disorders and complicated grief, and show that the system is well accepted and valued by both the patients and the therapists.

At this moment, an adaptation of EMMA's World for children and adolescent are being carried out and a study is being planned to test the efficacy of this adaptation in abused children and adolescent. We expect to apply this system for the treatment of other kind of problems in the future. We think that this is the first step in a line of research that will be very productive. We envision a time in the near future when each researchers and clinicians will have a versatile and complex virtual world available to them, to which they will be able to add elements, enriching it and exploring new utilities and possibilities. This will allow us to work with our patients in "virtual worlds" and "real worlds" that fit the specific needs of every patient in different situations.

Limitations of VR

In spite of the advantages we have highlighted so far, VR has shortcomings too. First of all, it is important to stand out the necessity of studying and specifying the possible harm of the wrong use of virtual worlds. One of the possible harmful effects are the collateral effects suffered by some users after the virtual exposure (Rizzo, Wierderhold, & Buckwalter, 1998). The symptoms include eye problems, disorientations, balance alteration, and sickness. These secondary effects are not frequent in users who do not suffer any physical nor neurological problems, affecting only to a short percentage of the population. In any case, and specifically the first times when people interact with the virtual world, some motion sickness and instability may appear. In these situations, the therapist has to calm the user down, explaining that these effects do not suppose a risk for his/her health. In addition, we should explain that the sensations are temporary and very similar to others experimented when people travel by boat or by car. The therapist has to make sure that the symptoms have disappeared before the user goes out of the therapist's office. Another limitation is related to individual differences in the sense of presence achieved by users. Some users do not feel immersed in the virtual world, and then the efficacy of this technology to overcome their problems may be limited. Further research is needed on how to improve virtual worlds to make achievements that are easily transferable to the real world.

Augmented Reality and the Treatment of Specific Phobias

An advance in VR technology, augmented reality (AR) has overcome some of the limitations of VR, contributing with mixed realities to increase realism and the patients' sense of presence, "filling" reality instead of substituting it. AR could also improve therapy duration and acceptance.

AR allows the user to see the real world, but in this real world, some objects and 3D elements are placed, as part of what the user can see. Then, it can be said that AR

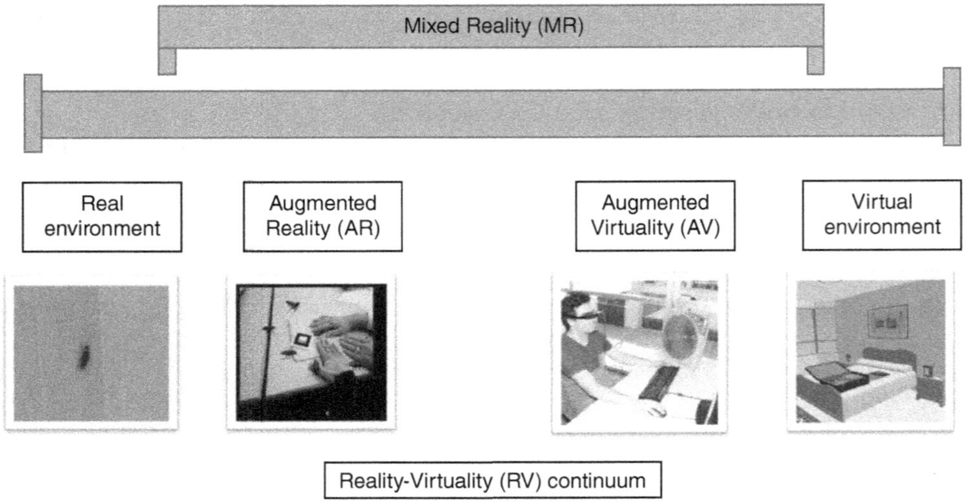

Figure 25.3 The reality-virtuality-continuum.

Source: Milgram, P., Takemura, H., Utsumi, A., & Kishino, F. (1994). Augmented reality: A class of displays on the reality-virtuality continuum. *Proceedings of Telemanipulator and Telepresence Technologies* (pp. 285–292). SPIE.

complement the reality instead of fully replacing it, as VR does. The central characteristic of AR is that virtual elements that are superimposed let additional and relevant information to the image of the system the user is visualizing, with the purpose of helping him/her.

The reality–virtuality continuum proposed by Milgram, Takemura, Utsumi, and Kishino (1994) can help us to better understand AR and the main differences with VR. As it can be observed in Figure 25.3, in an extreme of the continuum "the real environment" is located, which, obviously, is not modelled at all. In the other extreme, "the virtual environment" is positioned, which is totally modelled. AR would be located near from the real world, as the user can see an image composed by the visualization of the real world and some virtual elements superposed on it. "Augmented virtuality" consists of a virtual environment where real stimuli are included (e.g., the smell of the flowers in a virtual environment). Both AR and augmented virtuality have two main characteristics: they both mix realities and they are both worlds partially modelled but in a different degree, the AR being modelled in a lower degree.

AR is a technology that offers very promising possibilities in several fields, due to its characteristics, especially in the clinical psychology area. This technology offers additional advantages over VR. First, AR can provoke a greater feeling of presence and reality than VR because the environment and the tools the patient uses to interact with the application are real. Second, in AR the users can see their own bodies in context, interacting with the feared stimuli; the system allows patients to use real elements and their own hands and bodies to interact with the feared stimuli. Finally, AR is cheaper than VR, as it is only necessary to design and simulate the feared elements rather than an entire immersive environment.

Figure 25.4 AR system for the treatment of acrophobia.

AR is having several applications in engineering, medicine, entertainment, and in other disciplines and areas (Azuma et al., 2001). However, there are few AR applications in psychology. So far, the scarce available applications in the psychological treatment field have been carried out by our research group and have been focused on the treatment of specific phobias: acrophobia and small animal phobia (spiders and cockroaches).

The AR system for acrophobia uses immersive photography technique (Juan, Baños, Botella Pérez, Alcañiz, & Monserrat, 2006), consisting of photographed environments representing the real ones, where the user can move around. By using immersive photographs, a complete space is captured from a single point and it is worked in a digital way in order to create a photogram of 360°. The system includes different locations and levels that patients can visualize (see Figure 25.4): images taken from a window of a building (located on the first, second, third, fourth, fifth, and fifteenth floors), view of a stairwell from the second and third floors, view of a staircase from a terrace situated on the second floor, view of a dam (both sides), and images taken from a terrace located on the second and third floors. The therapist can modify each level with an option menu. The treatment includes AR exposure by using immersive photos, reinforcement practice, and cognitive challenge.

The AR system for small animal phobia uses a USB or FireWire camera incorporated in a Head-Mounted Display (see Figure 25.5). When the camera finds a marker (a white square with a black border containing symbols or letters) in the real world, the program recognizes it and activates the virtual elements (spiders and cockroaches in this case).

The therapist can monitor and control the virtual stimuli presented to participants during the exposure session. The system includes the following menu options: number of cockroaches (it is possible to increase or reduce the number of animals with 60 as the maximum number); movement of cockroaches (static or moving); state of the

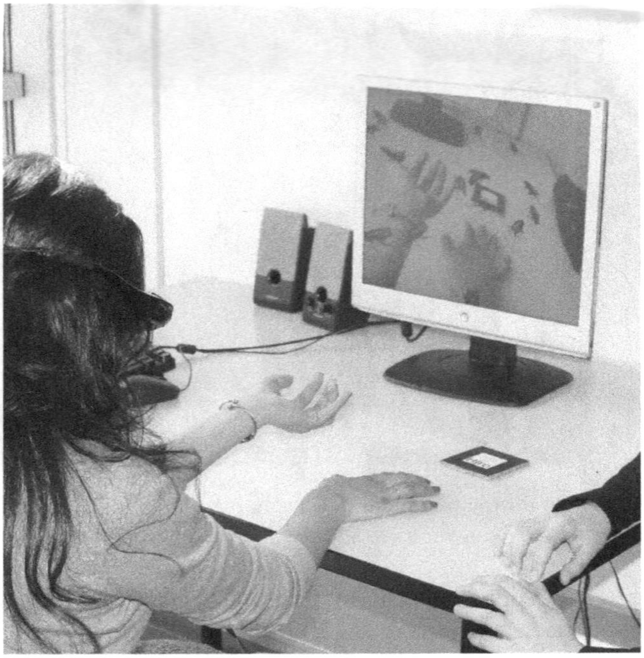

Figure 25.5 AR system for the treatment of small animal phobia.

animal (alive or dead: the participant can "kill" the animal using an insecticide or crushing it with some object); and kind of animals (there are three different kinds of spiders and cockroaches). The AR treatment is applied following the guidelines developed by Öst, Salkovskis, and Hellström (1991), which involve intensive exposure, lasting up to 3 hr and incorporate treatment components like modeling, exposure (in this case using AR), reinforced practice and cognitive challenge.

Several studies have offered data about the efficacy and utility of AR for the treatment of small animal phobia (e.g., Botella, Bretón-López, Quero, Baños, & García-Palacios, 2010). These studies have shown that AR exposure can reach an important reduction of the fear, avoidance, and beliefs linked to situations that involve confronting these feared elements (spiders and cockroaches). In addition, virtual objects (small animal) are assessed by patients as very realistic and useful to activate the anxiety, mainly when the animals are in movement. These positive results suggest AR's potential in psychology. It is needed to apply this treatment to larger samples and in a group design that includes a control group. Furthermore, some aspects could be improved. Firstly, as usual, participants came to the "one-session treatment" with high levels of fear. We hypothesize that asking the patients to complete homework assignments involving familiarization with the feared object (cockroach) could help make the "one-session" exposure treatment less aversive. Secondly, following Öst's recommendation (Öst et al., 1991), participants are advised to continue confronting cockroach-related phobic situations after therapy in order to completely surmount the problem. This aspect can be difficult, since the feared object (cockroach) is sometimes not present for the duration that the patient requires. Computer games could help make post-treatment self-exposure tasks more effective. We have developed a serious game for the treatment of cockroach phobia that uses a mobile phone as the application device. We have analyzed whether the use of this mobile game can facilitate treatment of this specific phobia preparing his or her for the AR exposure (Botella et al., 2011). Preliminary data show that the use of the game reduced fear and anxiety punctuations of participants before starting the AR session, and consolidated treatment gains reached by the therapy, reducing more the punctuations of fear, avoidance, and beliefs in irrational thoughts.

In the same way that VR began with the treatment of specific phobias and its use was extended in a quickly way to more complex disorders, AR can become a very useful tool in the treatment of several problems. Future research lines have to be focused on testing the utility of these systems, by controlled studies, large samples, and with follow-up data in order to confirm the efficacy and efficiency of AR as a therapeutic tool for the treatment not just of specific phobias but other anxiety disorders.

Conclusions

VR and AR are showing their versatility and utility in the psychological treatment application. Since the past years, we have witnessed an important development of applications for the treatment of different mental disorders, especially anxiety disorders. VR and AR are effective tools for delivering exposure technique as several controlled studies have stated, with short- and long-term results in specific phobias and panic disorders with agoraphobia. Regarding PTSD and the disorders related to stress, results are preliminary, but yet very promising. These technologies can be an effective alternative to patients who do not have a good response to the imaginal exposure, which could help increase the number of patients who benefit from a cognitive–behavioral treatment.

VR and AR can contribute to improve the clinical utility axis (*APA Task Force on Psychological Intervention Guidelines*, 1995), increasing the acceptation of the exposure technique or proving efficacy in a group of patient who do not respond to other ways of applications. The usefulness of VR and AR in clinical psychology lies on its ability to improve existing intervention techniques in order to better support to people with mental disorders.

Finally, we should remark that in addition to expanding the applications field, technological innovations include the use of other information and communication technologies (ICTs) such as the Internet, mobile devices, or a combination of different ICTs as pervasive computing, ubiquitous computing, or ambient intelligence. The possibilities of VR and other ICTs like AR are very enormous. Although currently most VR applications have been designed to be applied in one PC located in the therapist's office, however, it is important to be aware that the technology is changing very quickly. Recently we are witnessing a vertiginous development of the mobile communications, and an increase in the range, accessibility, and extent of communications, by the development of the multimedia technologies. Linked to this process, it is expected the emergency of the called "Virtual Immersive Telepresence". This consists of VR systems that combine "wireless" multimedia devices, streaming videos, biosensors, and brain-computers interfaces.

Rigorous research will be necessary to test the utility of these tools in the treatment of psychological disorders and the promotion of health. The ICTs have become a reality that is changing codes and forms of communication in our society. Psychological therapy has to adapt to these changes in order to offer its worthy techniques and procedures to the present and future generations. However, in order to reach an optimum in the use of these technologies, it will be necessary to work in a multidisciplinary way, collaborating together engineers, computer programmers, graphic designers, and experts in ergonomics, and also with clinicians and experts in other fields of psychology. Only from this joint work of all these experts, we will be able to develop innovative applications that use appropriate technologies, and which are useful for improving well-being and quality of life of people.

Finally, the necessity of obtaining more data regarding the therapeutic efficacy of VR and AR has to be emphasized, especially regarding the differential efficacy in comparison to other techniques widely used in therapy. VR and AR are still new tools; a lot of work has to be done. A fundamental feature in this work, not existing yet, is the necessity of structuring the theoretical framework to guide predictions and to organize the results. VR and AR have a great future and the available applications so far are only the starting point.

Acknowledgments

The research presented in this paper was funded in part by Generalitat Valenciana, Conselleria de Educación Programa de Investigación de Excelencia PROMETEO (2008/157), and CIBER Fisiopatología de la Obesidad y Nutrición (an initiative of ISCIII).

References

Anderson, P., Jacobs, C., & Rothbaum, B. O. (2004). Computer-supported cognitive behavioral treatment of anxiety disorders. *Journal of Clinical Psychology*, *60*(3), 253–267.

Andreu-Mateu, S., Botella, C., Quero, S., Guillén, V., & Baños, R. M. (2012). La utilización de la realidad virtual & estrategias de psicología positiva en el tratamiento de los trastornos adaptativos. *Behavioral Psychology/Psicología Conductual*, *20*, 323–348.

Azuma, R., Baillot, Y., Behringer, R., Feiner, S., Julier, S., & MacIntyre, B. (2001). Recent advaces in augmented reality. *Computers & Graphics*, *25*, 1–15.

Baños, R. M., Botella, C., Perpiñá, C., Alcañiz, M., Lozano, J. A., Osma, J., & Gallardo, M. (2002) Virtual reality treatment of flying phobia. *IEEE- Transactions on Information Technology in BioMedicine*, *6*(3), 206–212.

Baños, R. M., Guillén, V., Quero, S., García-Palacios, A., Alcañiz, M., & Botella, C. (2011). A virtual reality system for the treatment of stress-related disorders: A preliminary analysis of efficacy compared to a standard cognitive behavioral program. *International Journal of Human Computer Studies*, *69*(9), 602–613.

Becker, C. B., Zayfert, C., & Anderson, E. (2004). A survey of psychologists′ attitudes towards and utilization of exposure therapy for PTSD. *Behaviour Research and Therapy*, *42*, 277–292.

Botella, C., Baños, R.M., Guerrero, B., García-Palacios, A., Quero, S., Alcañíz, M. (2006) Using a flexible Virtual Environment for Treating a Storm Phobia. *PsychNology Journal*, 4(2) 129–144.

Botella, C., Baños, R. M., Perpiñá, C., Villa, H., Alcañiz, M., & Rey, A. (1998). Virtual reality treatment of claustrophobia: A case report. *Behaviour Research and Therapy*, *36*, 239–246.

Botella, C., Baños, R. M., Villa, H., Perpiñá, C., & García-Palacios, A. (2000). Virtual reality in the treatment of claustrophobic fear: A controlled multiple baseline design. *Behavior Therapy*, 31, 583–595.

Botella, C., Bretón-López, J., Quero, S., Baños, R., García-Palacios, A., Zaragoza, I., & Alcaniz, M. (2011). Treating cockroach phobia using a serious game on a mobile phone and augmented reality exposure: A single case study. *Computers in Human Behavior*, *27*, 217–227.

Botella, C., Bretón-López, J., Quero, S., Baños, R., & García-Palacios, A. (2010). Treating cockroach phobia with augmented reality. *Behavior Therapy*, *41*, 401–413.

Botella, C., García-Palacios, A. Baños, R.M., Quero S. (2009) Cybertherapy: Advantages, Limitations, and Ethical Issues. *PsychNology Journal*, 7 (1), 77–100.

Botella, C., García-Palacios, A., Villa, H., Baños, R. M., Quero, S., Alcañiz, M., & Riva, G. (2007). Virtual reality exposure in the treatment of panic disorder and agoraphobia: A controlled study. *Clinical Psychology & Psychotherapy*, *14*(3), 164–175.

Botella, C., Osma, J., García-Palacios, A., Guillén, V., & Baños, R. M. (2008). Treatment of complicated grief using virtual reality: A case report. *Death Studies*, 32, 674–692.

Botella, C. Osma, J., García-Palacios, A., Quero, S., & Baños, R. (2004). Treatment of flying phobia using virtual reality: Data from a year follow-up using a multiple baseline design. *Clinical Psychology & Psychotherapy*, *11*(5), 311–323.

Difede, J., Cukor, J., Jayasinghel, N., Patt, I., Jedel, S., Spielman, L., & Hoffman, H. (2007). Virtual reality exposure therapy for the treatment of posttraumatic stress disorder following September 11, 2001. *The Journal of Clinical Psychiatry*, *68*, 1639–1647.

Foa, E. B., & Rothbaum, B. O. (1998). *Treating the trauma of rape: Cognitive-behavior therapy for PTSD.* New York, NY: Guildford Press.

Juan, C., Baños, R. M., Botella, C., Pérez, D., Alcañiz, M., & Monserrat, C. (2006). An augmented reality system for acrophobia: The sense of presence using immersive photography. *Presence: Teleoperators and Virtual Environments, 15,* 393–402.

Kamphuis, J. H., Emmelkamp, P. M. G., & Krijn, M. (2002). Specific phobia. In M. Hersen (Ed.), *Clinical behavior therapy, adults and children.* New York, NY: Wiley.

Meyerbröker, K., & Emmelkamp, P. (2010). Virtual reality exposure therapy in anxiety disorders: A systematic review of process-and-outcome studies. *Depression and Anxiety*, *27*(10), 933–944.

Milgram, P., Takemura, H., Utsumi, A., & Kishino, F. (1994). Augmented reality: A class of displays on the reality-virtuality continuum. In Hari Das (Ed.), *Proceedings of Telemanipulator and Telepresence Technologies* (pp. 285–292). Bellingham, WA: SPIE. Digital Library.

Neimeyer, R. A. (2001). Mourning and meaning. *American Behavioral Scientist*, *46*, 235–251.

Öst, L. G., Salkovskis, P., & Hellström, K. (1991). One-session therapist directed exposure vs. self-exposure in the treatment of spider phobia. *Behavior Therapy*, *22*, 407–422.

Pérez-Ara, M., Quero, S., Botella, C., Baños, R., Andreu-Mateu, S., García-Palacios, A., & Bretón-López, J. (2010). Virtual reality interoceptive exposure for the treatment of panic disorder and agoraphobia. In B. Wiederhold, G. Riva, & S. Kim (Eds.), *Annual review of cybertherapy and telemedicine* (pp. 61–64). San Diego, CA: Interactive Media Institute.

Powers, M., & Emmelkamp, P. (2008). Virtual reality exposure therapy for anxiety disorders: A meta-analysis. *Journal of Anxiety Disorders*, *22*(3), 561–569.

Rizzo, A. A., Wiederhold, M. D., & Buckwalter, J. G. (1998). Basic issues in the use of virtual environments for mental health applications. In G. Riva, B. K. Widerhold, & E. Molinari (Eds.), *Virtual environments in clinical psychology and neuroscience* (pp. 21–42). Amsterdam, The Netherlands: IOS Press.

Rothbaum, V. O., Hodges, L. F., Kooper, R., Opdyke, D., Williford, J. S., & North, M. (1995). Virtual-reality graded exposure in the treatment of acrophobia—A case report. *Behaviour Therapy*, *26*, 547–554.

Rothbaum, B. O., Hodges, L. F., Ready, D. J., Graap, K., & Alarcon, R. D. (2001). Virtual reality exposure therapy for Vietnam veterans with posttraumatic stress disorder. *Journal of Clinical Psychiatry*, *62*, 617–622.

Rothbaum, B. O., Hodges, L., Watson, B. A., Kessler, G. D., & Opdyke, D. (1996). Virtual reality exposure therapy in the treatment of fear of flying: A case report. *Behavior Research and Therapy*, *34*, 477–481.

Tortella-Feliu, M., Botella, C., Llabrés, J., Bretón-López, J. M., Riera del Amo, A., Baños, R. M., & Gelabert, J. M. (2011). Virtual reality versus computer-aided exposure treatments for fear of flying. *Behavior Modification*, *35*(1), 3–30.

Index

The Handbook of the Psychology of Communication Technology, First Edition. Edited by S. Shyam Sundar.

Printed in the USA
CPSIA information can be obtained
at www.ICGtesting.com
CBHW050854191223
R14744900001B/R147449PG2629CBX00002B/1

9 781118 413364